ORGANIC REACTION MECHANISMS

Selected Problems and Solutions

William C. Groutas
Athri D. Rathnayake

First edition published 2023
by CRC Press
6000 Broken Sound Parkway NW, Suite 300, Boca Raton, FL 33487-2742

and by CRC Press
4 Park Square, Milton Park, Abingdon, Oxon, OX14 4RN

CRC Press is an imprint of Taylor & Francis Group, LLC
First edition published by Wiley, 1996

ISBN: 978-1-032-48343-6 (HB)
ISBN: 978-1-032-48825-7 (PB)
ISBN: 978-1-003-39096-1 (EB)

DOI: 10.1201/9781003390961

Typeset in Arial
by William C. Groutas and Athri D. Rathnayake

To Susan, Mark, Christopher, Kalyn, Toby, Kindsey

and the memory of Mom and Dad

To Mom, Dad and Uncle

PREFACE TO THE FIRST EDITION

The primary goal of this book is to use organic reaction mechanisms as a means of facilitating the mastery and understanding of the fundamental principles of organic chemistry, while at the same time sharpening a student's reasoning ability and critical thinking. This is achieved through the judicious selection and use of a large number of problems selected from the chemical literature. Each question is meant to illustrate one or more fundamental principles of mechanistic organic chemistry. The level of difficulty and suitability of the questions in this book have been tested by including them in exams and/or as homework assignments in the undergraduate organic chemistry course (Part A), or the first year graduate-level course in organic chemistry (Part B).

Special emphasis has been placed on the organization of the book. Part A contains questions geared toward students taking the sophomore-level organic chemistry course. The questions and principles illustrated thereof, are organized in the same sequence as they are normally discussed in a standard textbook of organic chemistry. A series of minireviews that summarize and reinforce fundamental principles that underlie a particular set of related problems have been included at the beginning of each set of questions. Thus, Part A can serve as a supplement to a standard textbook used in the first year organic chemistry course, and is intended to meet an existing need, since only a token number of end-of-chapter mechanism questions is included in most textbooks of organic chemistry.

The questions included in Part B are suitable for students in an honors course in organic chemistry, and beginning graduate students in chemistry, medicinal chemistry, biochemistry and related disciplines. A limited number of applied problems has been included (Part C) to demonstrate how a knowledge of basic organic reaction mechanisms can be used to understand problems related to everyday life.

I am deeply indebted to professor Richard A Bunce (Oklahoma State University) for his diligence in reviewing the manuscript, and his many valuable comments. I am also grateful to Jennifer Yee for her editorial assistance, guidance and constant encouragement throughout the preparation of the workbook. Any errors that may have crept in is the responsibility of the author.

PREFACE TO THE SECOND EDITION

As in the first edition, our overarching goal is to use organic reaction mechanisms to help students master the fundamental principles of organic chemistry and sharpen their thinking and reasoning skills. In doing so, we have focused on mechanisms that involve Lewis acid/Lewis base reactions. The revised book is aimed at undergraduates and beginning graduate students in organic and medicinal chemistry, biochemistry, and related disciplines, and the questions are organized in the same general sequence as in a standard textbook of organic chemistry.

The second edition contains several important changes. The scope of each minireview has been expanded and the number of problems has been increased considerably (from 210 to 500 problems). The questions are graded in difficulty with part A containing questions aimed at students taking the sophomore-level organic chemistry class while part B contains questions of somewhat greater difficulty suitable for students taking an honors course in organic chemistry and beginning graduate students. Special emphasis has been placed on the selection of questions to ensure that each question illustrates one or more fundamental principles of organic chemistry.

We hope that the second edition will enhance the understanding of organic reaction mechanisms and minimize rote learning. Finally, we would like to thank Hilary LaFoe for her editorial assistance and encouragement, and Sukirti Singh and the CRC press production team for their help and support.

TABLE OF CONTENTS

GLOSSARY

Name	Abbreviation	Structure
p-Toluenesulfonic acid	p-TSA/p-TsOH	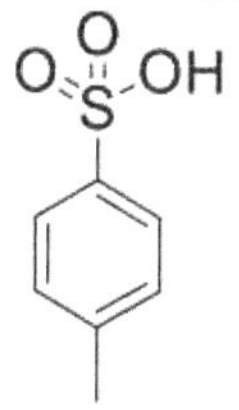
4-Dimethylaminopyridine	DMAP	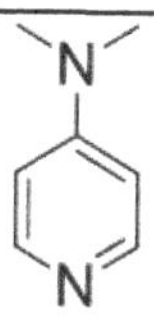
Polyphosphoric acid	PPA	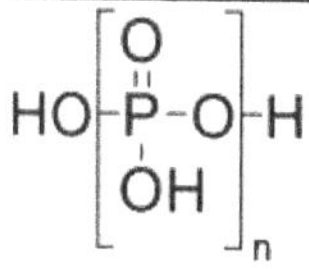
1,8-Diazabicyclo[5.4.0]undec-7-ene	DBU	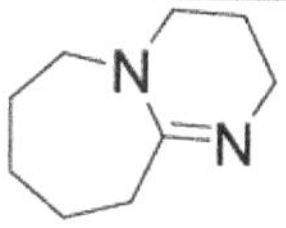
Tetra n-butyl ammonium fluoride	TABF	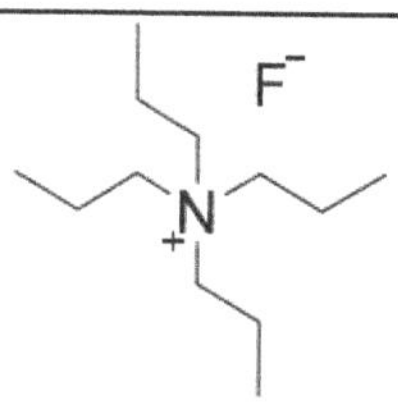
1,4-Diazabicyclo[2.2.2]octane	DABCO	
Lithium diisopropylamide	LDA	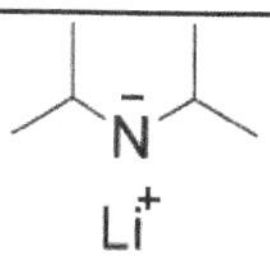
N-Chlorosuccinimide	NCS	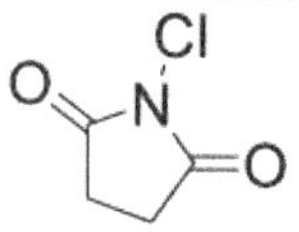
2,4-Bis(4-methoxyphenyl)-1,3,2,4-dithiadiphosphetane-2,4-disulfide	Lawesson's reagent	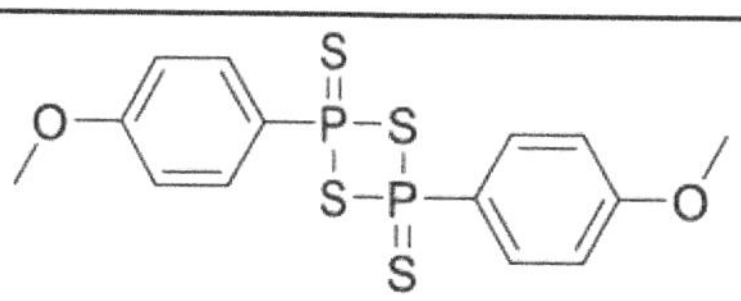

Sulfonic acid polymer resin (strongly acidic cation exchanger)	Amberlyst-15	
Lithium bis(trimethylsilyl)amide	LHMDS	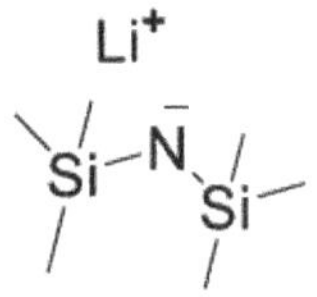
N,N′-Dicyclohexylcarbodiimide	DCC	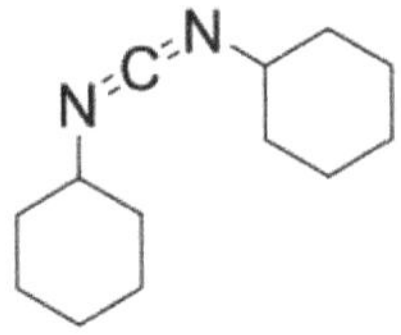
(7,7-dimethyl-2-oxobicyclo[2.2.1]heptan-1-yl)methanesulfonic acid	CSA	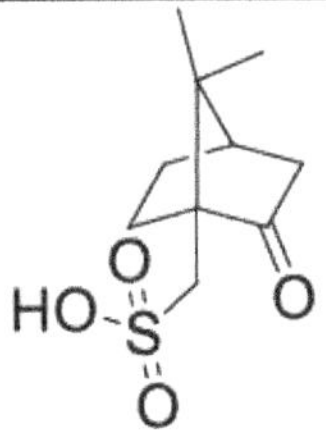
Diethyl azodicarboxylate	DEAD	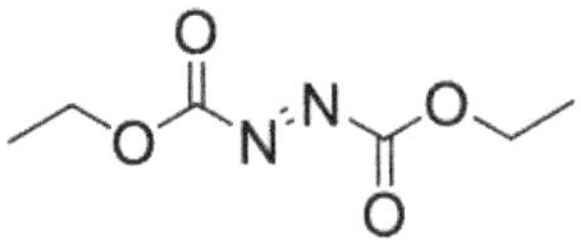
1-methoxy-N-(triethylazaniumyl)sulfonylmethanimidate	Burgess reagent	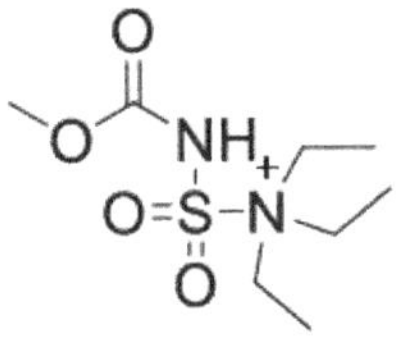
Tetrabutylammonium iodide	TBAI	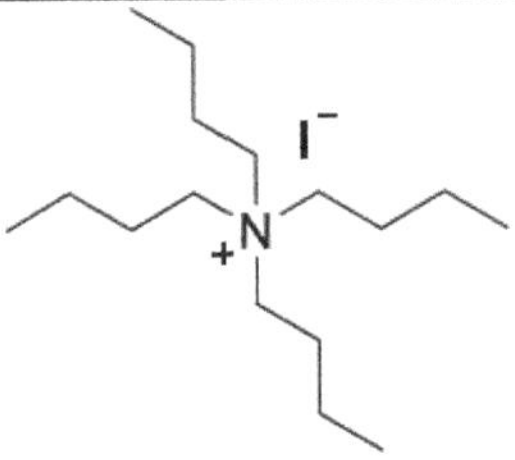

A NOTE ON WRITING MECHANISMS

The majority of organic reactions can be viewed as being Lewis acid/Lewis base reactions. Recall that a *Lewis base* (LB) is any substance that can donate a pair of non-bonded or pi electrons to a *Lewis acid* (LA) to form a covalent bond, and a *Lewis acid* is any substance that can accept a pair of electrons to form a covalent bond (see Minireviews 1 and 2 for a general discussion of Lewis structures, Lewis acids and Lewis bases, and Lewis acid/ Lewis base reactions).

In writing mechanisms, the following general approach should be followed:

(a) *add a sufficient number of non-bonded electrons pairs on any heteroatoms* (atoms other than carbon, such as, O, N, S, etc.) *to complete their octets*, since chemical structures are customarily drawn without the non-bonded electron pairs shown. By doing so, you will immediately identify the atom(s) in a given reactant that, in principle, can donate a pair of electrons to a Lewis acid. With just a little practice this will turn out to be a trivial task and, furthermore, in many cases you'll only be needing to add non-bonded electron pairs to the atoms that are *directly* involved in the reaction.

EXAMPLES

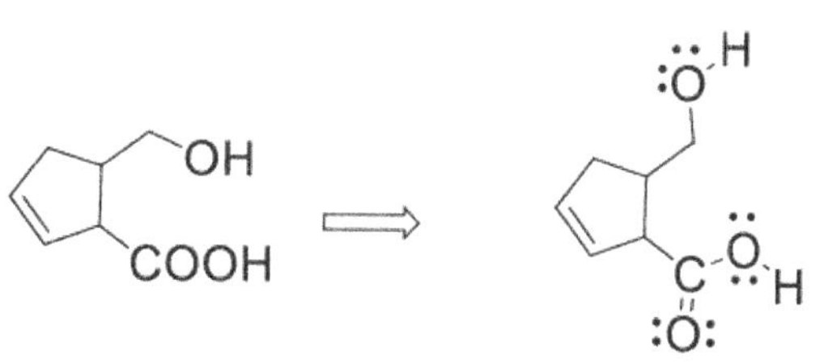

(b) *Identify the reactant that functions as a Lewis base, and the reactant that functions as a Lewis acid in a given reaction and use curved arrows to indicate the flow of electrons from the atom that donates the pair of non-bonded or pi electrons to the receiving electron deficient atom*, as illustrated below. The reactant that acts as a Lewis acid (LA) typically has an atom that is electron deficient, namely, it may have a positive charge, or a partial positive charge (δ^+) if it is bonded to one or more electronegative atoms, or it may have an atom that lacks an octet of electrons.

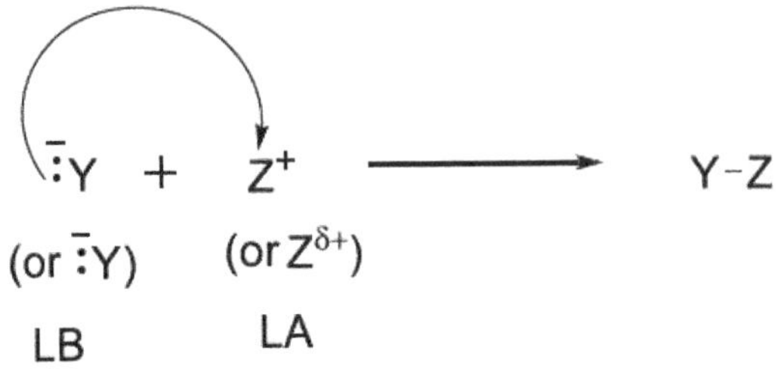

Y-Z
LB
LA

Examples illustrating these ideas are shown below.

EXAMPLE 1

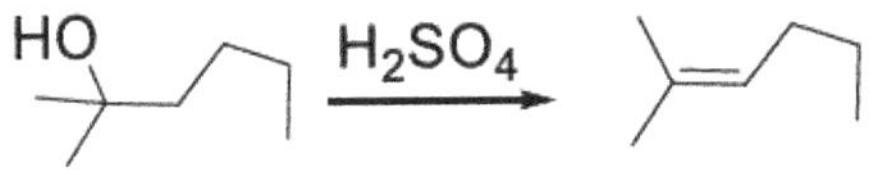

HO
H2SO4

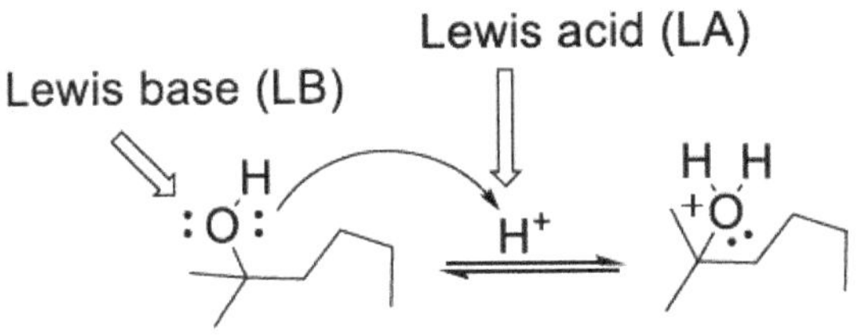

Lewis acid (LA)
Lewis base (LB)
H+

EXAMPLE 2

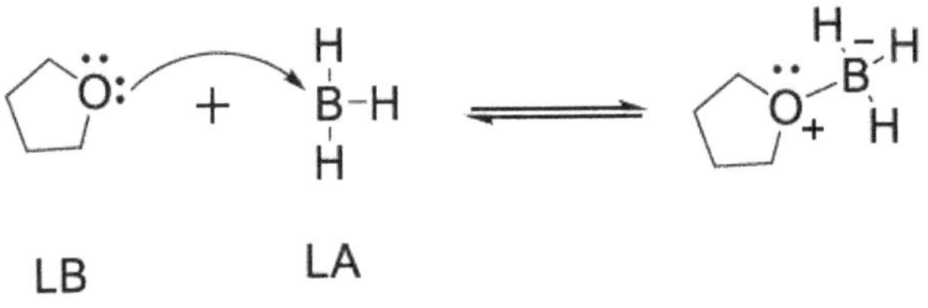

LB
LA

Note In general, compounds of group IIIA elements (B, Al, etc.) typically have an incomplete octet of electrons and invariably function as Lewis acids. Examples include boron trifluoride (BF_3), aluminum chloride ($AlCl_3$), boron tribromide (BBr_3), etc.

EXAMPLE 3

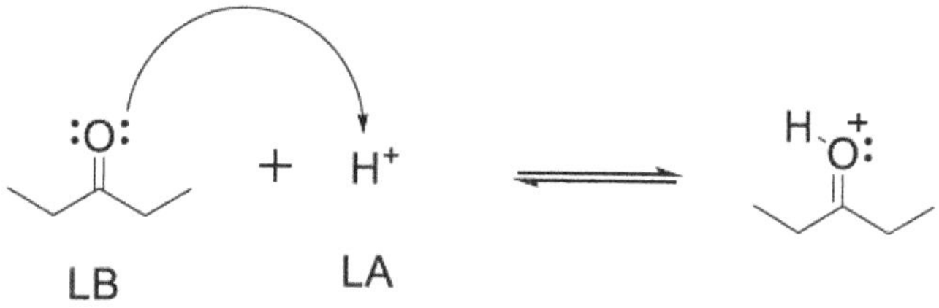
LB
LA

EXAMPLE 4

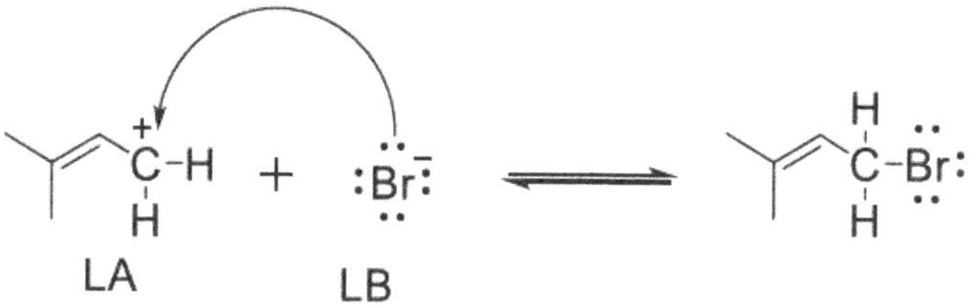
LA
LB

EXAMPLE 5

It is helpful to remember that strong mineral acids (HCl, HBr, HNO_3, etc.) or strong organic acids (for example RSO_3H, CF_3COOH) are fully or predominantly ionized in solution. Thus, the reaction shown above can be viewed as shown below, making it easier to identify the Lewis acid(s) and Lewis base(s).

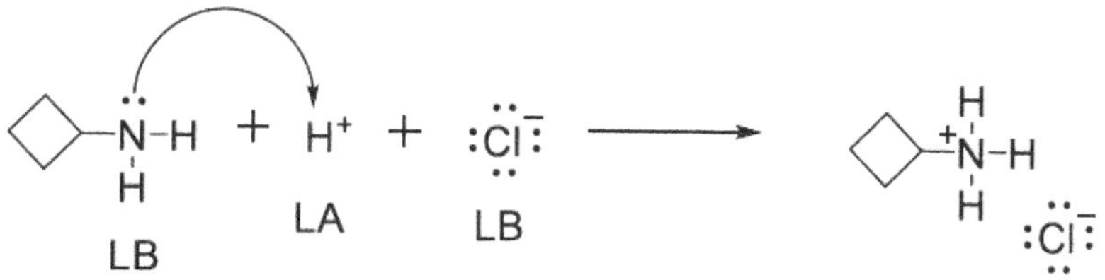
LB
LA
LB

EXAMPLE 6

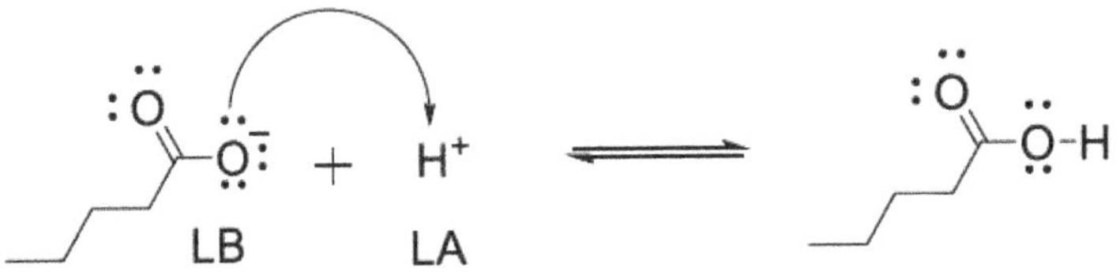

LB
LA

EXAMPLE 7

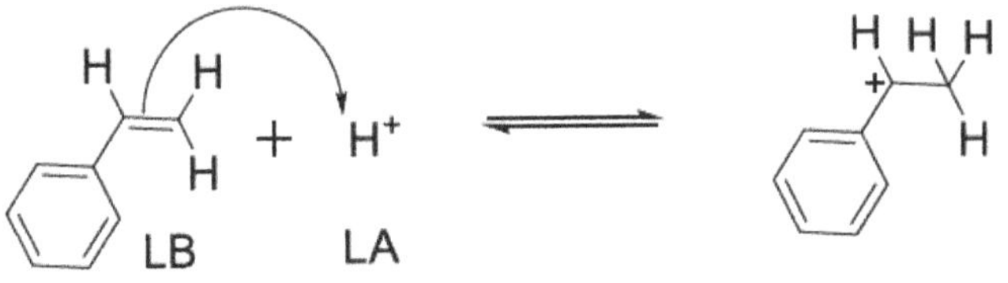

LB
LA

EXAMPLE 8

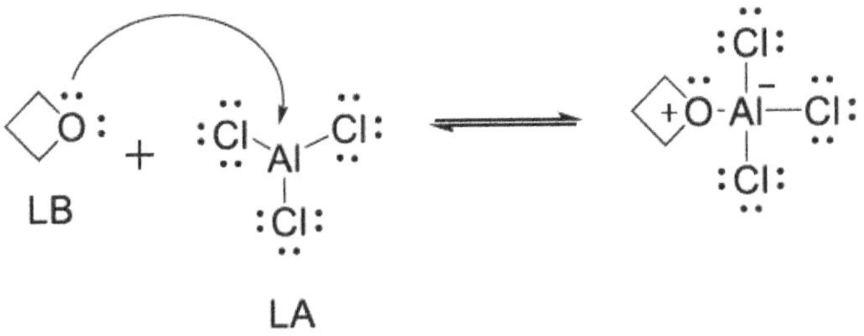

LB
LA

(c) *Assign formal charges to the atoms directly involved in the reaction* (the formal charges on atoms that are not *directly* involved in the reaction do not change). Recall that the formal charge can be readily determined as follows:

Formal charge on an atom = X - Y - Z

where X = number of valence electrons

Y = number of non-bonded electrons

Z = half the number of bonded electrons

The number of valence electrons for an atom corresponds to its group number in the periodic table. For example, sulfur is in group VI in the periodic table, therefore it has six valence electrons (see also Minireview 1).

(d) If a reactant has more than one functional group which can donate a pair of non-bonded or pi electrons, then inspection of the structures of the reactant and product will usually reveal which functional group will react initially. Other considerations coming into play in deciding which functional group in a reactant is involved in the first step of a reaction include, for example, the stability of the initial carbocation. *The more stable a carbocation is, the easier it is to form* (see Minireview 4 for a discussion of carbocation chemistry). In the example shown below, while a pair of non-bonded electrons can be donated from either hydroxyl group, reaction takes place preferentially at the one that leads to the formation of the more stable benzylic carbocation, as opposed to the less stable primary carbocation.

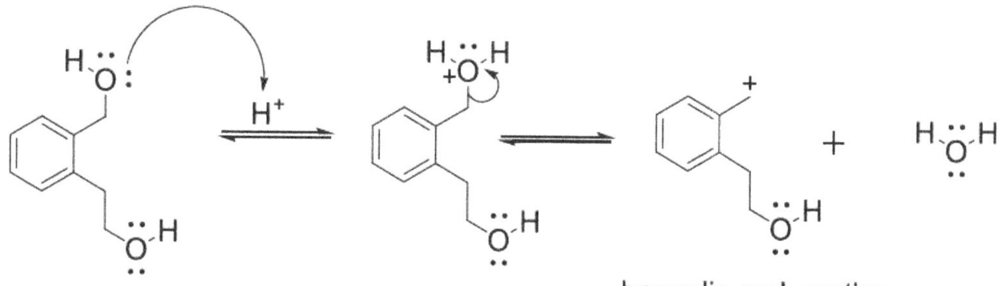
H⁺
benzylic carbocation

(e) Most mechanisms, particularly those involving skeletal rearrangements, may involve several sequential steps. Most of the steps are *consecutive Lewis acid/Lewis base reactions which can be intermolecular* (reaction involves two separate molecules that can be the same or different) or *intramolecular* (reaction takes place within the same molecule) in nature. The *driving force* behind these steps is the *resulting gain in stability* in going from one transient species to another such as, for example, from a less stable carbocation to a more stable carbocation, relief of ring strain, etc.

The following examples are meant to serve as a guide on how to write a reasonable mechanism for a reaction you may have never seen before using the Lewis acid/Lewis base approach. The same approach can be used to understand and explain how a reagent mediates a particular transformation. The thought processes underlining the general approach used in writing mechanisms are explicitly stated.

EXAMPLE 1 Write a mechanism for the following reaction.

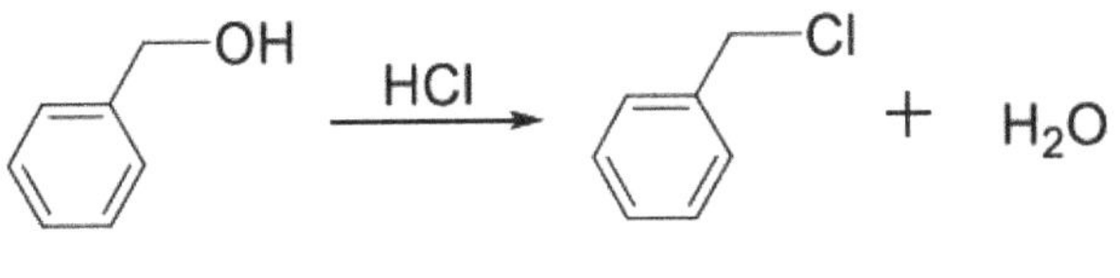

OH
HCl
Cl
+ H_2O

Answer

The first step in writing a mechanism for any chemical reaction is the *identification of the Lewis acid and Lewis base.* This can be readily accomplished by placing a sufficient number of non-bonded electron pairs on the heteroatom (oxygen in this case) in a reactant to complete its octet. This immediately reveals that the alcohol will function as a Lewis base and the oxygen atom of the hydroxyl group will donate a pair of non-bonded electrons to the Lewis acid. Since mineral assets are fully ionized in solution, the second reactant can be viewed as existing as H^+ (a Lewis acid) and Cl^- (a Lewis base) ions. Therefore, the first step of the mechanism for this reaction involves a Lewis base /Lewis acid reaction to yield a protonated alcohol. The second step involves the loss of a molecule of water via the cleavage of the C-O bond (the pair of electrons ends up on the oxygen atom of water as a pair of non-bonded electrons), at the same time relieving the positive charge on the electronegative atom. The second step in this mechanism is the rate-determining step, namely, the step with the highest free energy of activation (see Minireview 4 for a full discussion of carbocation chemistry). The third step completes the mechanism of the reaction. It involves a straightforward Lewis acid/Lewis base reaction between the carbocation (LA) and the chloride ion (LB).

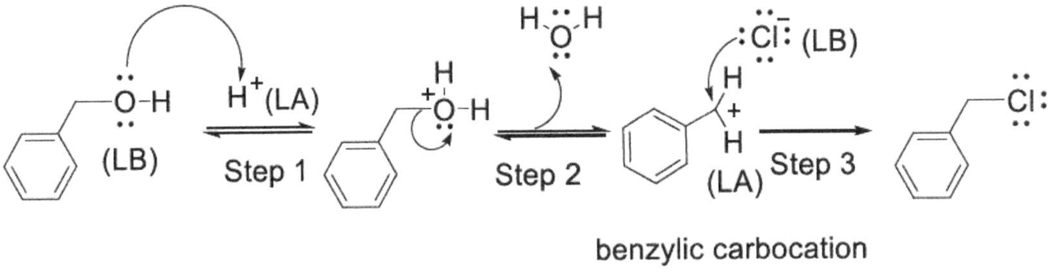

O-H
(LB)
H^+(LA)
Step 1
H
O-H
H
O
H
Step 2
Cl (LB)
H
H
(LA)
Step 3
Cl
benzylic carbocation

EXAMPLE 2 Write a mechanism for the following reaction.

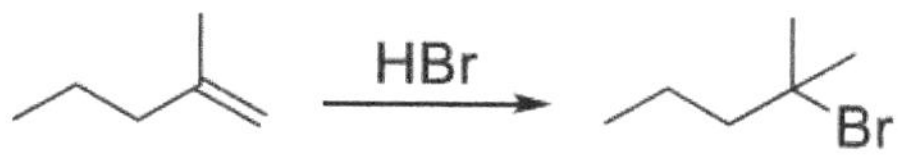

HBr
Br

Answer

By definition, any substance that can donate a pair of *non-bonded* or *pi electrons* is a Lewis base. Thus, alkenes (as well as alkynes) invariably function as Lewis bases by donating a pair of pi electrons to a Lewis acid. In the first step of this reaction, the pair of pi electrons in the C=C bond is donated to H^+ (Lewis acid) to form a covalent bond yielding a carbocation (a Lewis acid). Notice that in this instance the pair of pi electrons can potentially be used to form a covalent bond with either one of the carbon atoms of the C=C bond. Recall that, *the reaction of an alkene with H^+ always leads to the initial formation of the most stable carbocation. In order to form the most stable carbocation, the H^+ will have to bond to the carbon of the C=C bond that bears the greater number of hydrogens (Markovnikov's rule)*. The second step of the mechanism involves the reaction of a carbocation (LA) with a Lewis base (bromide ion).

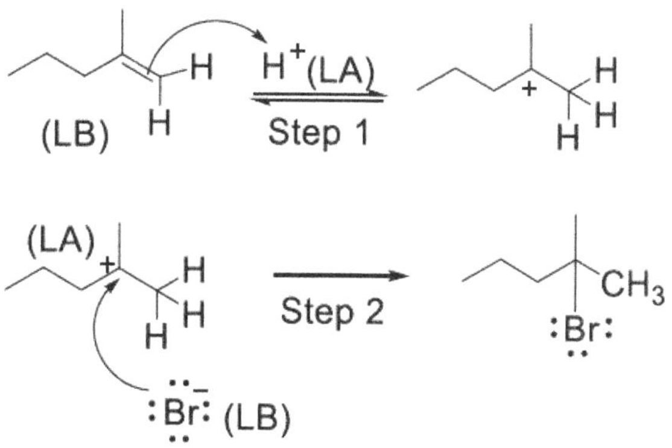

(LB)
H
H
H⁺(LA)
Step 1
H
H
H
(LA)
H
H
H
Step 2
CH₃
:Br:
:Br: (LB)

EXAMPLE 3 Write a mechanism for the following reaction.

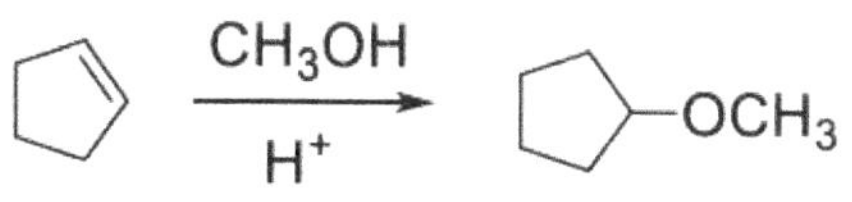

CH₃OH
H⁺
OCH₃

Answer

Unlike examples 1 and 2 which involve the reaction of either an alcohol or an alkene with a Lewis acid, this example involves the participation of two Lewis Bases (an alkene and an alcohol) and a Lewis acid (H^+), leading to the formation of an ether. Prior to writing a mechanism for this reaction the following question must be addressed: which one of the two Lewis bases will react with H^+ and how do you go about deciding that? While the alcohol (LB) can potentially react with H^+ (LA), this is an unproductive process (it cannot lead to the formation of the observed product). Note also that methyl carbocations are highly unstable (see Minireview 4). Thus, the reaction is initiated via a Lewis base/Lewis acid reaction between the alkene and H^+ forming a secondary carbocation. Further reaction of the carbocation (a Lewis acid) with the alcohol (a Lewis base) leads to the formation of a protonated ether (Step 2). Loss of a H^+ yields the observed product (step 3). The hydrogen in the protonated ether is *acidic*, consequently step 3 is essentially an ionization step.

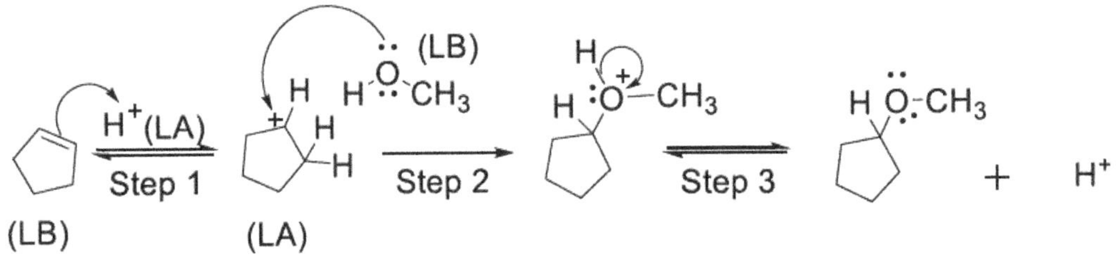

(LB)
H
H
O
CH3
H+(LA)
H
H
Step 1
Step 2
H
H
O
CH3
Step 3
H O CH3
+
H+
(LB)
(LA)

<u>Exercise</u> Write a mechanism for the following reaction

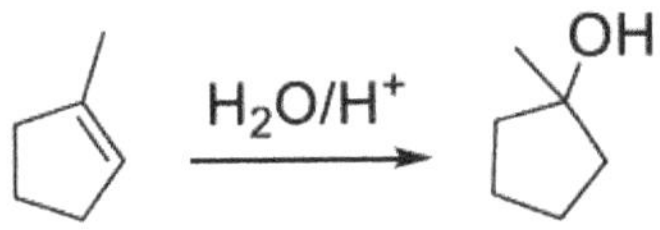

H2O/H+
OH

<u>EXAMPLE 4</u> Write a mechanism for the following reaction.

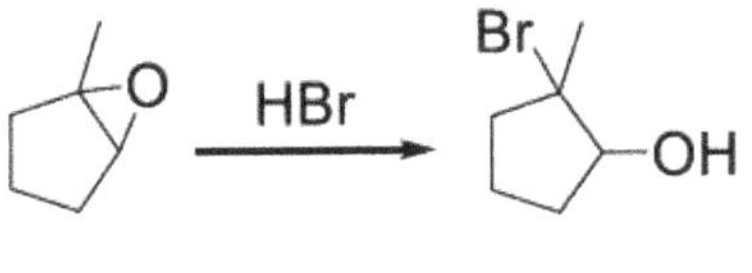

O
HBr
Br
OH

<u>Answer</u>

Once the appropriate number of pairs of non-bonded electrons are placed on the oxygen to complete its octet, it can be readily seen that ethers (like alcohols), can function as Lewis bases. Thus, the first step of this reaction is a Lewis acid/Lewis base reaction leading to the formation of a protonated ether. This is followed by cleavage of the C-O bond *in a way that yields the most stable carbocation* (tertiary vs secondary). A Lewis acid/Lewis base reaction between the carbocation and bromide ion completes the mechanism for this reaction. Note Three and four-membered cyclic ethers readily undergo ring-opening reactions in the presence of Lewis acids. The relief of ring strain serves as the driving force for these reactions (see Minireview 7).

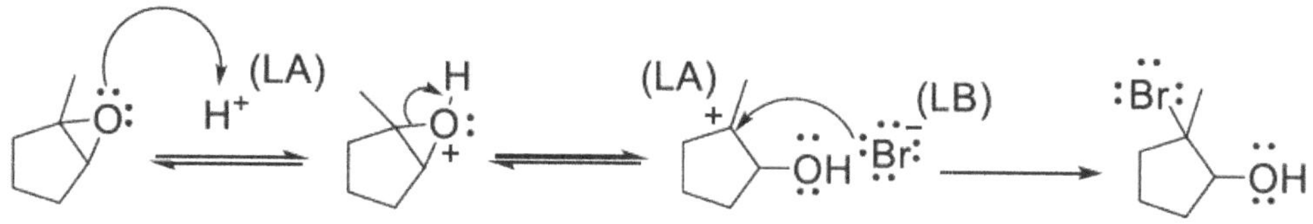

(LA)
H+
(LB)
H
O
(LA)
(LB)
OH
Br
Br
OH

EXAMPLE 5 Write a mechanism for the following reaction.

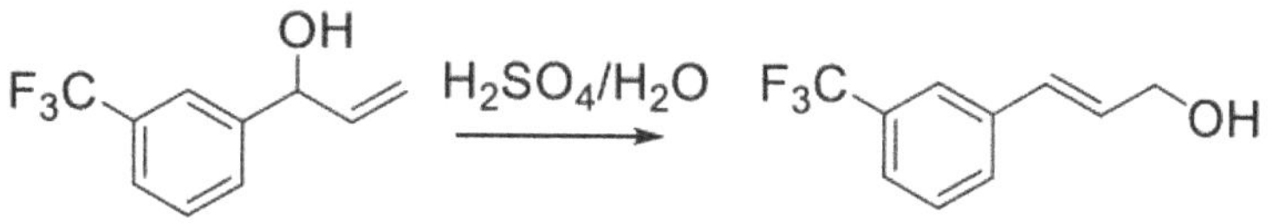

OH
F3C
H2SO4/H2O
F3C
OH

Answer

Once the Lewis acid and the Lewis base are identified, casual inspection of the Lewis base indicates that, in principle, three different functional groups (OH, alkene C=C bond and aromatic ring C=C bonds) can act as a Lewis base, namely, donate a pair of non-bonded or pi electrons to the Lewis acid (H^+). At this point a decision has to be made as to which of these will react with the Lewis acid. *As a general rule, it is useful to remember that (a) non-bonded electrons are more available for donation than pi electrons (since pi electrons are held by two nuclei) and, (b) an alkene C=C bond is more reactive than the C=C bond of an aromatic ring,* since reaction of an aromatic ring C=C bond results in the loss of the aromatic character of the ring (30 kcal/mol loss in resonance stabilization energy).

Thus, in step 1 of this reaction the hydroxyl group of the Lewis base donates a pair of non-bonded electrons to the Lewis acid, which is then followed by cleavage of the C-

O bond and the loss of a molecule of water (step 2), leading to the formation of a carbocation (notice that when the C-O bond is cleaved, the pair of electrons making up the bond goes with the oxygen). Step 1 (LB/LA reaction) and step 2 (loss of water) are common to all reactions involving an alcohol and acid. The carbocation formed in step 2 is resonance-stabilized, thus, reaction of the carbocation (LA) with water (LB) gives rise to a protonated alcohol (step 3). The final step (Step 4) in this reaction involves the loss of hydrogen ion (H^+) to give the observed product. Step 4 is a simple ionization step, analogous to the ionization of the hydronium ion ($H_3O^+ = H_2O + H^+$).

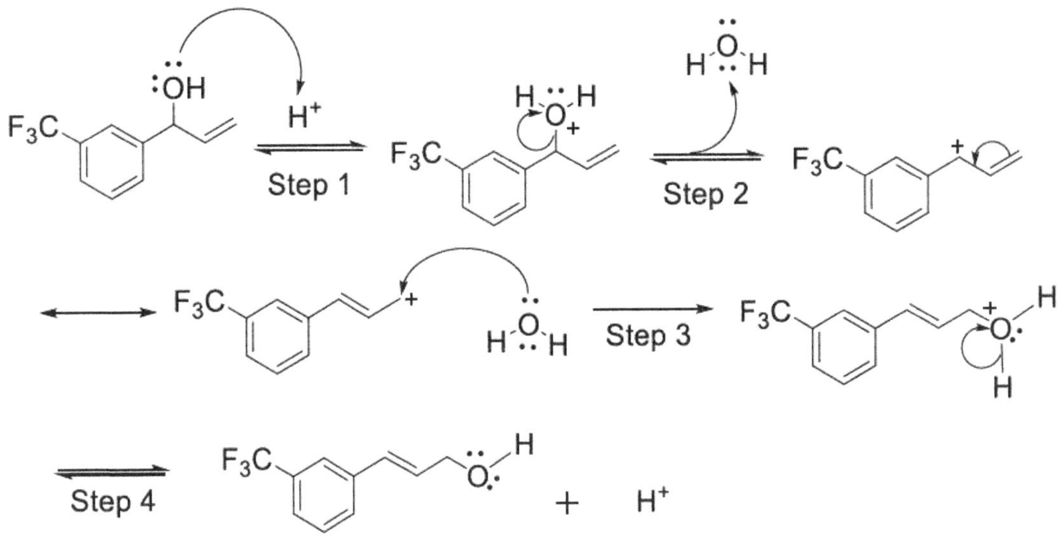

H+
Step 1
Step 2
Step 3
Step 4

EXAMPLE 6 Write a mechanism for the following reaction.

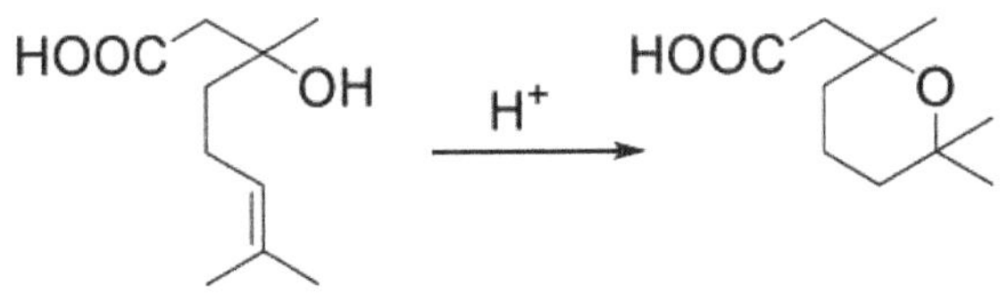

HOOC
OH
H+
HOOC
O

Answer

It is always a good idea to inspect the structures of the starting material and product, since in many instances this will quickly reveal which functional group(s) are involved in the reaction. In this instance, it is evident that it's the hydroxyl group and C=C bond, and not the carboxyl group. With the non-bonded electrons added on the heteroatoms, and recalling the fundamental definition of a Lewis base, it is apparent that either the hydroxyl group or the C=C bond can donate a pair of electrons to the Lewis acid. In example 5, it was stated that, as a general rule, a pair of non-bonded electrons is *more* available for donation than a pair of pi electrons. This example was chosen to (a) demonstrate that this is not always true, and (b) emphasize the need for you to keep an open (flexible) mind as you consider plausible mechanistic pathways. In other words, organic reactions frequently follow an unpredictable course, and the task on hand is to use fundamental principles to account for the formation of the observed product. Indeed, herein lies the pedagogical value of writing mechanisms. In so doing, you will be forced to look at a situation in many ways and consider plausible pathways within a framework of principles. Thus, in step 1 of this reaction, a Lewis base (C=C)/Lewis acid (H^+) reaction gives rise to the 3^0 carbocation (Markovnikov's rule). In step 3, the carbocation (LA) accepts a pair of non-bonding electrons from the Lewis base to form a product that ionizes to give the observed product.

Notice that an initial reaction between the hydroxyl group and H^+, followed by loss of a molecule of water *also* leads to the formation of 3^0 carbocation, however, this pathway cannot account for the observed product (unproductive pathway).

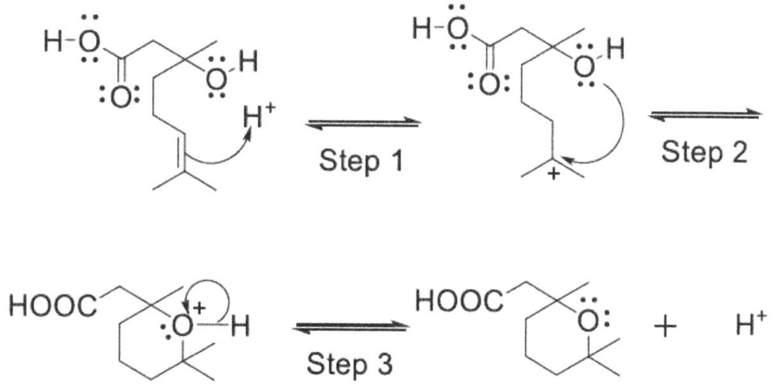
H-O
:O:
O
H
H+
Step 1
H-O
:O:
O
H
+
Step 2
HOOC
+
O
H
Step 3
HOOC
O:
+
H+

EXAMPLE 7 Write a mechanism for the following reaction.

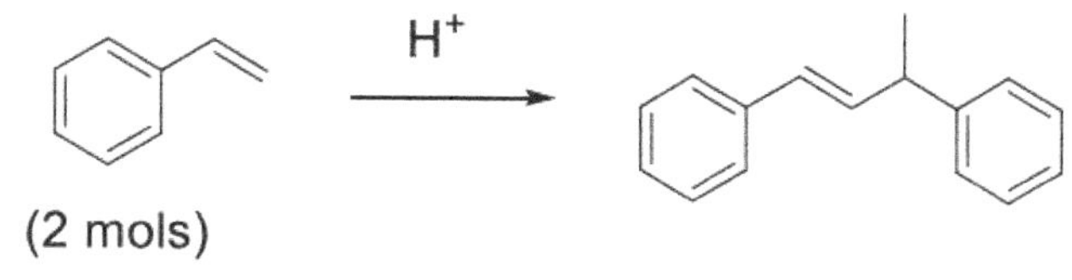
H+
(2 mols)

Answer

As stated previously (example 2), the C=C bond of an alkene functions as a Lewis base by donating a pair of pi electrons to a Lewis acid. However, in this instance the pi bonds in the phenyl (aromatic) ring could potentially behave the same way. Since the reaction of a ring C=C with a Lewis acid would result in the formation of a much less stable non-aromatic species, the alkene C=C bond reacts preferentially (step 1). The carbocation formed is a Lewis acid which then reacts with a second molecule of the alkene (a Lewis base) to give rise to a second benzylic carbocation (step 2). Besides reacting with a Lewis base, a carbocation can lose a H^+ from an adjacent carbon atom to form an alkene (Minireview 3). Thus, step 3 leads to the formation of a new alkene.

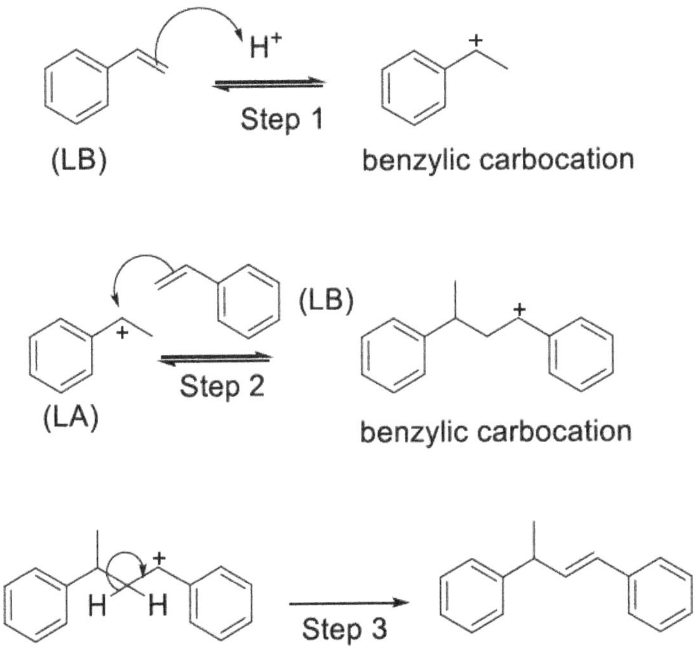

H+
Step 1
(LB)
benzylic carbocation
(LB)
Step 2
(LA)
benzylic carbocation
H
H
Step 3

EXAMPLE 8 Write a mechanism for the following reaction.

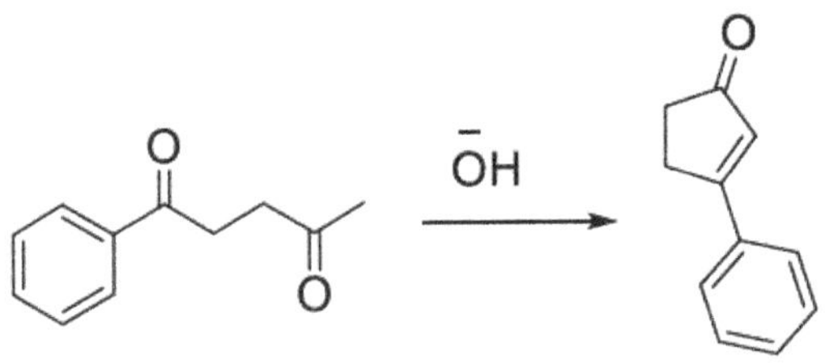

O
O
OH
O

Answer

A mechanism question must always be approached from first principles, namely, it is not necessary for you to be able to realize that this particular reaction is an example of an *aldol condensation reaction*, in order to write a reasonable mechanism. Thus, the approach is always the same: first add non-bonded electron pairs to the two reactants, and classify each as a Lewis base and a Lewis acid. The stronger Lewis base (^{-}OH, since it has a negative charge) is going to react with the second reactant which, by necessity, must function as a Lewis acid (LA). The carbon of the C=O group is electron-deficient (oxygen is more electronegative than carbon, consequently the electrons connecting the carbon and oxygen are not equally shared, and hence the carbon has a partial positive charge and oxygen a partial negative charge, $^{\delta+}C{=}O^{\delta-}$). Thus, one possibility is for the hydroxide ion to donate a pair of electrons to the electron deficient carbon (*nucleophilic addition*) or, since the hydrogens on the alpha carbon of a ketone are acidic, an acid-base reaction can take place instead, yielding an anion (a Lewis base or nucleophile). *As a general rule, Bronsted acid-base reactions are faster than most other types of organic reactions*. Thus, a Bronsted acid-base reaction in step 1 yields an anion, which then reacts in a Lewis base/Lewis acid reaction (step 2) to form the product. This product is the conjugate base of an alcohol and can be viewed as being in equilibrium with the acid (step 3). Step 4 is a β-elimination reaction that leads to the formation of the product. Step 4 is facile because it leads to the formation of a highly stable *conjugated system* (a system that consists of an array of *alternating* double and single bonds).

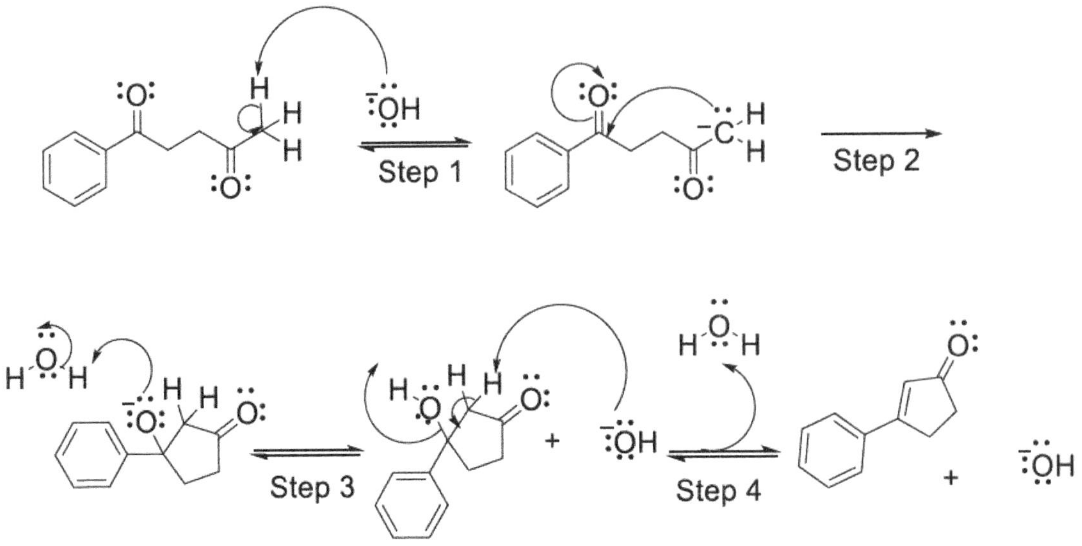
Step 1
Step 2
Step 3
Step 4

EXAMPLE 9 Write a mechanism for the following reaction.

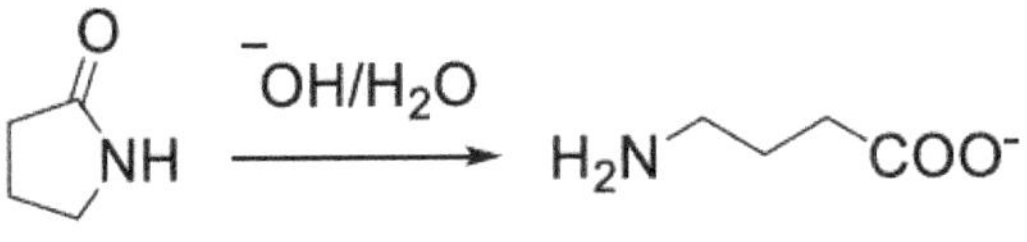
$^{-}OH/H_2O$
H_2N
COO^-

Answer

The carbonyl carbon is electron deficient, i.e., has a partial positive charge, because it is bonded to two electronegative atoms. Any atom that bears either a full positive charge (such as a carbocation, for example) or partial positive charge, is capable of accepting a pair of electrons from a Lewis base to form a covalent bond. Thus, an initial LA/LB reaction leads to the formation of a tetrahedral intermediate (step 1). Subsequent collapse of this intermediate leads to a ring-opened product (step 2). The ring-opened product has an acidic group (COOH) and a basic group, thus a fast Lewis acid/base reaction (H^+ transfer) takes place, leading to the formation of the final product (step 3).

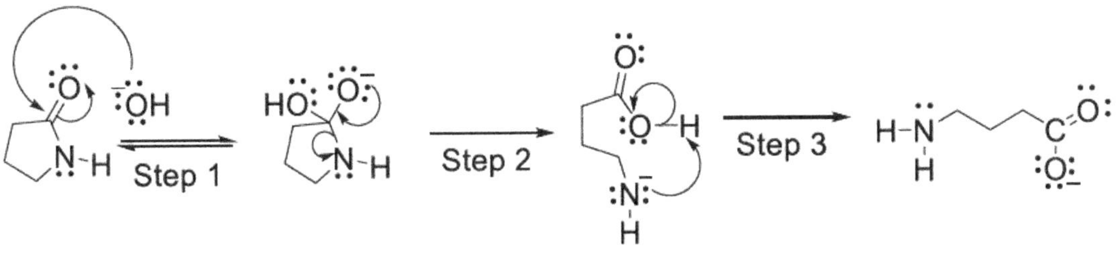
Step 1
Step 2
Step 3

EXAMPLE 10 Write a mechanism for the following reaction.

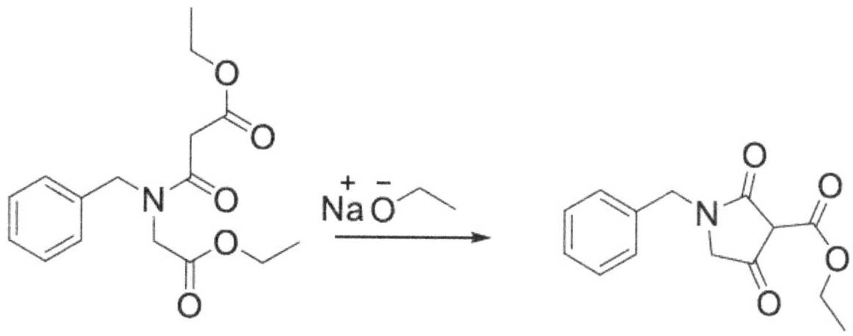

Answer

The Lewis base (ethoxide ion, $CH_3CH_2O^-$) reacts with the reactant to generate an anion (step 1). Notice that the ethoxide ion reacts with the *most acidic hydrogen* (see Minireview 5 for a discussion of acidity). The pKa of the reactant acid is ~11, while that of the product acid ($HOCH_2CH_3$) is ~16. Thus, the equilibrium lies to the right, i.e., favors the formation of the anion derived from the stronger acid. Once the anion (nucleophile) is formed, an intramolecular LB/LA reaction (*nucleophilic acyl substitution*) takes place (Steps 2 and 3), forming the product.

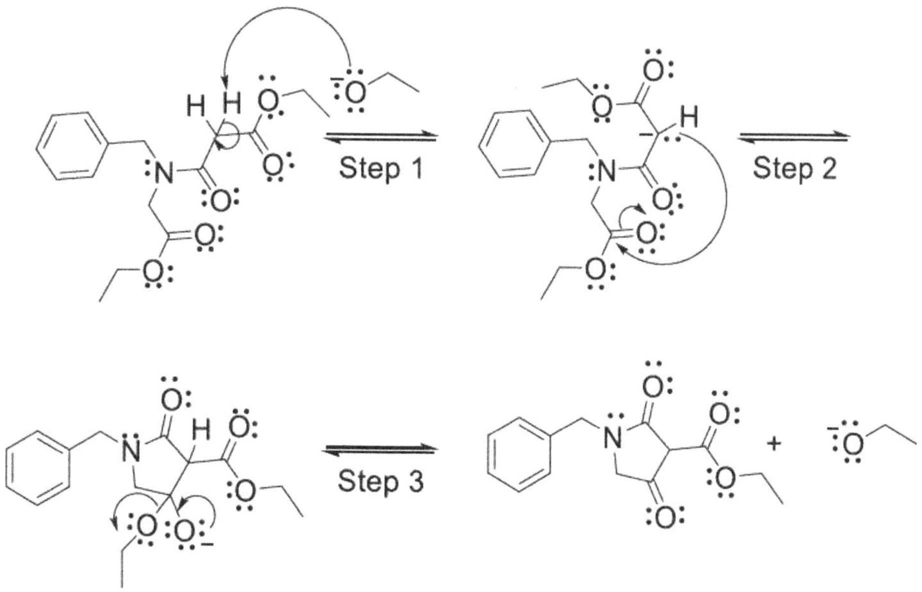
Step 1
Step 2
Step 3

Minireviews 1-4 are intended to provide a quick review of the fundamental principles related to Lewis structures, Lewis acid/Lewis base reactions, and carbocation chemistry. These should be studied prior to attempting questions 1-34, Part A.

PART A

Minireview 1

1. Lewis Structures

A sound understanding of mechanistic organic chemistry requires a proficiency in writing Lewis structures. Without the ability to draw Lewis structures correctly and with facility, a student is so severely handicapped that he or she will ultimately resort to learning organic chemistry by rote (a tedious, frustrating, and minimally-successful endeavor). The importance of this will become apparent momentarily.

A *Lewis structure* is a type of structural formula that shows the way in which the atoms are bonded together and depicts the bonding between atoms using pairs of non-bonded electrons (shown as dots) and bonded electrons (shown as dashes). In writing Lewis structures, the general approach outlined below should be followed.

1. ***Determine the total number of valence electrons***.

For neutral molecules, this is simply accomplished by adding up the valence electrons of the individual atoms. In the case of ions, an electron is added for each negative charge (anions), and an electron is subtracted for each positive charge (cations). *Recall that the number of valence electrons for an element corresponds to the group number of that element in the periodic table.* For example, nitrogen has five valence electrons (nitrogen is located in group five of the periodic table), fluorine has seven valence electrons (fluorine is located in group seven), etc.

2. ***Connect the atoms in the given molecular formula using single lines (dashes)***.

It's helpful to remember that in the case of polyatomic molecules or ions, the atom of lower electronegativity is typically the central atom. Recall that electronegativity follows the order F > O > Cl, N > Br > C, H.

3. ***Place a sufficient number of non-bonded electron pairs on each atom to give each atom an octet of electrons (octet rule)*** (keep in mind that hydrogen can only share a pair of electrons). If at this point the total number of valence electrons used is greater than that computed in step 1 above, use double or triple bonds or rings to arrive at a Lewis structure that has the correct total number of valence electrons and all the atoms have an octet electrons.

4. **Determine the formal charge on each atom**.

As stated earlier, the formal charge can be readily determined as follows (see sidebar).

Minireview 1

Formal charge = X -Y- Z

where,

X= number of valence electrons (of atom under consideration)

Y= number of non-bonded electrons, and

Z= half the number of bonded electrons

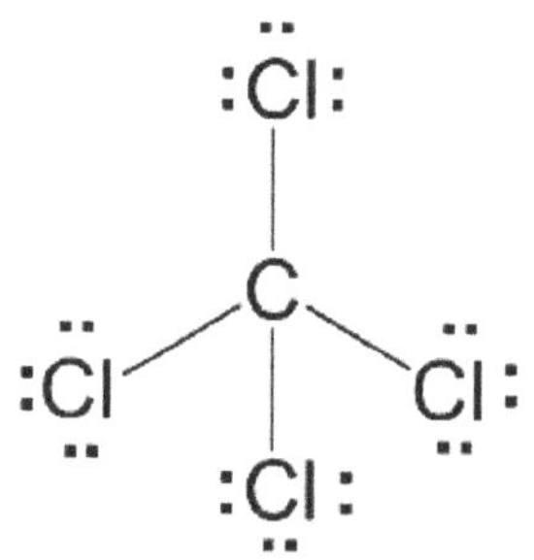
:Cl:
C
:Cl
Cl:
:Cl:

CCl_4

1 C	1 X 4 = 4	
4 Cl	4 X 7= 28	
	32 Total number of valence electrons	

Minireview 1

Summary

Species that lack an octet of electrons (electron-deficient species) have two distinct characteristics: they function as Lewis acids and are highly reactive (vide infra).

Notes

a) The Lewis structures of compounds derived from group IIIA elements (B, Al, etc.), have an incomplete octet (six electrons). As expected, these compounds invariably function as Lewis acids and are also highly reactive (because of their incomplete octet, they tend to readily accept a pair of electrons from a Lewis base, thereby acquiring an octet of electrons).

b) Many chemical reactions proceed through the transient formation of highly reactive species. For example, carbocations, as well as other species that lack an octet of electrons, have high energy (low stability) and, consequently, are highly reactive.

c) Lewis structures in which all the atoms have an octet of electrons cannot be written for molecules and ions that have an *odd* total number of valence electrons (for example, nitric oxide, NO). As might be expected, such species also exhibit high chemical reactivity.

d) The atoms of elements that have empty d orbitals can expand their octets, namely, they can accommodate more than eight electrons. Sulfur and phosphorus are the two elements most commonly encountered in organic chemistry capable of accommodating more than eight electrons.

Examples

PH_3 P = 1 X 3 = 5
H = 3 X 1 = 3
8

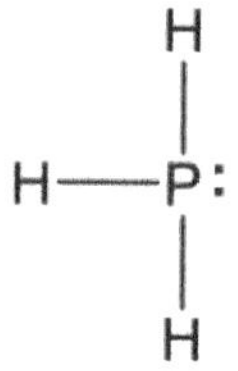

C_2H_2 C = 2 X 4 = 8
H = 4 X 1 = 4
12

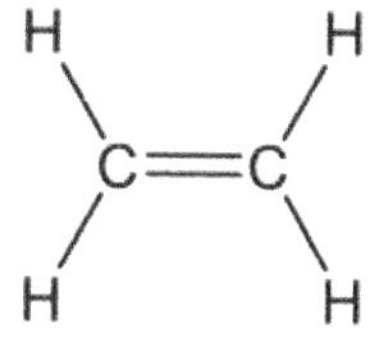

$SOCl_2$ S = 1 X 6 = 6
O = 1 X 6 = 6
Cl = 2 X 7 = 14
26

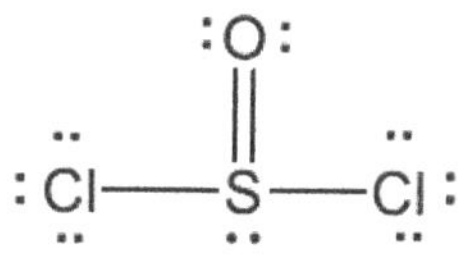

or

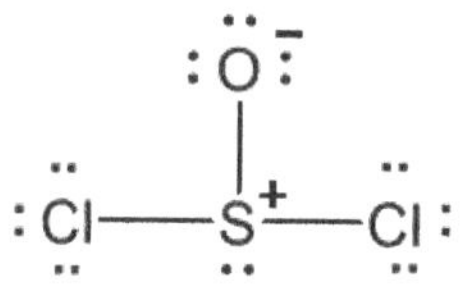

BF_3 B = 1 X 3 = 3
F = 3 X 7 = 21
24

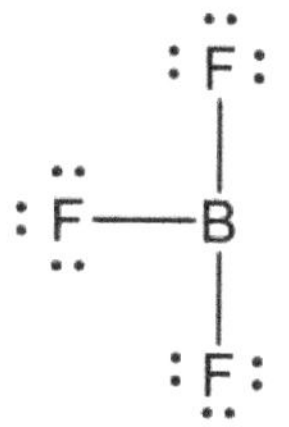

NOCl N = 1 X 5 = 5
O = 1 X 6 = 6
Cl = 1 X 7 = 7
18

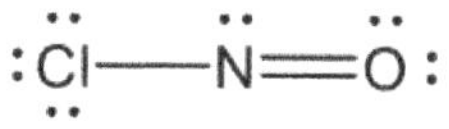

CH_3NO_2 C 1 X 4 = 4
H 3 X 1 = 3
N 1 X 5 = 5
O 2 X 6 = 12
24

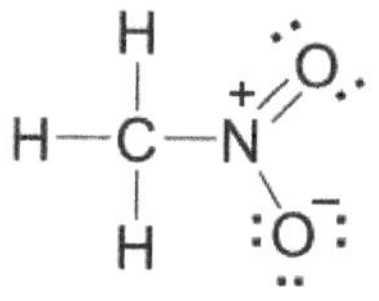

or

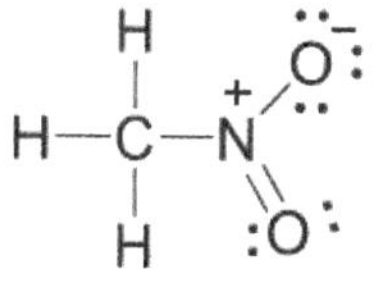

CH_2O C 1 X 4 = 4
H 2 X 1 = 2
O 1 X 6 = 6
12

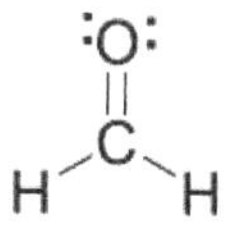

NO_2^- N 1 X 5 = 5
O 2 X 6 = 12
= 1 (one negative charge)
18

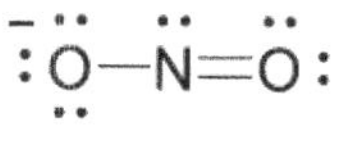

CN^-

C 1 X 4 = 4
N 1 X 5 = 5
= 1 (one negative charge) 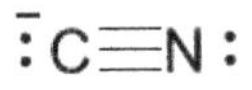
10

CO_3^{2-}

C 1 X 4 = 4
O 3 X 6 = 18
= 2 (two negative charges) 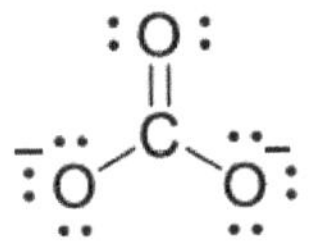
24

CH_3O^-

C 1 X 4 = 4
H 3 X 1 = 3
O 1 X 6 = 6
= 1 (one negative charge) 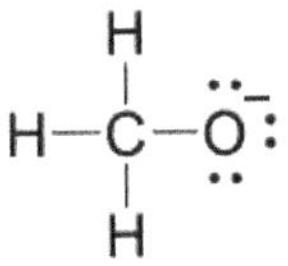
14

$NaBH_4$ This is the same as Na^+ BH_4^-. Thus, the Lewis structure that we want is that of BH_4^-

B 1 X 3 = 3
H 4 X 1 = 4
1 (one negative charge) 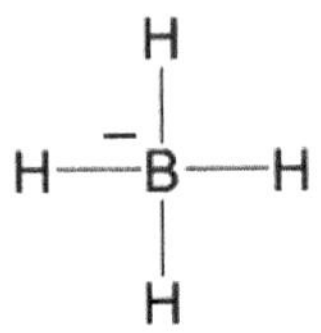
8

NO^+

N 1 X 5 = 5
O 1 X 6 = 6
11
-1 (subtract one electron for each positive charge)
10

Formal charge on O
6 - 2 - 3 = +1

$C_2H_5^+$

C 2 X 4 = 8
H 1 X 5 = 5
13
-1 (subtract one electron for each positive charge)
12

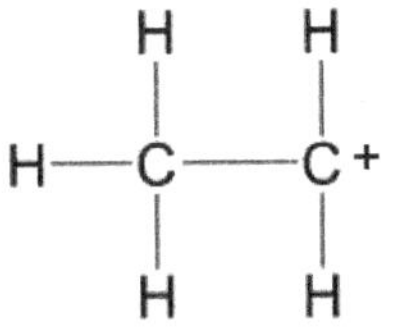

Notes

1) Compounds having a molecular formula that includes a group I or group II element are ionic compounds, namely, they consist of a cation and an anion. For example, $NaBH_4$, NaCN, $BaSO_4$, etc. In drawing a Lewis structure for an ionic compound, write the cation and anion first, and then draw the Lewis structure of the non-metallic component (usually the anion). By way of illustration, $BaSO_4$ consists of Ba^{2+} and SO_4^{2-} and you simply draw the Lewis structure of the sulfate ion.
2) On occasion, instead of being asked to draw the Lewis structure of a compound, you may be asked to deduce its relative stability, in which case you draw the Lewis structure and deduce its stability using the octet rule.

Minireview 2

1. Lewis acids
2. Lewis bases
3. Lewis acid/Lewis base reactions

A Lewis acid/Lewis base reaction can be generally described as shown below.

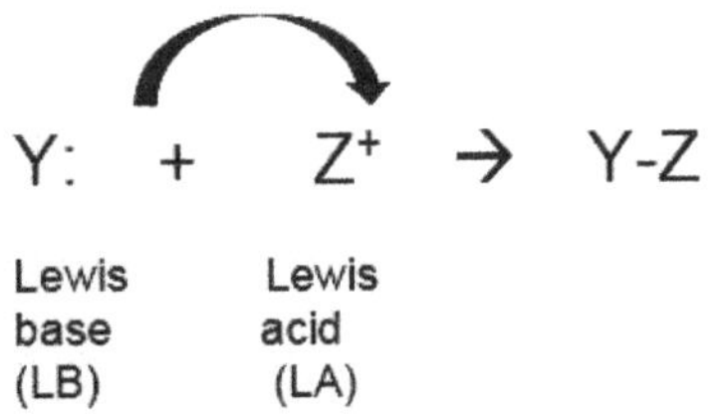
Y: + Z+ → Y-Z
Lewis base (LB)
Lewis acid (LA)

When organic reactions are described this way, the product of an organic reaction can be readily predicted, *without recourse to memorization*.

A Lewis base is substance that can donate a pair of non-bonded or pi electrons to a Lewis acid to form a covalent bond. Lewis bases that have a negative charge are stronger Lewis bases than those without a negative charge. For example, CH_3O^- is a stronger Lewis base than CH_3OH and, consequently, will react faster with a Lewis acid.

Furthermore, Lewis basicity is directly related to the availability of the pair of non-bonded electrons. A pair of non-bonded electrons on a *less* electronegative atom is more available for donation than a pair of non-bonded electrons on a *more* electronegative atom. This is the reason, for example, why amines are stronger Lewis bases than alcohols and ethers. Likewise, aliphatic amines are stronger Lewis bases than aromatic amines because the pair of non-bonded electrons on the nitrogen in aromatic amines is delocalized over the aromatic ring via resonance and, consequently, is not as available for donation.

In Lewis acid/Lewis base reactions where a Lewis acid can bond to either of two atoms that differ in electronegativity, the stability of the product determines the site of the reaction. For example, the reaction of an amide ($R(C{=}O)NH_2$) with H^+ takes place on the O (more electronegative atom) than nitrogen (less electronegative atom) because the product is resonance-stabilized (vide infra).

A Lewis acid is a substance that can accept a pair of non-bonded or pi electrons from a Lewis base to form a covalent bond. Lewis acids are electron deficient, namely, an atom in a Lewis acid may have a positive charge, or it may have a *partial positive charge* because it is bonded to one or more electronegative atoms. An atom with an incomplete octet of electrons also acts as a Lewis acid, readily accepting a pair of electrons from a Lewis base. Recall that group IIIA elements are *trivalent*, namely, form compounds with three covalent bonds and, consequently, they lack an octet of electrons.

Example 1

$$BH_3 \quad + \quad NH_3 \longrightarrow H_3\overset{-}{B}\text{-}\overset{+}{N}H_3$$

or

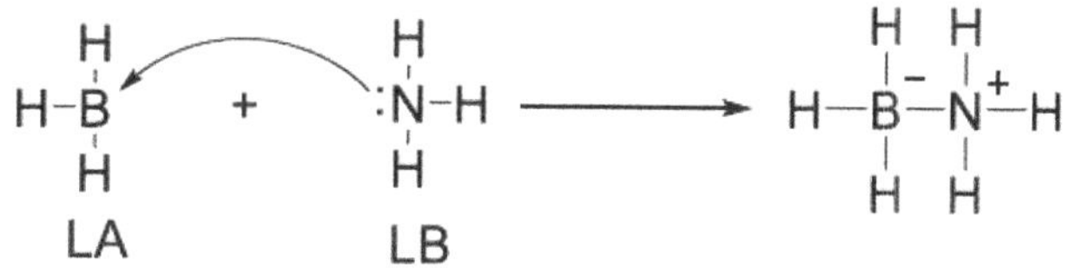

LA
LB

Note that the overall process involved in a Lewis acid/Lewis base reaction entails

(a) writing the Lewis structures of the two reactants and identifying the reactant which functions as a Lewis acid and the reactant that functions as a Lewis base;

(b) using a curved arrow to show how the non-bonded or pi electron pair of electrons in the Lewis base is used to form a covalent bond with the electron-deficient atom in the Lewis acid and

(c) determining the formal charges on the two atoms involved in the formation of the new covalent bond.

Example 2

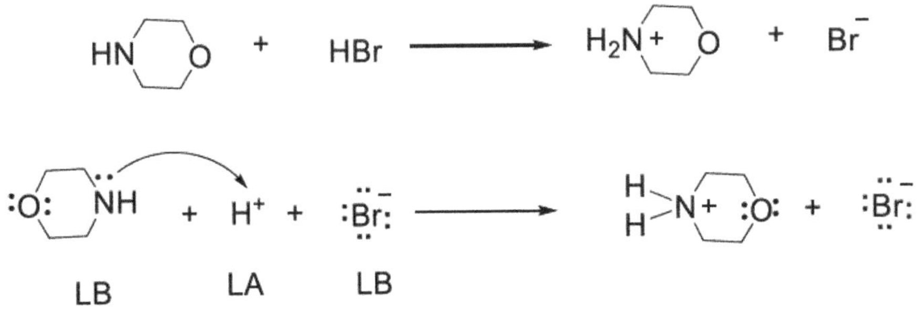
LB
LA
LB

Notes

(a) As mentioned earlier, strong mineral acids (HCl, HBr, HNO_3, etc.) and strong organic acids (RSO_3H, RCOOH, etc.) are *ionized* in solution, therefore it's best to represent them as ions (*for the sake of simplicity, H^+ instead of H_3O^+ is used throughout the book*). When written that way, it can be readily ascertained which is the Lewis acid and which is the Lewis base. Strong bases and ionic compounds should be treated the same way.

(b) Recall that a Lewis base will only react with a Lewis acid, but *not* with another Lewis base.

Example 3

$$CH_3NH_2 + HCl \longrightarrow CH_3NH_3^+ + Cl^-$$

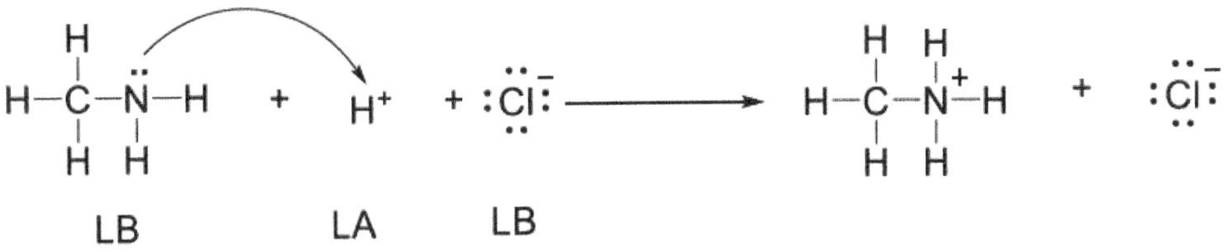
LB
LA
LB

Example 4

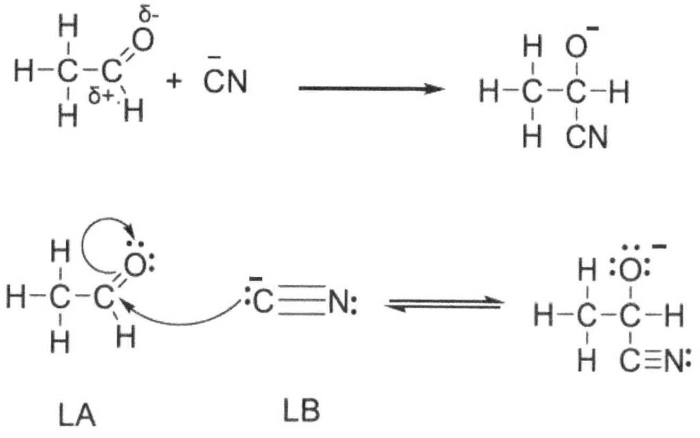
LA
LB

Notice that in this example the carbonyl carbon is electron-deficient because it is bonded to an electronegative atom and, consequently, has a partial positive charge; Thus, the molecule can behave as a Lewis acid and accept a pair of electrons from the Lewis base. In general, an atom that is bonded to one or more electronegative atoms will have a partial positive charge and can, in principle, function as a Lewis acid by accepting a pair of electrons from a Lewis base.

Minireview 3

Resonance

Resonance involves the alternate placement of non-bonded or pi electrons over the same atomic skeleton, without any change in the position of the atoms.

A general familiarity with the concept of resonance and a facility in writing resonance structures is very helpful in

a) assessing the *stability* of an individual species (anion, cation, or radical)

b) assessing the *relative stability* of two similar species, thereby enabling one to predict the pathway that a reaction will likely to follow and/or which species is likely to form (since the greater the stability of a species is, the easier it is to form) and,

c) predicting the *site* of reaction in a molecule.

When a molecule or ion can be represented by two or more Lewis structures that differ only in the position of the electrons,

a) none of those Lewis structures represents the *actual* structure of the molecule or ion and,

b) the actual structure of the molecule or ion is best represented by a hybrid (called *resonance hybrid*) of these resonance structures.

In writing resonance structures, the following *general rules* should be followed:

a) the greater the number of resonance structures that can be written for a species, the more stable the species is;

b) resonance structures in which all the atoms have an octet of electrons are more stable;

c) other things being equal, a resonance structure with a negative charge on the most electronegative atom will have greater stability. Conversely, a resonance structure with a positive charge on the least electronegative atom will be more stable;

d) maximum stabilization of a species (anion or cation) is achieved when the contributing resonance structures are *equivalent*, namely, they have the same energy.

Examples illustrating these rules are given below.

Resonance structures involving anions

Example 1

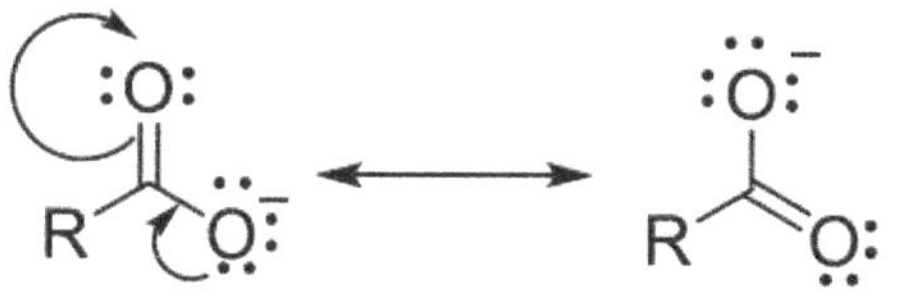

The two resonance structures that can be written for the carboxyl anion ($RCOO^-$) are shown here *(in writing resonance structures for anions, the movement of the electrons is initiated from the atom bearing the negative charge toward the pi bond and the other electronegative atom)*.

Note that these structures are *equivalent*, namely, they have the same energy (they both have a negative charge on the same kind of atom) [Rule 4]. Notice also that the delocalization of the negative charge over the two oxygen atoms via resonance stabilizes the anion and is the reason for the acidity of carboxylic acids. In other words, the hydrogen ion in a carboxylic acid (RCOOH) is acidic and readily donated to a base because a resonance-stabilized anion is formed in the process (see also Minireview 5 for a full discussion of the relationship between acid strength and resonance stabilization of the conjugate base).

Example 2

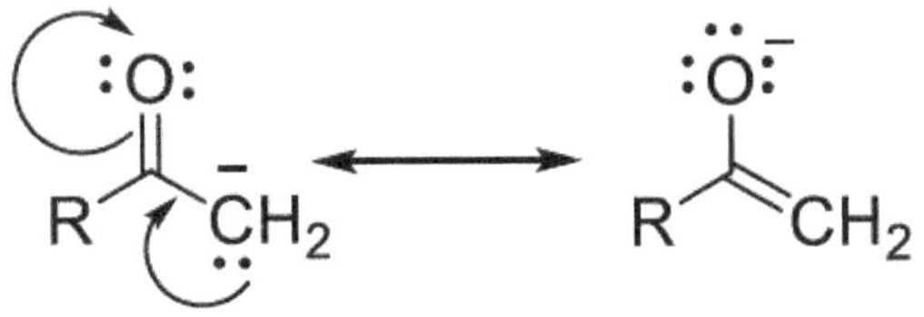

The resonance structures shown here are *non-equivalent*, since the negative charge is on two different atoms (carbon and oxygen). The resonance structure with the negative charge on the more electronegative atom is more stable [Rule 3].

Recall that electronegativity follows the order F > O > N, Br > Cl > C, H.

Example 3

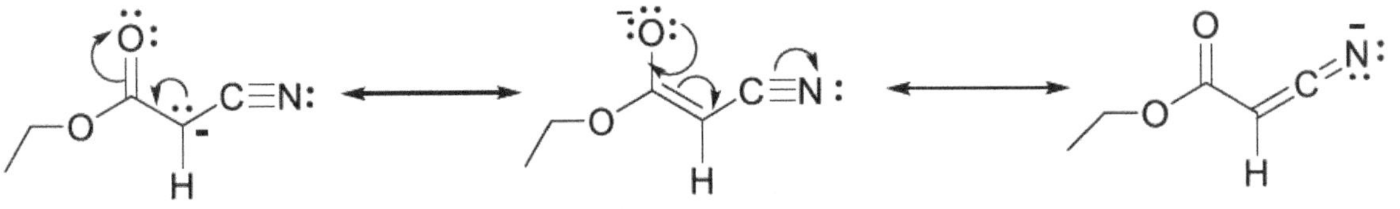

(I) (II) (III)

The order of stability in resonance structures (I) to (III) parallels the order of electronegativity of the three atoms bearing the negative charge [Rule 3]. Thus,

(II) > (III) > (I)

⟵

Increasing stability

O > N > C

⟵

Increasing electronegativity

As stated earlier, the greater the number of resonance structures that can be written for an anion, the greater the stability of the anion [Rule 1]. Put differently, *the greater the delocalization of the negative charge, the greater the stability of an anion.* Consequently, in comparing the relative stability of two anions, the more stable anion will be the one for which a greater number of resonance structures can be written.

Corollary

Resonance-stabilized anions have higher stability and, therefore, are easier to form. Thus, the relative acidity of a acid is determined by the stability of the corresponding anion.

Example 4

When ethyl acetoacetate is treated with base it readily forms the resonance-stabilized anion shown below:

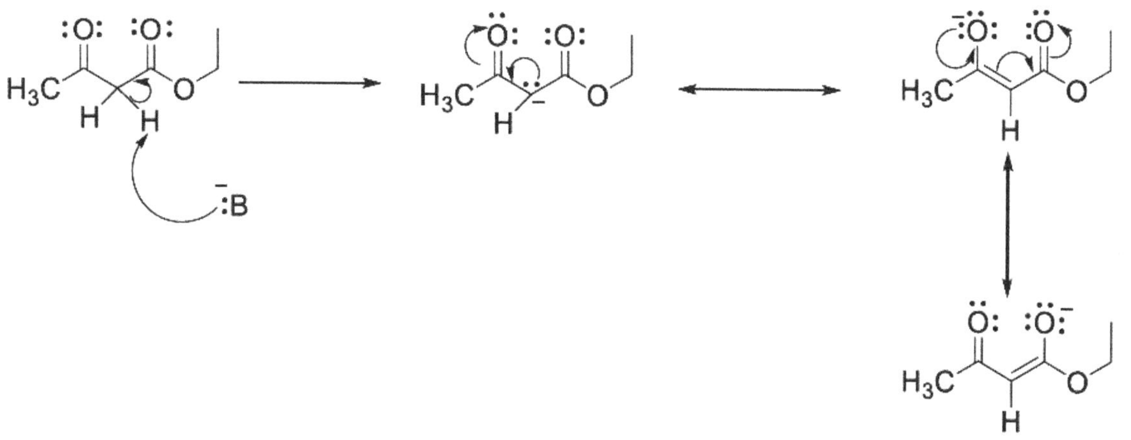
H3C
O
H
H
:B

Exercise

a) Rank the resonance structures in example 4 in order of decreasing stability (most stable first).

b) The three methyl hydrogens in ethyl acetoacetate are less acidic than the two methylene hydrogens. Why?

Example 5 Assess the relative stability of anions A and B below:

A

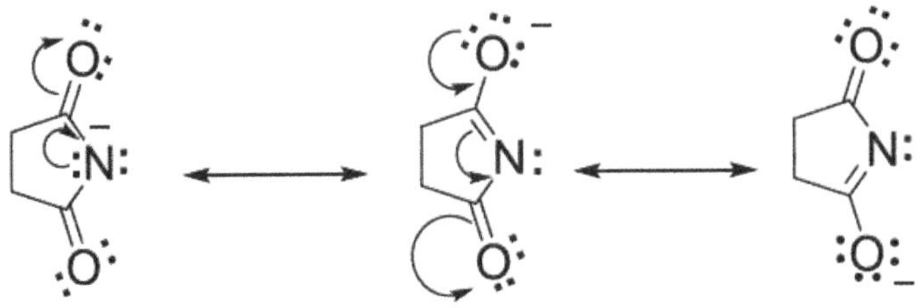

B

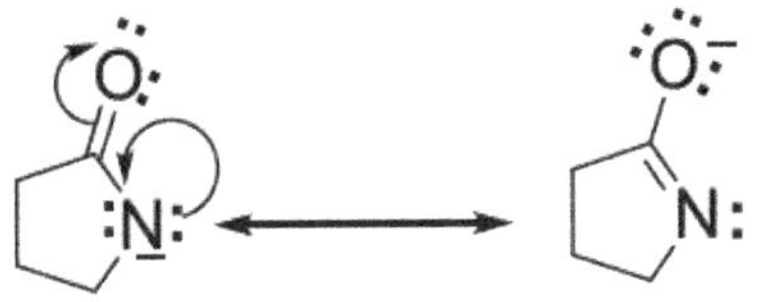

Three resonance structures can be written for anion A versus two resonance structures for anion B. Hence, anion A is more stable than anion B [Rule 1].

Exercise Write all the major resonance structures for the following anions.

a)

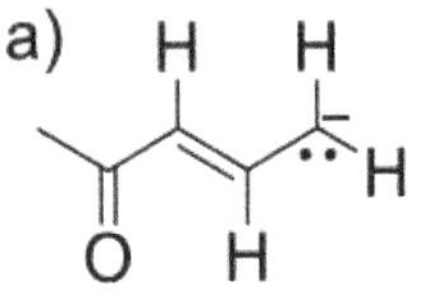

b)

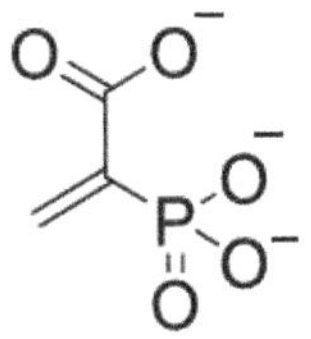

Resonance structures involving cations

Allylic and benzylic carbocations, as well as carbocations having a heteroatom directly bonded to the carbon bearing the positive charge, are stabilized by resonance. When writing resonance structures for cations, *the movement of the non-bonded or pi electrons is toward the positively charged atom* (electrons are negatively-charged and are therefore attracted to the positive charge).

Example 1

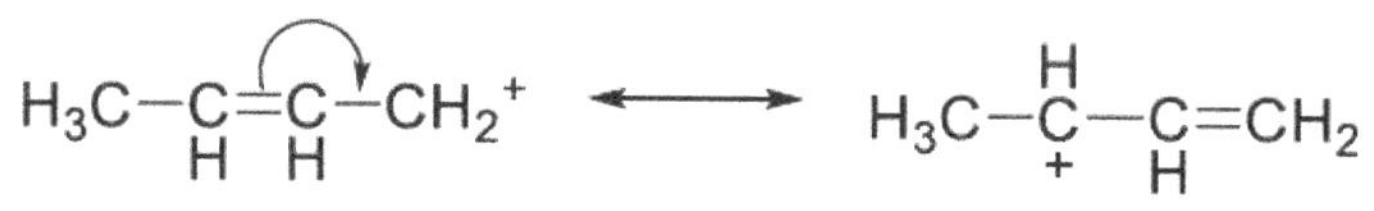

Example 2

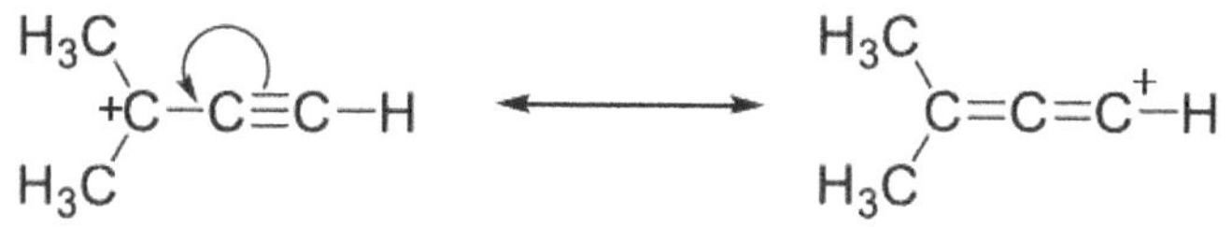

Example 3

Example 4

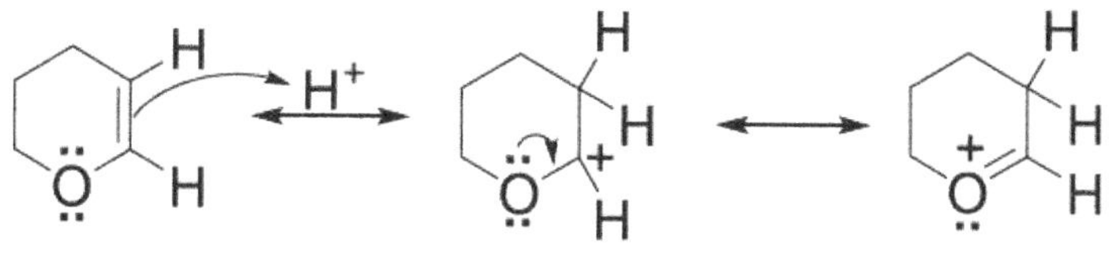

Example 5

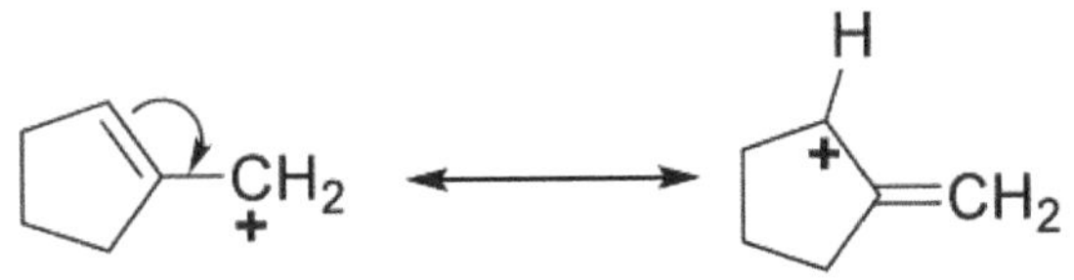

Example 6

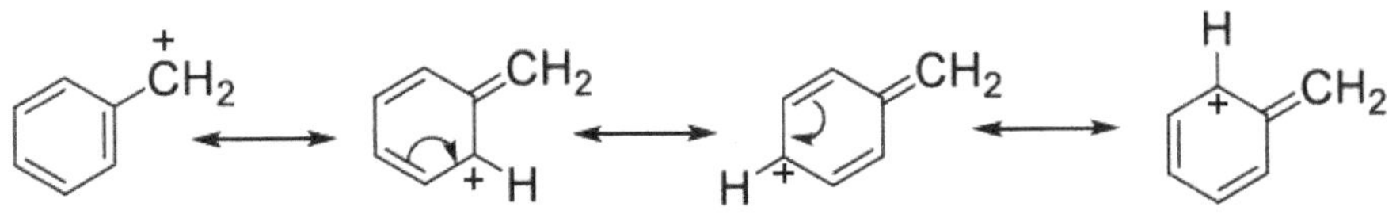

In this example, the first resonance structure is more stable than the others because the ring is aromatic, which is not the case with the other three structures.

Example 7

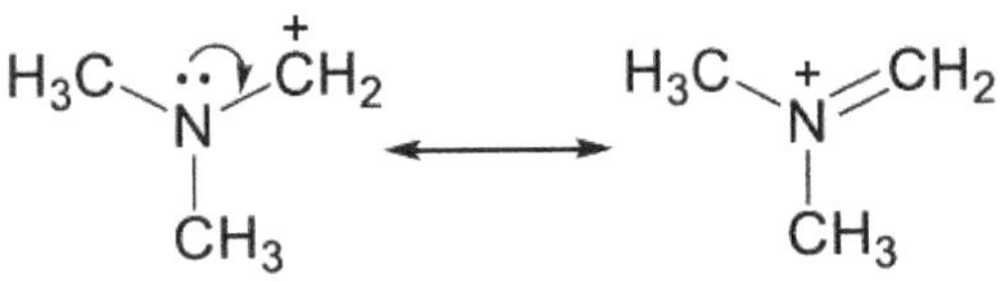

Example 8

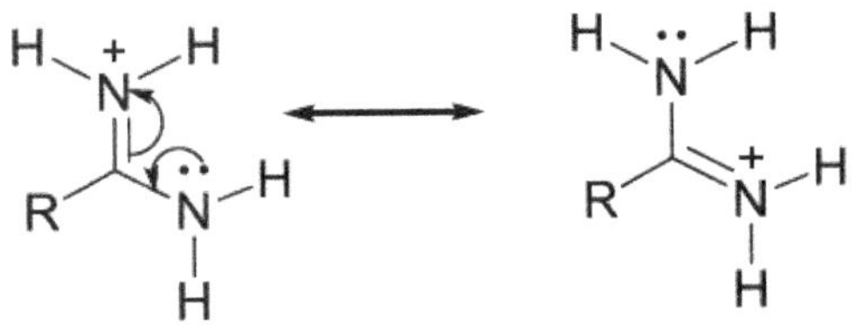

Resonance structures involving neutral molecules

Example 1

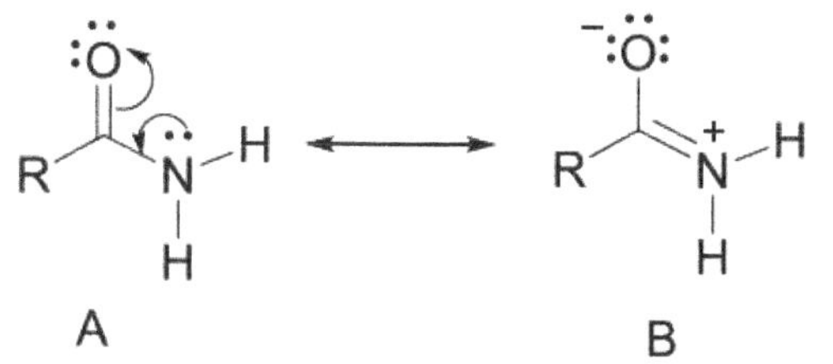

Structure A is more stable than structure B, however, structure B is a significant contributor that accounts for a) the observed *restricted rotation* around the C-N bond in amides and peptides and, b) the *decreased Lewis basicity* of amides verses amines (arising from the lower availability of the non-bonded electron pair on the nitrogen in amides).

In thiourea and thioamides, SN2 reactions occur at sulfur because of the contribution of the resonance structures shown below and the high nucleophilicity of sulfur (see Minireview 6 for a discussion of nucleophilicity).

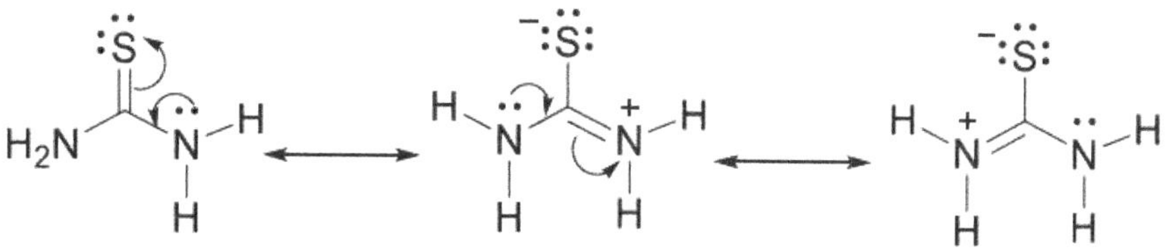

Resonance structures can be used to predict the *site* of reaction in neutral molecules. For example, the course of the reaction of α,β-unsaturated compounds with a Lewis base (*Michael addition reaction*) can be predicted by considering the resonance structure shown below. The resonance structure clearly indicates that the β carbon is electron-deficient (Lewis acid) and can accept a pair of electrons from a nucleophile (Lewis base).

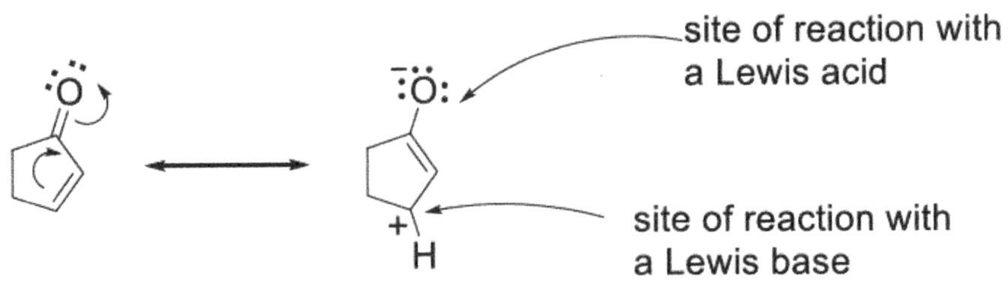

site of reaction with
a Lewis acid
site of reaction with
a Lewis base

Likewise, the following resonance structures identify the sites of reaction of vinyl ethers and enamines with Lewis acids and provide a better understanding of the chemical behavior of these classes of compounds.

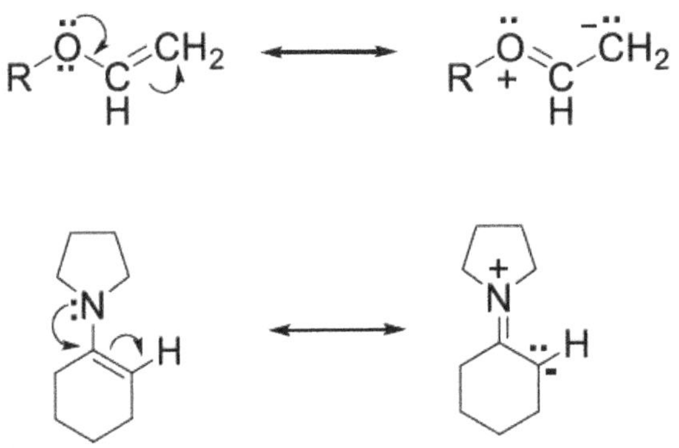

Minireview 4

Carbocation Chemistry

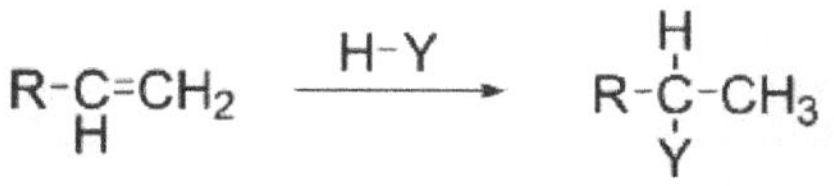

The reaction of alkenes, alkynes, or alcohols (all Lewis bases) with H^+ (a strong Lewis acid) leads to the initial formation of a carbocation. *Carbocations are transient, electron-deficient, and highly-reactive species.* Once formed, they function as Lewis acids that react rapidly with Lewis bases (in the process the carbon atom bearing the positive charge completes its octet).

Some noteworthy characteristics of carbocations are the following:

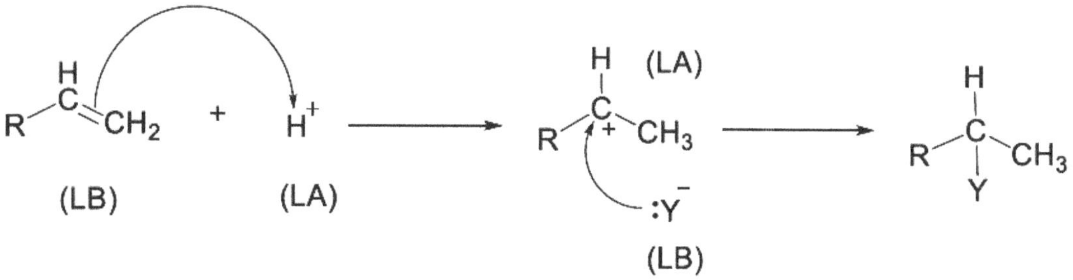

1) Carbocations vary in stability depending on their structure. *The greater the stability of a carbocation is, the easier it is to form.* Thus, the initial carbocation formed in a given reaction is invariably the one with the highest stability. The *order of stability of* carbocations is as follows:

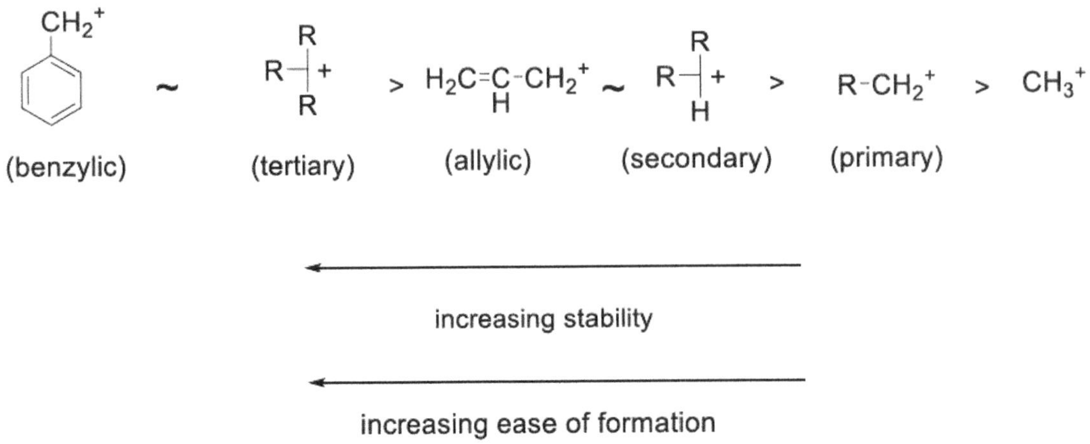
CH2+
R
R
R
H2C=C-CH2+
H
R
R
H
R-CH2+
CH3+
(benzylic)
(tertiary)
(allylic)
(secondary)
(primary)
increasing stability
increasing ease of formation

Methyl, vinyl and phenyl carbocations are highly unstable. Consequently, although the hydroxyl group of a phenol can, in principle, function as a Lewis base (just like the hydroxyl group of an alcohol), it never yields a phenyl carbocation.

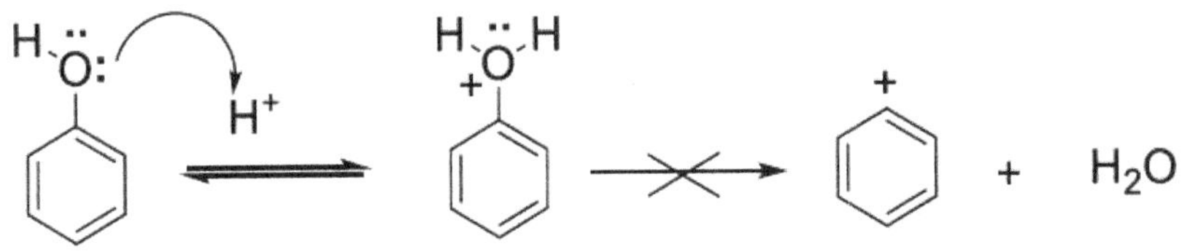
H
O
H+
H
H
O
H2O

As mentioned earlier, carbocations that have a positive charge on a carbon atom which is bonded to a heteroatom (O, N, S) are stabilized by resonance.

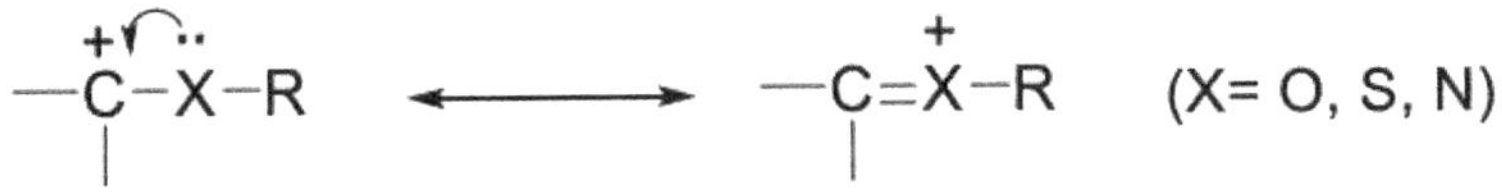
C-X-R
C=X-R
(X= O, S, N)

2) *Carbocations frequently undergo rearrangements via 1,2-hydride or 1,2-alkyl shifts to form carbocations of equal or greater stability*. These carbocations may on occasion arise via consecutive 1,2-hydride and/or 1,2-alkyl shifts. Further reaction with a Lewis base, or loss of a H^+ from an adjacent carbon (E_1), ultimately yields the observed product(s).

In certain cases, initial formation of a carbocation is followed by *ring expansion* to form a new carbocation. The driving force behind the observed ring expansion is the relief of ring strain and the formation of a more stable species. *Recall that as ring size decreases, ring strain increases*. Thus, 3- and 4-membered rings have considerable ring strain and tend to undergo ring expansion. *Recall also that as ring strain increases, chemical reactivity increases.*

An interesting variation of this theme involves the cyclopropylmethyl carbocation.

(relief of ring strain via ring expansion)

(homoallylic carbocation)

(ring expansion)

3) *Carbocations are planar (flat), since the carbon bearing the positive charge is sp*^{2} *hybridized*. Consequently, *a carbocation derived from an optically active reactant will ordinarily yield an optically inactive product (a racemic mixture) upon reaction with a Lewis base*. Since attack on an sp^2 hybridized carbon by a Lewis base is equally likely from either side, this leads to a 1:1 mixture of the R and S isomers.

<u>Example</u>

When (S) 2-butanol is treated with a trace amount of acid, it undergoes racemization. Write a plausible mechanism that accounts for this observation.

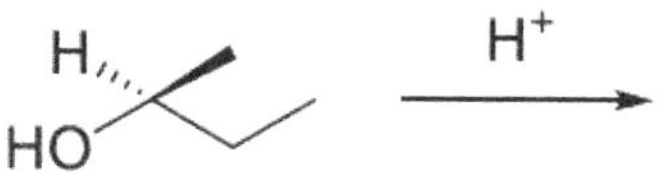
H,,,
HO
H+

(R)

(RS) 2-butanol

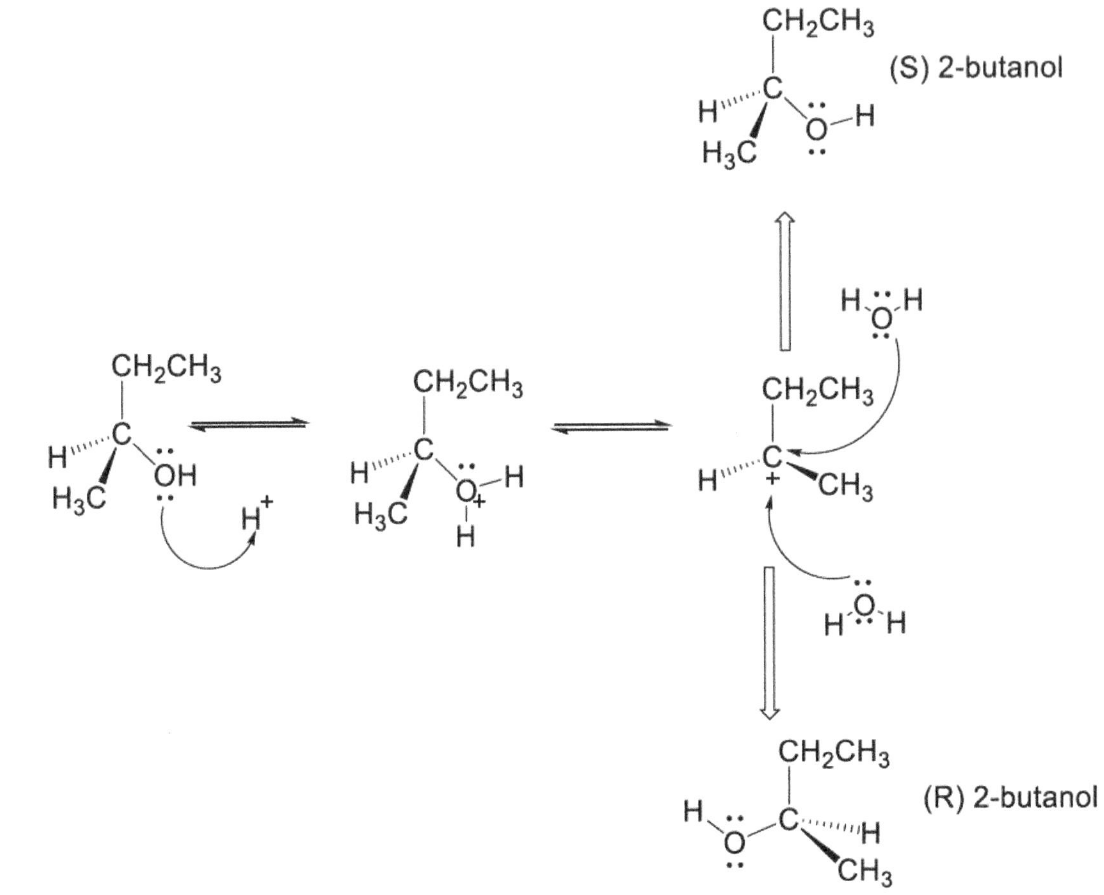
(S) 2-butanol
(R) 2-butanol

4) An *electrophilic aromatic substitution* reaction can be viewed as a two-step Lewis base/Lewis acid reaction involving an aromatic compound (acting as a Lewis base) and a transiently-generated Lewis acid (also called an *electrophile*). Recall that the first step in an electrophilic aromatic substitution reaction is the *rate-determining* step (has the highest free energy of activation). Typical transient Lewis acids include carbocations and other electron-deficient species such as NO_2^+, R^+, etc.

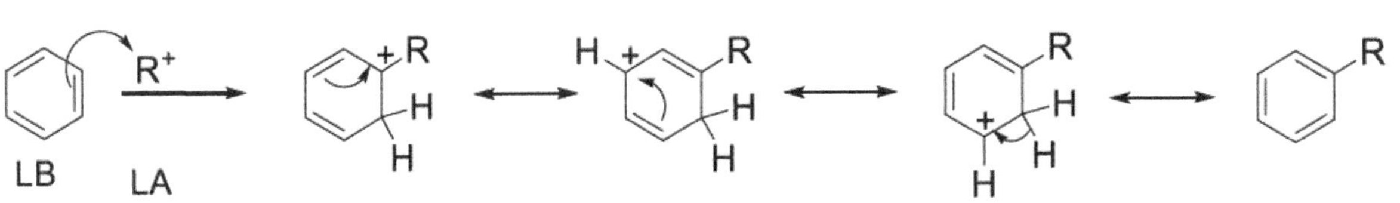
LB
LA

In the familiar *Friedel-Craft alkylation reaction* a carbocation is generated by mixing an alkyl halide (RX) with a Lewis acid (AlX_3), or by mixing an alcohol (ROH) or alkene with acid. As expected, the formation of a carbocation in this reaction may lead to the formation of rearranged products.

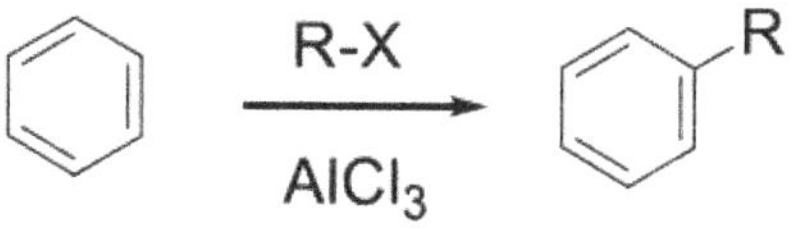
R-X
AlCl3
R

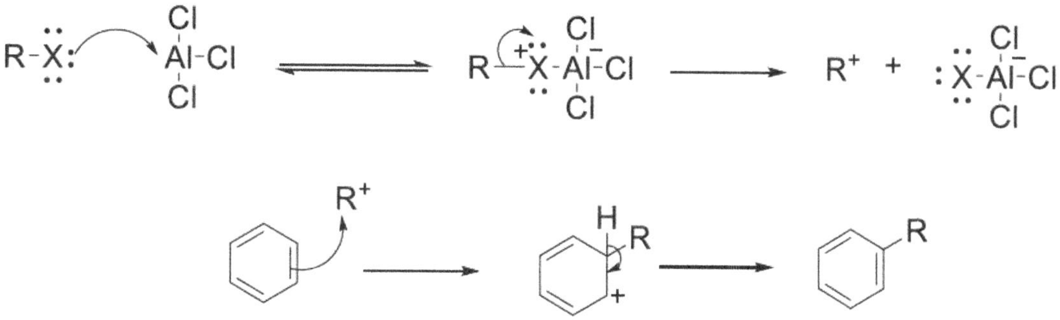
R-X:
Cl
Al-Cl
Cl
R
X-Al-Cl
R+ +
:X-Al-Cl
H
R
+

The reaction of an aromatic compound with an acid chloride (RCOCl) in the presence of a Lewis acid is referred to as the *Friedel-Craft acylation reaction*. An initial Lewis acid/ Lewis base reaction leads to the transient formation of a resonance-stabilized carbocation (also called an *acylium ion*) which then reacts with the aromatic compound (illustrated below). In contrast to ordinary carbocations, *acylium ions do not undergo rearrangements*.

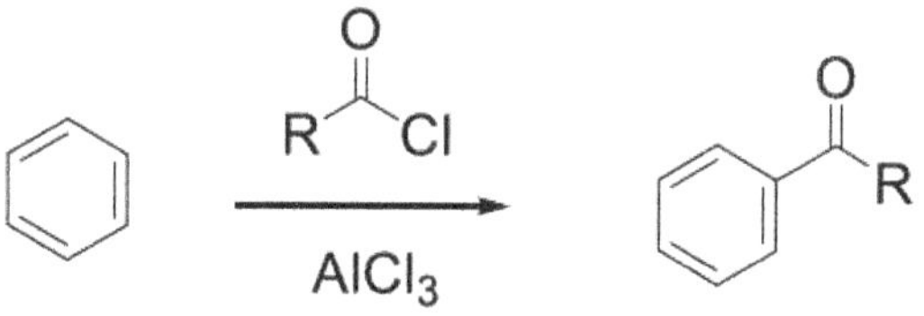
O
R
Cl
AlCl3
O
R

Mechanism

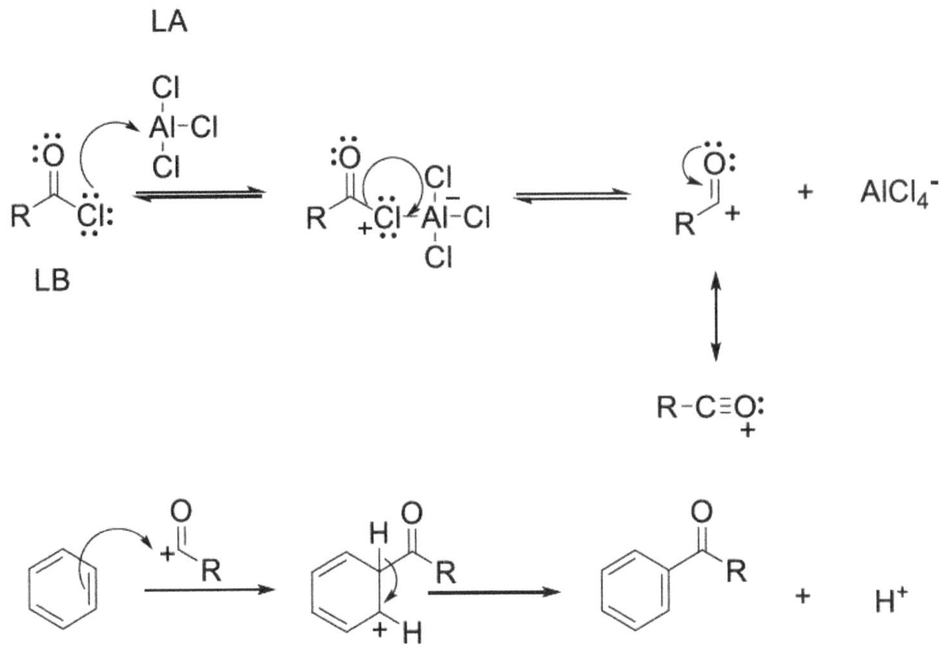
LA
LB
AlCl4-
R-C≡O:
H+

In the nitration and halogenation reactions the Lewis acid (electrophile) is transiently generated via a sequence of Lewis acid/ Lewis base reactions using concentrated HNO_3/H_2SO_4 and Br_2 (or Cl_2) with $FeBr_3$ (or $FeCl_3$), respectively. Recall that the presence and nature of substituent on the aromatic ring has a profound effect on *reactivity* and *orientation*. *Electron-donating groups* (R, OR, NHCOR, OH, NH_2, etc.) enhance Lewis basicity and, hence, the reactivity of an aromatic compound. Furthermore, these groups direct the Lewis acid (electrophile) to the *ortho* and *para* positions. In contrast, *electron-withdrawing groups* (NO_2, CN, CHO, COR, COOR, etc.) decrease reactivity and direct the electrophile to the *meta* position. Halogens are deactivating, but *ortho* and *para*-directing. *Activating groups have one or more pairs of non-bonded elctrons on the attom that is directly bonded to an aromatic carbon.*

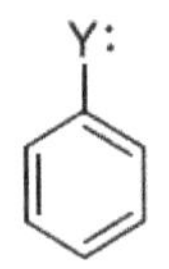
Y:

Exercise 1 Is the nitroso group (-NO) an activating or deactivating group?

Exercise 2 Predict the product of the following reaction

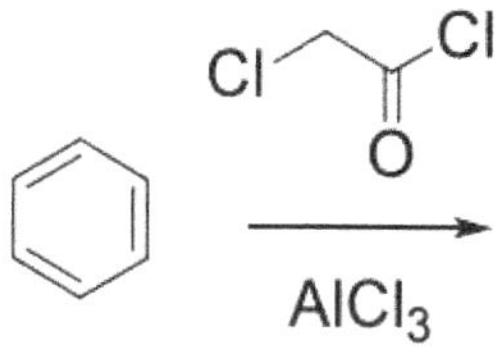
Cl
Cl
O
$AlCl_3$

Questions 1 – 34

Questions 1-34 aim at reviewing and gaining a better understanding of the following topics.

- Lewis structures
- Lewis acids and Lewis bases
- Lewis acid/ Lewis base reactions
- Resonance (rules of resonance, writing resonance structures, assessment of relative stability of anions and cations using resonance)
- Carbocation chemistry (formation, stability, rearrangements, stereochemistry, and reactions of carbocations)
- Lewis acid/ Lewis base reactions of alcohols, alkenes, alkynes, and epoxides
- Lewis acid/ Lewis base reactions involving aromatic rings (electrophilic aromatic substitution reactions, including Friedel-Craft alkylation and acylation reactions)

1

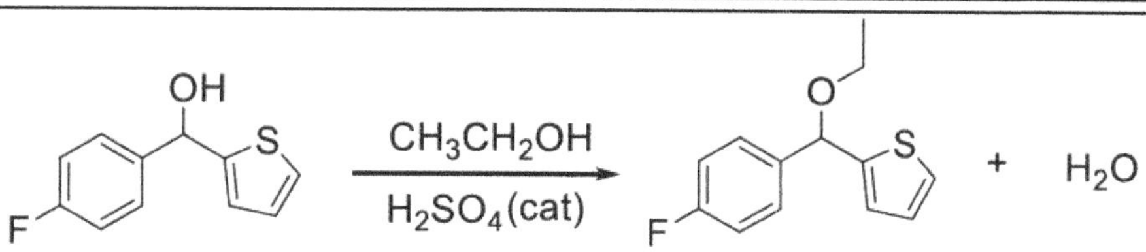
OH
S
F
CH3CH2OH
H2SO4(cat)
O
S
F
+
H2O

2

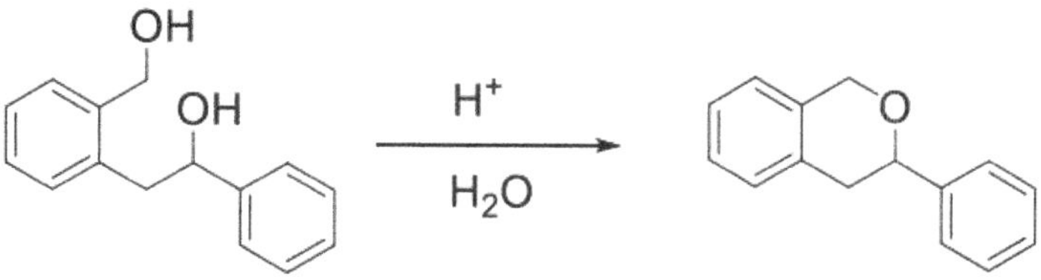
OH
OH
H+
H2O
O

3

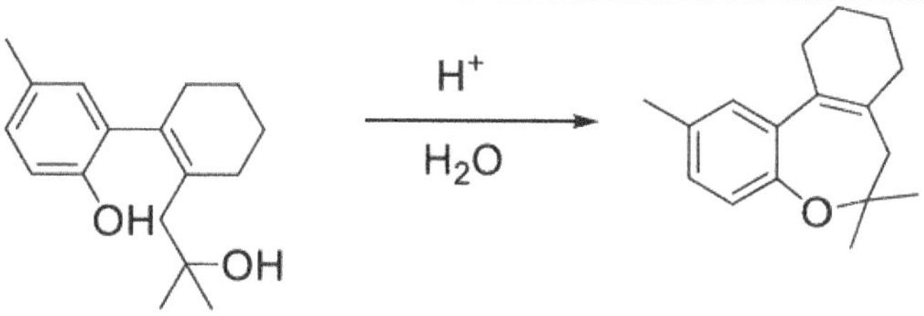
OH
OH
H+
H2O
O

4

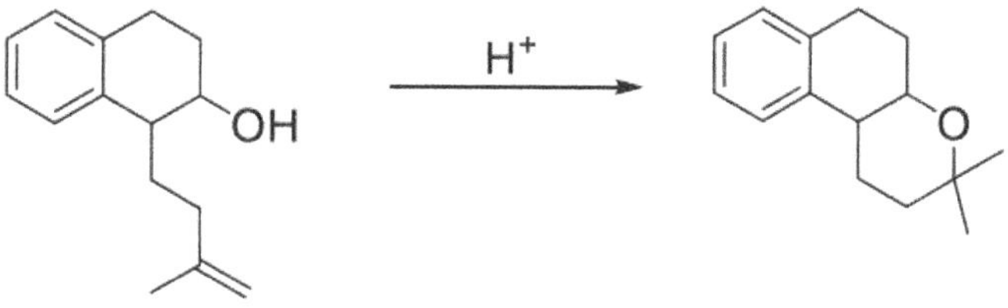
OH
H+
O

5

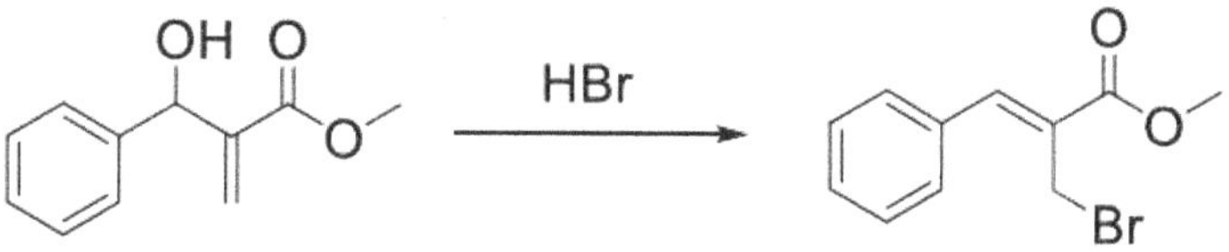
OH
O
O
HBr
O
O
Br

6

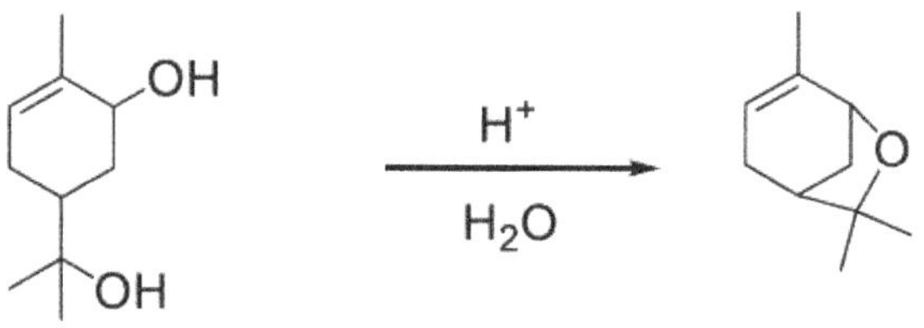
OH
OH
H+
H2O
O

7

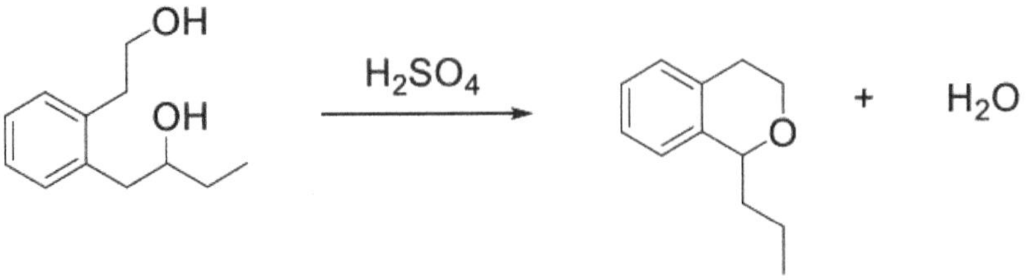
OH
OH
H2SO4
O
+
H2O

8

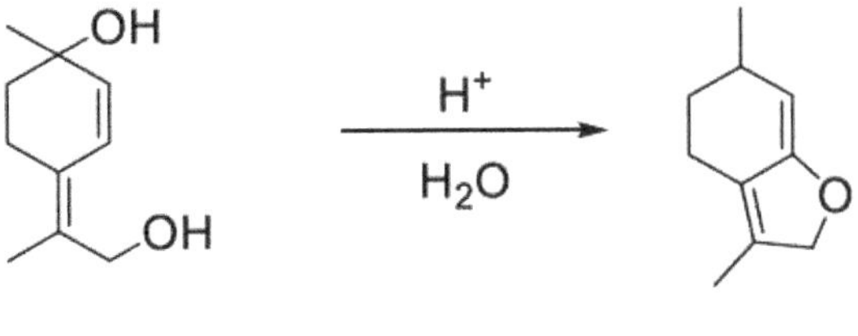
OH
OH
H+
H2O
O

9

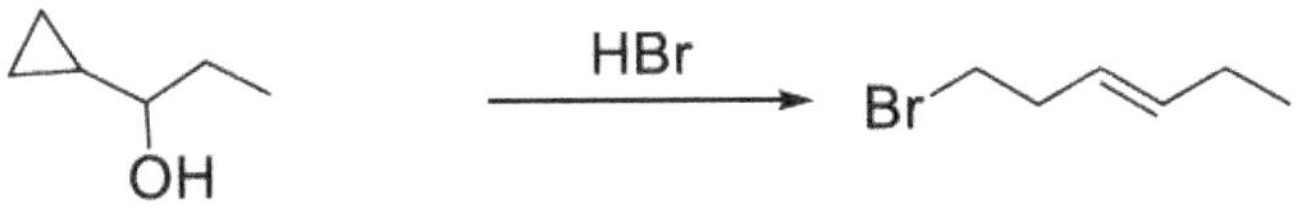
OH
HBr
Br

10

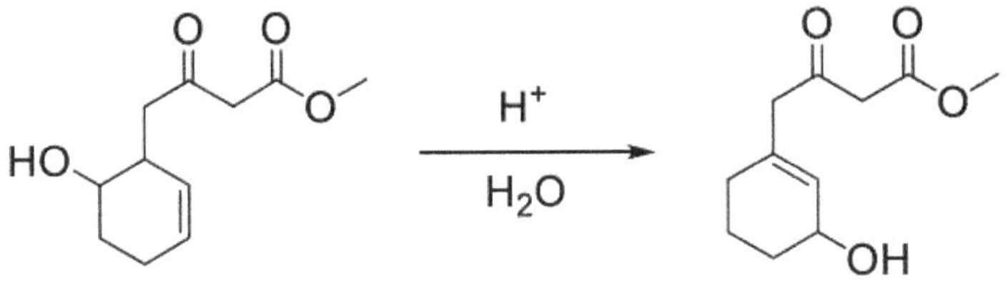
O
O
O
HO
H+
H2O
O
O
O
OH

11

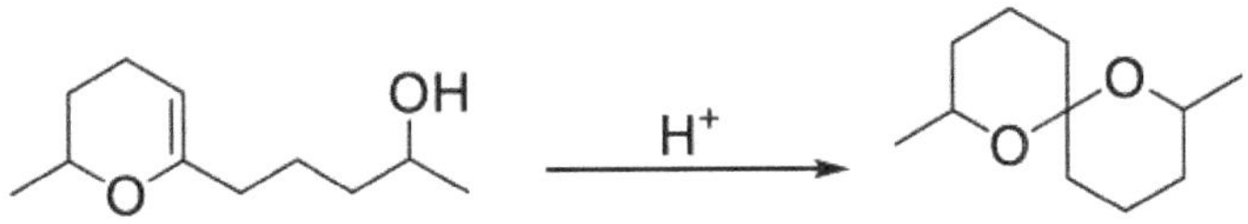
O
OH
H+
O
O

12

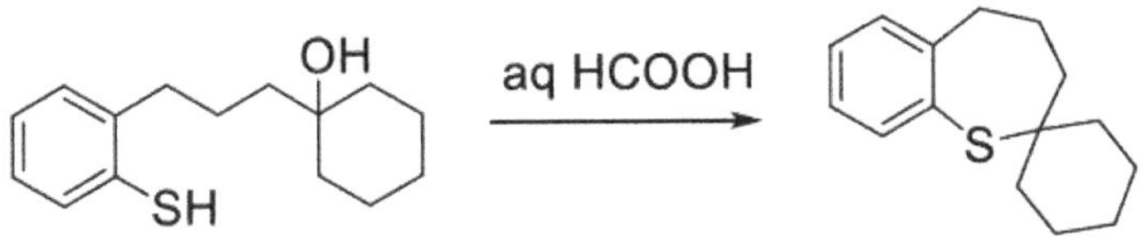
OH
SH
aq HCOOH
S

13	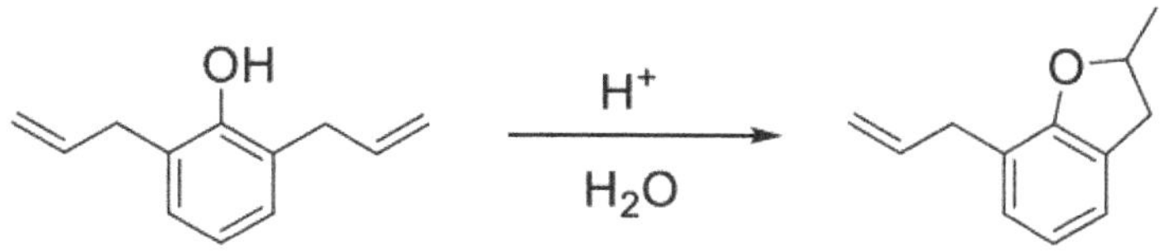 OH H+ H2O O
14	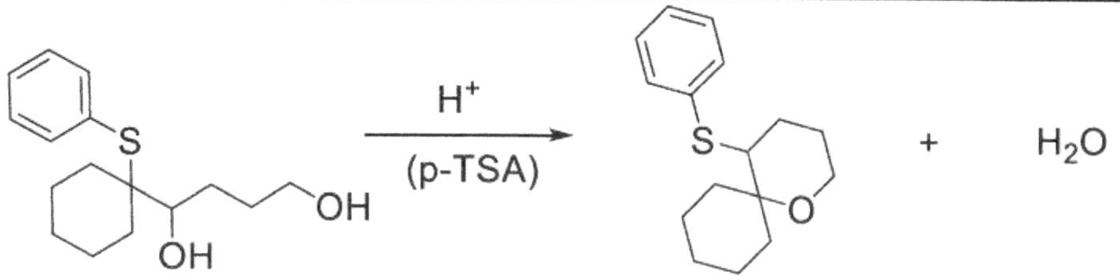 S OH OH H+ (p-TSA) S O + H2O
15	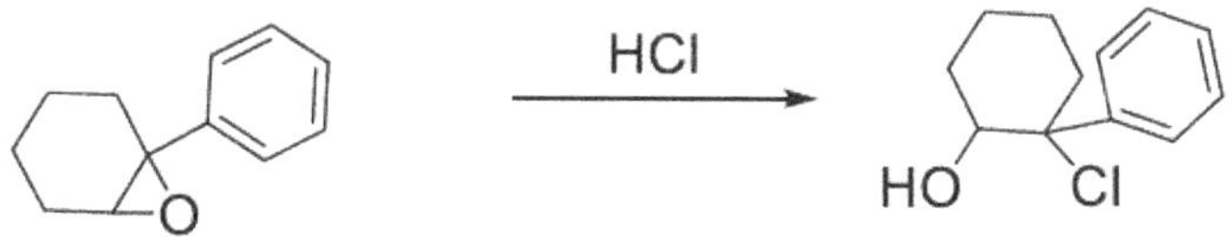 O HCl HO Cl
16	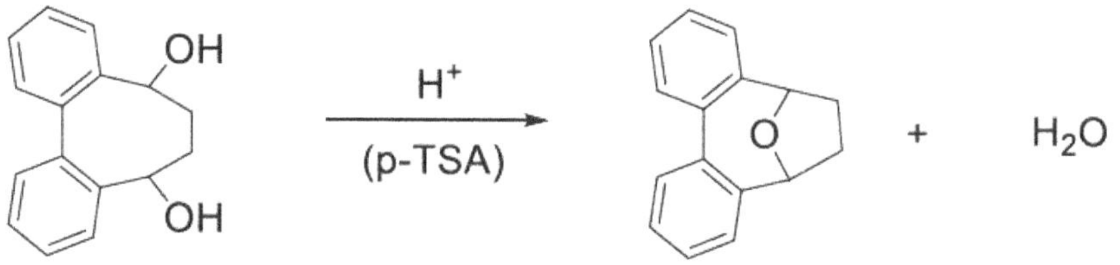 OH OH H+ (p-TSA) O + H2O
17	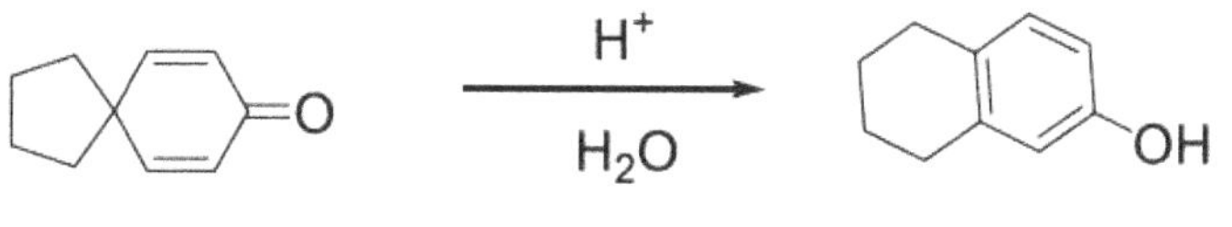 O H+ H2O OH
18	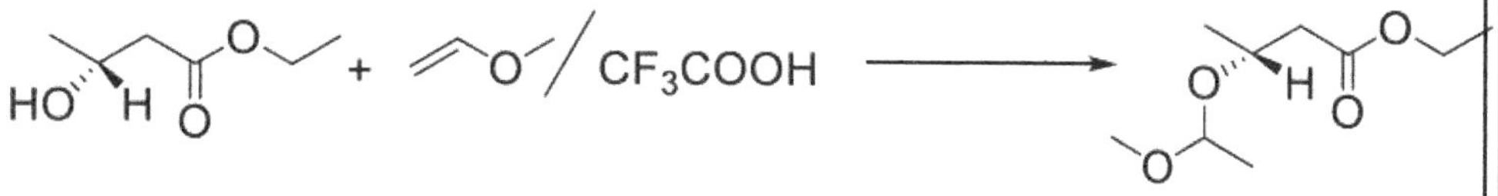 HO H O O + O CF3COOH O H O O O

19	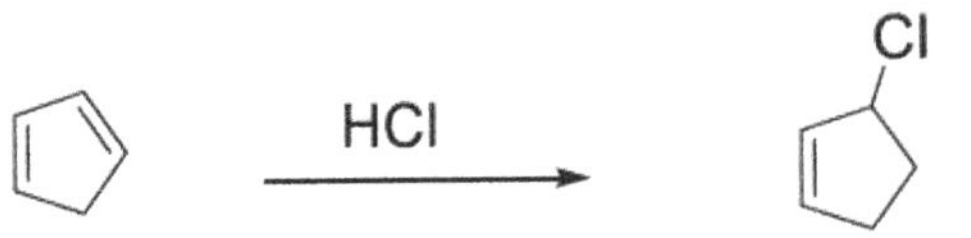 HCl Cl
20	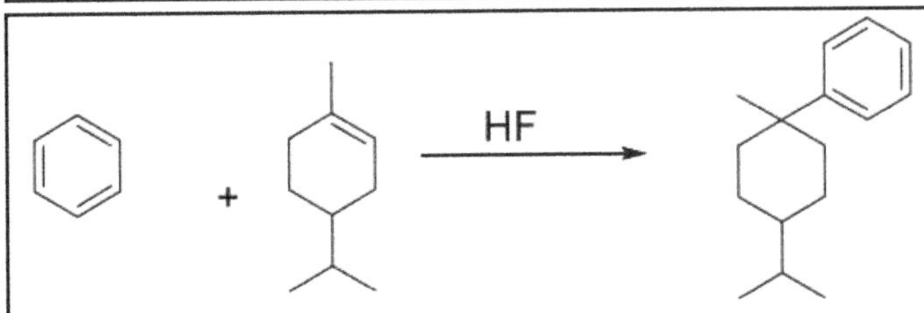 + HF
21	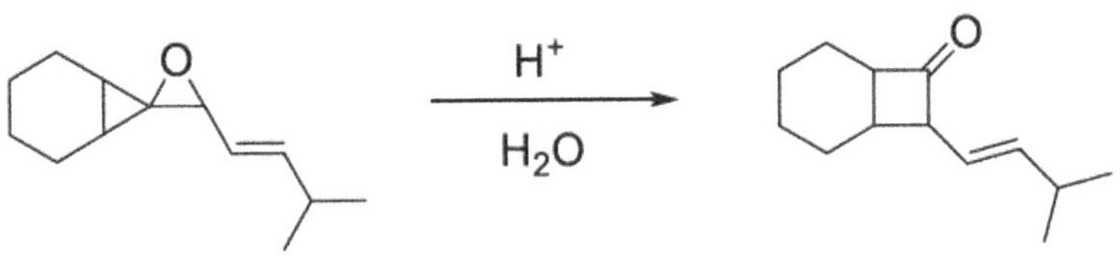 O H+ H2O O
22	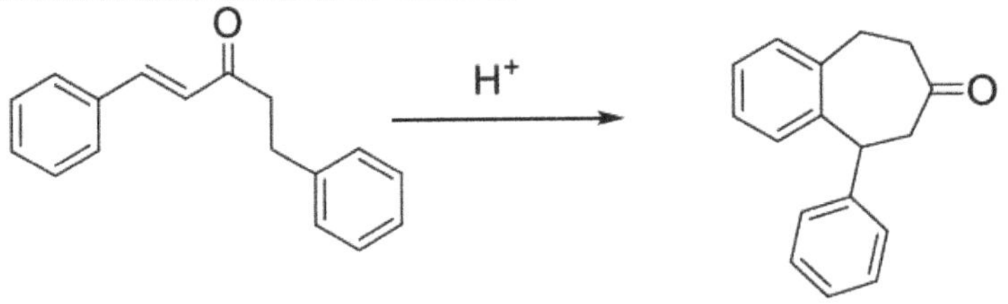 O H+ O
23	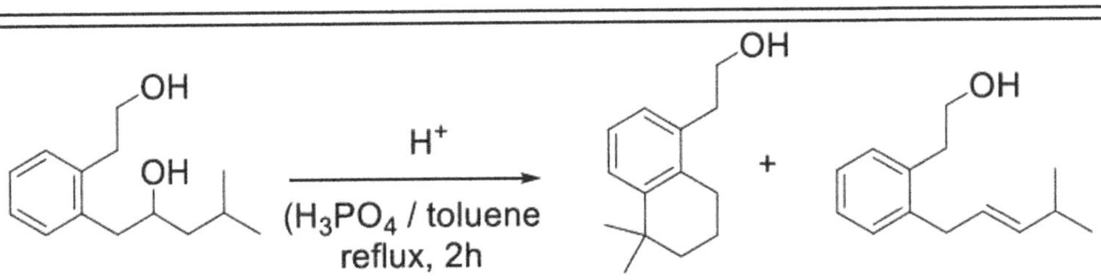 OH OH H+ (H3PO4 / toluene reflux, 2h OH + OH
24	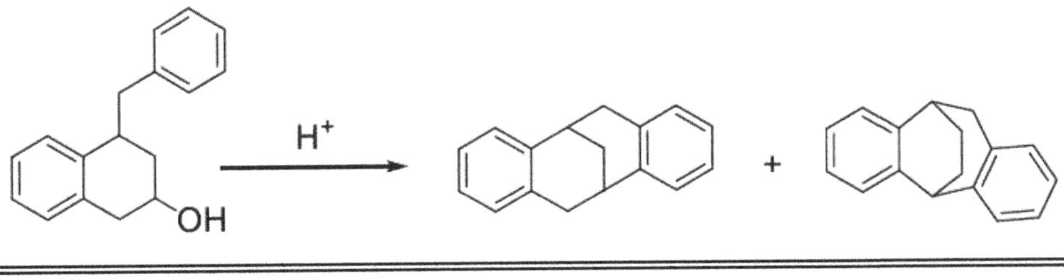 OH H+ +

25	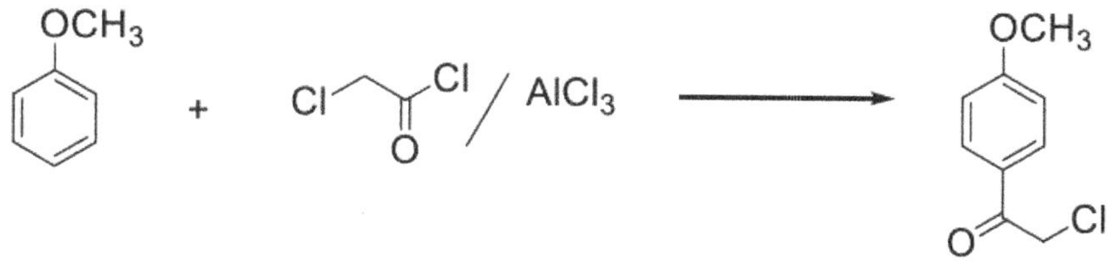OCH3 + Cl Cl O / AlCl3 OCH3 O Cl
26	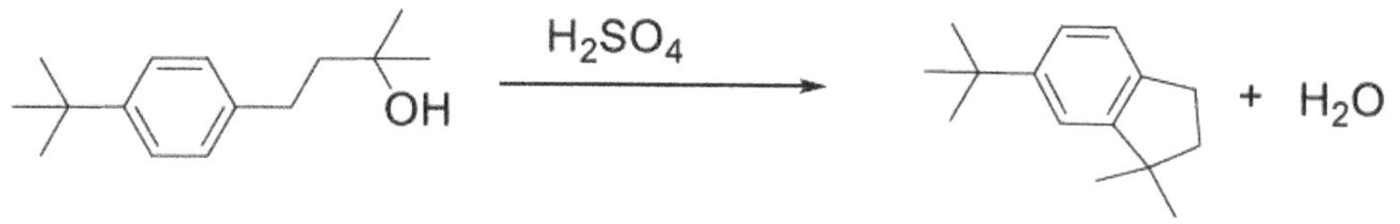OH H2SO4 + H2O
27	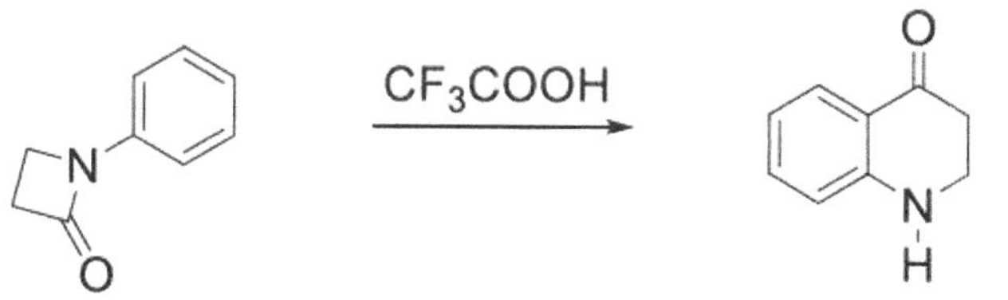N O CF3COOH O N H
28	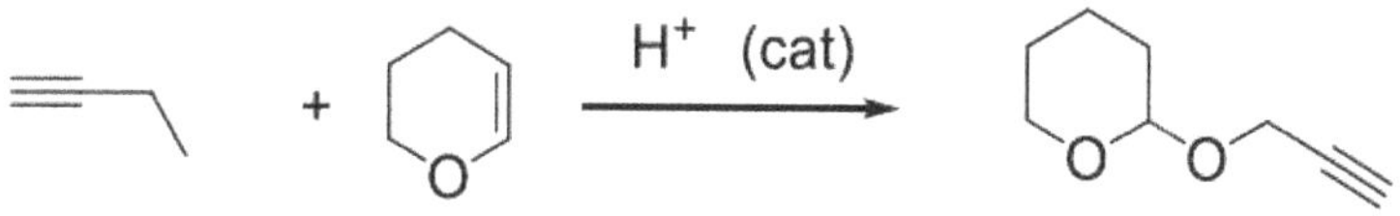+ O H+ (cat) O O
29	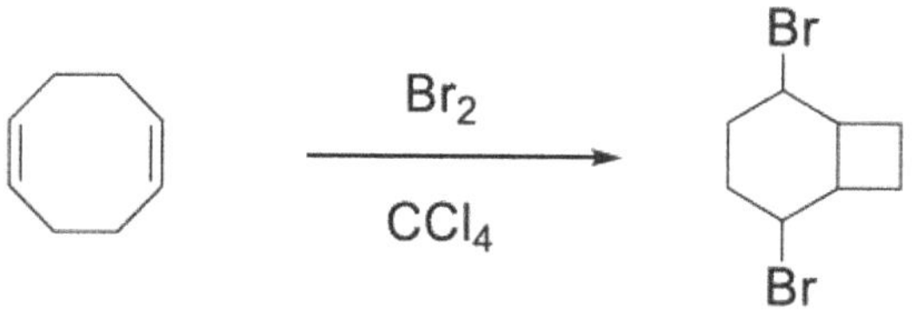Br2 CCl4 Br Br
30	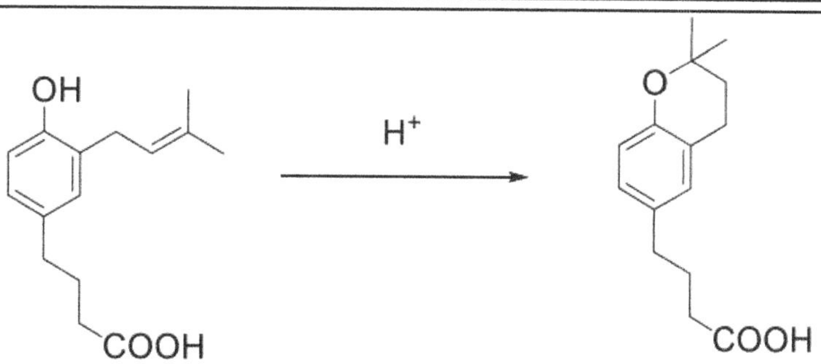OH COOH H+ O COOH

31	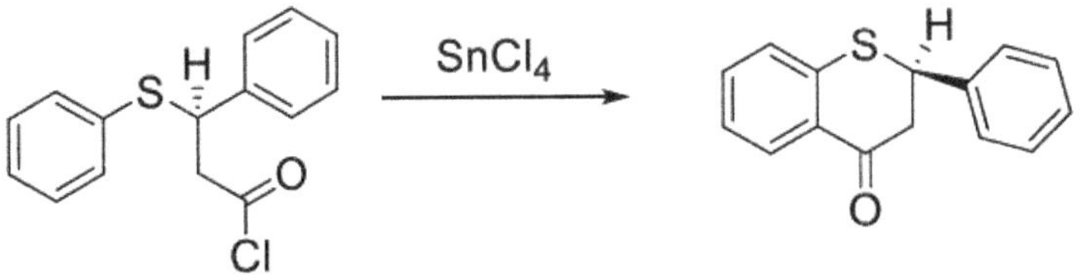 S H O Cl SnCl4 S H O
32	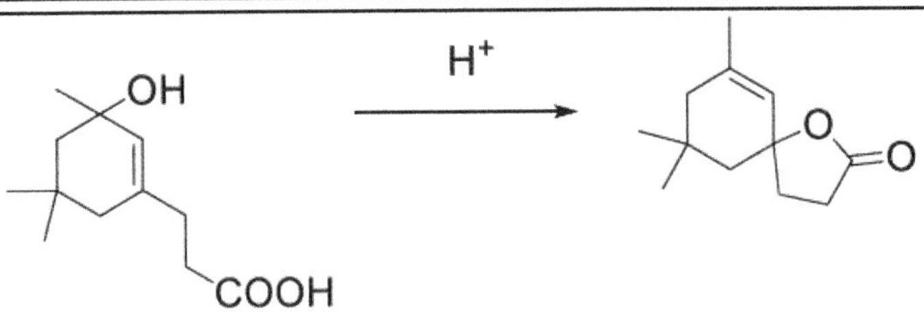 OH COOH H+ O O
33	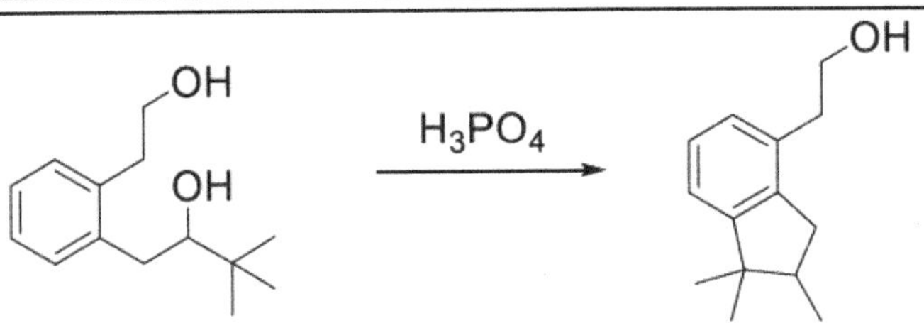 OH OH H3PO4 OH
34	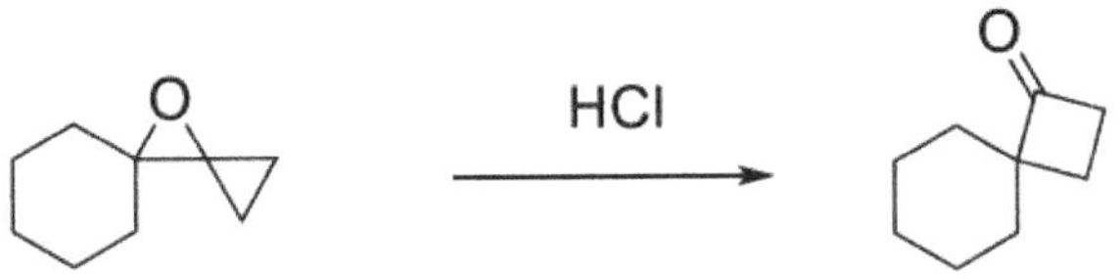 O HCl O

Answers to Questions 1-34

1

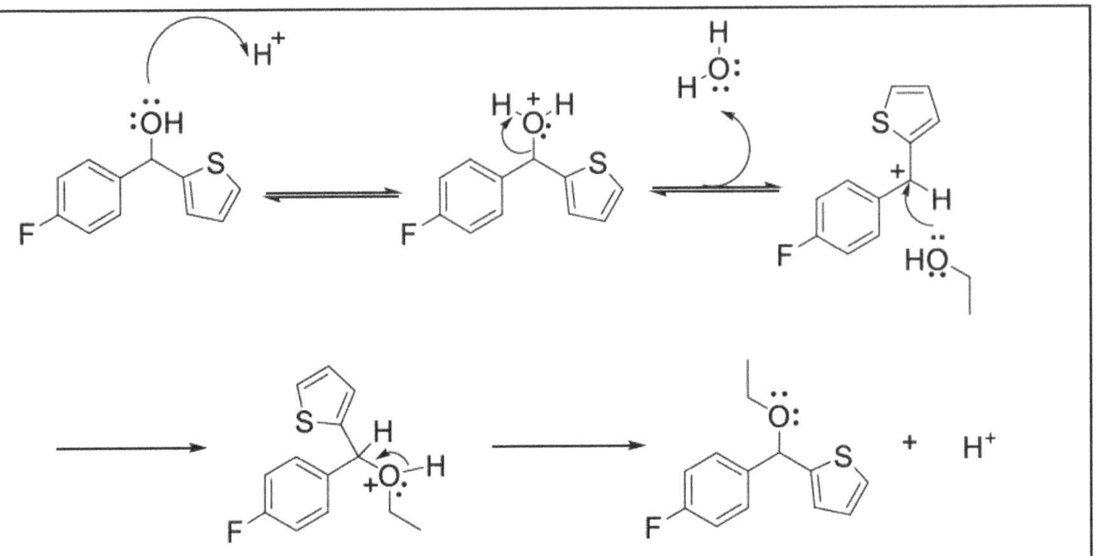

An initial LA/LB reaction leads to the formation of a resonance stabilized benzylic carbocation (a Lewis acid), which then reacts with ethyl alcohol (a Lewis base) to form the product. Recall that (a) non-bonded electrons are more available for donation than bonded electrons (pi electrons); (b) reaction of the LA with either aromatic ring would result in the formation of a non-aromatic species and the loss of resonance stabilization energy.

2

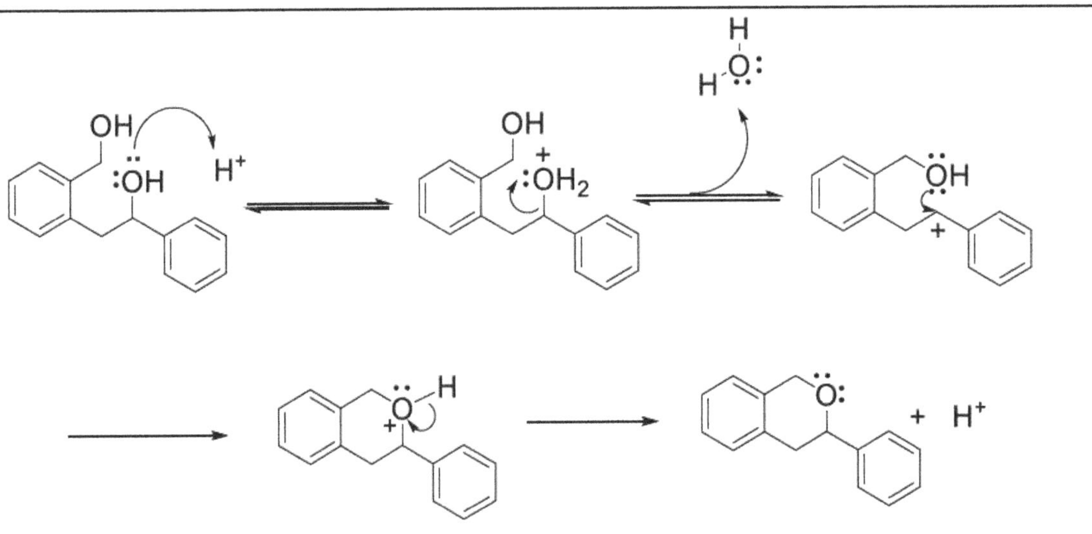

Although either oxygen atom can donate a pair of electrons to the Lewis acid, initial formation of the more stable species (a 2° benzylic carbocation) is preferred. A subsequent intramolecular LA/LB reaction leads to the formation of the product.

3

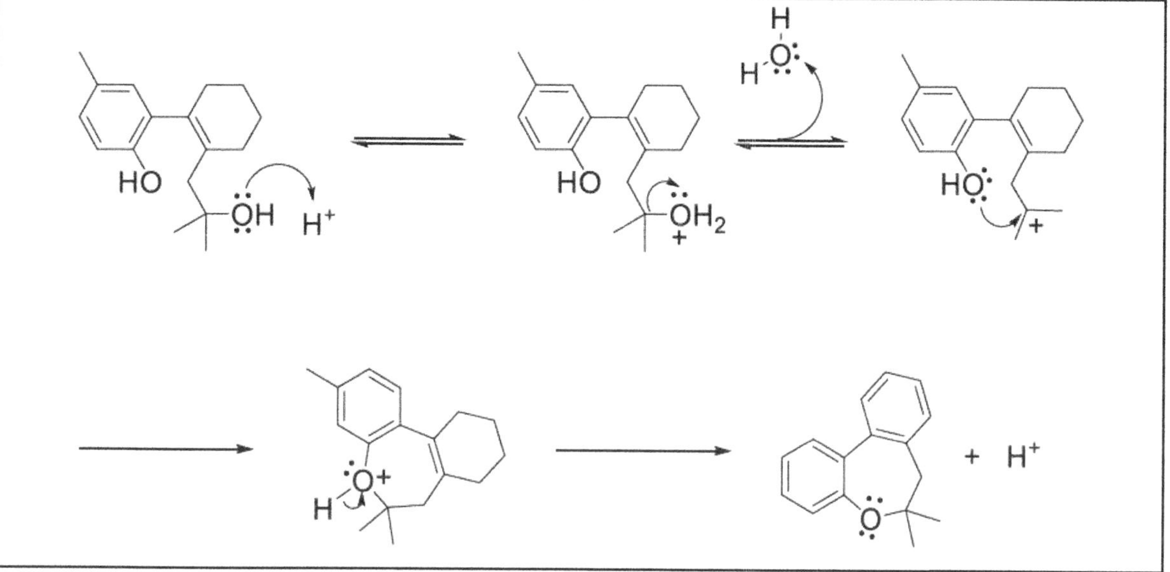
HO
OH
H+
HO
OH2
+
H
H O
HO
+
O+
H
+ H+

Note that although the phenolic-OH can, in principle, react with the Lewis acid, formation of a phenyl carbocation does not take place because of the low stability of this carbocation. Consequently, only the much more stable tertiary carbocation is formed. This is then followed by an intramolecular LA/LB reaction to form the product.

4

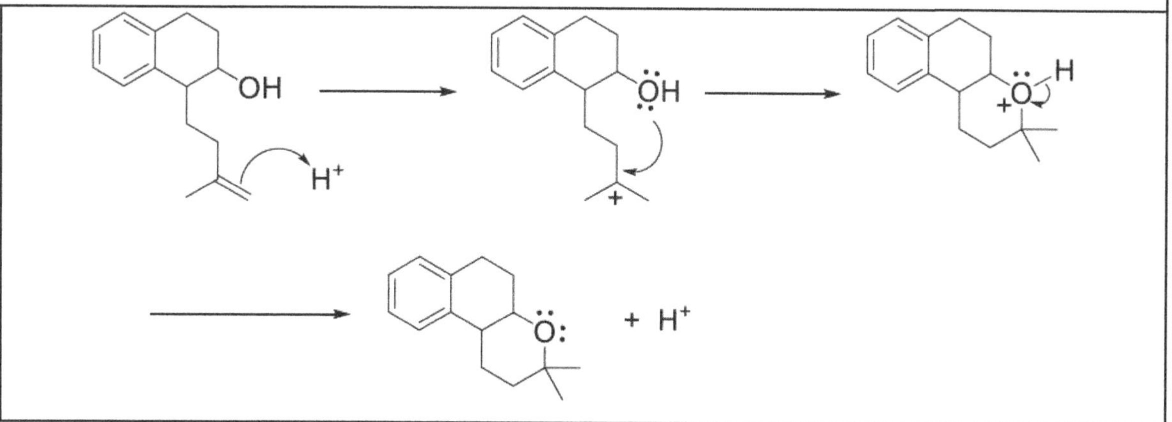
OH
H+
OH
+
O
H
+
O
+ H+

A LA/LB reaction between the pi-bond and the hydrogen ion (LA) yields a tertiary carbocation. This is followed by an intramolecular LA/LB reaction, leading to the formation of the observed product. As stated earlier, non-bonded electrons are more available for donation than pi electrons, however, in this instance donation of a pair of pi electrons results in the formation of a highly stable 3° carbocation.

5

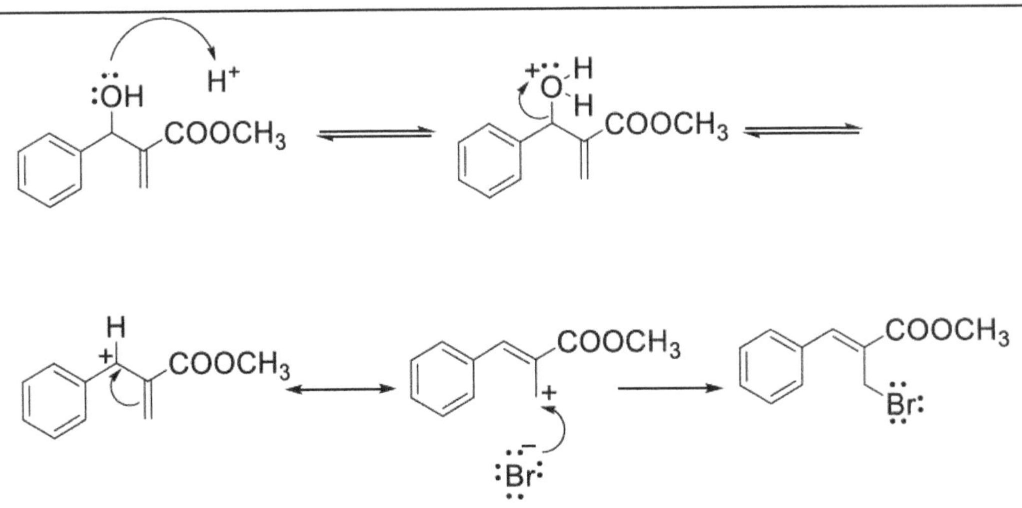

:ÖH
H+
COOCH3
COOCH3
COOCH3
COOCH3
COOCH3
Br:
:Br:

Recall that strong acids are fully ionized in solution. Formation of the resonance-stabilized benzylic and allylic carbocation is followed by a LA/LB reaction to form the product.

6

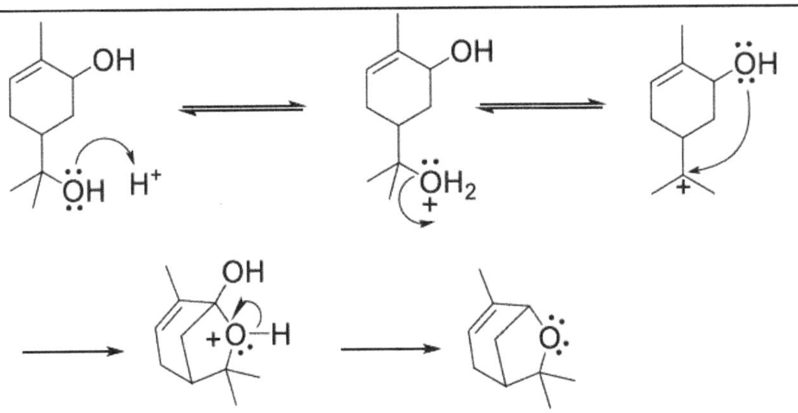

OH
OH
H+
OH
OH2
OH
OH
H
O

Although either -OH group can react in a LA/LB reaction, the one that leads to formation of the more stable carbocation reacts preferentially (General principle to remember: most organic reactions proceed through the formation of the more stable intermediate species and, ultimately, the more stable final product). A subsequent intramolecular LA/LB reaction leads to the formation of the product

7

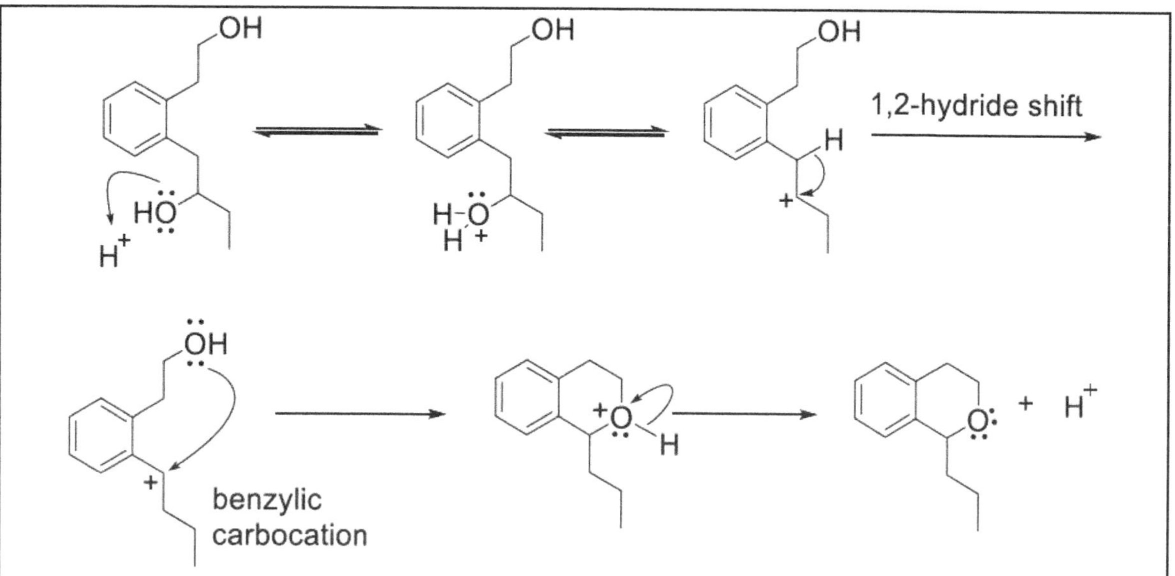

1,2-hydride shift
benzylic
carbocation

Initial formation of the more stable 2° carbocation is followed by a 1,2-hydride shift to yield a benzylic carbocation. This is followed by an intramolecular LA/LB reaction to yield the product.

8

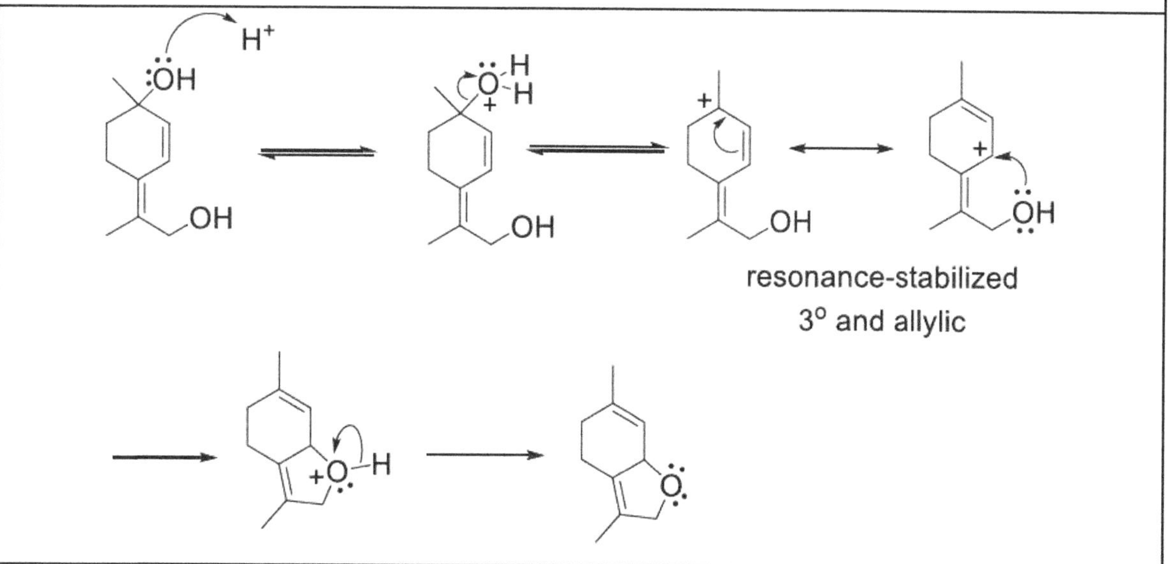

resonance-stabilized
3° and allylic

Initial formation of a resonance-stabilized 3° and allylic carbocation is followed by an intramolecular LA/LB reaction to yield the product.

9

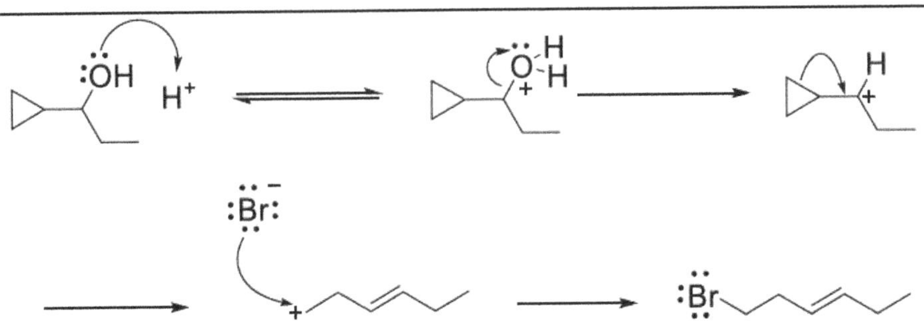

See Minireview 4 for a discussion of cyclopropyl methyl carbocations.

10

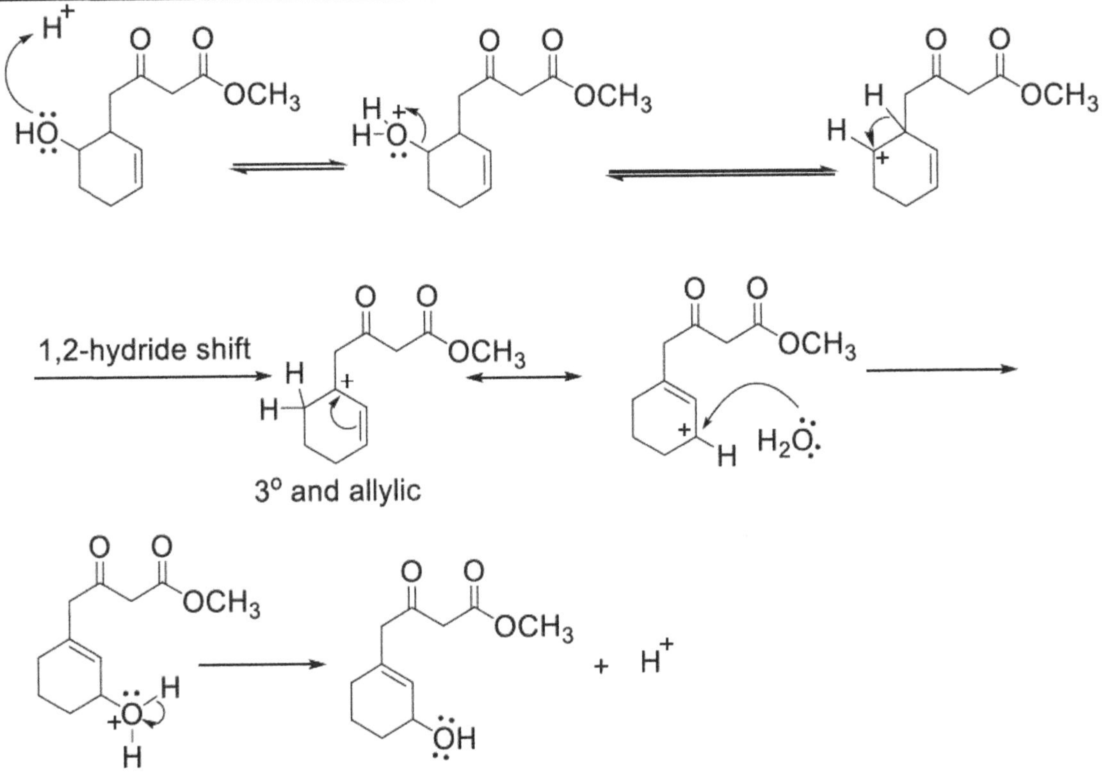
1,2-hydride shift
3° and allylic

While the initial LA/LB reaction can also take place at the carbonyl oxygens, that's an unproductive process, namely, it does not lead to the formation of the observed product

11

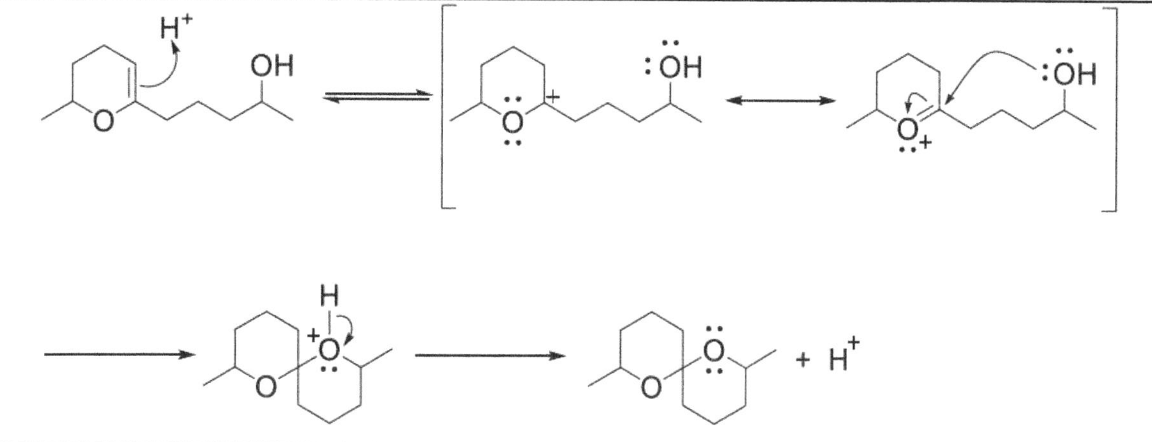

Although the starting material has three functional groups or reactive sites where a LA/LB reaction can take place, the more stable resonance-stabilized cation is initially formed preferentially.

12

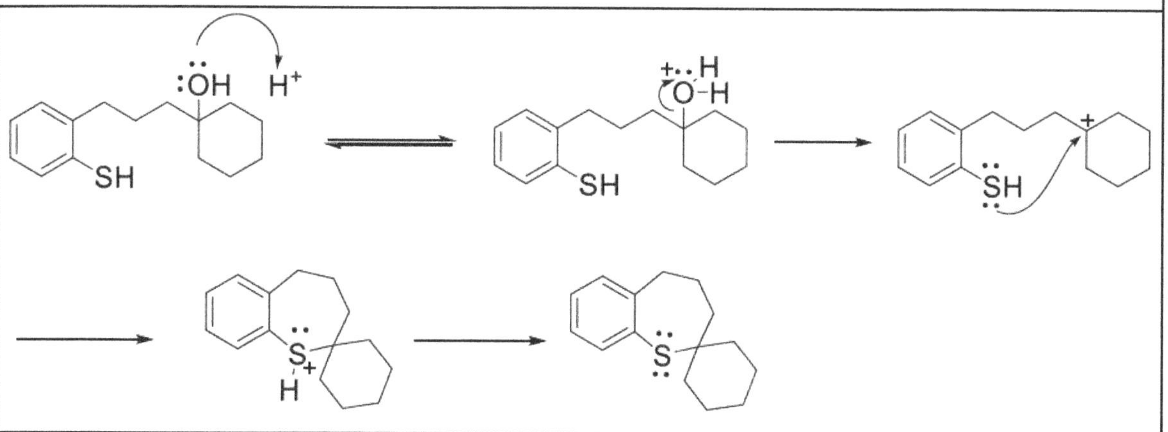

Formation of the 3° carbocation is followed by an intramolecular LA/LB reaction to yield the product.

13	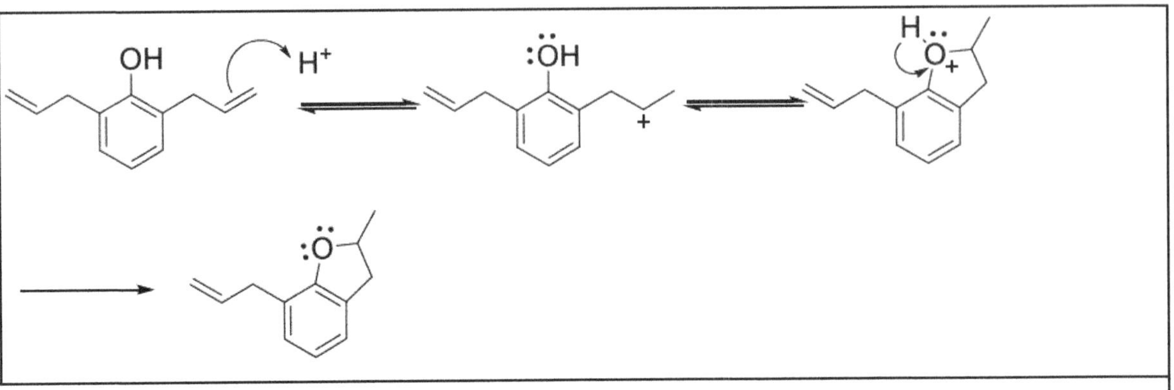 OH H+ :ÖH + H O+ :Ö
	Either one of the two alkene double bonds can react to form a secondary carbocation (notice that simultaneous reaction of both double bonds would give rise to a highly unstable species having two positive charges, therefore, it never happens). As noted earlier, while the aromatic ring can react in a LA/LB reaction, this is energetically unfavorable (loss of aromaticity).
14	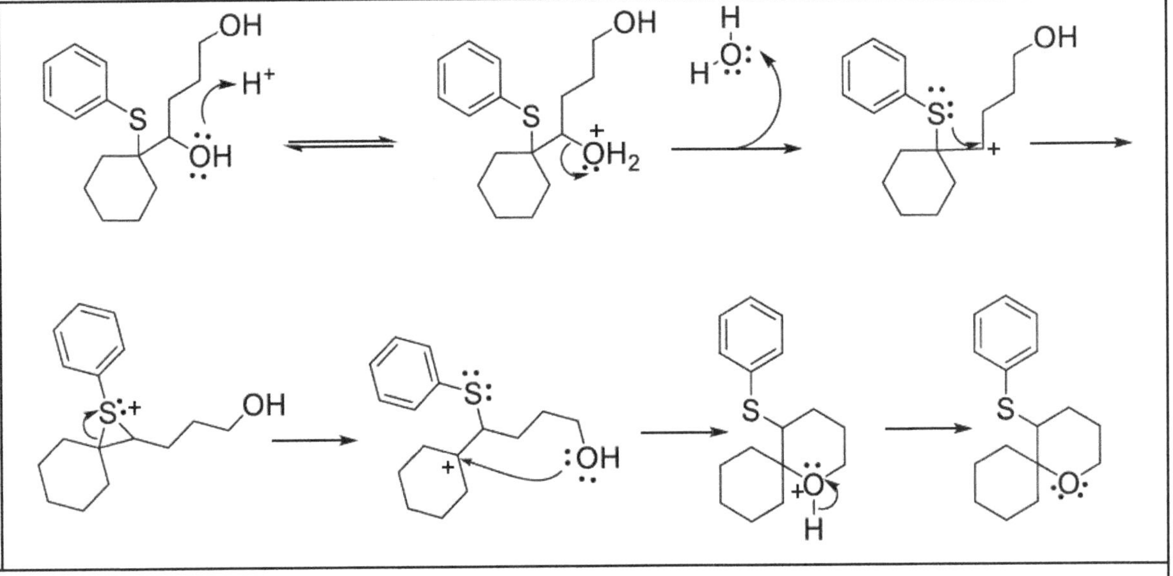 OH H+ S ÖH OH S + OH2 H H O: OH S: + S:+ OH S: + :OH S +O H S O:
	Initial formation of the 2° carbocation is followed by a LA/LB reaction to form a highly reactive episulfonium ion. Ring opening leads to the formation of a 3° carbocation which subsequently undergoes an intramolecular LA/LB reaction to form the observed product.

15

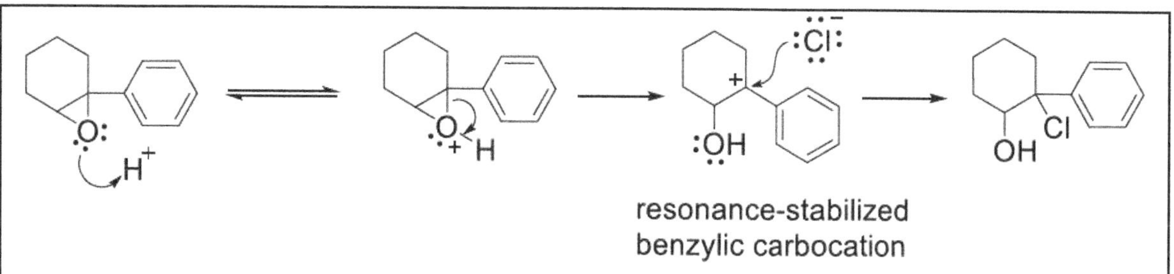
:Cl:
:O:
H+
H
:OH
OH
Cl
resonance-stabilized
benzylic carbocation

This is an example of acid-catalyzed ring opening of an epoxide. A LA/LB reaction is followed by ring-opening to yield the most stable carbocation (benzylic instead of the less stable secondary carbocation). A second LA/LB reaction yields the observed product.

16

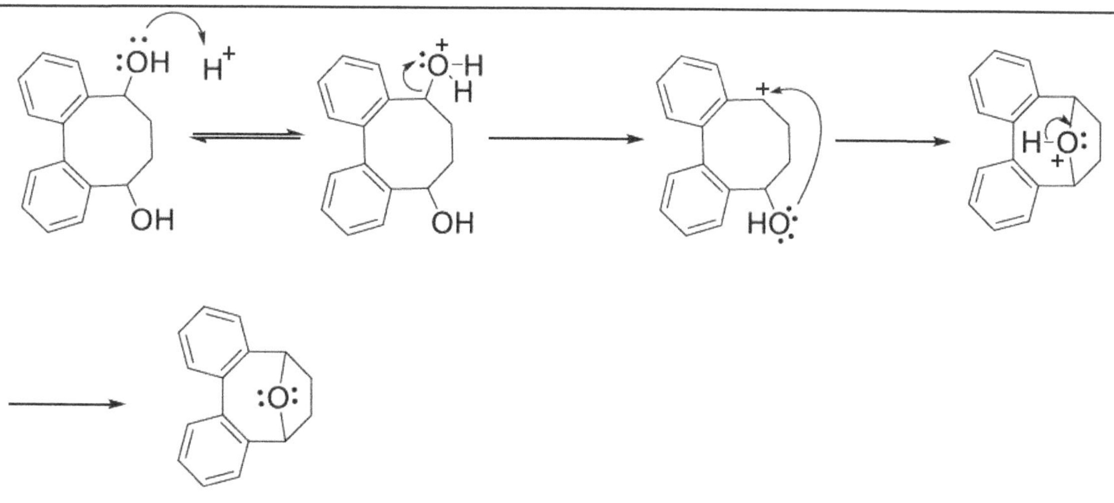
:OH
H+
OH
:O-H
H
HO:
H-O:
:O:

Both alcohols are benzylic, consequently, either one can form a benzylic carbocation which can form the product following an intramolecular LA/LB reaction.

17

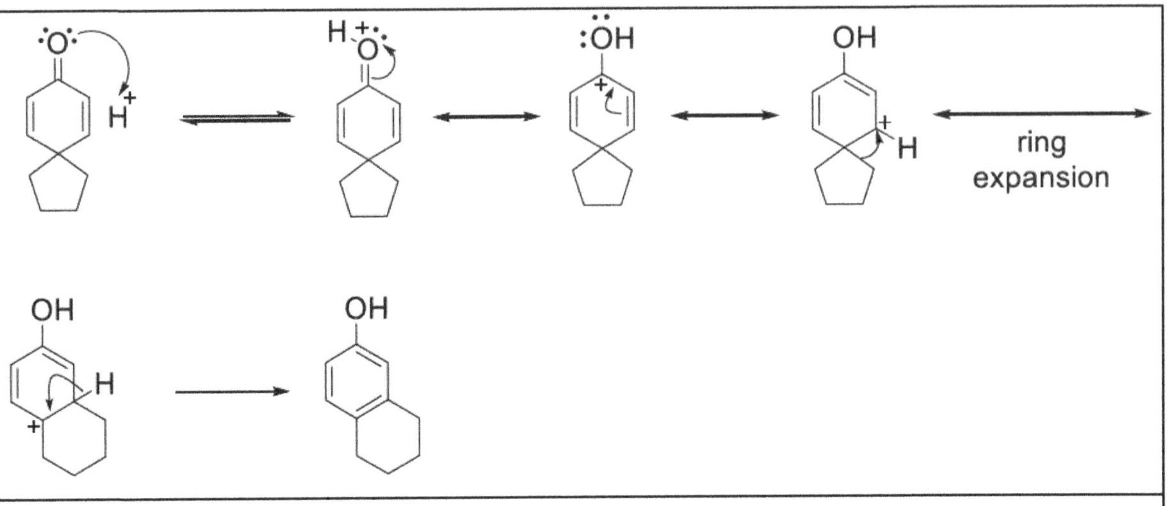

ring expansion

The driving force for this reaction is primarily due to the formation of the aromatic ring. Recall that aromatic compounds are cyclic, planar, satisfy Huckel's 4n+2 rule, where n is the number of pi and non-bonded electron pairs, and can sustain a ring current in a magnetic field.

18

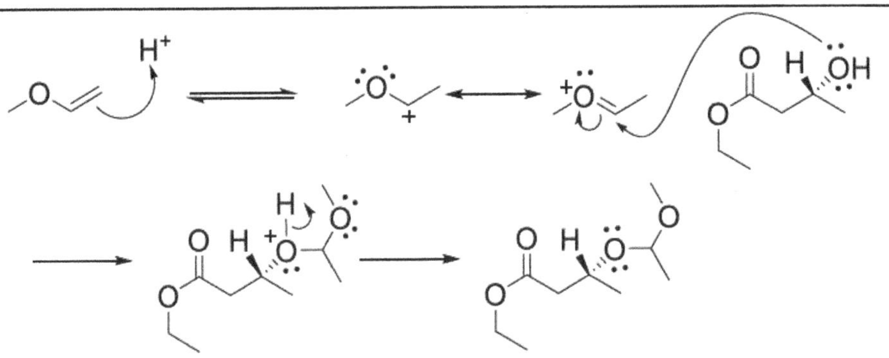

A LA/LB reaction yields the most stable carbocation, which then reacts in a second LA/LB reaction to furnish the product. Note that this reaction proceeds with *retention of configuration*, namely, the atoms or groups bonded directly to the stereogenic (chiral) center are oriented in space the same way in the product, as in the starting material, since none of the bonds to the stereogenic center is broken during the course of the reaction.

19

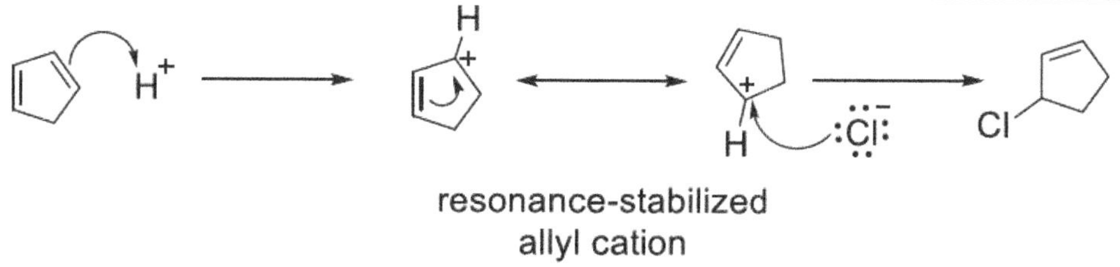
H
H+
H
:Cl:
Cl
resonance-stabilized
allyl cation

Conjugated dienes undergo electrophilic addition reactions to yield 1,2- and 1,4-addition products. In this example, the 1,2- and 1,4-addition products are the same.

20

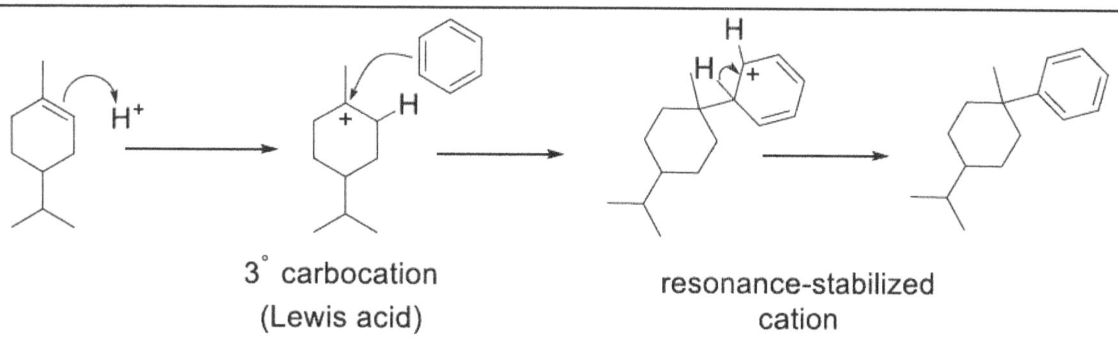
H+
H
H
H
3° carbocation
(Lewis acid)
resonance-stabilized
cation

A LA/LB reaction leads to the formation of the most stable carbocation. This is followed by an electrophilic aromatic substitution reaction, specifically, a Friedel-Craft's alkylation reaction (see Minireview 4).

21

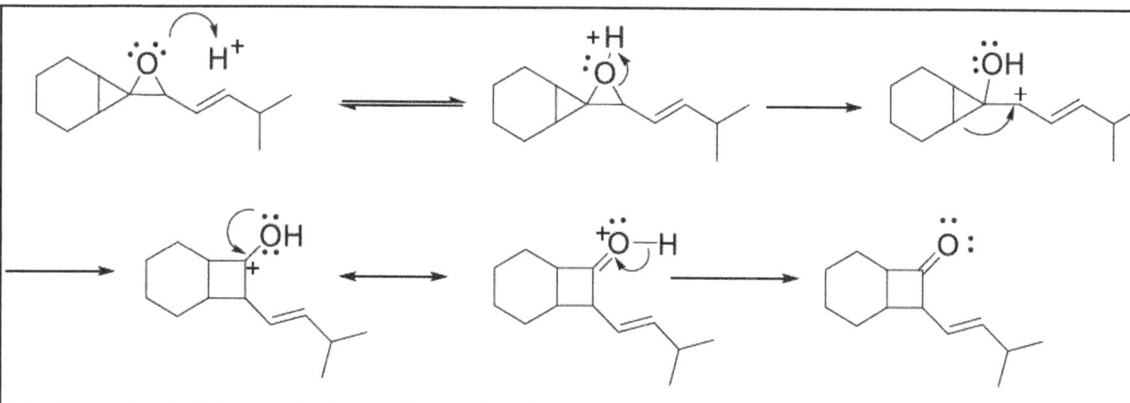
H+
+H
:OH
OH
O—H
O:

A LA/LB reaction is followed by ring opening of the epoxide (oxirane) ring to form a resonance-stabilized allyl carbocation. Ring expansion results in the relief of ring strain and the formation of a resonance-stabilized carbocation.

22

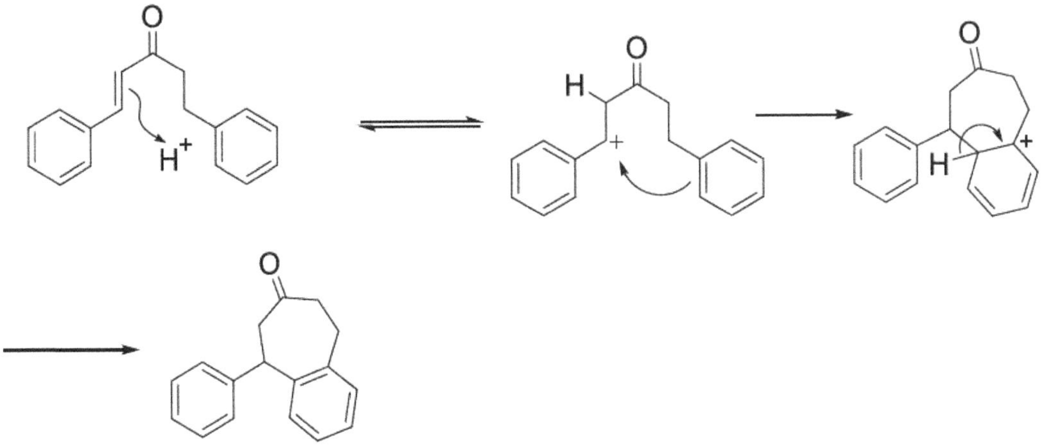

O
H+
O
H
+
O
H
+
O

An initial LA/LB reaction yields the most stable carbocation (note that carbocations with a positive charge on a carbon alpha to a carbonyl group are highly unstable because the positive charge is next to the partial positive charge on the carbonyl carbon. Repulsion of the adjacent positive charges raises the energy). This is followed by an intramolecular Friedel-Craft alkylation reaction.

23

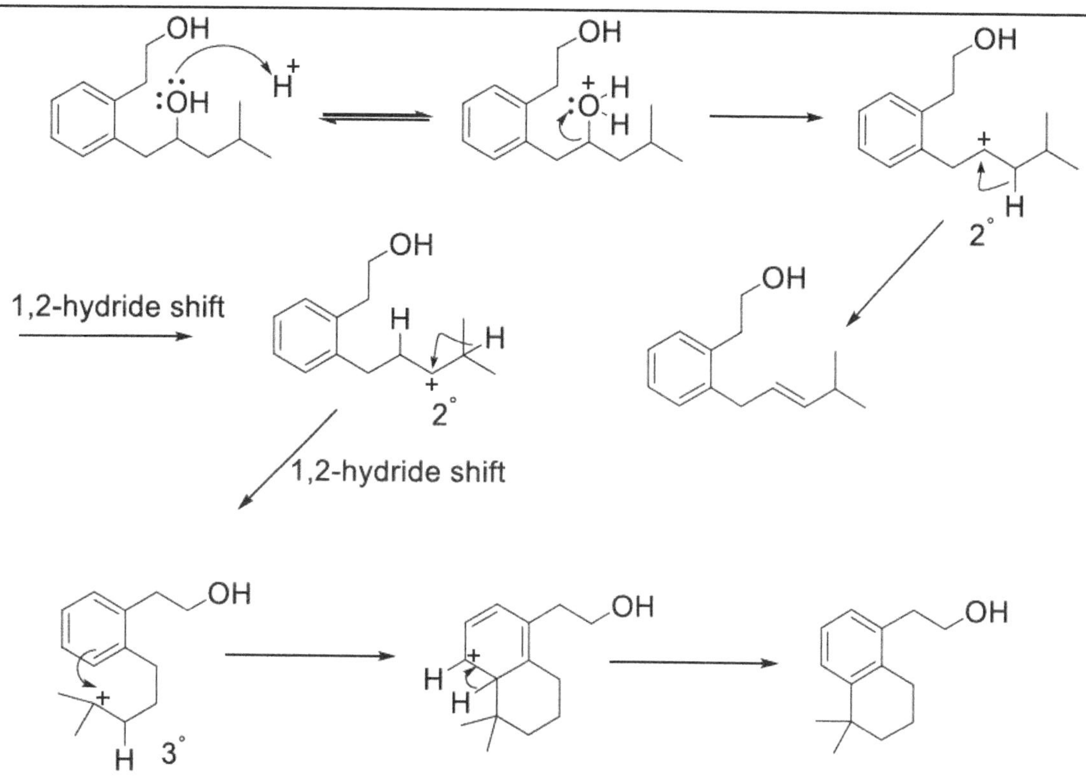
OH
:ÖH
H+
OH
OH
2°
1,2-hydride shift
OH
H
H
2°
OH
1,2-hydride shift
OH
H
3°
OH
H
H
OH

Formation of the 2° carbocation is followed by a 1,2-hydride shift to form a more stable 3° carbocation which undergoes an *intramolecular* Friedel-Craft alkylation reaction.

24	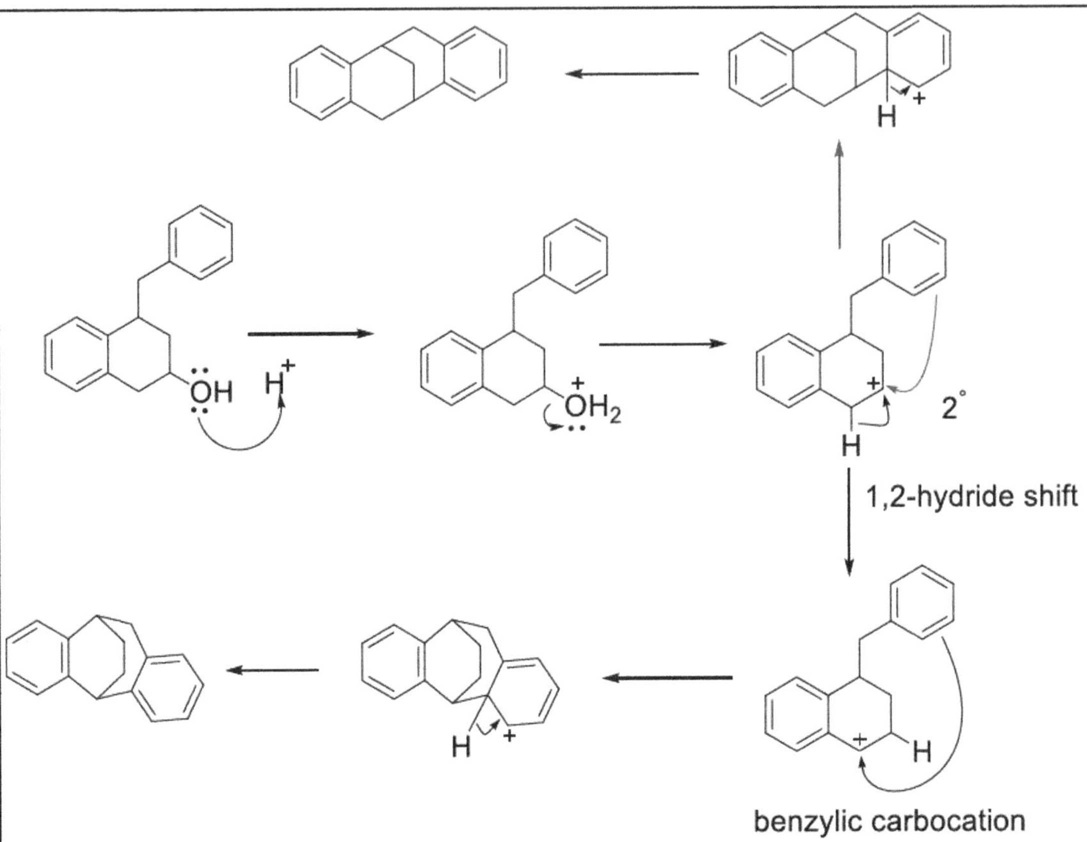OH H+ OH2 2° H 1,2-hydride shift H benzylic carbocation H H
	The initial 2° carbocation follows two pathways in forming the observed products.

25

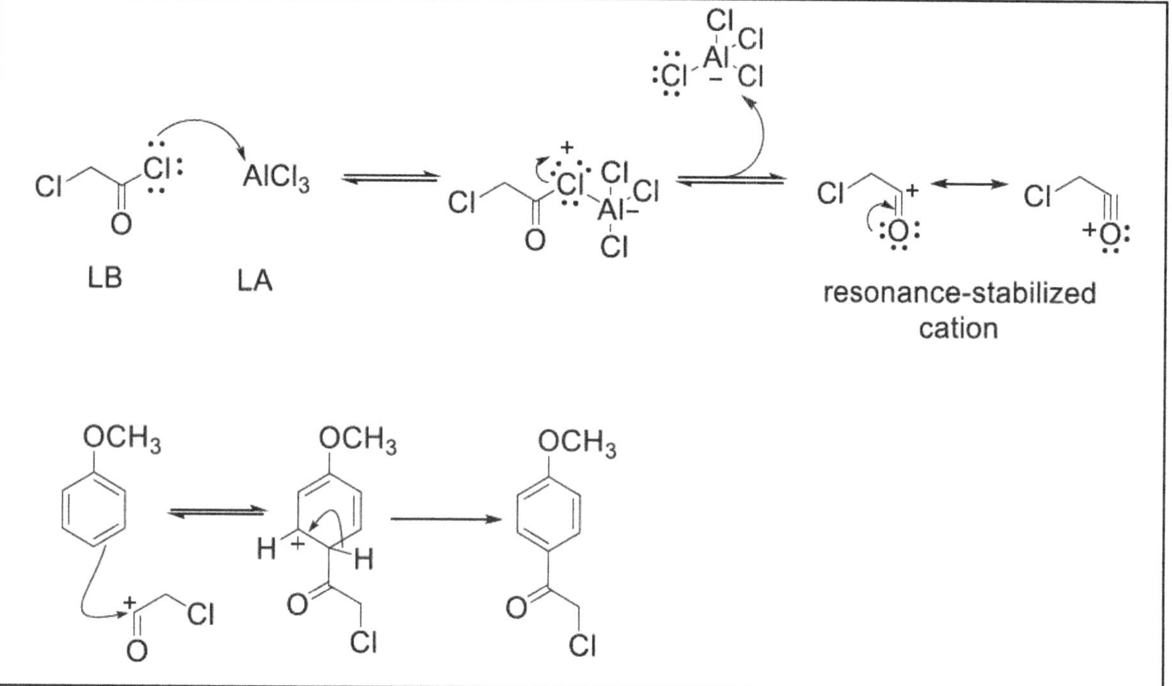
LB
LA
resonance-stabilized cation

This is an example of the *Friedel-Craft acylation reaction* (see Minireview 4). Since the methoxy group is an activating, *o*- and *p*-directing group, the major product formed is that shown above. Note also that while either chlorine can function as Lewis base by donating a pair of electrons to the Lewis acid, the more stable species is formed (a resonance-stabilized acylium ion).

26

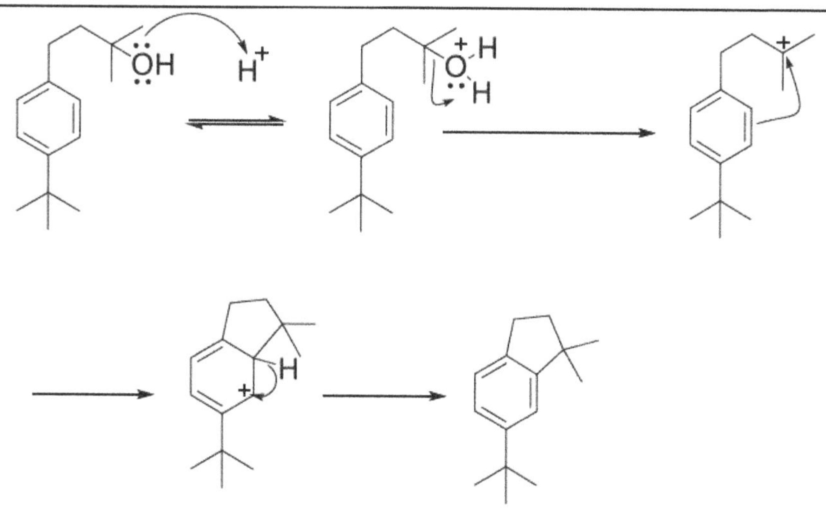

A LA/LB reaction leads to the formation of a 3° carbocation, which is followed by an intramolecular Friedel-Craft alkylation reaction.

27

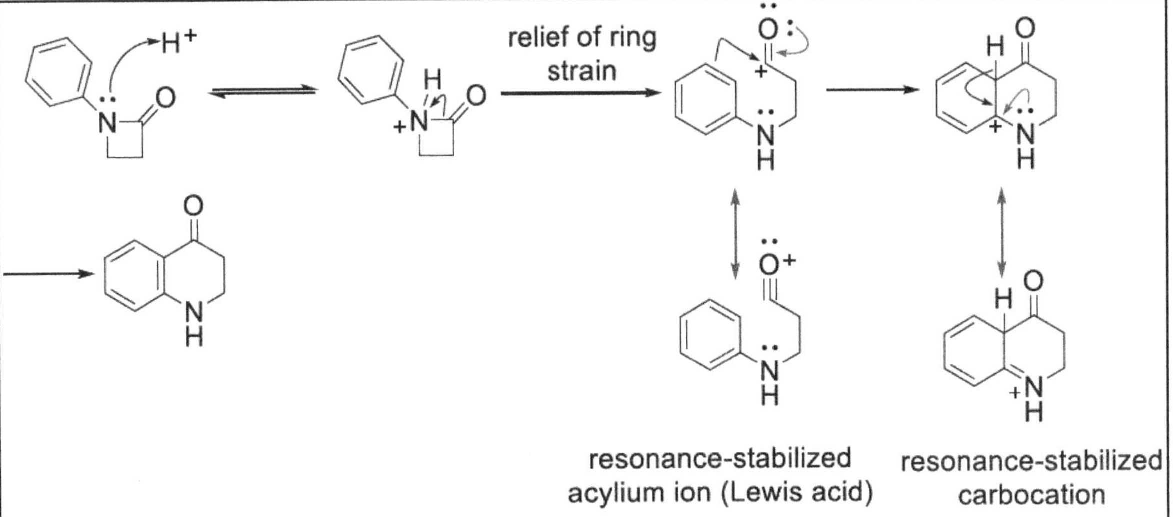

H+
relief of ring strain
resonance-stabilized acylium ion (Lewis acid)
resonance-stabilized carbocation

The pair of non-bonded electrons on the nitrogen is more available for donation than in acyclic amides because of ring strain.

28

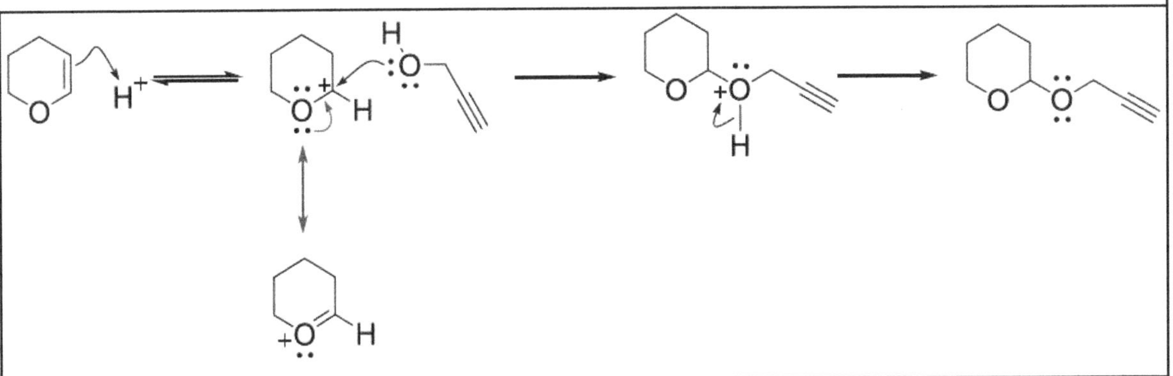

Two consecutive LA/LB reactions lead to the formation of the product. Note that a LA/LB reaction at the oxygen atom of the starting material is an unproductive process.

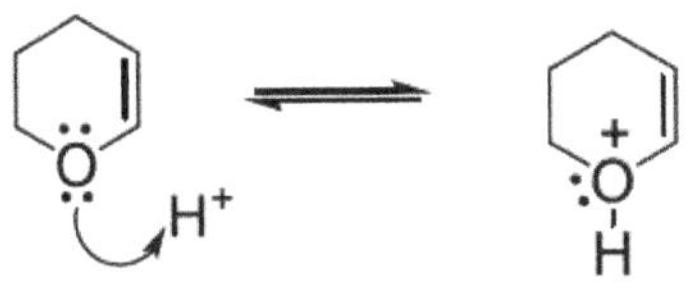

29

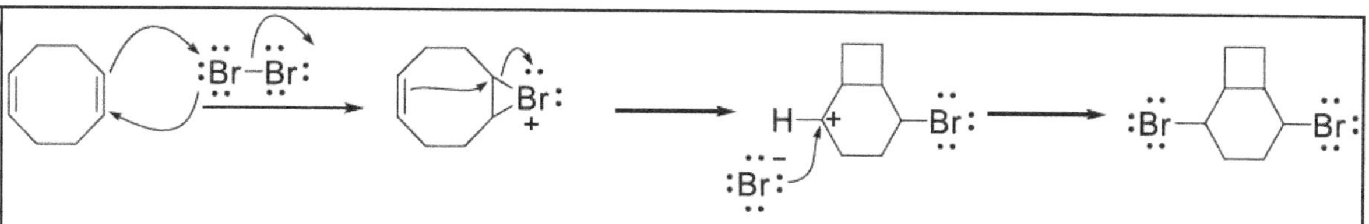

Formation of the highly reactive bromonium ion is followed by a transannular LA/LB reaction to form a 2° carbocation. A subsequent LA/LB reaction yields the observed product.

30

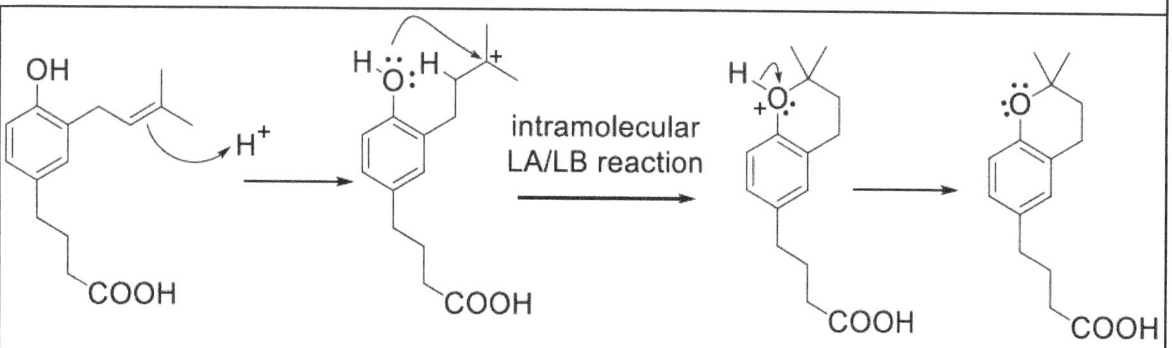

OH
H+
COOH
intramolecular
LA/LB reaction

31

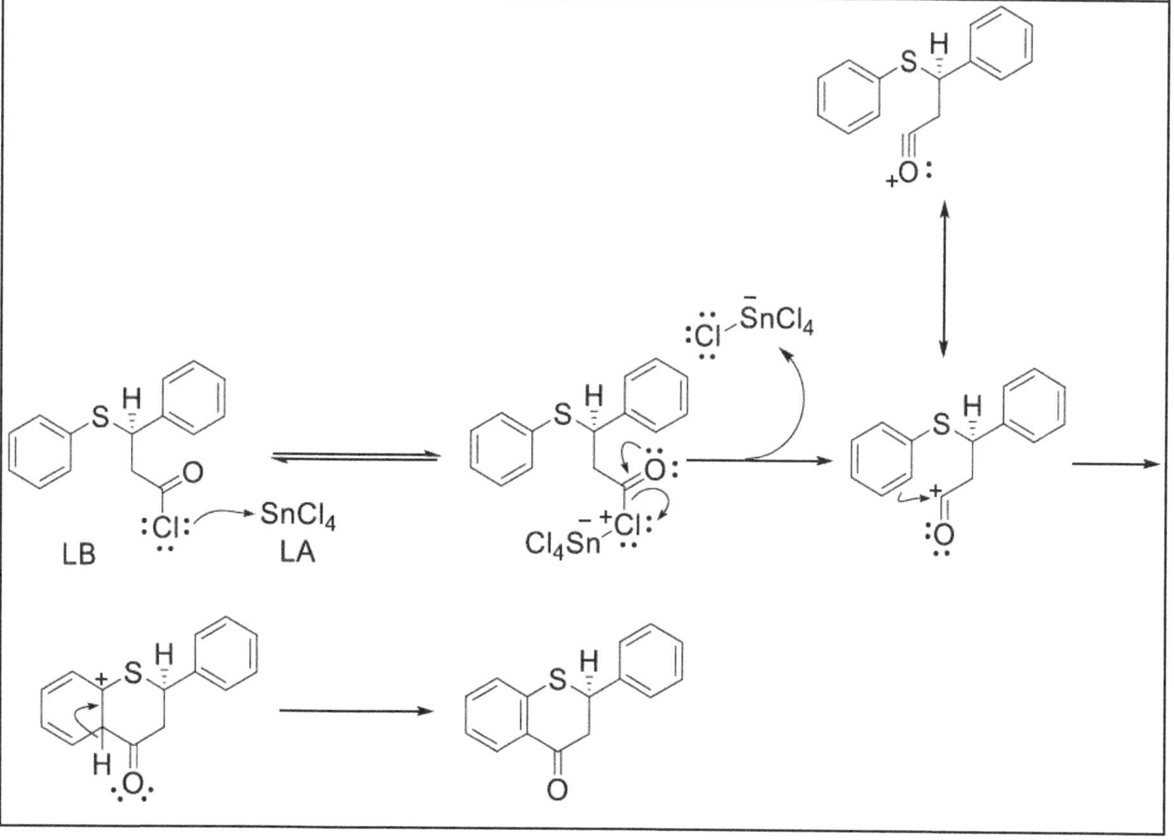

LB
SnCl4
LA

SnCl4 behaves as a Lewis acid (Sn has a partial positive charge), like AlCl3 and related compounds, initiating the formation of an acylium ion that subsequently undergoes an intramolecular Friedel-Craft acylation reaction.

32

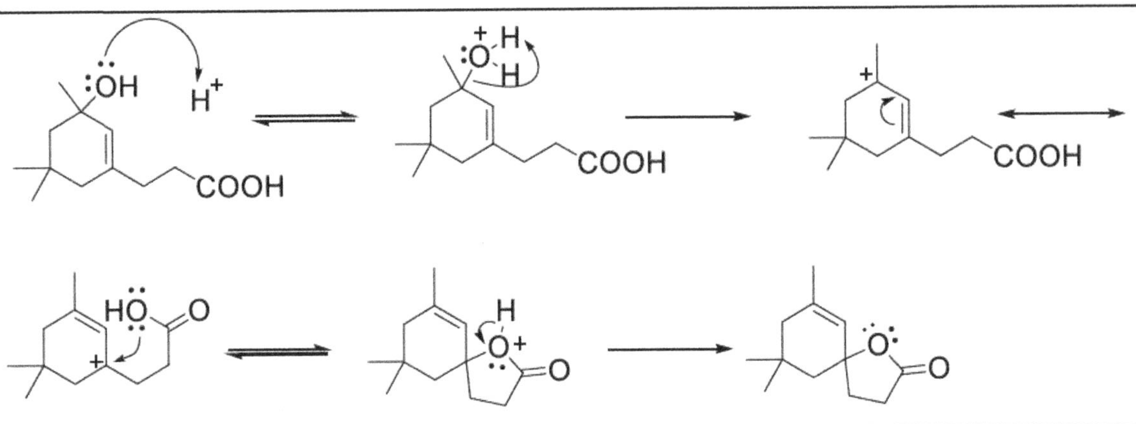

:OH
H+
+ H
:O
H
+
COOH
COOH
COOH
HO
O
+
H
O+
O
O
O

Formation of the 3° and allylic carbocation is followed by a LA/LB reaction to form the product.

33

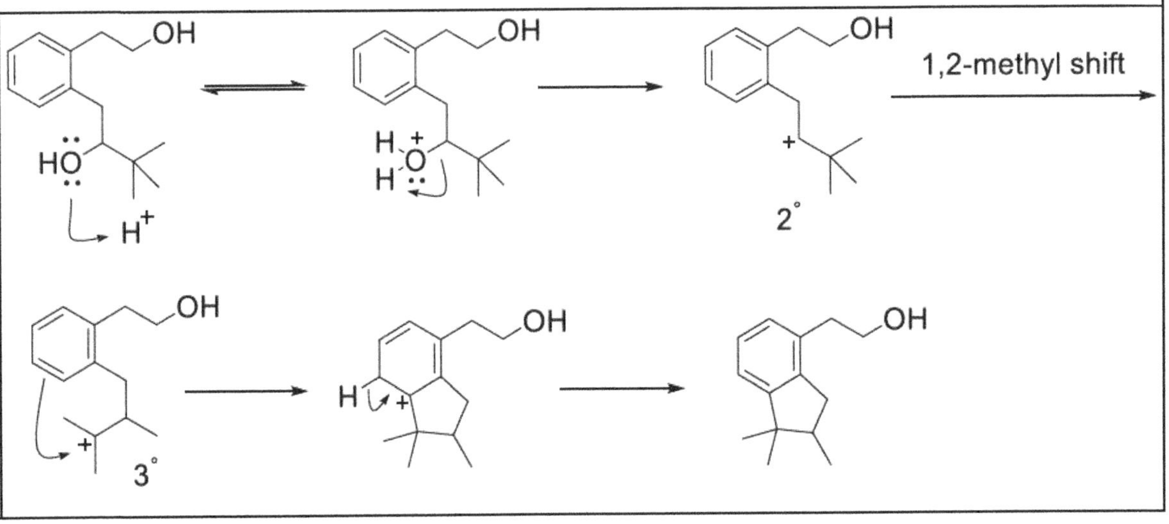

OH
HO
H+
H
O
H
+
2°
1,2-methyl shift
3°
H

34

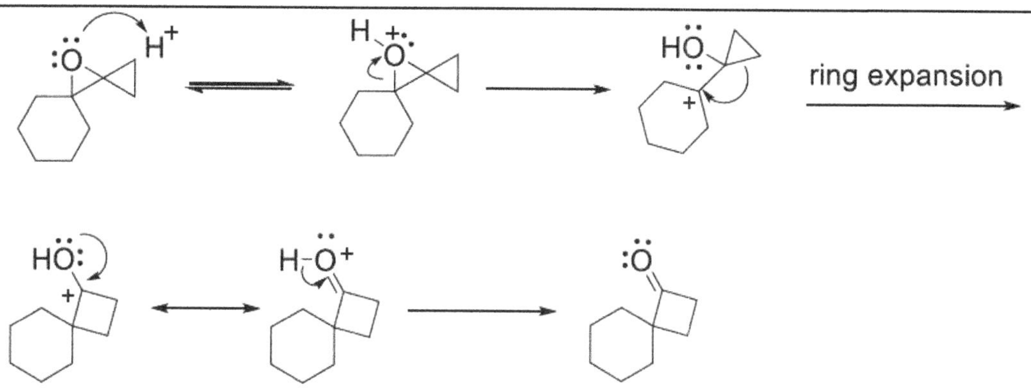

H+
ring expansion

Minireviews 5-7 discuss briefly some basic principles related to the formation of anions (nucleophiles), nucleophilic substitution (S_N2) and elimination (E_2) reactions, chemical reactivity and ring strain. These should be studied prior to attempting questions 35-50.

Minireview 5

Formation of Anions (Nucleophiles)

Before discussing anions (Lewis bases/ nucleophiles), it is helpful to remember that the chemistry of alkyl halides, aldehydes and ketones, carboxylic acid derivatives (acid halides, anhydrides, thioesters, esters, amides and epoxide) can be simply described and best understood by equations (1-4) below, without recourse to memorization. Thus, the typical reaction of alkyl halides is *nucleophilic substitution* (S_N2) (eq 1) (see Minireviews 6 and 7 for more details and some variations of the same theme), the typical reaction of aldehydes and ketones is *nucleophilic addition* (eq 2) (Minireview 8), and the typical reaction of carboxylic acid derivatives is *nucleophilic acyl substitution* (eq 3) (Minireview 9). Nucleophilic ring opening reactions of epoxides are illustrated by eq 4. These four general reactions can be viewed as Lewis base/Lewis acid reactions and are discussed in greater detail in Minireviews 6-9. Recall that S_N2 reactions proceed most readily when the Lewis base (nucleophile) has a negative charge. In the case of S and P nucleophiles and amines, because of their large size and high basicity, respectively, they can participate in S_N2 reactions as such. *Anions (Y:⁻) are typically generated by reacting a molecule with an acidic hydrogen (Y-H) with an appropriate base (B:⁻) for subsequent use in reactions of the type shown below.*

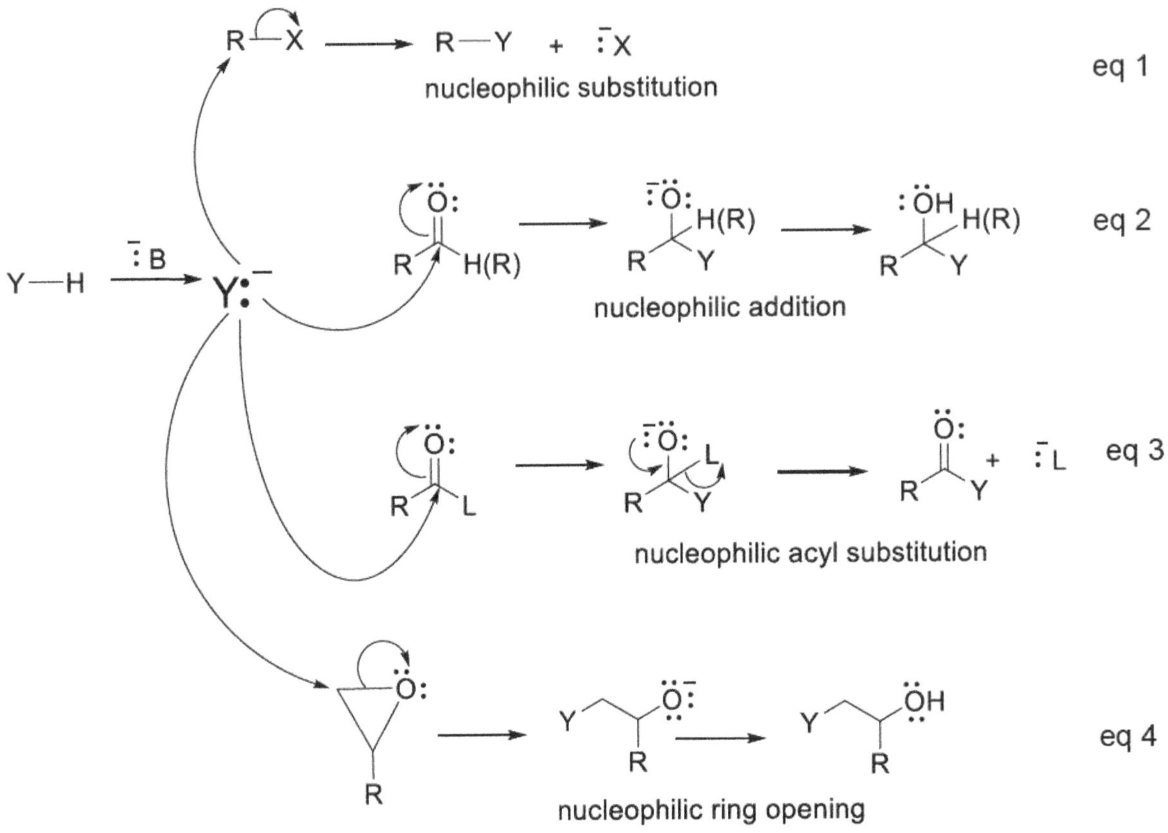

nucleophilic substitution
eq 1
nucleophilic addition
eq 2
nucleophilic acyl substitution
eq 3
nucleophilic ring opening
eq 4

Factors to consider prior to generating an anion include the relative acidity of the starting material (HA) and type of base (B^-) to use. Acid strength is dependent on the following factors: a) ***electronic effects: resonance and inductive effects***. Anions (Y:-) that are stabilized by resonance and/or inductive effects are more stable and easier to form; b) ***size***. The greater the size of the atom (Y) to which the hydrogen is bonded, the higher the acidity of a hydrogen; c) ***electronegativity***. The higher the electronegativity of Y, the higher the acidity of a hydrogen. The following examples illustrate the aforementioned factors and acid strength.

Exercise 1 Rank the hydrogens in the structure shown below in order of decreasing acidity.

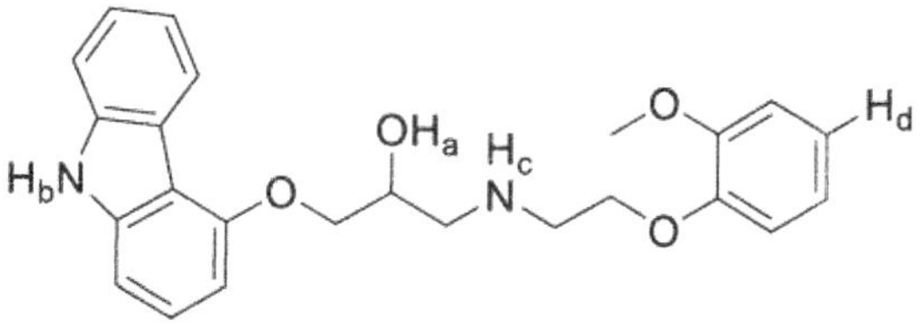

H_bN
O
OH_a
H_c
N
O
O
H_d

It can be readily shown that the equilibrium constant (K_{eq}) for the general acid/base reaction shown below (equation 5) is given by equation 6, where pK_a is the acidity constant of the reactant acid (HY) and pK_a' is the acidity constant of the product acid (HB). Since the pK_as of a large number of acids have been determined and are readily available, an approximate K_{eq} for any acid-base reaction can be readily calculated using equation 6 (*vide infra*). In order for the aforementioned reactions to proceed readily, a high concentration of the anion (nucleophile) is desirable. Hence, *a sufficiently strong base is typically selected to generate (Y^-) rapidly and quantitatively.*

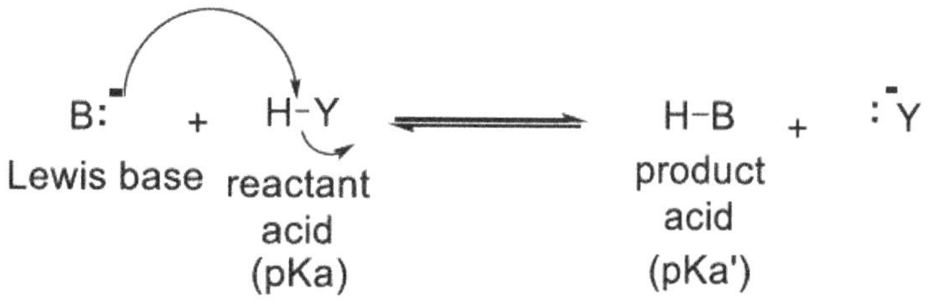

B:
+
H-Y
H-B
+
:Y
Lewis base
reactant acid (pKa)
product acid (pKa')

eq 5

$$K_{eq} = 10^{pKa' - pKa}$$ eq 6

Recall that the pK_a of an acid is an index of the relative strength of the acid. *The lower the pK_a is, the stronger the acid.* Table 1 lists the pK_a values of some common types of organic compounds

Table 1: pK_a values of some common classes of organic compounds *

	pK_a
RSO_3H	-6.5
RCOOH	4-6
ArSH	6-8
RSH	10-11
ArOH	8-11
ROH	16-18
$RCONH_2$	17
RCH_2CHO	19-20
RCH_2COR	19-20
RCH_2COOR	25

*values are approximate

The acidity of substances is greatly affected by the presence of one or more functional groups capable of delocalizing the negative charge in the corresponding anion via resonance (Minireview 3). Common functional groups that enhance acidity include NO_2, COR, CN, COOR, SO_2R and phenyl (Ph). These groups differ in their ability to delocalize the negative charge in the anion, hence the extent to which acidity is affected is dependent on the *nature* and *number* of such functional groups. For example, in a series of compounds of the type RCH_2X, the pK_a varies as follows: X= NO_2 (10), (C=O)R (20), CN (25), COOR (25) and SO_2R (29). The presence of two such groups reduces the pK_a further.

Selection of Base

In selecting a base to be used in the generation of an anion, the *strength* of the base and its *compatibility* with any other functional groups that might present in the molecule, are factors of paramount importance. In general, a sufficiently strong base is selected so that the desired anion is generated rapidly and quantitatively. Secondly, depending on the structure of the acidic molecule (HY), the base used may have to be *non-nucleophilic* (vide infra). Common bases used to generate anions include the following:

a) Alkoxide ions (RO^-M^+, where $M^+ = Na^+$ or K^+)

Alkoxide ions are conveniently generated by reacting dry methanol (pK_a ~15.5), ethanol (pK_a ~16), or t-butyl alcohol (pK_a ~18) with sodium or potassium metal under a nitrogen atmosphere. The generated base is then used <u>in situ</u> (in the flask, without isolation). *Note that alkoxides can function as both bases and nucleophiles.*

$$\underset{pK_a\ 13}{EtOO\text{-}CH_2\text{-}COOEt} + CH_3CH_2O^- \rightleftharpoons EtOO\text{-}\ddot{C}H^-\text{-}COOEt + \underset{pK_a\ 16}{CH_3CH_2OH}$$

Using equation (6), $K_{eq} = 10^{16-13} = 10^3$, therefore the formation of the anion ($:Y^-$) is greatly favored (equilibrium lies to the right).

In cases where a reactant having more than one acidic hydrogen is treated with base, *the most acidic hydrogen will react faster, since it leads to the formation of the more stable anion*. The following examples illustrate this concept.

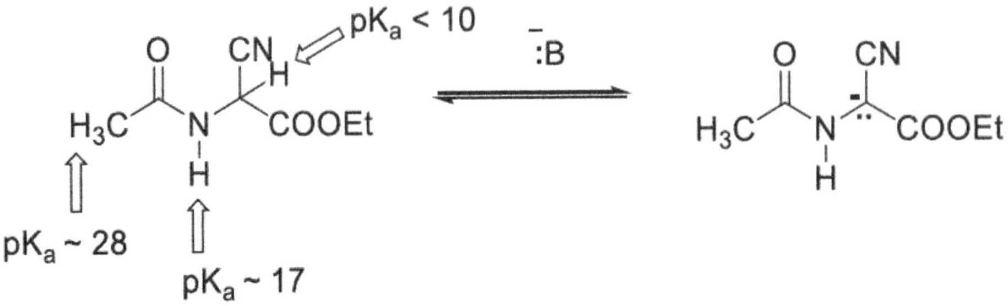
pKa < 10
:B
pKa ~ 28
pKa ~ 17
COOEt
CN

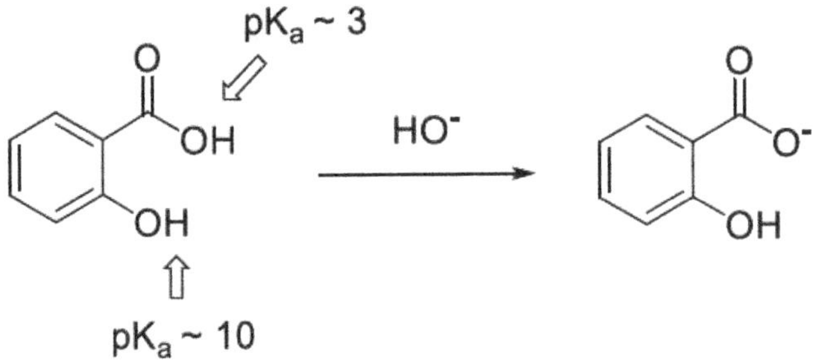
pKa ~ 3
HO-
pKa ~ 10

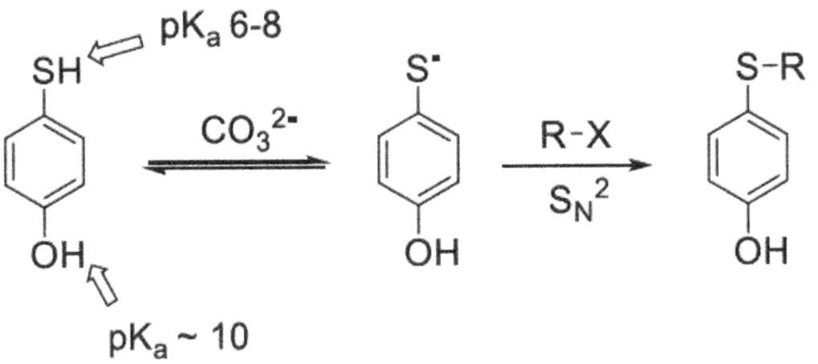
pKa 6-8
CO3 2-
R-X
SN2
pKa ~ 10

b) Alkyl lithium reagents (:R⁻ Li⁺)

Alkyl lithium bases, such as methyl lithium, n-butyl lithium, and t-butyl lithium, are *strongly basic* and *nucleophilic*. Because alkyl lithium reagents are the conjugate bases of the corresponding hydrocarbons (pKa ~ 50), their high basicity and reactivity requires the use of an inert atmosphere and anhydrous conditions. Recall that the lower the pKa is, the stronger the acid and the weaker the conjugate base. Conversely, the higher the pKa is, the weaker the acid and the stronger the conjugate base.

c) If a much stronger, non-nucleophilic base is needed, then use of one of the following bases is preferred:

➢ Lithium diisopropyl amide (LDA)

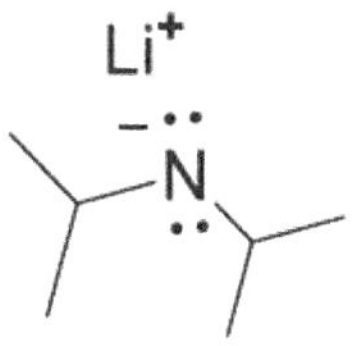
Li+
N

LDA can be generated *in situ* by treating diisopropyl amine with n-butyl lithium (the pK_a of diisopropyl amine is ~ 36), however, it is also commercially available.

➢ Lithium hexamethyldisilane (LHMDS)

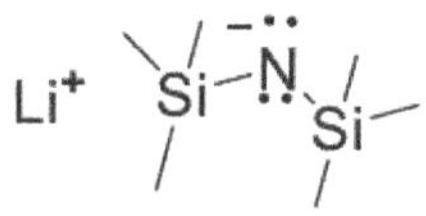
Li+
Si
N
Si

➢ Sodium or potassium hydride NaH or KH ($:H^-$)

➢ Sodium amide ($NaNH_2$)

d) Amines

Recall that the basicity of amines follows the order

$3^0 > 2^0 > 1^0 > NH_3 > Ar\text{-}NH_2$

⟵

Increasing base strength

(in other words, aliphatic amines are more basic than aromatic amines and, secondly, 3° aliphatic amines are more basic than 2° aliphatic amines which in turn are more basic than 1° amines).

Triethylamine (TEA), pyridine and N-methylmorpholine (NMM) are organic bases that are widely used in organic synthesis.

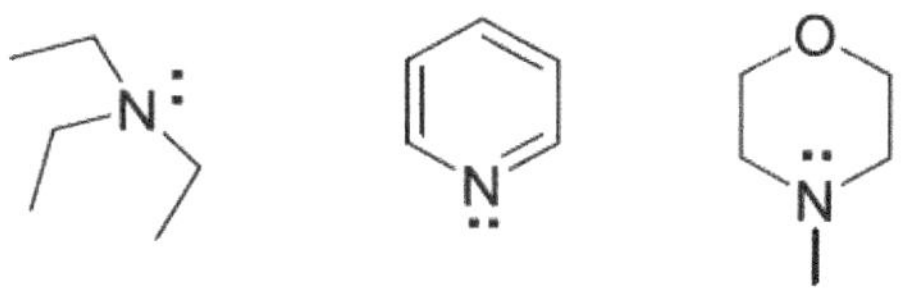
N
N
O
N

When the objective is to generate an anion via an acid-base reaction, or to induce an E2 elimination reaction (vide infra), the use of a *sterically hindered, non-nucleophilic base*, such as diisopropylethyl amine (DIEA), 1,8-diazabicyclo[5.4.0]undecen-7-ene (DBU), or 1,5-diazabicyclo[4.3.0]non-5-ene (DBN) should be considered.

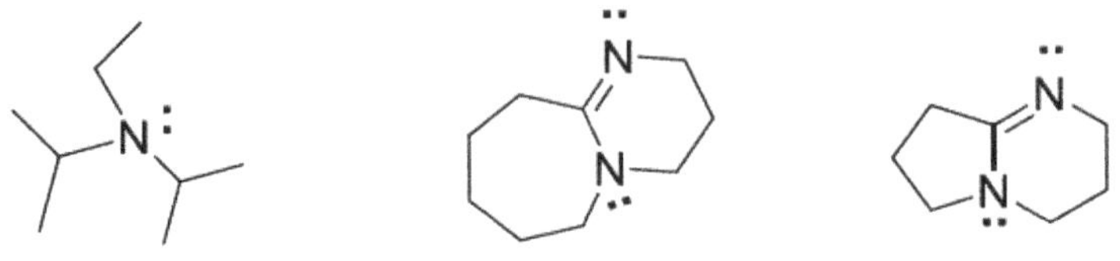

DIEA DBU DBN

Example 1

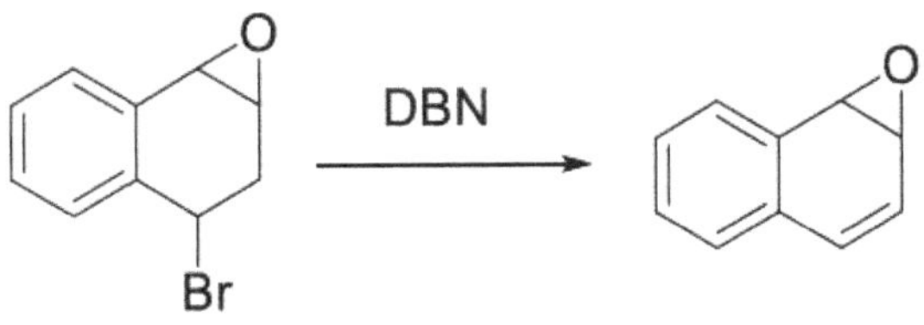

DBN

The following transformation was successfully achieved using a bulky and non-nucleophilic base, greatly minimizing competing S_N2 and ring-opening reactions between the base and the highly reactive functionalities in the starting material and product.

Example 2

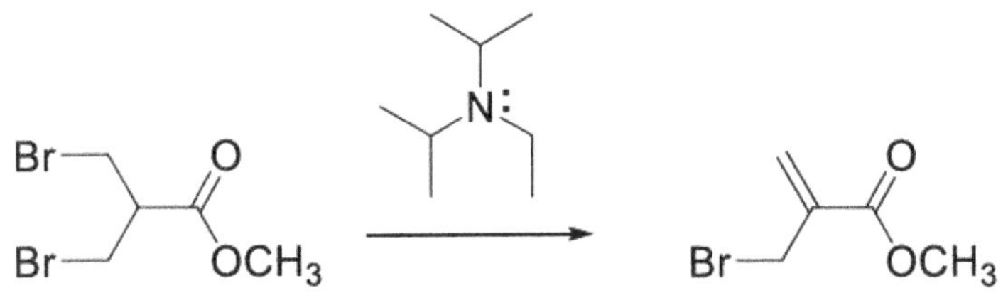

The objective here was to induce an E_2 elimination only, leading to the formation of the desired product. The selection of base was critical for the success of this reaction because of the high reactivity of the product. The latter is an allylic bromide (highly reactive in S_N2 reactions), thereby necessitating the use of a bulky and non-nucleophilic base.

e) Miscellaneous bases

These include bases such as sodium or potassium bicarbonate, carbonate or hydroxide. The pK_a values of carbonic acid, bicarbonate and water are 6.4, 10.3 and 15.7, respectively. Consequently, base strength follows the order $HO^- > CO_3^{2-} > HCO_3^-$.

Examples illustrating the *rationale* underlying the selection of a particular base are given below. See Minireview 5 for additional examples. It would be very beneficial to consider the logic behind the use of a particular base, or reaction conditions, as you work through the problems in this workbook.

Minireview 6

- Nucleophilic Substitution Reactions (S_N2)
- Elimination Reactions (E_2)

The S_N2 reaction is *a concerted, one-step reaction involving backside attack by a strong Lewis base (nucleophile) on an sp^3-hybridized carbon bearing a partial positive charge (δ+) (Lewis acid) and a leaving group (L). The reaction proceeds with inversion of configuration* (eq 1).

S_N2 reactions are greatly influenced by the nature of the alkylating agent (R-L), nucleophile ($^-$:Y), leaving group (L), and solvent.

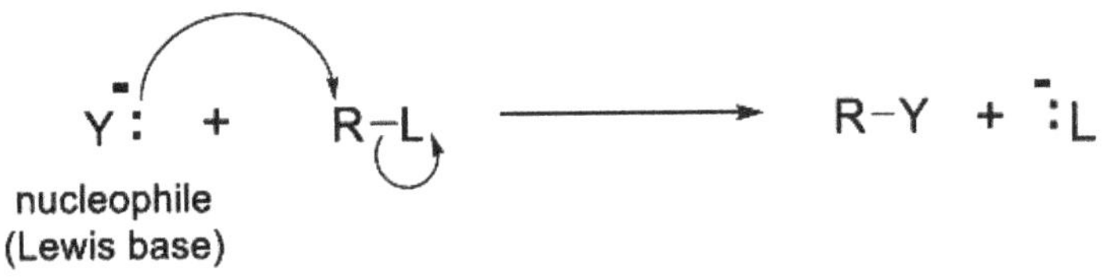

Y:
+
R–L
R–Y + :L
nucleophile
(Lewis base)

eq 1

a) Alkylating agent (R-L)

The order of reactivity of halides in S_N2 reactions is

benzyl halide > CH_3X ~ allyl > 1^0 > 2^0 > 3^0 > vinyl halide, aryl halide, R-F (Unreactive)

← increasing reactivity

Note:

1) Tertiary alkyl halides undergo E2 reactions, while vinyl, aryl halides, and alkyl fluoride do *not* undergo S_N2 reactions.
2) The reactivity profile of α-haloesters, α-haloethers and α-haloketones is similar to that of benzylic halides.

b) Strength and Nature of Lewis Base (Nucleophile)

In general, *nucleophilicity increases with increasing basicity, provided nucleophiles with a common nucleophilic atom are compared.* Thus, a nucleophile with a negative charge ($Y:^-$) will always be more basic and, therefore, more nucleophilic than a neutral nucleophile ($Y:$). With the exception of amine, phosphorus, and sulfur nucleophiles, nucleophiles used in S_N2 reactions are anions, namely, they have a negative charge ($Y:^-$). As pointed out earlier, the lower the pK_a is, the stronger the acid and the weaker the conjugate base.

$R\text{-}O^- \; > \; HO^- \; > \; Ar\text{-}O^- \; >$ 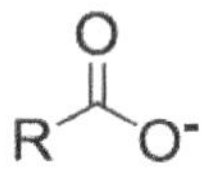

$\longleftarrow$ increasing base strength

The *size* of the atom that donates the pair of electrons has a profound effect on its nucleophilicity. *The larger the atom is, the greater its nucleophilicity.* Thus,

$:PH_3 \; > \; :NH_3$

$RSe^- \; > \; R\text{-}S^- \; > \; R\text{-}O^-$

$I^- \; > \; Br^- \; > \; Cl^- \; > \; F^-$

$\longleftarrow$ increasing size

$\longleftarrow$ increasing nucleophilicity

Nucleophiles that are capable of reacting at more than one atom, thereby giving rise to two different products, are referred to as *ambident nucleophiles*. The site of reaction in ambident nucleophiles is primarily determined by the nature of the solvent used in an S_N2 reaction, and the size of the atom. Some examples of ambident anions are shown below.

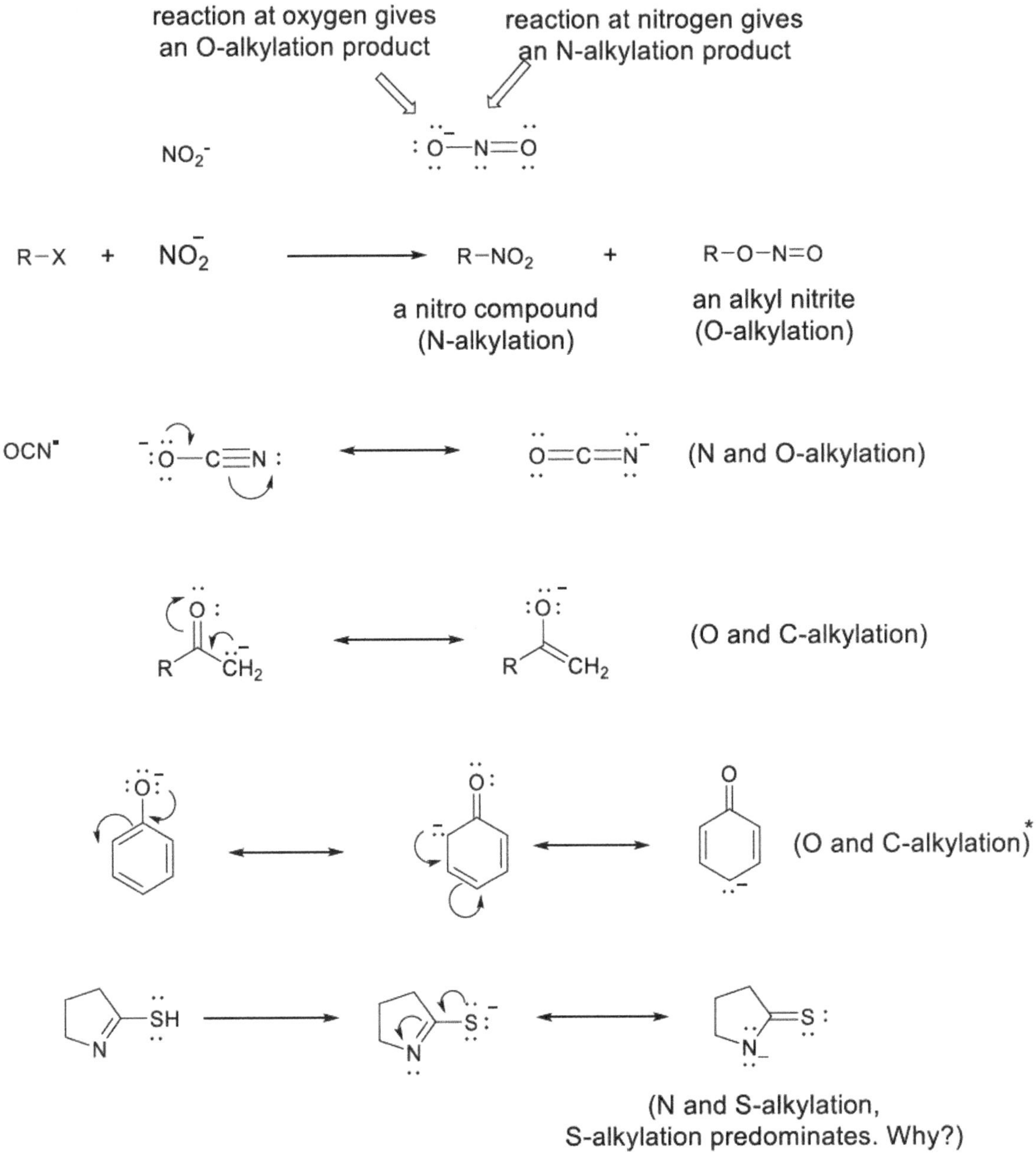
reaction at oxygen gives an O-alkylation product
reaction at nitrogen gives an N-alkylation product
NO2-
R-X + NO2 → R-NO2 + R-O-N=O
a nitro compound (N-alkylation)
an alkyl nitrite (O-alkylation)
OCN-
(N and O-alkylation)
(O and C-alkylation)
(O and C-alkylation)*
(N and S-alkylation, S-alkylation predominates. Why?)

* Note that C-alkylation leads to loss of aromaticity

c) Nature of the Leaving Group (L)

The group that is displaced by the nucleophile (Y:$^-$) in an SN2 reaction is referred to as the *leaving group. Good leaving groups are stable anions, namely, they are conjugate bases derived from strong acids*. For example, halide ions (I^-, Br^-, Cl^-, except F^-) are good leaving groups since they are the conjugate bases of strong acids. Many other good leaving groups are derived from strong organic acids and are stabilized by resonance (Minireview 3). These include the triflate, mesylate and tosylate groups (vide infra). The hydroxyl group of an alcohol (a poor leaving group that is not displaced by any nucleophile) is frequently transformed into one of these leaving groups prior to carrying out an S_N2 reaction (Scheme I below).

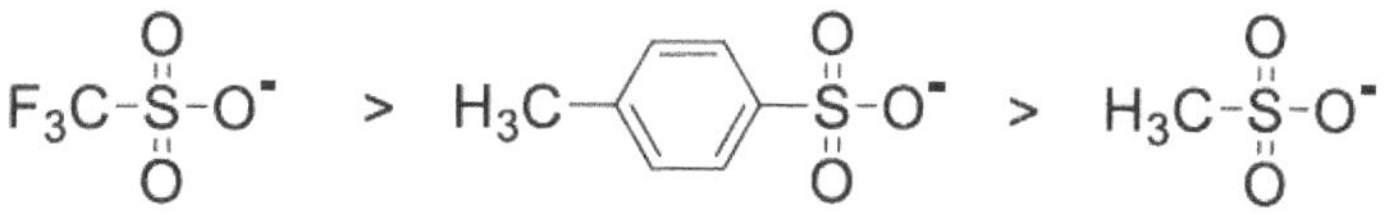

triflate tosylate mesylate

$CF_3COO^- > CH_3COO^-$

$CF_3SO_3^- > CH_3SO_3^-$

$I^- > Br^- > Cl^- > F^-$

← increasing leaving group ability

Scheme I

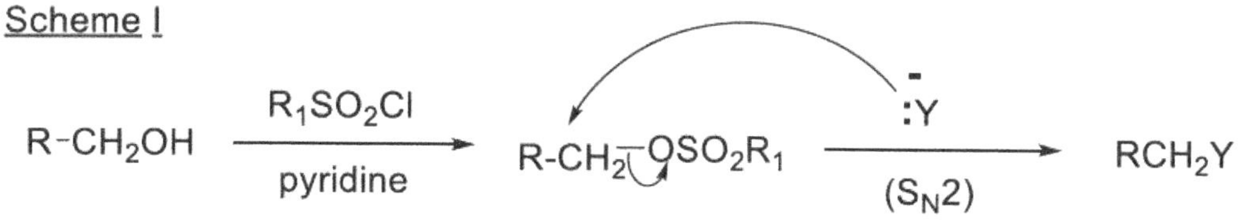

R-CH2OH
R1SO2Cl
pyridine
R-CH2-OSO2R1
:Y
(SN2)
RCH2Y

d) Nature of the Solvent

The nature of the solvent used in carrying out an S_N2 reaction can have a dramatic effect on the rate and success of the reaction. S_N2 reactions are greatly facilitated by dipolar aprotic solvents such as dimethyl sulfoxide (DMSO), dimethyl fornamide (DMF) and acetonitrile. These solvents enhance the nucleophilicity of anions by solvating the cation (solvation refers to the clustering of solvent molecules around the anion or cation).

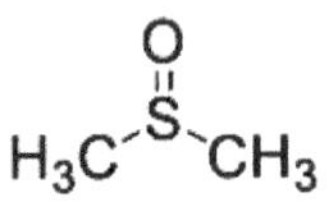

DMSO

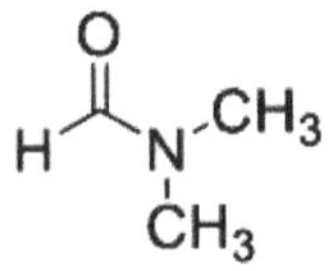

DMF

CH_3CN

acetonitrile

The nucleophilicity of an anion can also be enhanced by using a *crown ether* that complexes with the cation, as shown below. The size of the cavity in a polyether determines the type of cation that is complexed. Inorganic reagents, such as potassium permanganate, which are ordinarily insoluble in organic solvents, become soluble in organic solvents in the presence of a crown ether.

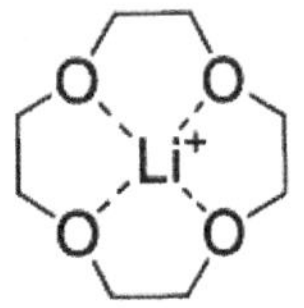

12-crown-4

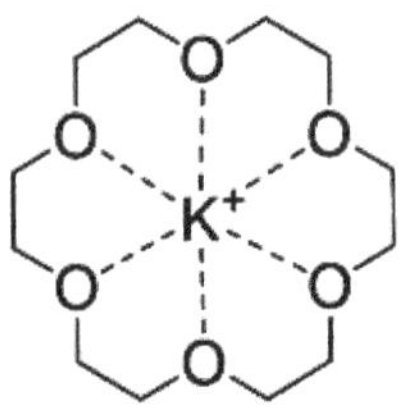

18-crown-6

Note:

S_N2 reactions can be *intermolecular* (eq 2) or *intramolecular* (eq 3). Intramolecular S_N2 reactions take place within the same molecule. Reactions in which the nucleophile and reactive center are tethered together, are favored entropically and proceed at a much higher rate than the corresponding intermolecular reactions.

Note:

Intramolecular reactions lead to the formation of cyclic structures, consequently, 5- and 6-membered ring structures are favored since they are free of ring strain.

Y:
+
R-L
R-Y
+
:L

eq 2

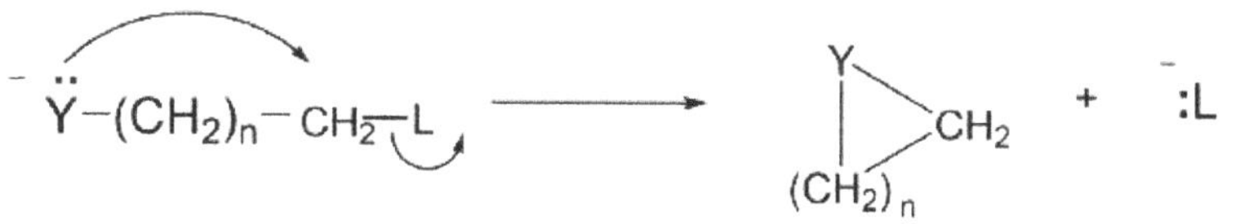

Y-(CH2)n-CH2-L
Y
CH2
(CH2)n
+
:L

eq 3

Elimination (E_2) Reactions

As mentioned earlier, when an alkyl halide or similar type of compound (R-L, where L= halogen or other good leaving group) is treated with a strong base ($:Y^-$), an S_N2 reaction take place. This is the course that an S_N2 reaction typically follows when a methyl, allylic, or benzylic halide is used. However, when a 3° R-L is used, the reaction follows a different cause due to steric reasons. An E_2 elimination reaction takes place, leading to a formation of a C=C bond (eq 4). *An E_2 reaction is a concerted 1,2- or β-anti elimination reaction* (the term *anti* means that the β-hydrogen removed by the base and the leaving group are opposite to each other). Secondary alkyl halides undergo competing substitution (S_N2) and elimination (E_2) reactions. The rate of an E_2 reaction increases with increasing base strength, leaving group ability, and the relative stability of the product(s). Typical bases used in E_2 reactions include HO^-, RO^-, and bulky organic bases such as diisopropyl ethyl amine, DBU, and DBN.

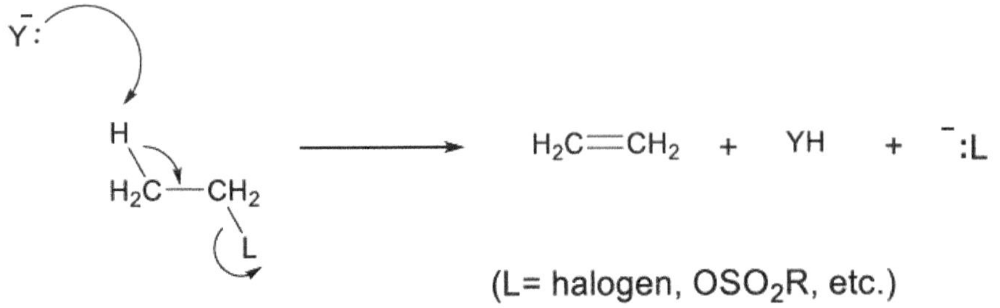
Y:
H
H2C—CH2
L
H2C=CH2 + YH + :L

eq 4

(L= halogen, OSO_2R, etc.)

Thermal Syn Elimination Reaction

Thermal *syn elimination* reactions constitute another somewhat less common type of β-elimination reaction. This type of reaction proceeds thermally through a cyclic transition state, and does not require the use of base (the term syn means that the β-hydrogen and the leaving group are on the same side). Syn eliminations involving sulfoxides and selenoxides are widely used in the facile formation of C=C bonds. While other leaving groups have been used in the past (acetate, amine oxide, etc.), elimination occurs at a much higher temperature.

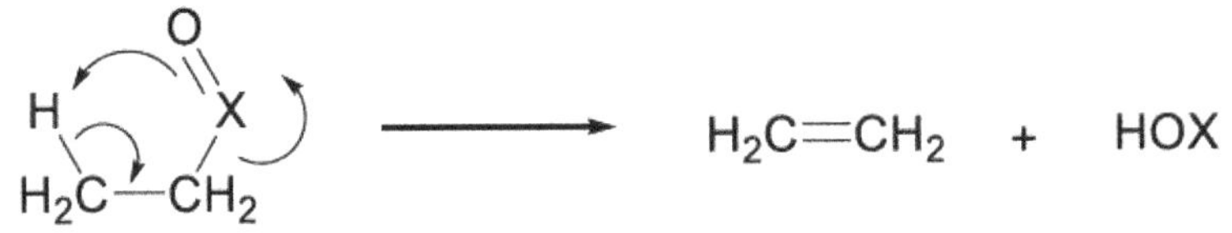
O
H
X
H2C—CH2
H2C=CH2 + HOX

eq 5

(X= SR or SeR)

Minireview 7

➢ Chemical Reactivity and Ring Strain

Three and four membered rings have considerable *ring strain*, namely, the ring bonds are weak because of poor overlap of the atomic orbitals and, consequently, exhibit high reactivity. They are highly reactive toward Lewis bases (nucleophiles), giving rise to ring-opened products. Note that *nucleophilic ring opening of an epoxide, for example, involves back-side attack at the least substituted (least congested) carbon, and proceeds with inversion of configuration* (eq 1). Transient 3-membered species, such as halonium and selinarium ions, behave similarly (eq 2 and 3).

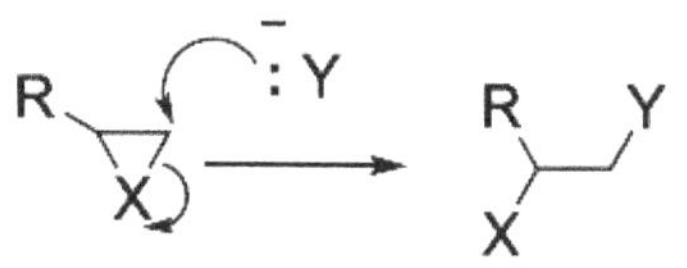

X= O (epoxide), S (episulfide)

eq 1

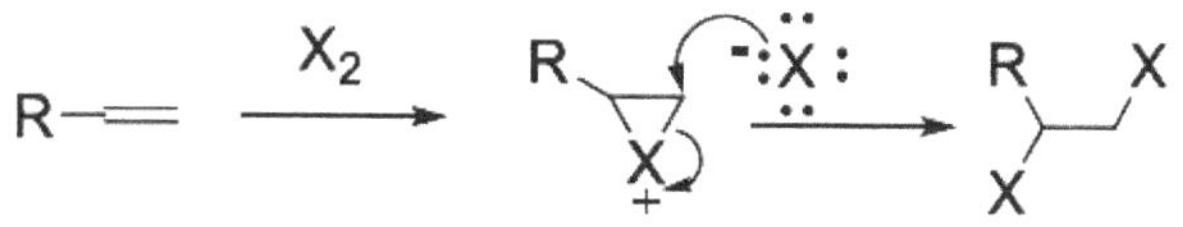

(X_2= Cl_2, Br_2, I_2)

eq 2

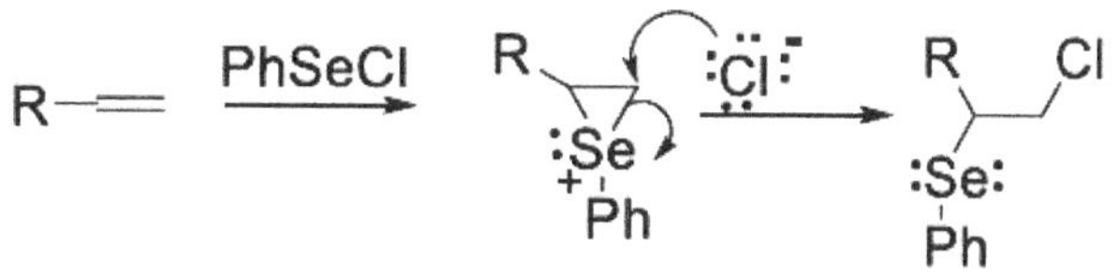

eq 3

3-and 4- membered cyclic esters (lactones) and cyclic amides (lactams) also react rapidly with nucleophiles (eq 4).

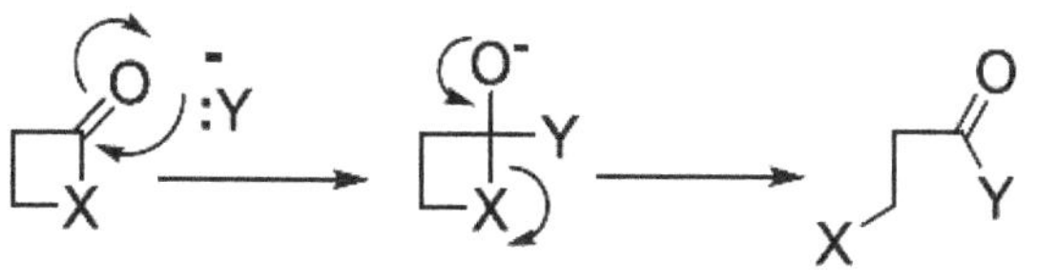

(X= O, NH, NR)

eq 4

Example 1

The highly reactive β-lactam ring of penicillins and related antibacterial agents undergoes ring opening rapidly. Hydrolysis of the β—lactam ring in penicillins results in total loss of antibacterial activity (in practical terms, suspensions of penicillins, such as pivampicillin, are kept refrigerated in order to slow down the hydrolysis reaction. They are thrown away after a week or so since they contain mostly inactive penicillin).

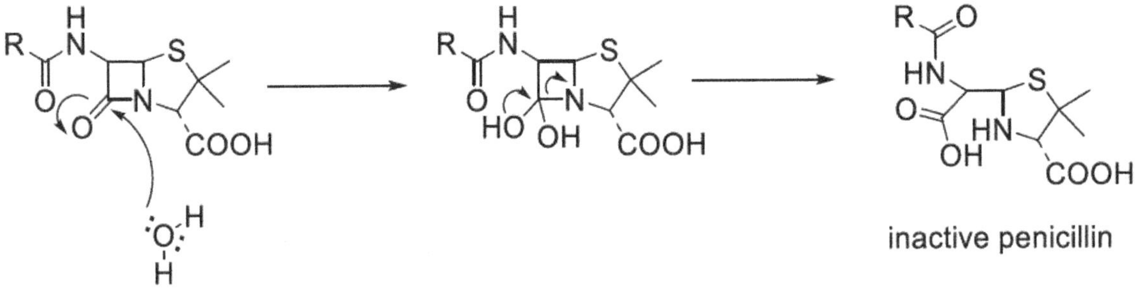

inactive penicillin

Example 2

The toxicity associated with a drug or substance is frequently due to the formation of a highly reactive metabolite. For example, the toxicity and carcinogenicity of polycyclic aromatic hydrocarbons and other aromatic compounds is attributed to the formation of highly reactive epoxides by liver-metabolizing enzymes, and their subsequent reaction with nucleophilic cellular components, including DNA.

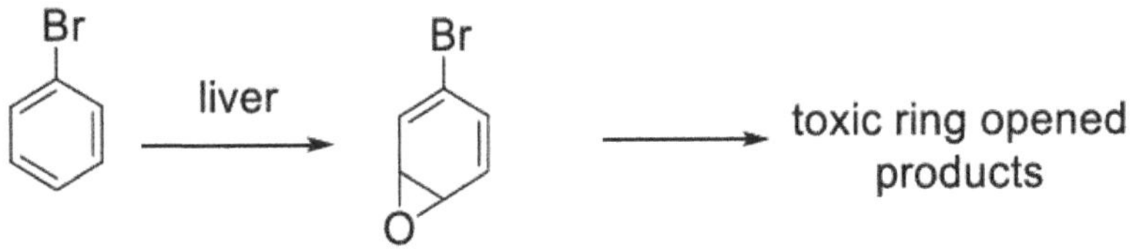
liver
toxic ring opened products

Questions 35 - 50

Questions 35-50 are meant to help you gain a better understanding of following topics.

- Generation of anions/acidity
- S_N2 reactions (mechanism, nucleophilicity, ambident nucleophiles, leaving groups, stereochemistry, solvent effects, etc.)
- E_2 reactions (anti and thermal syn elimination reactions)
- Nucleophilic ring-opening reactions

35	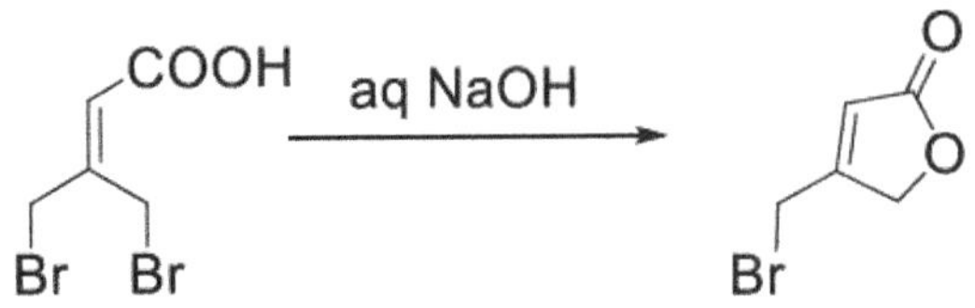COOH aq NaOH Br Br O O Br
36	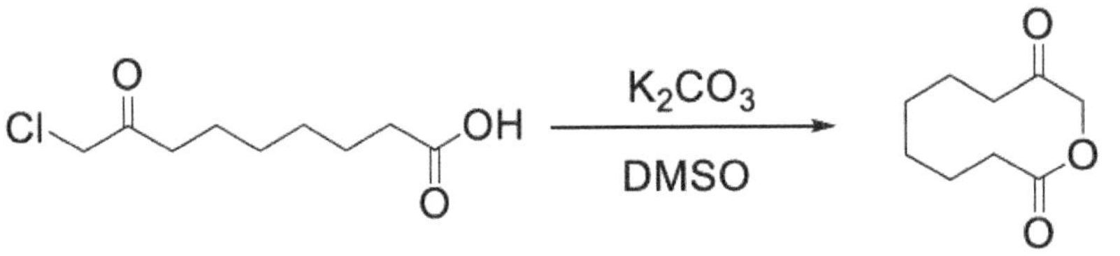Cl O OH O K_2CO_3 DMSO O O O
37	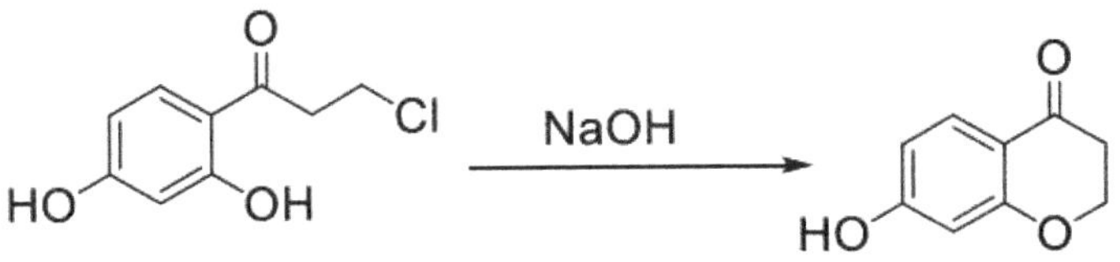O Cl HO OH NaOH O HO O
38	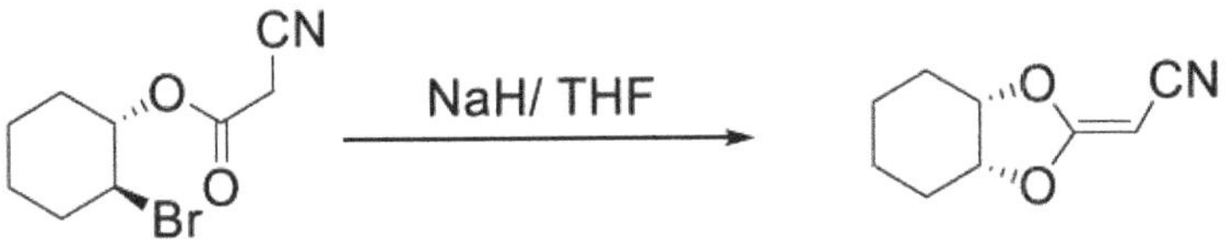CN O O Br NaH/ THF O O CN
39	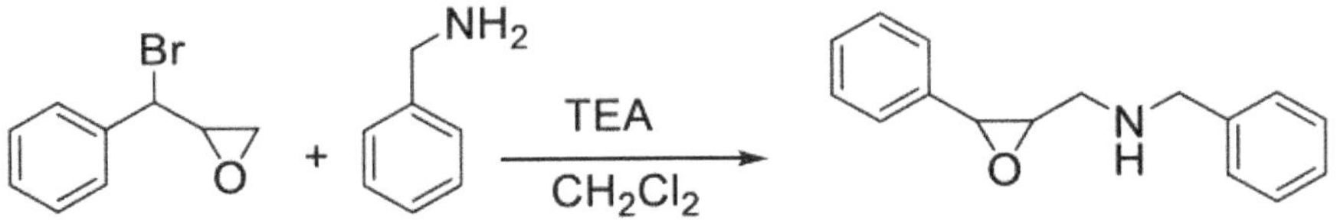Br O + NH_2 TEA CH_2Cl_2 O N H
40	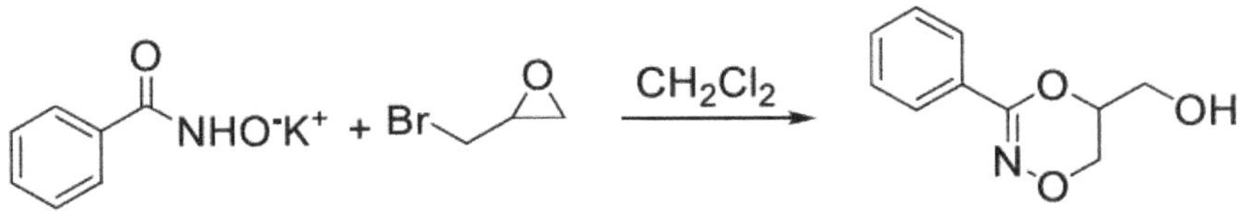O NHO^-K^+ + Br O CH_2Cl_2 O N O OH

41

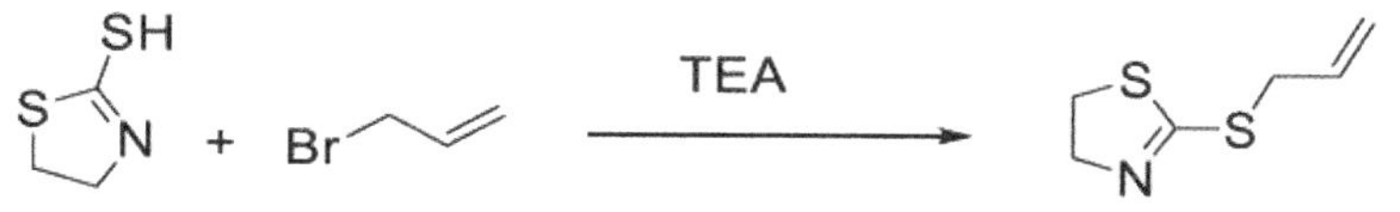
SH
S
N
+
Br
TEA
S
N
S

42

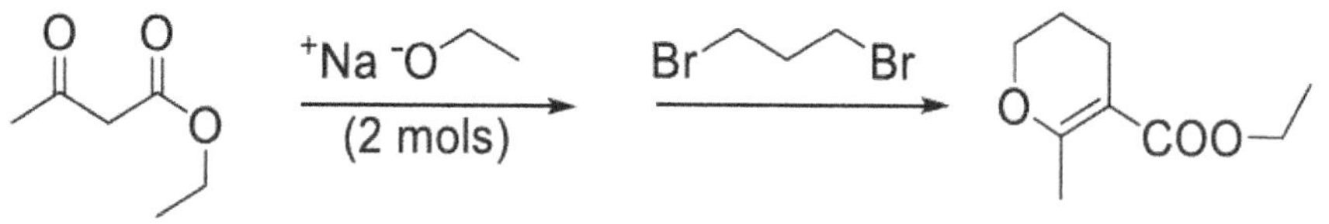
O
O
O
+Na -O
(2 mols)
Br
Br
O
COO

43

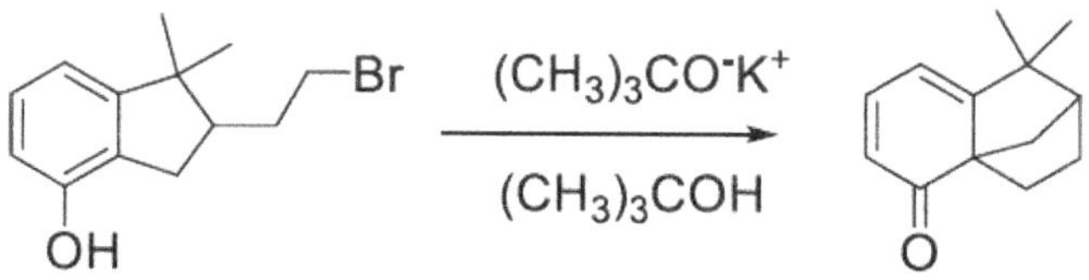
Br
OH
(CH3)3CO-K+
(CH3)3COH
O

44

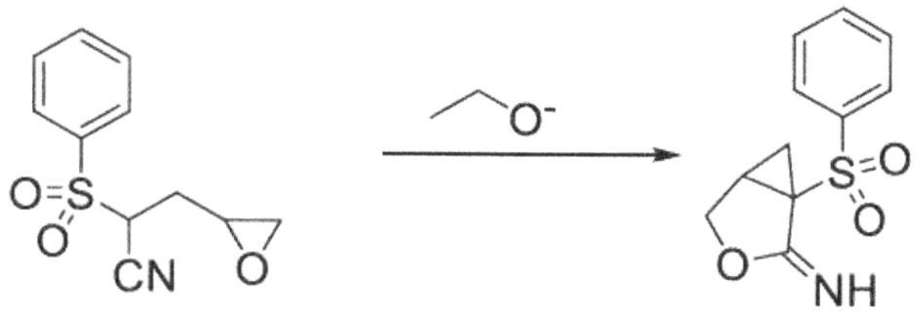
O=S
O
CN
O
O-
S=O
O
O
NH

45

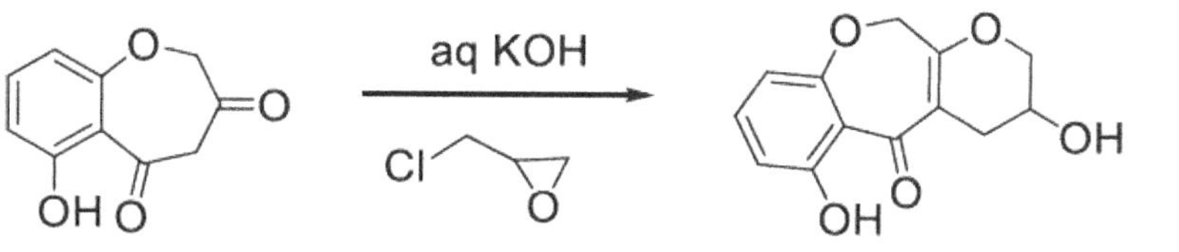
O
O
OH O
aq KOH
Cl
O
O
O
OH
O
OH

46

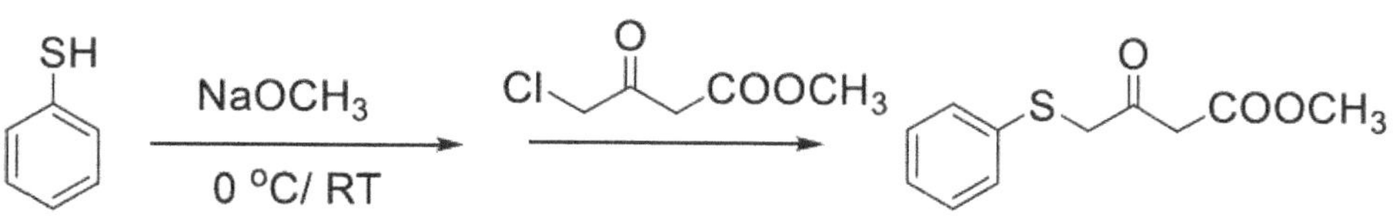
SH
NaOCH3
0 °C/ RT
O
Cl
COOCH3
S
O
COOCH3

47	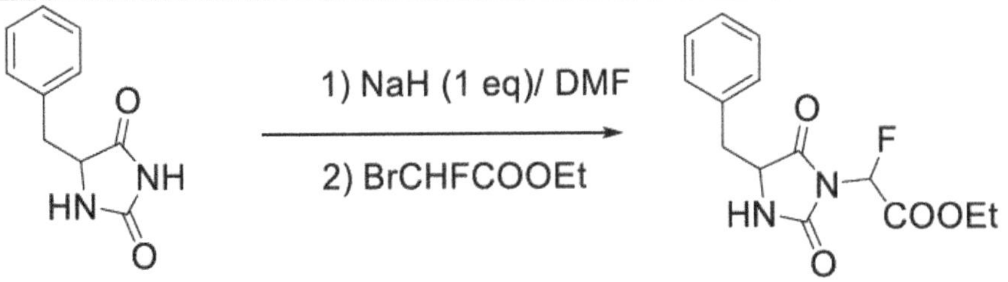O NH HN O 1) NaH (1 eq)/ DMF 2) BrCHFCOOEt O F N HN O COOEt
48	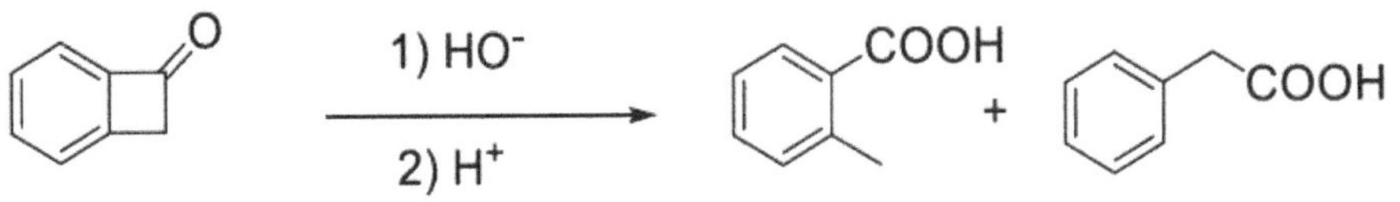O 1) HO- 2) H+ COOH + COOH
49	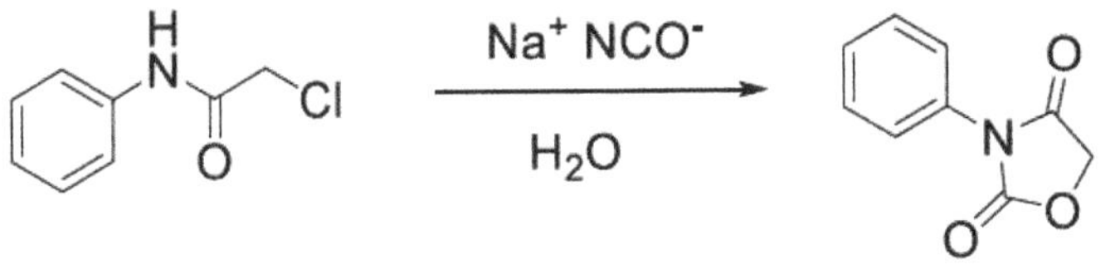H N O Cl Na+ NCO- H2O O N O O
50	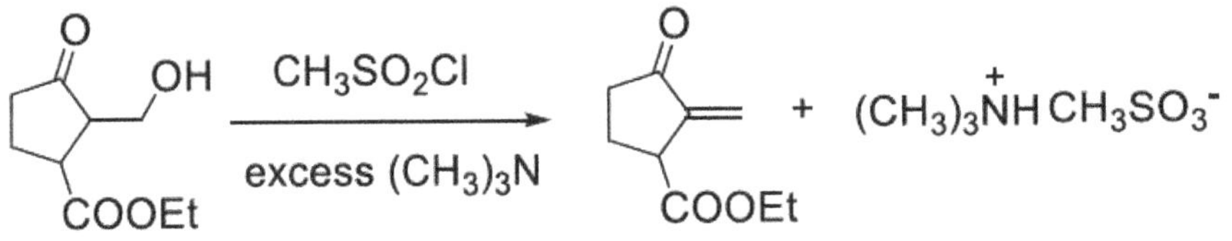O OH COOEt CH3SO2Cl excess (CH3)3N O COOEt + (CH3)3NH CH3SO3-

Answers to Questions 35-50

35

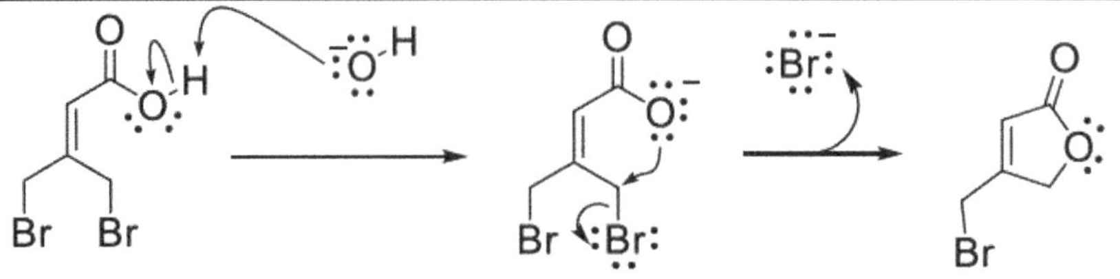

An initial fast acid-base reaction yields the anion (LB or nucleophile). This is followed by an intramolecular S_N2 reaction. The reaction proceeds readily at room temperature despite the fact that the carboxylate anion is a weak nucleophile because (a) the reaction is intramolecular (reaction takes place within the same molecule) and is therefore favored entropically; (b) the reacting groups are arranged in a favorable geometric alignment and, (c) the reaction involves a highly reactive allyl bromide (Minireview 6).

36

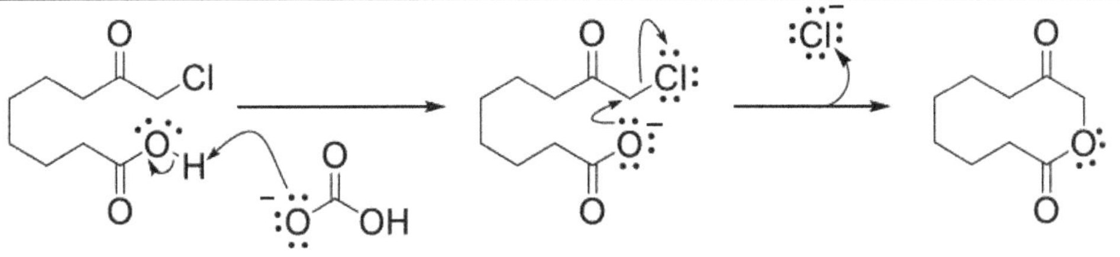

Formation of the anion is followed by an intramolecular S_N2 reaction. In general, intramolecular ring formation (cyclization) is favored when the reaction is carried out in *high dilution*. The relative ease of ring formation is dependent on many factors, including the stability of the ring to be formed, with lowest yields obtained with medium size (8 to 11-membered) rings.

37

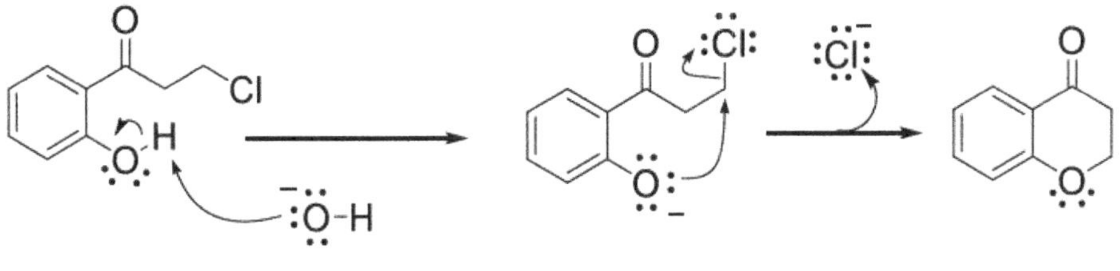

Formation of the anion of phenol (pKa~10) is rapid and quantitative. This is followed by an intramolecular S_N2 reaction. Note that the hydroxide ion could, in principle, displace the chloride first. It should be remembered, however, that the rates of acid/base reactions are extremely fast

38

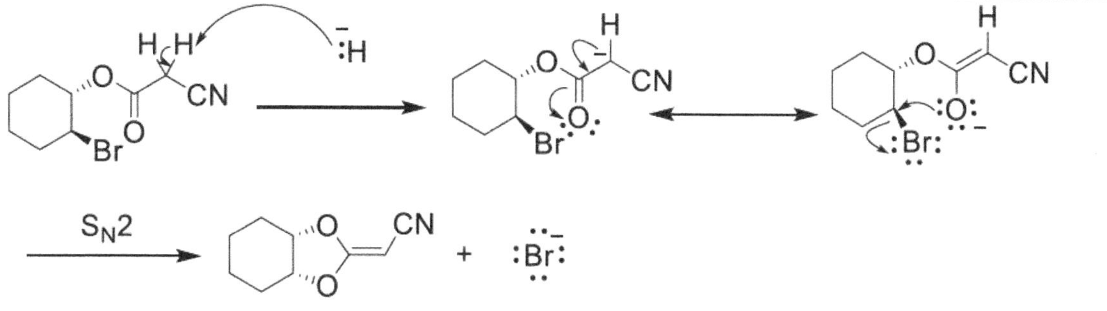

H
H
:H
O
CN
O
Br
S$_N$2
CN
O
:Br:

Rapid and quantitative formation of the anion using sodium hydride is followed by an intramolecular S$_N$2 reaction. As expected, the reaction proceeds with inversion of configuration, accounting for the observed stereochemistry. See Minireview 5 for a discussion of acidity and anion formation.

39

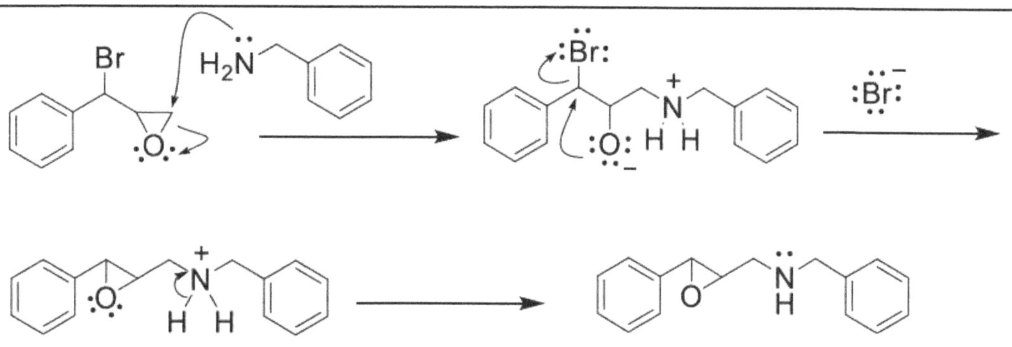

Br
H$_2$N
:O:
:Br:
N
H H

Epoxide (oxiranes) are highly reactive because of ring strain and, consequently, undergo ring opening reactions with the nucleophile attacking the least substituted (least congested) carbon.

40

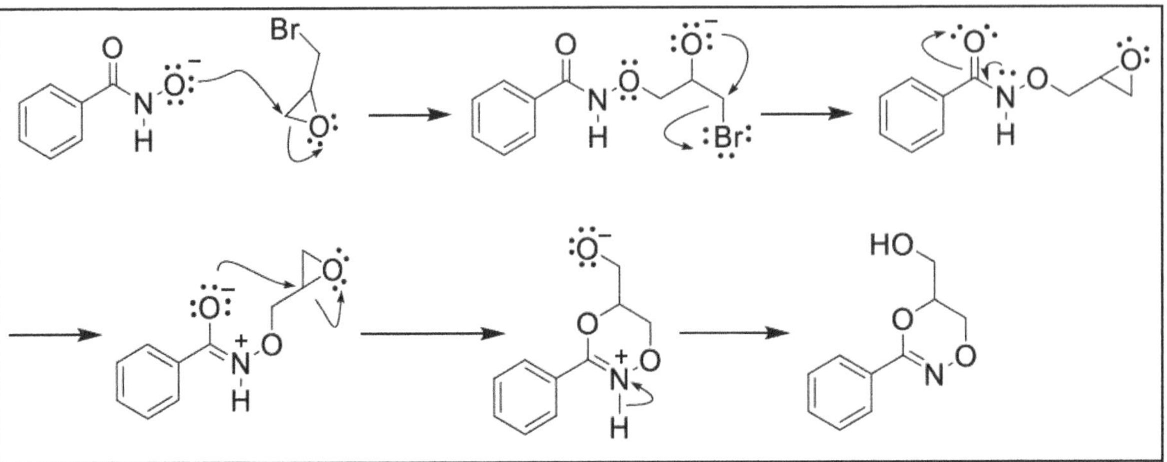

Epoxides are more reactive toward nucleophiles than 1° halides. Thus, nucleophilic ring opening is followed by epoxide formation which undergoes ring opening to form the observed product.

41

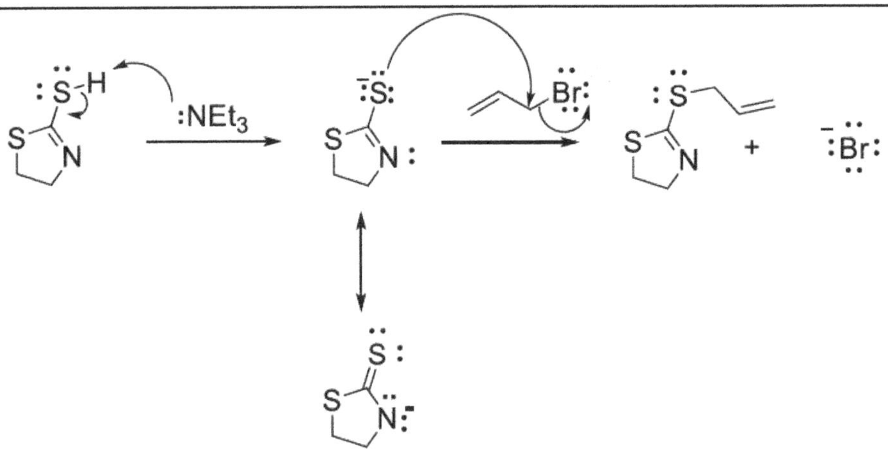
:NEt3

The anion generated in the first step (see Minireview 5) is an ambident anion, i.e., it can react at either the N or S atom. The S_N2 reaction takes place on S (the larger atom), attesting to the greater nucleophilicity of sulfur (see Minireview 6). Recall that size trumps basicity in S_N2 reactions.

42

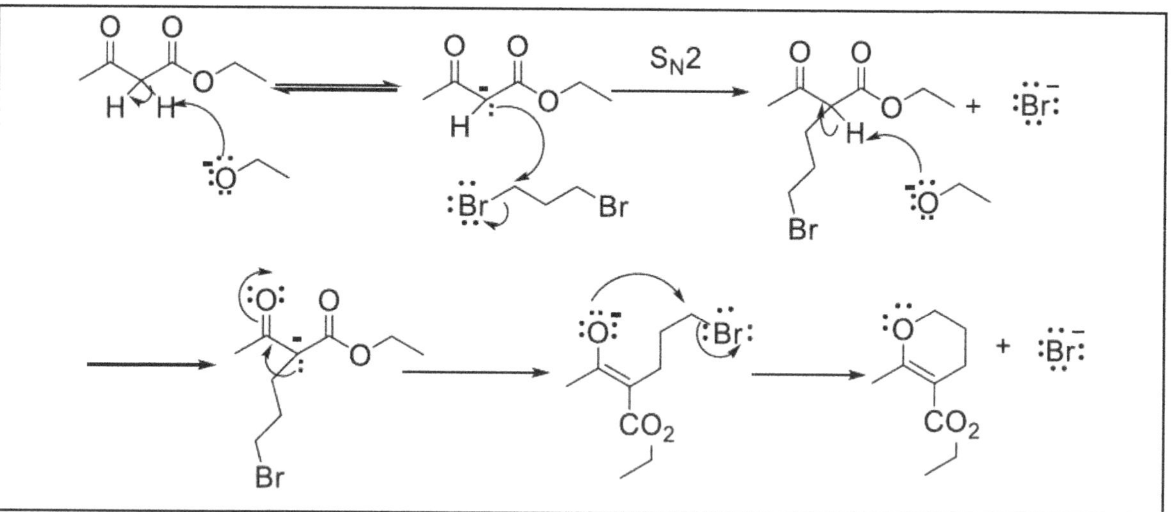

Formation of the ambident anion is followed by C-alkylation. A subsequent acid/base reaction leads to a second ambident anion which undergoes O-alkylation, leading to the formation of the observed cyclic product. Note that C-alkylation would be less favorable since it would result in the formation of a highly-strained 4-membered ring.

43

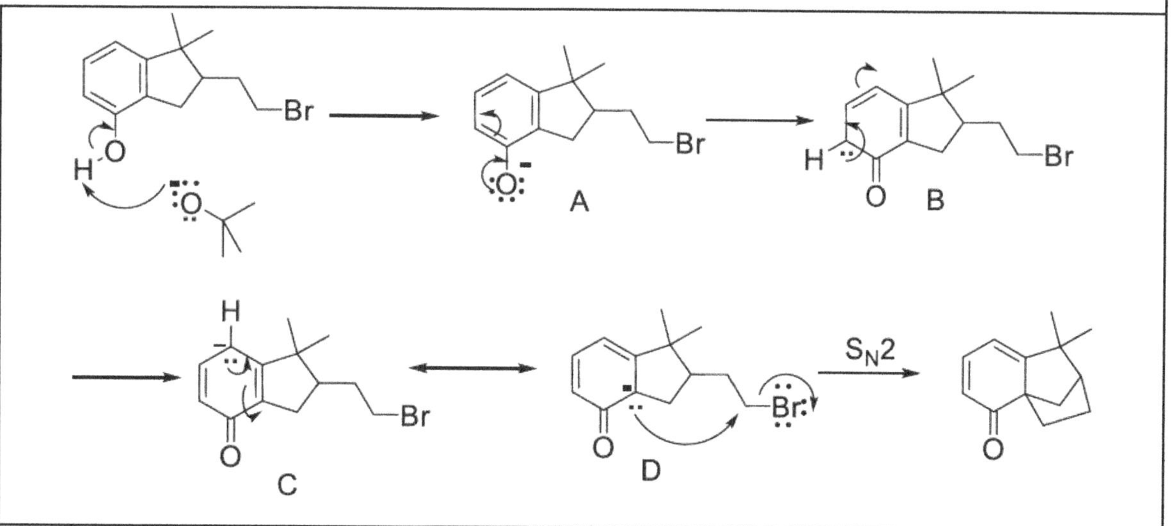

The first step involves a rapid and quantitative formation of a resonance-stabilized ambident anion. A subsequent intramolecular S_N2 reaction (C-alkylation) leads to the observed tricyclic non-aromatic product. As pointed out earlier, the nature of the product formed in reactions involving ambident anions is dependent on a range of factors.

44	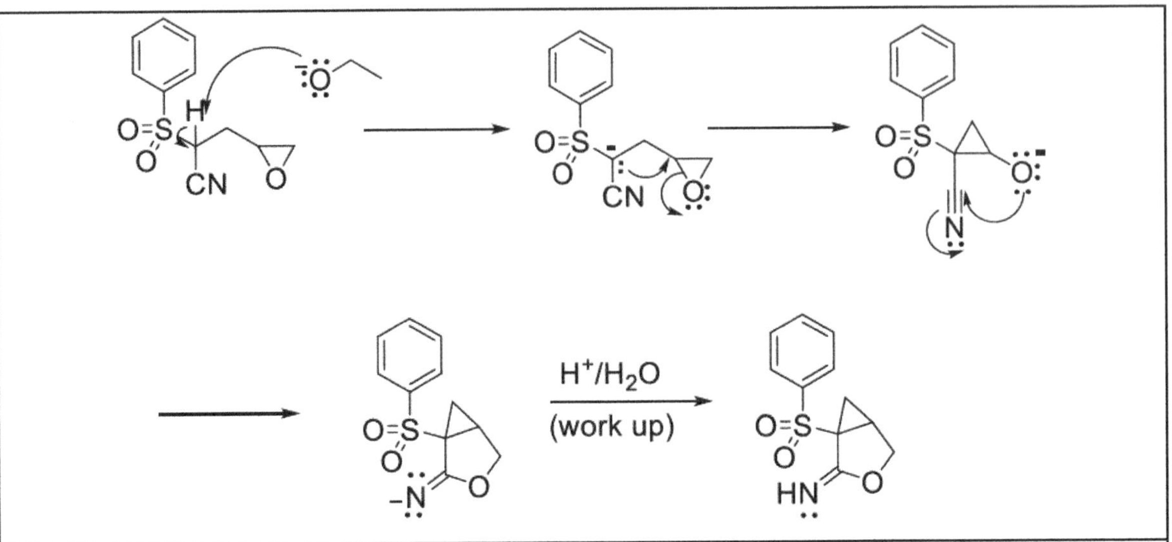
45	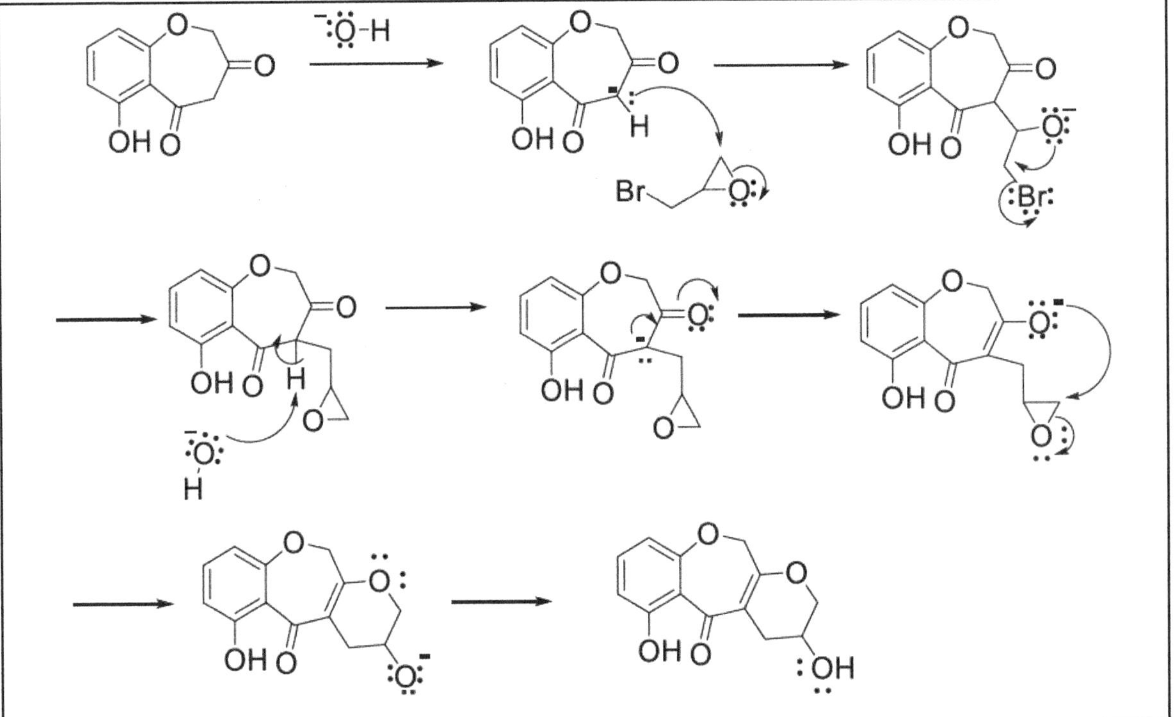
46	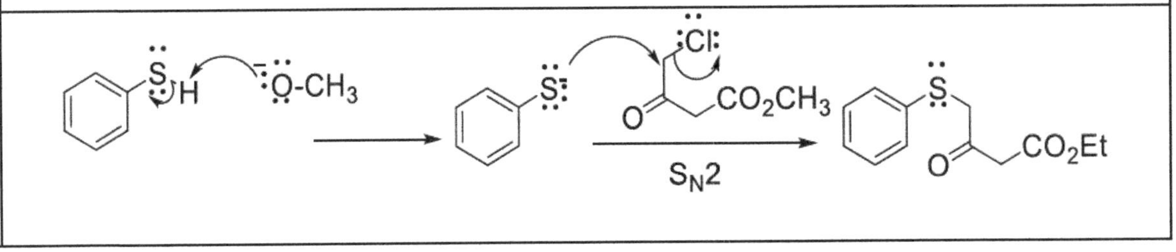

H⁺/H₂O
(work up)

S_N2
CO₂CH₃
CO₂Et

47

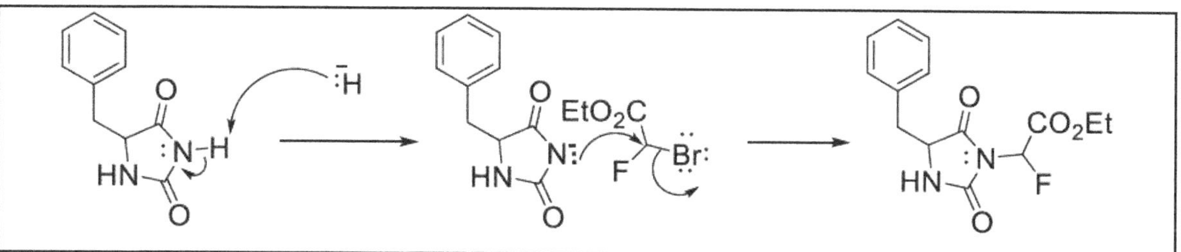

EtO2C
Br
CO2Et
F
HN

The base reacts preferentially with the more acidic hydrogen leading to the formation of the more stable anion (3 resonance structures). Removal of the less acidic hydrogen would result in the formation of a less stable anion (2 resonance structures). Note also that the nucleophile displaces the bromide ion (better leaving group) instead of the fluoride anion (a poor leaving group).

48

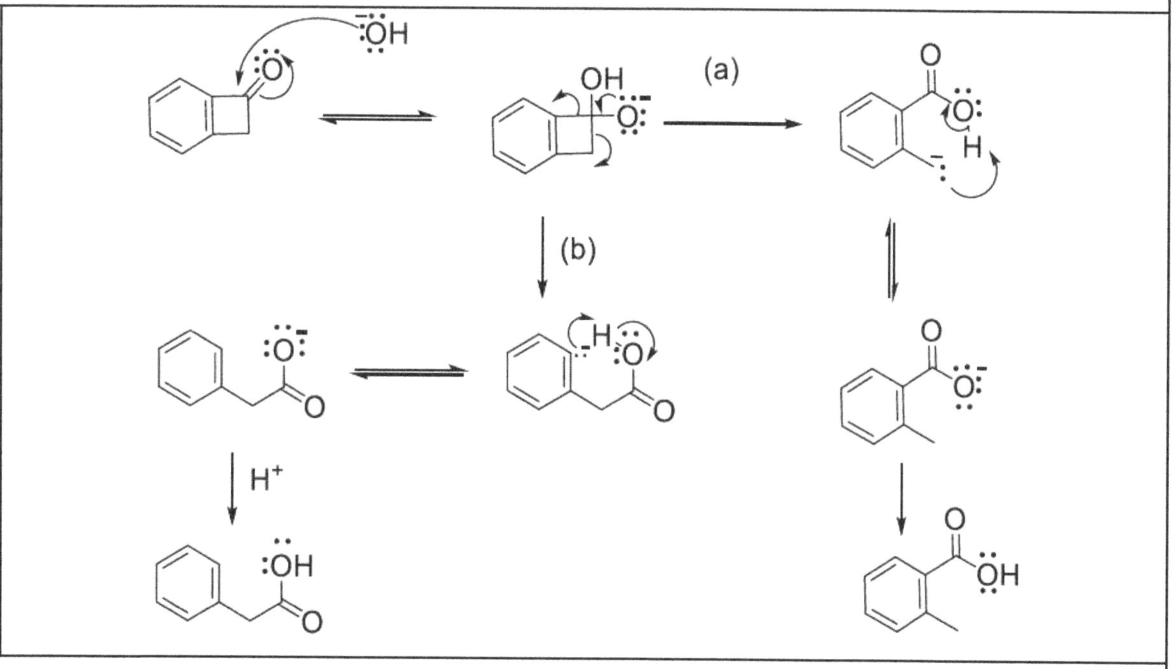

(a)
(b)
H+

The driving force for the reaction is the ring opening of the cyclobutane ring and the relief of ring strain.

49

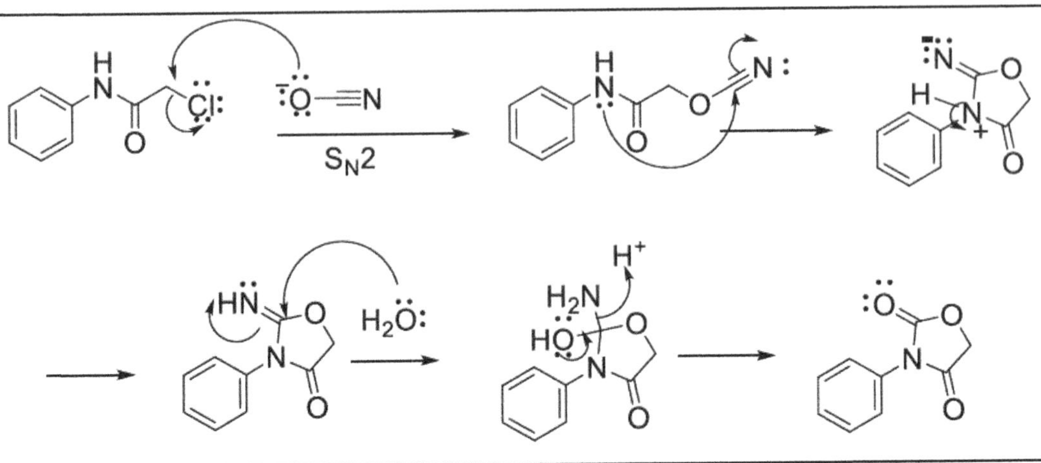
S$_N$2
H$^+$
H$_2$O:
H$_2$N
HO

50

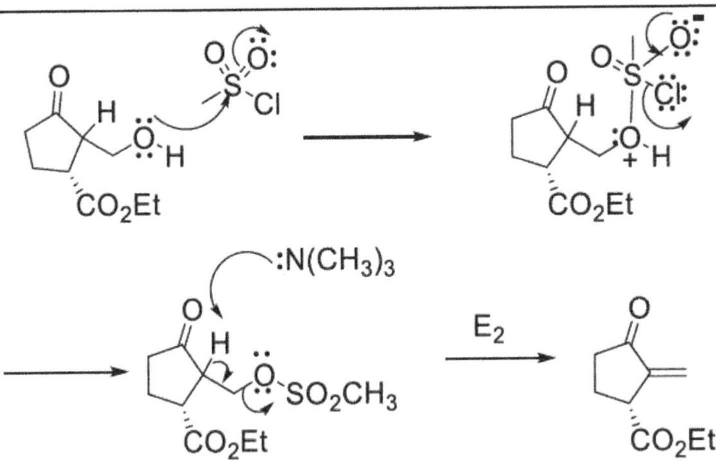
Cl
CO$_2$Et
:N(CH$_3$)$_3$
SO$_2$CH$_3$
E$_2$

Formation of the sulfonate ester transforms the hydroxyl group from a poor leaving group into a good leaving group (see also Minireview 6).

Minireview 8 focuses on nucleophilic addition reactions (the typical reaction of aldehydes and ketones). This should be studied prior to attempting questions 51-70.

Minireview 8

- Nucleophilic Addition Reactions of Aldehydes and Ketones

The typical reaction of aldehydes and ketones is *nucleophilic addition*. Notice that the nucleophilic addition reaction is essentially a Lewis acid/Lewis base reaction involving a Lewis base ($:Y^-$ or :Y) with the electron-deficient carbon (Lewis acid) of the carbonyl group. The addition product arising from strong nucleophiles is then protonated (typically during the workup of the reaction) and the final product isolated (eq 1). Typical strong nucleophiles including Grignard reagents (RMgX or $R:^-\ Mg^{2+}X^-$), alkyl lithium reagents (RLi or $R:^-\ Li^+$), cyanide ion, alkyne anions ($RC{\equiv}C:^-$), reducing agents that deliver hydride ($H:^-$) ions, etc.

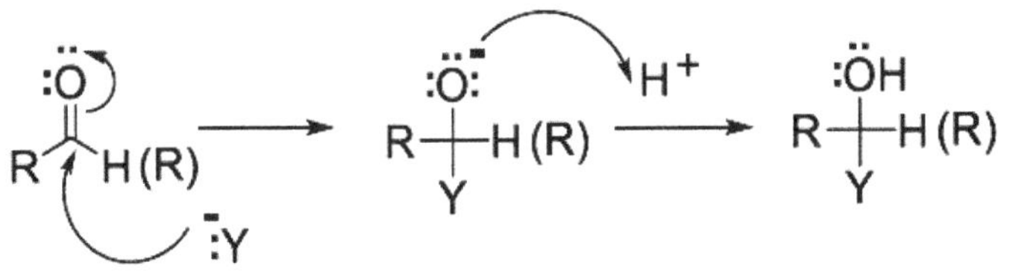

eq 1

In the familiar aldol *condensation reaction*, an aldehyde or ketone having an acidic α-hydrogen reacts with dilute base to yield a carbon anion (Lewis base), which then adds to the carbonyl carbon of a second molecule of the aldehyde or ketone, leading to the formation of a β-hydroxy aldehyde or ketone (an α,β-unsaturated aldehyde or ketone may also form depending on the reaction conditions) (Scheme I).

Scheme I

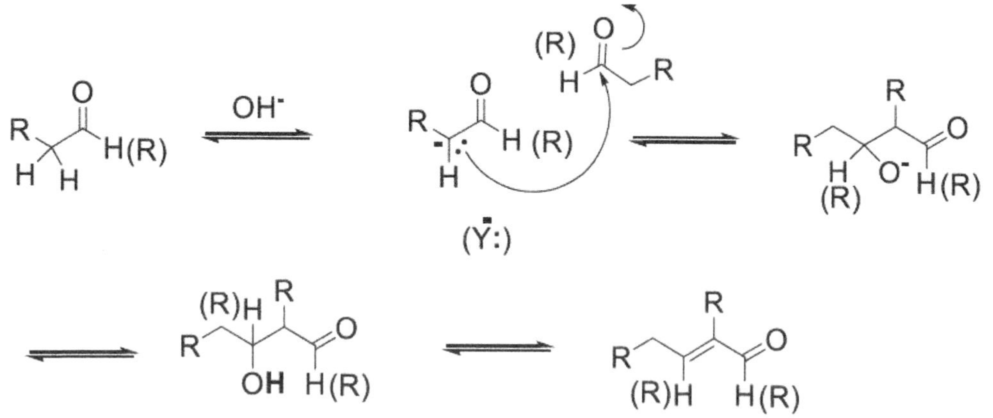

In the case of some nucleophilic addition reactions, the initial nucleophilic addition product reacts further to yield a more stable product. For example, the reaction of an aldehyde or ketone with a *Wittig reagent* involves an initial nucleophilic addition reaction. An intramolecular Lewis acid/base reaction yields a 4-membered intermediate which collapses, forming an alkene (Scheme II).

Scheme II

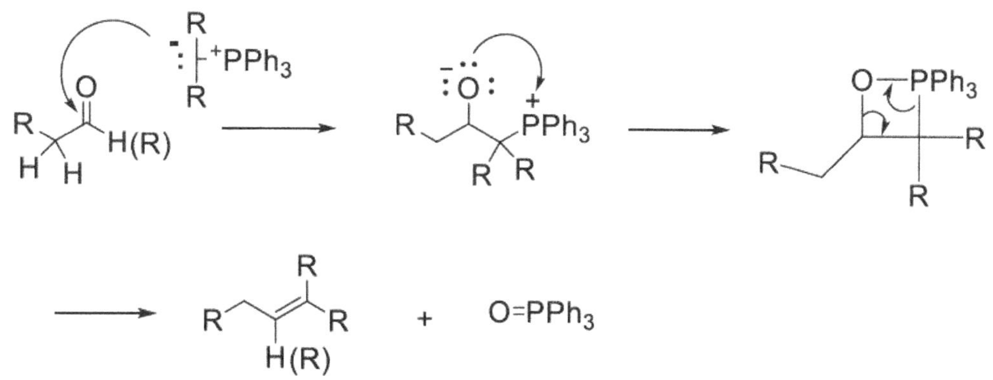

Likewise, *primary* amines react rapidly with aldehydes and ketones, however, the initial adduct loses a molecule of water to form an *imine* (Scheme III). Imines undergo addition reactions with Lewis bases. These reactions proceed readily when the imine is protonated (Scheme IV). *Secondary* amines react with aldehydes and ketones to form enamines (Scheme V).

Scheme III

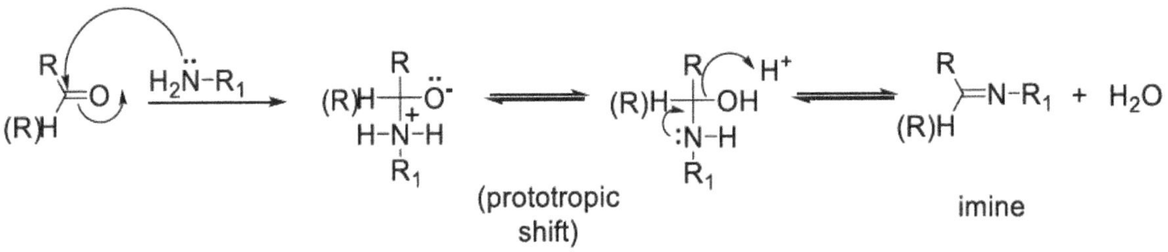
(prototropic shift)
imine

Scheme IV

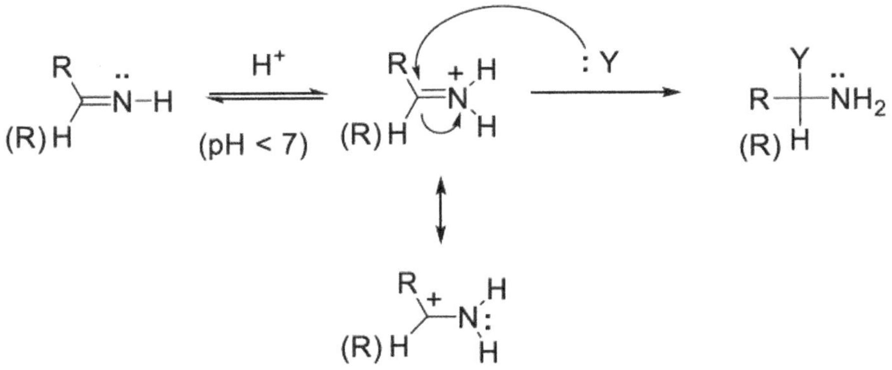
(pH < 7)

Scheme V

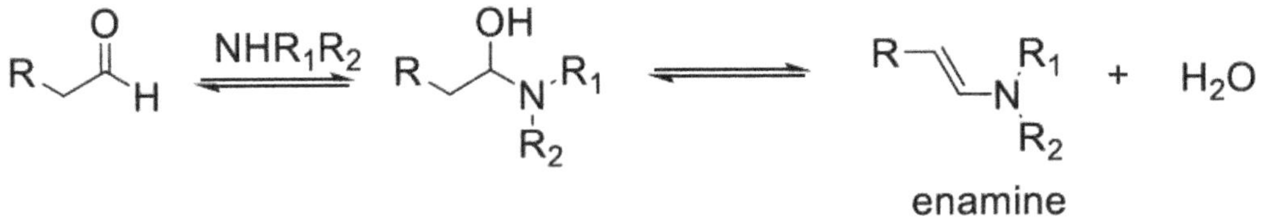
enamine

The reaction of aldehydes and ketones with weak nucleophiles (:Y), such as alcohols, requires acid catalysis (Scheme VI). Recall that the reaction of two mols of an alcohol with an aldehyde or ketone in the presence of acid (dry HCl) leads to the formation of an acetal or ketal, respectively (Scheme VI).

Scheme VI

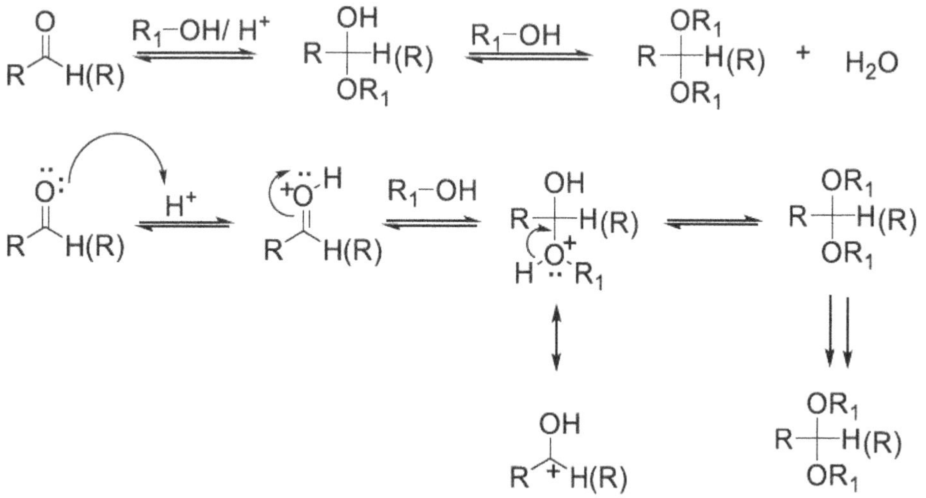

α,β-Unsaturated aldehydes and ketones undergo the *Michael addition reaction* with Lewis bases (nucleophiles) (Scheme VII). Related conjugated systems, R-CH=CH-X, where X=COOR, CN, SO_2R and NO_2, behave similarly.

Scheme VII

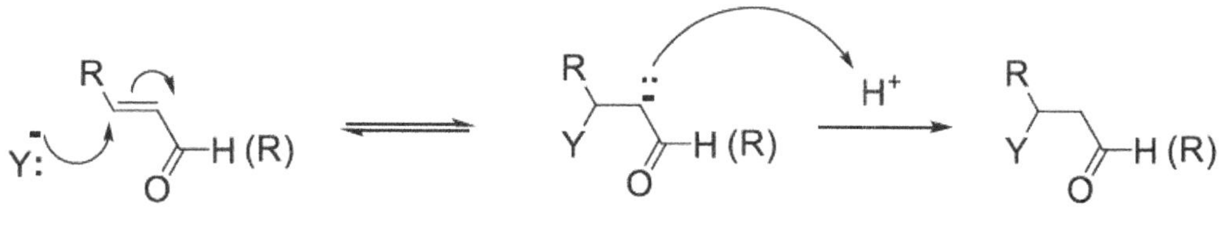

Lastly, the pK_a of the α-hydrogens in aldehydes and ketones is about 20. Consequently, treatment with a strong base leads to the formation of the corresponding anion (a Lewis base or nucleophile). The anion can then participate in nucleophilic displacement reactions (S_N2), nucleophilic addition reactions, or nucleophilic substitution reactions (see Minireview 6).

Questions 51 - 70

Questions 51-70 are meant to help you gain a better understanding of following topics.

- Acid and base-catalyzed addition reactions
- Aldol condensation reactions
- Michael addition reactions
- Keto-enol tautomeric equilibria

51	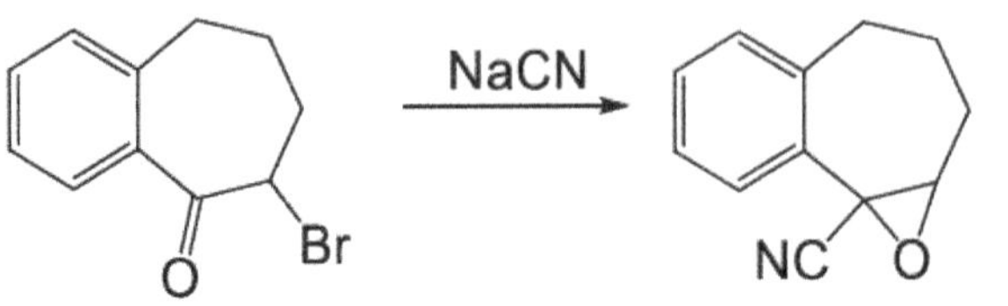 NaCN Br O NC O
52	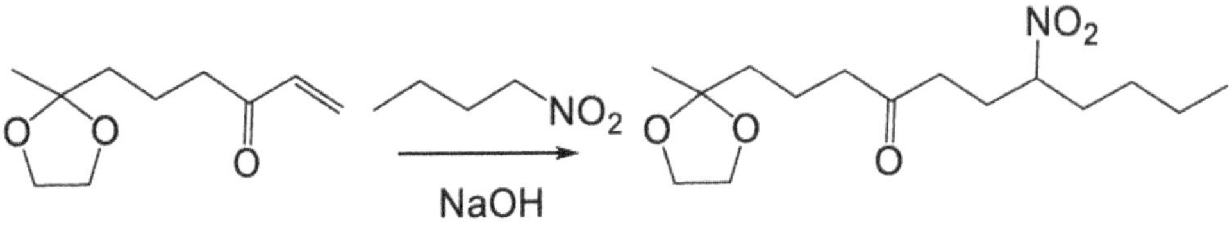 O O O NO2 NaOH O O NO2 O
53	 O O O dil. NaOH O O
54	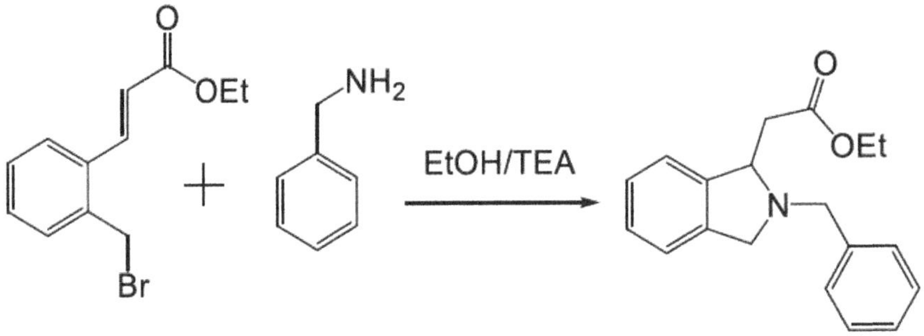 O OEt Br NH2 EtOH/TEA O OEt N
55	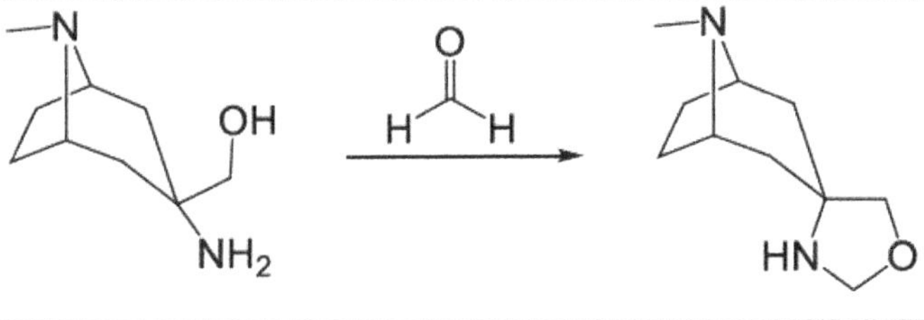 N OH NH2 O H H N HN O
56	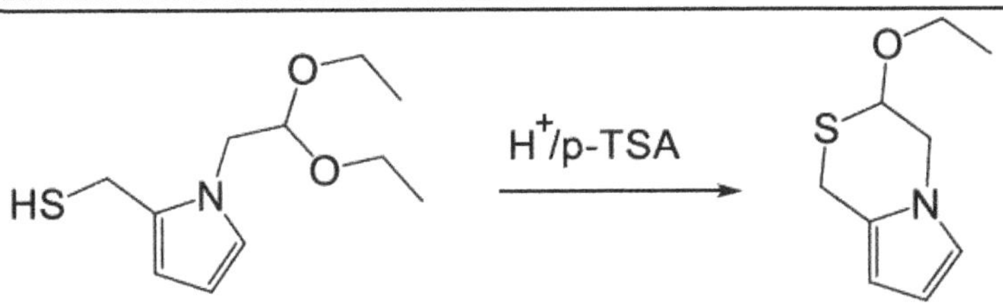 O O HS N H+/p-TSA O S N

57	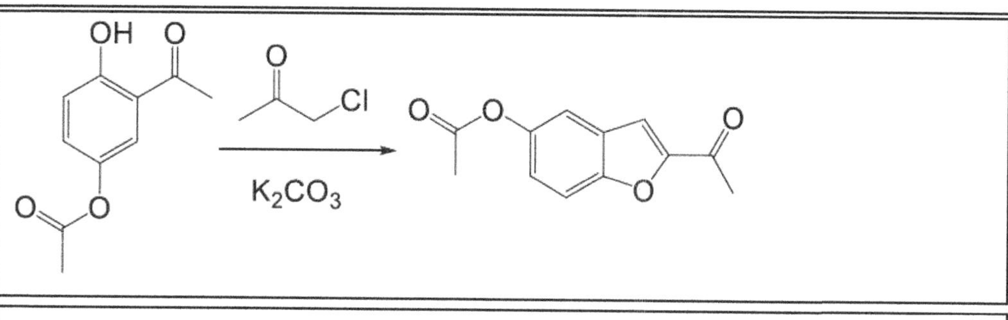OH O Cl K2CO3 O
58	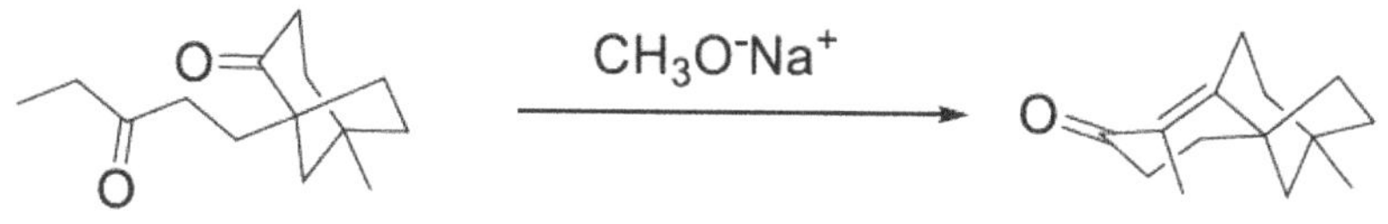O CH3O-Na+
59	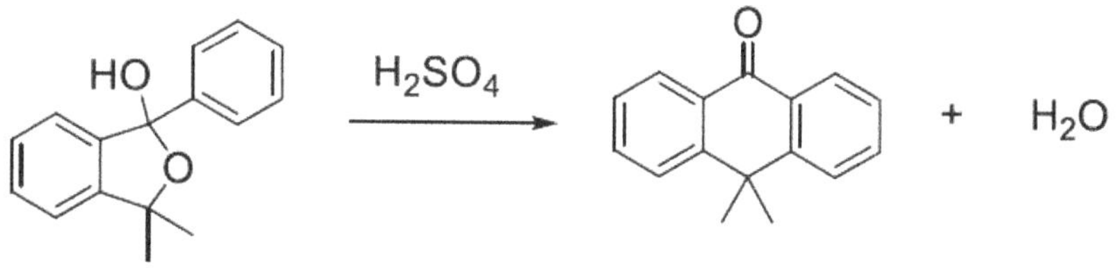HO O H2SO4 + H2O
60	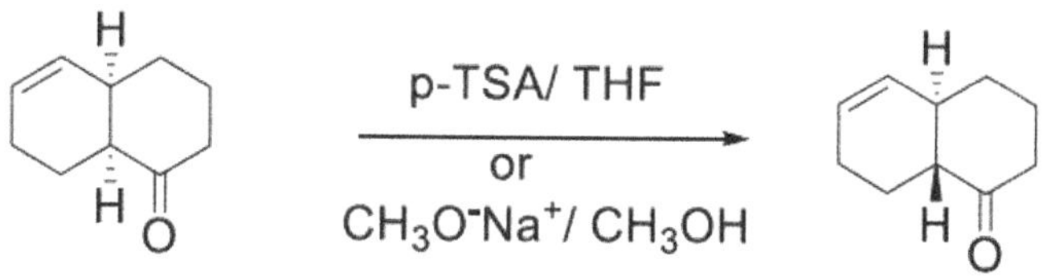H O p-TSA/ THF or CH3O-Na+/ CH3OH
61	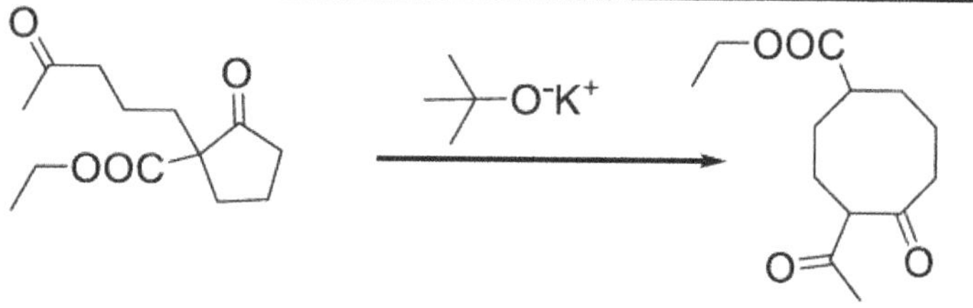O OOC O-K+
62	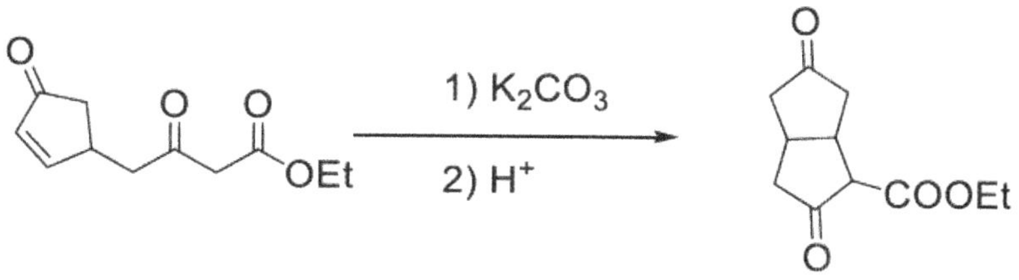O OEt 1) K2CO3 2) H+ COOEt

63	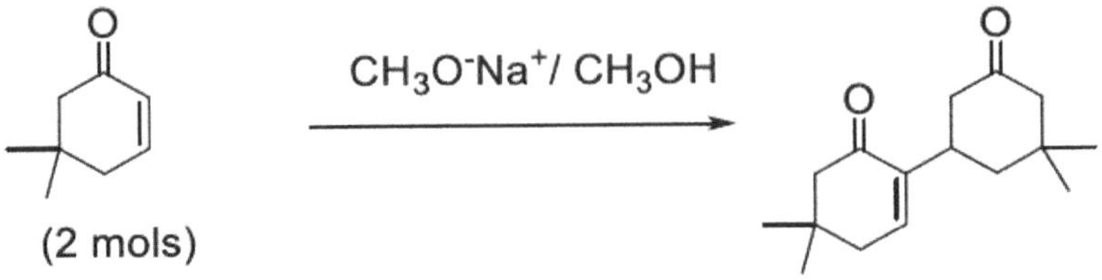CH3O-Na+/ CH3OH (2 mols)
64	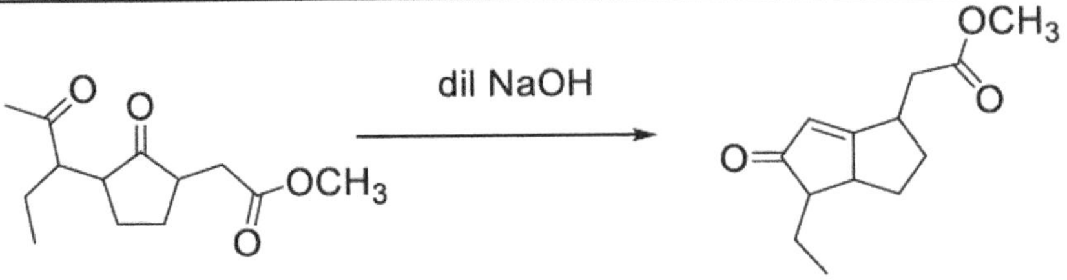dil NaOH OCH3 OCH3
65	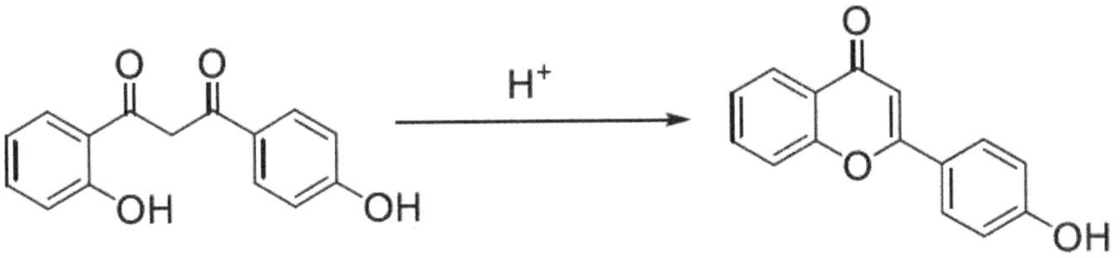H+ OH OH OH
66	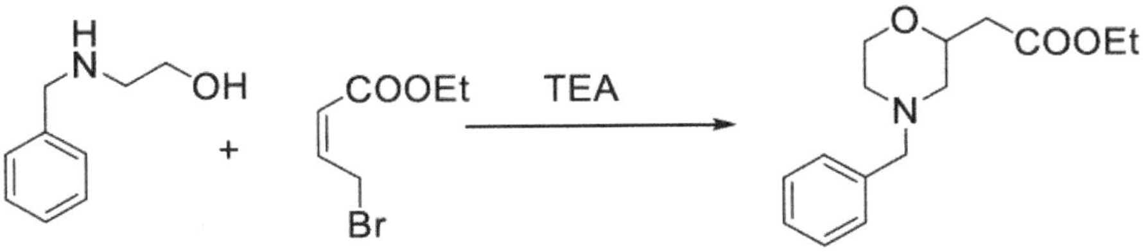OH + COOEt Br TEA COOEt
67	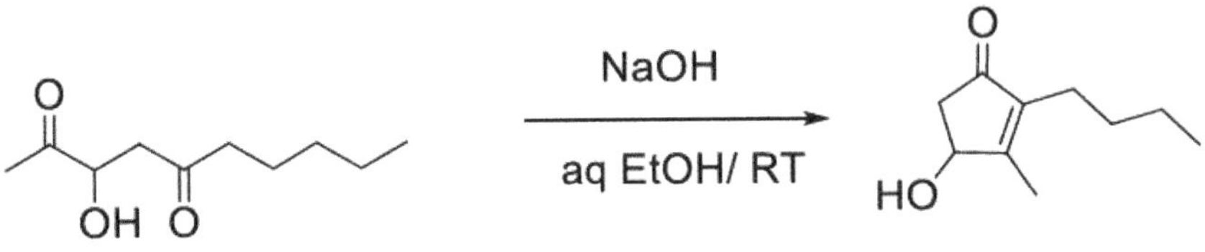NaOH aq EtOH/ RT OH HO
68	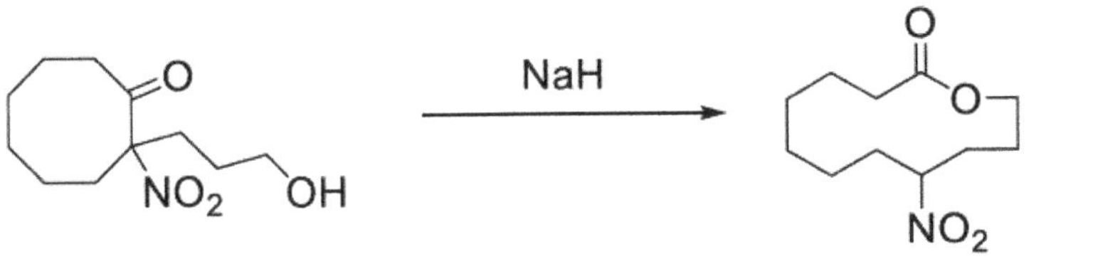NO2 OH NaH NO2

69	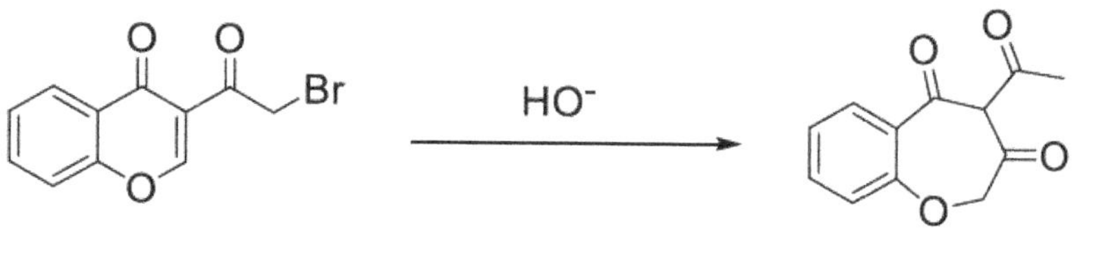 Br HO^-
70	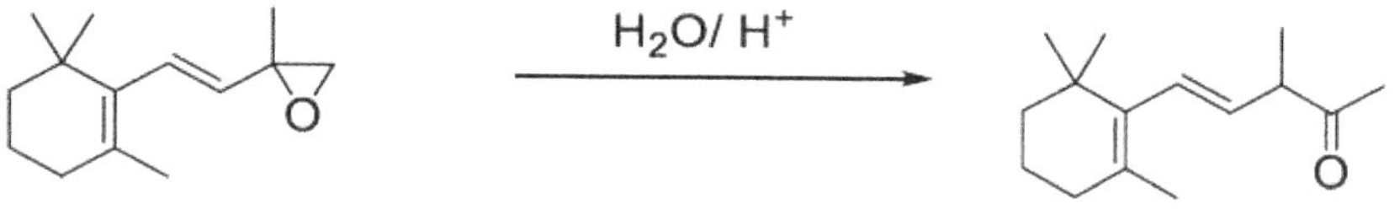 H_2O/ H^+

Answers to Questions 51-70

51

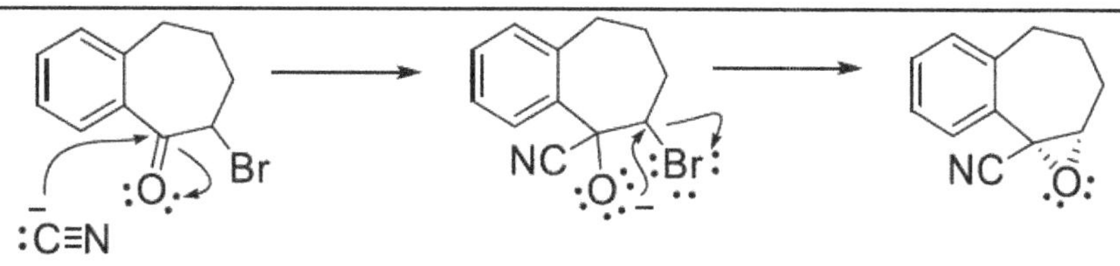

A straightforward nucleophilic addition is followed by an intramolecular S_N2 reaction.

52

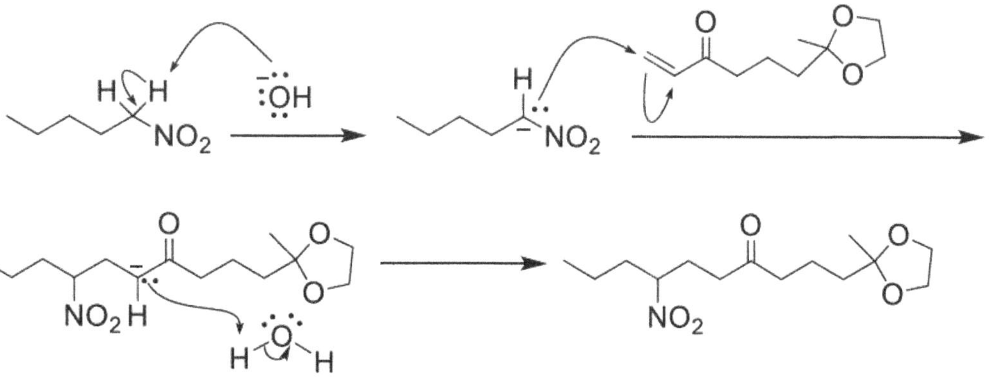

Note that once the anion is generated, what happens next is determined by the nature of the other reactant (see Minireview 5). Since the reactant in this case is a conjugated carbonyl compound, a Michael addition reaction takes place.

53

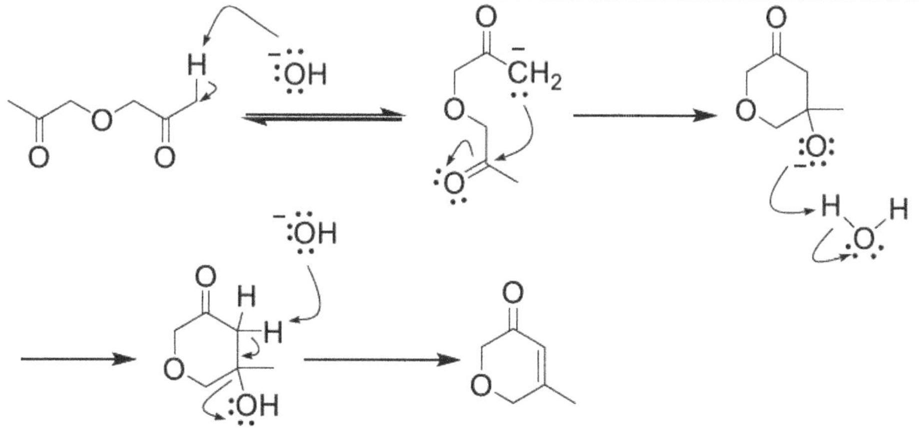

This is an example of an intramolecular aldol condensation reactio

54	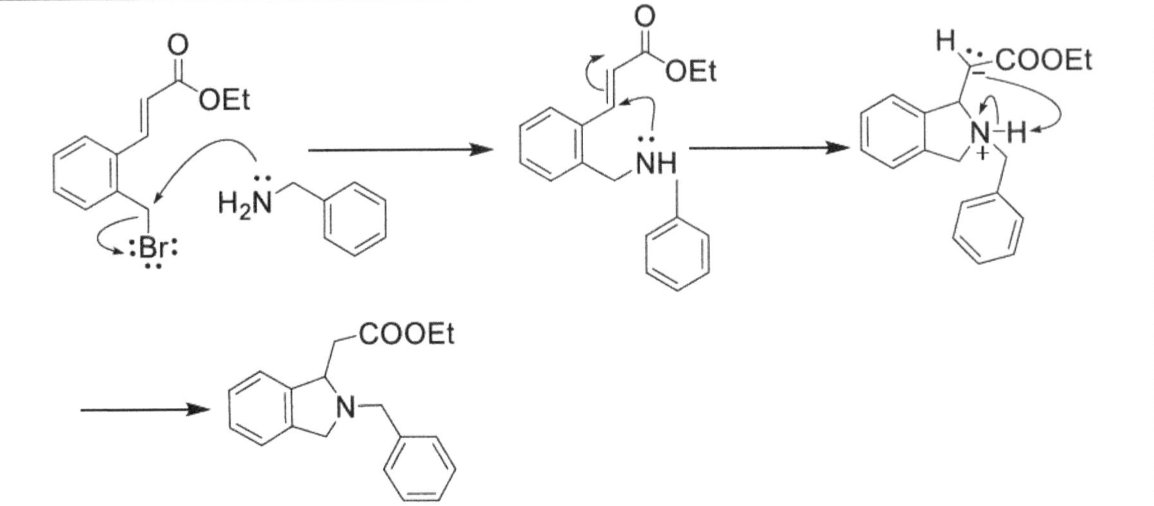
	An S_N2 reaction is followed by an intramolecular Michael addition reaction.
55	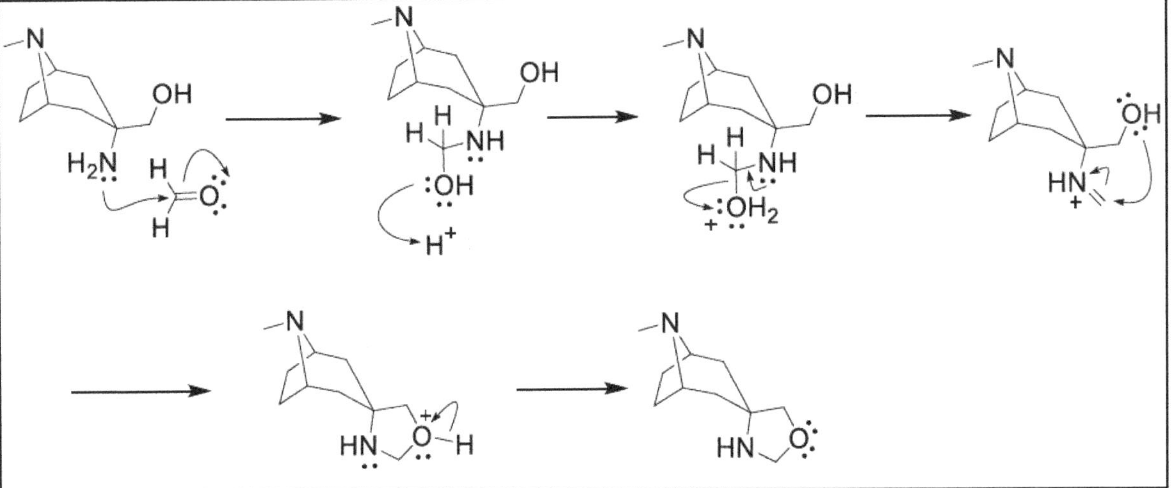
56	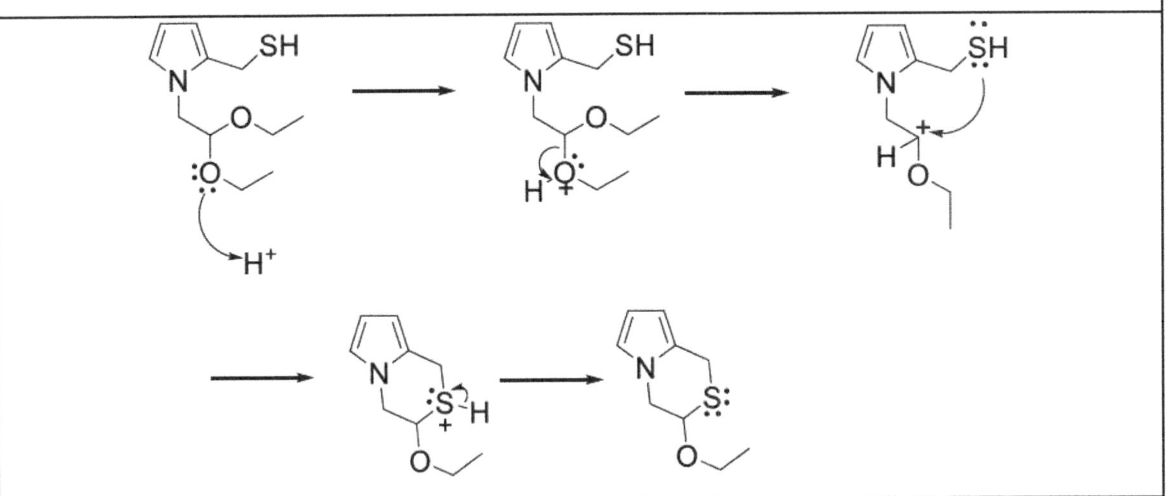

57	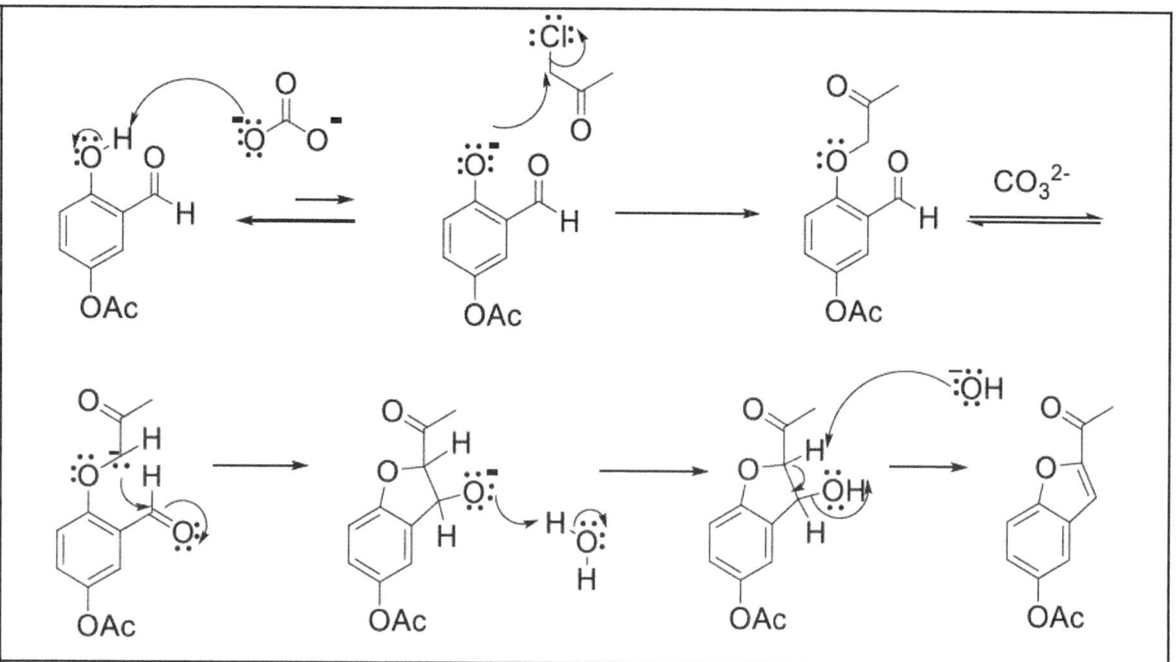 CO_3^{2-}
58	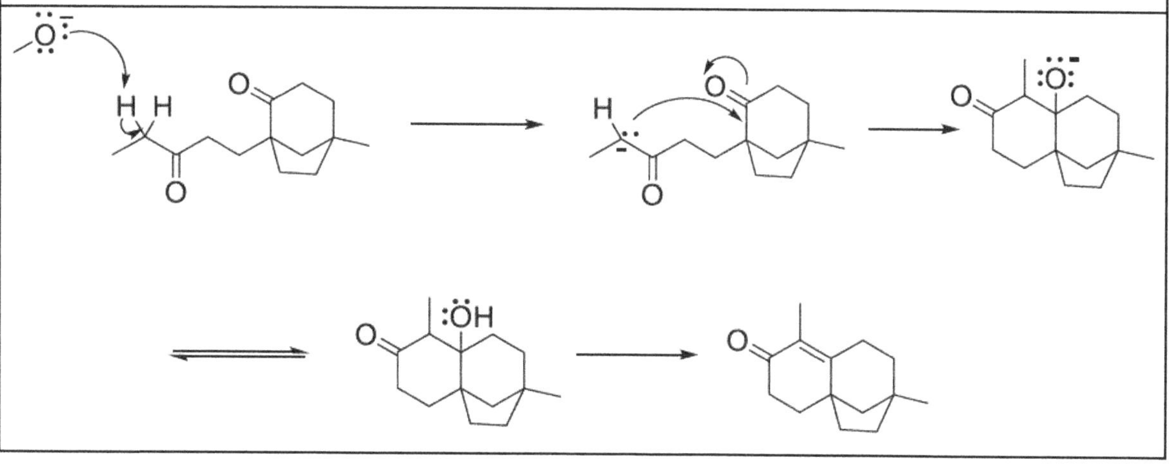

59	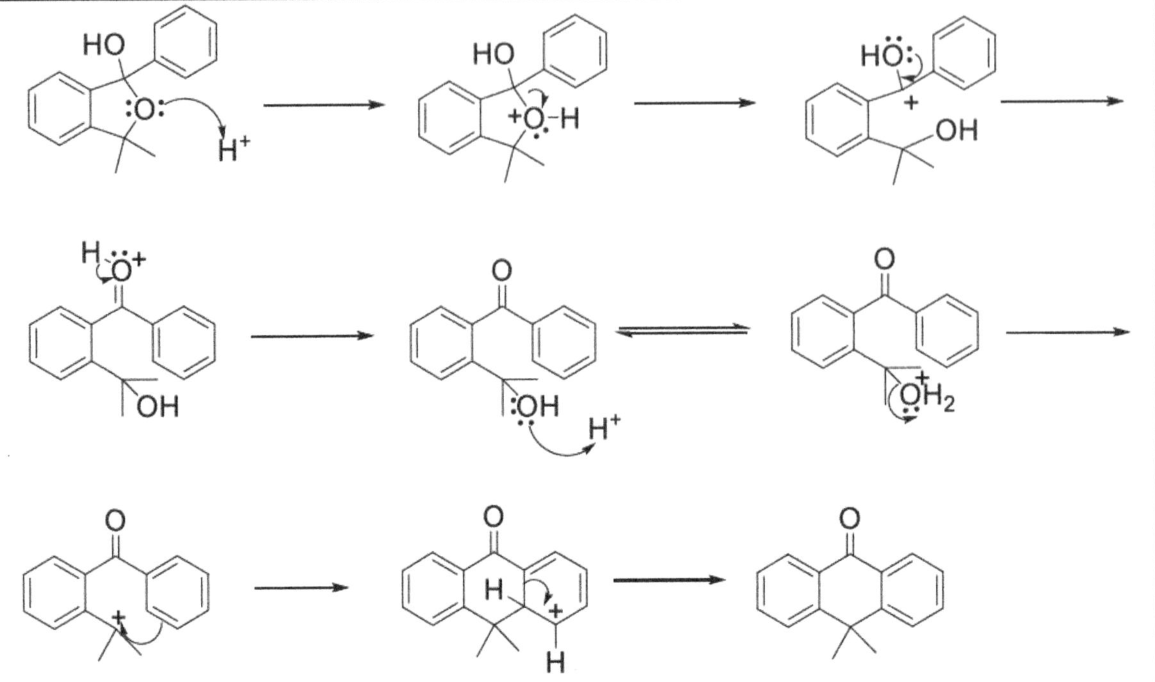
60	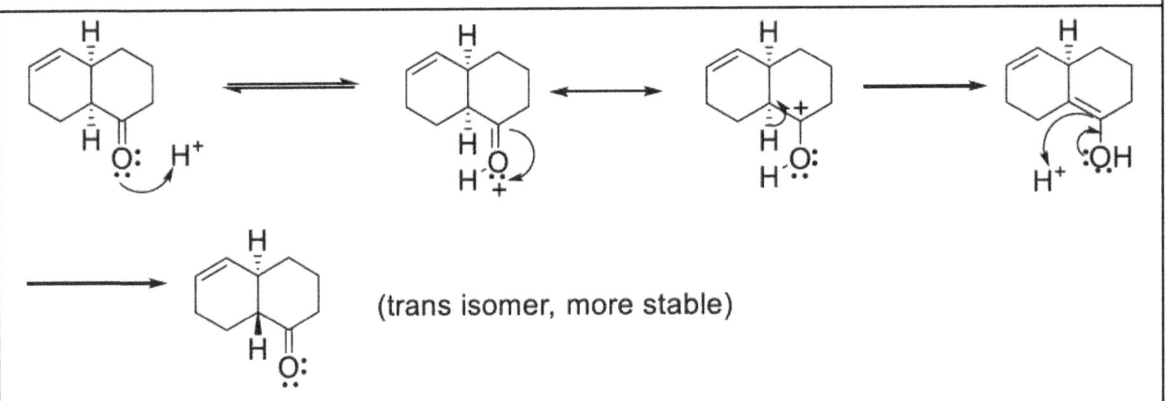 (trans isomer, more stable)
61	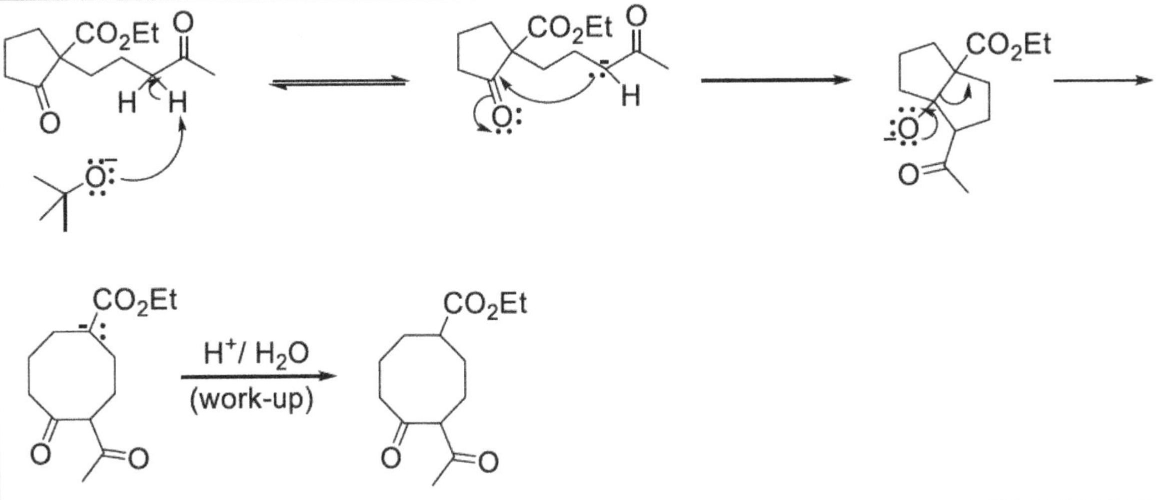 H+/ H2O (work-up)

62

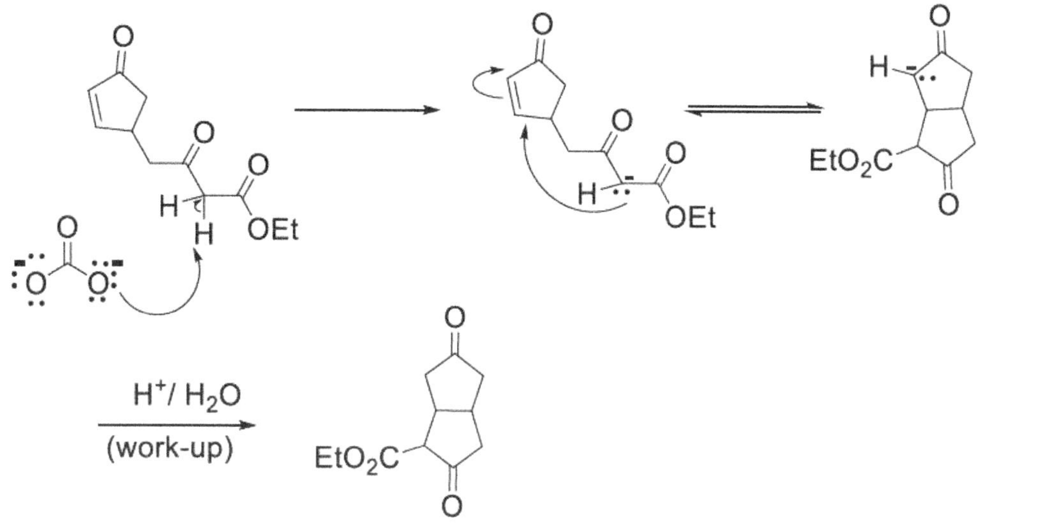

OEt
OEt
EtO2C
H+/ H2O
(work-up)
EtO2C

63

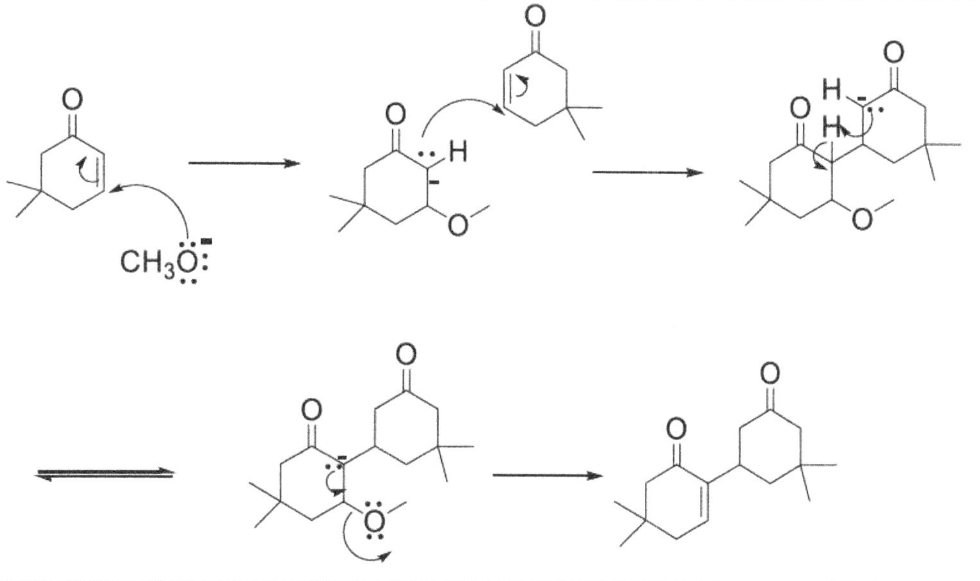

CH3O

Two consecutive Michael addition reactions are followed by a prototropic shift (an acid-base reaction) and elimination of the methoxide ion to yield the product

64

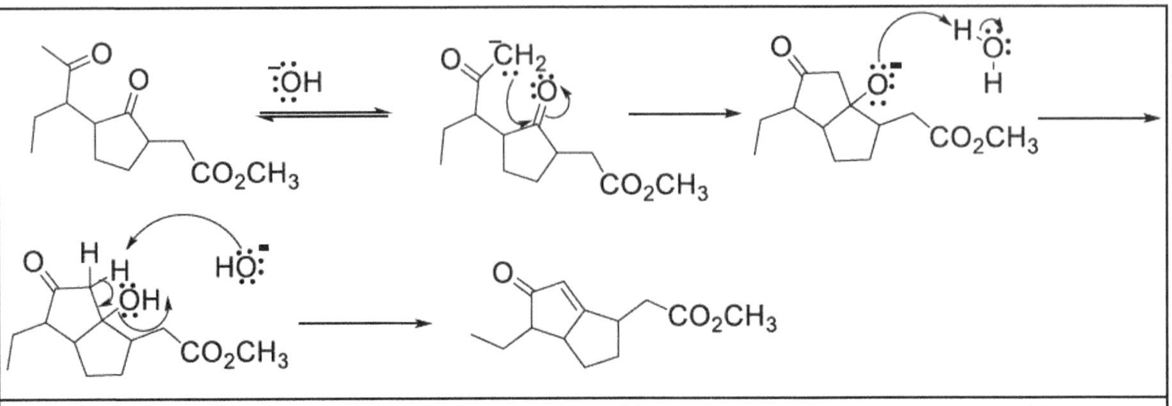

O
O
CO2CH3
:ÖH
CH2
CO2CH3
H
O
H
CO2CH3
H
H
HO
OH
CO2CH3
O
CO2CH3

65

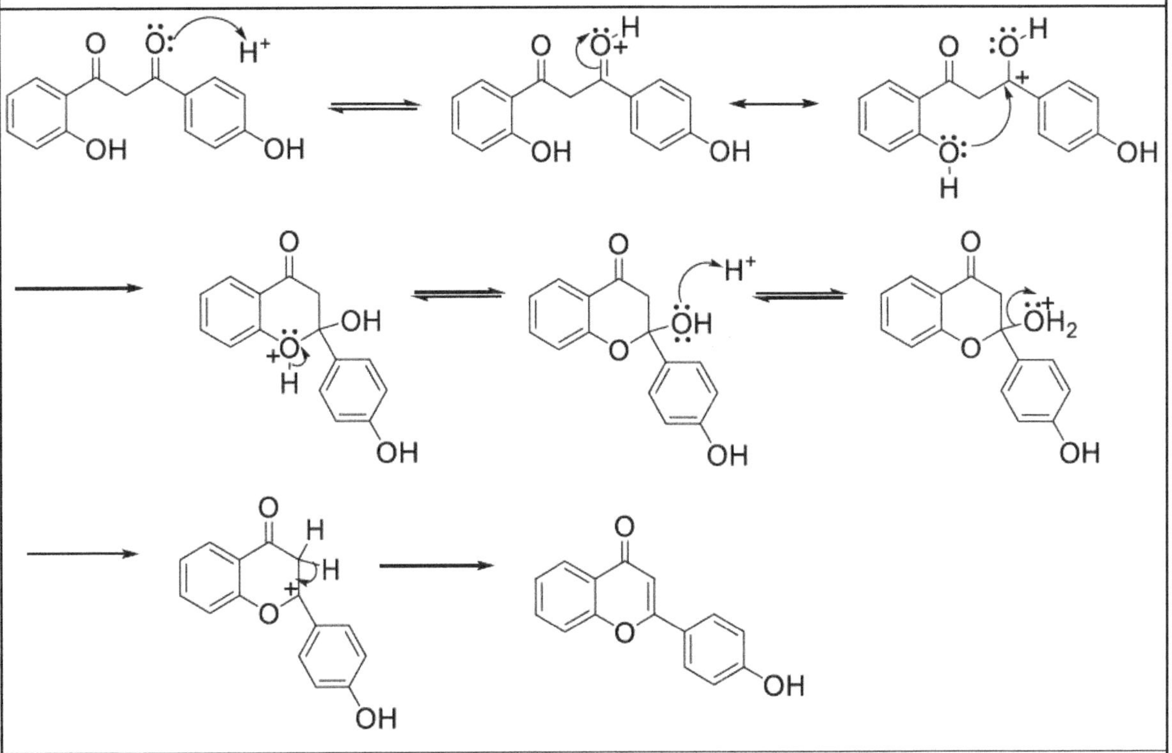

O
O
H+
OH
OH
H
OH
OH
H
+
OH
OH
H
O
OH
O
H
OH
O
H+
OH
O
OH
O
OH2
OH
O
H
H
O
OH
O
O
OH

66

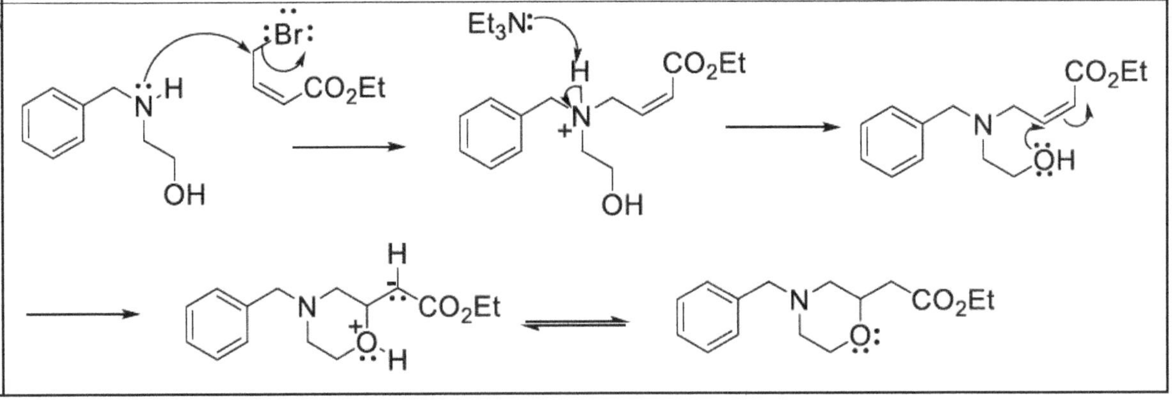

Br
N
H
CO2Et
OH
Et3N
H
N
CO2Et
OH
N
CO2Et
OH
N
H
CO2Et
O
H
N
CO2Et
O

67

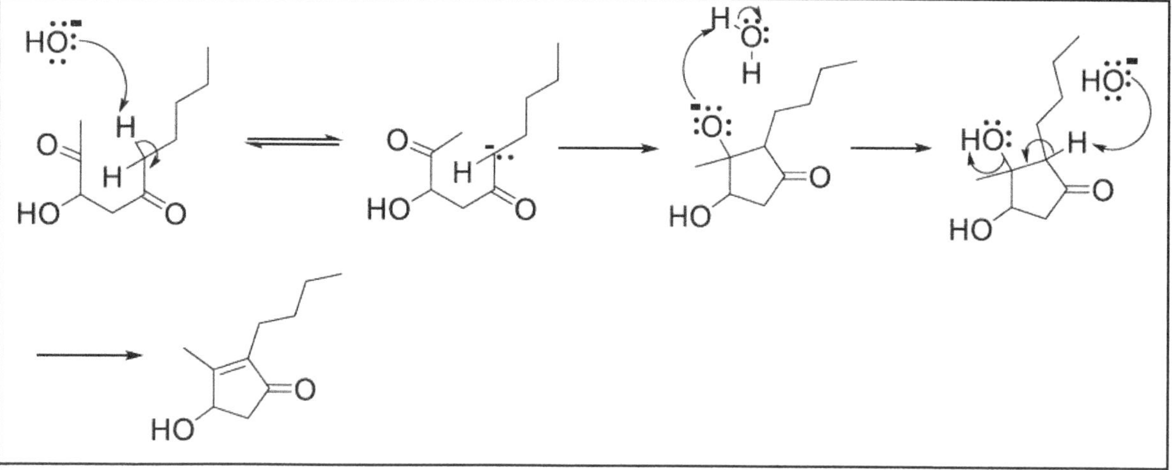

68

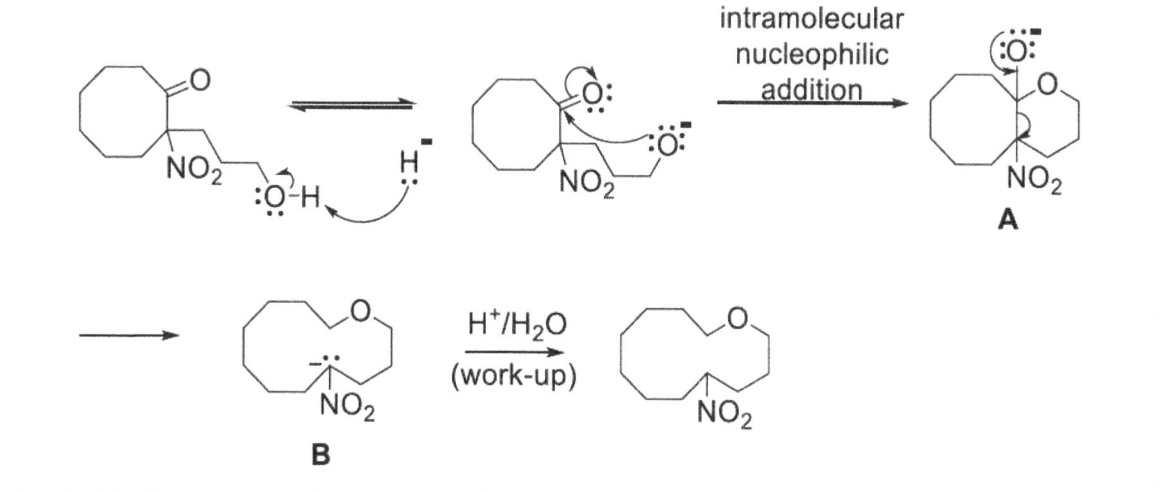
intramolecular nucleophilic addition
A
B
H^+/H_2O (work-up)

Species A is the expected nucleophilic addition product, however, instead of becoming protonated, A reforms the strong carbonyl bond, yielding a highly stable anion B.

<u>Exercise:</u> Write all the resonance structures that can be written for anion B.

69	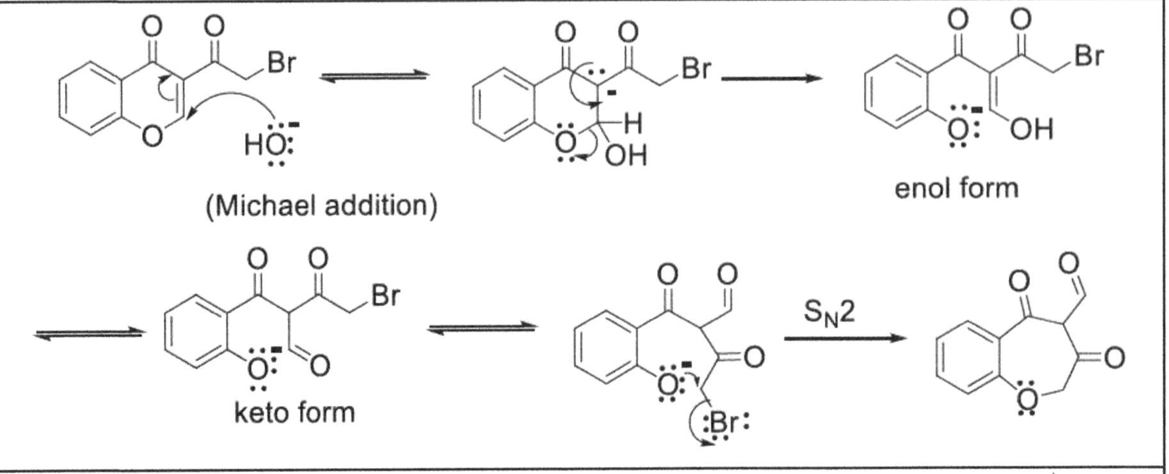 (Michael addition) enol form keto form S_N2
70	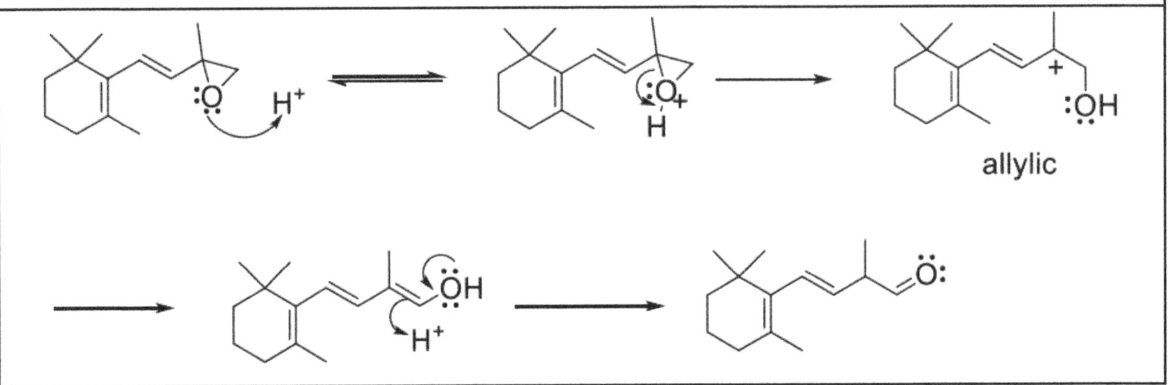 allylic

Minireview 9 focuses on nucleophilic acyl substitution reactions, the typical reaction of carboxylic acid derivatives. This should be studied prior to attempting questions 71-200.

Minireview 9

- Nucleophilic Acyl Substitution Reactions

The typical reaction of carboxylic acid derivatives (acid halides, anhydrides, esters, thioesters, and amides) with Lewis bases is *nucleophilic acyl substitution*. The mechanism of this reaction is shown below (eq 1).

Mechanism

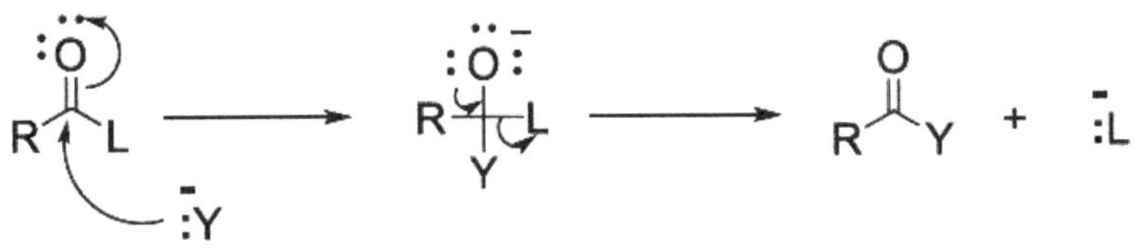

R
L
:Y
Y
+
:L

eq 1

Reactions involving weak nucleophiles (:Y) can be speeded up by acid catalysis (eq 2). With the exception of amines and phosphorus nucleophiles, nucleophiles that do not bear a negative charge (:Y) are classified as weak, while those having a negative charge are classified as strong nucleophiles.

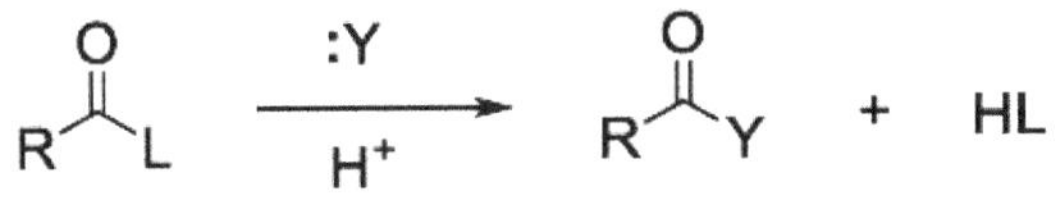

O
R
L
:Y
H+
Y
+
HL

eq 2

Mechanism

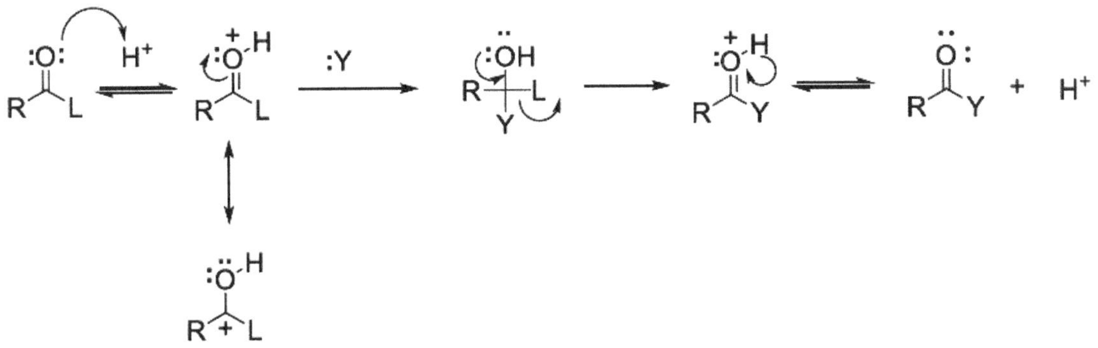

H+
:Y
+
H+

Exercise

1) Using general mechanistic principles, write a mechanism for each of the following reactions

a) 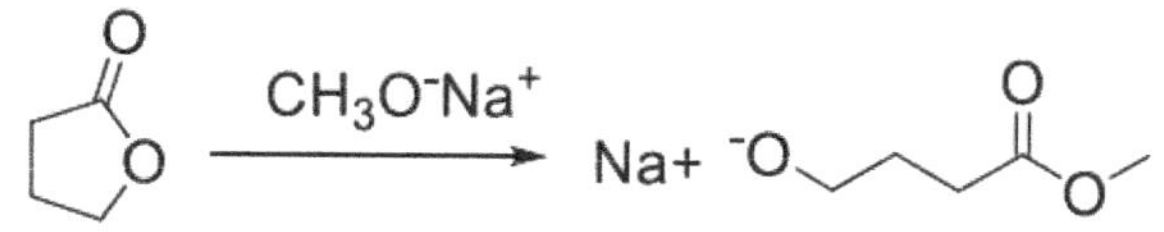

b) 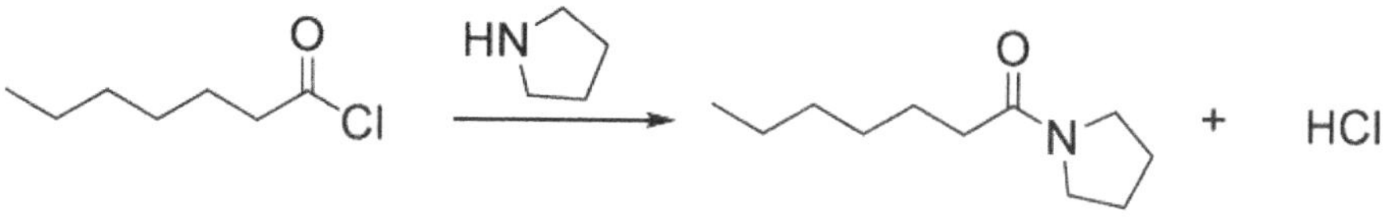

c) 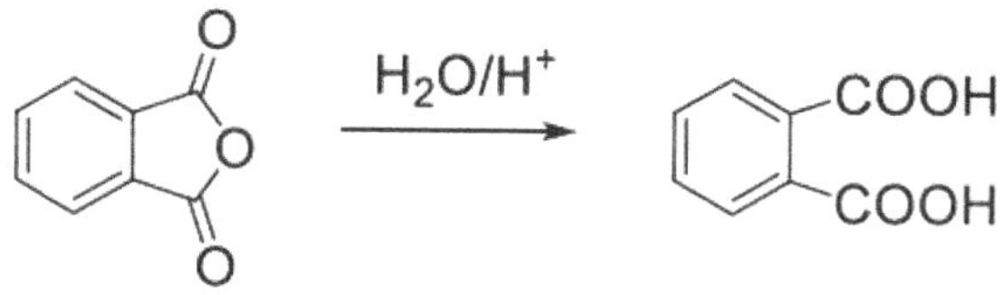

2) Predict the product of the following reaction:

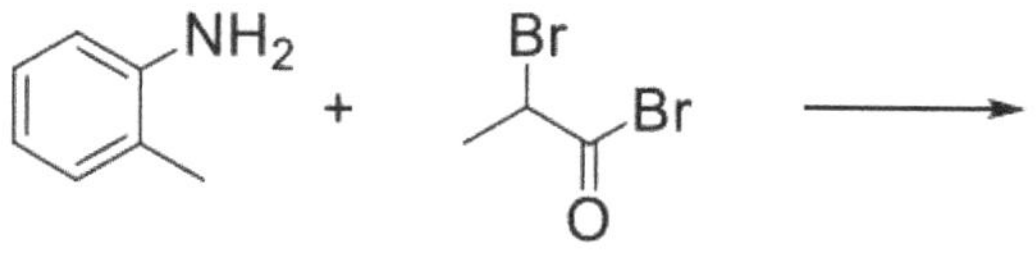

Questions 71-200

71

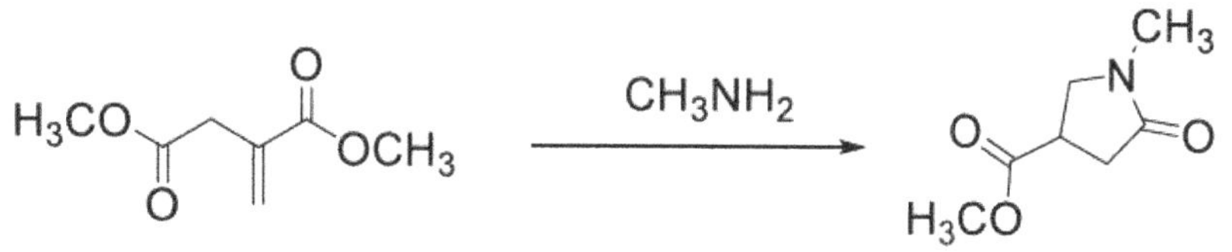
H3CO
O
O
OCH3
CH3NH2
CH3
N
O
O
H3CO

72

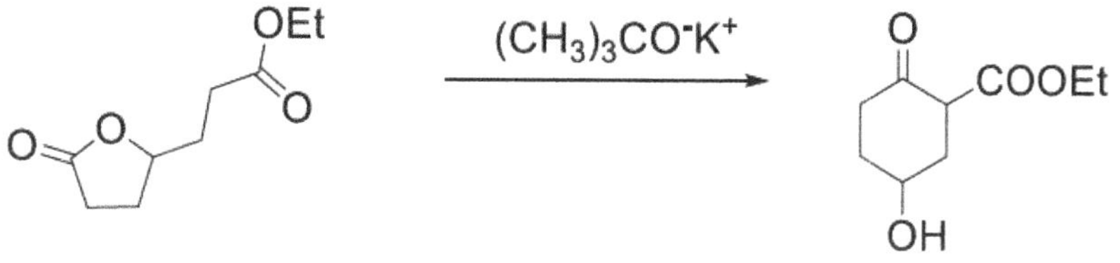
OEt
O
O
O
(CH3)3CO-K+
O
COOEt
OH

73

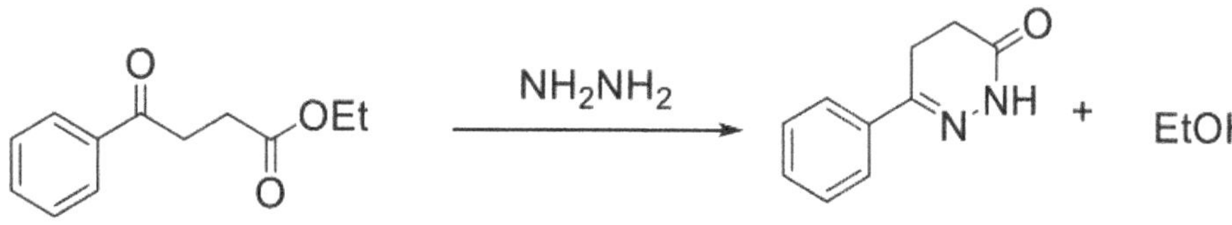
O
OEt
O
NH2NH2
O
NH
N
+
EtOH

74

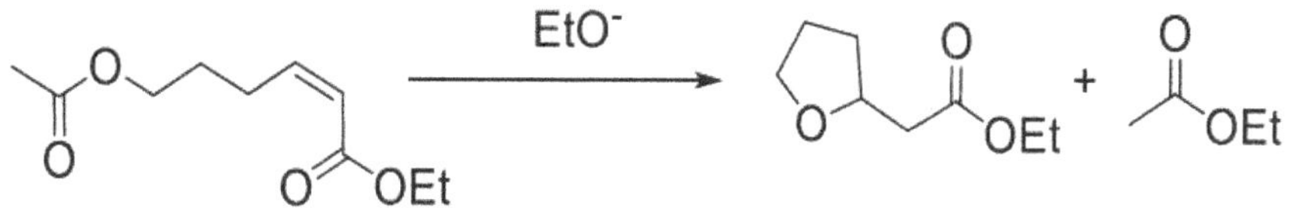
O
O
O
OEt
EtO-
O
O
OEt
+
O
OEt

75

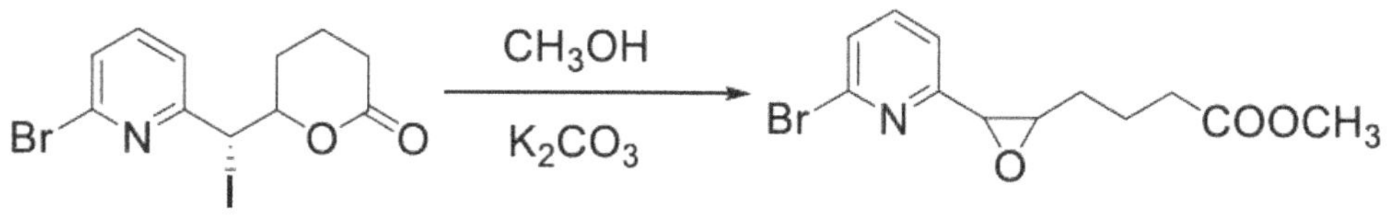
Br
N
I
O
O
CH3OH
K2CO3
Br
N
O
COOCH3

76

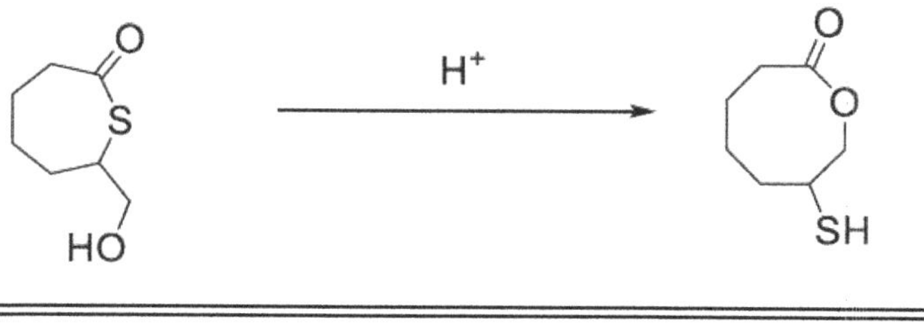
O
S
HO
H+
O
O
SH

77

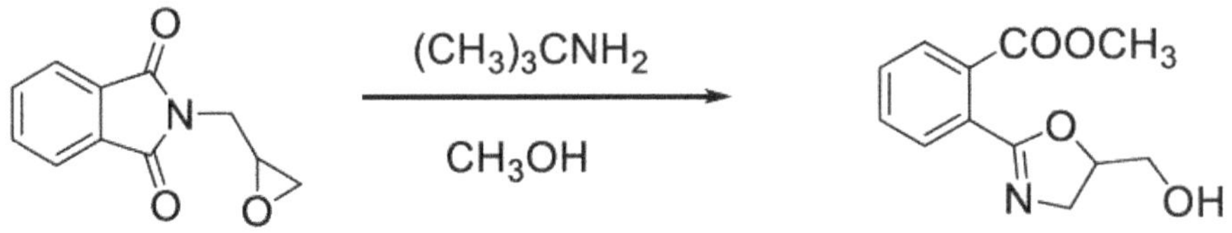

(CH3)3CNH2
CH3OH
COOCH3
OH

78

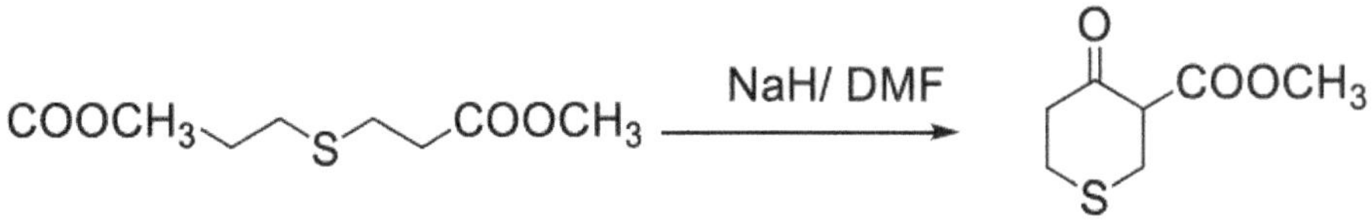

COOCH3
COOCH3
NaH/ DMF
COOCH3

79

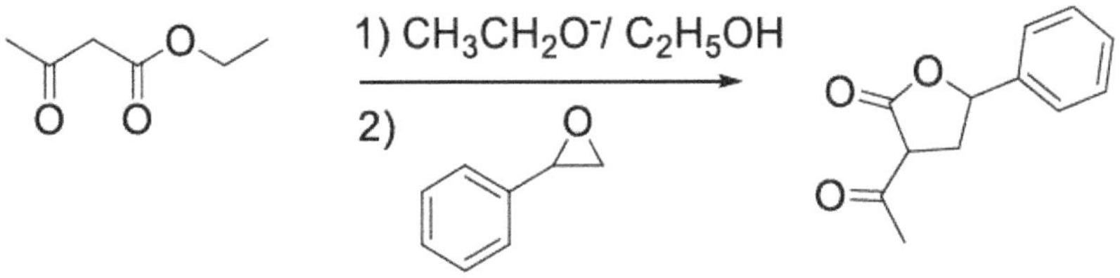

1) CH3CH2O-/ C2H5OH
2)

80

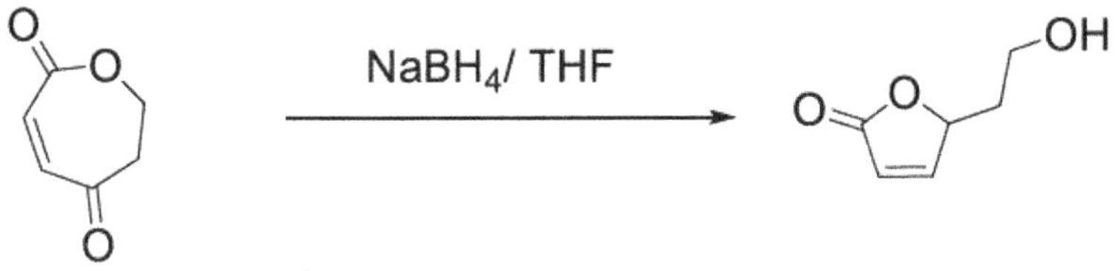

NaBH4/ THF
OH

81

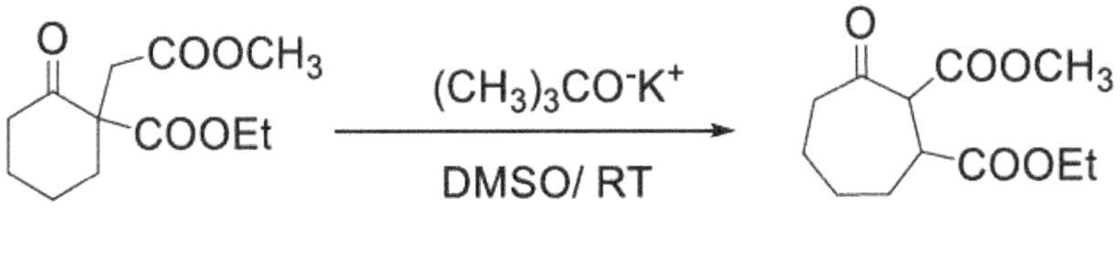

COOCH3
COOEt
(CH3)3CO-K+
DMSO/ RT
COOCH3
COOEt

82

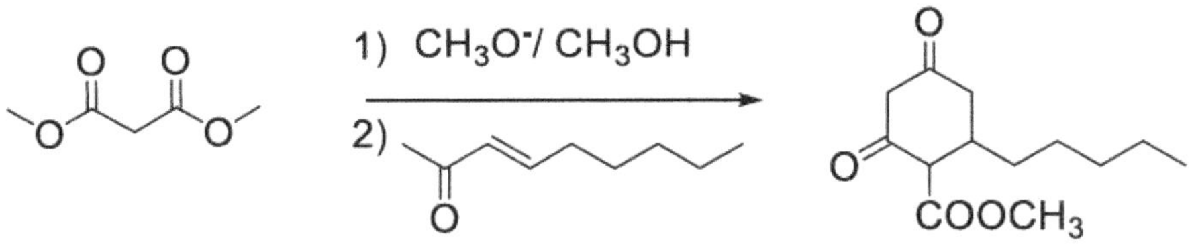

1) CH3O-/ CH3OH
2)
COOCH3

83	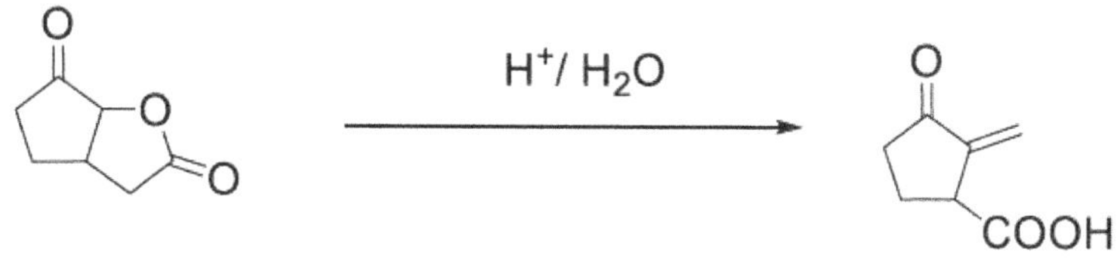H^+/ H_2O COOH
84	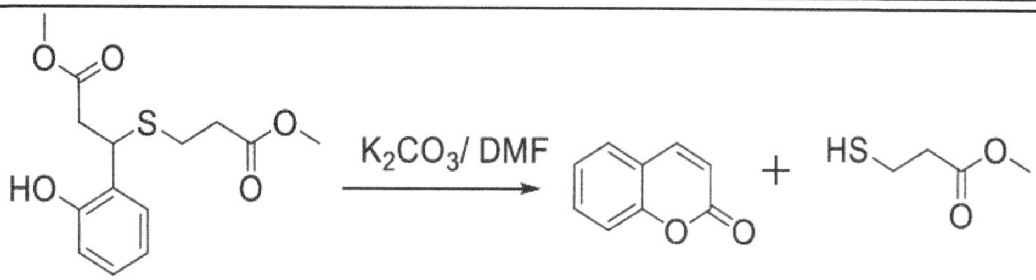HO S K_2CO_3/ DMF HS
85	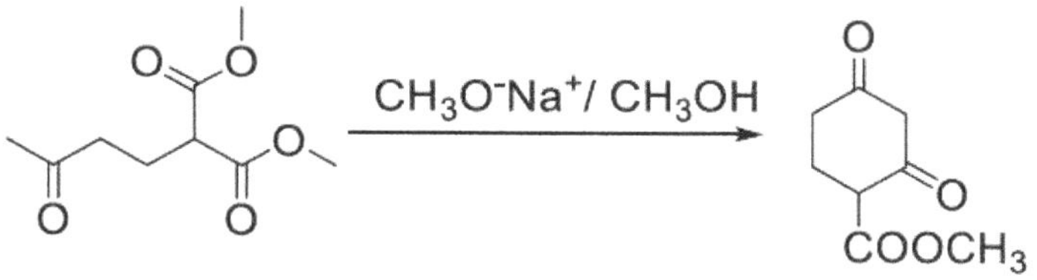$CH_3O^-Na^+$/ CH_3OH $COOCH_3$
86	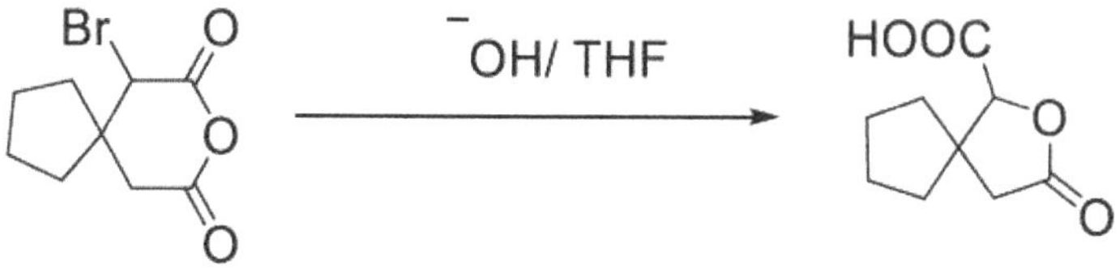Br ^-OH/ THF HOOC
87	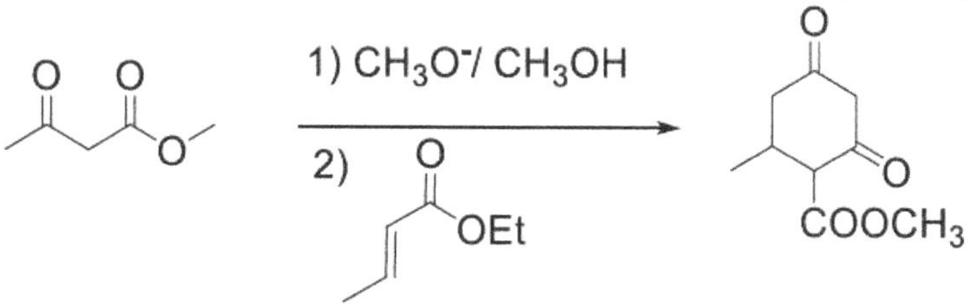1) CH_3O^-/ CH_3OH 2) OEt $COOCH_3$
88	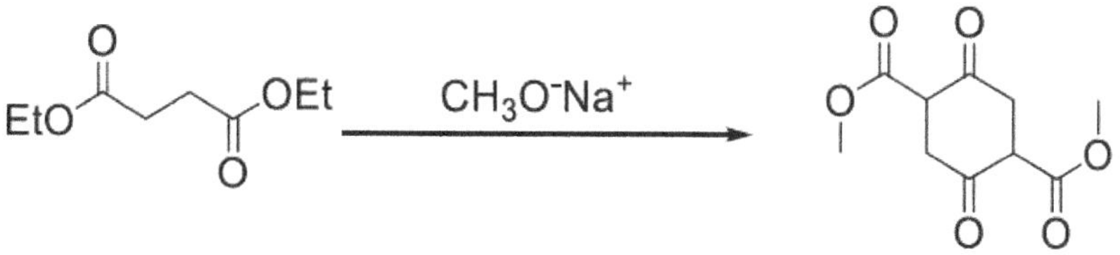EtO OEt $CH_3O^-Na^+$

89

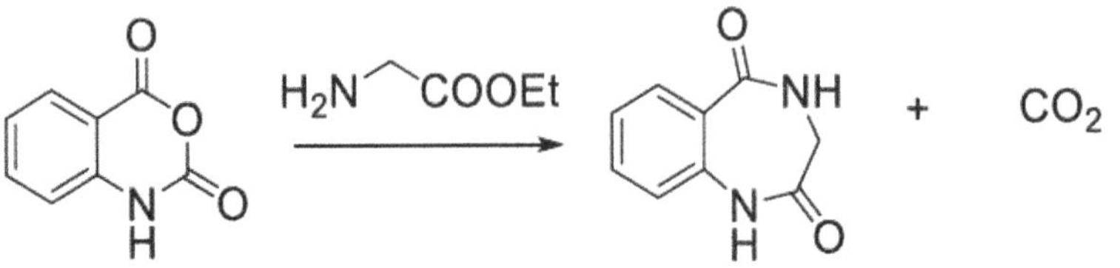
H_2N COOEt
+
CO_2

90

The following decarboxylation occurs spontaneously. Explain.

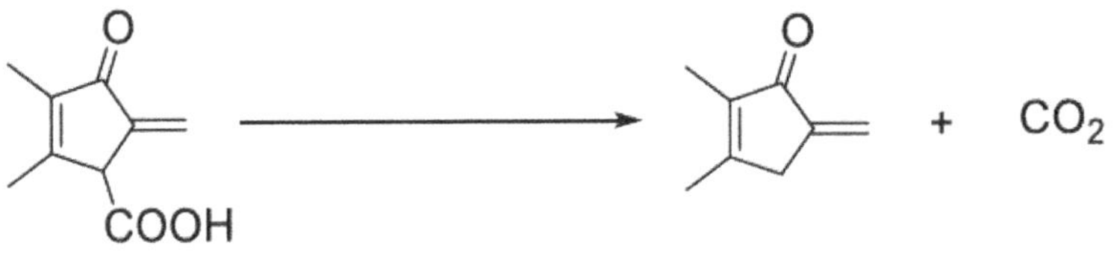
COOH
+
CO_2

91

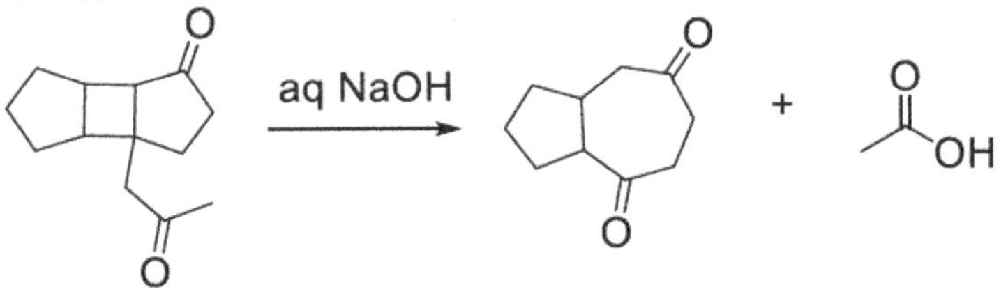
aq NaOH
+
OH

92

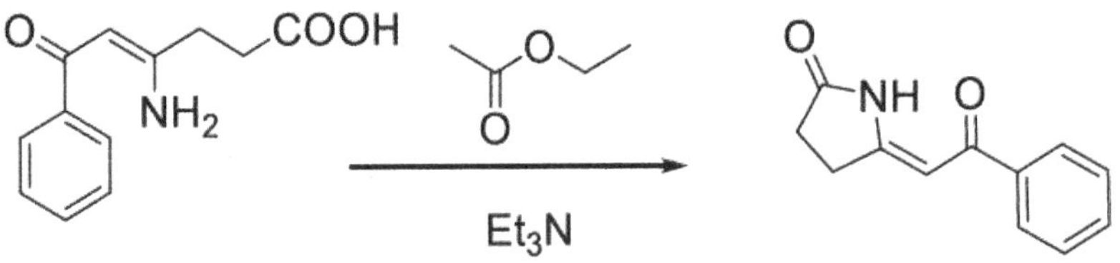
COOH
NH_2
Et_3N
NH

93

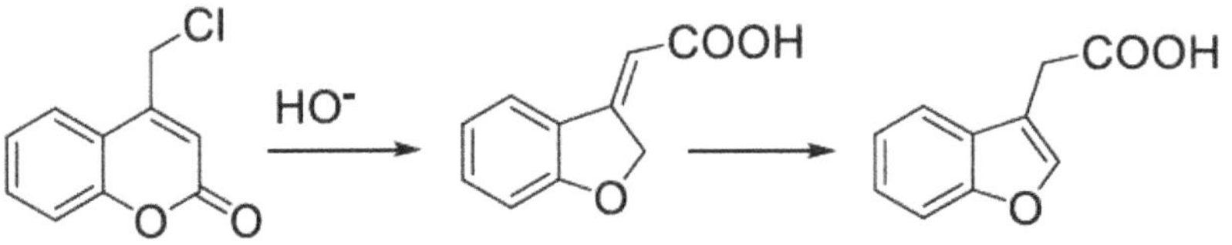
Cl
HO^-
COOH
COOH

94

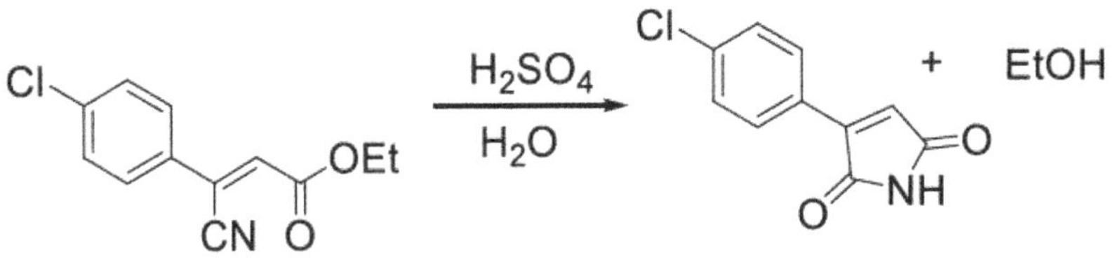
Cl
OEt
CN
H_2SO_4
H_2O
Cl
NH
+
EtOH

95

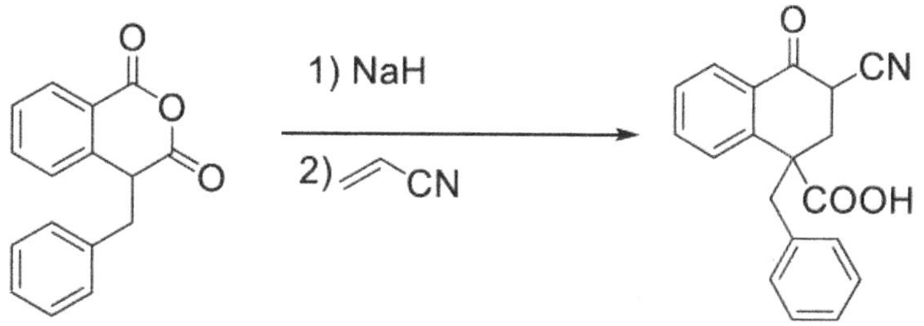
1) NaH
2) CN
CN
COOH

96

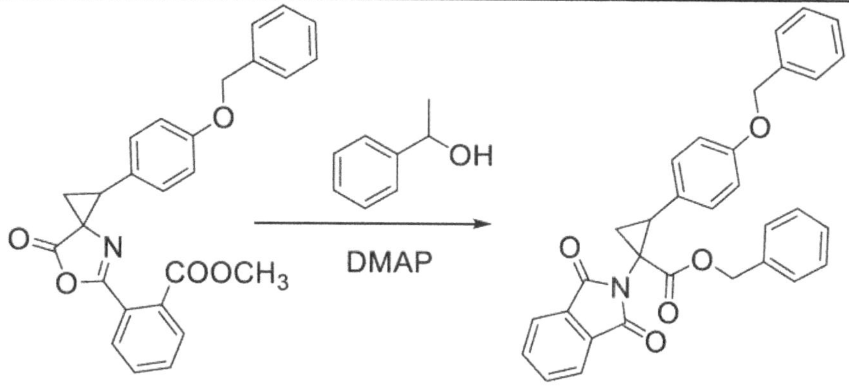
OH
DMAP
COOCH3

97

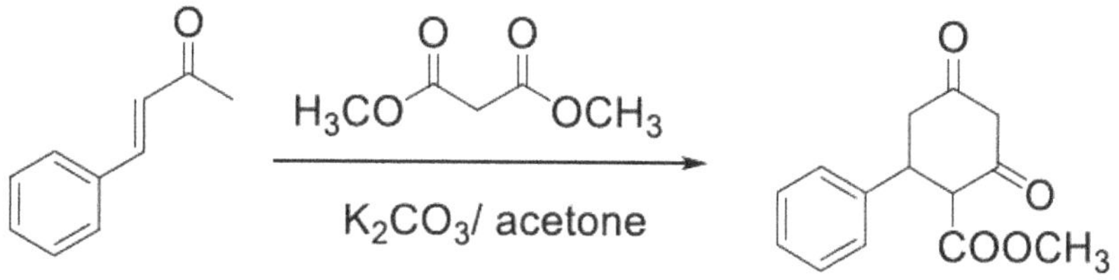
H3CO
OCH3
K2CO3/ acetone
COOCH3

98

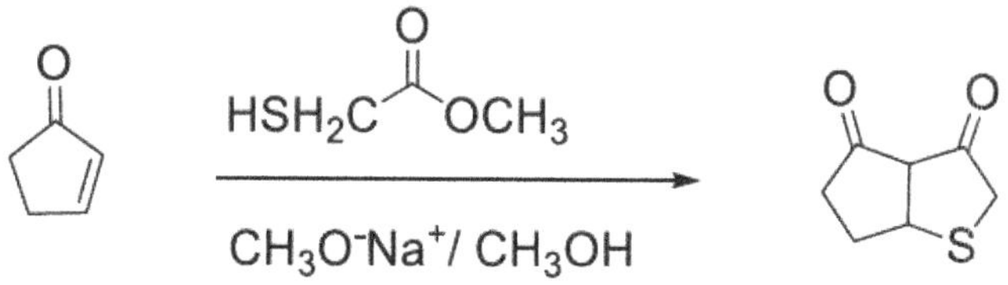
HSH2C
OCH3
CH3O-Na+/ CH3OH

99

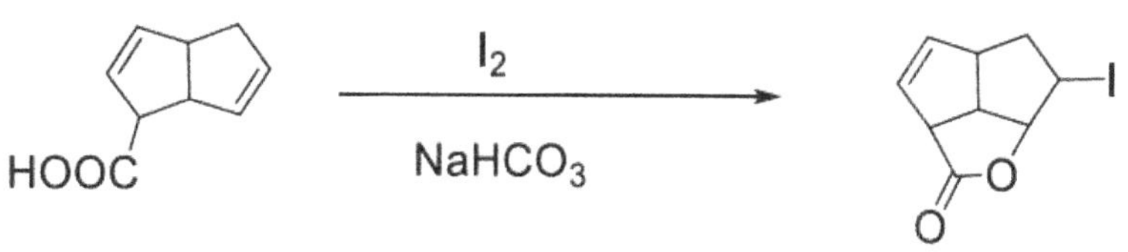
I2
NaHCO3
HOOC

100

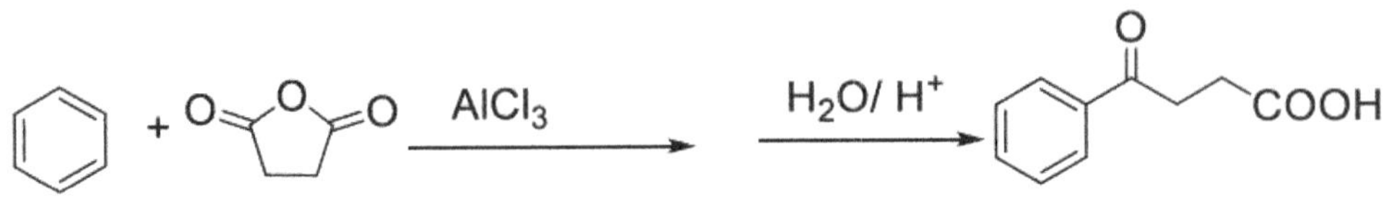
AlCl$_3$
H$_2$O/ H$^+$
COOH

101

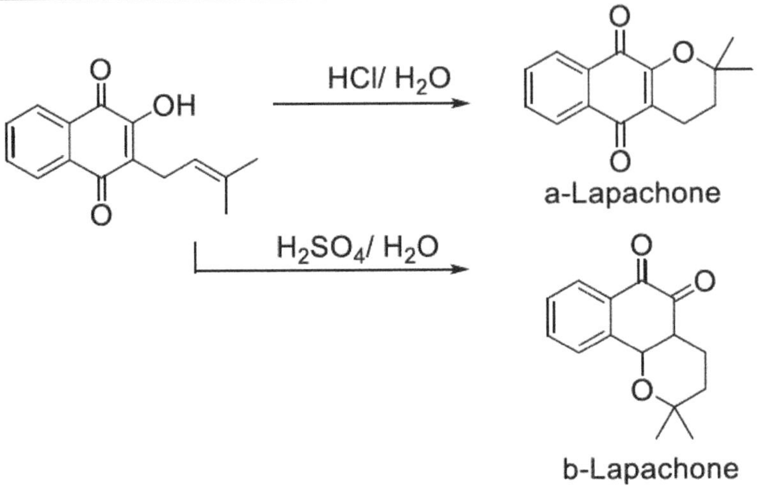
OH
HCl/ H$_2$O
H$_2$SO$_4$/ H$_2$O
a-Lapachone
b-Lapachone

102

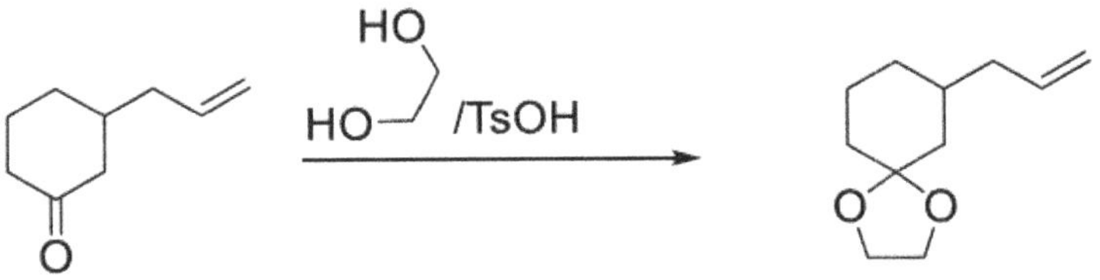
HO
HO
/TsOH

103

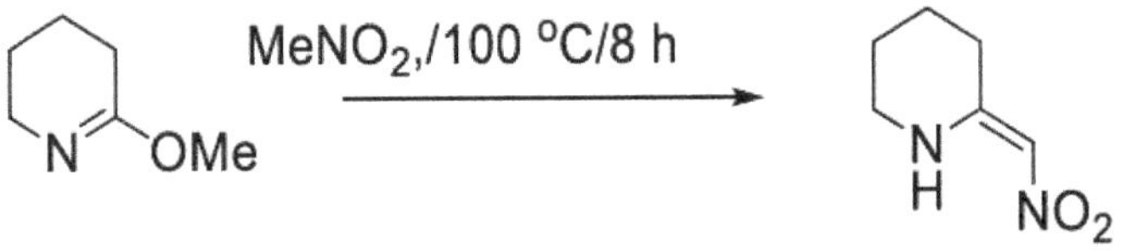
N
OMe
MeNO$_2$,/100 °C/8 h
N
H
NO$_2$

104

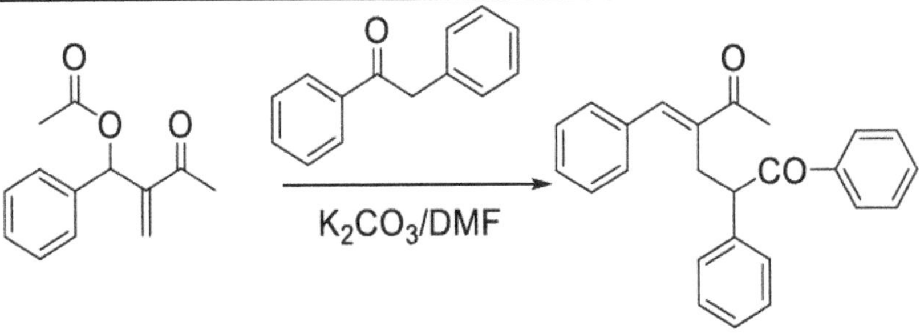
K$_2$CO$_3$/DMF
CO

105	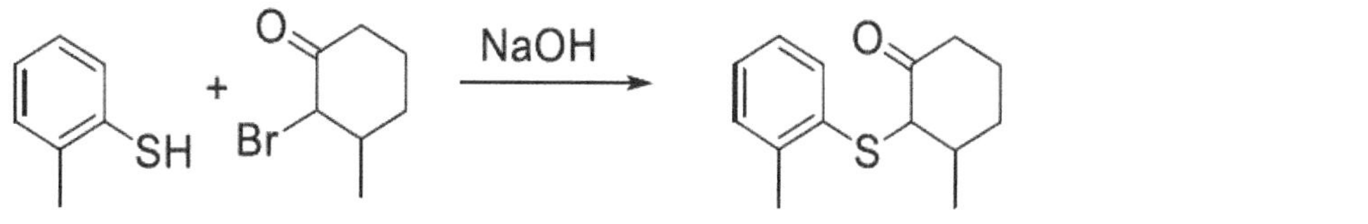 SH O Br NaOH O S
106	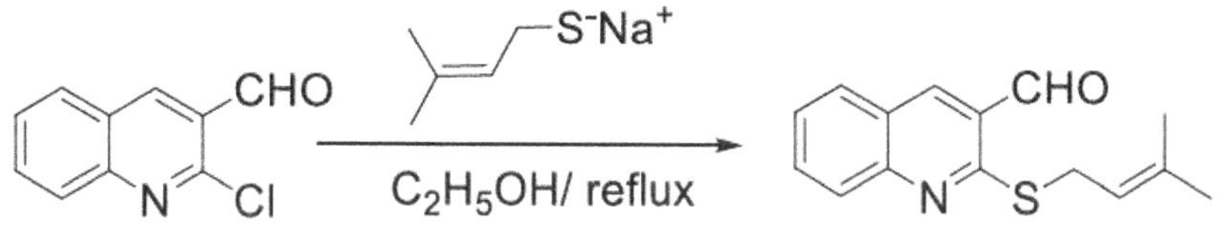 CHO N Cl S⁻Na⁺ C2H5OH/ reflux CHO N S
107	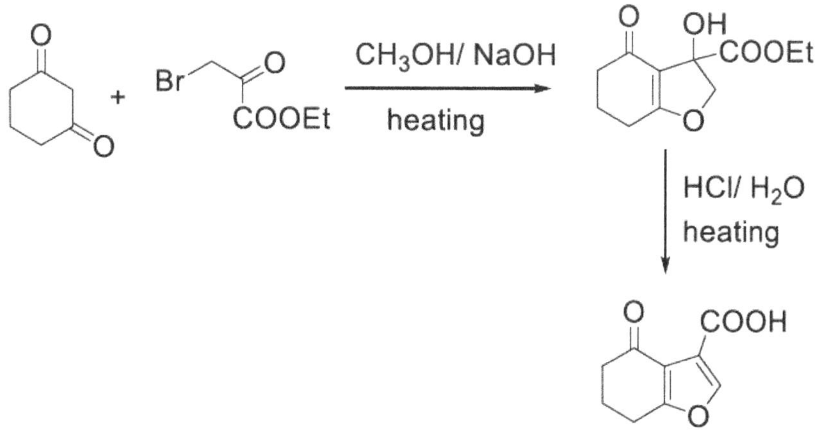 O O Br O COOEt CH3OH/ NaOH heating O OH COOEt O HCl/ H2O heating O COOH O
108	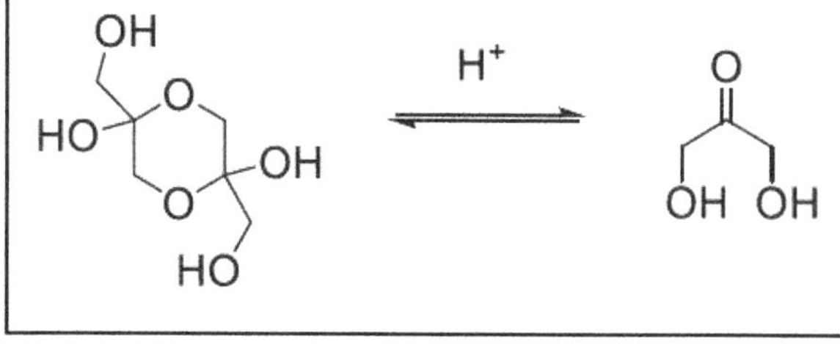 OH HO O OH O HO H⁺ O OH OH
109	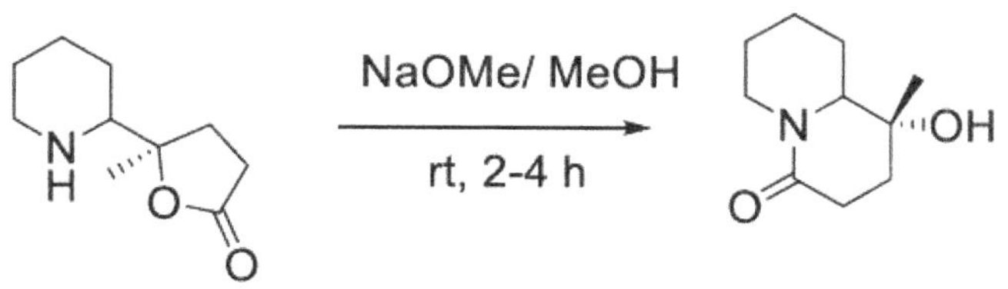 N H O O NaOMe/ MeOH rt, 2-4 h N OH O

110	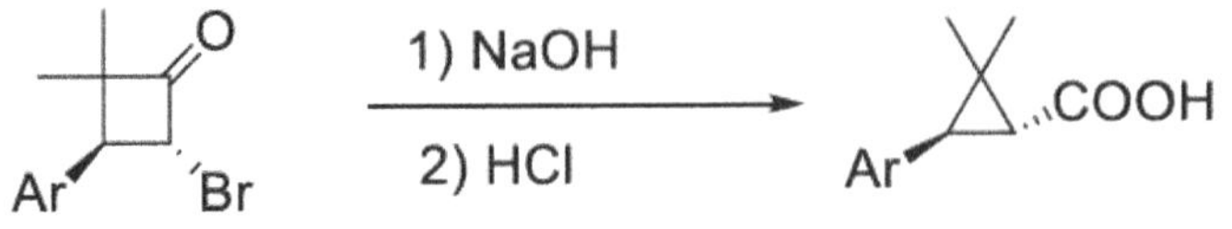 O Ar Br 1) NaOH 2) HCl Ar COOH
111	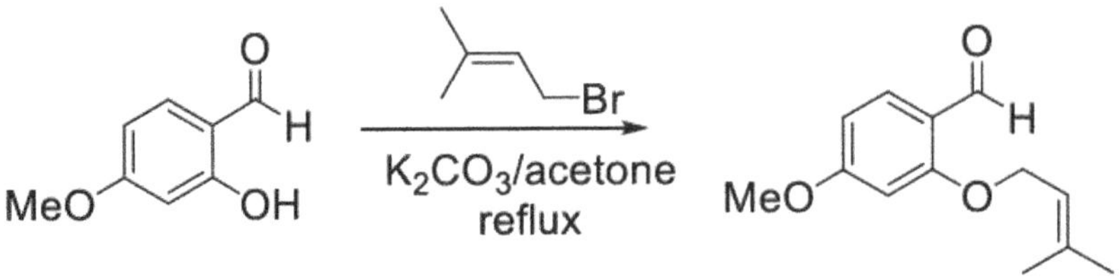 O H MeO OH Br K_2CO_3/acetone reflux O H MeO O
112	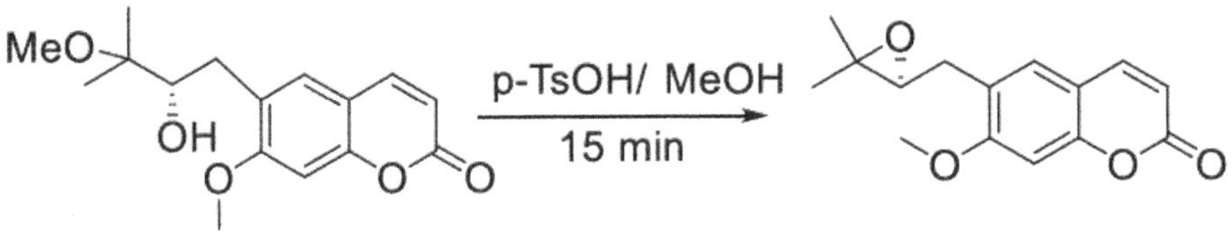 MeO OH O O O p-TsOH/ MeOH 15 min O O O O
113	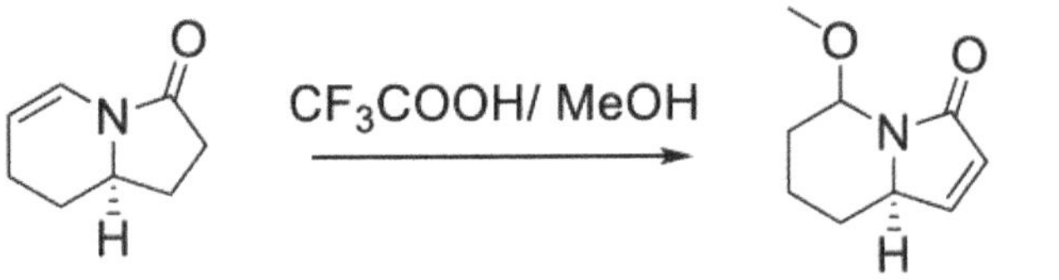 O N H CF_3COOH/ MeOH O O N H
114	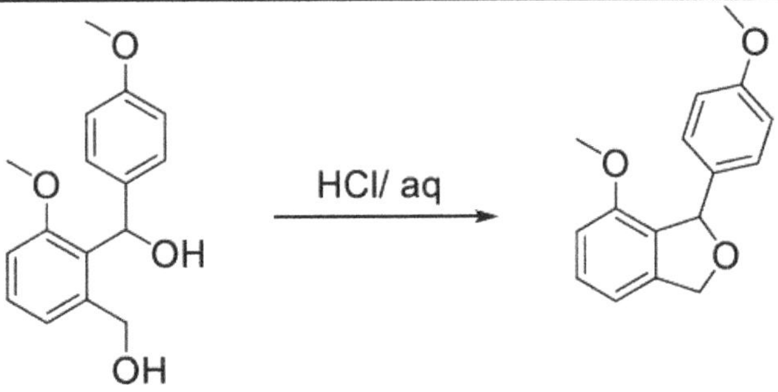 O O OH OH HCl/ aq O O O
115	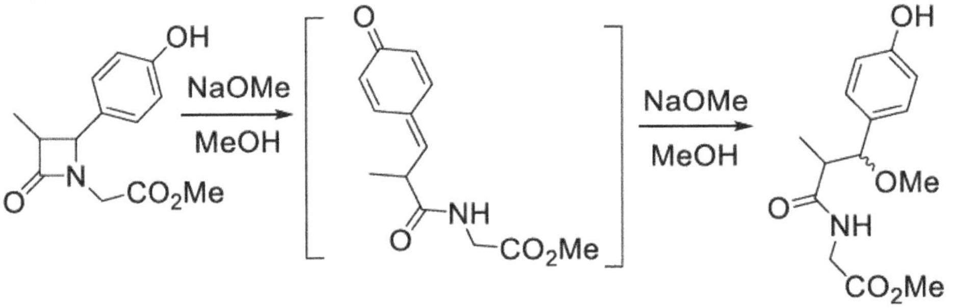 OH O N CO_2Me NaOMe MeOH O O NH CO_2Me NaOMe MeOH OH OMe O NH CO_2Me

116	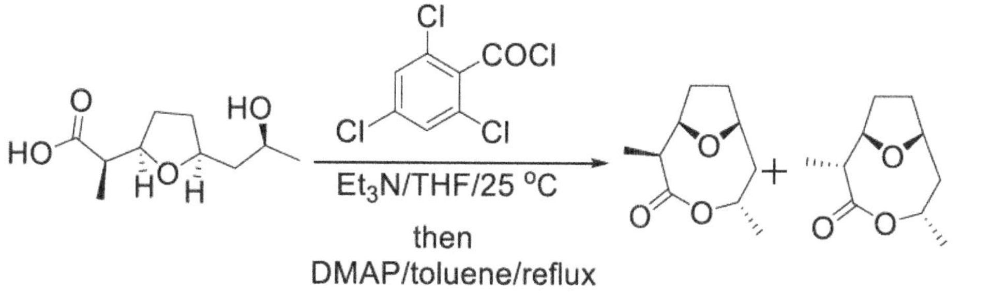COCl Cl Et3N/THF/25 °C then DMAP/toluene/reflux
117	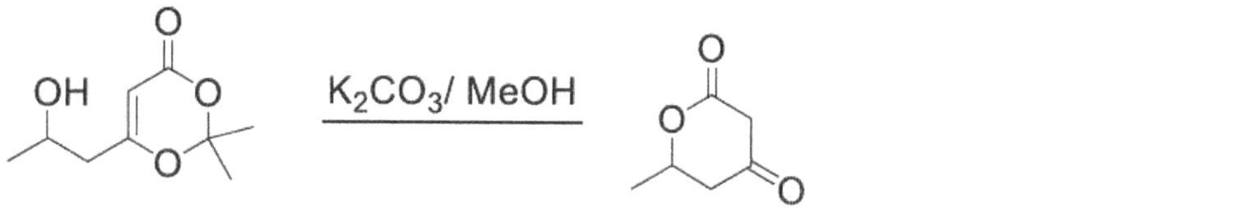K2CO3/ MeOH
118	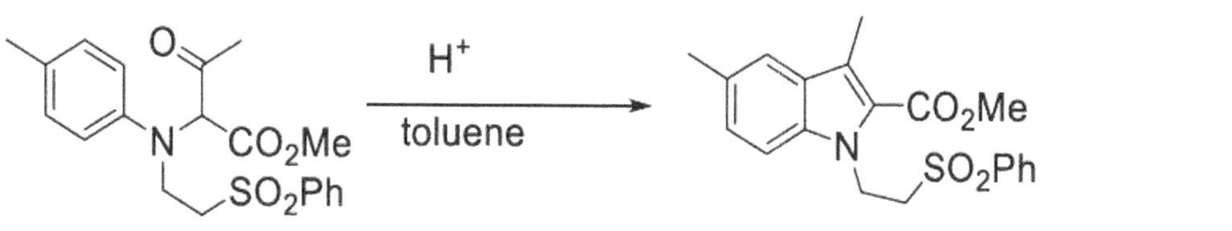H+ toluene CO2Me SO2Ph
119	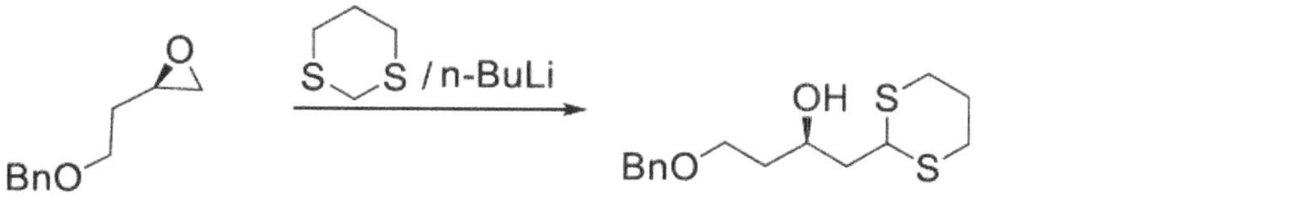/ n-BuLi BnO OH
120	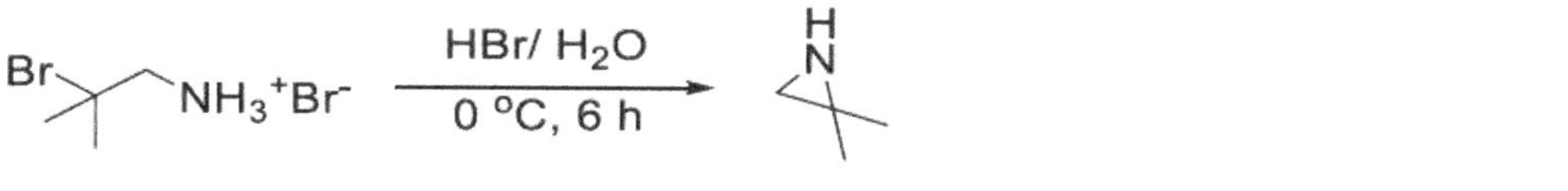HBr/ H2O 0 °C, 6 h NH3+Br-
121	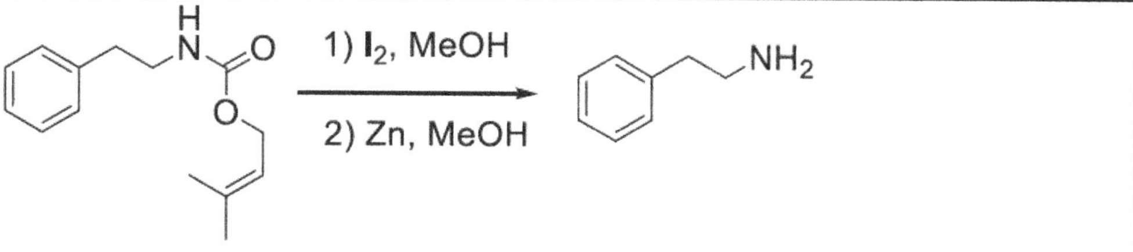1) I2, MeOH 2) Zn, MeOH NH2

122	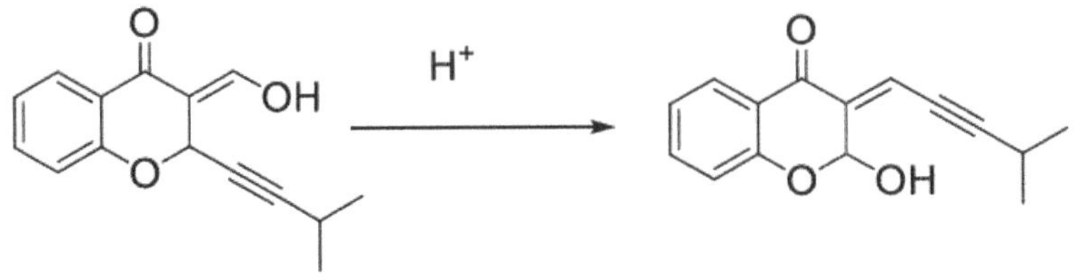 H+
123	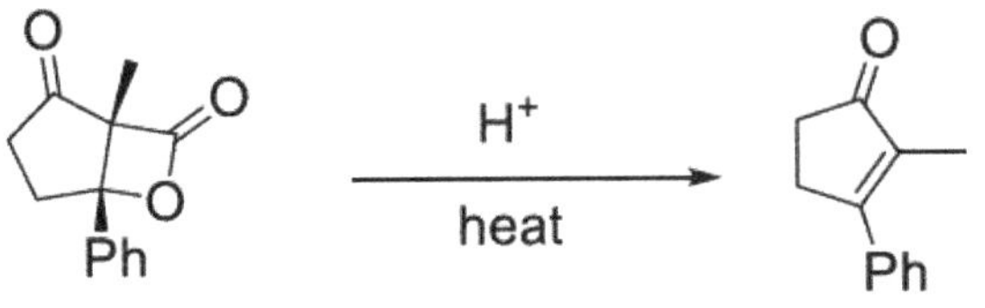 H+ heat
124	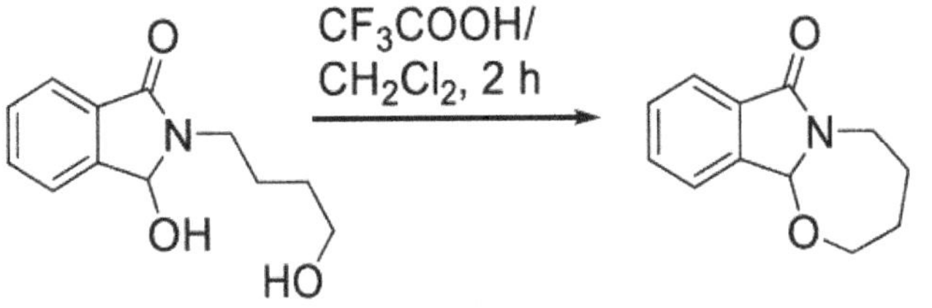 CF3COOH/ CH2Cl2, 2 h
125	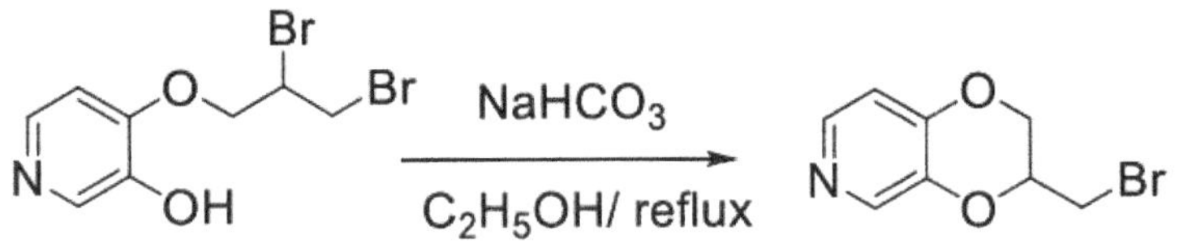 NaHCO3 C2H5OH/ reflux
126	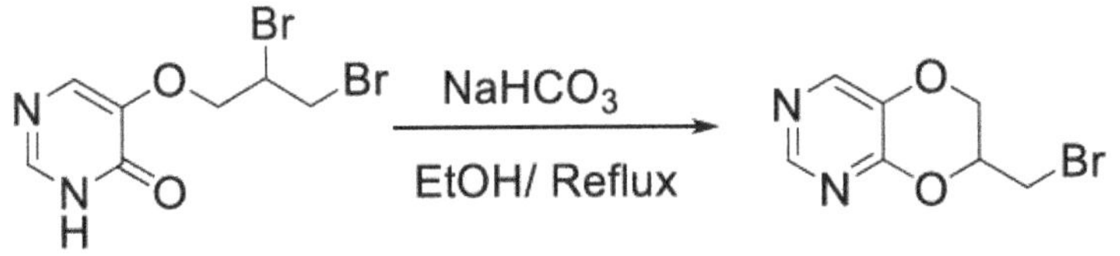 NaHCO3 EtOH/ Reflux
127	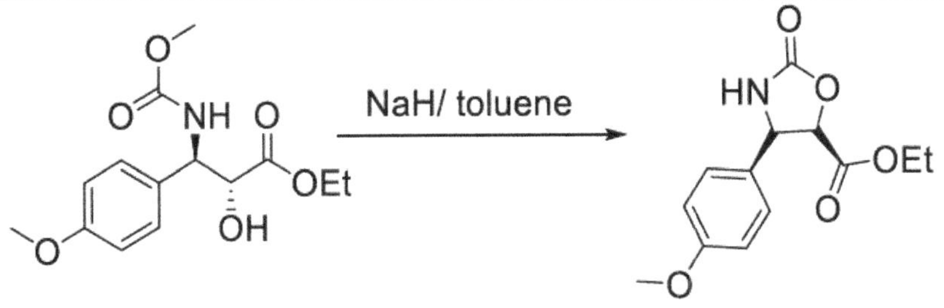 NaH/ toluene

128

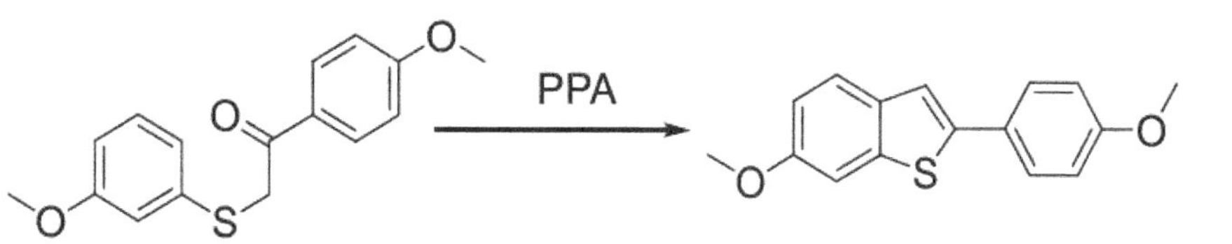

O
O
S
PPA
O
S
O

129

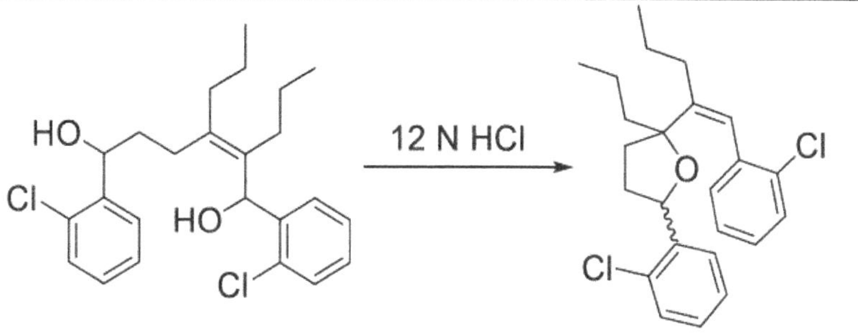

HO
Cl
HO
Cl
12 N HCl
O
Cl
Cl

130

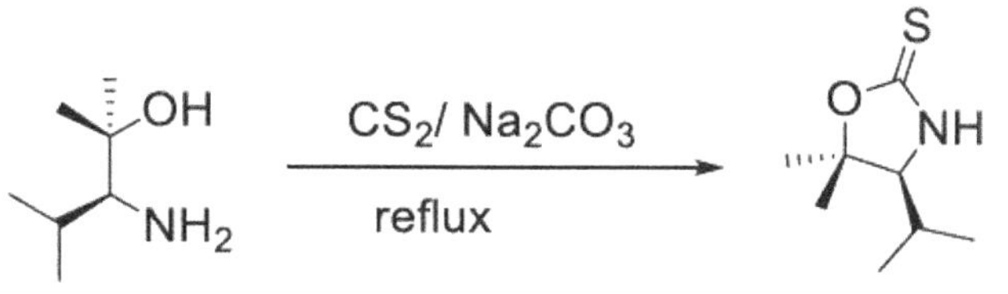

OH
NH2
CS2/ Na2CO3
reflux
S
O
NH

131

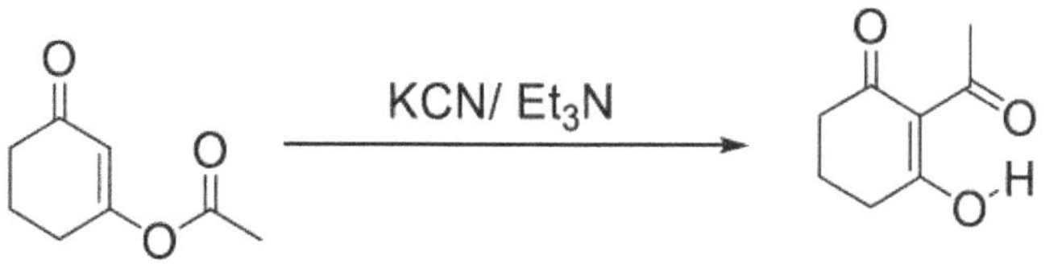

O
O
O
KCN/ Et3N
O
O
O
H

132

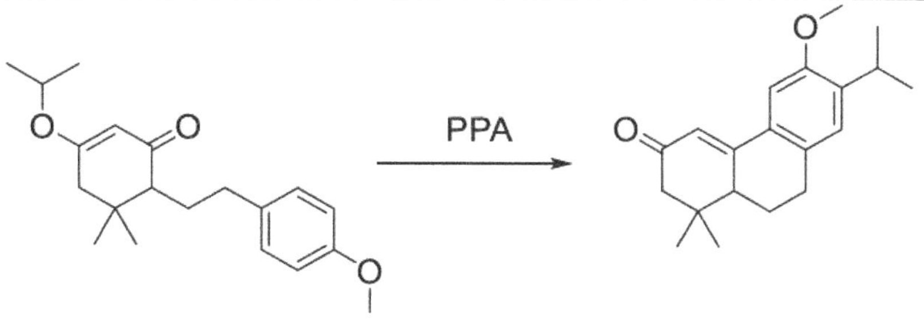

O
O
O
PPA
O
O

133

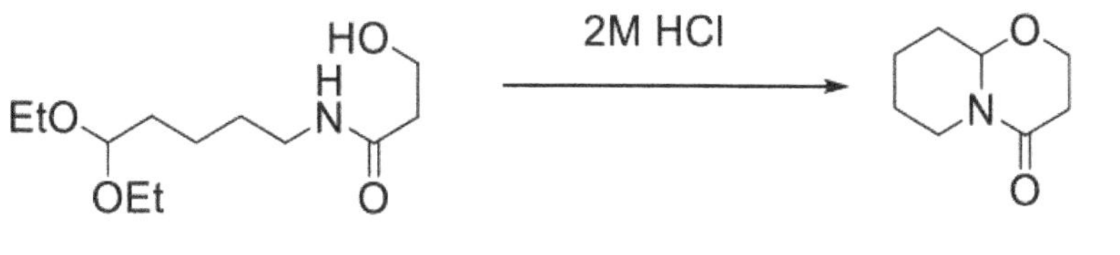

EtO
OEt
HO
H
N
O
2M HCl
O
N
O

134	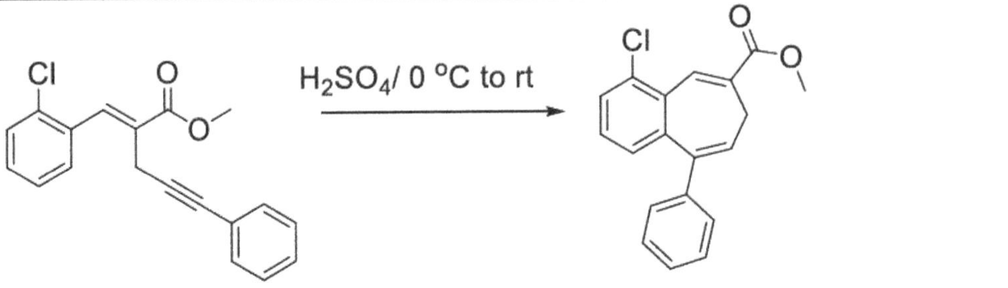 Cl O O H2SO4/ 0 °C to rt Cl O O
135	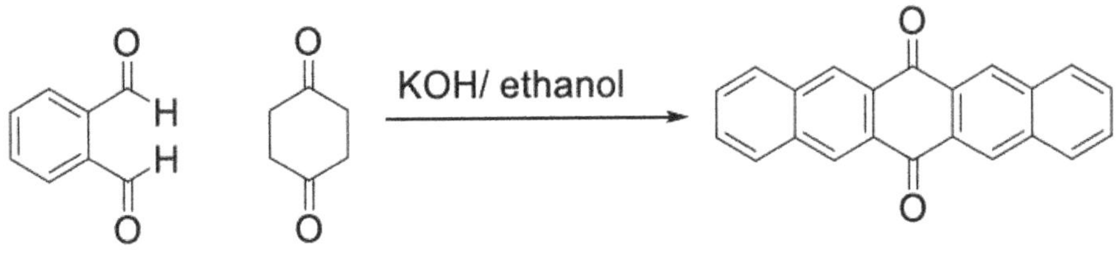 O H H O O O KOH/ ethanol O O
136	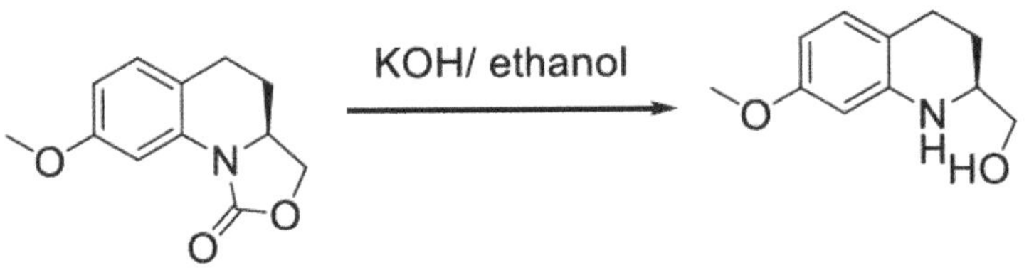 O N O O KOH/ ethanol O N H HO
137	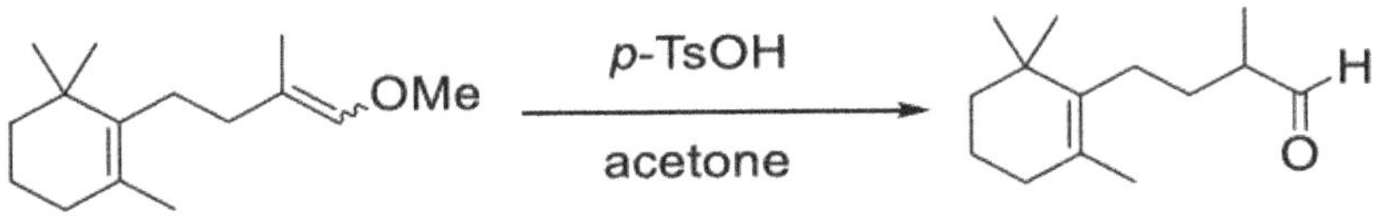 OMe p-TsOH acetone H O
138	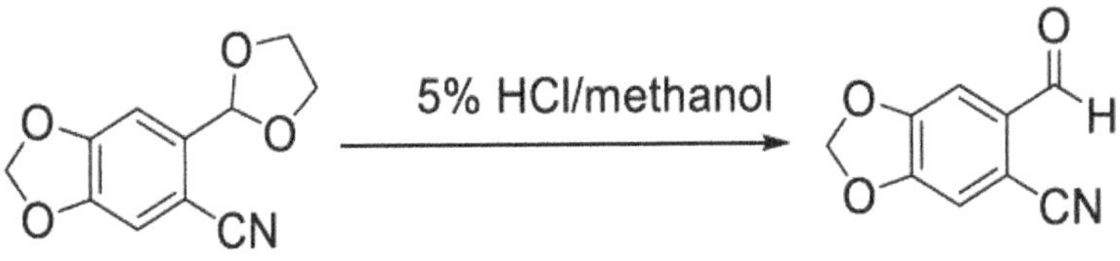 O O O O CN 5% HCl/methanol O H O O CN

139	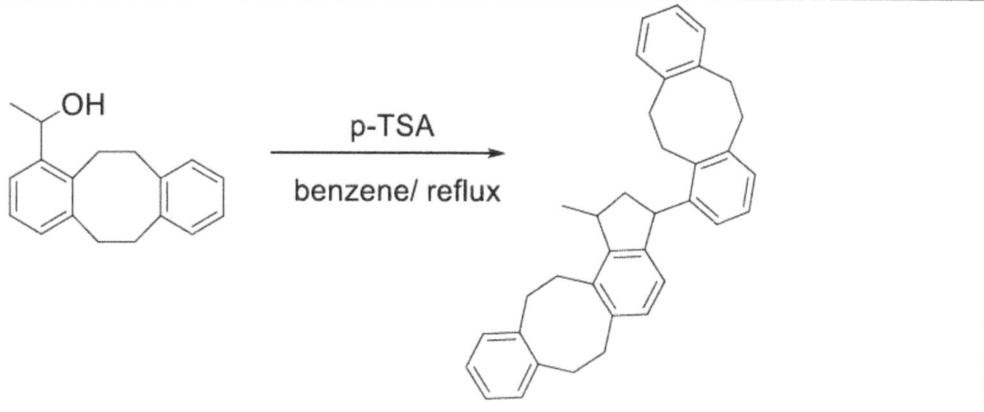OH p-TSA benzene/ reflux
140	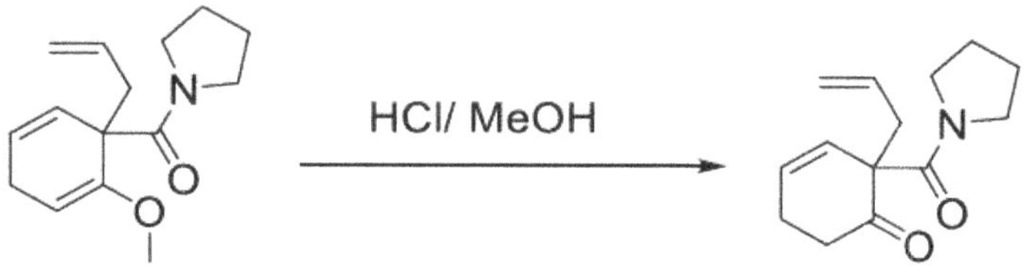N O O HCl/ MeOH N O O
141	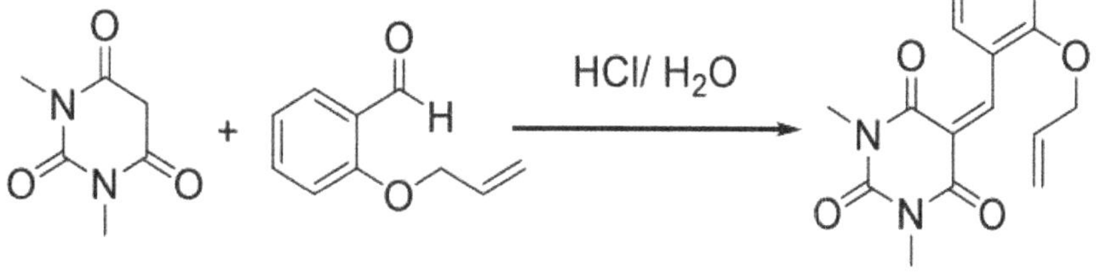O N O N O + O H O HCl/ H_2O O O N O N O
142	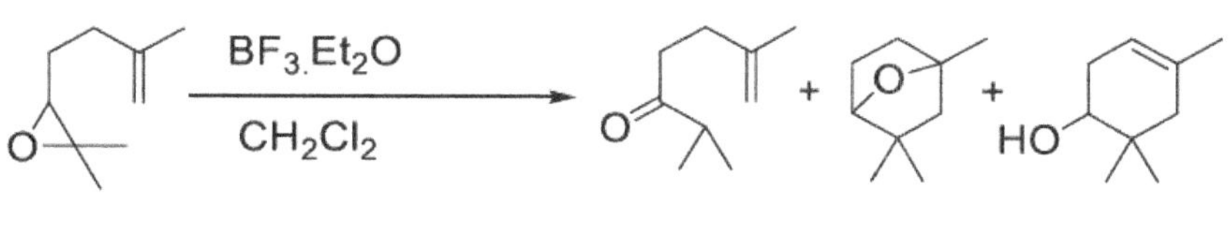O $BF_3.Et_2O$ CH_2Cl_2 O + O + HO
143	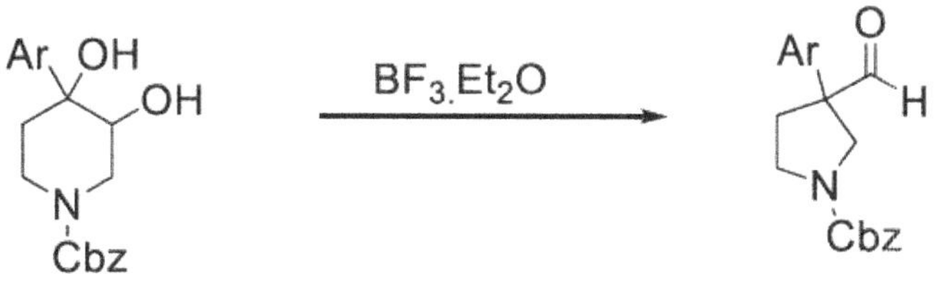Ar OH OH N Cbz $BF_3.Et_2O$ Ar O H N Cbz

144	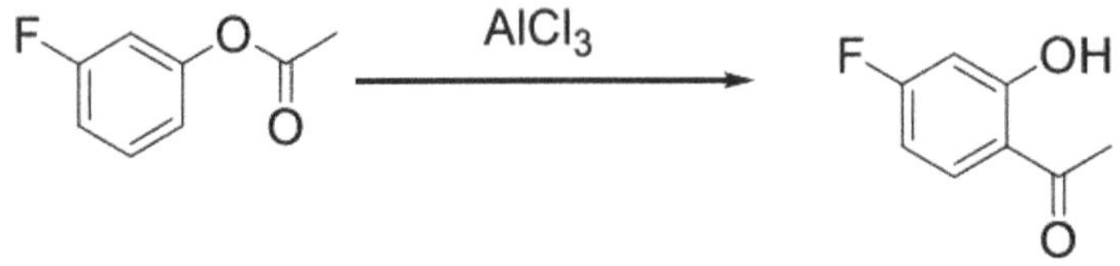F O O AlCl3 F OH O
145	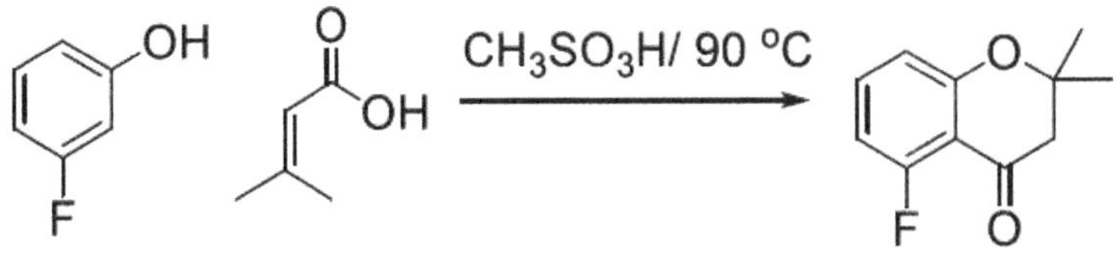OH F O OH CH3SO3H/ 90 °C O F O
146	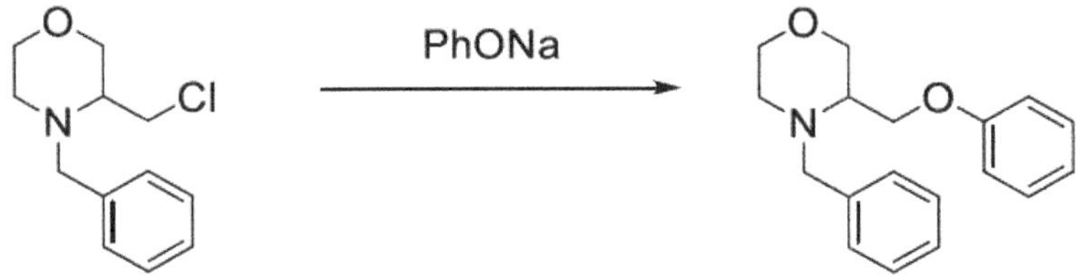O N Cl PhONa O N O
147	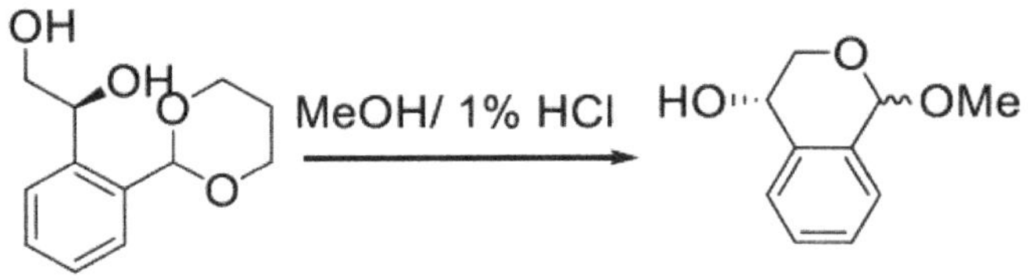OH OH O O MeOH/ 1% HCl HO O OMe
148	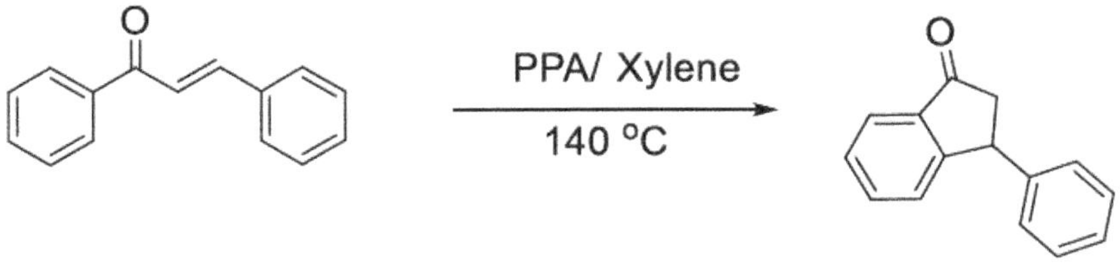O PPA/ Xylene 140 °C O
149	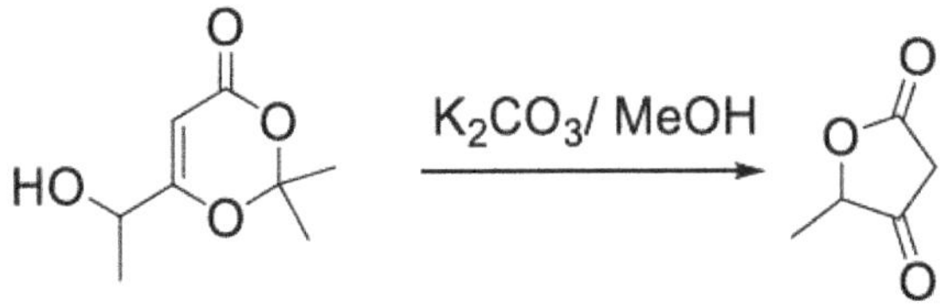O O HO O K2CO3/ MeOH O O O

150	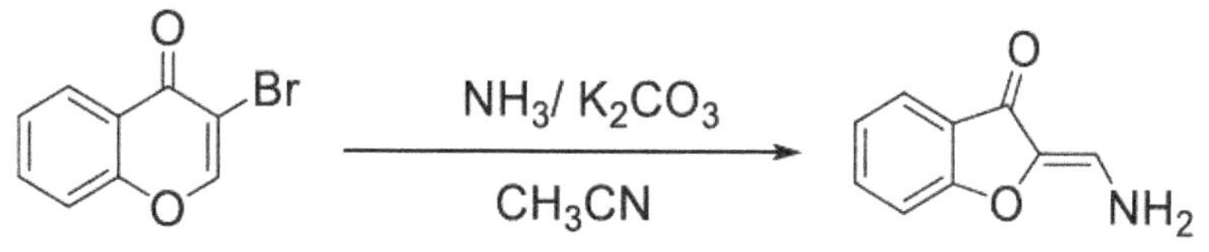NH3/ K2CO3 CH3CN
151	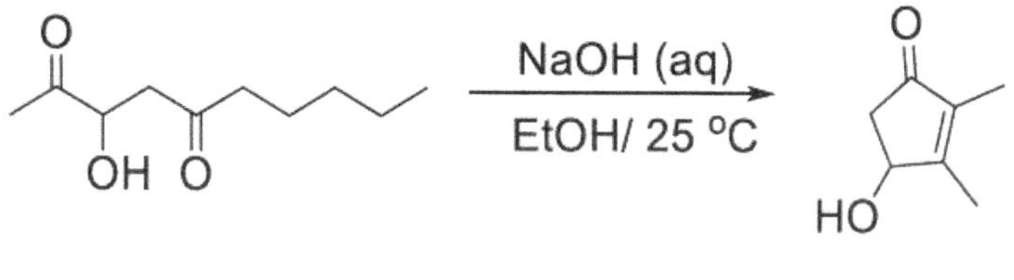NaOH (aq) EtOH/ 25 °C
152	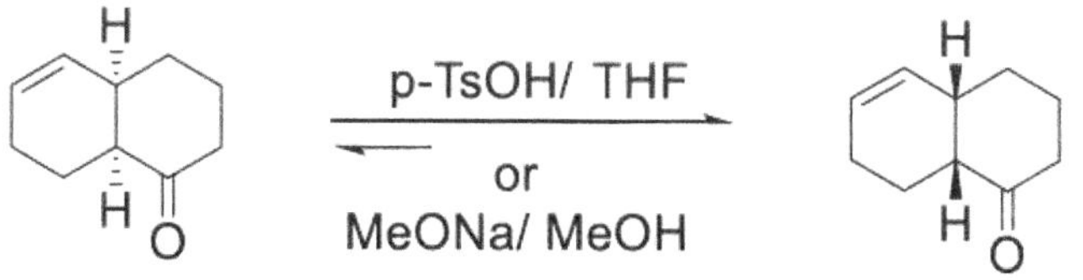p-TsOH/ THF or MeONa/ MeOH
153	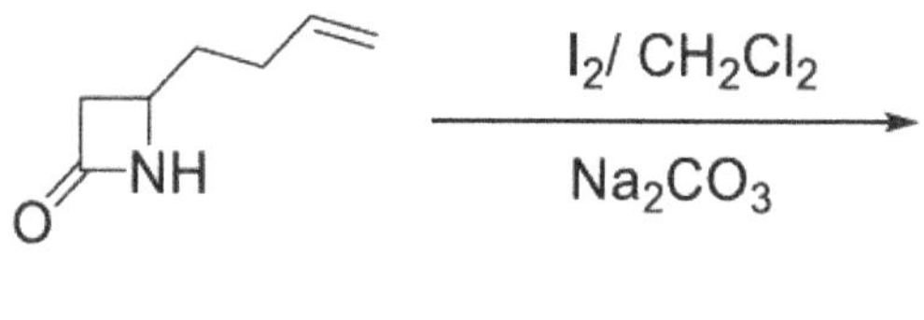I2/ CH2Cl2 Na2CO3
154	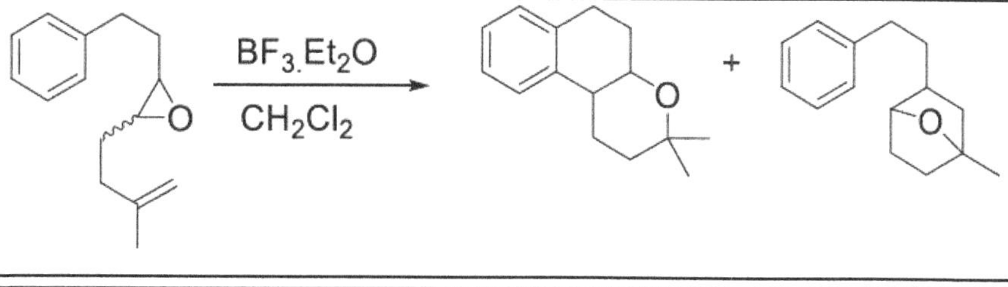BF3.Et2O CH2Cl2
155	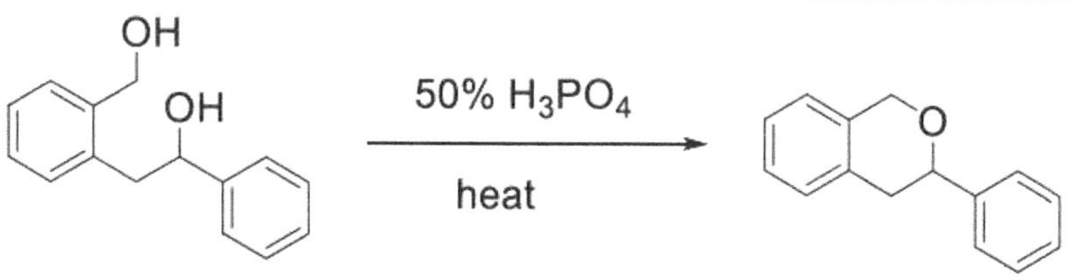50% H3PO4 heat

156	Simple cyclopropanes do not react with nucleophilic reagents. In contrast, "activated" cyclopropanes undergo nucleophilic cleavage. Write a mechanism that accounts for this observation. 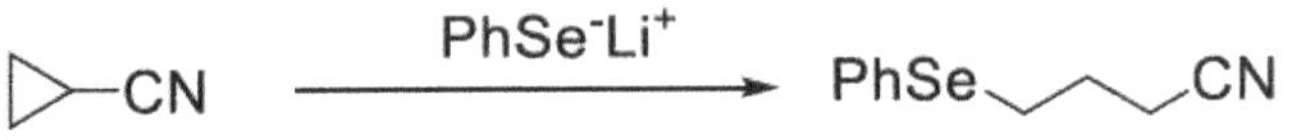 CN PhSe⁻Li⁺ PhSe CN
157	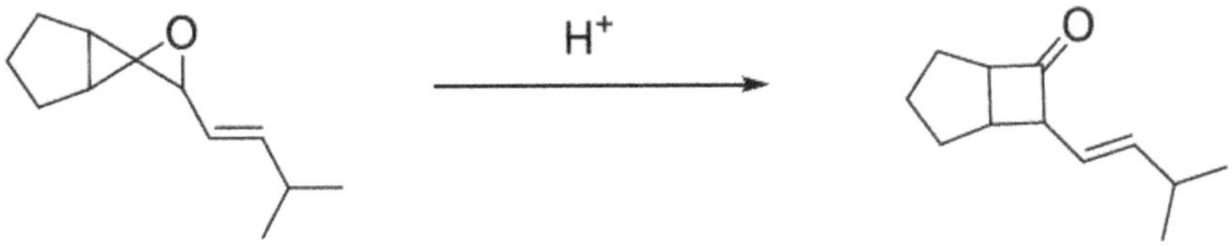 O H⁺ O
158	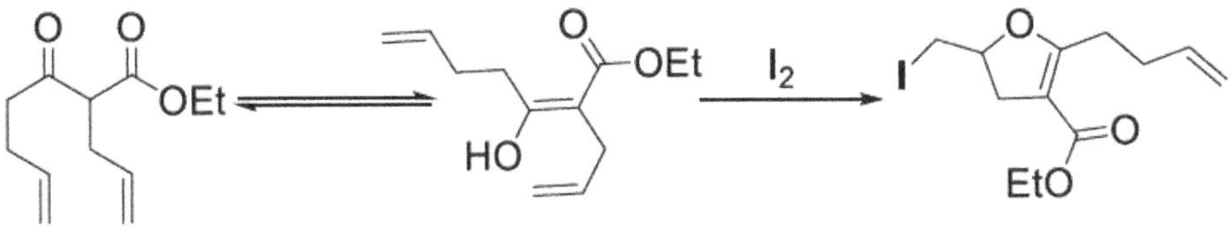 O O OEt O OEt HO I2 I O O EtO
159	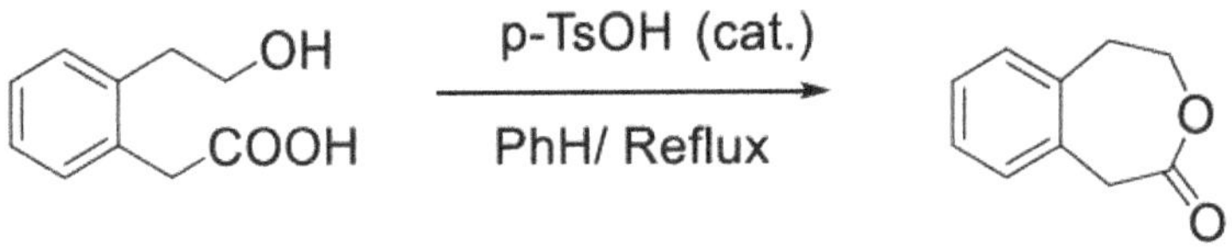 OH COOH p-TsOH (cat.) PhH/ Reflux O O
160	 O OEt O O tBuOK/DMSO CO2Et O O
161	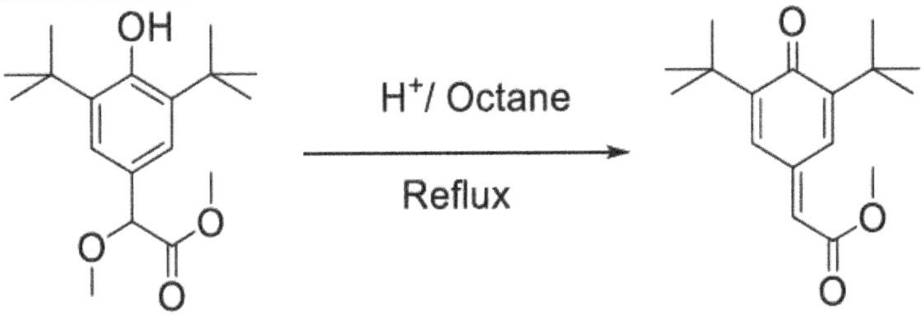 OH O O O H⁺/ Octane Reflux O O O

162

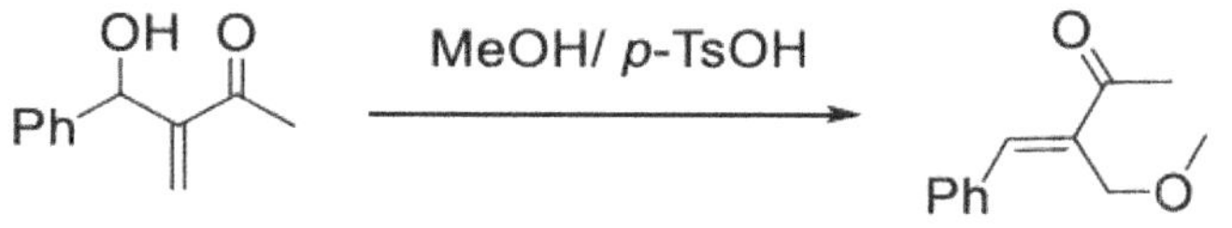

OH
O
Ph
MeOH/ p-TsOH
O
Ph
O

163

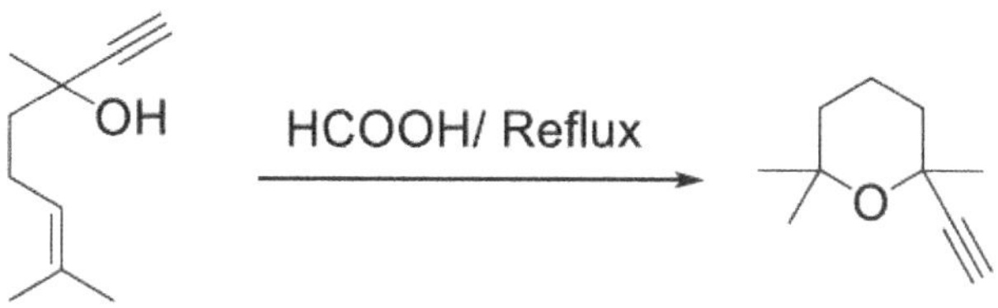

OH
HCOOH/ Reflux
O

164

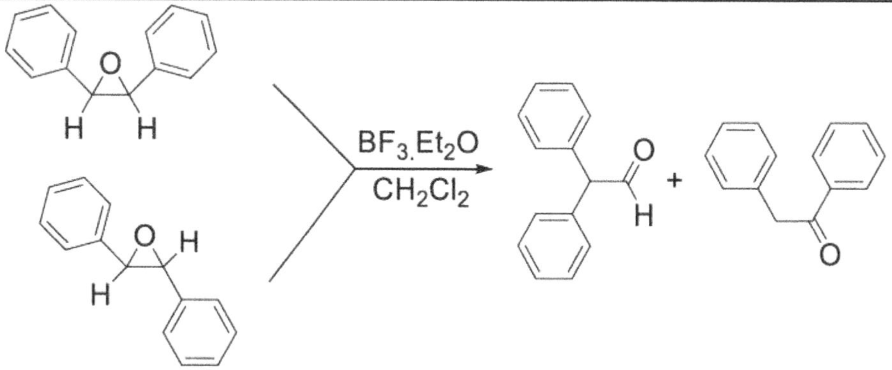

O
H
H
O
H
H
BF3.Et2O
CH2Cl2
O
H
+
O

165

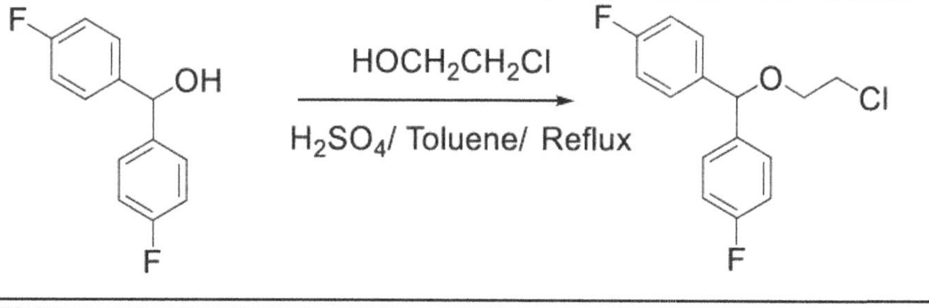

F
OH
F
HOCH2CH2Cl
H2SO4/ Toluene/ Reflux
F
O
Cl
F

166

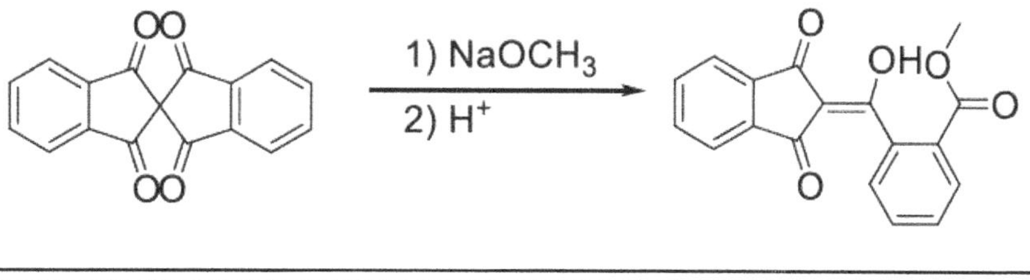

OO
OO
1) NaOCH3
2) H+
O
OH
O
O
O

167

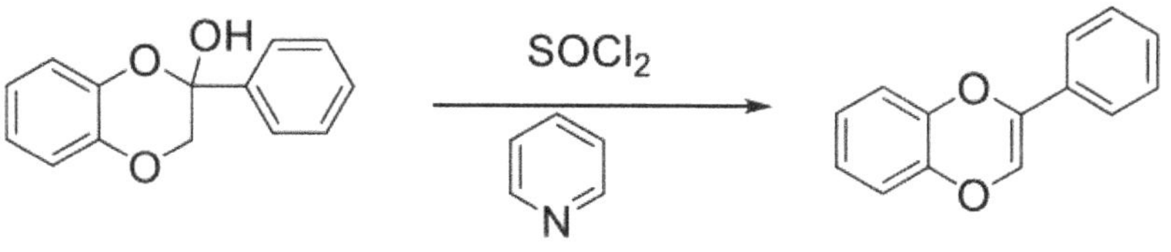

OH
O
O
SOCl2
N
O
O

168	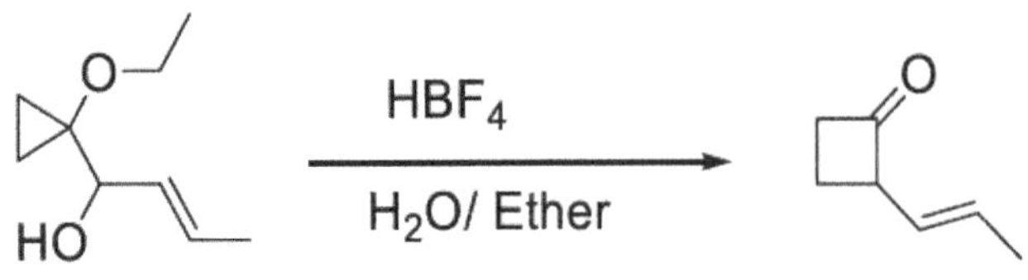HBF4 H2O/ Ether
169	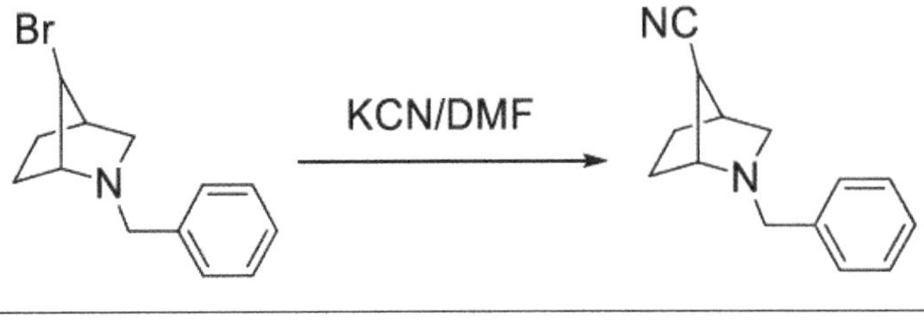Br KCN/DMF NC
170	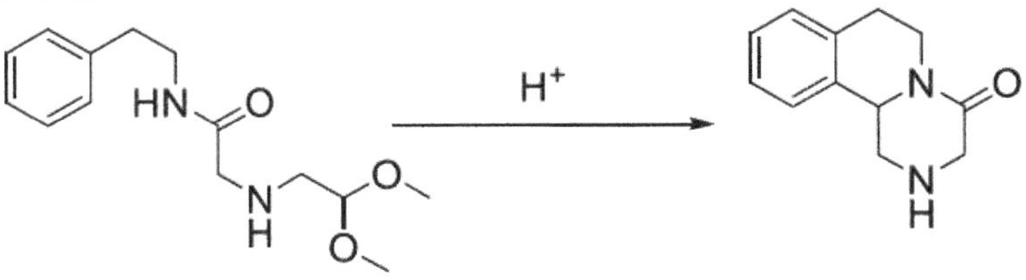H+
171	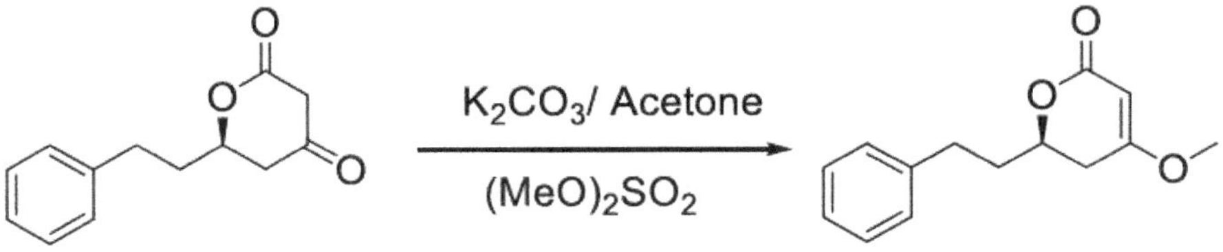K2CO3/ Acetone (MeO)2SO2
172	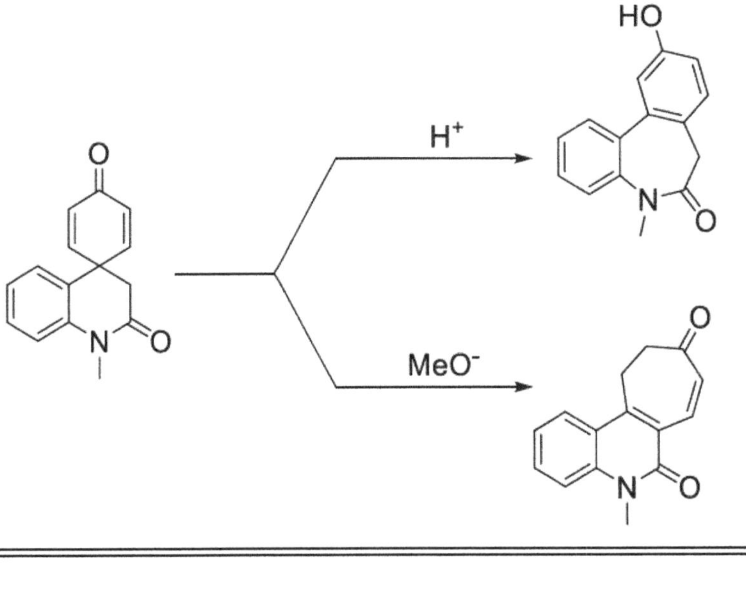H+ MeO- HO

173	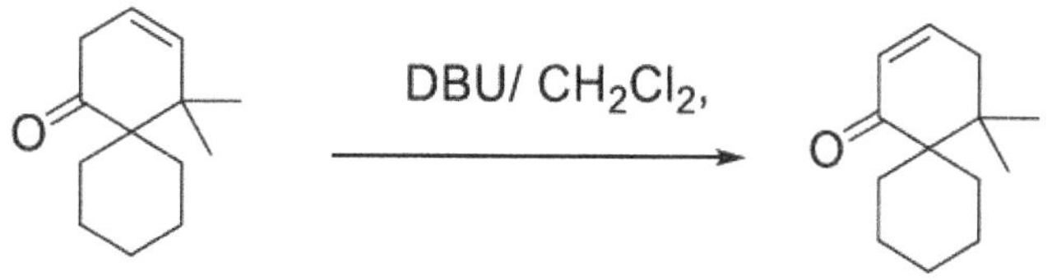 DBU/ CH2Cl2,
174	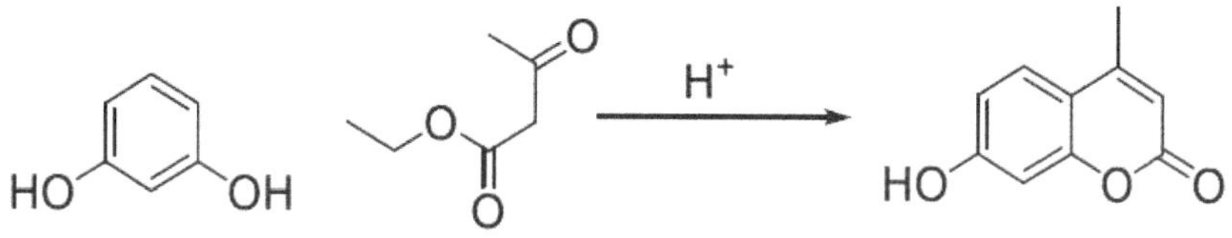 HO OH O O O H+ HO O O
175	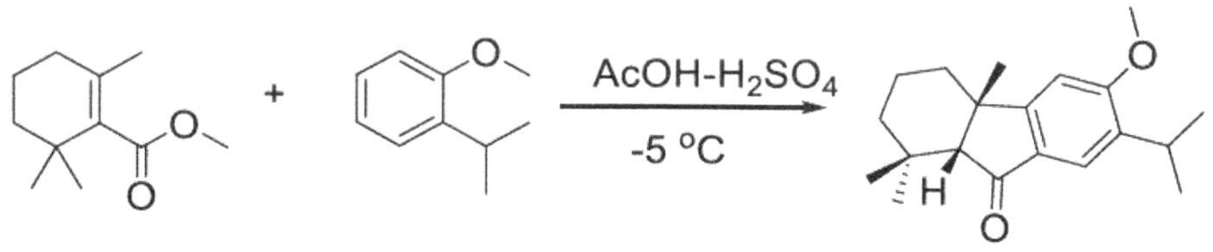 + AcOH-H2SO4 -5 °C O O O H O
176	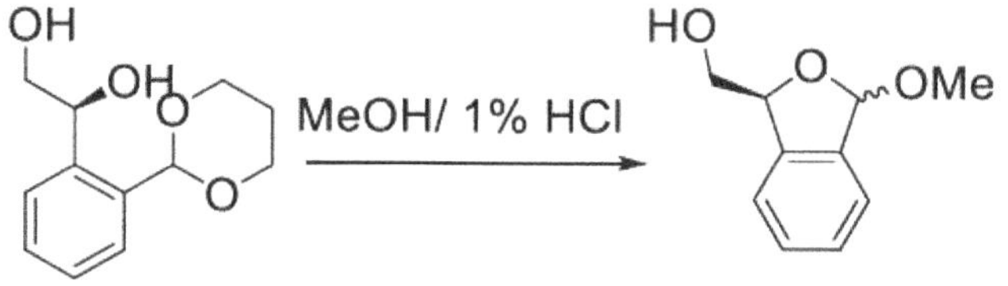 OH OH O O MeOH/ 1% HCl HO O OMe
177	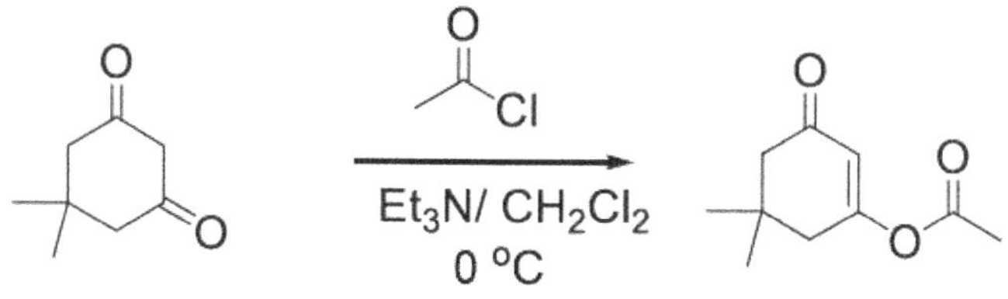 O O O Cl Et3N/ CH2Cl2 0 °C O O O
178	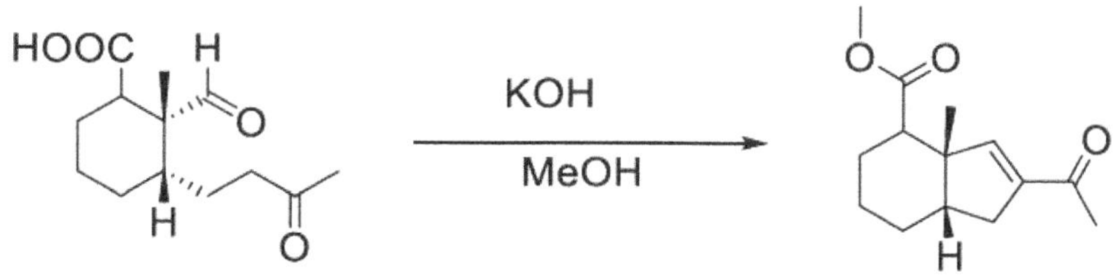 HOOC H O H O KOH MeOH O O O H

179	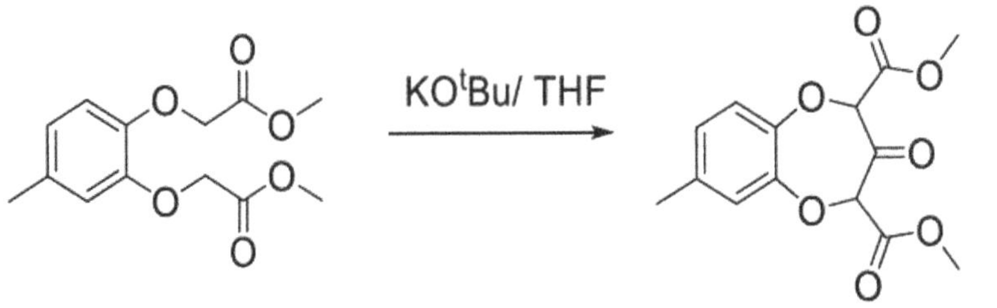KO^tBu/ THF
180	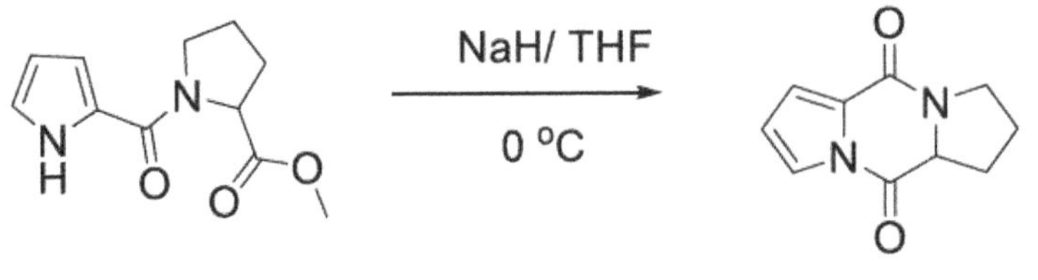NaH/ THF 0 °C
181	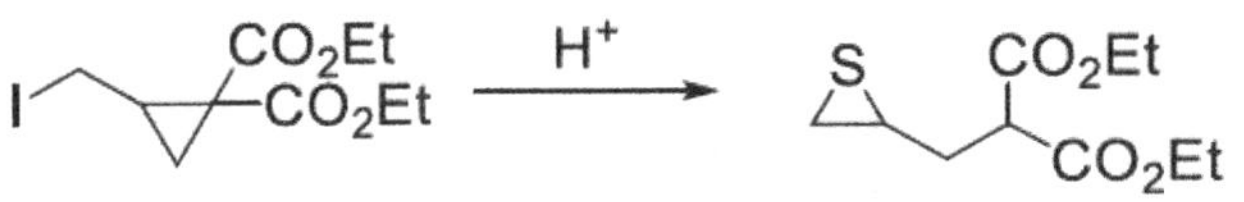I CO2Et CO2Et H+ S CO2Et CO2Et
182	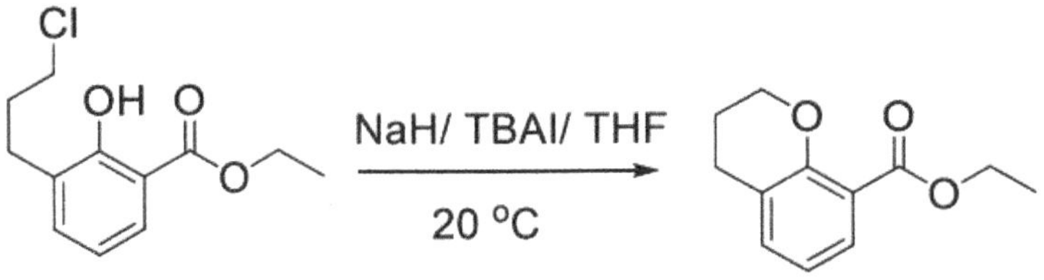Cl OH NaH/ TBAI/ THF 20 °C
183	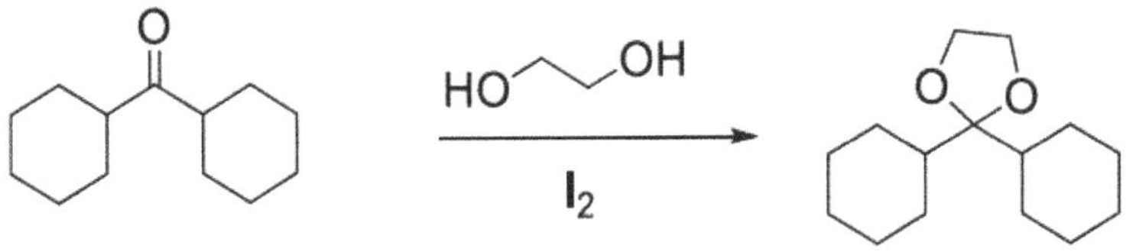HO OH I2
184	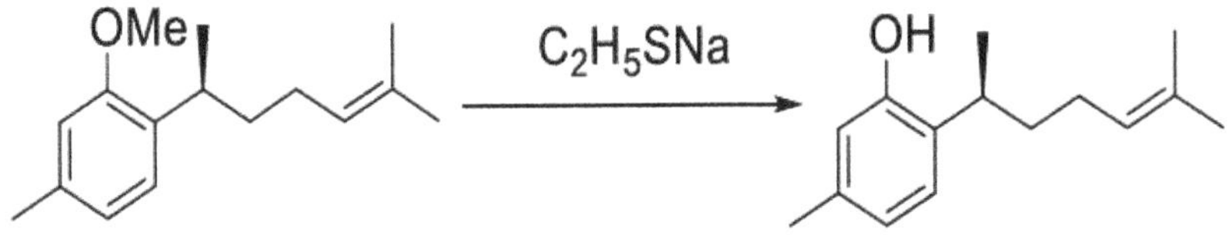OMe C2H5SNa OH

185

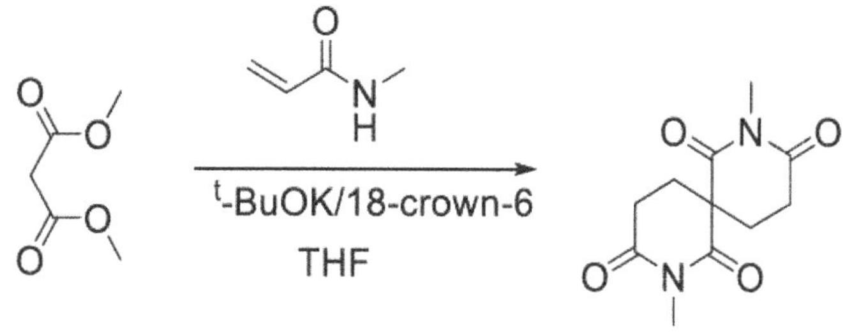

t-BuOK/18-crown-6
THF

186

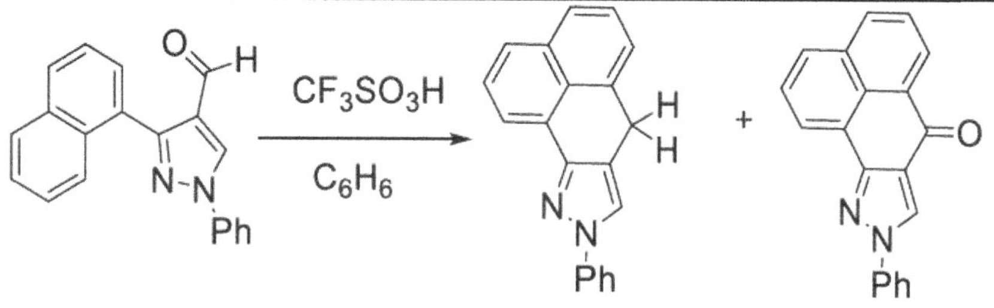

CF3SO3H
C6H6

187

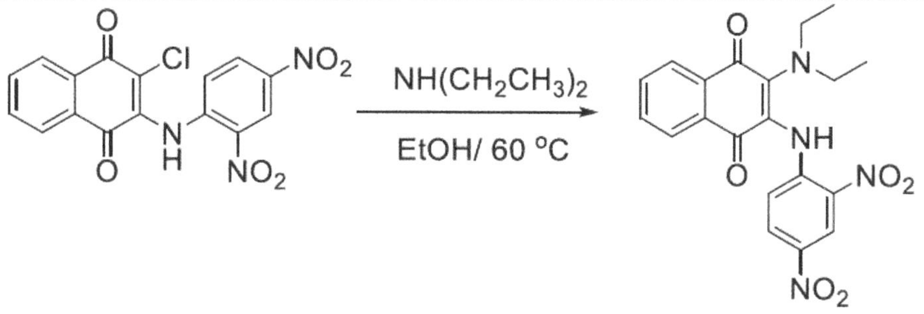

NH(CH2CH3)2
EtOH/ 60 °C

188

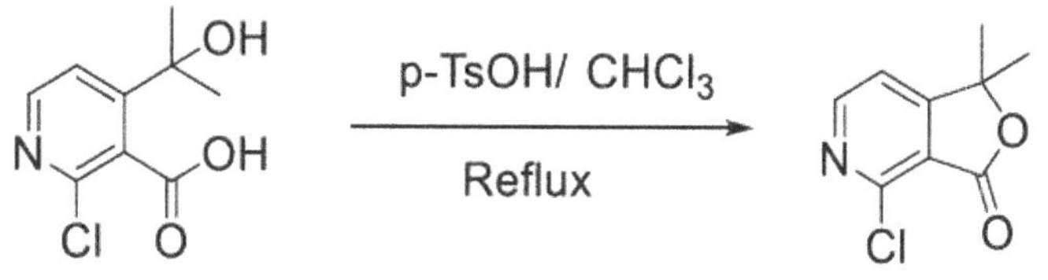

p-TsOH/ CHCl3
Reflux

189

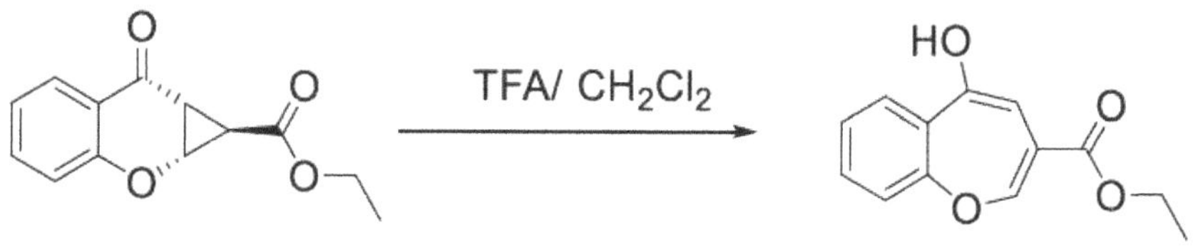

TFA/ CH2Cl2

190

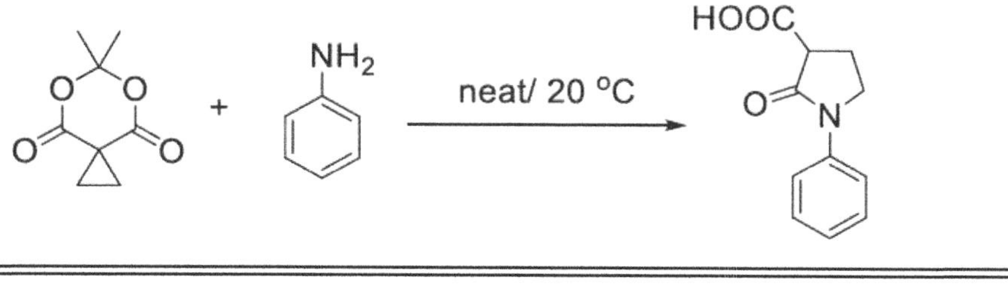

neat/ 20 °C

191	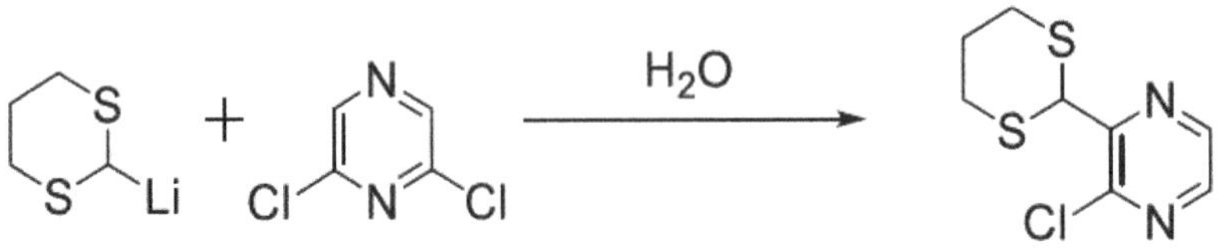 Li Cl N H2O S
192	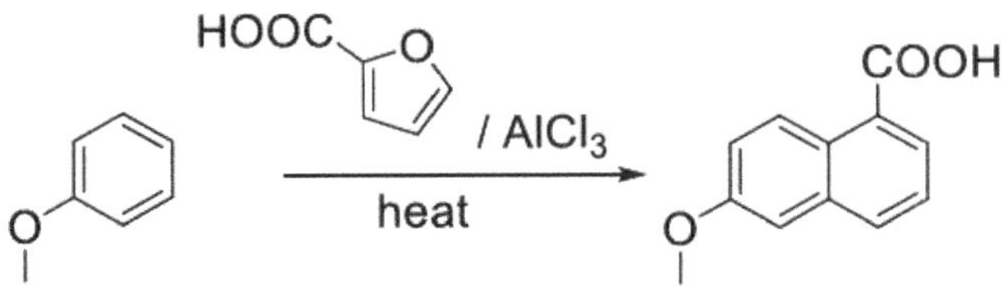 HOOC O / AlCl3 heat COOH
193	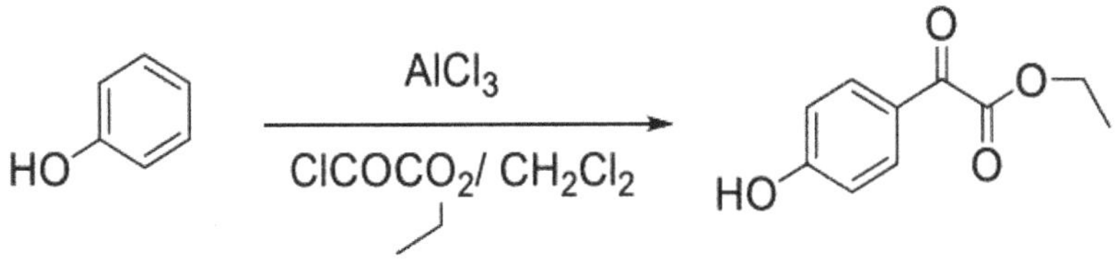 HO AlCl3 ClCOCO2/ CH2Cl2 O
194	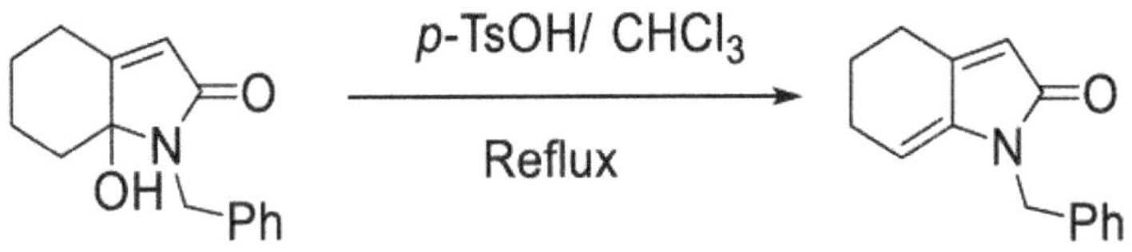 OH N Ph O p-TsOH/ CHCl3 Reflux
195	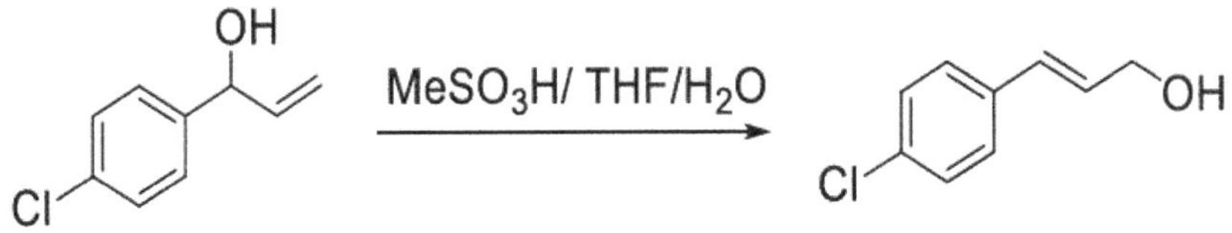 OH Cl MeSO3H/ THF/H2O
196	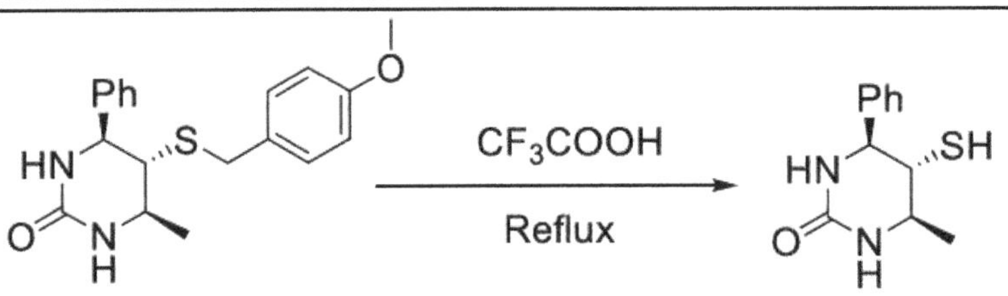 Ph HN S O N H CF3COOH Reflux SH

197

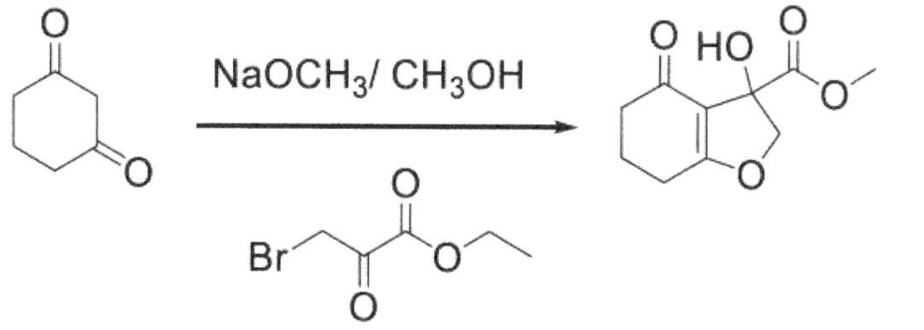
NaOCH3/ CH3OH
HO
Br

198

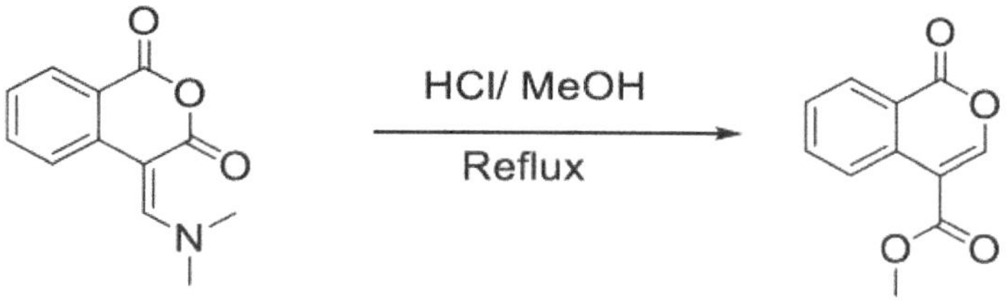
HCl/ MeOH
Reflux

199

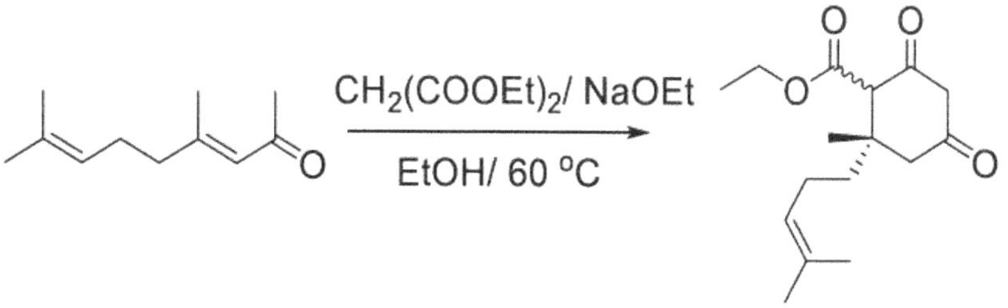
CH2(COOEt)2/ NaOEt
EtOH/ 60 °C

200

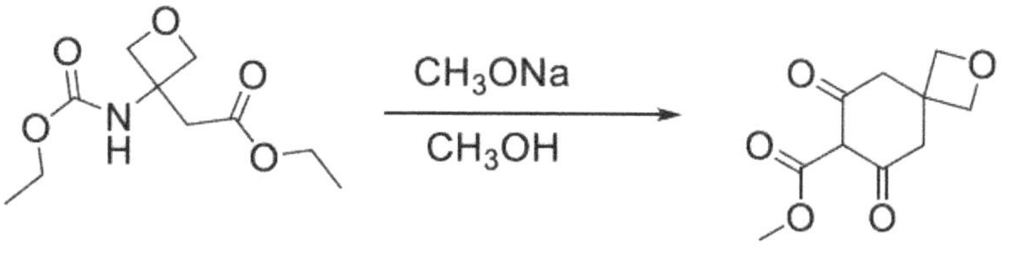
CH3ONa
CH3OH
N
H

Answers to Questions 71-200

71

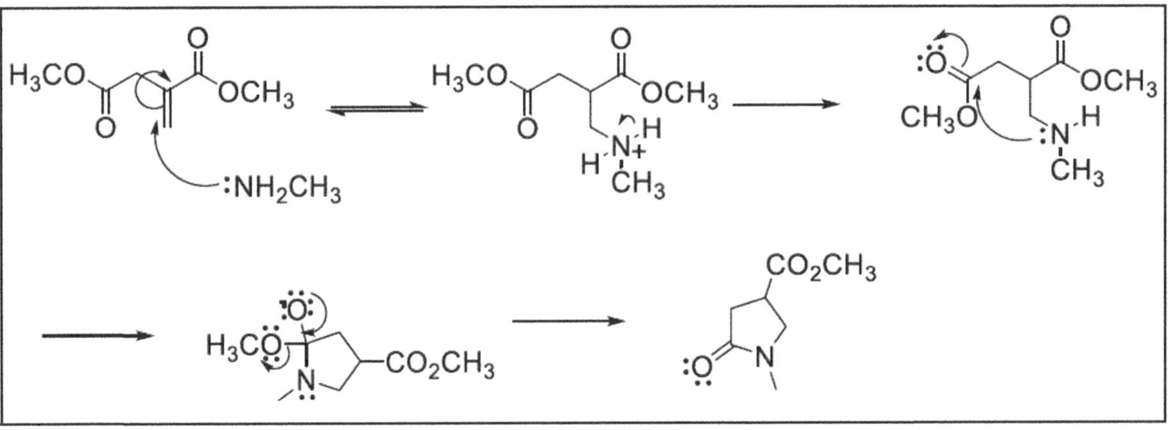

H3CO
O
OCH3
O
:NH2CH3
H3CO
O
OCH3
O
H
H N+
CH3
:O
O
OCH3
CH3O
:N H
CH3
:O:
H3CO
N
CO2CH3
CO2CH3
:O
N

A Michael addition reaction is followed by an intramolecular nucleophilic acyl substitution reaction to yield the product.

72 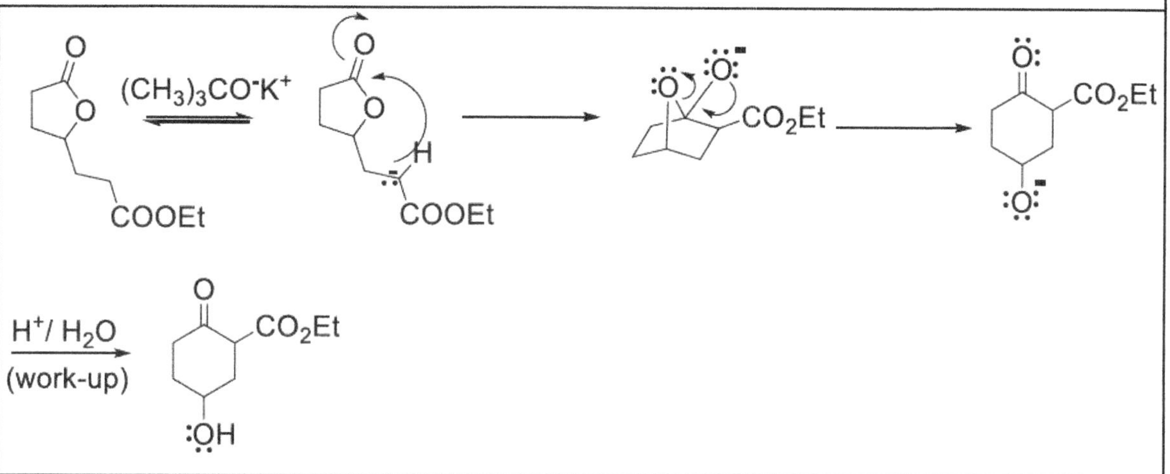

O
O
(CH3)3CO-K+
COOEt
O
O
H
COOEt
:O:
:O:
CO2Et
Ö:
CO2Et
:O:
H+/ H2O
(work-up)
O
CO2Et
:OH

Rapid formation of the anion is followed by an intramolecular acyl substitution reaction.

73 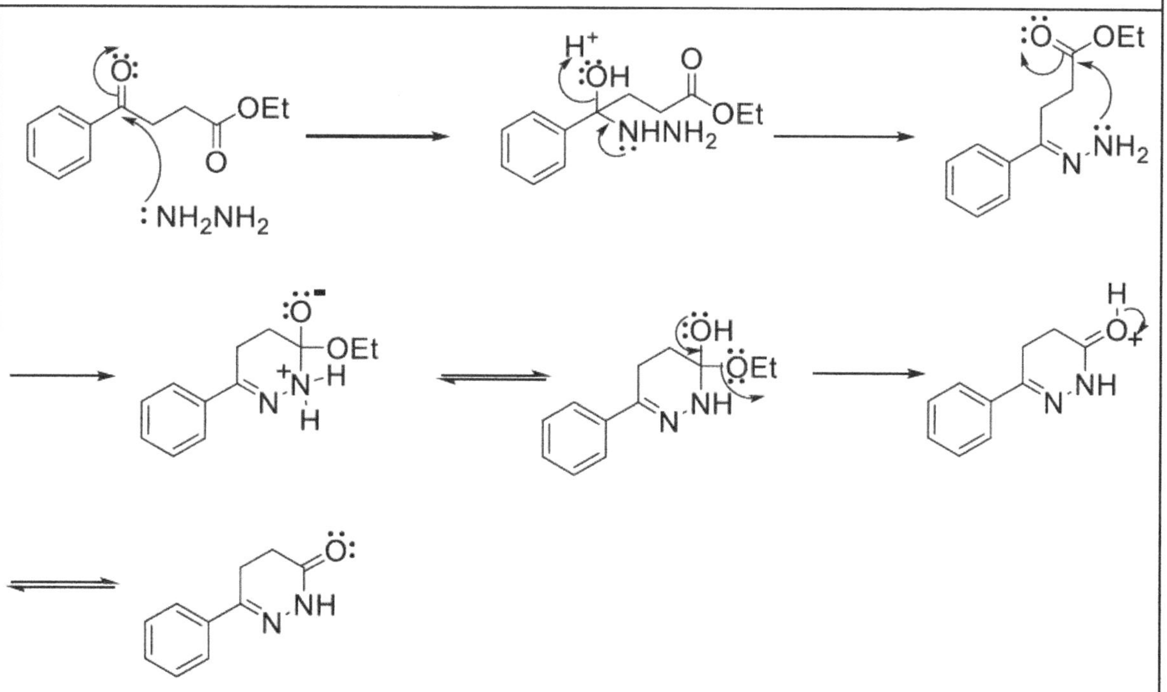

:Ö:
OEt
O
: NH2NH2
H+
:ÖH
O
OEt
NHNH2
:Ö
OEt
N
NH2
:O:
OEt
N
N-H
H
:ÖH
ÖEt
N
NH
H
O
N
NH
:Ö:
N
NH

In the first step of this reaction, the nucleophile attacks the ketone carbonyl instead of the ester carbonyl. Why?

74

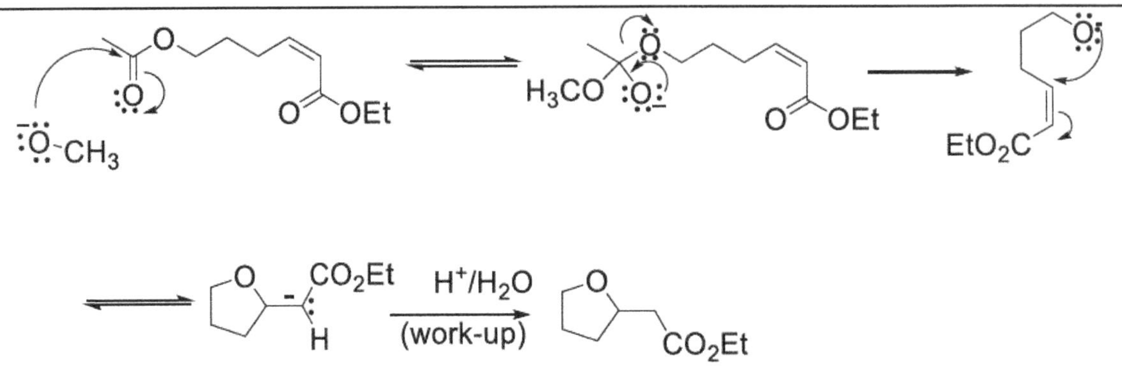

O
:O:
OEt
:O-CH3
H3CO
OEt
EtO2C
CO2Et
H
H+/H2O
(work-up)
CO2Et

A nucleophilic acyl substitution reaction is followed by an intramolecular Michael addition reaction.

75

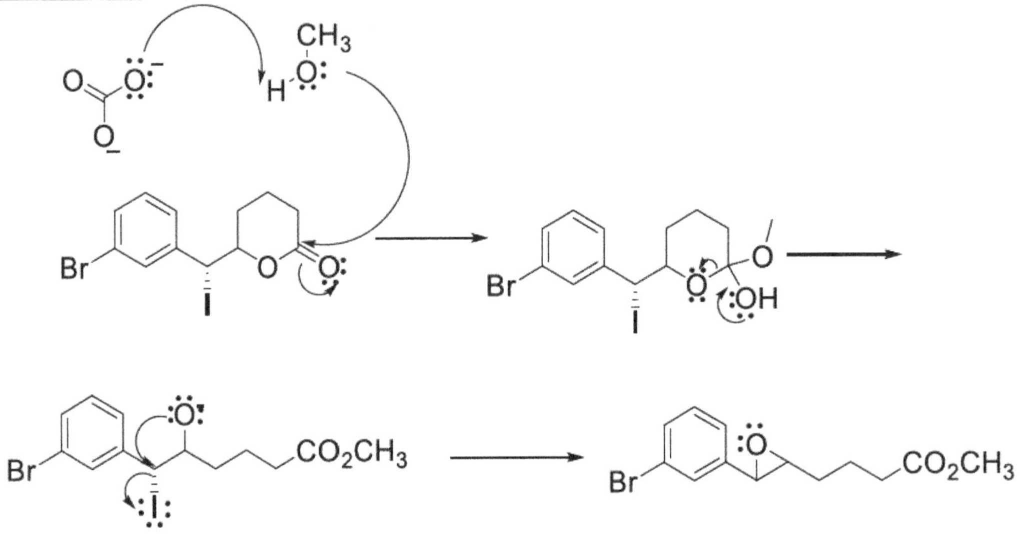

CH3
H
Br
I
O
:OH
CO2CH3
CO2CH3

An acid-catalyzed acyl substitution reaction is followed by epoxide formation.

76

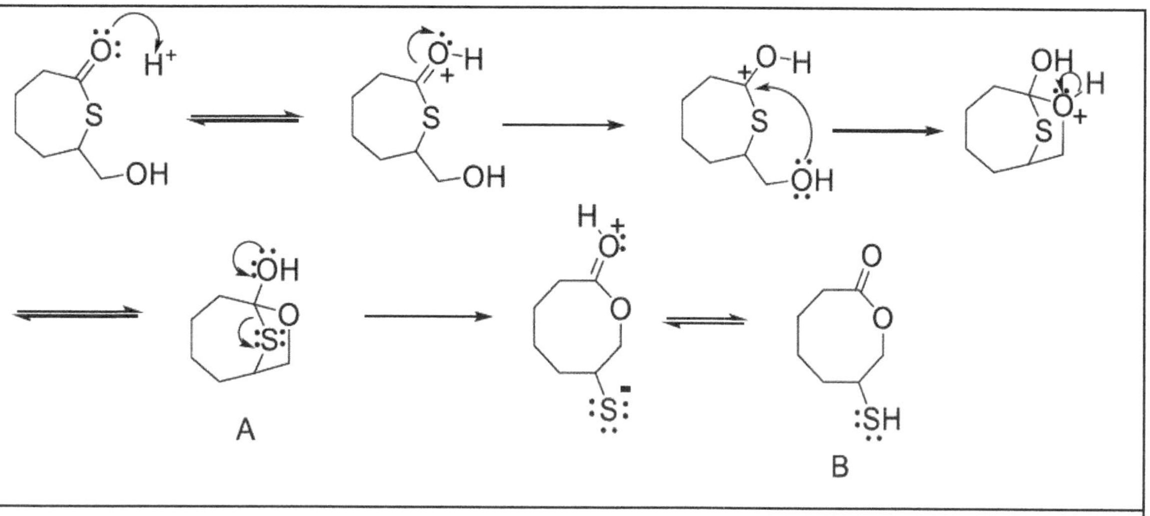

O:
H+
S
OH
O-H
+
S
OH
+
O-H
S
OH
OH
H
O+
S
:OH
O
:S:
A
H
+
O:
O
:S:
O
O
:SH
B

When tetrahedral intermediate A reforms the strong C=O bond, it yields B instead of the compound shown below. Why?

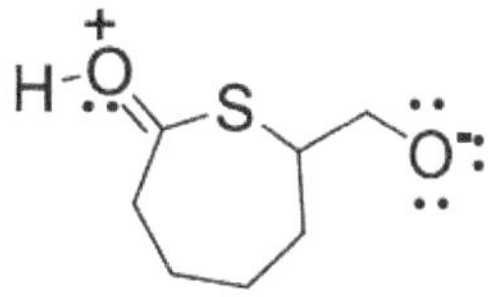

+
H-O
S
O:

77 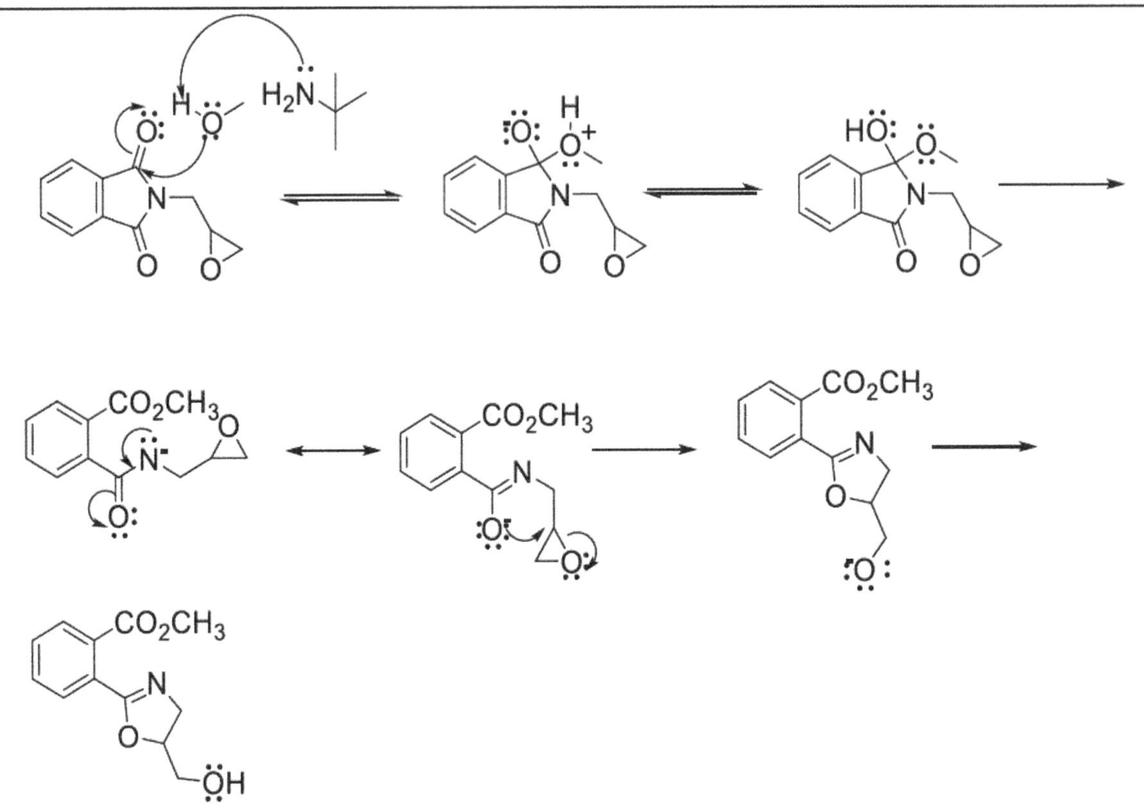

An acyl substitution reaction is followed nucleophilic ring opening of the epoxide to form an oxazolidine.

78 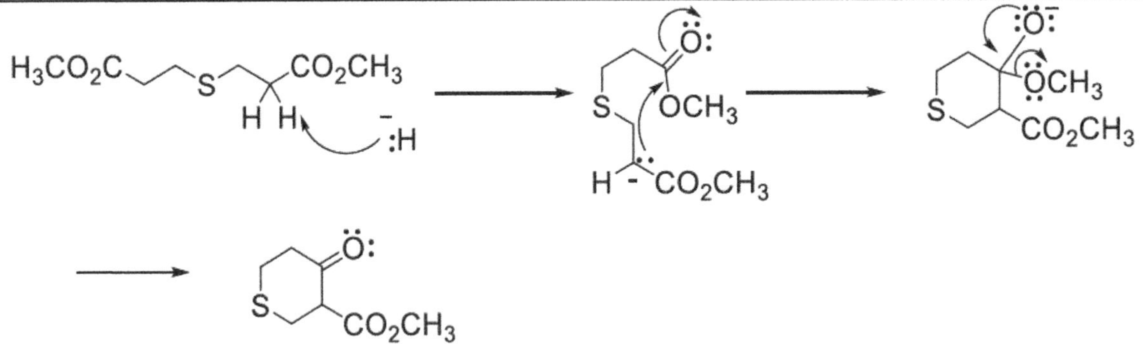

Formation of the anion is followed by an intramolecular acyl substitution reaction.

79

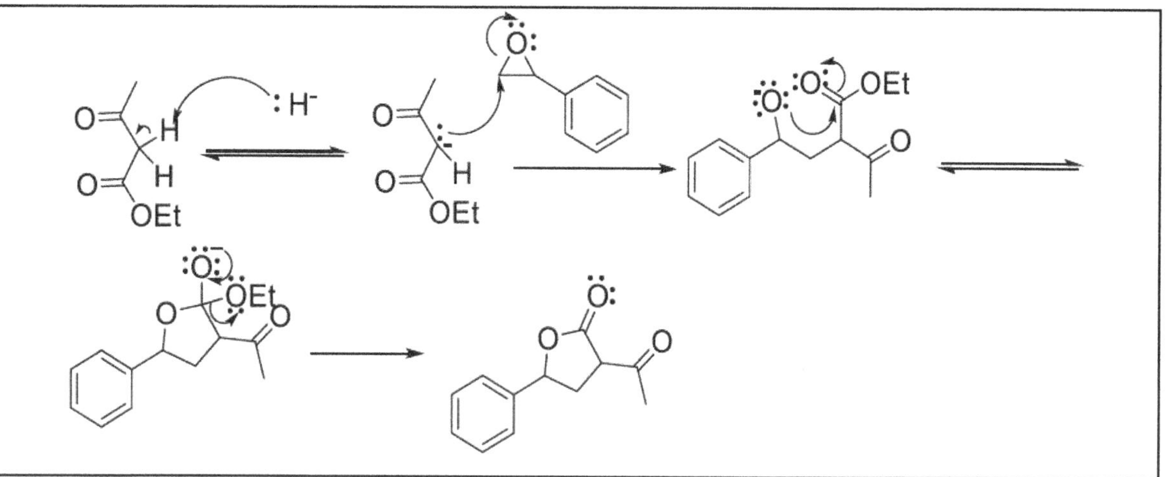

Formation of the anion (nucleophile) is followed by nucleophilic ring opening of the epoxide (note that attack is on the least congested carbon of the epoxide ring) and a subsequent intramolecular acyl substitution reaction.

80

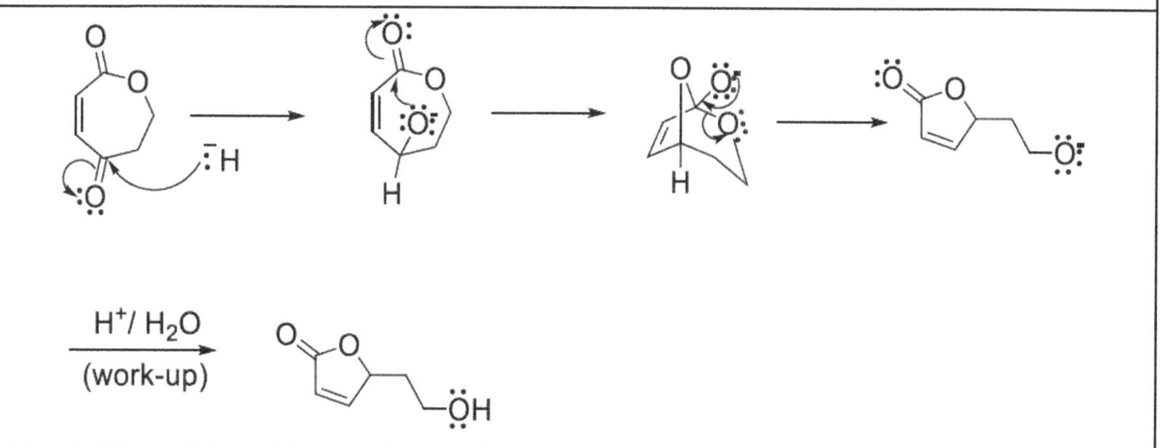
H⁺/ H₂O
(work-up)

Reducing agents, such as $NaBH_4$, $LiAlH_4$, etc. function by delivering hydride ($^-$:H) ions. Addition of the hydride to the ketone carbonyl is followed by a transannular nucleophilic acyl substitution reaction.

81	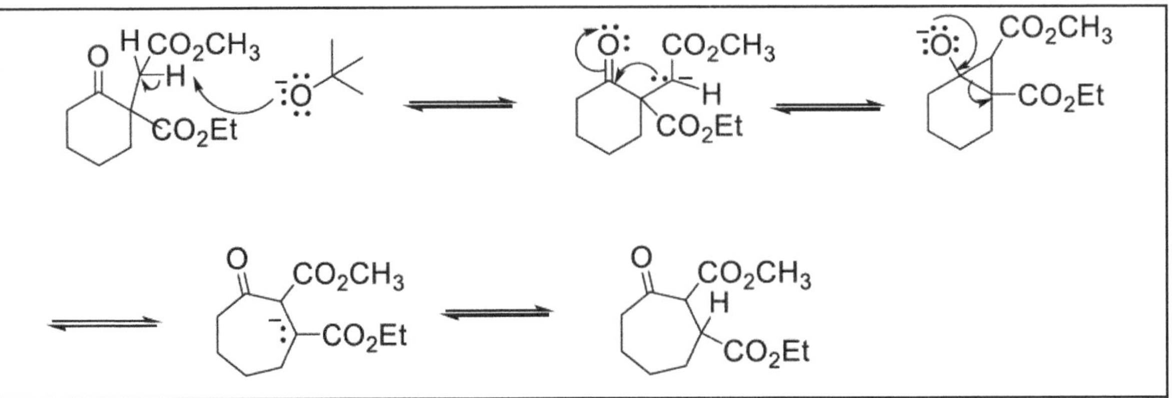
82	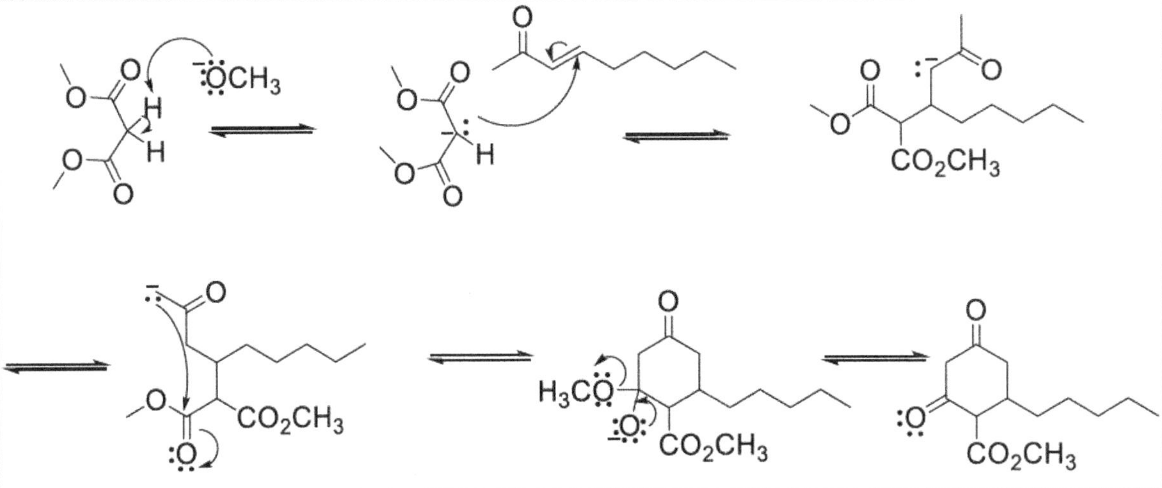

Initial formation of the anion is followed by a Michael addition reaction. A prototropic shift yields a new anion that undergoes an intramolecular acyl substitution reaction .

83

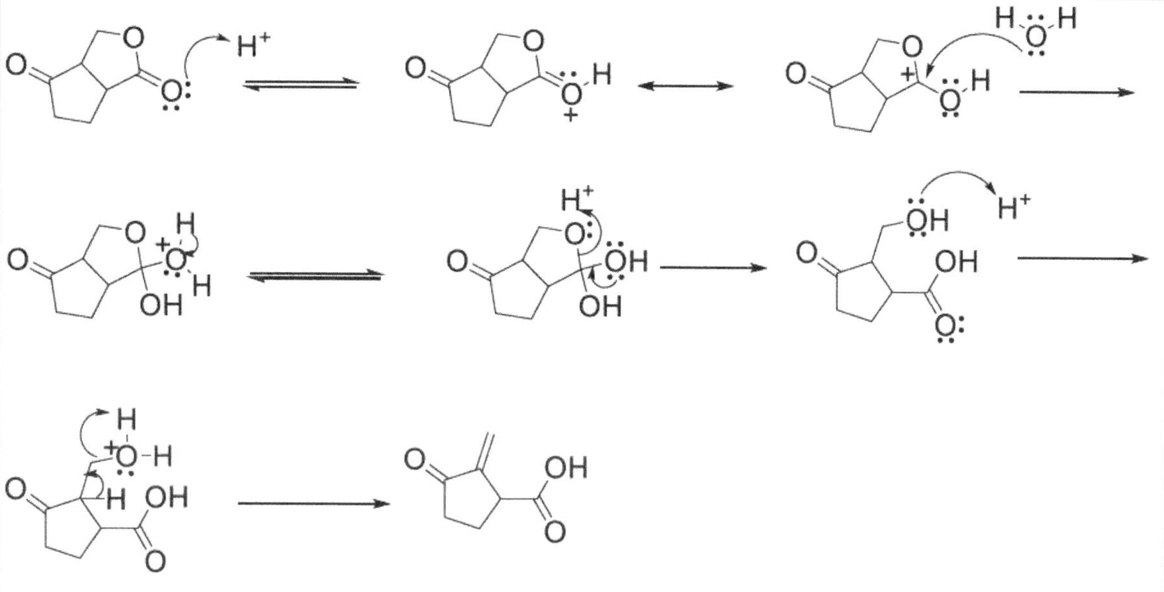

Acid-catalyzed hydrolysis of the lactone yields an α-hydroxymethyl ketone which undergoes elimination of water to form the α-methylene ketone.

84

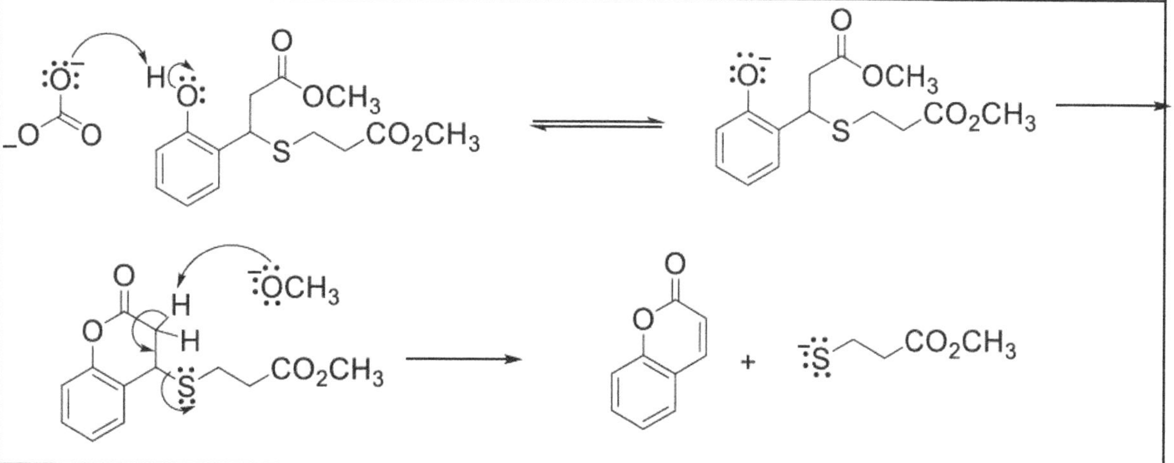

Reaction of the most acidic hydrogen yields an anion that participates in a nucleophilic acyl substitution reaction. This is followed by a base-catalyzed β-elimination reaction to form the α,β-unsaturated lactone.

85

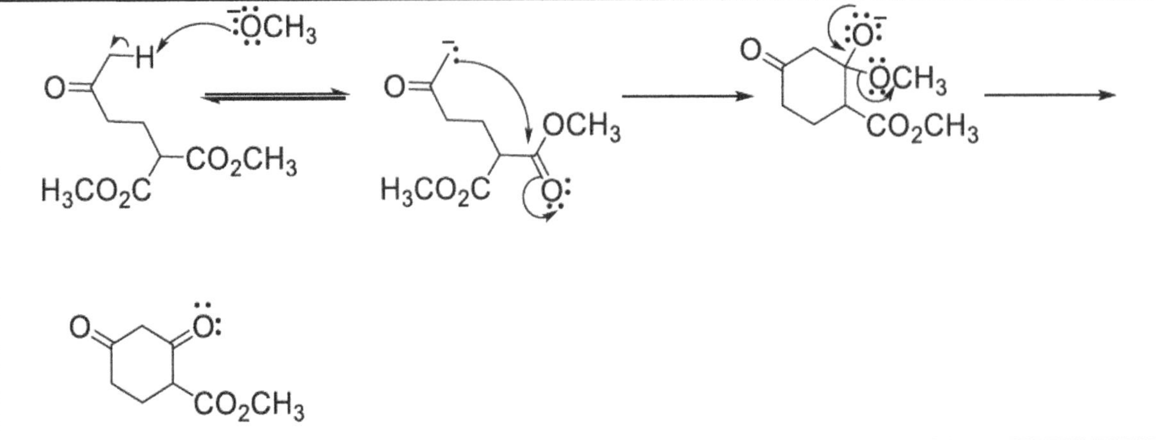

H
:ÖCH3
O
H3CO2C
CO2CH3
OCH3
:O:
ÖCH3
CO2CH3
Ö:

The pK_a of malonic ester is ~13 and the α-H of the ketone is ~18. Why do you think the reaction proceeds as shown?

86

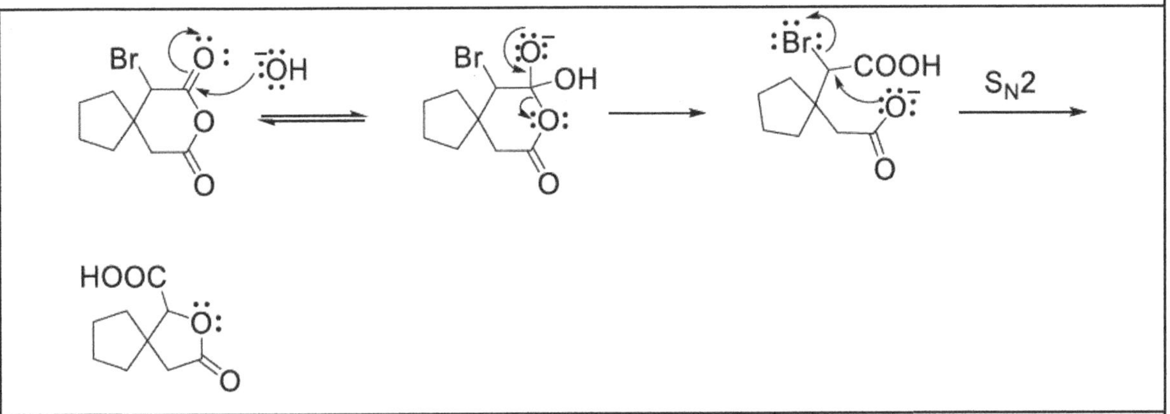

Br
O
:ÖH
OH
:Br:
COOH
SN2
HOOC
Ö:

Note that in the first step of the reaction the Lewis base (nucleophile) attacks the most electron-deficient (most electrophilic) carbonyl carbon. The electron-withdrawing inductive effect of the electronegative atom (Br) renders the top carbonyl group more electrophilic (higher partial positive charge).

87

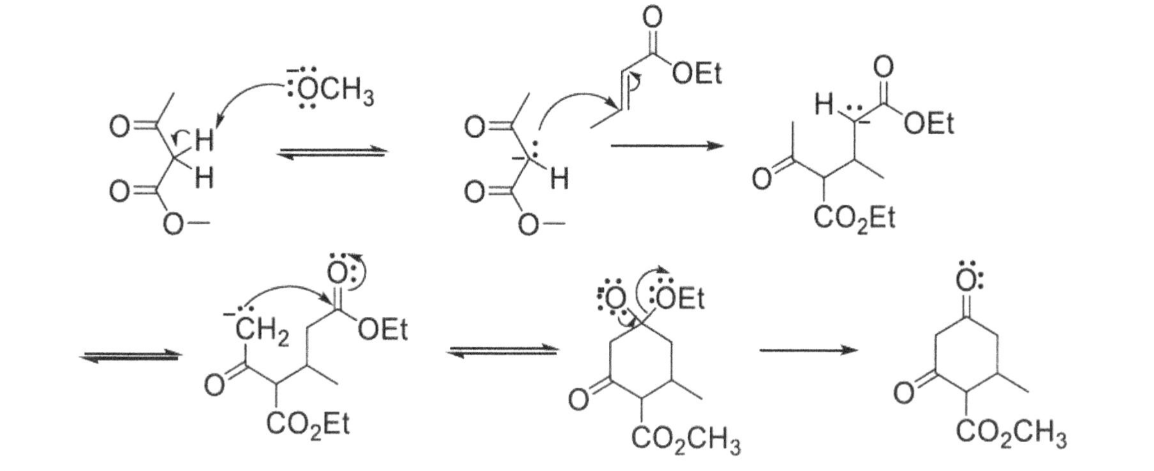

Formation of the anion is followed by a Michael addition reaction, a prototropic shift and an intramolecular acyl substitution reaction.

88

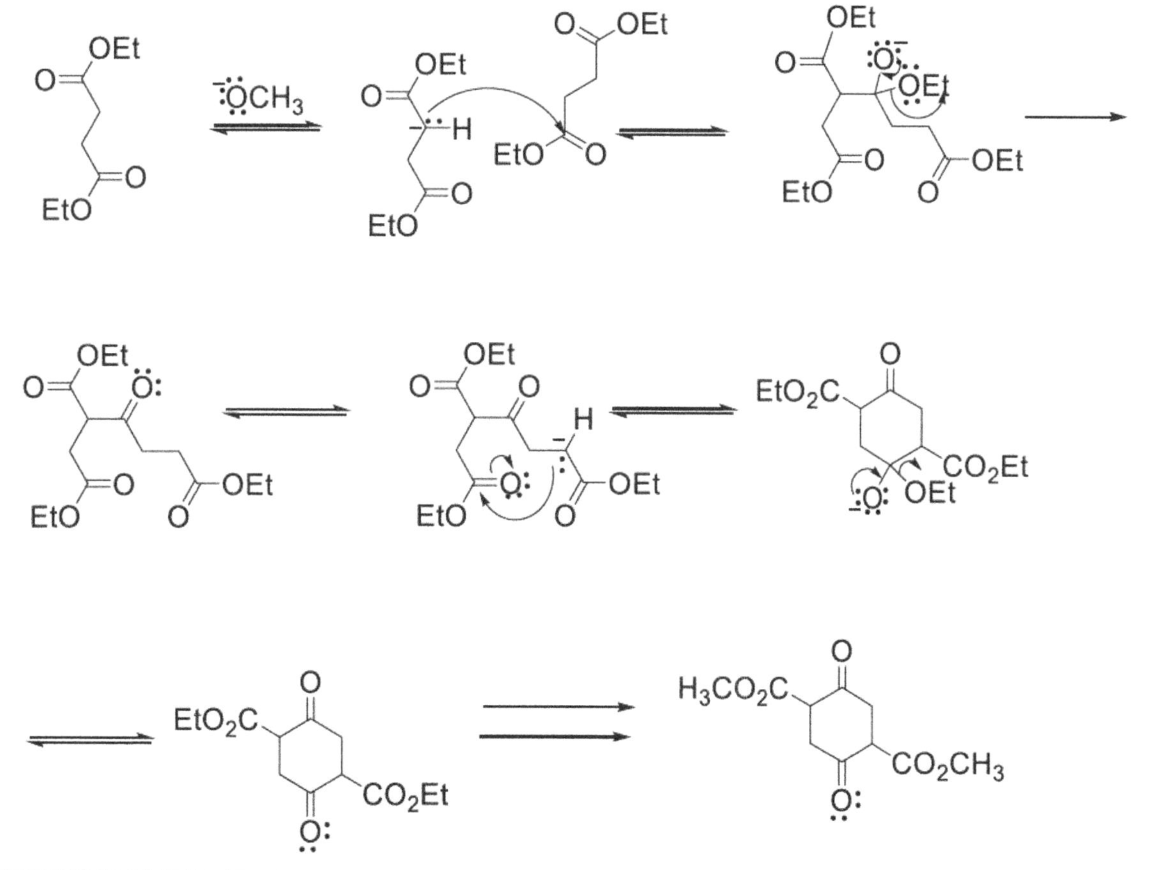

An intermolecular Claisen condensation reaction is followed by an intramolecular condensation reaction (*a Dieckman condensation*). A double transesterification reaction yields the final product.

89

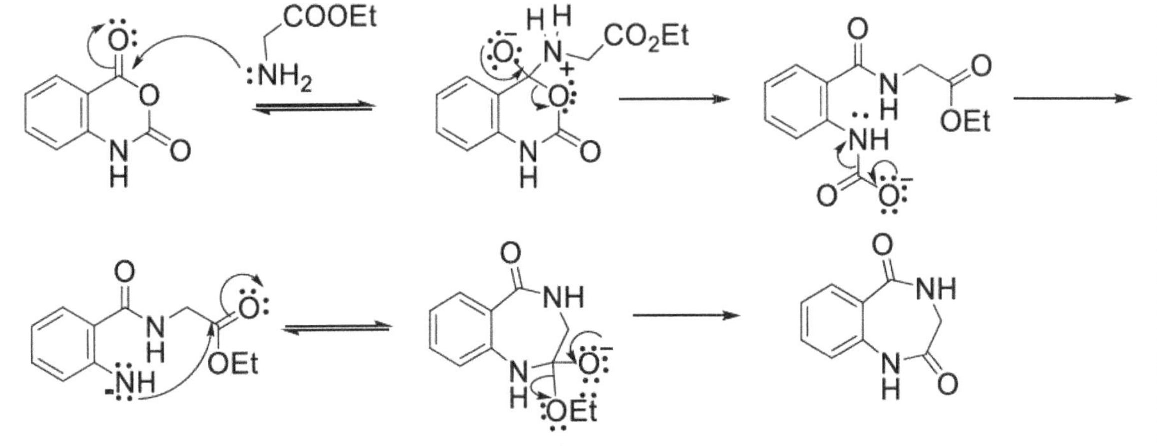
COOEt
NH2
CO2Et
OEt
NH
OEt

Exercise The top carbonyl carbon in this mixed anhydride is more reactive than the bottom one. Explain.

90

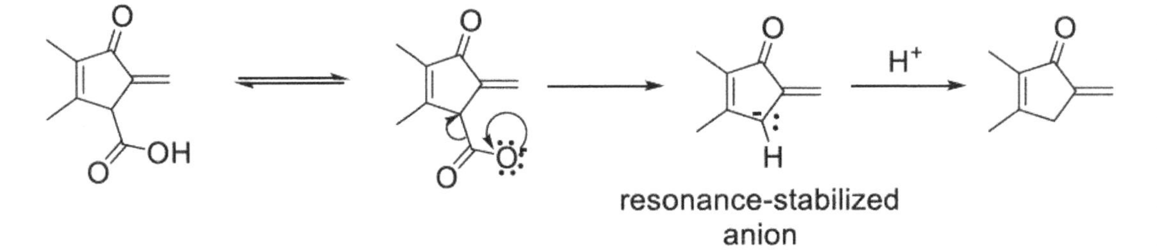
OH
H+
H
resonance-stabilized anion

Recall that, unlike ordinary carboxylic acids, β-ketoacids and *gem*-dicarboxylic acids (malonic acids) undergo facile loss of CO_2 to form a stable product (a resonance-stabilized anion). This unusual keto acid loses CO_2 for the same reason.

Exercise Draw all the resonance structures of the anion shown above.

91	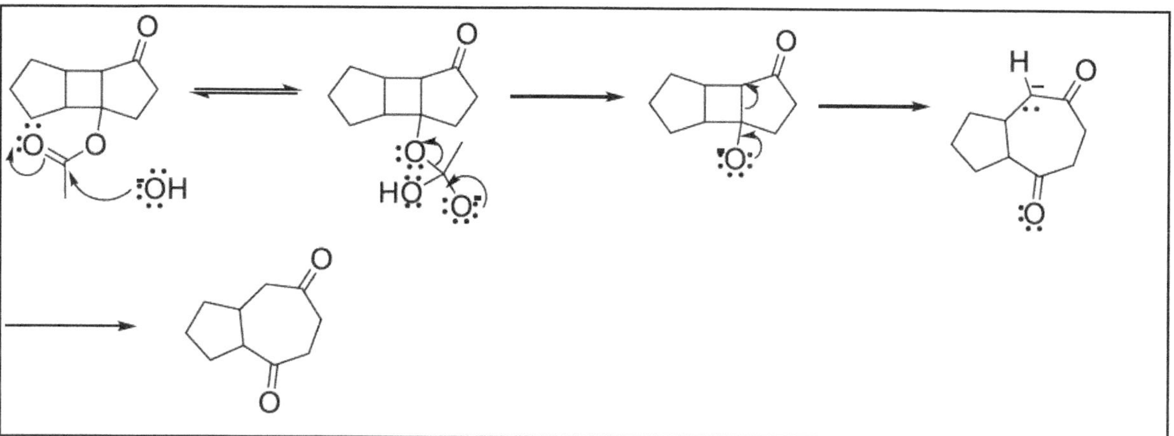
Base-catalyzed hydrolysis of the ester yields an anion that undergoes a retro aldol reaction (recall that the aldol condensation reaction is reversible).	
92	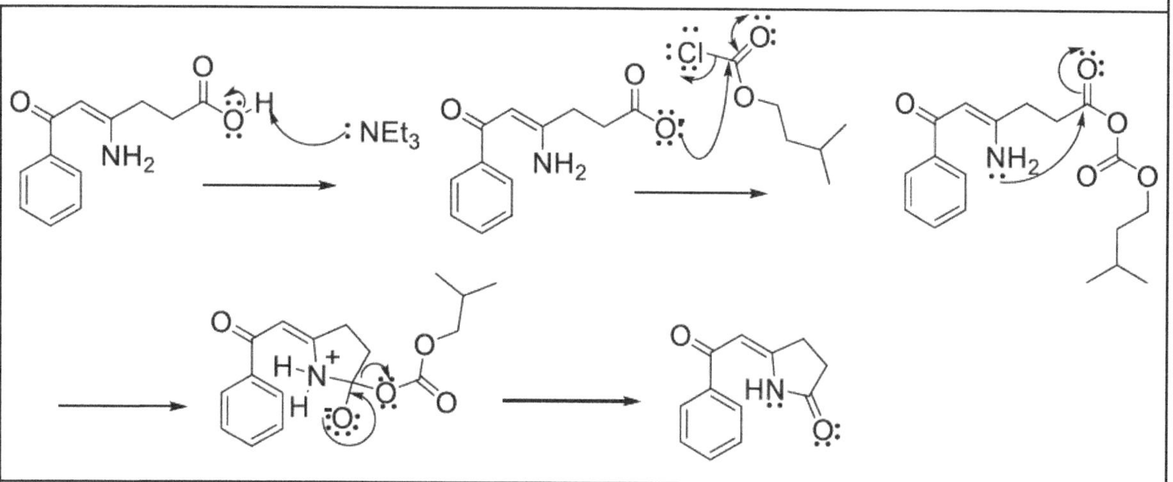
Activation by the acid (changing the hydroxyl group from a poor leaving group to a good leaving) is followed by an intramolecular acyl substitution reaction.	

93

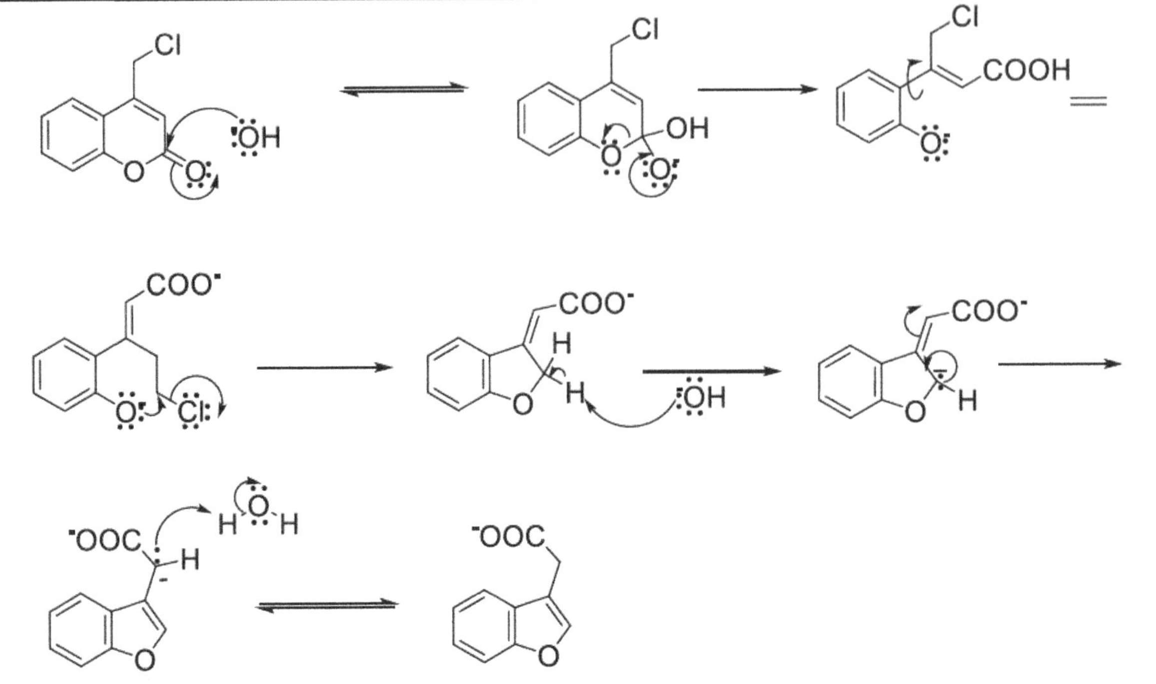

Cl
OH
COOH
COO-
H
-OOC

Base-catalyzed hydrolysis of the lactone yields an oxygen nucleophile that undergoes an S_N2 reaction with the allyl chloride. This is followed by a base-catalyzed isomerization to form the product.

94

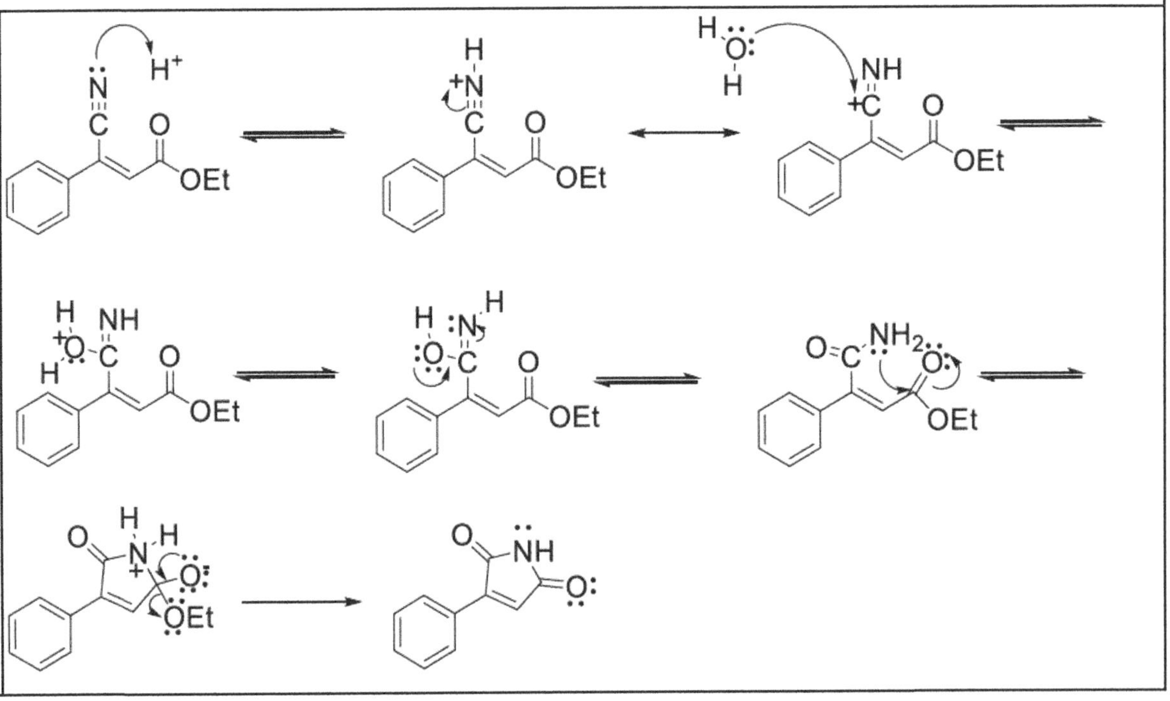

H+
OEt
NH
NH2

95	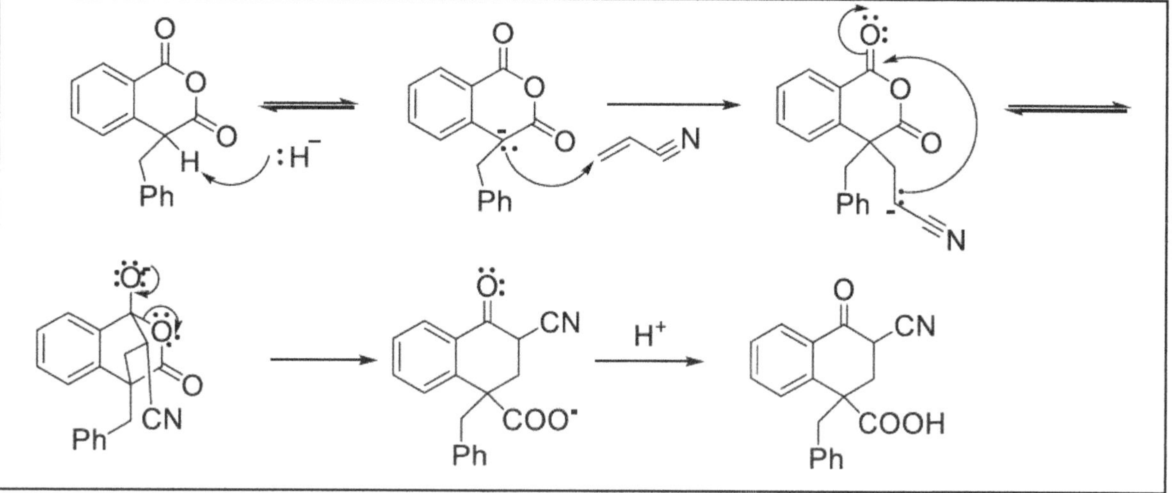
	Formation of the anion is followed by a Michael addition reaction and trapping of the anion via reaction with the mixed anhydride.
96	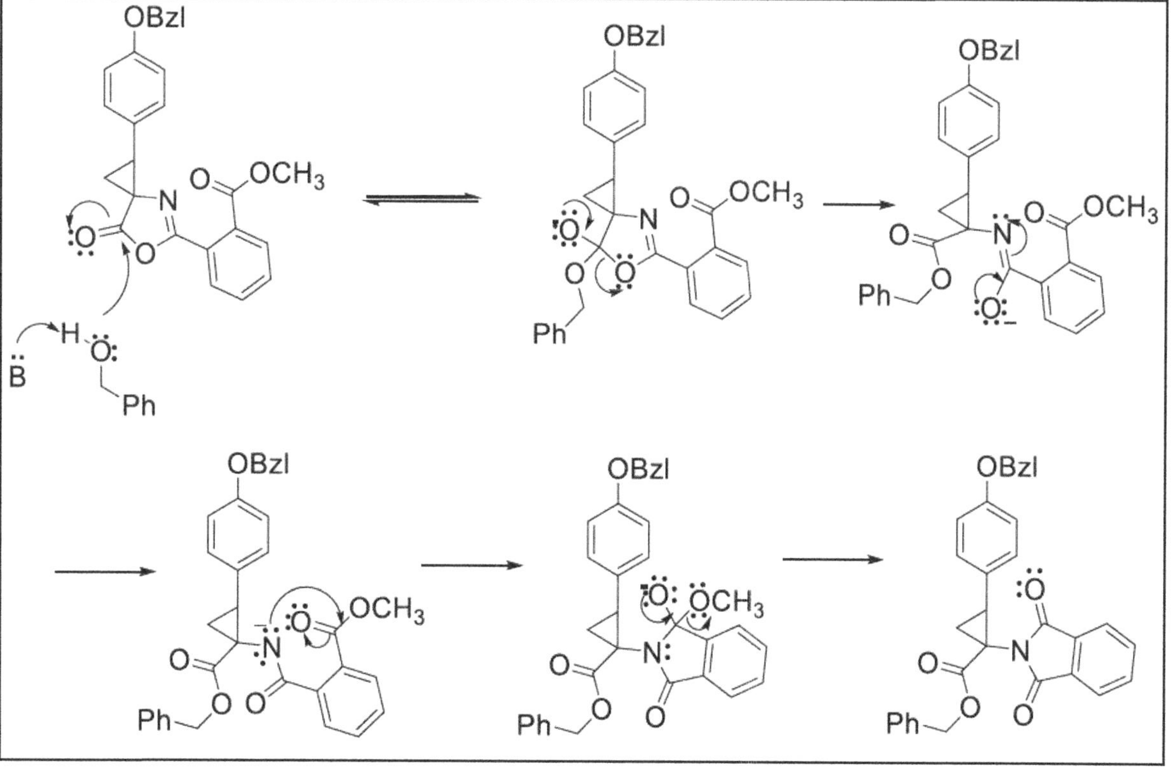

97	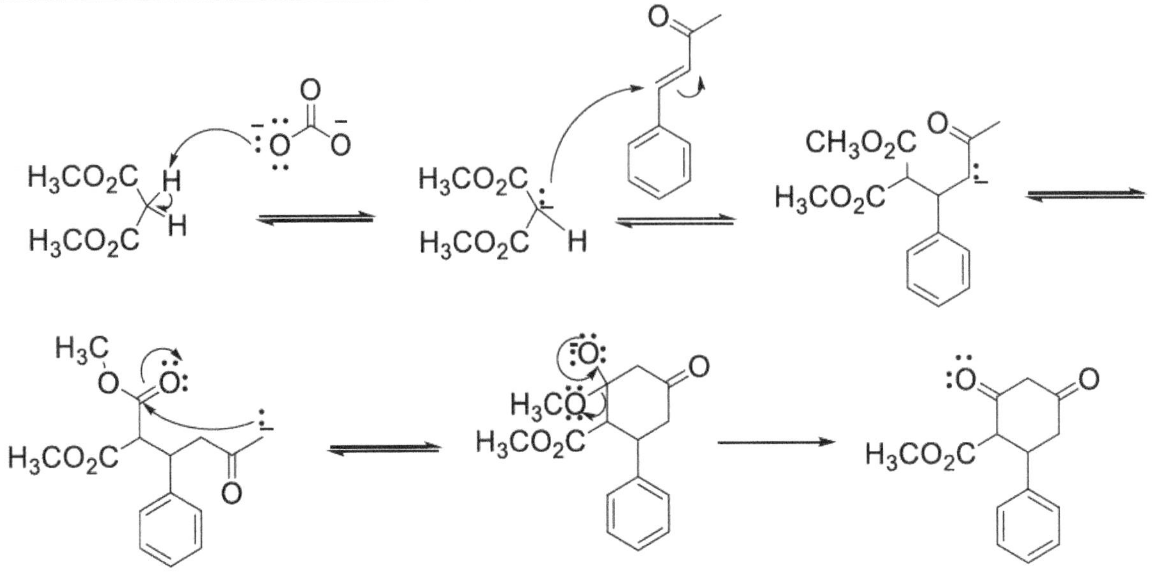
98	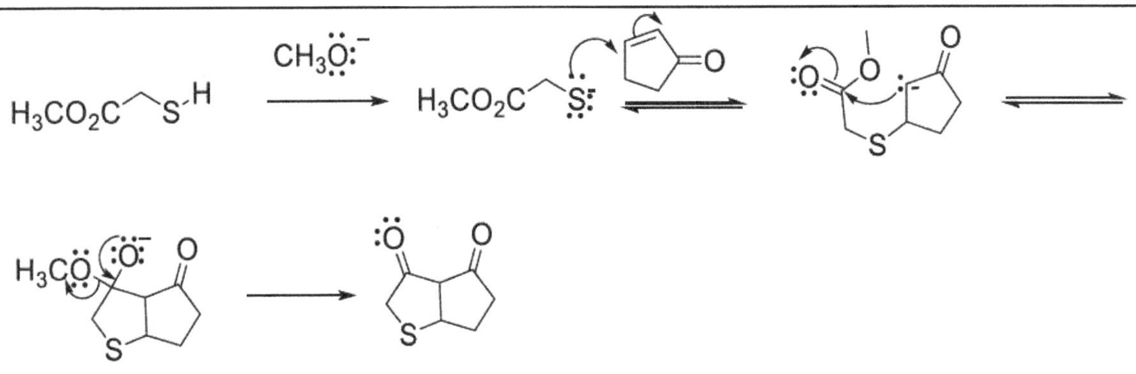
99	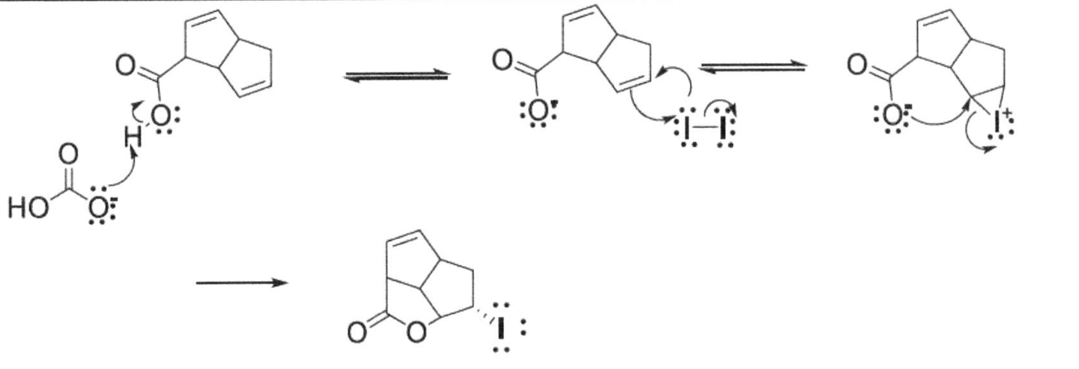
	This is an example of the *iodolactonization reaction*.
100	

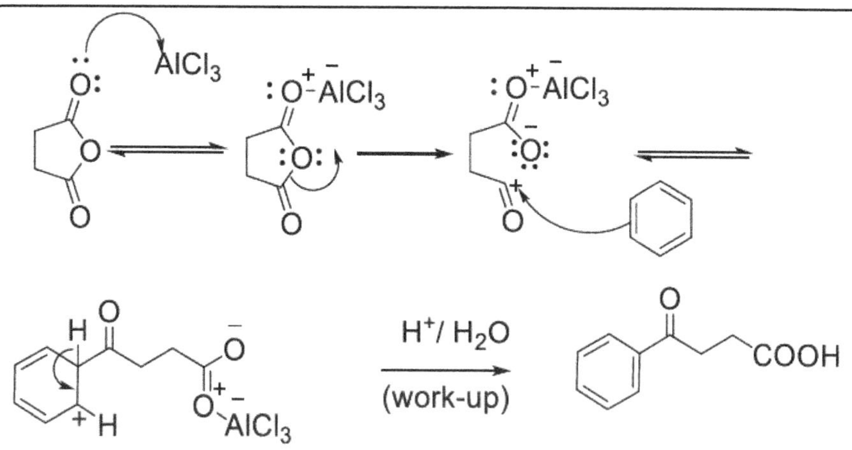

AlCl3
O
AlCl3
O
H⁺/ H2O
(work-up)
COOH

This is a variant of the *Friedel-Craft acylation reaction*.

101

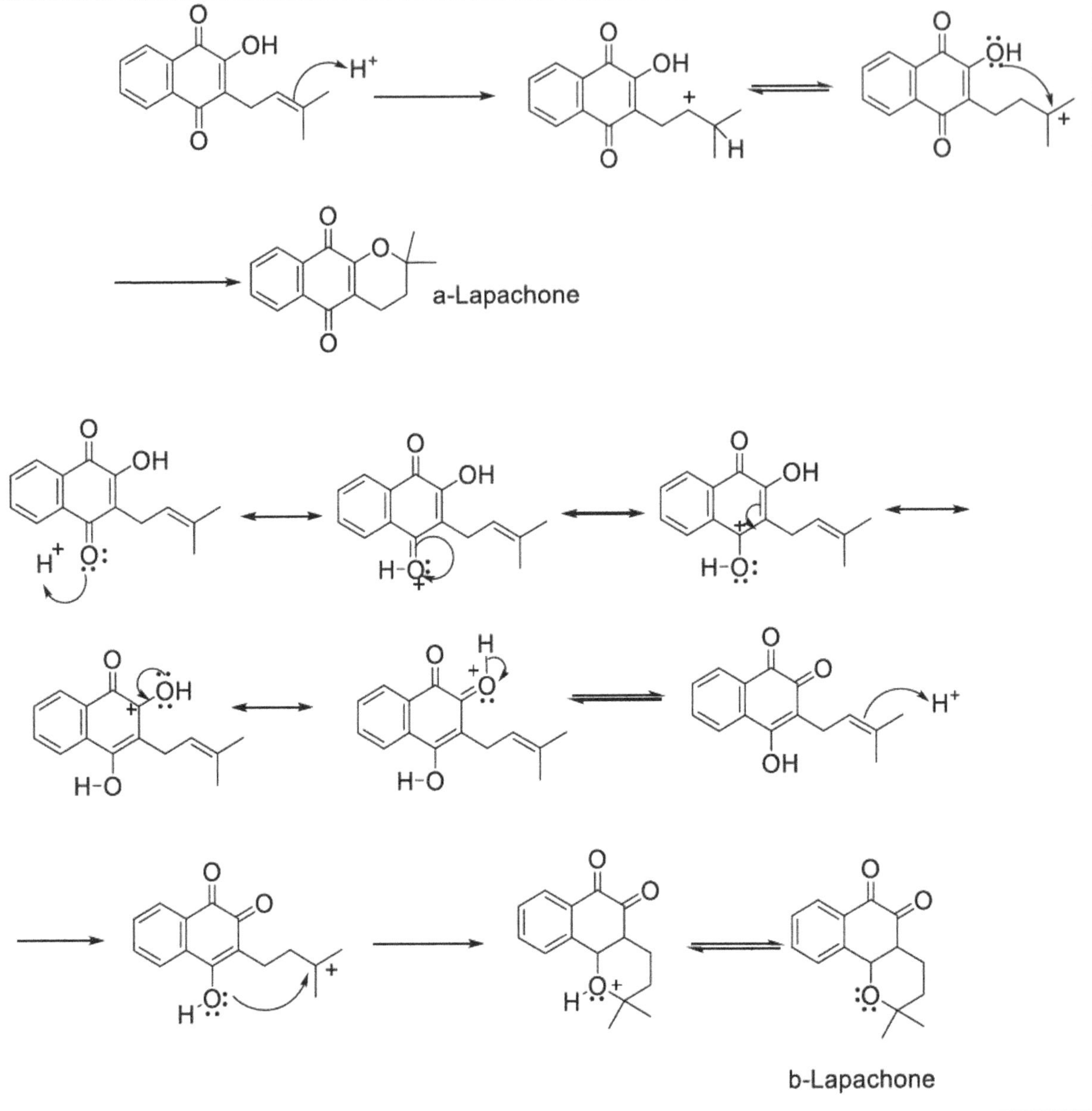
a-Lapachone
b-Lapachone

The starting material can exist in two enol forms each capable of reacting with the 3° carbocation in a LA/LB reaction to form two different products.

102

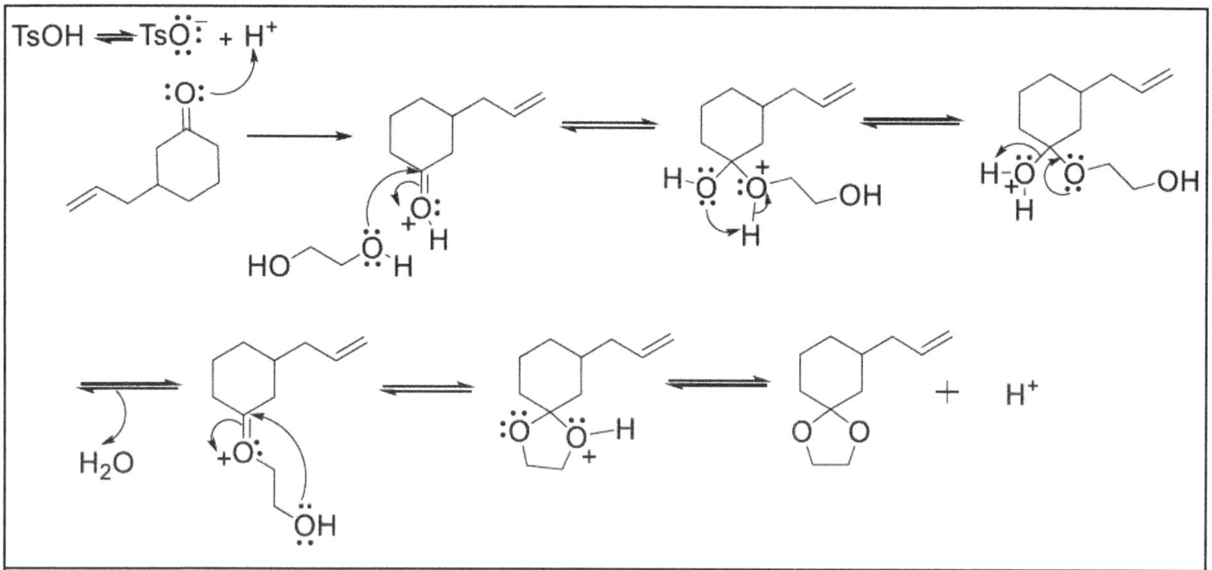

103

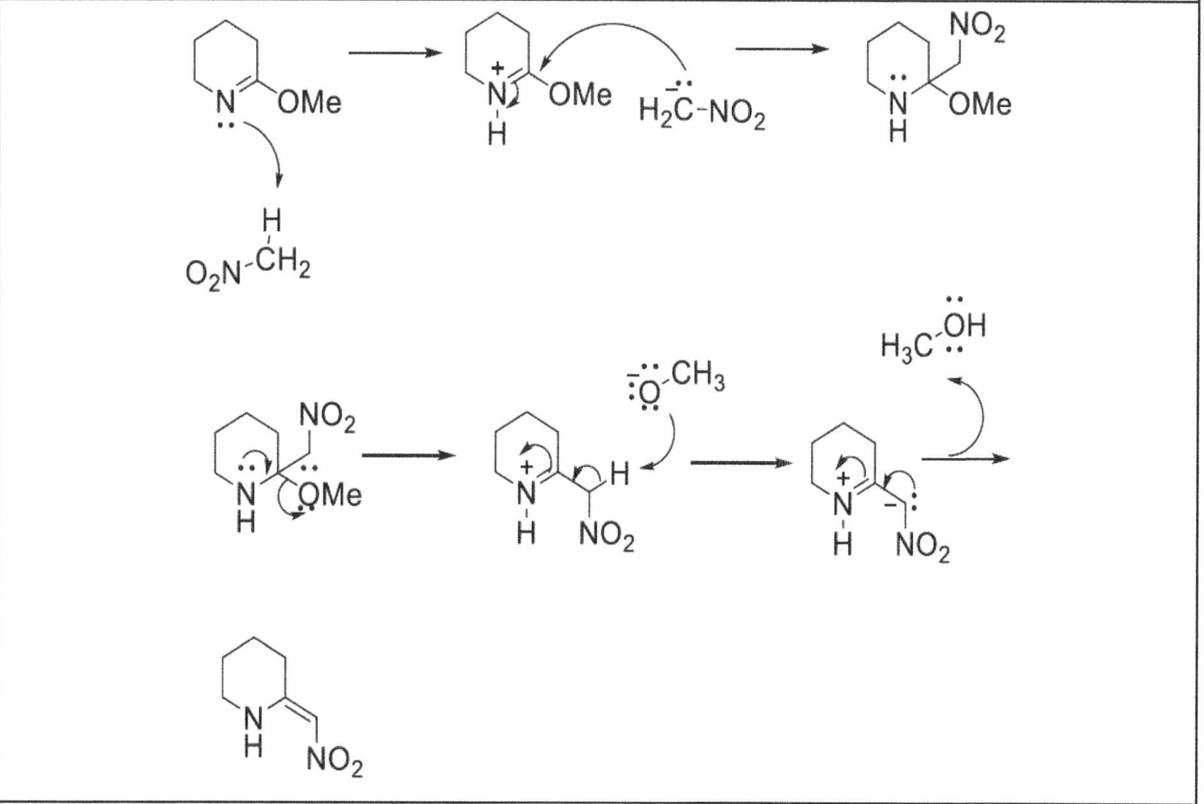

104

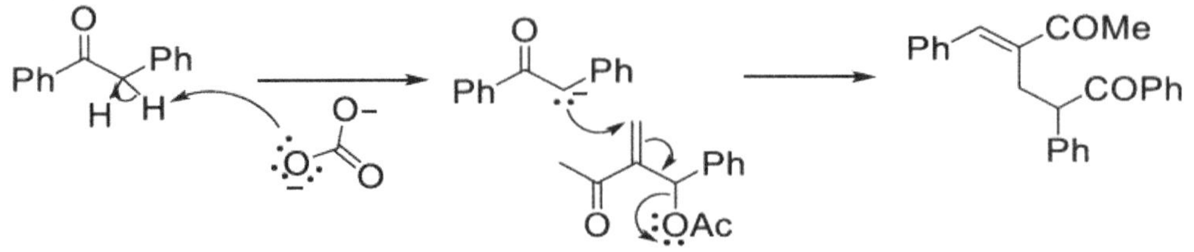

O
Ph
Ph
H
H
O−
O
O
O
Ph
Ph
Ph
O
OAc
Ph
COMe
COPh
Ph

Formation of the anion is followed by a Michael addition reaction and tandem elimination.

105

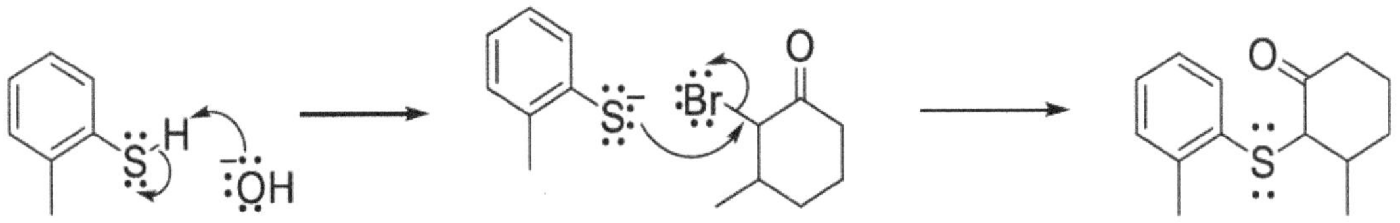

S
H
OH
S
Br
O
O
S

106

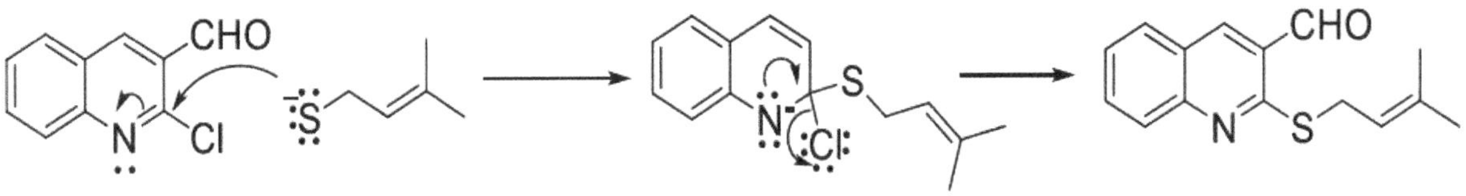

CHO
N
Cl
S
N
S
Cl
CHO
N
S

This is a two-step substitution reaction at an electron-deficient sp^2 hybridized carbon.

107

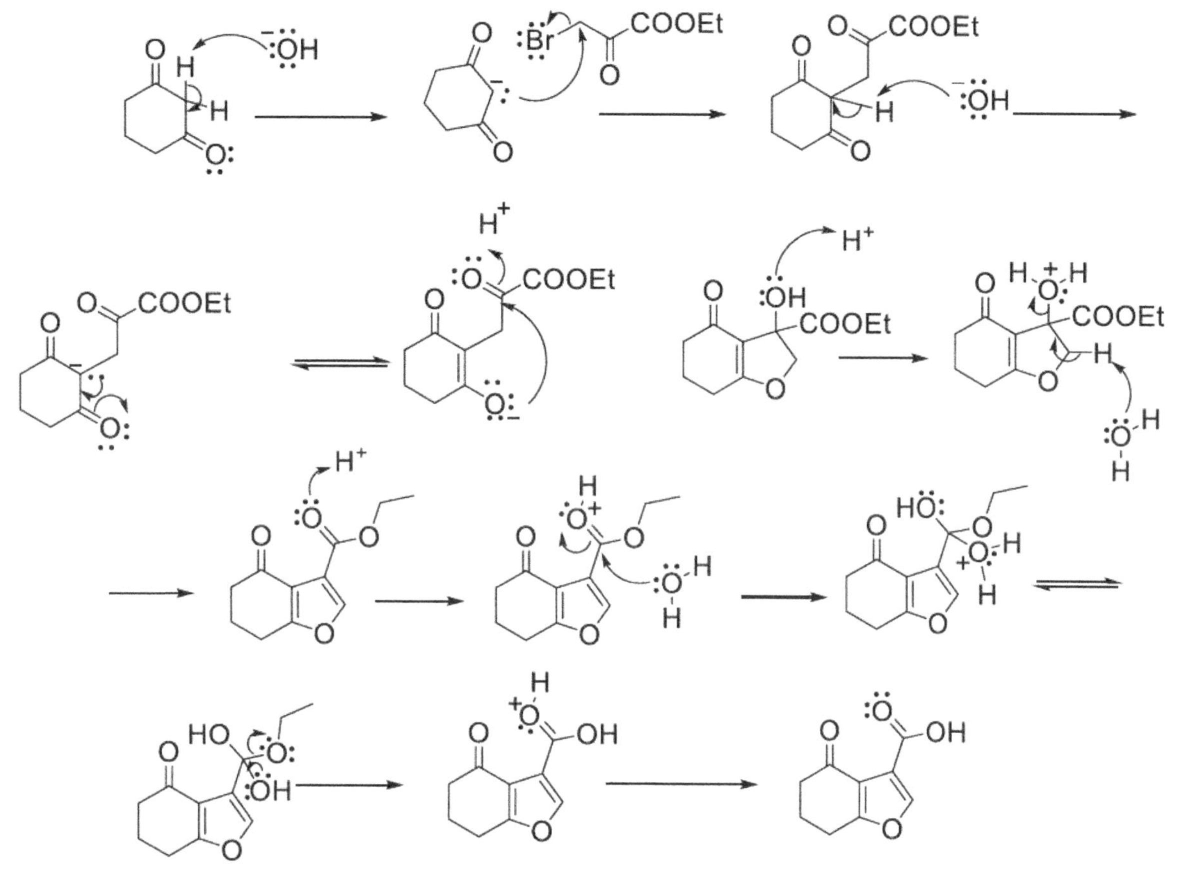

108

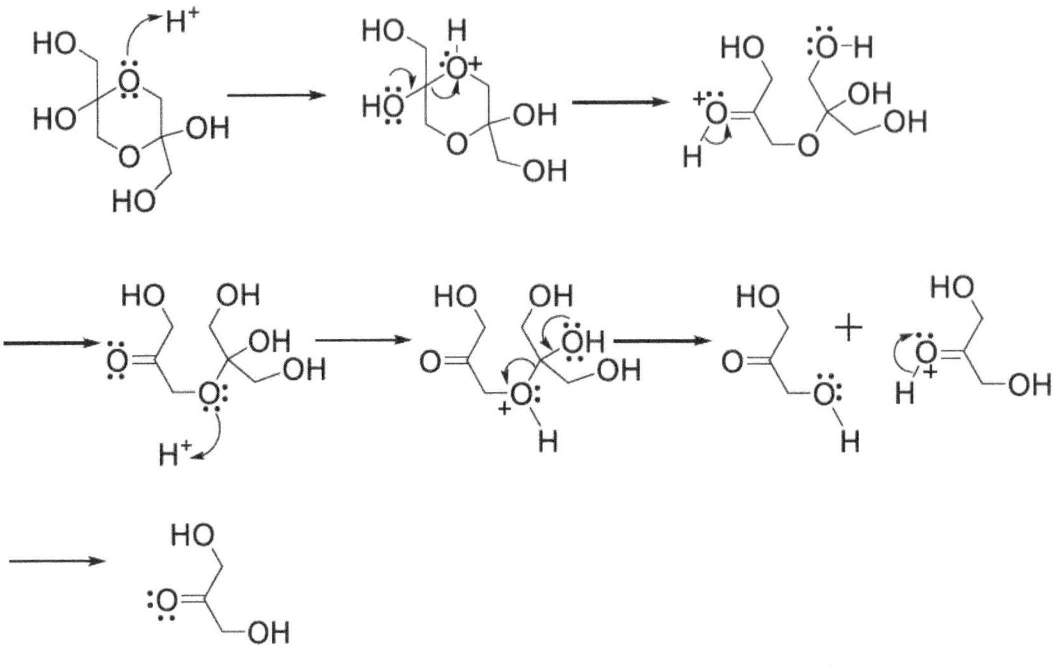

109

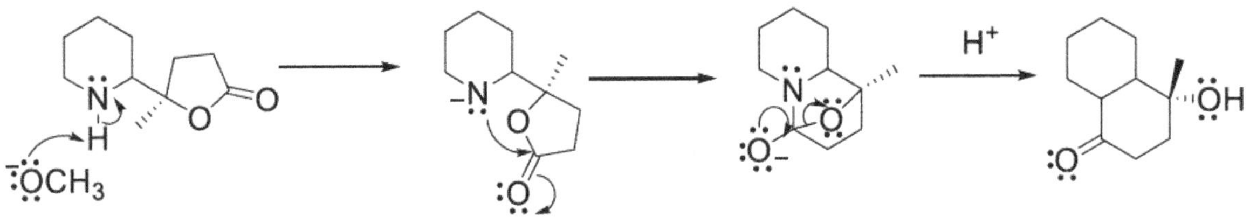

110

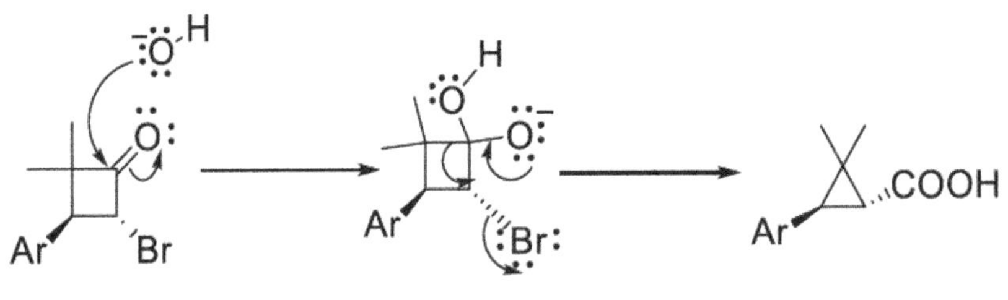

111

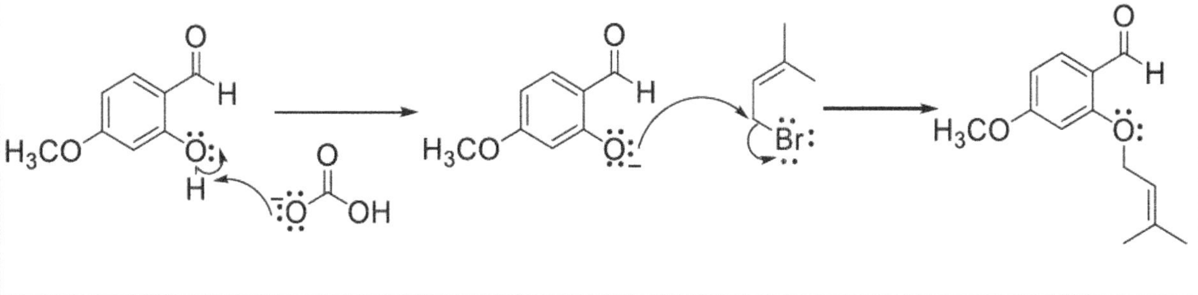

112

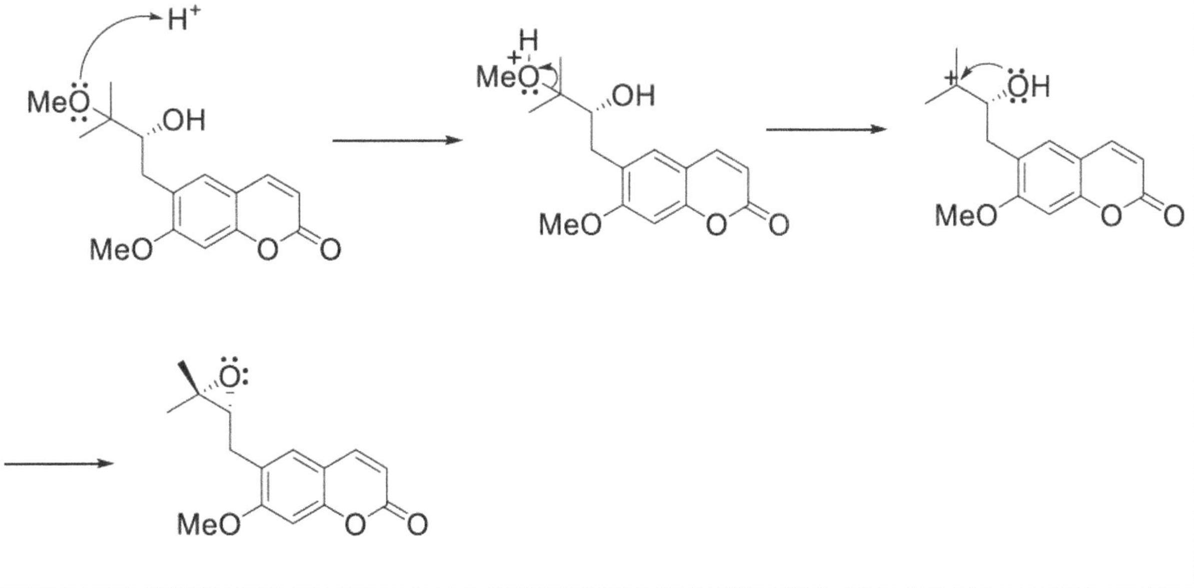

113

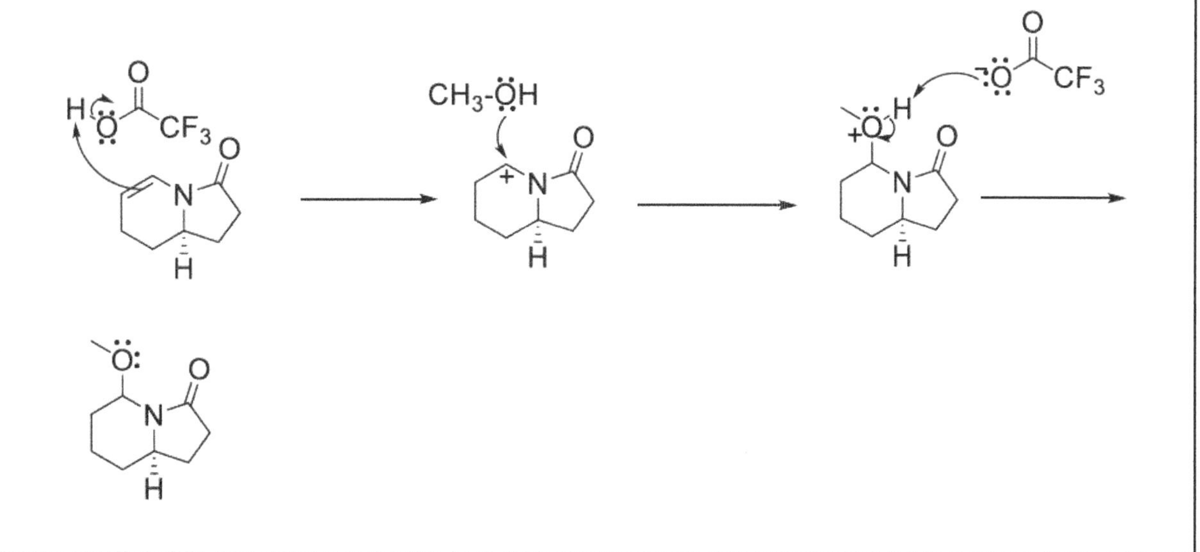

114

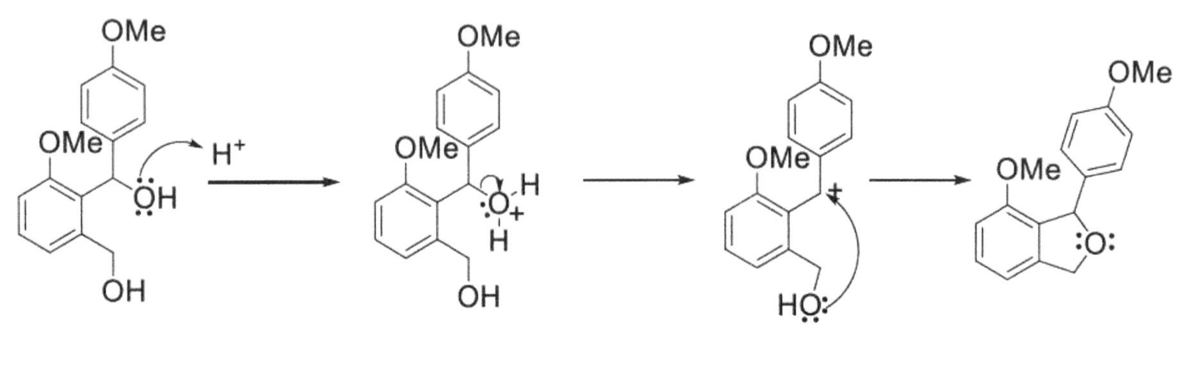
OMe
OMe
H+
OH
OH
OMe
OMe
H
O+
H
OH
OMe
OMe
+
HO
OMe
OMe
O

115

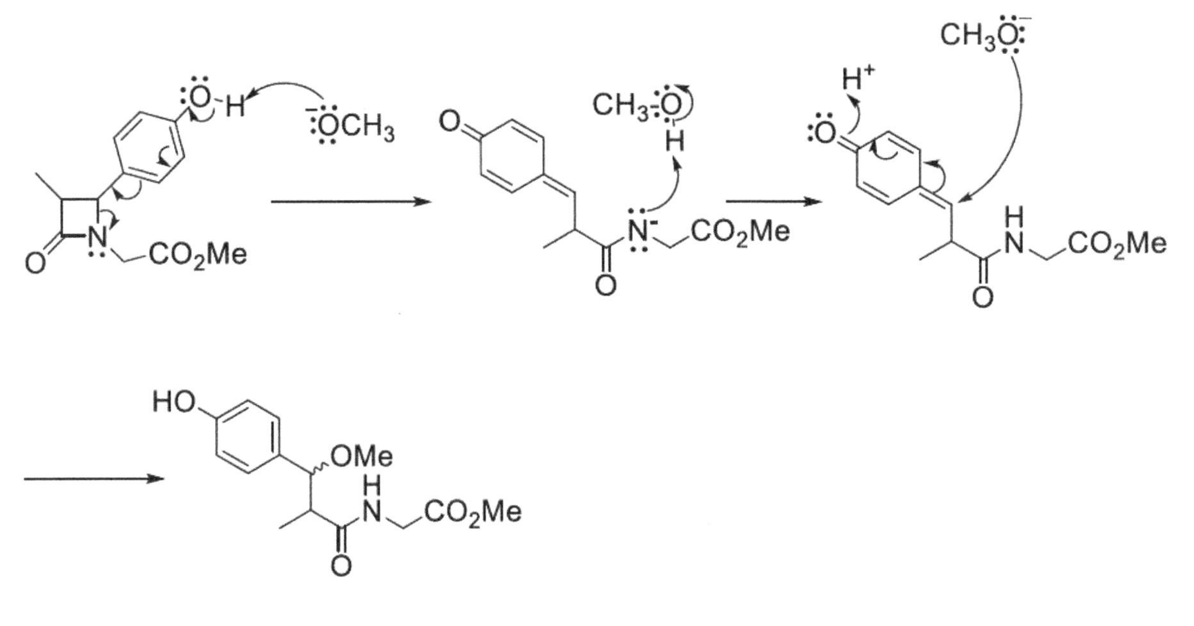
O
H
-OCH3
O
N
CO2Me
O
CH3-O
H
N-
CO2Me
O
CH3O-
H+
O
H
N
CO2Me
O
HO
OMe
H
N
CO2Me
O

116

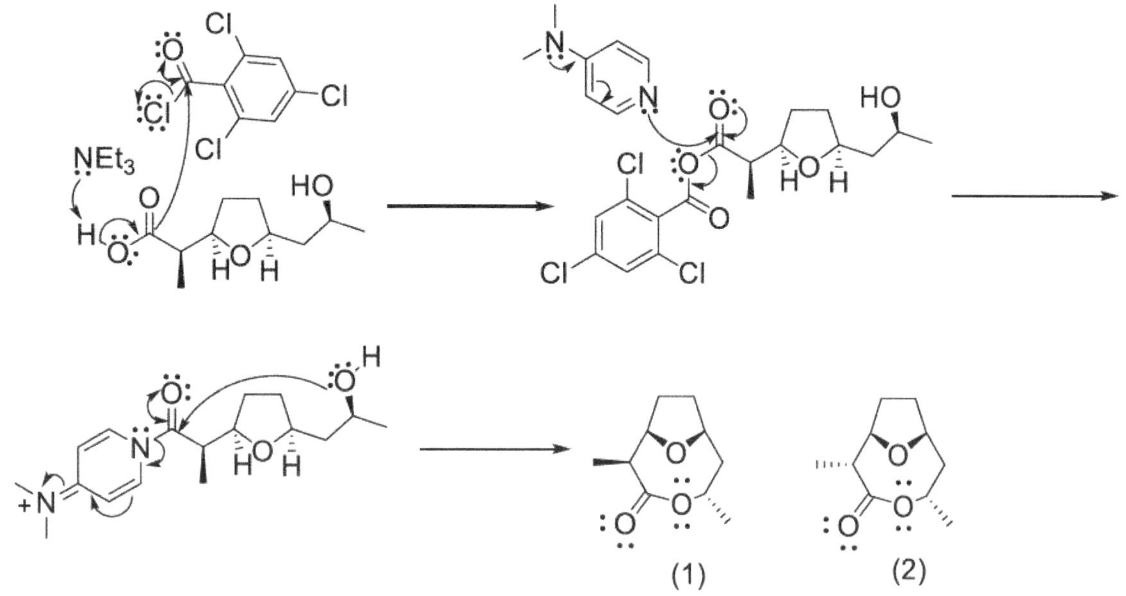

1

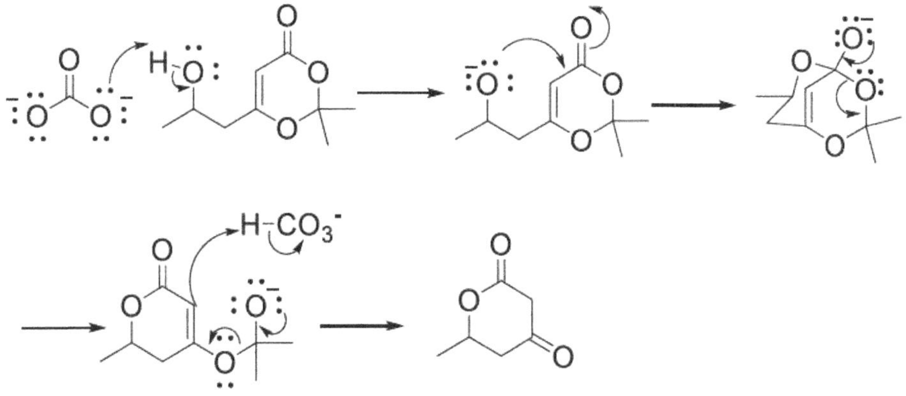

118

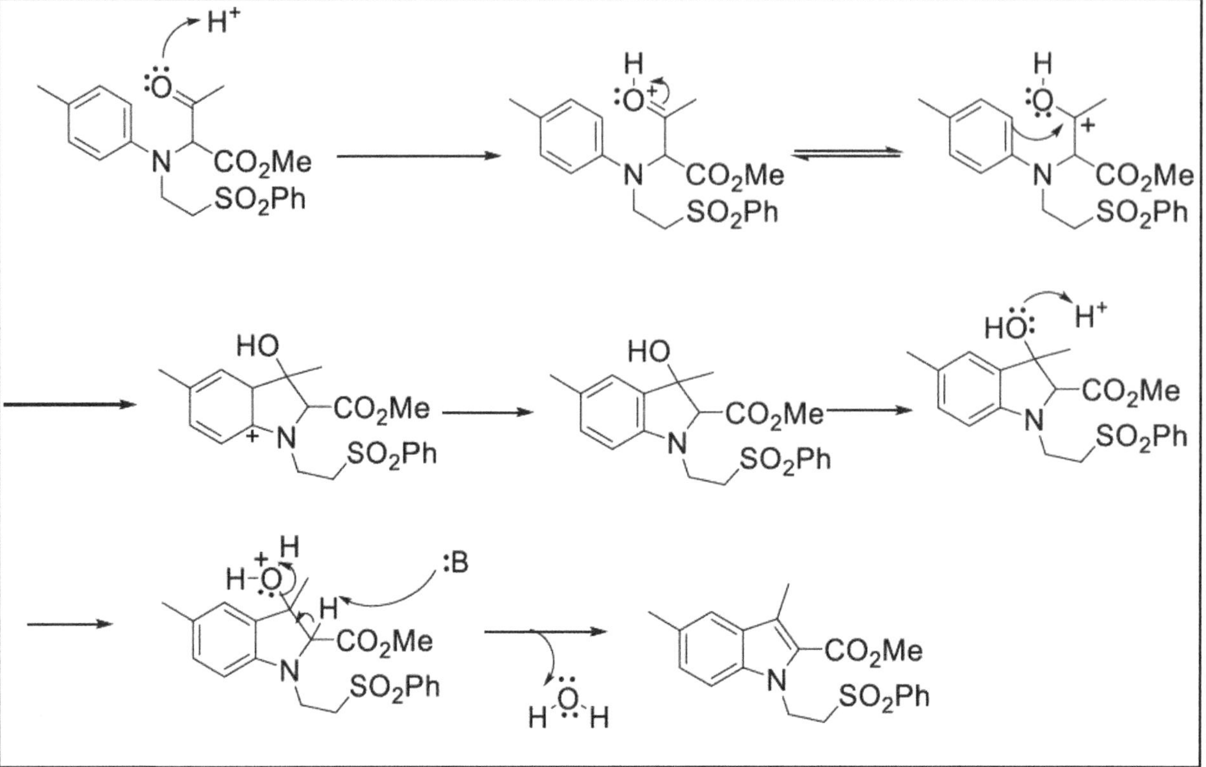
H+
CO2Me
SO2Ph
HO
:B
H

119

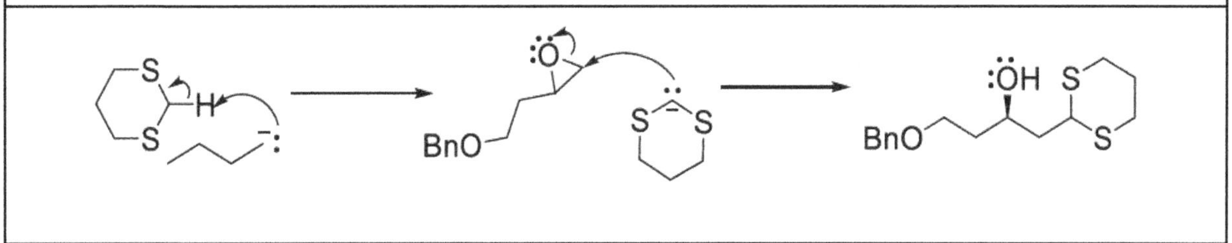
S
H
BnO
:OH

120

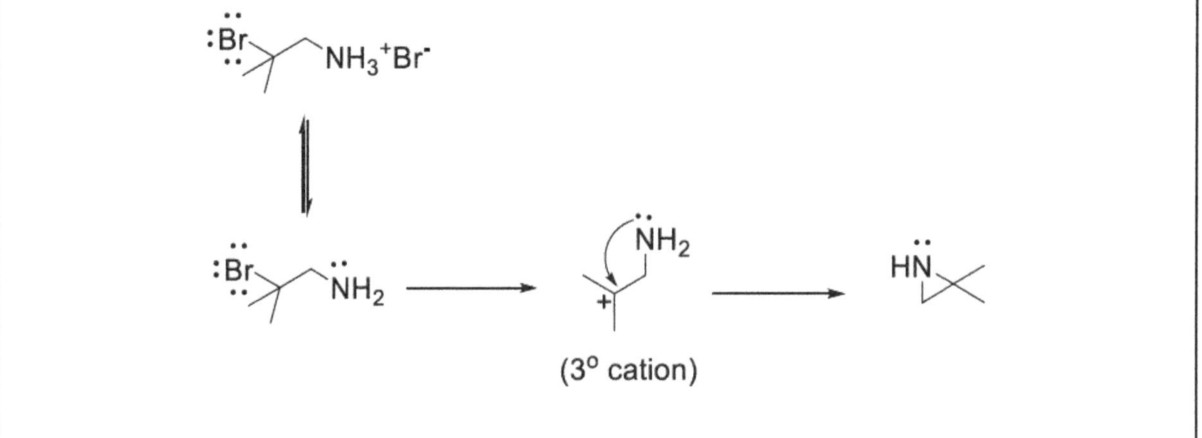
:Br
NH3+Br-
NH2
HN
(3° cation)

121

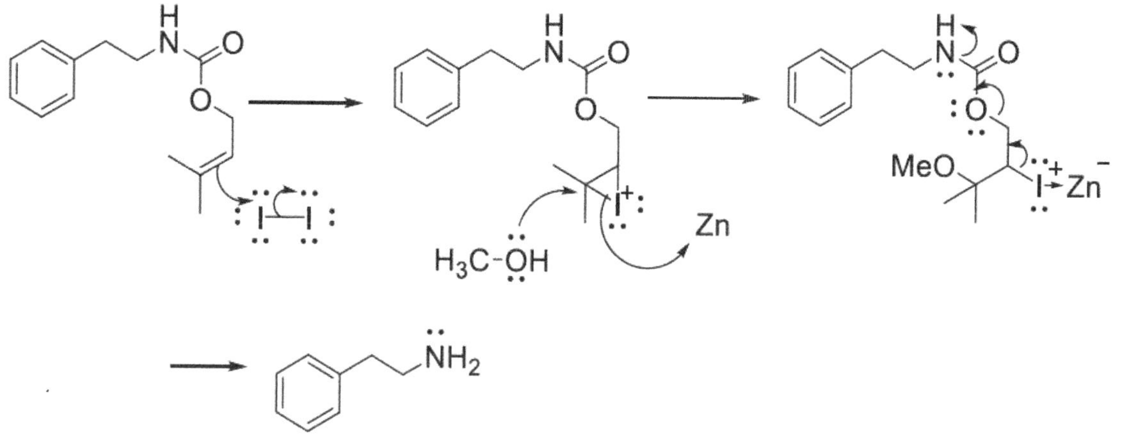
H
N
O
O
I
I
H
N
O
O
I
H3C-OH
Zn
H
N
O
O
MeO
I
Zn
NH2

122

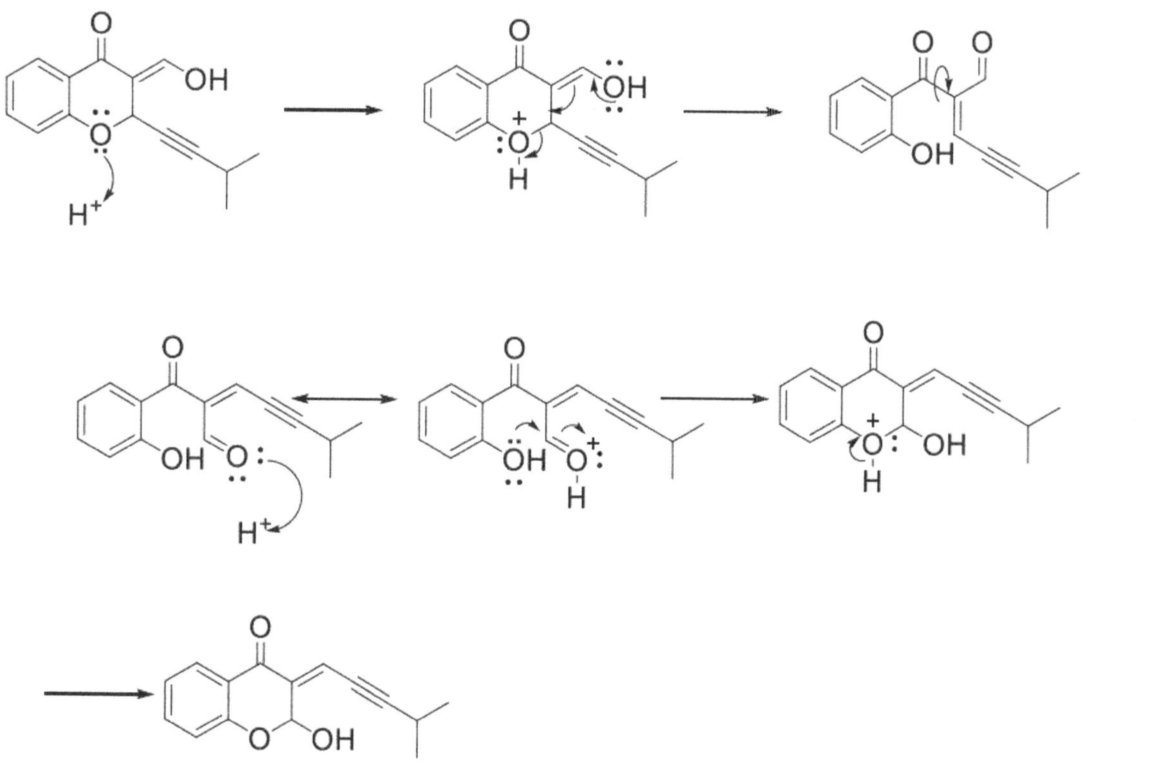
O
OH
O
H+
O
OH
O
H
O
O
OH
O
OH
O
H+
O
OH
O
H
O
O
OH
H
O
O
OH

123

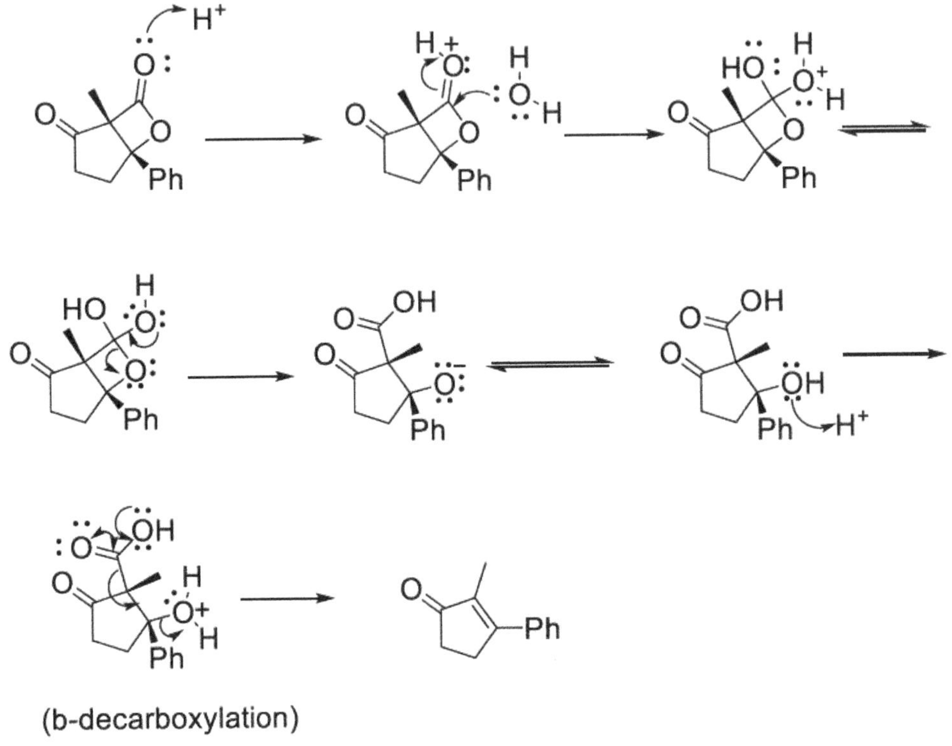

H+
Ph
HO
OH
(b-decarboxylation)

124

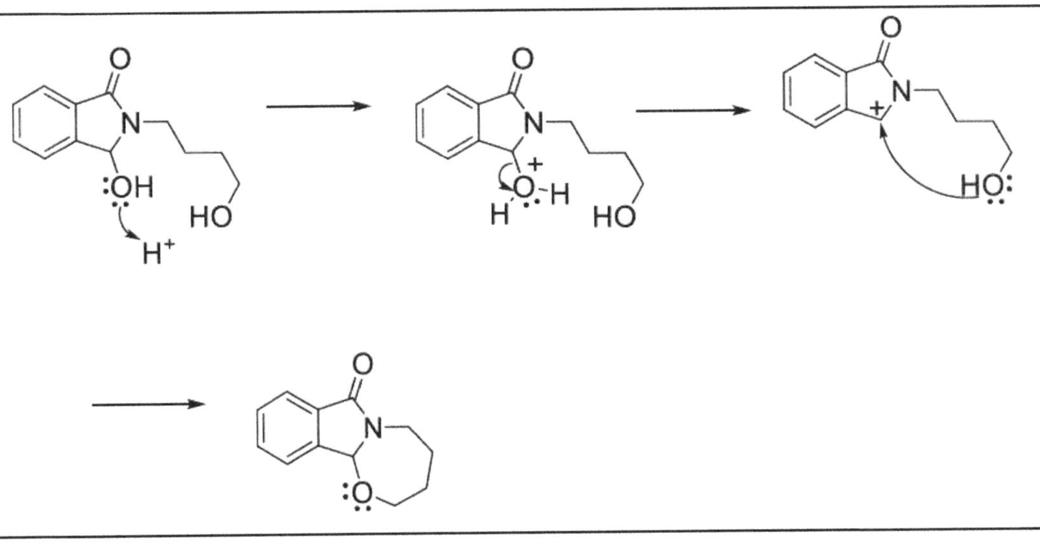

N
OH
HO
H+

125

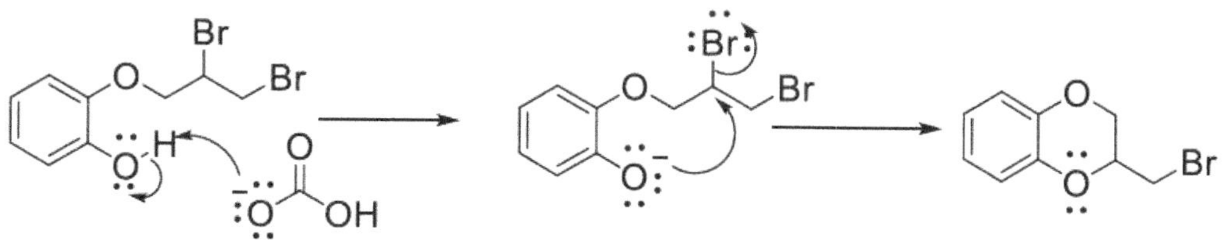

126

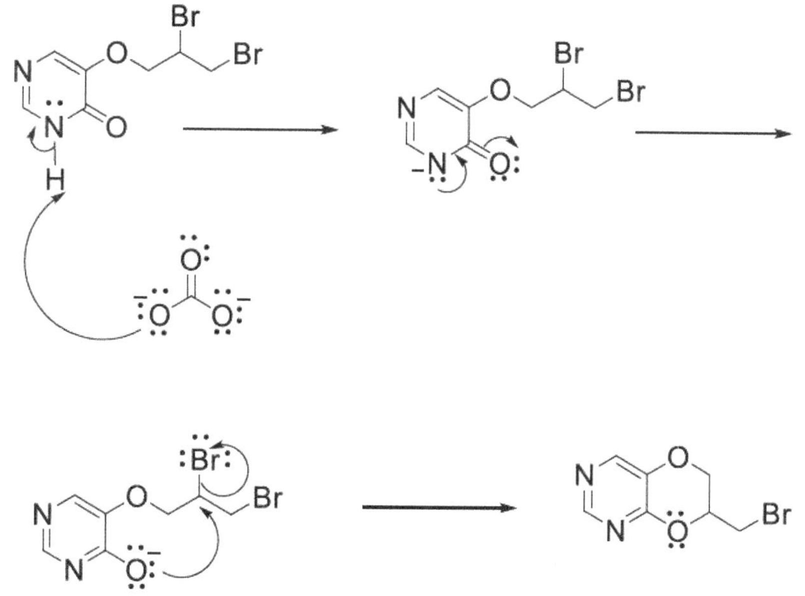

127

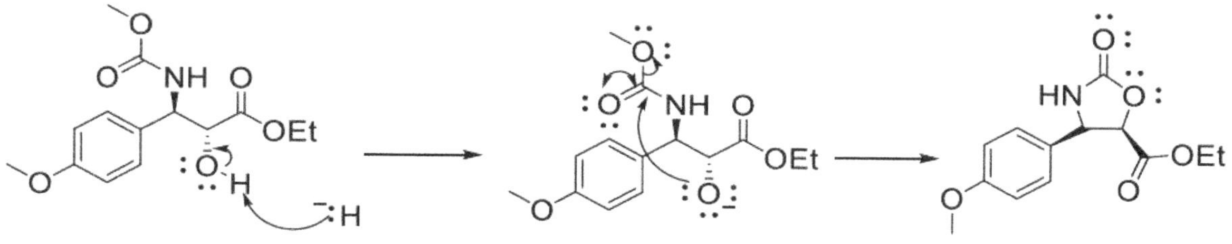

128

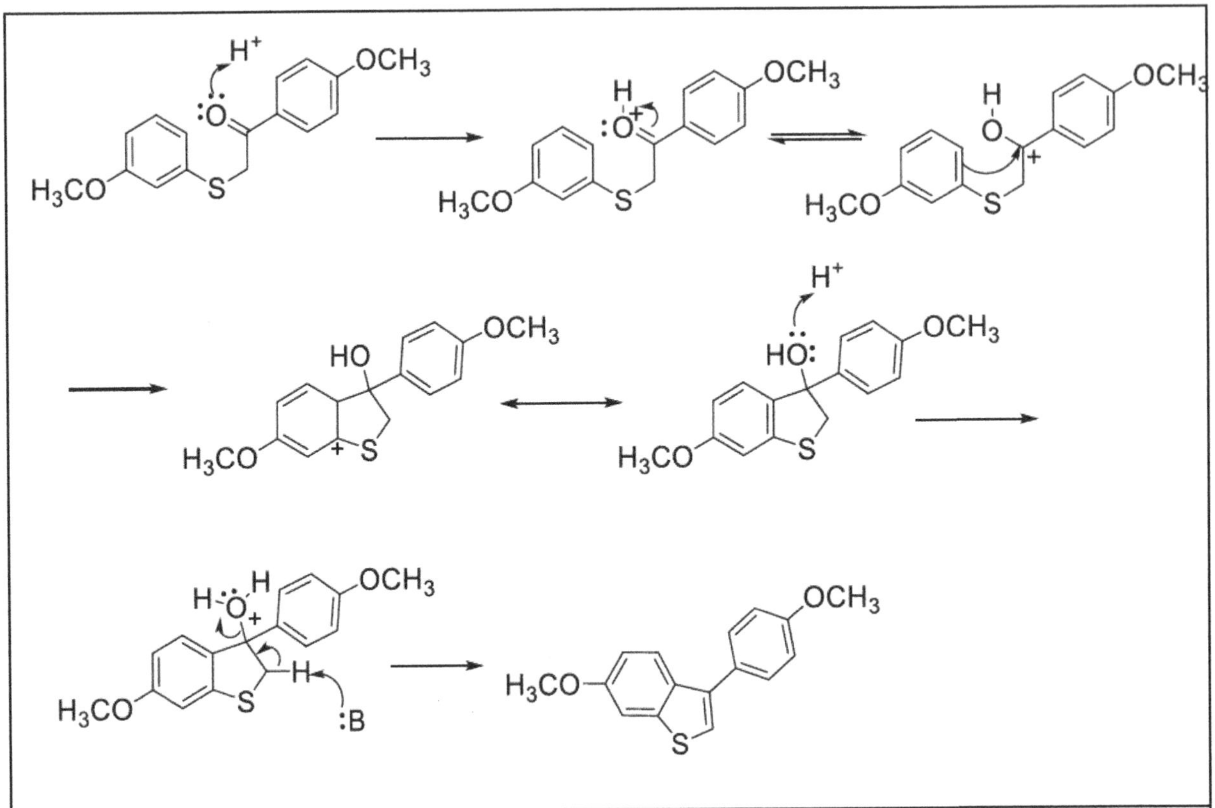

H⁺
OCH3
H3CO
S
:B

129

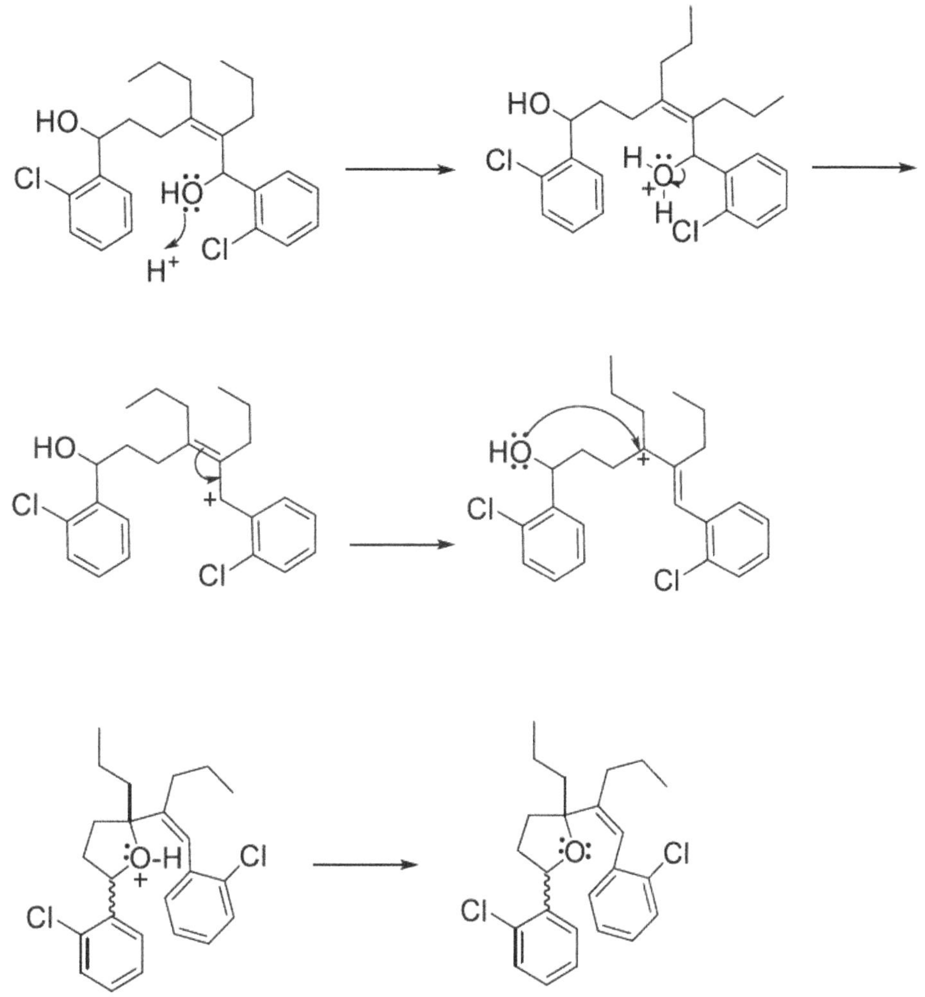
HO
Cl
HÖ
H+
H
Ö
+
Cl
HÖ
:O-H
+
:O:

130

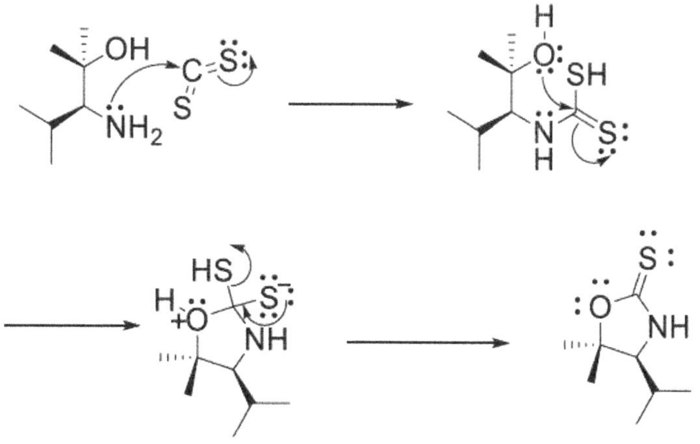
OH
C
S
NH2
H
Ö
SH
NH
H
S:
HS
S−
H
+O
NH
:O
S :

131

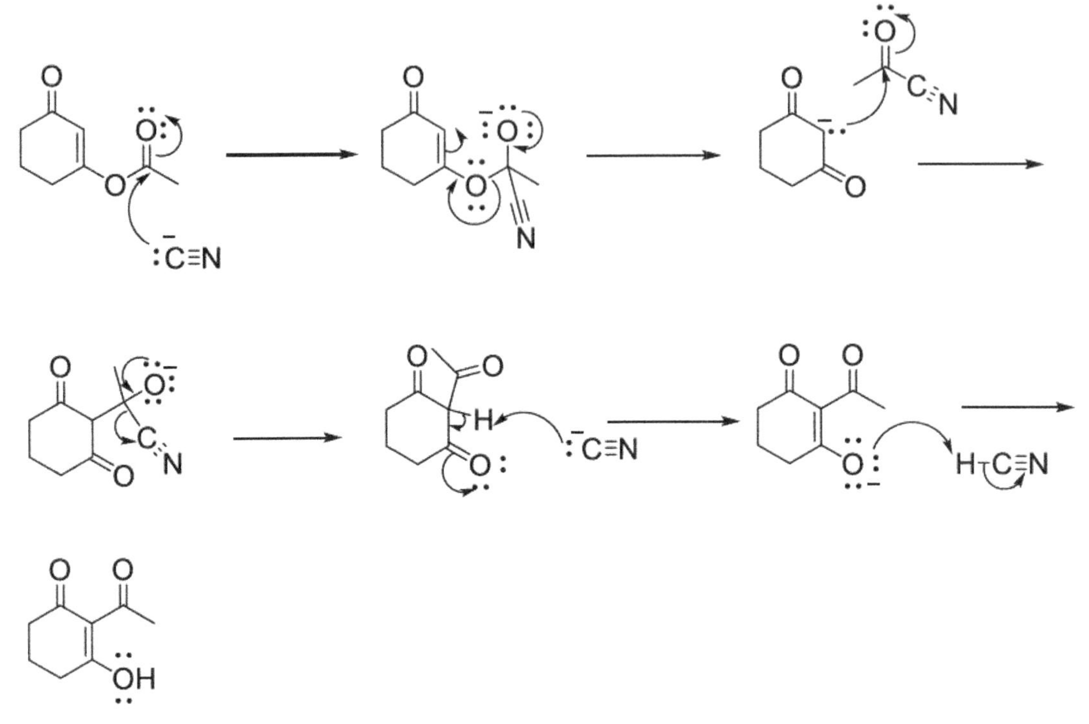

O
C≡N
H
OH

132

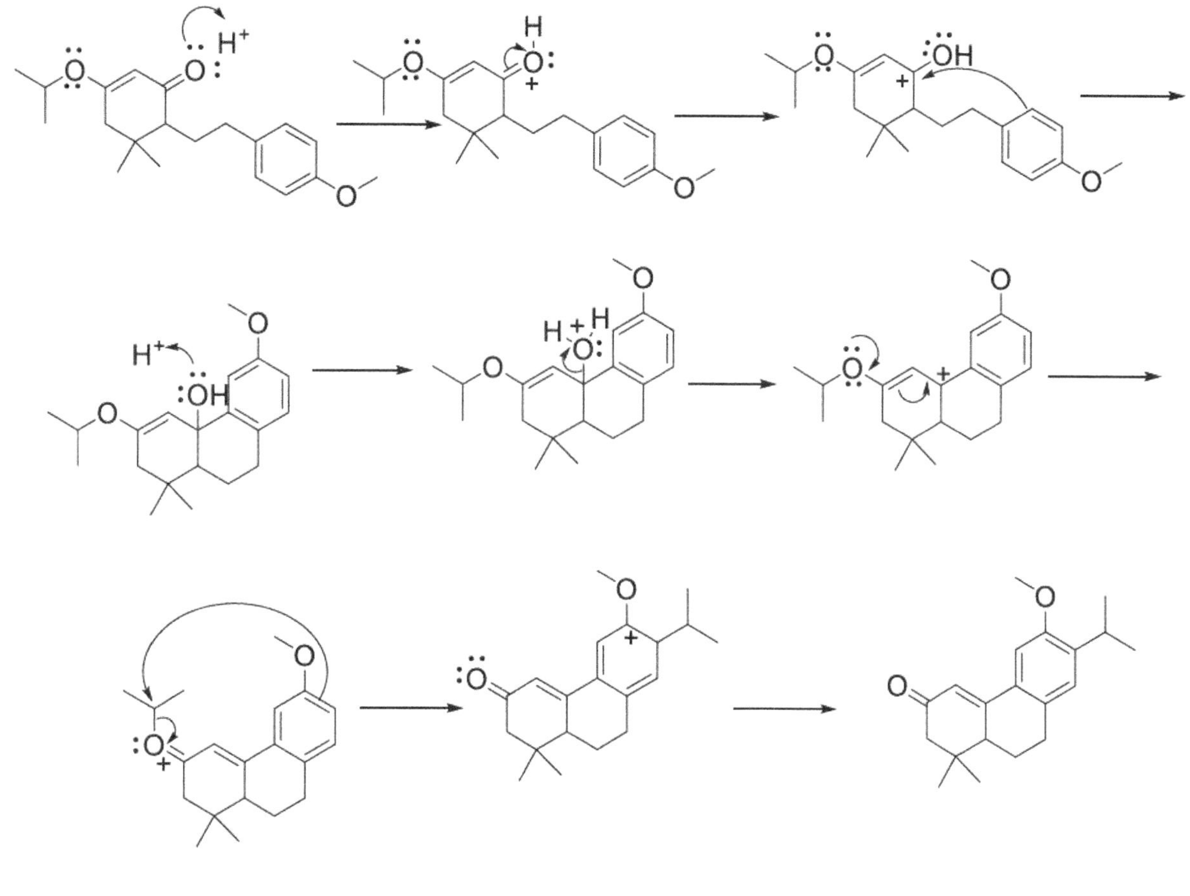

H⁺
H
OH
+
H

133

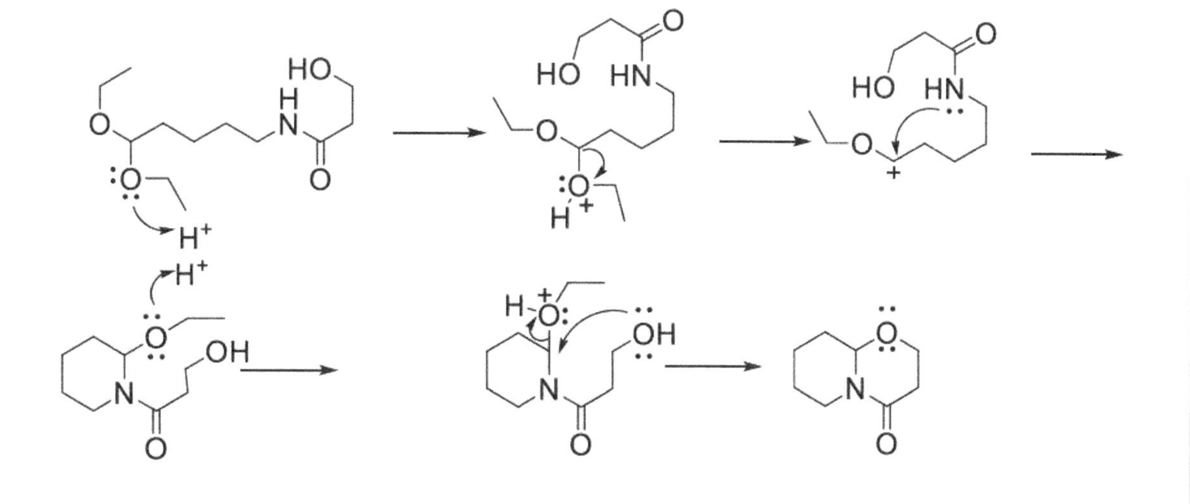

HO
H
N
O
H⁺
HN
O
N
OH

134

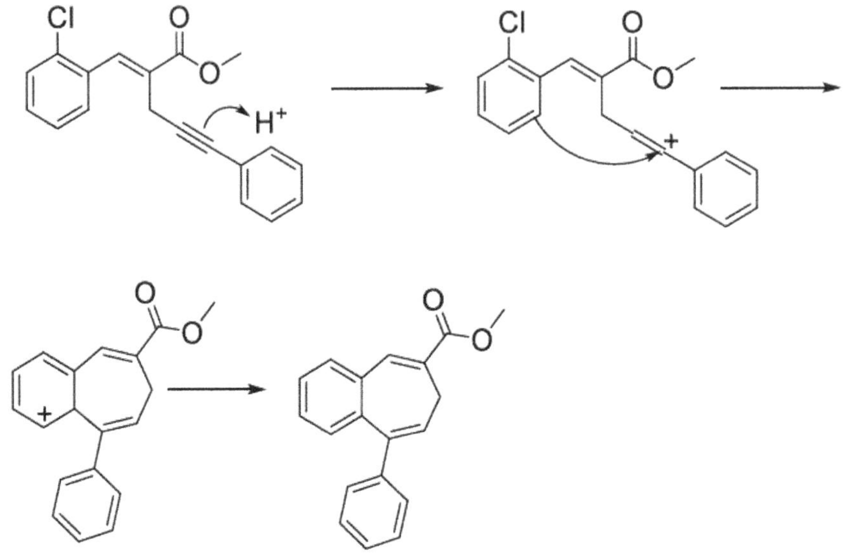
Cl
O
O
H+
Cl
O
O
+
O
O
+
O
O

135

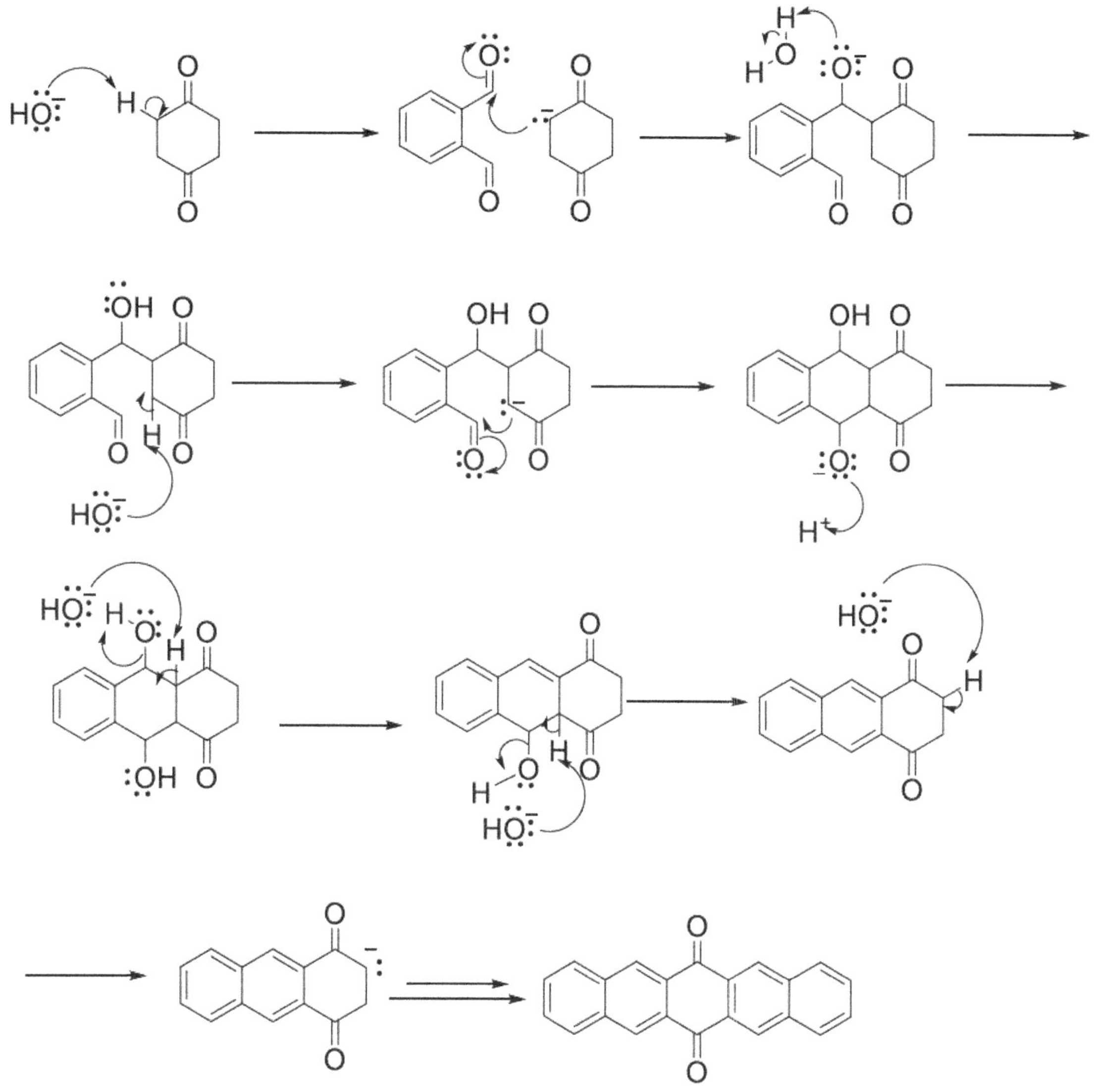
HO:
H
O
O
:O:
:O:
O
:O:
H
O
O
O
O
O
:OH
O
H
O
O
HO:
OH
O
:O
O
OH
O
:O:
O
H+
HO:
H
:O:
H
O
O
:OH
O
O
O
H
O
H
O
HO:
HO:
O
H
O
O
O
O
O
O

136

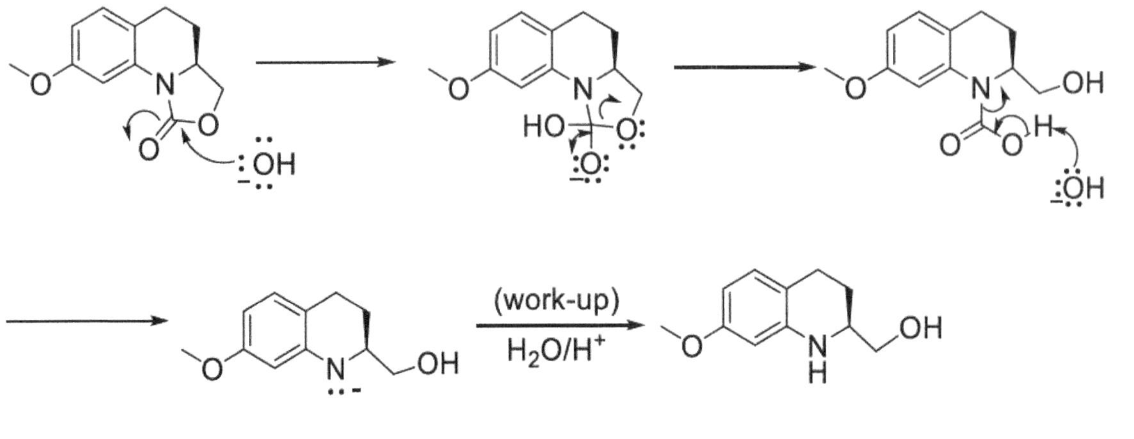

:ÖH
HO
:Ö:
OH
(work-up)
H_2O/H^+
N
H

137

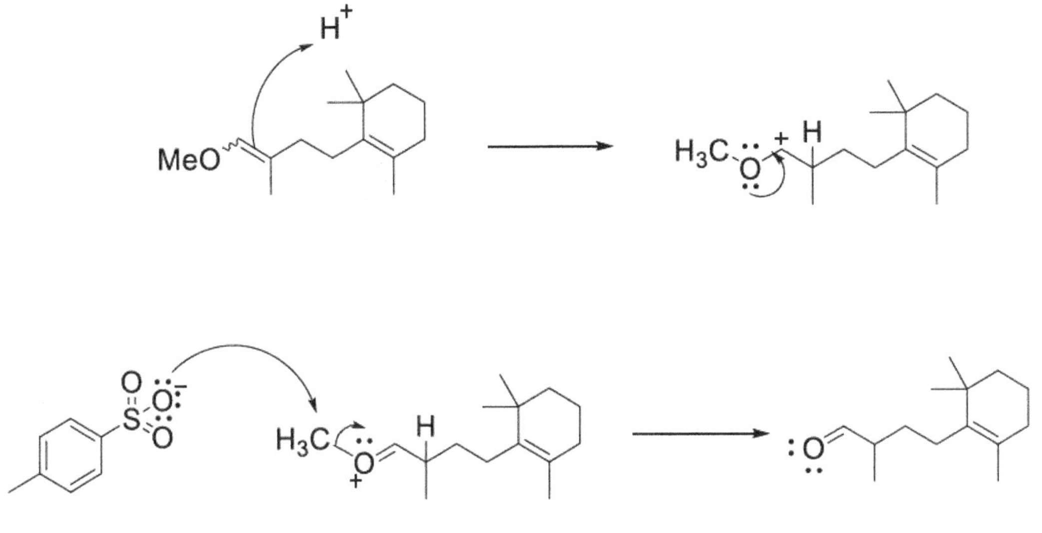

H^+
MeO
H_3C
H
O
S

138

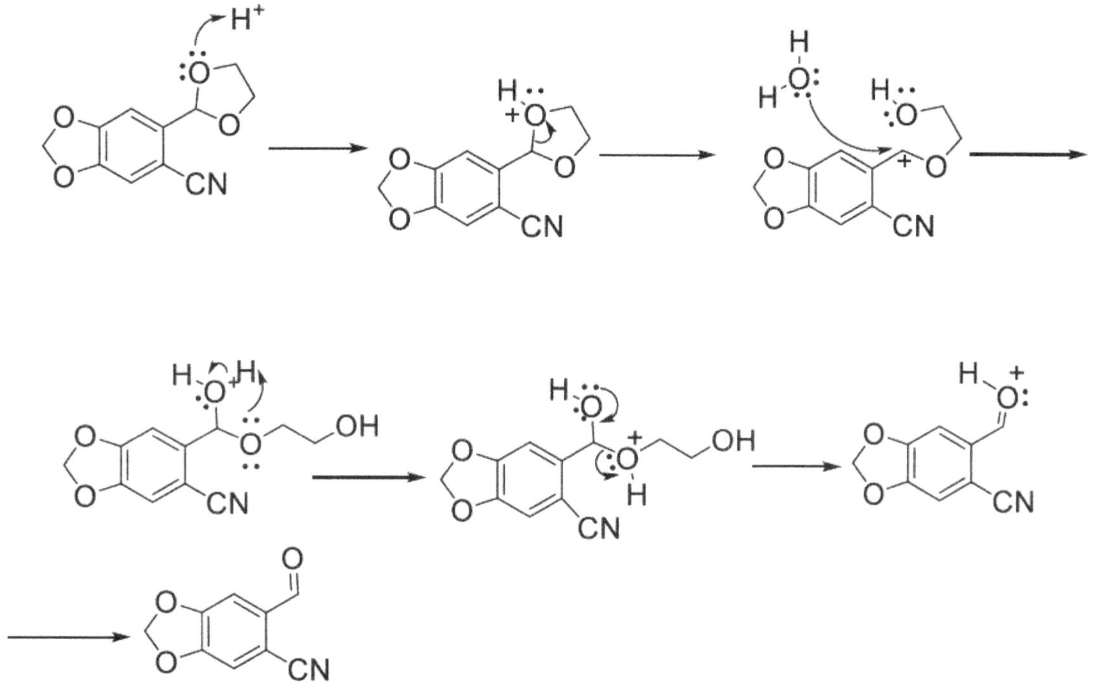
H+
O
O
O
O
CN
H
+O
O
O
O
CN
H
H O
H
O
O
O
+
O
CN
H
O+
H
O
OH
O
O
CN
H
O
O+
OH
H
O
O
CN
H
O+
O
O
CN
O
O
O
CN

139

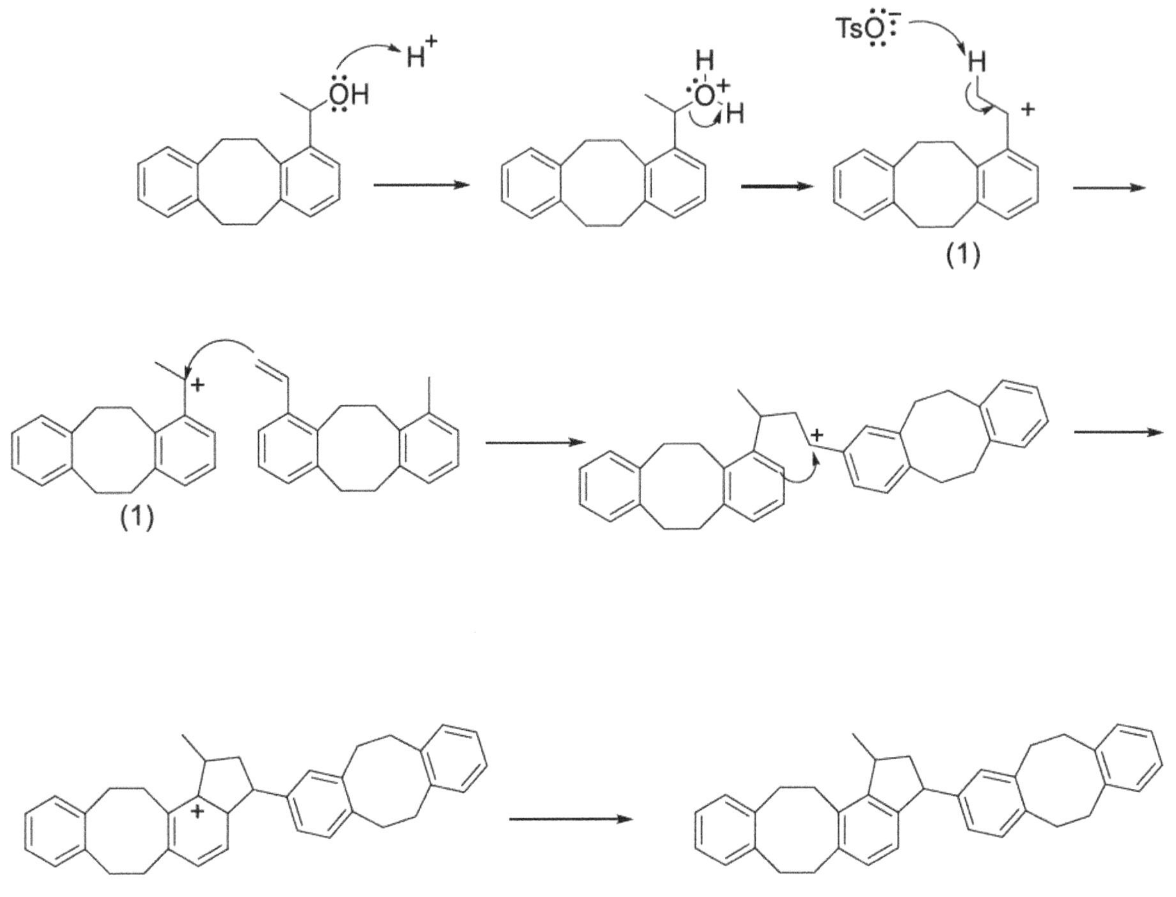

H+
OH
H
O+
H
TsO−
H
+
(1)
+
(1)
+
+

140

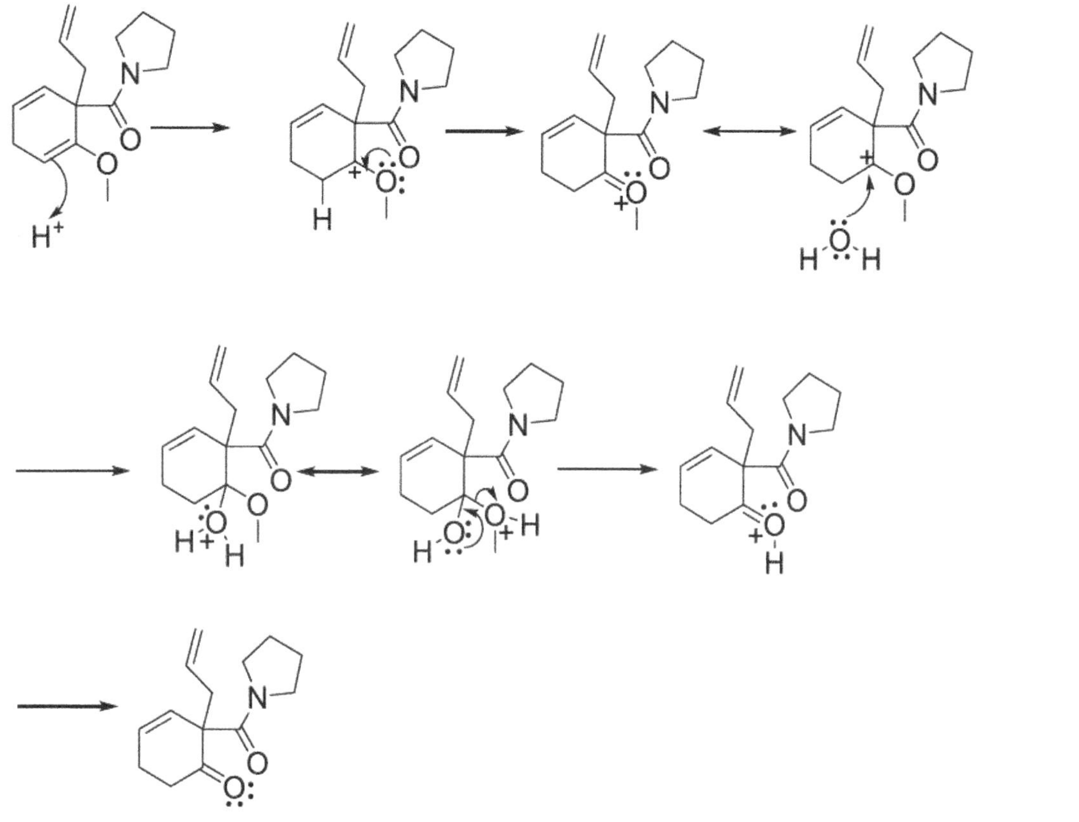

141

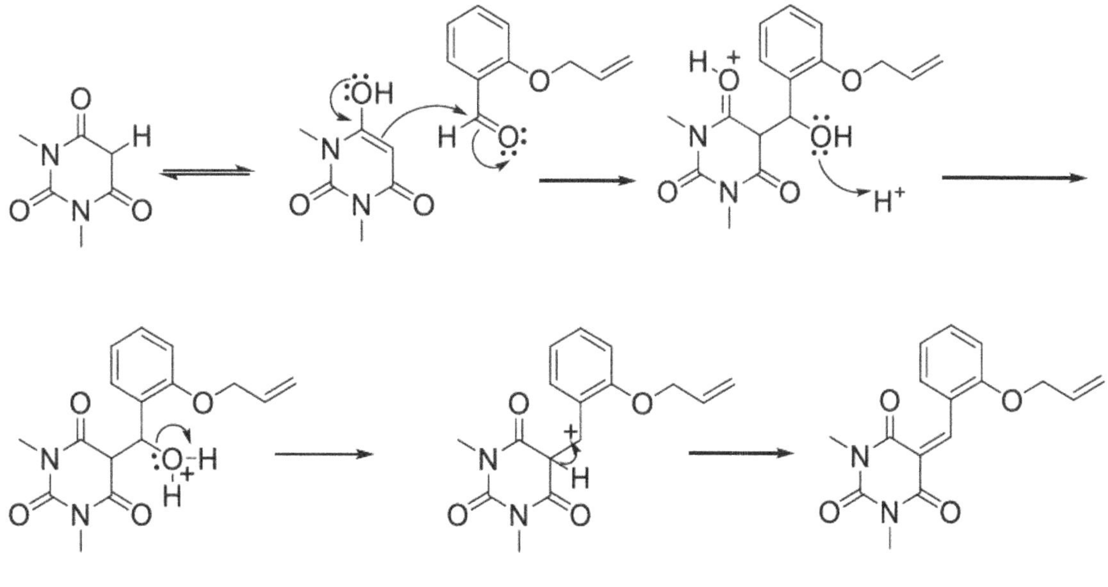

142

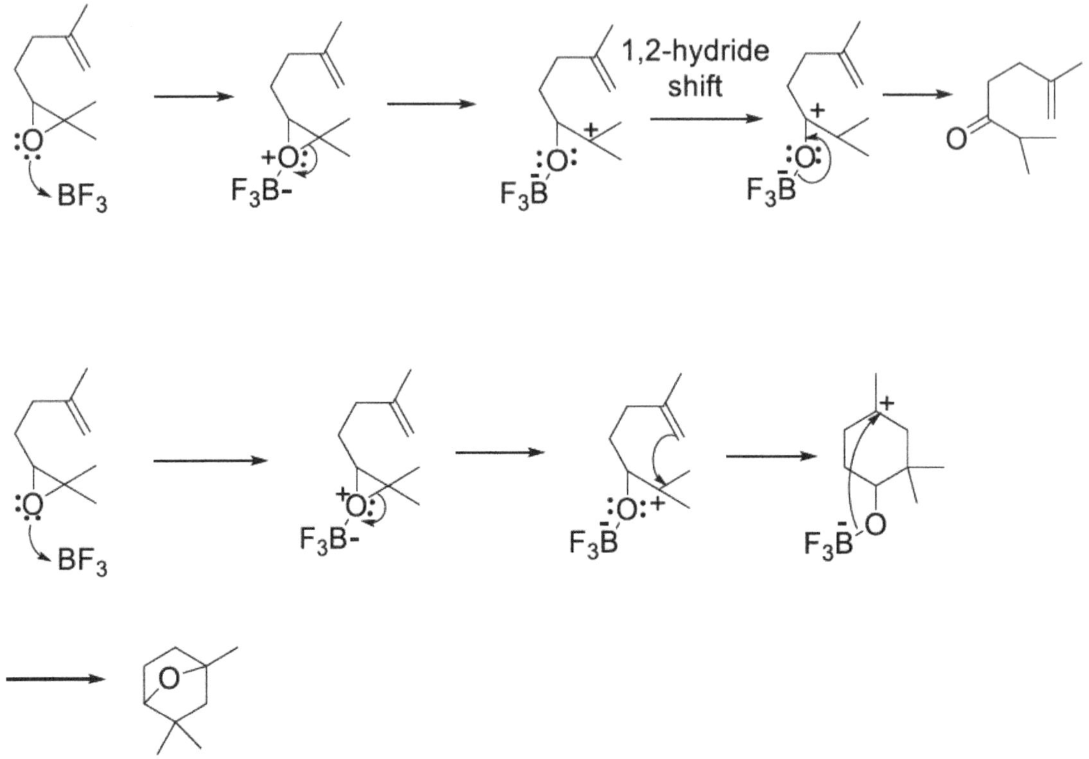

BF3
F3B-
F3B
1,2-hydride shift
F3B
O

143

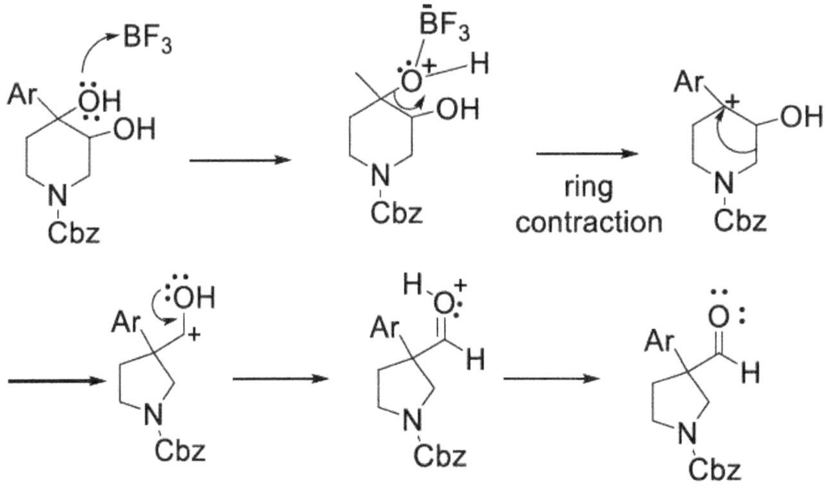

BF3
Ar
OH
OH
N
Cbz
ring contraction
H

144

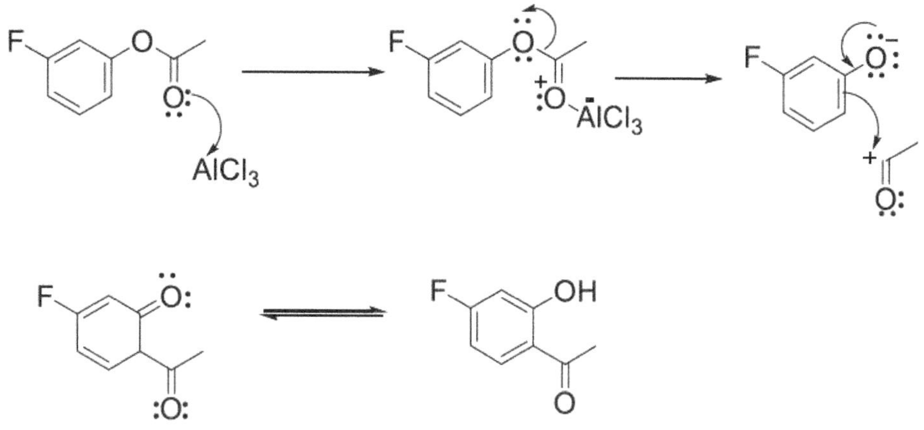
F
O
O:
AlCl3
F
O
+
:O
AlCl3
F
O:
+
O:
F
O:
:O:
F
OH
O

This is an example of the *Fries rearrangement*

145

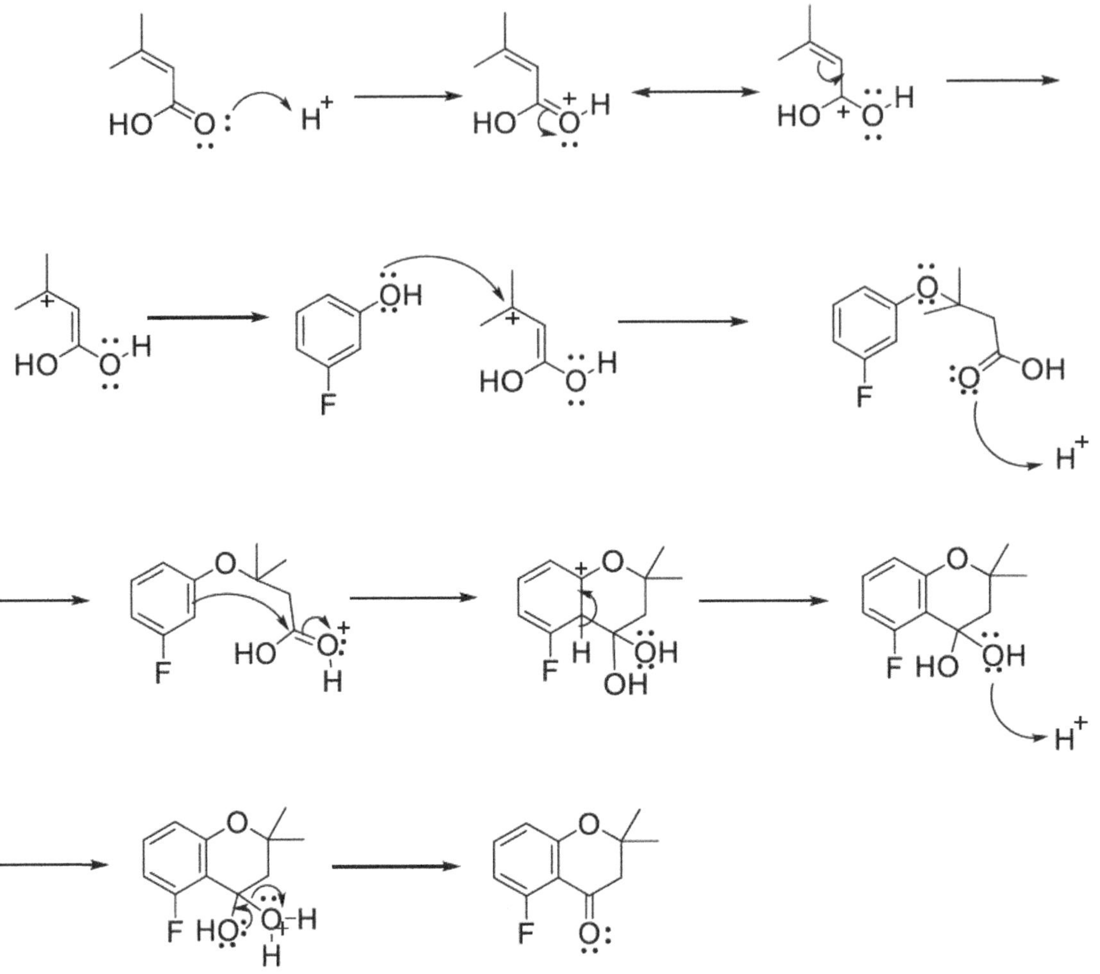

HO
O
H+
HO
O
H
HO
O
H
HO
O
H
OH
F
HO
O
H
O
F
O
OH
H+
O
F
HO
O
H
O
F
H
OH
OH
O
F
HO
OH
H+
O
F
HO
O
H
H
O
F
O

146

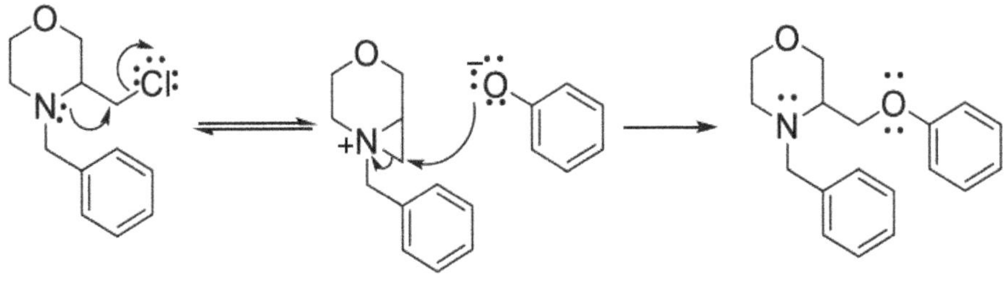

O
N
Cl
O
N
O
O
N
O

147

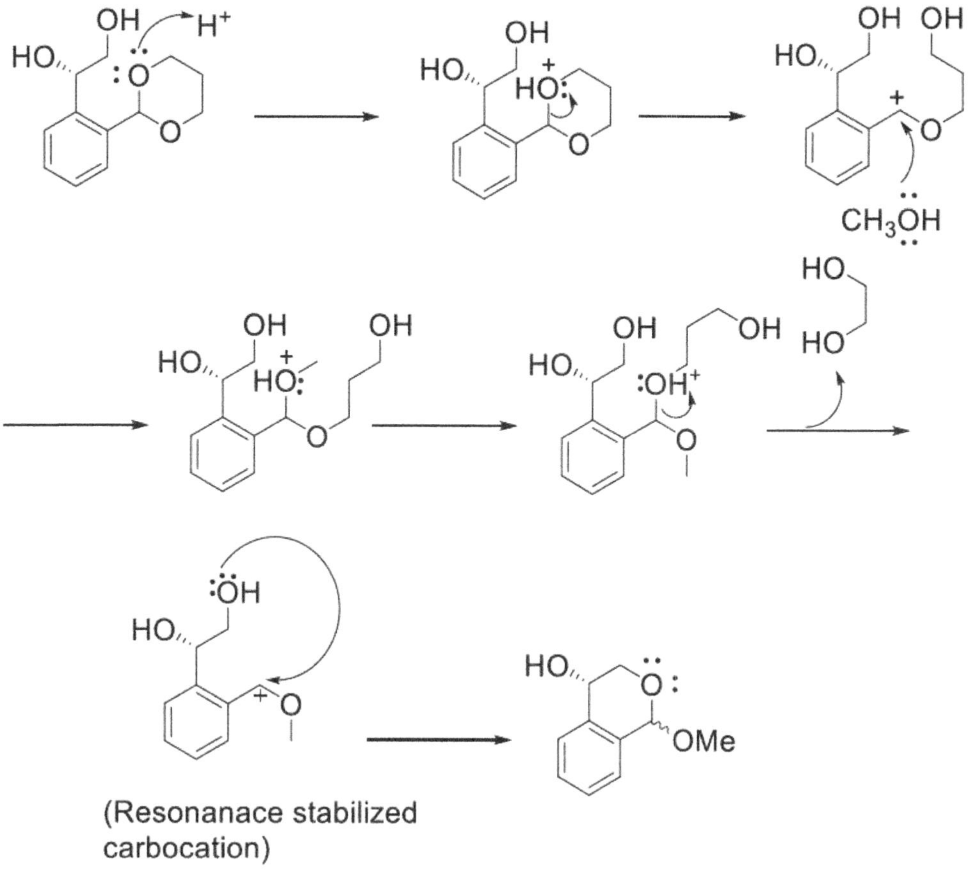

OH
H+
HO
O
O
HO
HO
HO
CH3OH
OMe
(Resonanace stabilized
carbocation)

148	

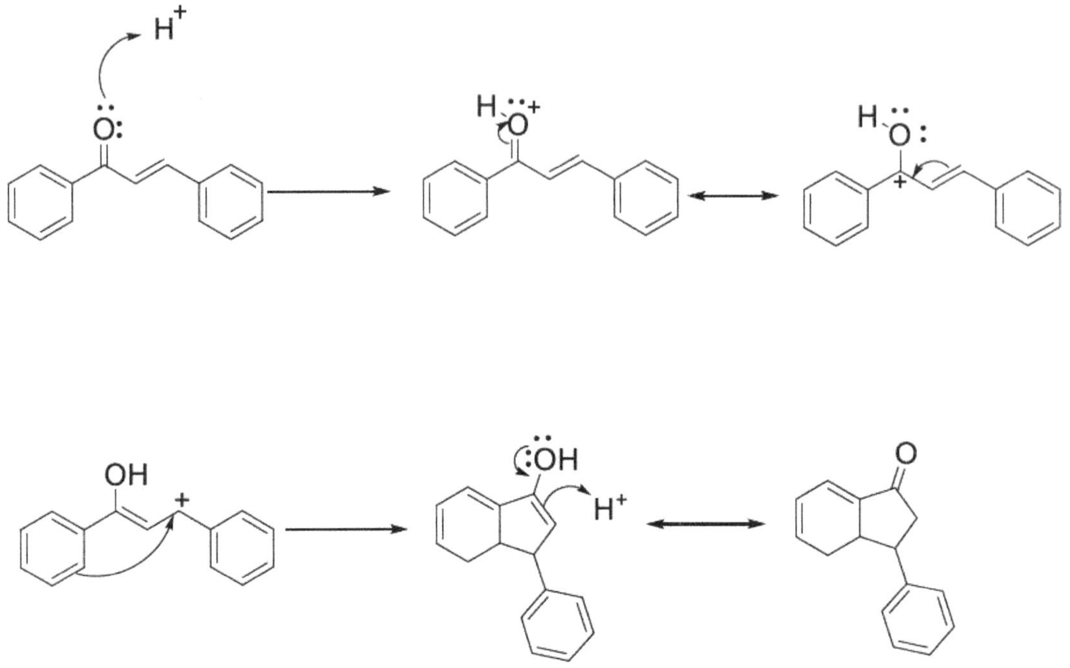
H+
OH
H+

149

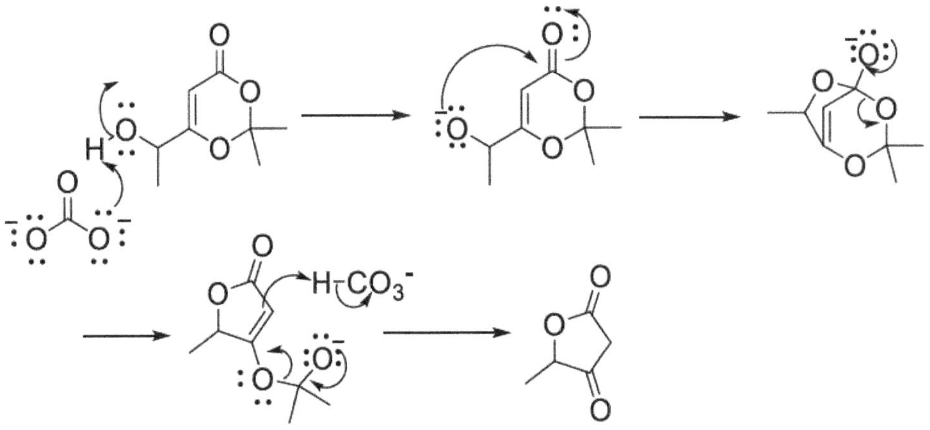
H-CO3-

150

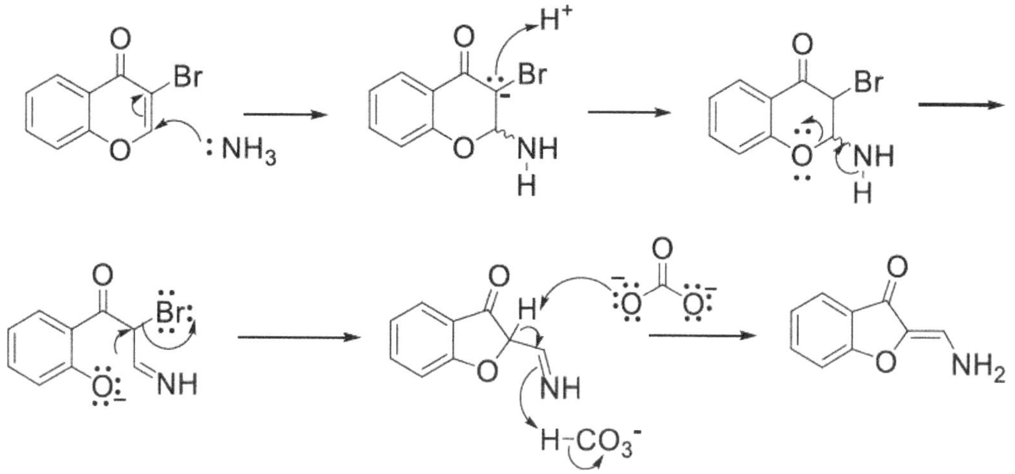
Br
NH3
H+
NH
H
Br
NH2
NH
H-CO3-

151

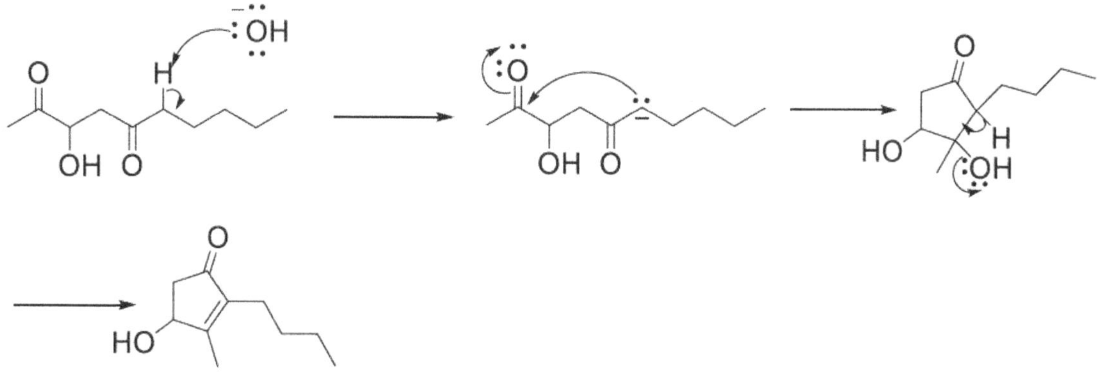
OH
HO
H

152

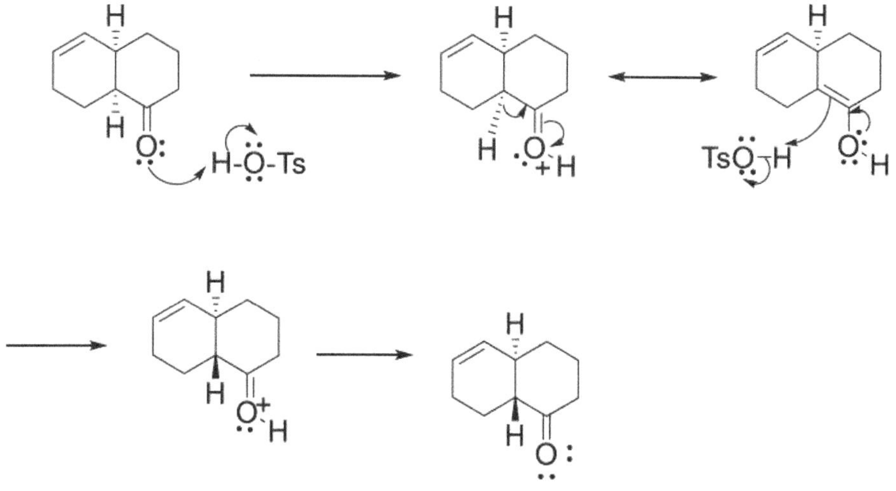
H
H-O-Ts
TsO-H

Acid-catalyzed isomerization to the more stable trans isomer
153
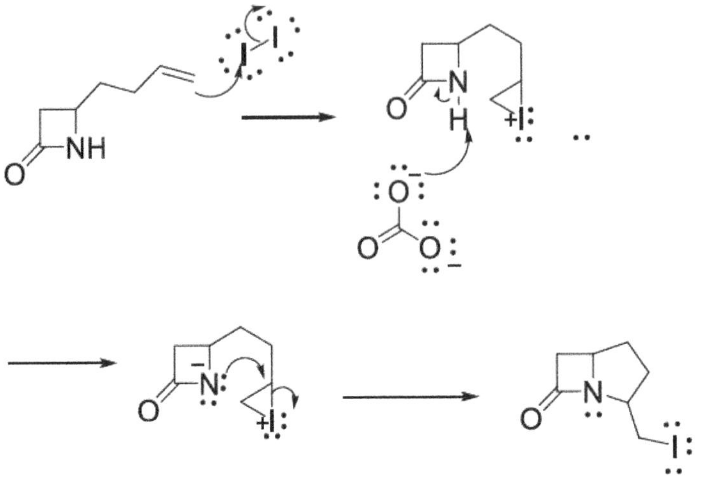
154

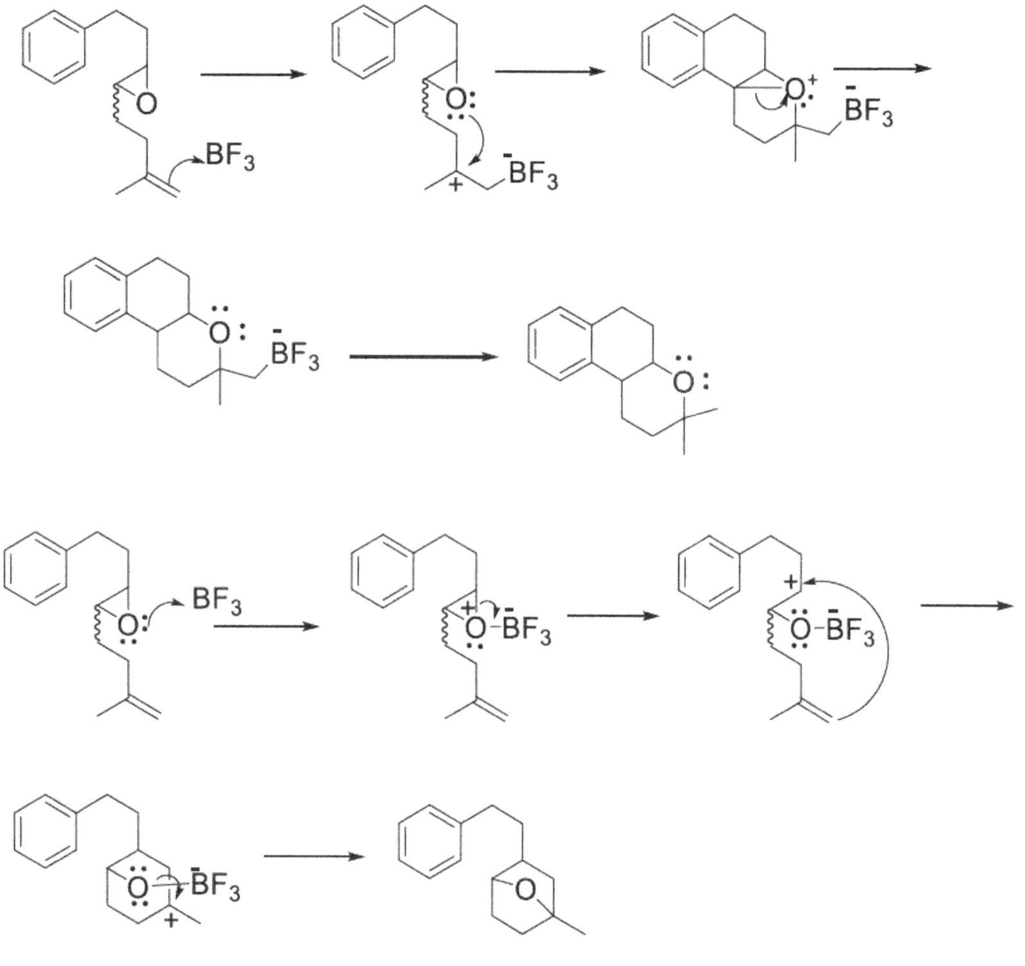

155

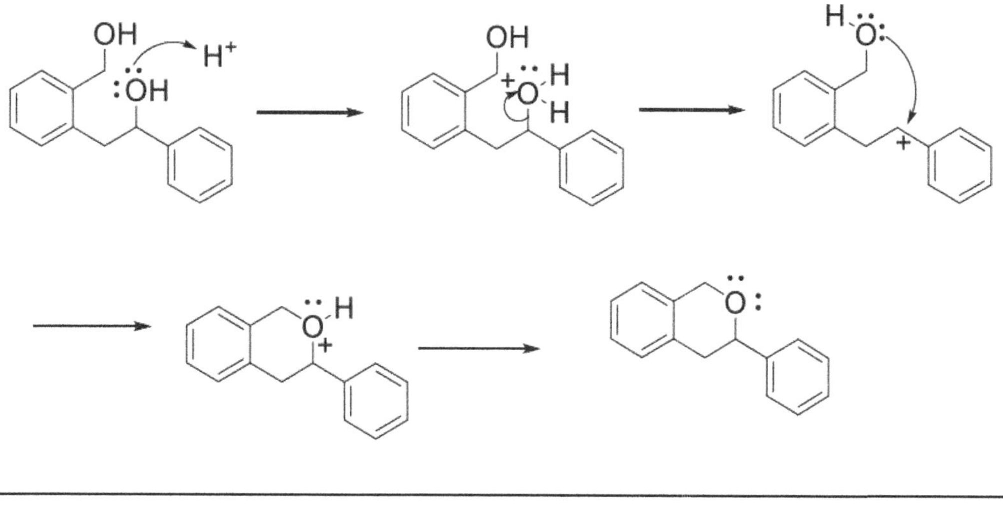

156

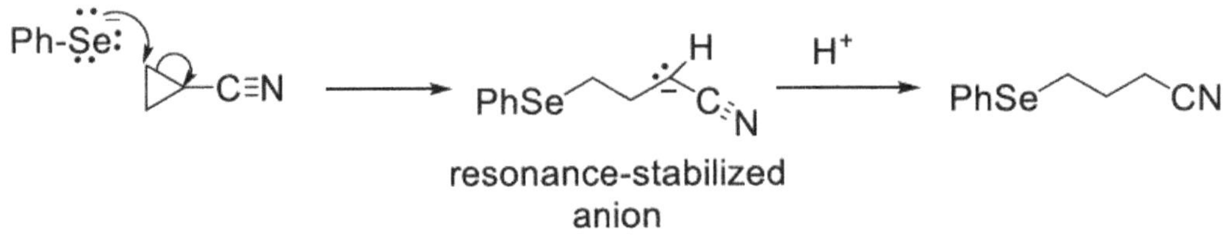

Ph-Se:
C≡N
PhSe
H
C
N
H+
PhSe
CN
resonance-stabilized
anion

157

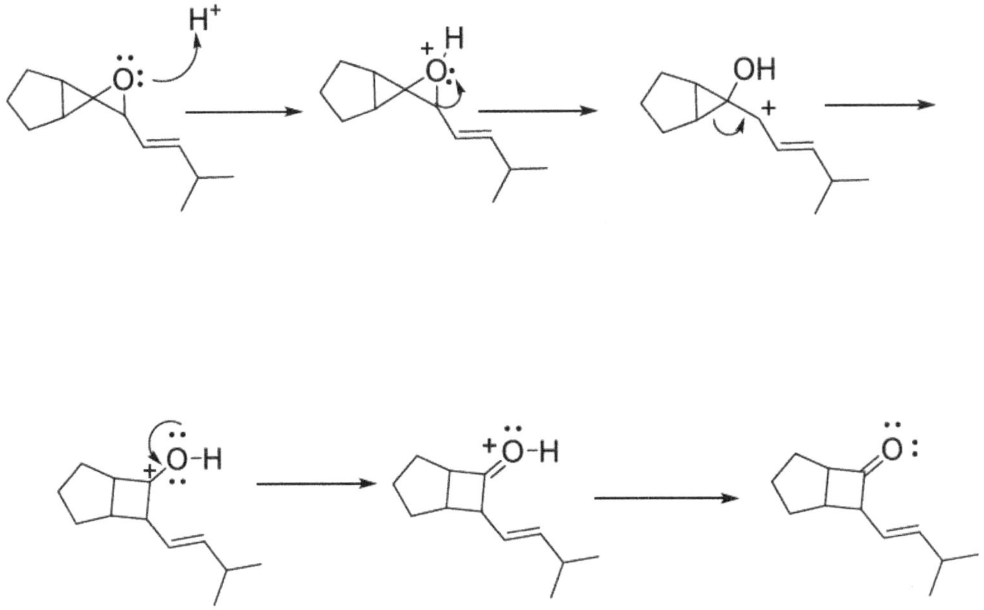

H+
O:
+ H
O
OH
+
O-H
+O-H
O:

158

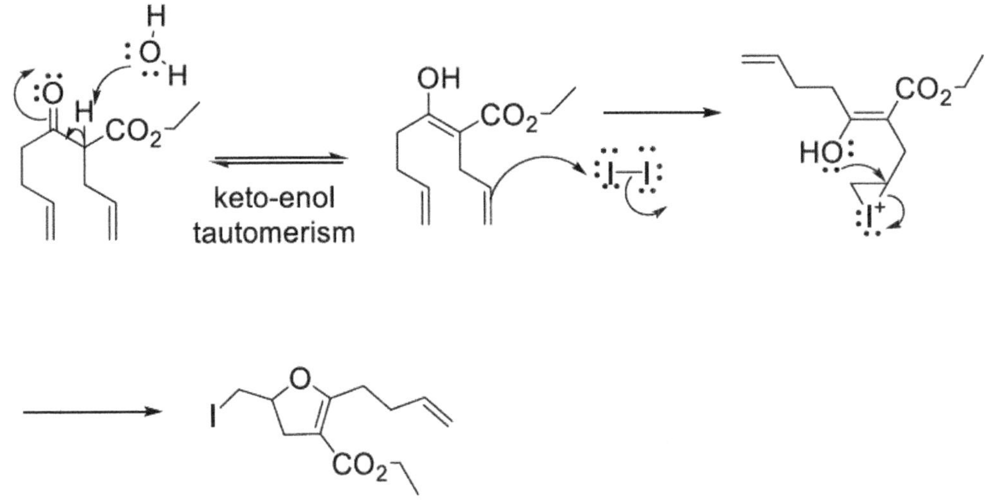

H
:O
H
:O
H
CO2
keto-enol
tautomerism
OH
CO2
:I—I:
CO2
HO:
I+
I
O
CO2

159

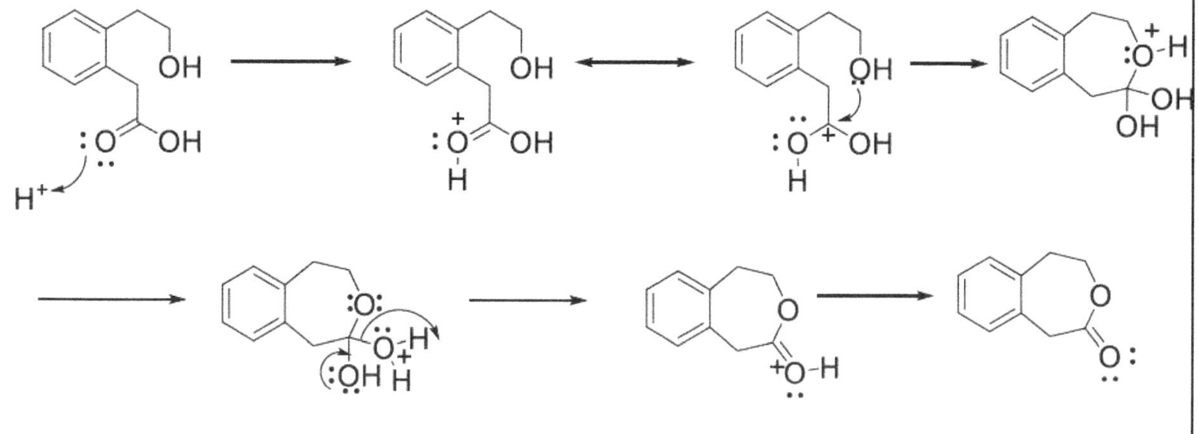

160

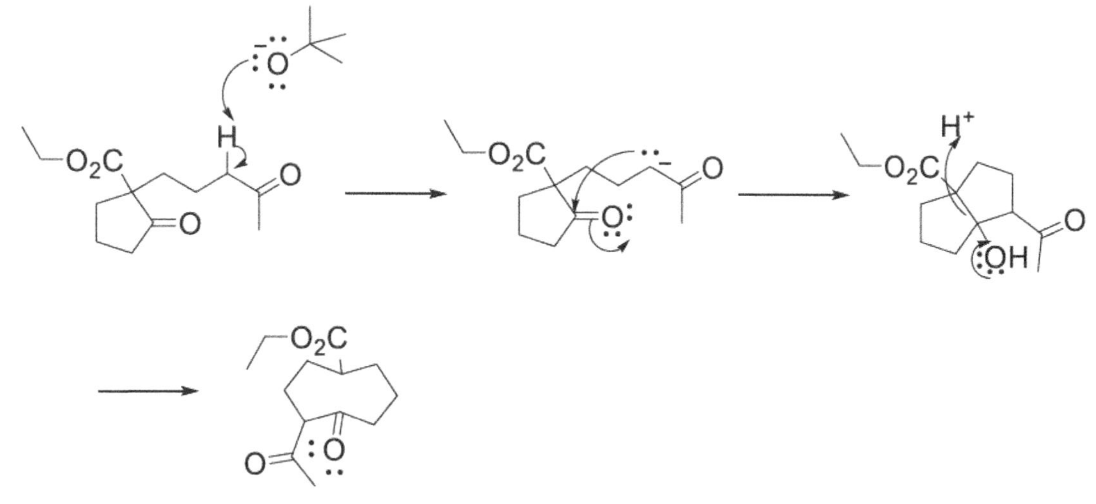

161

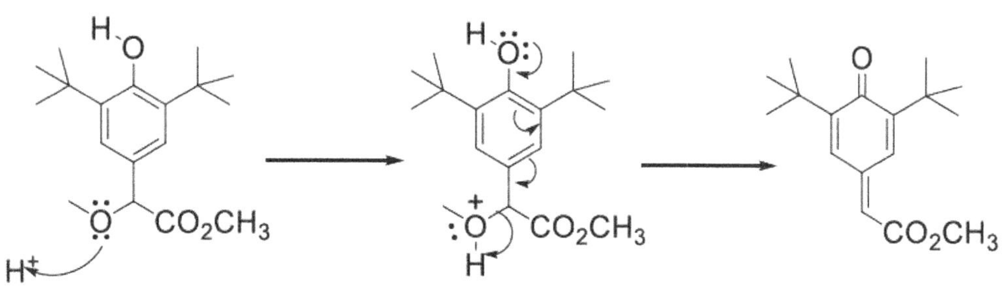

162

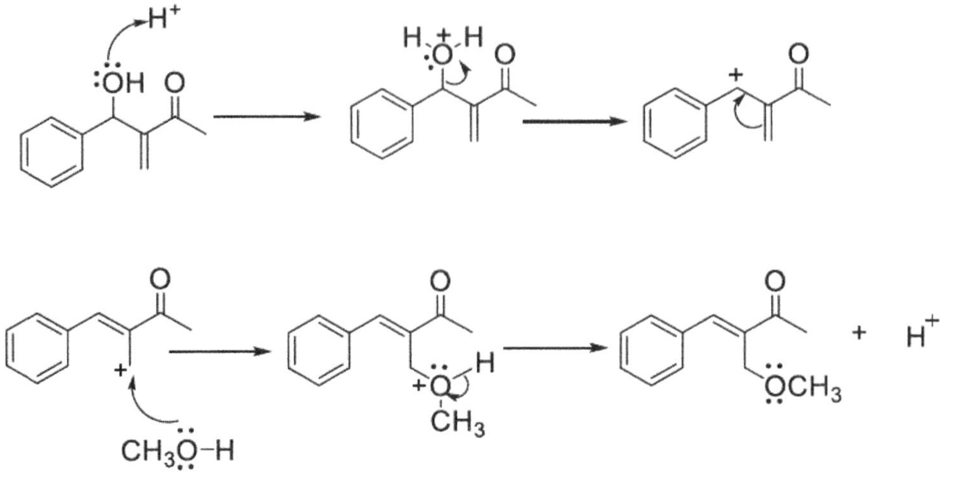
H+
OH
O
H + H
O
O
+
O
O
O
O
+
CH3O-H
+O
H
CH3
O
OCH3
+
H+

163

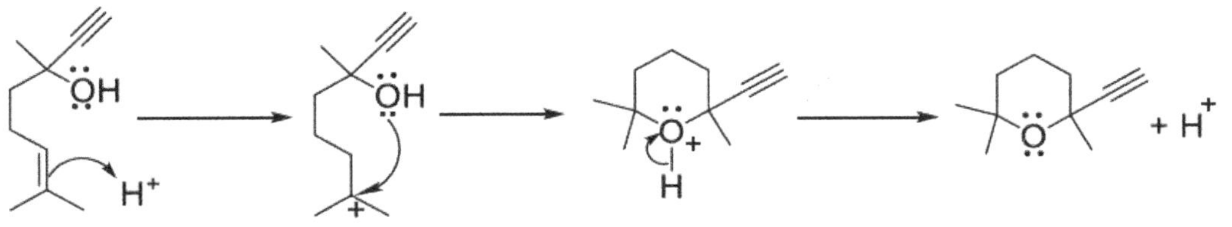
OH
H+
OH
+
O+
H
O
+ H+

164

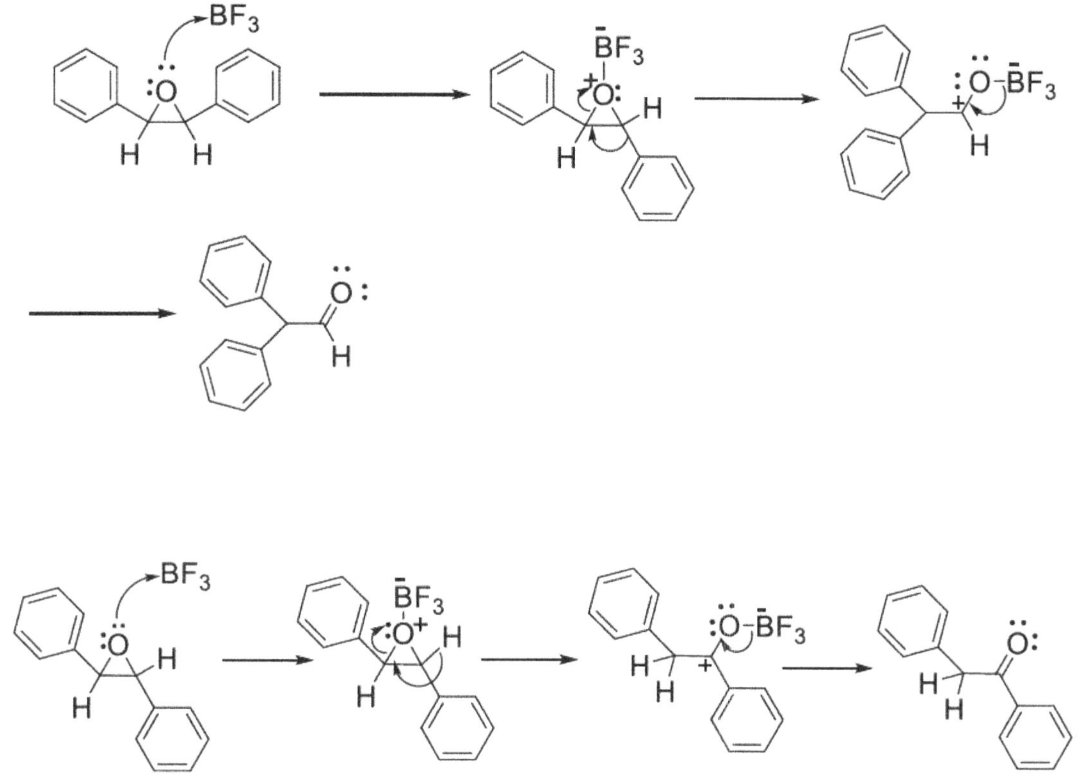

165

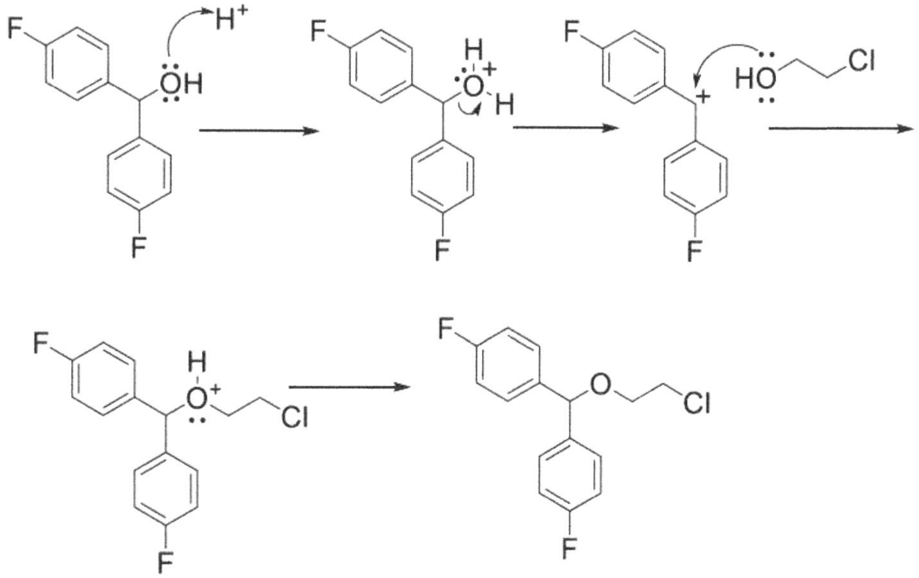

166

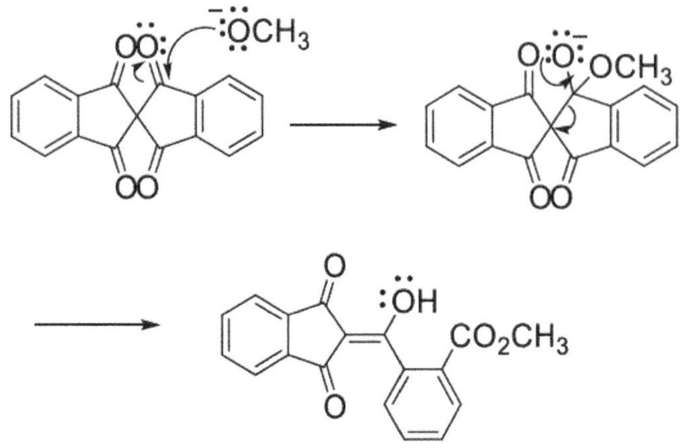

OCH3
H+
:OH
CO2CH3

167

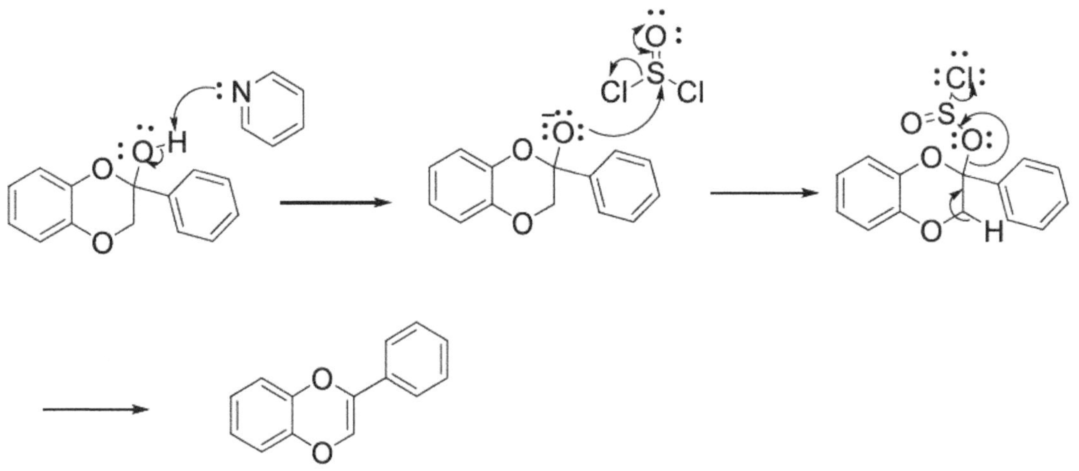

N
H
O
Cl
S

168

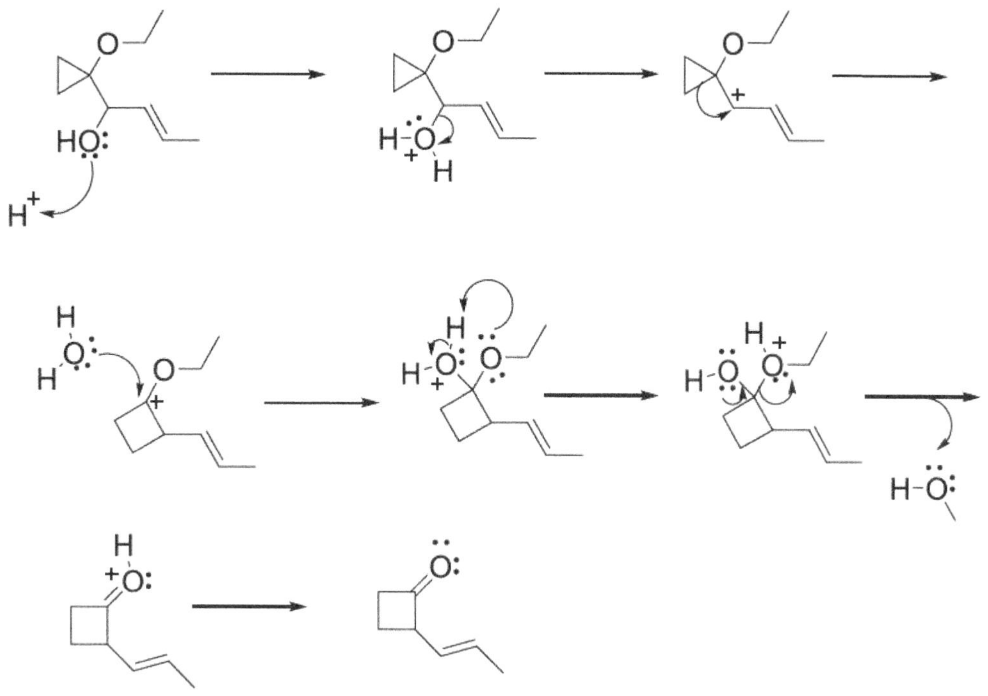
HO
H+
H-O
H
+
H
O
H
H-O
H
+
H
H-O
O
H-O
H
+O
O

169

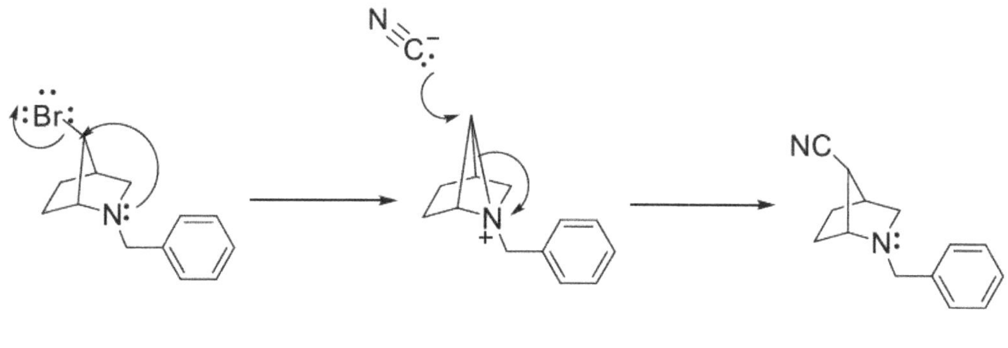
N
C
Br
N
N
+
NC
N

170

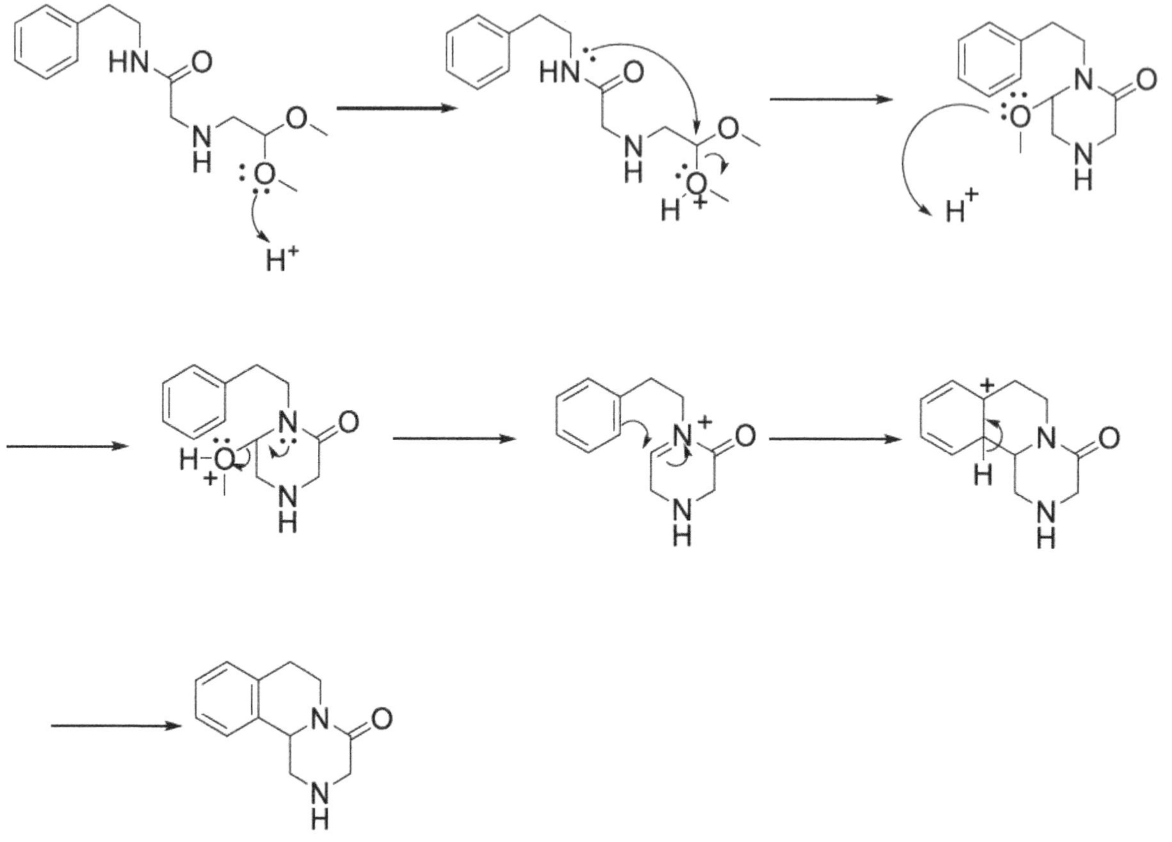

171

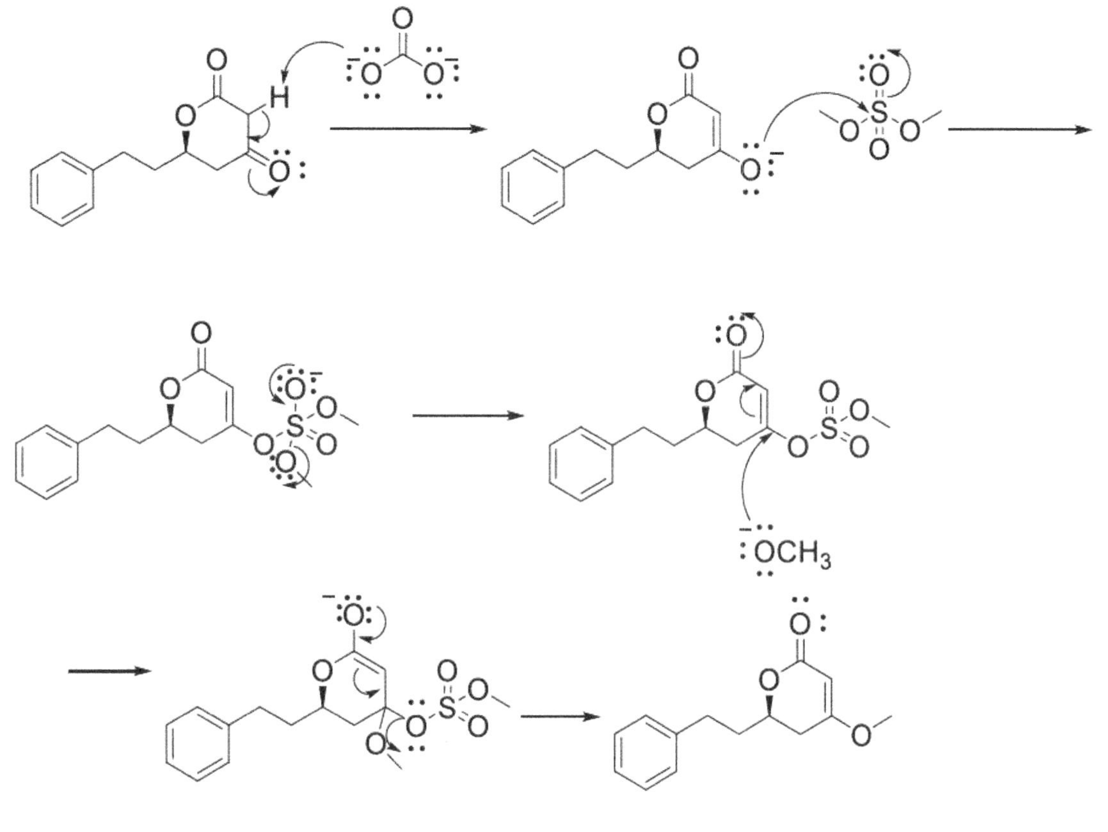

172

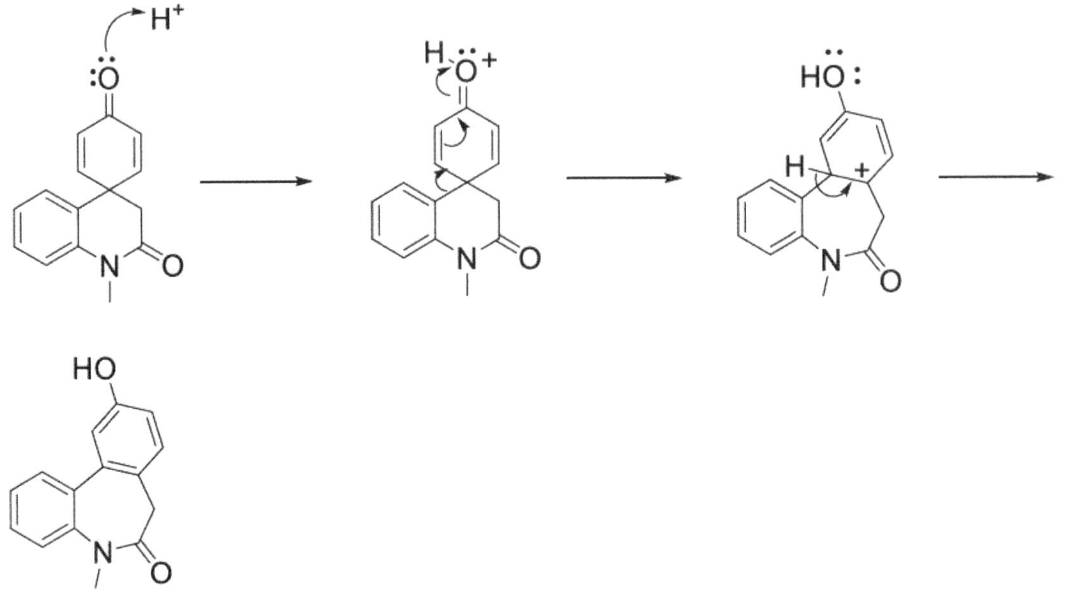

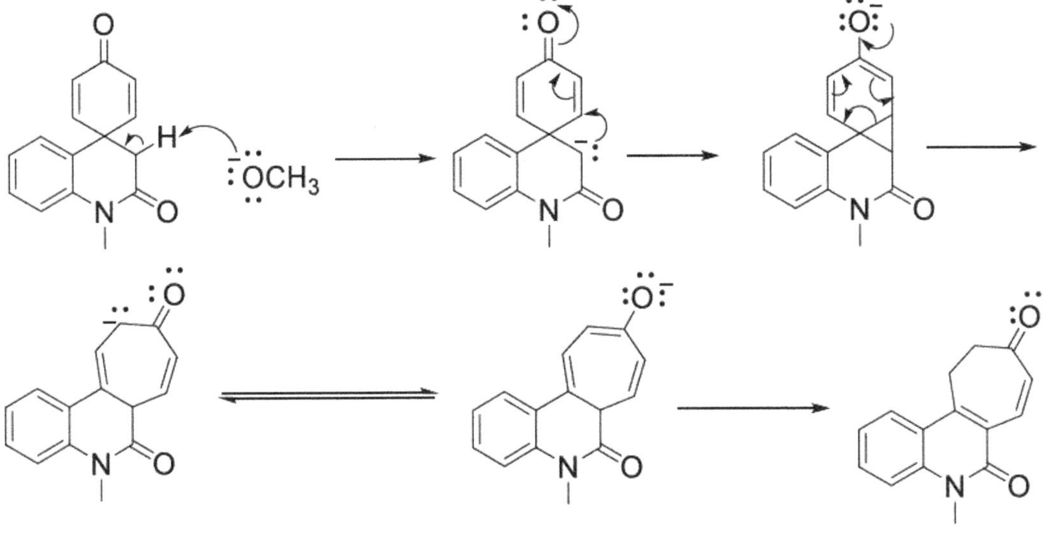

173

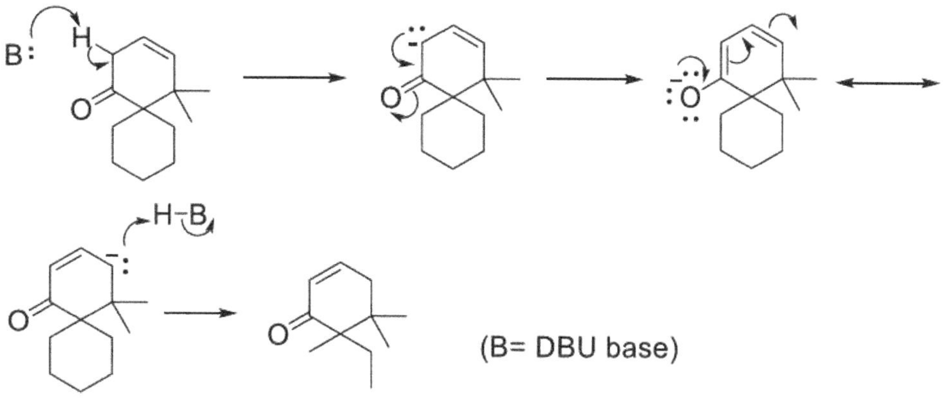

B:
H
O
H–B

(B= DBU base)

Base-catalyzed isomerization to form the more stable conjugated product.

174

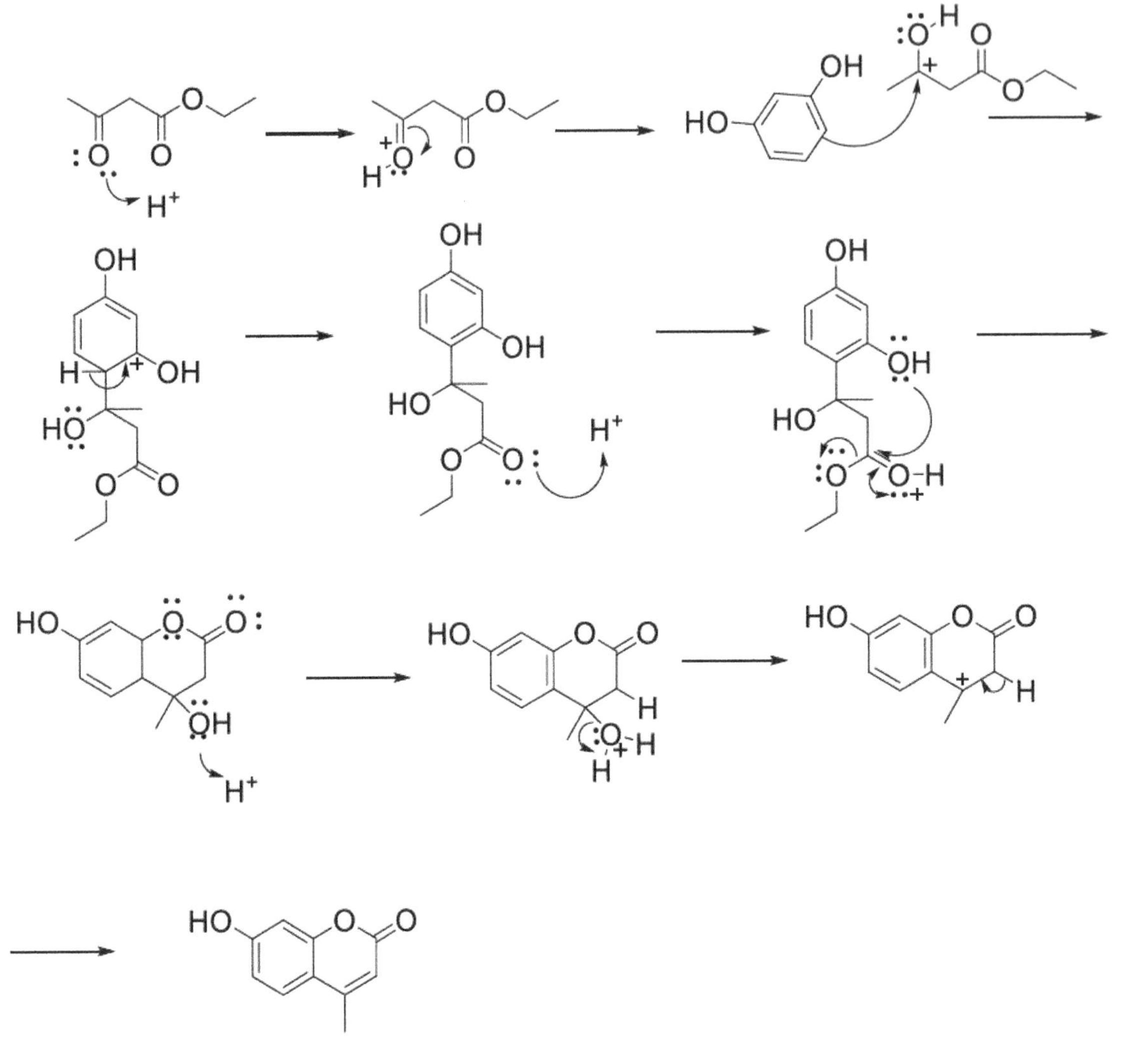
H+
OH
HO
O
H

175

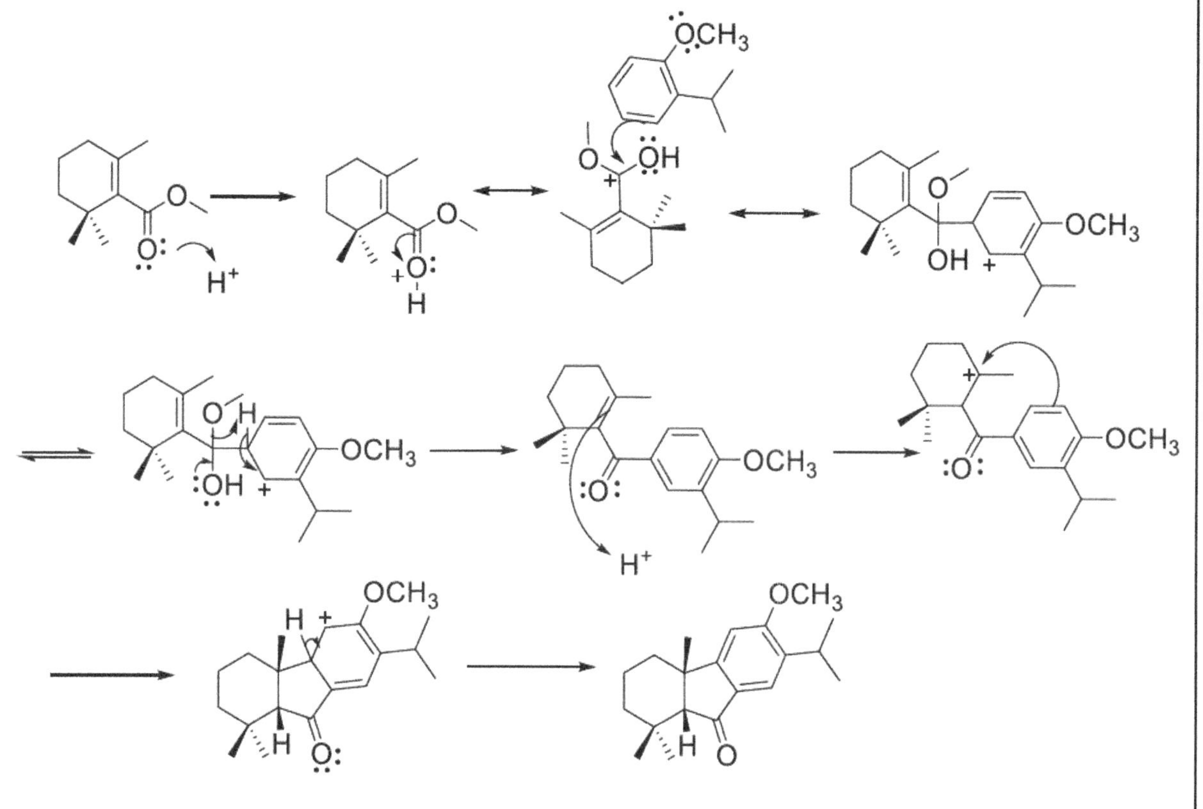

OCH3
OH
H+
OCH3
OH
OCH3
H
OCH3
H+
OCH3
OCH3
H
H
OCH3
H

176

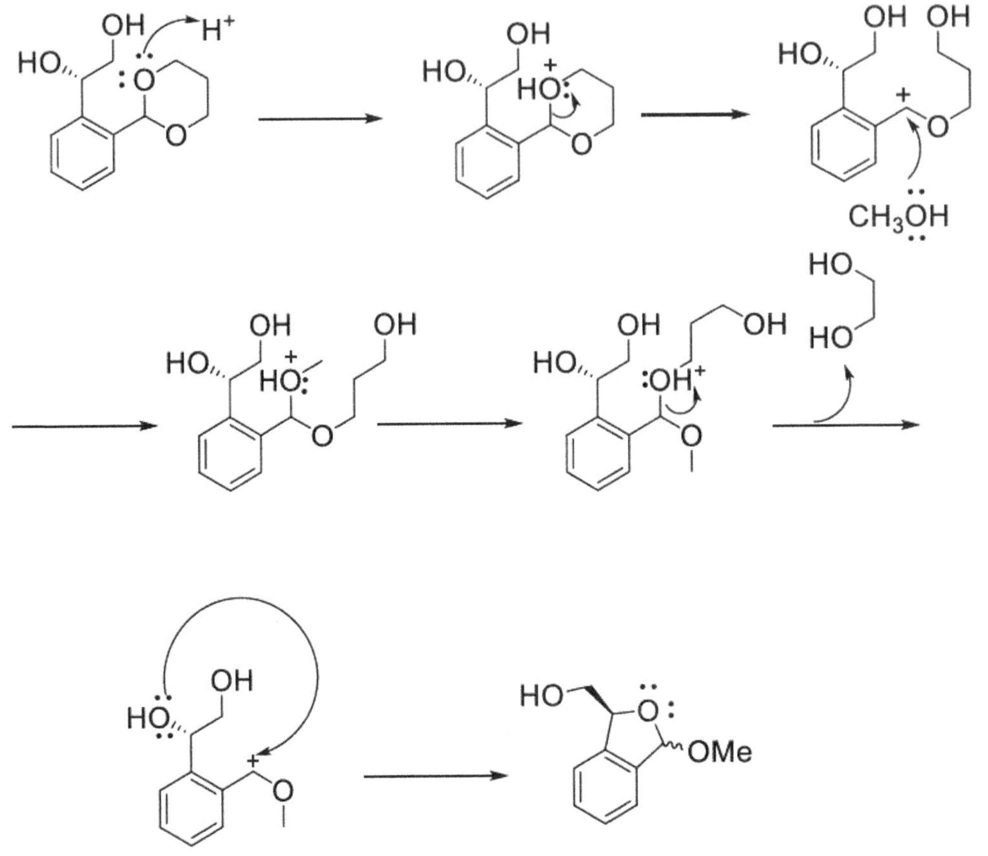

(Resonanace stabilized carbocation)

177

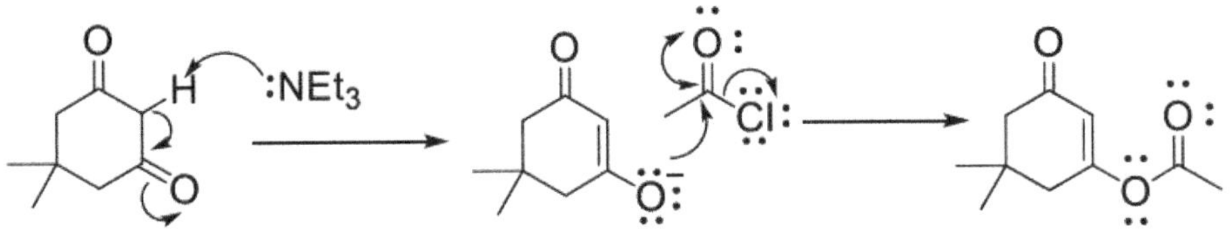

Formation of the ambident anion and subsequent reaction with the acid chloride yields the product.

178

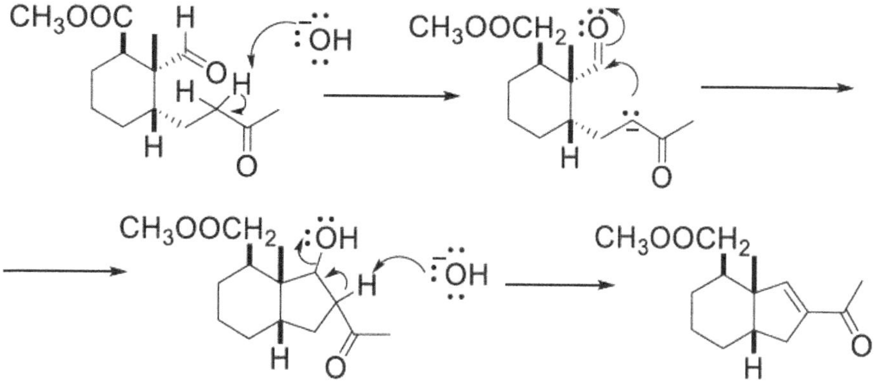
CH3OOC
H
OH
H
H
H
O
:OH
CH3OOCH2
:O
H
O
CH3OOCH2
:OH
H
H
O
:OH
CH3OOCH2
H
O

This is an example of an intramolecular aldol condensation reaction.

179

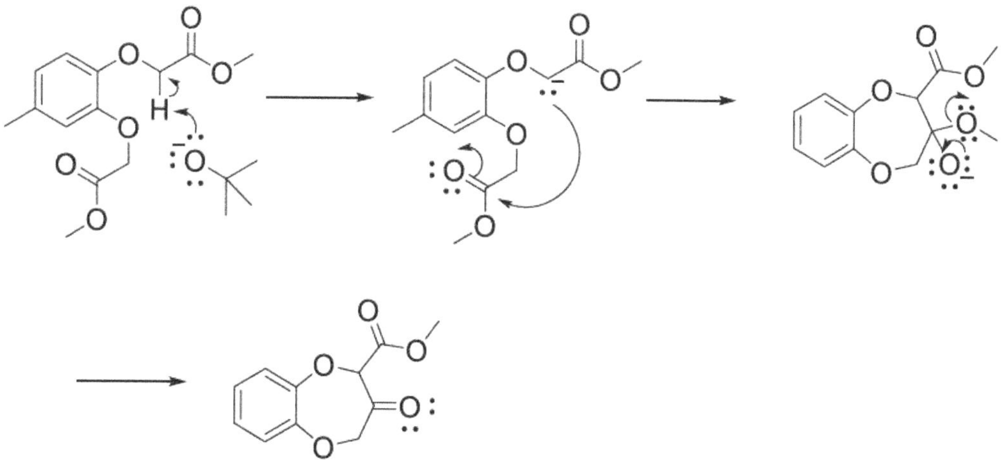

This is a *Dieckmann condensation reaction* (the intramolecular counterpart of the *intermolecular Claisen condensation reaction*)

180

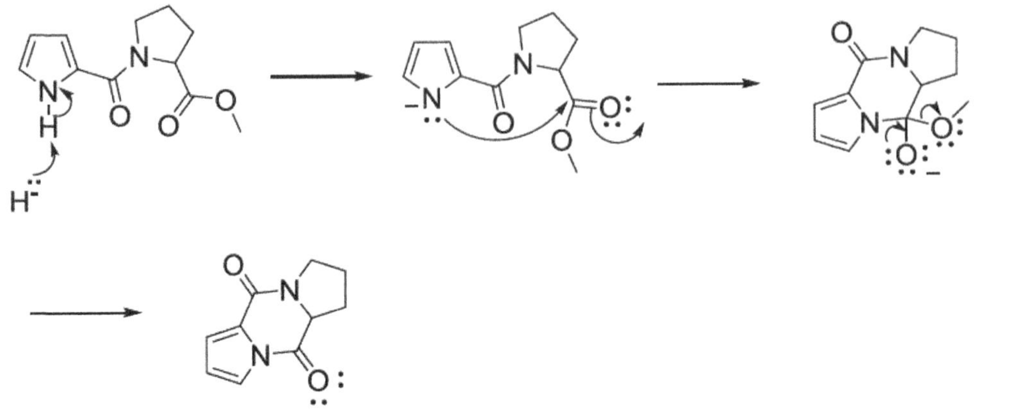

181

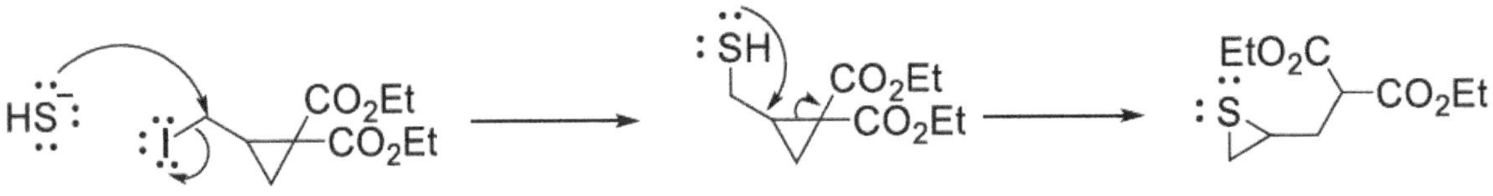

182

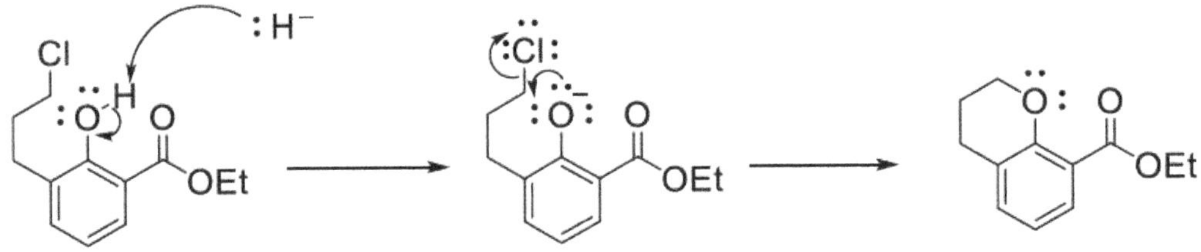

183

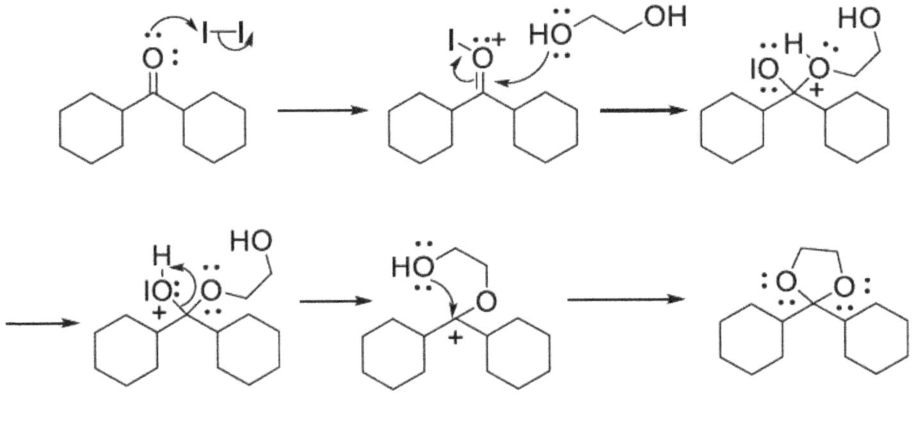

184

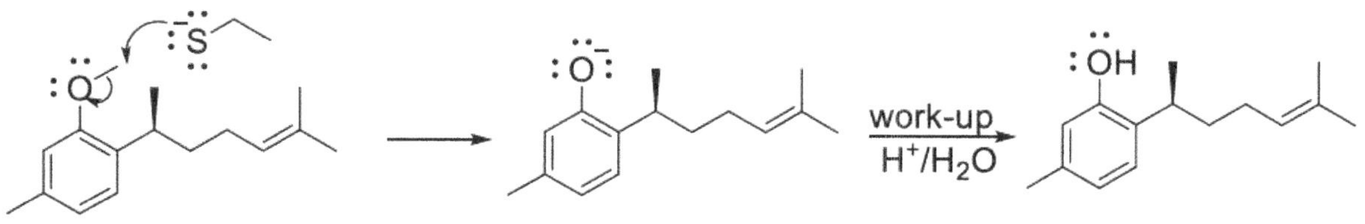
work-up
H+/H2O
OH

This is a general reaction involving the cleavage of an aryl alkyl ether.

185

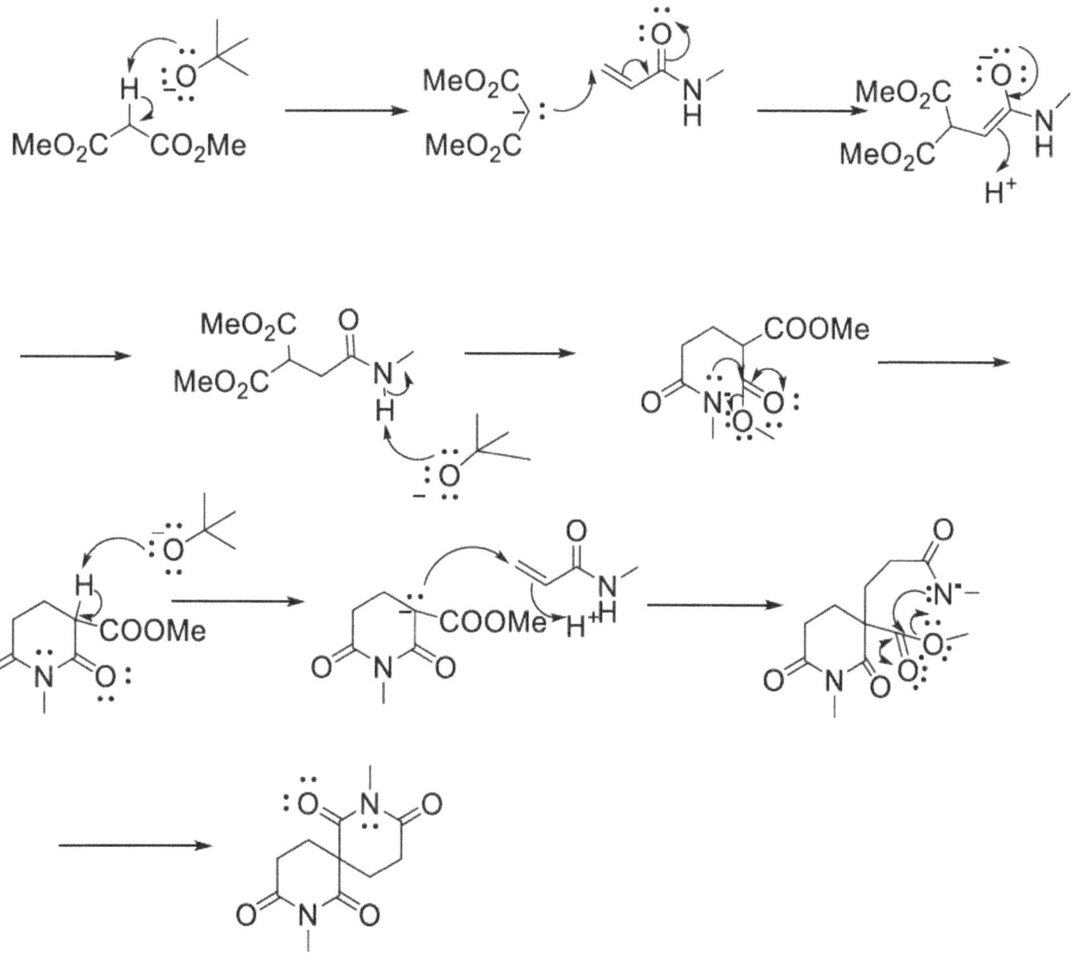
MeO2C
CO2Me
MeO2C
COOMe
H+

186

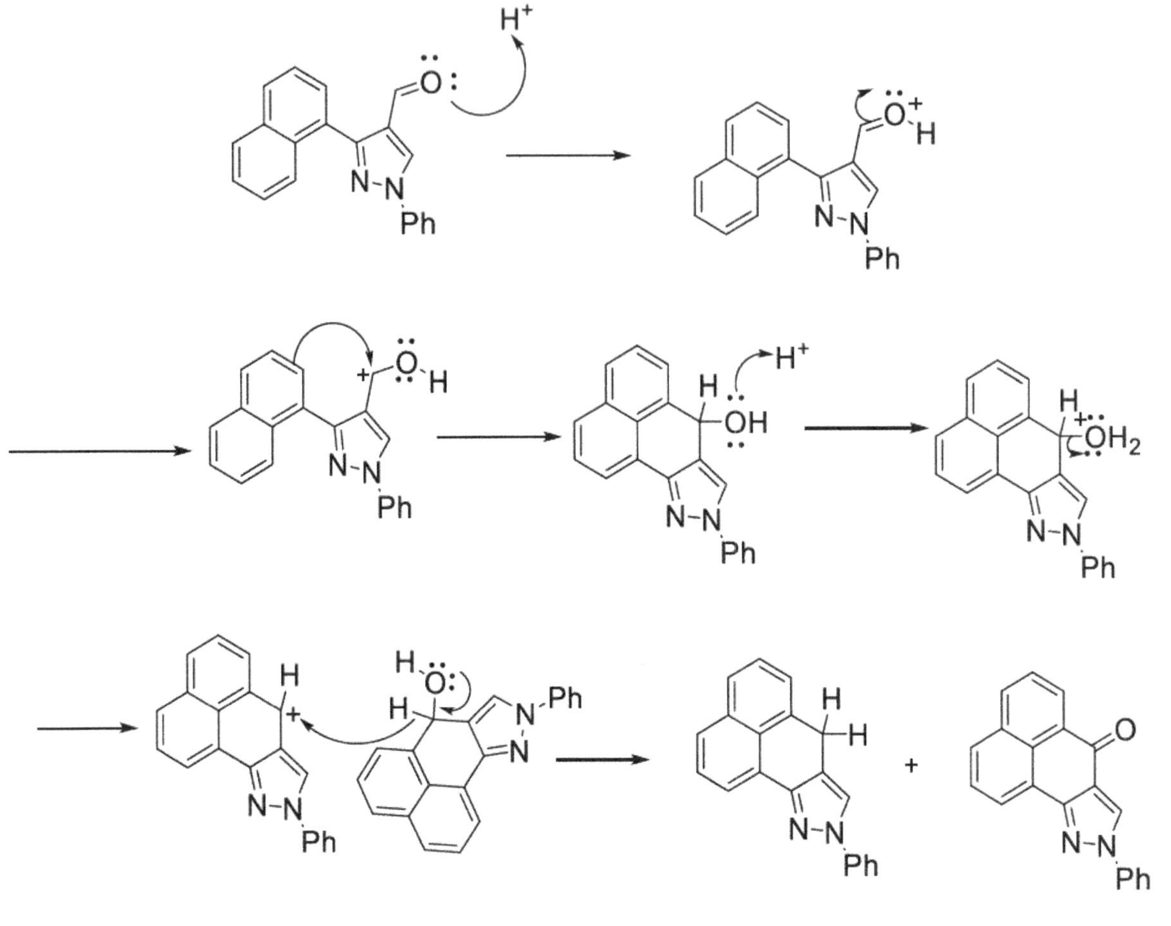

H+
O
H
N
Ph
OH
OH2

187

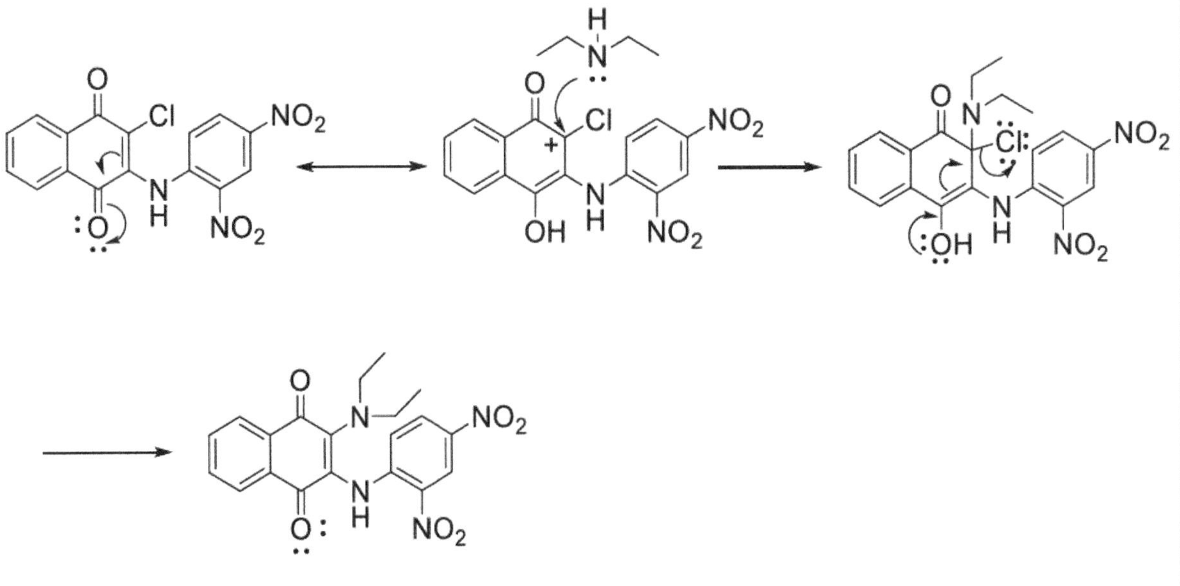

O
Cl
NO2
NH
H
N
OH

188

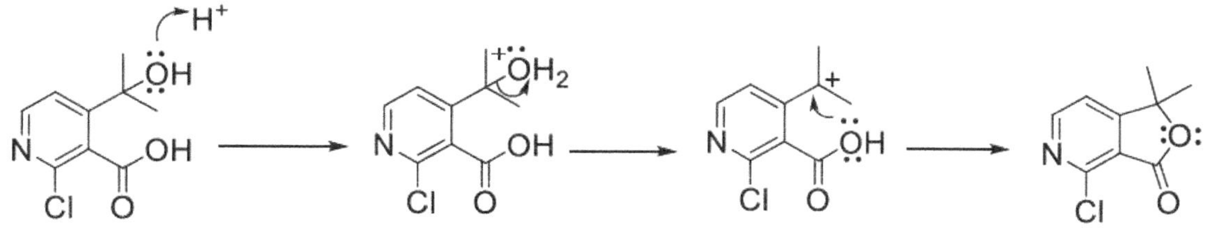

H+
OH
N
OH
Cl
O
OH2
+
O

189

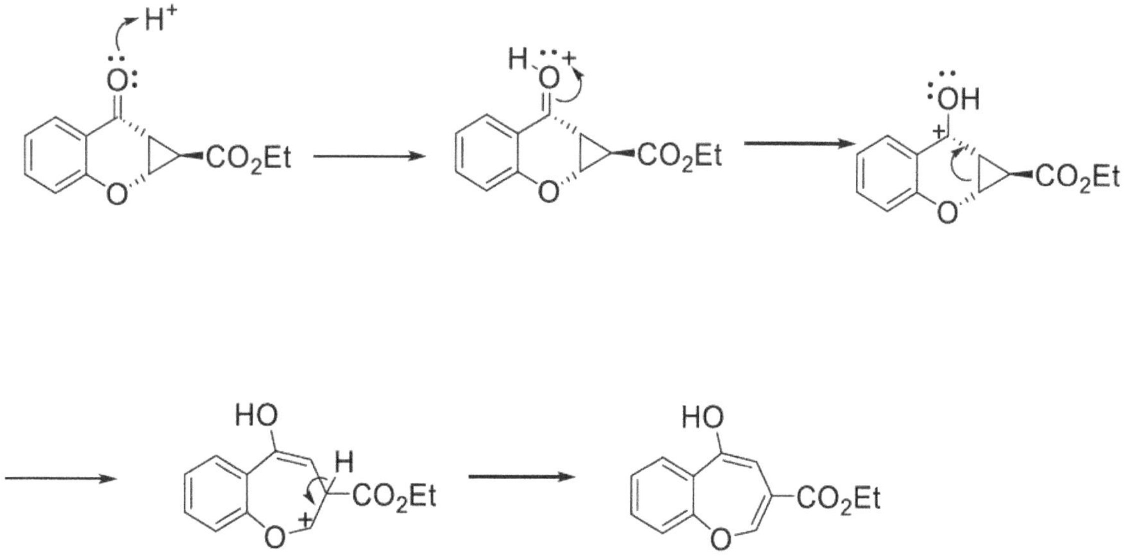

H+
O
CO2Et
H
+
OH
HO
H

190

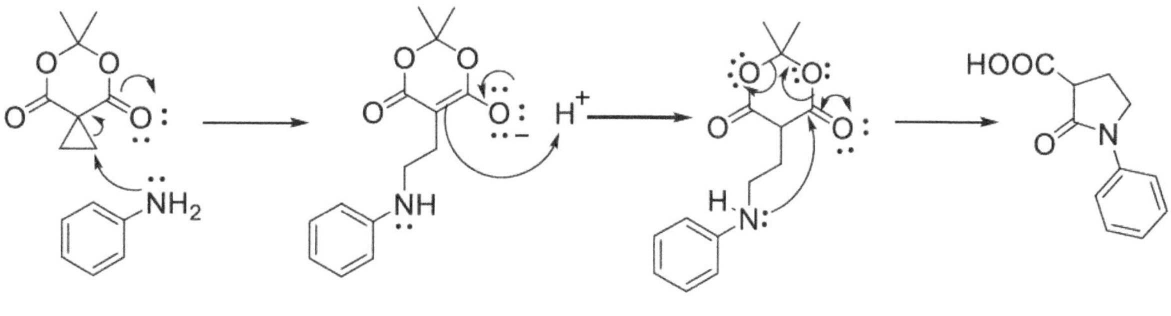

O
NH2
NH
H+
N
HOOC

191

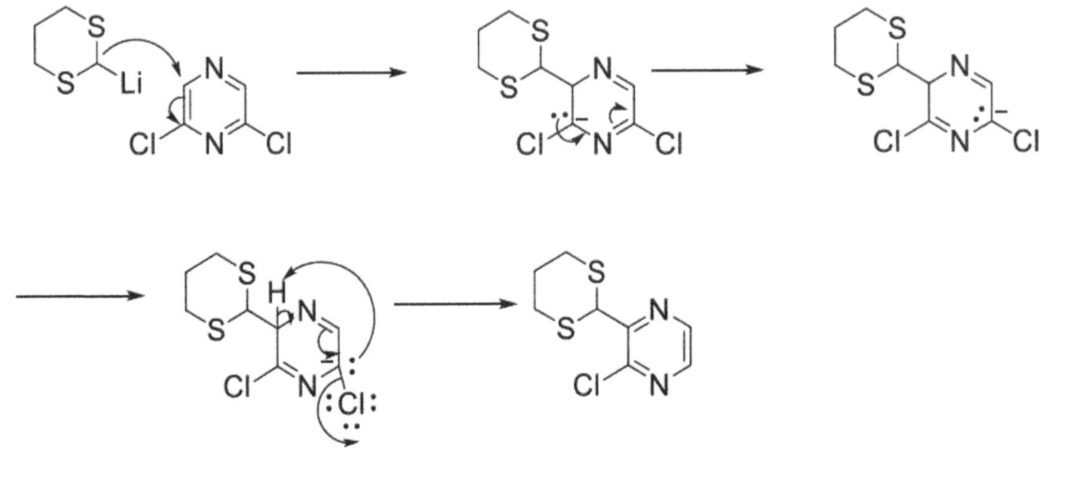

S
S
Li
N
Cl
N
Cl
S
S
N
Cl
N
Cl
S
S
N
Cl
N
Cl
S
S
H
N
Cl
N
Cl
S
S
N
Cl
N

192

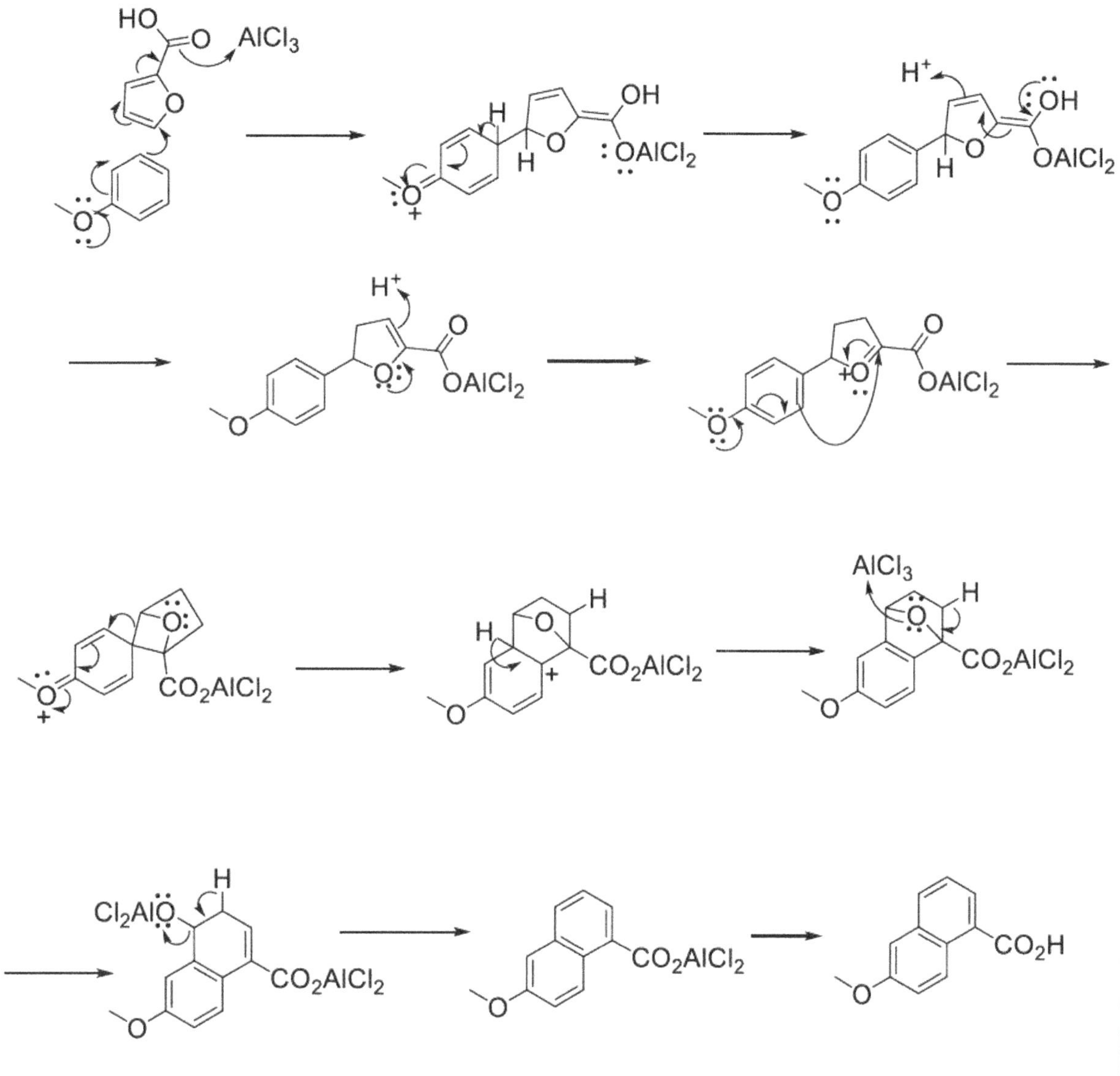

HO
O
AlCl$_3$
O
H
H
OH
OAlCl$_2$
H$^+$
OH
OAlCl$_2$
O
OAlCl$_2$
CO$_2$AlCl$_2$
AlCl$_3$
Cl$_2$AlO
CO$_2$H

193

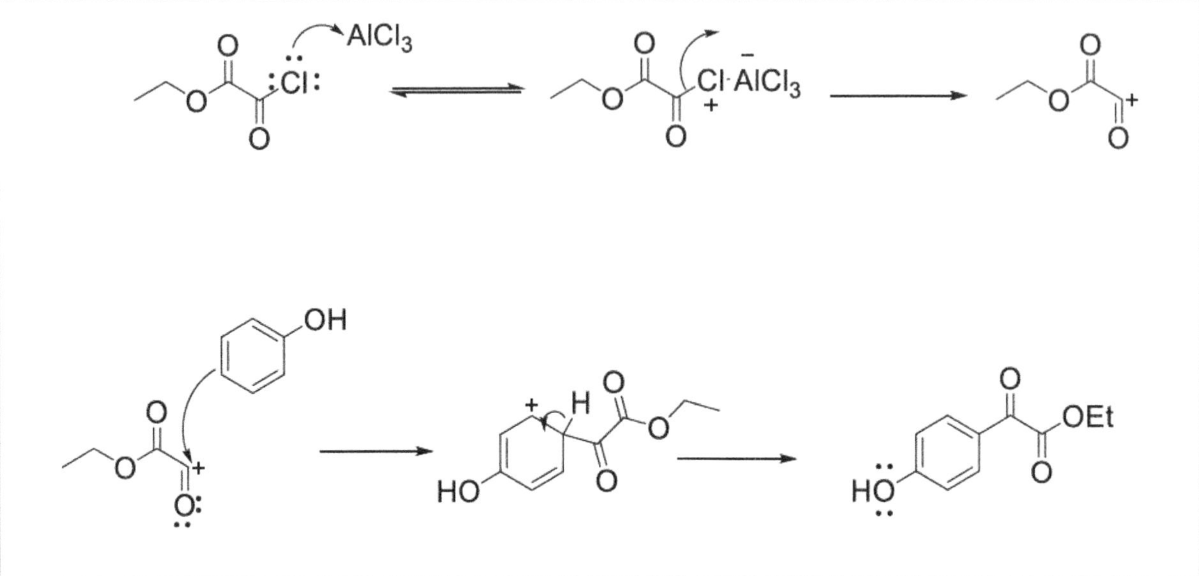
AlCl3
OEt
HO
OH

194

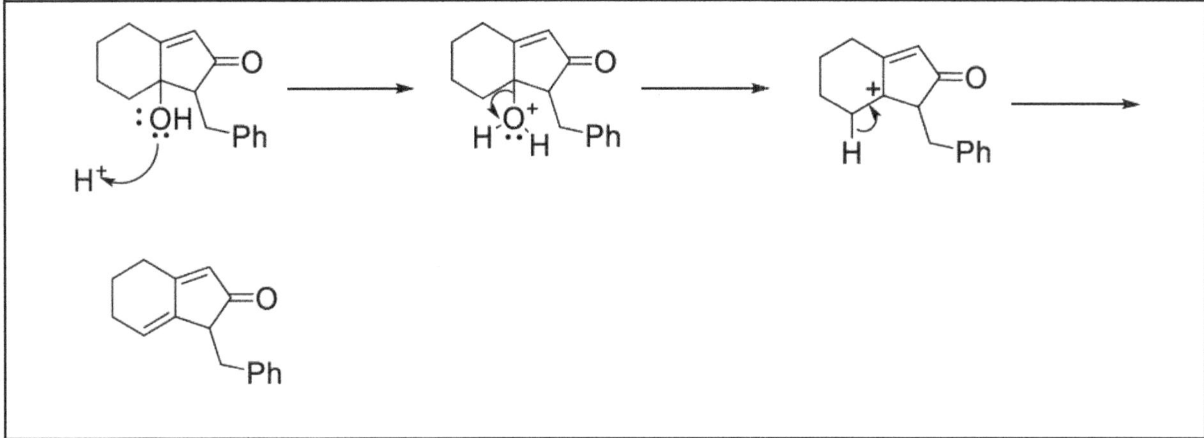
OH
H+
Ph

195

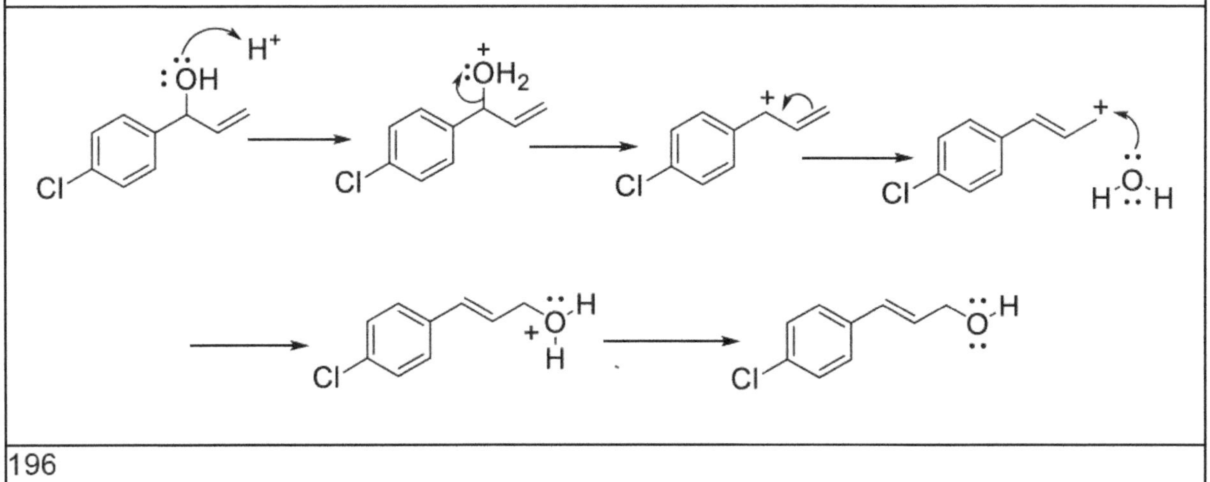
OH
H+
OH2
Cl

196

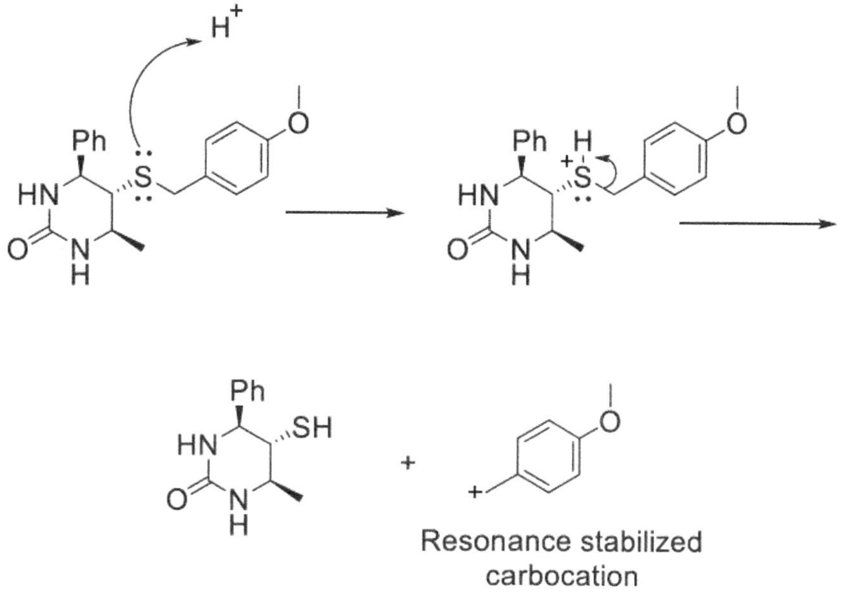
H+
Ph
HN
O
N
H
S
Ph
H
+
SH
Resonance stabilized carbocation

Exercise Draw all the resonance structuers of the *p*-methoxybenzyl carbocation.

197

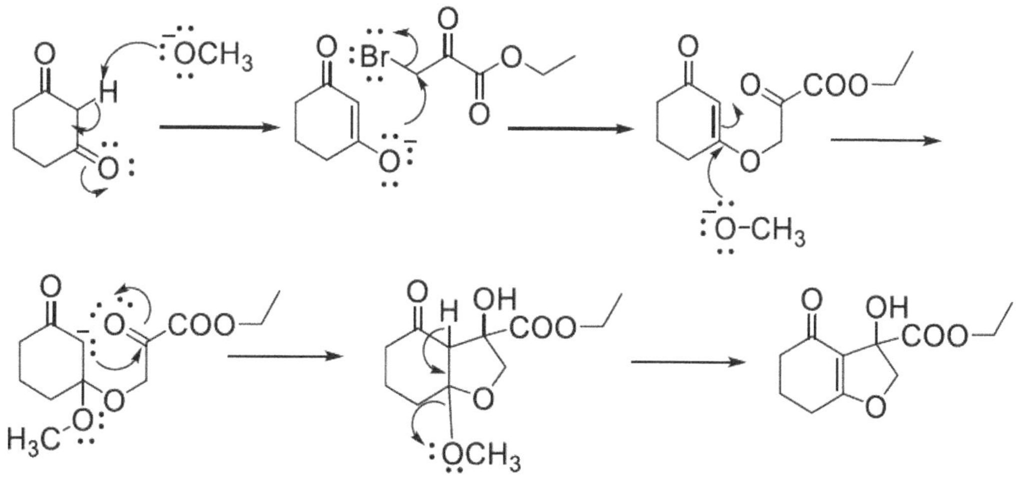
OCH3
H
Br
COO
O-CH3
OH
H3C
OCH3

198

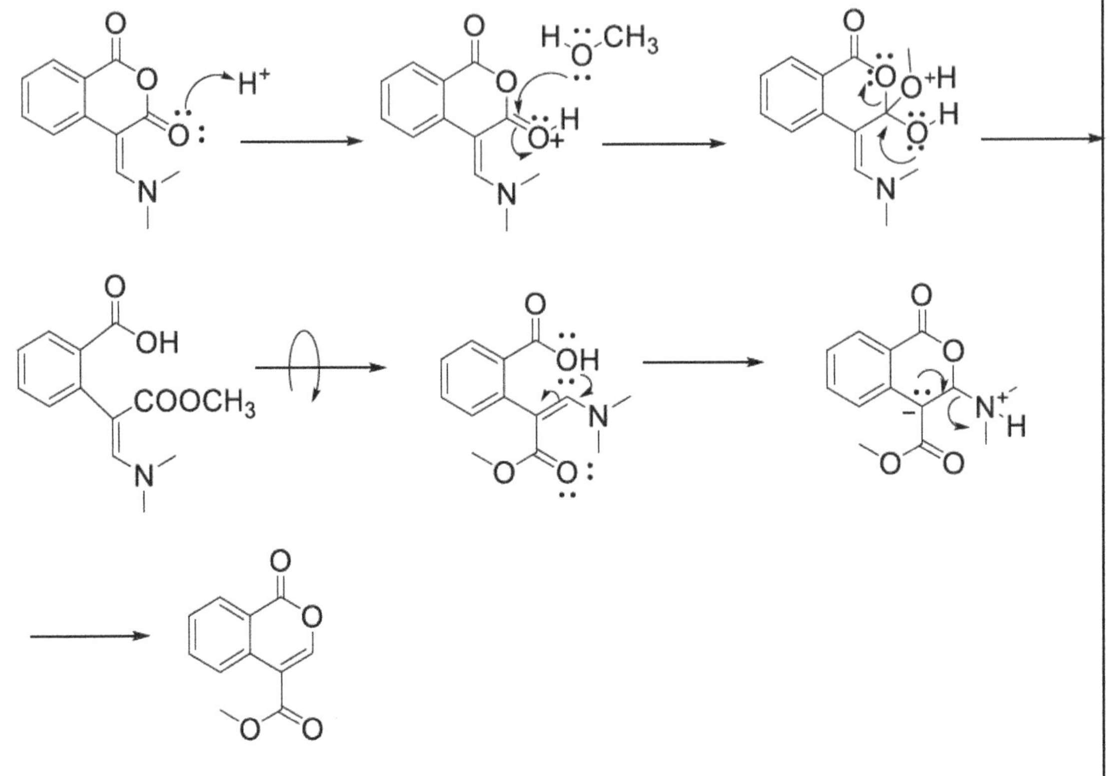

199

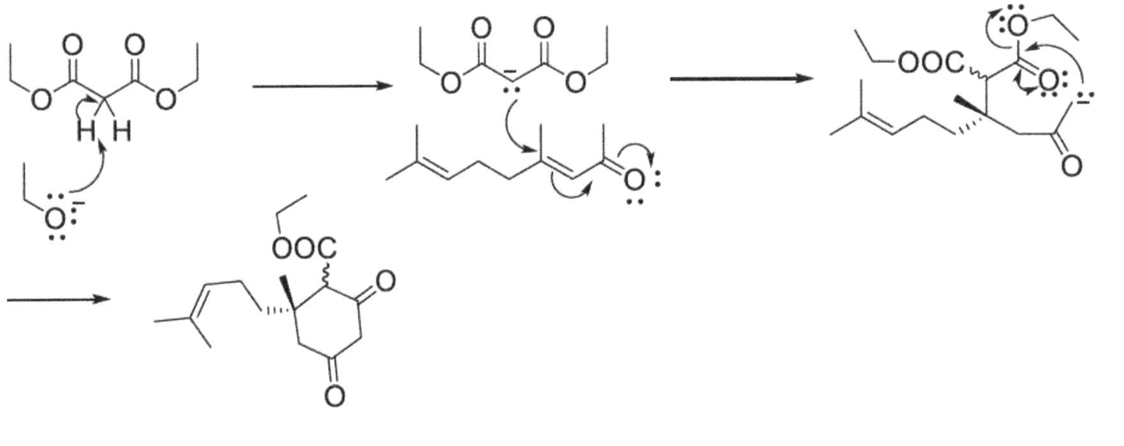

200

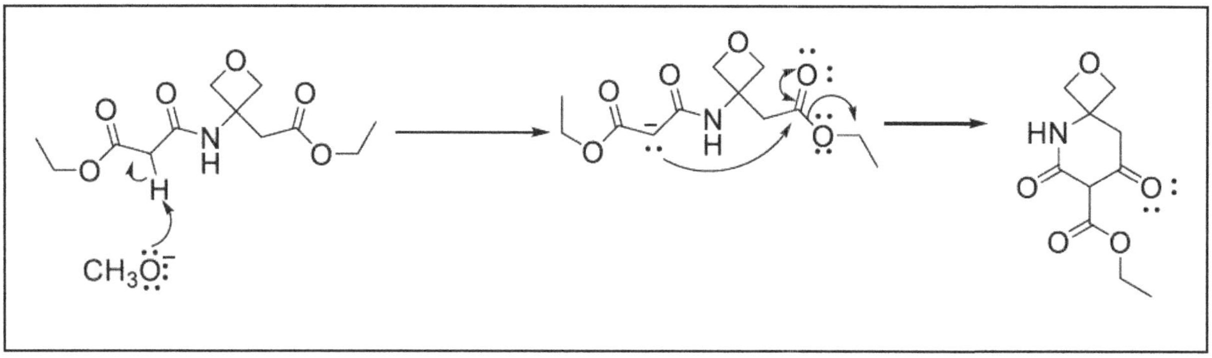
CH3O

PART B

Questions 201-490

201	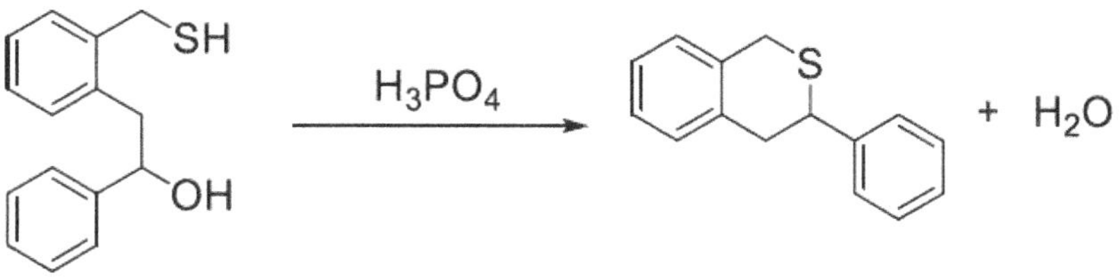SH OH H3PO4 S + H2O
202	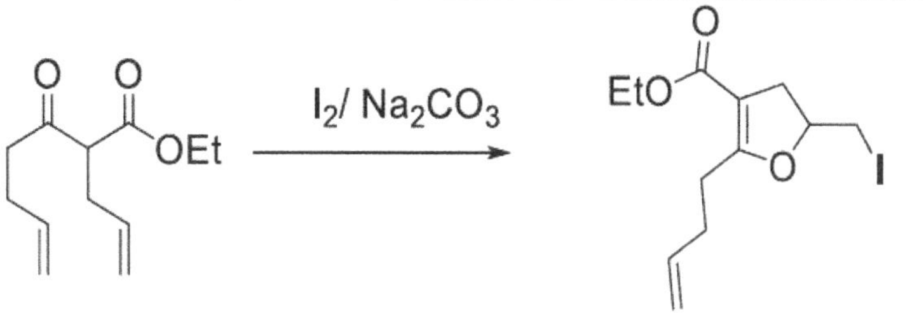O O OEt I2/ Na2CO3 O EtO O I
203	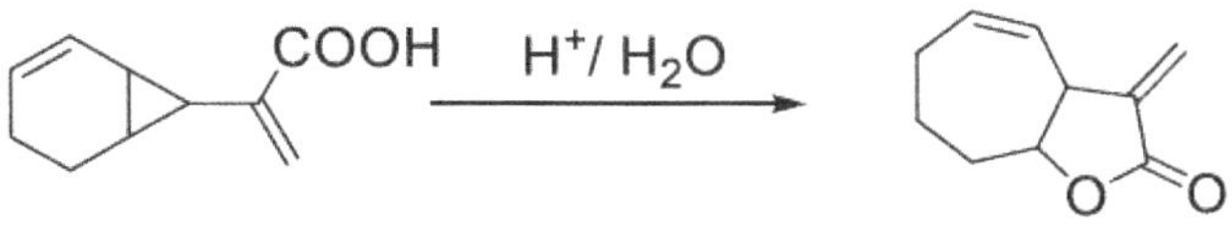COOH H+/ H2O O O
204	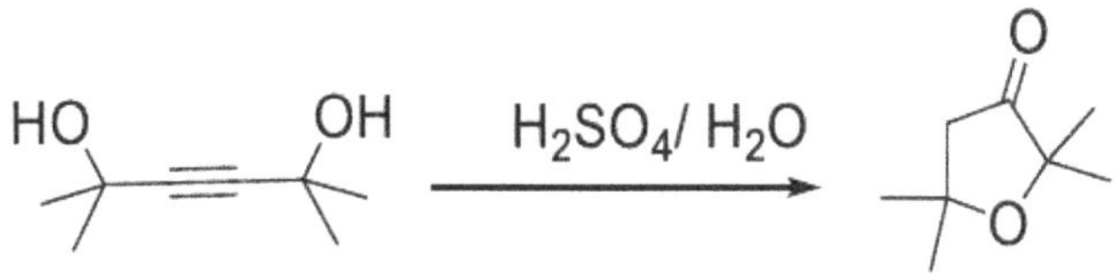HO OH H2SO4/ H2O O O
205	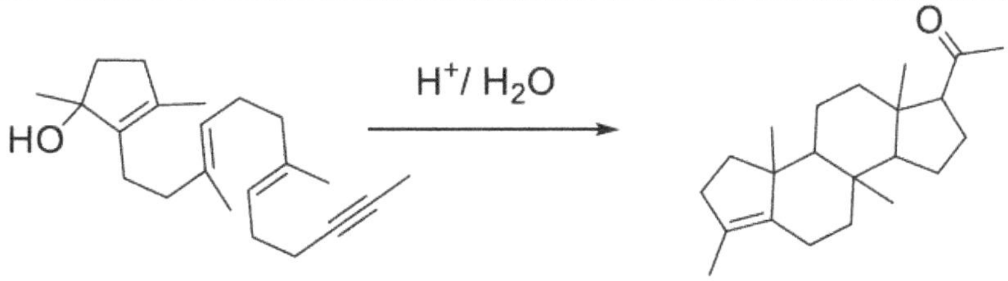HO H+/ H2O O
206	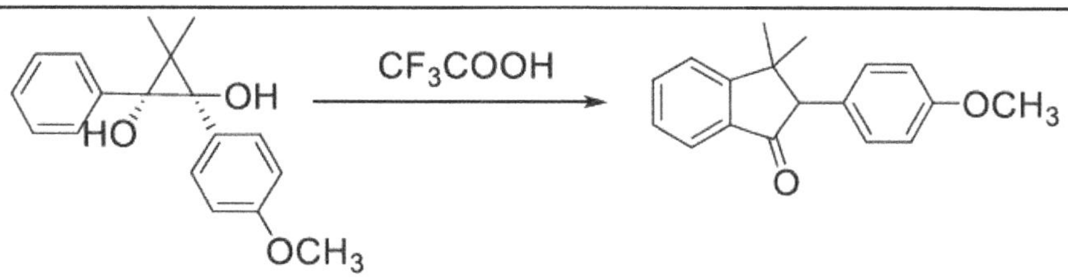OH HO OCH3 CF3COOH OCH3 O

207	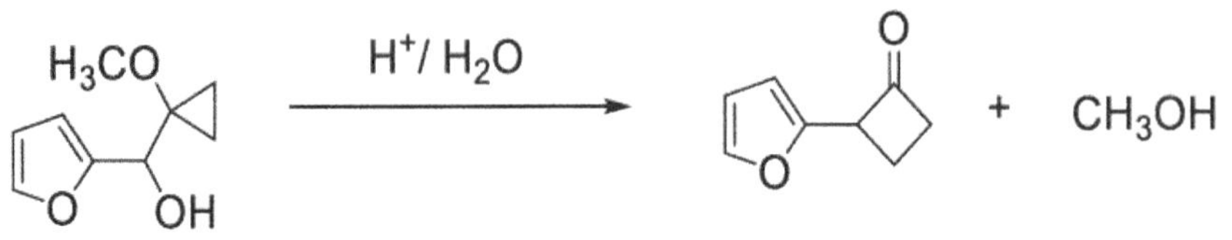
208	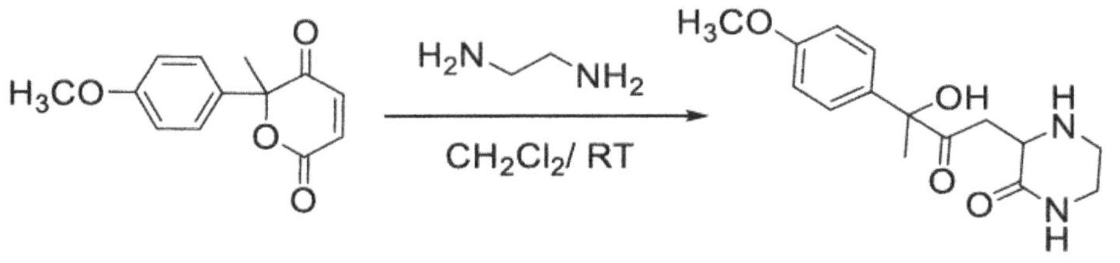
209	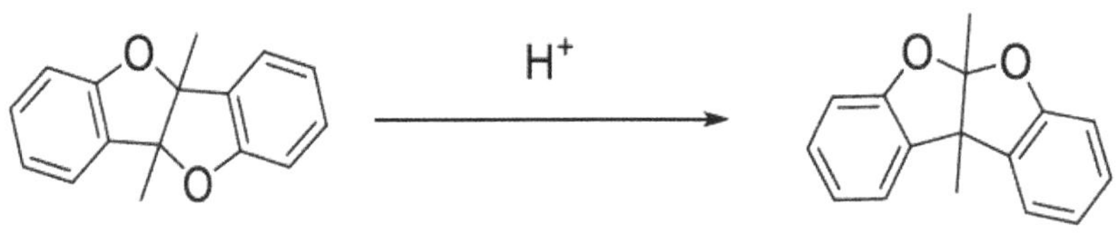
210	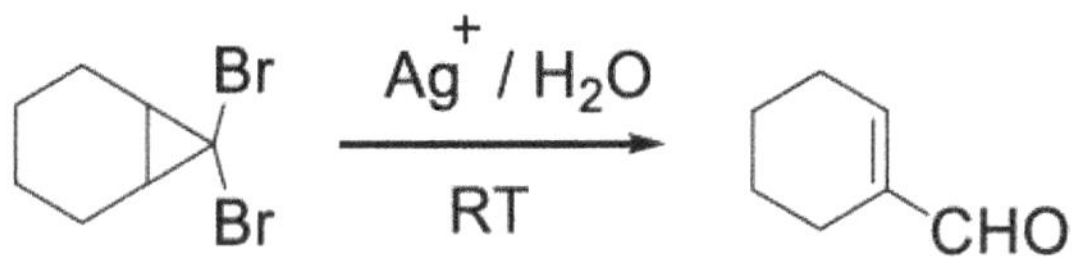
211	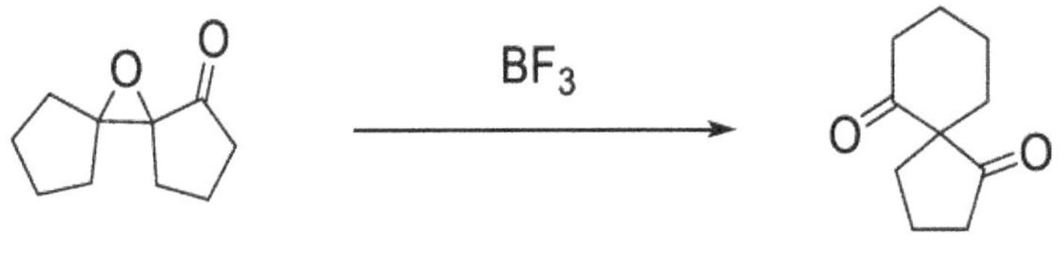
212	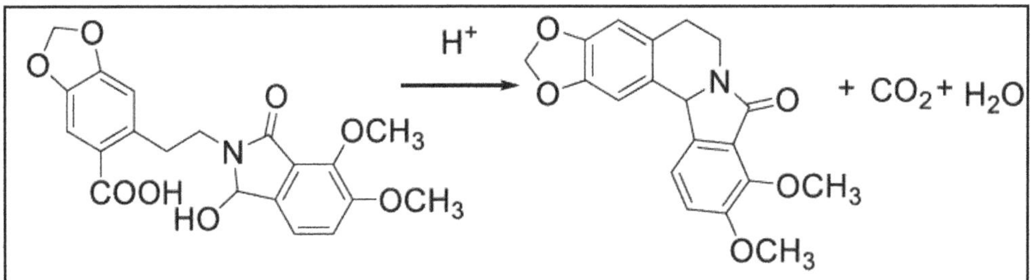

213	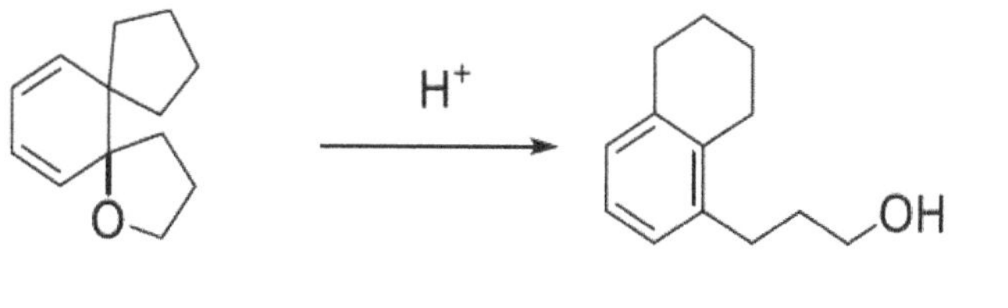H+ OH
214	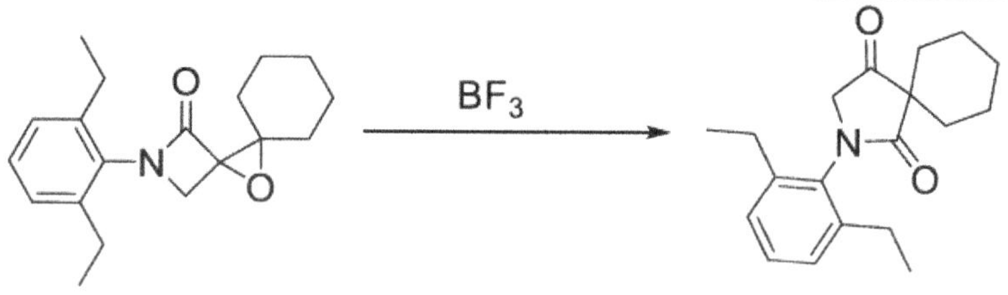O N O BF3 O N O
215	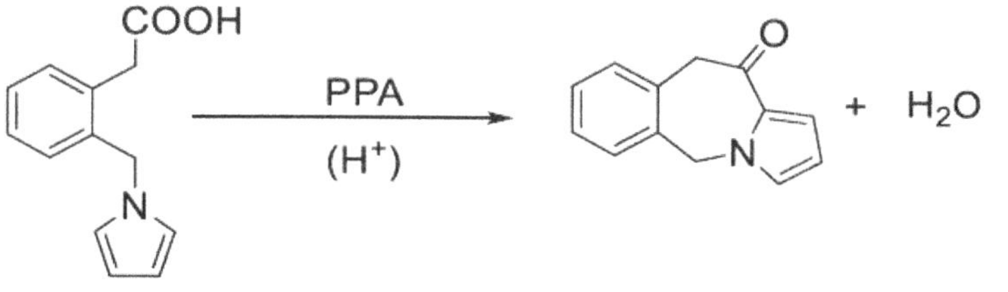COOH N PPA (H+) O N + H2O
216	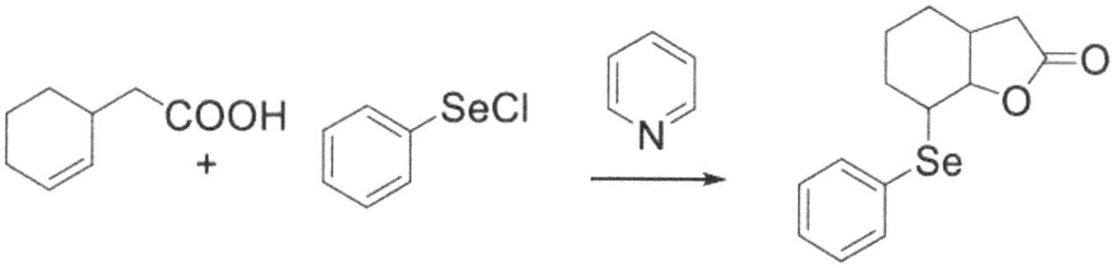COOH + SeCl N O O Se
217	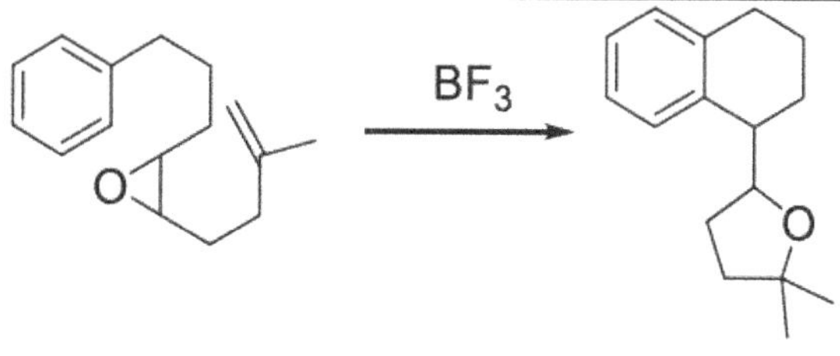O BF3 O
218	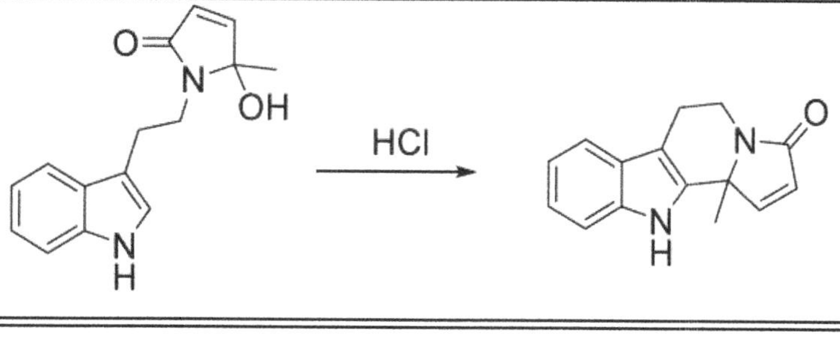O N OH N H HCl N O N H

219	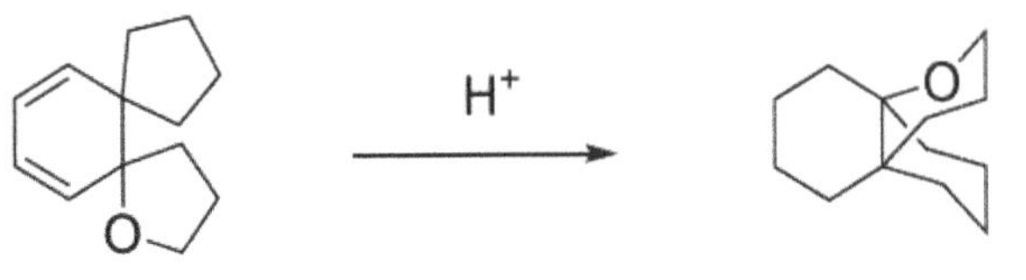H+ O
220	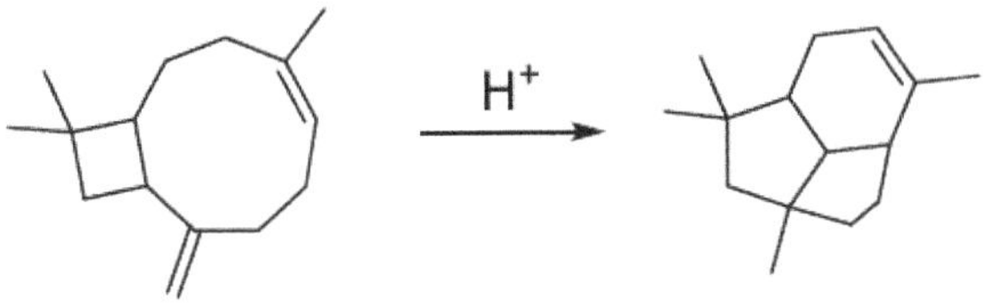H+
221	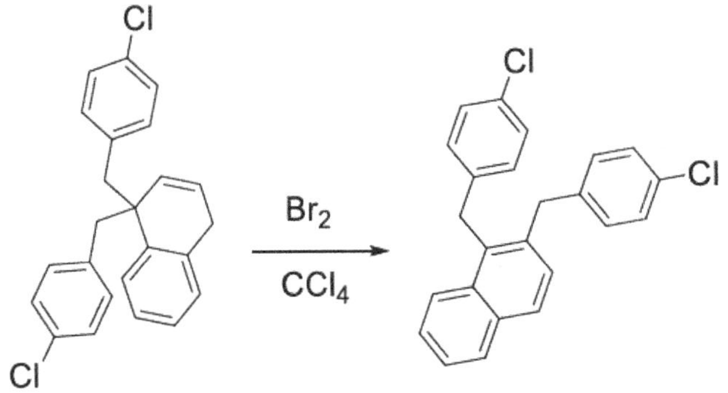Cl Br2 CCl4
222	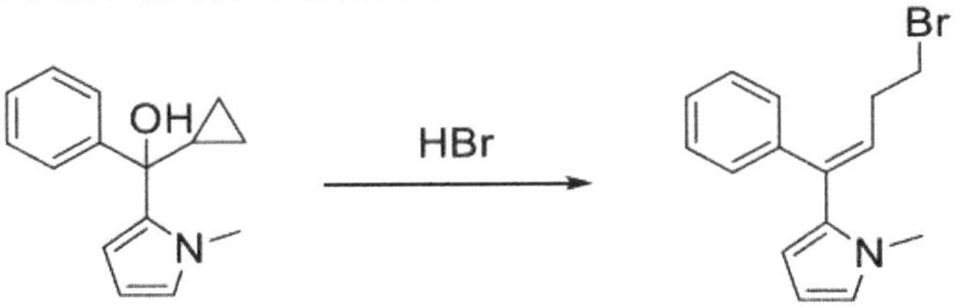OH N HBr Br
223	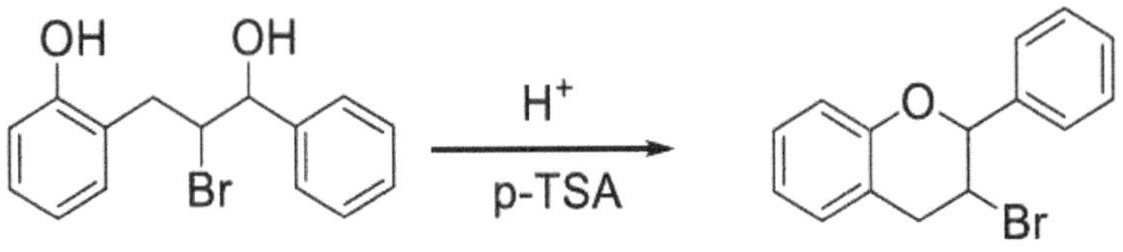OH Br H+ p-TSA O
224	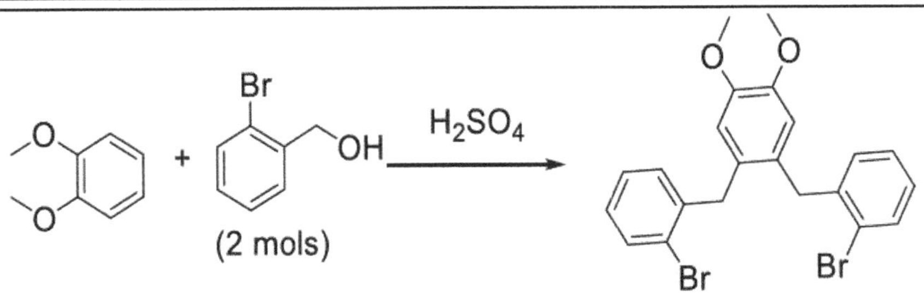O + Br OH (2 mols) H2SO4

225	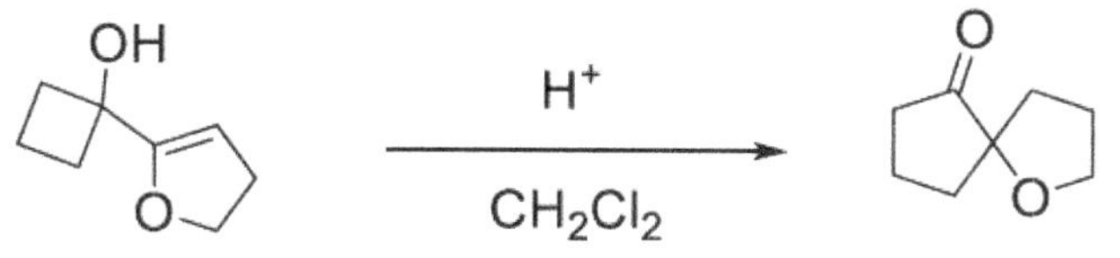OH H+ CH2Cl2 O O O
226	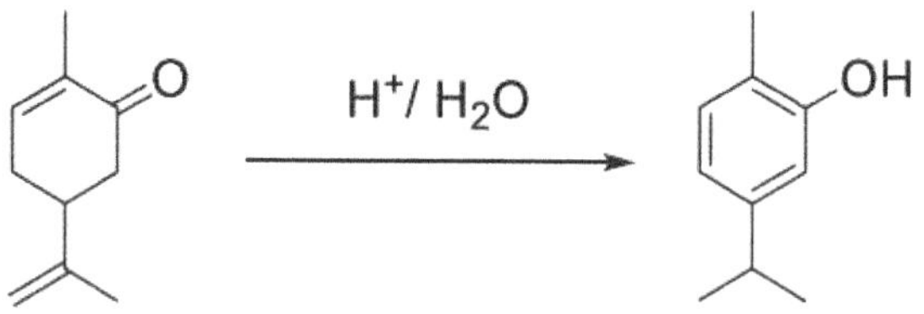O H+/ H2O OH
227	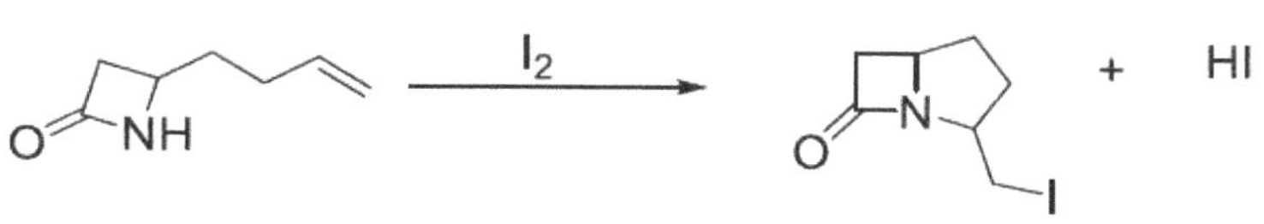O NH I2 O N I + HI
228	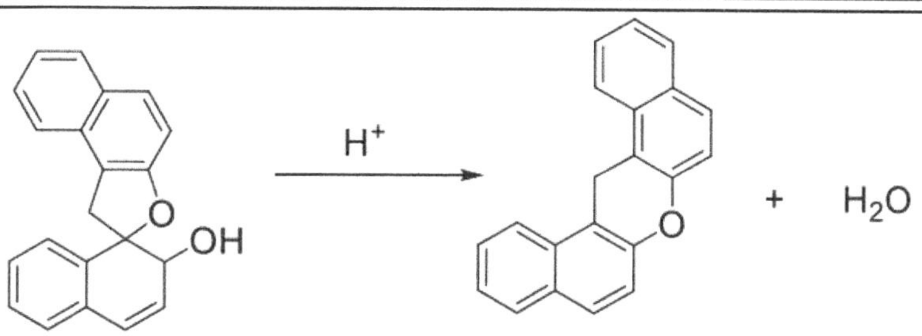O OH H+ O + H2O
229	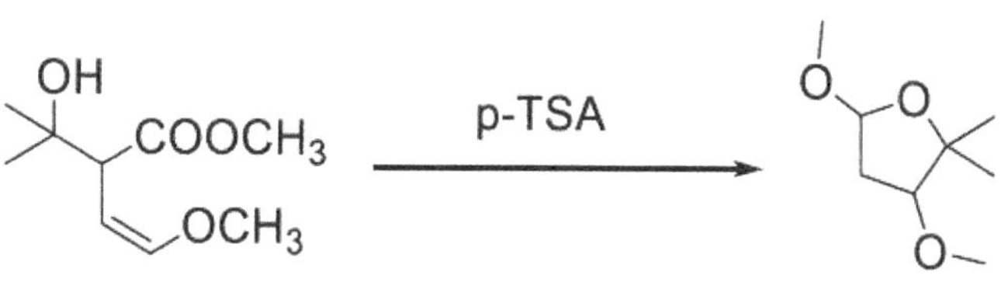OH COOCH3 OCH3 p-TSA O O O
230	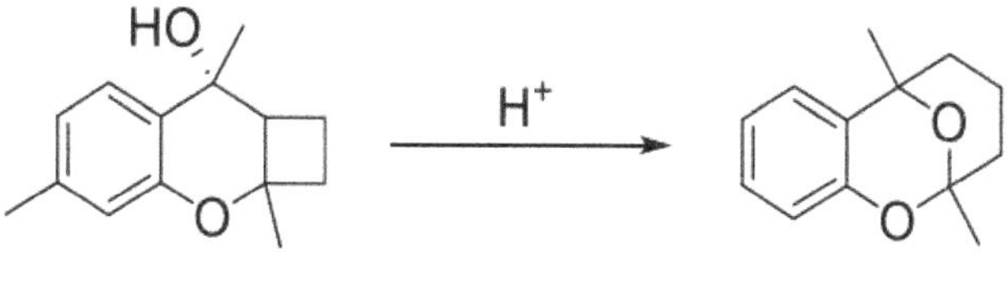HO O H+ O O

231	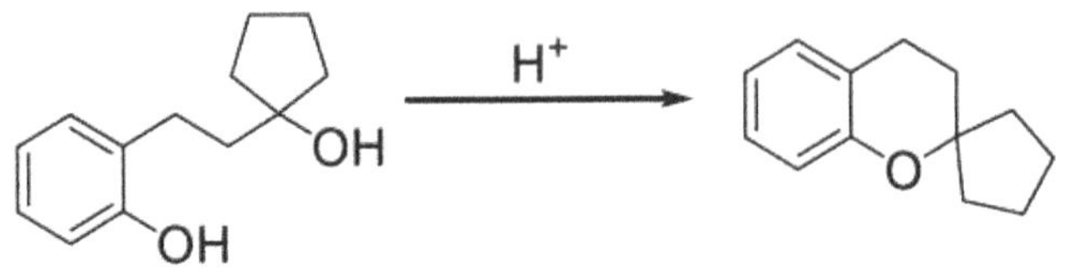OH OH H+ O
232	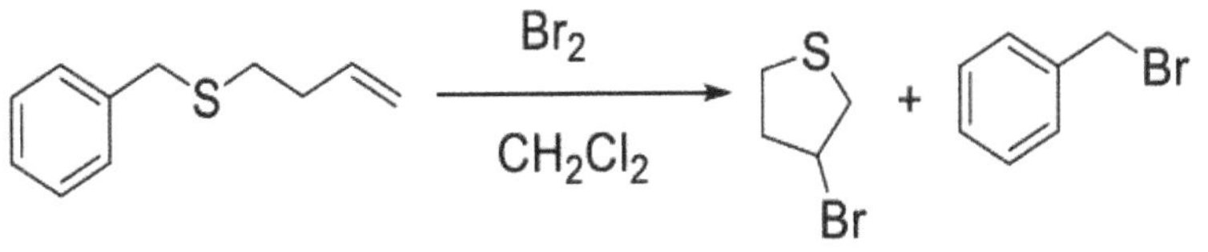S Br2 CH2Cl2 S Br + Br
233	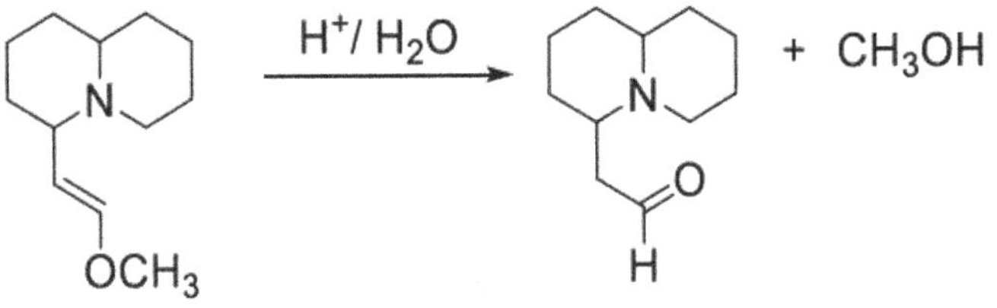N OCH3 H+/ H2O N O H + CH3OH
234	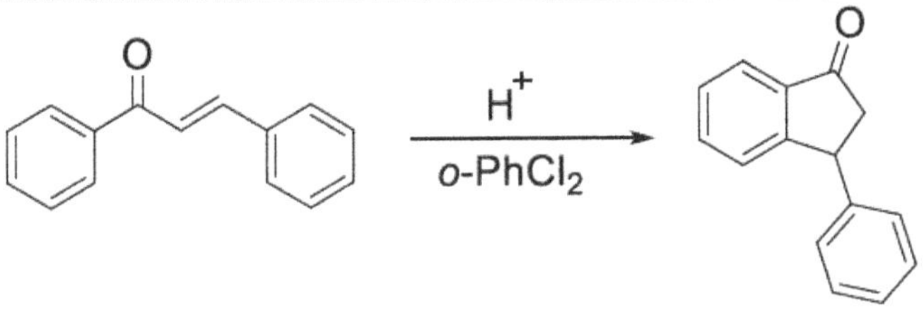O H+ o-PhCl2 O
235	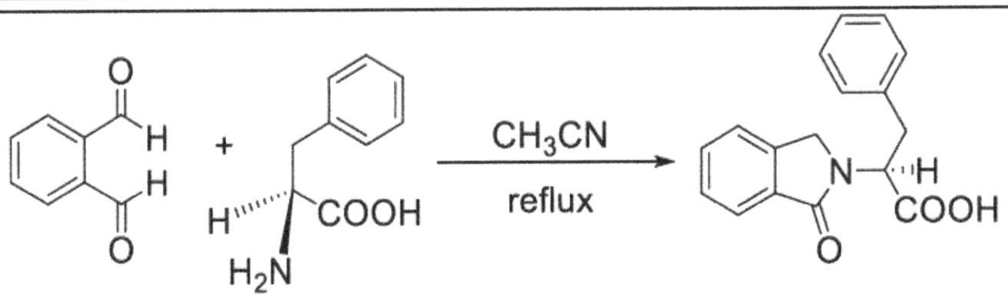O H H O + H COOH H2N CH3CN reflux N H COOH O
236	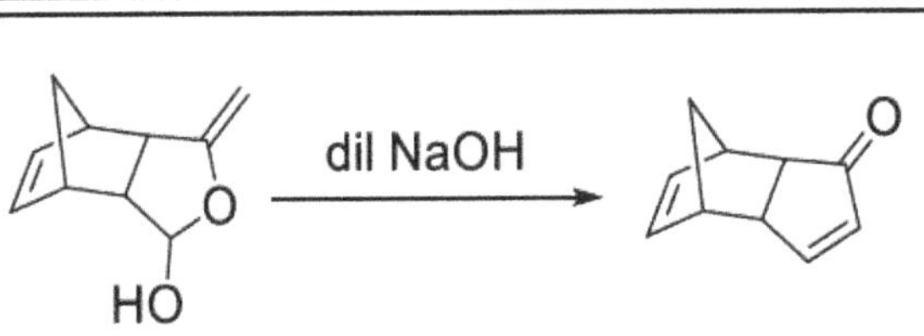O HO dil NaOH O

237

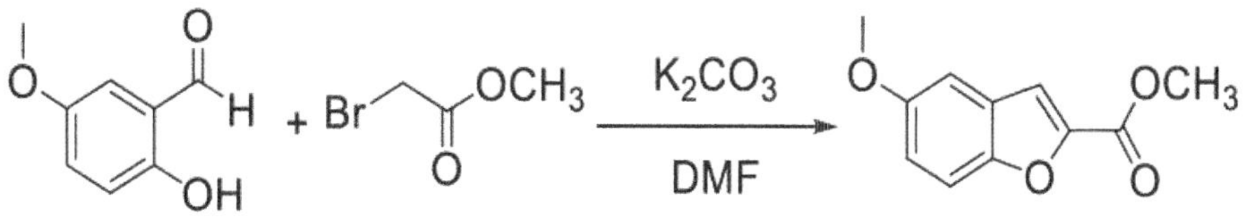

OCH3
K2CO3
DMF
OH
OCH3

238

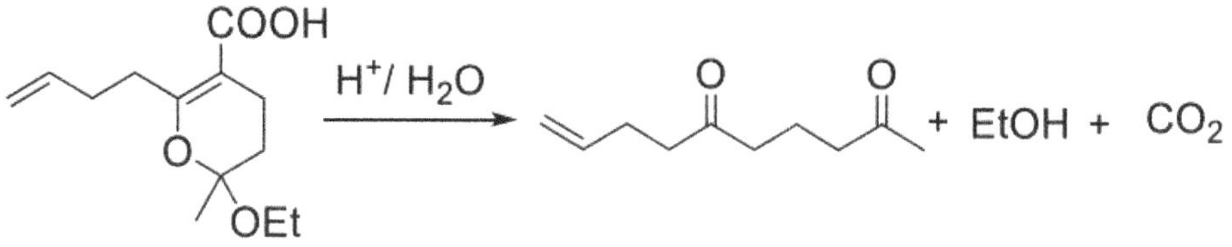

COOH
H+/ H2O
OEt
+ EtOH + CO2

239

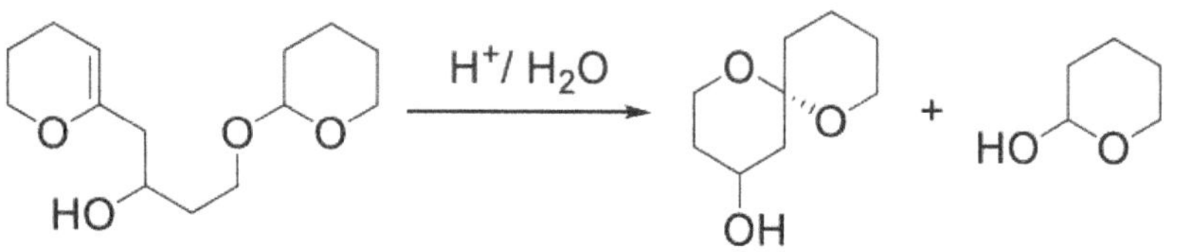

H+/ H2O
HO
OH

240

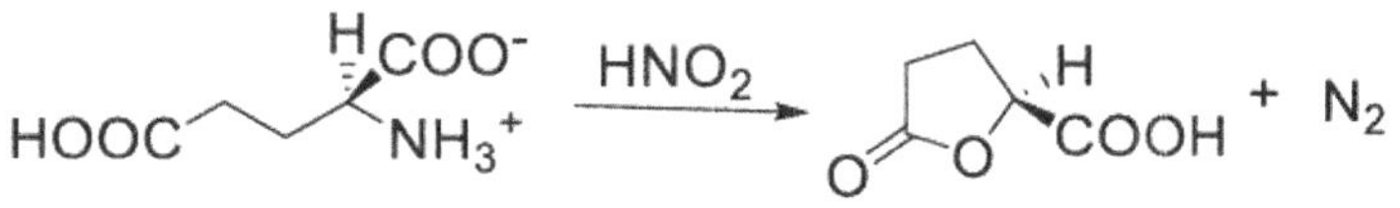

HNO2
HOOC
COO-
NH3+
COOH
+ N2

241

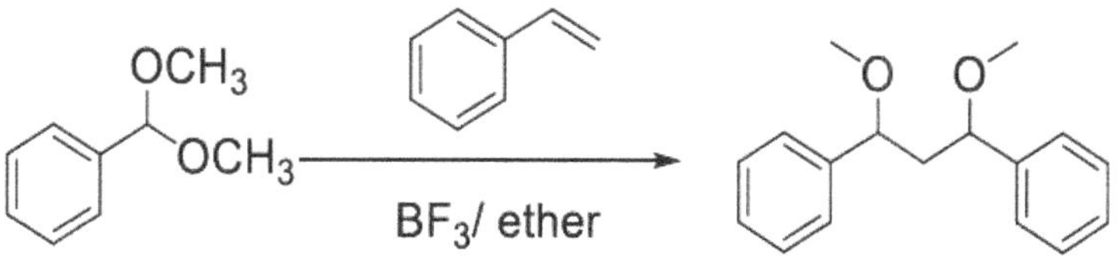

OCH3
OCH3
BF3/ ether

242

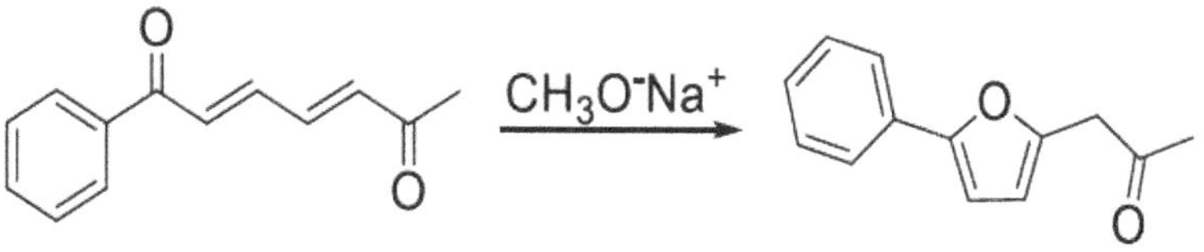

CH3O-Na+

243	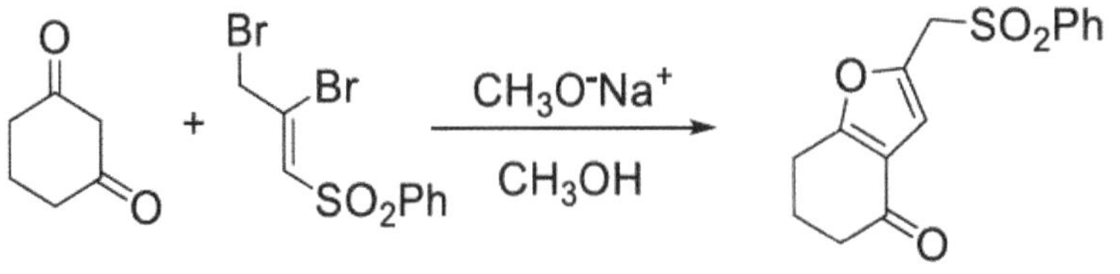 Br Br SO2Ph CH3O-Na+ CH3OH SO2Ph O
244	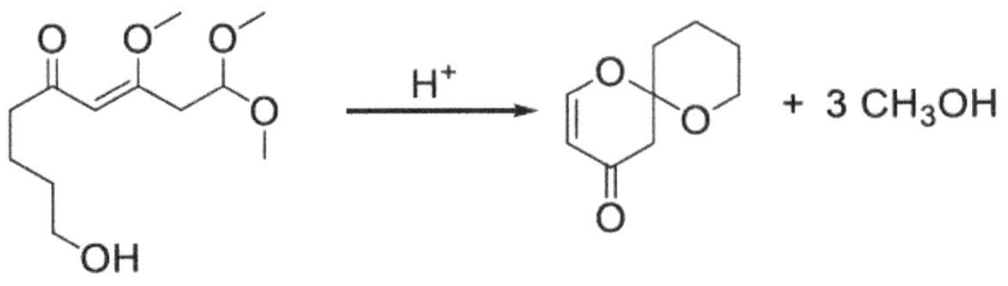 O O O O OH H+ O O O + 3 CH3OH
245	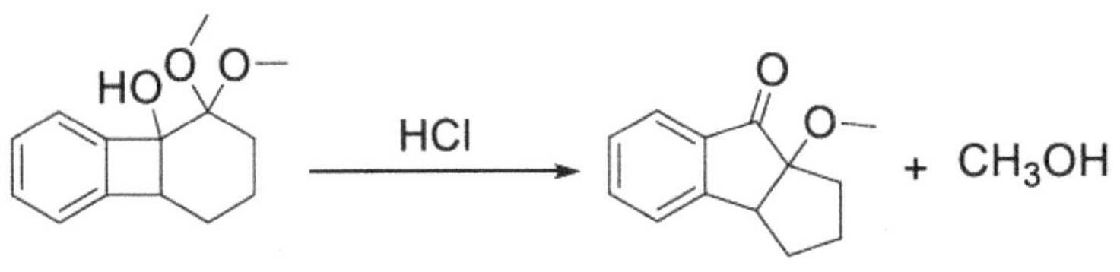 HO O O HCl O O + CH3OH
246	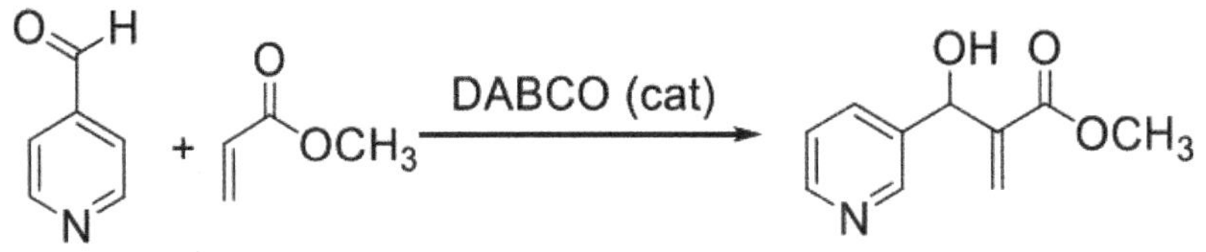 O H N + O OCH3 DABCO (cat) OH O OCH3 N
247	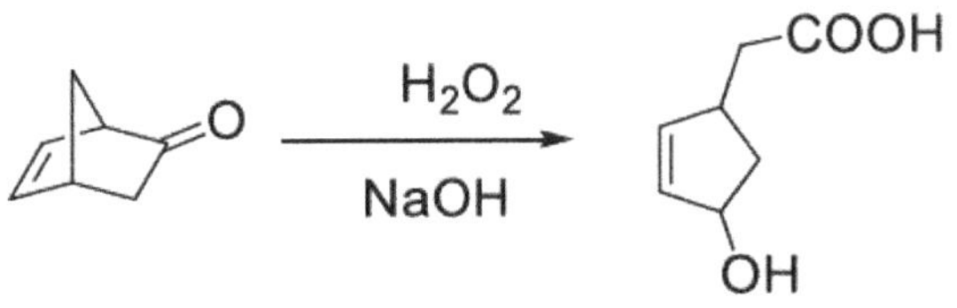 O H2O2 NaOH COOH OH
248	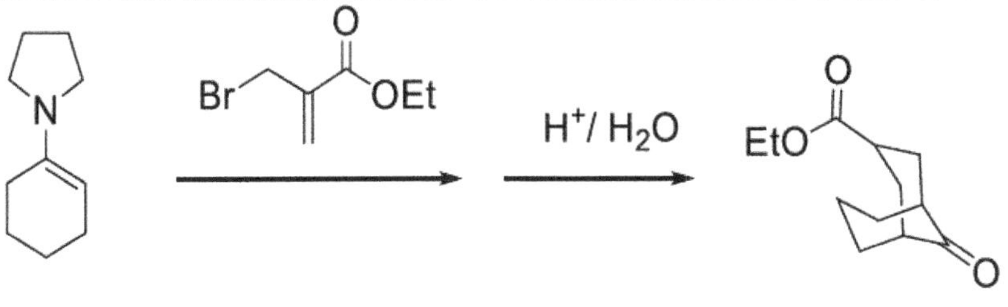 N O Br OEt H+/ H2O O EtO O

249	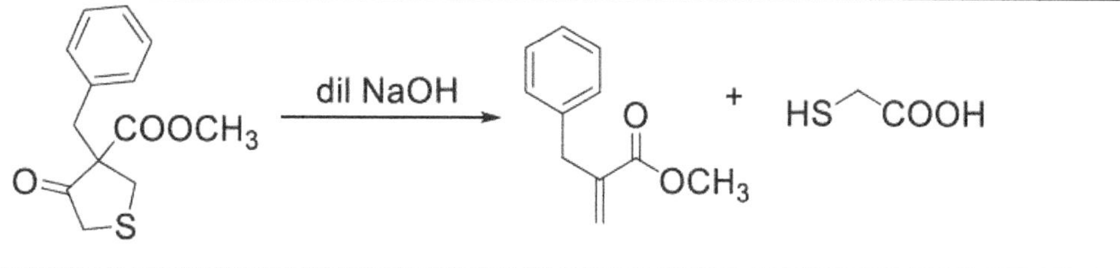dil NaOH
250	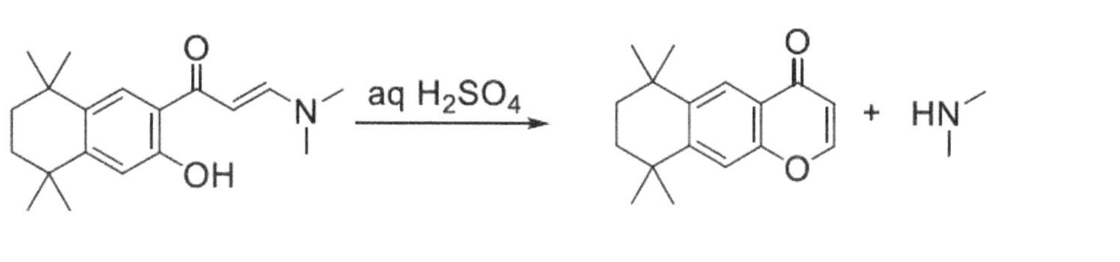aq H_2SO_4
251	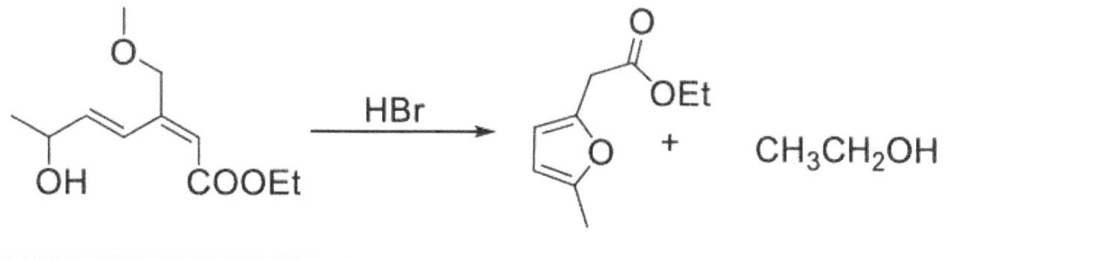HBr
252	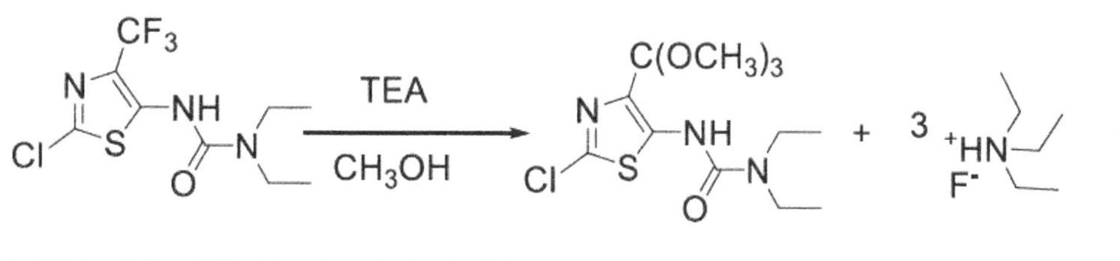TEA CH_3OH
253	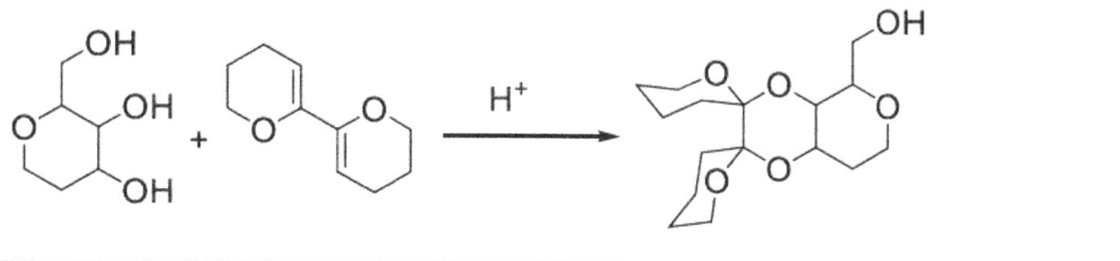H^+
254	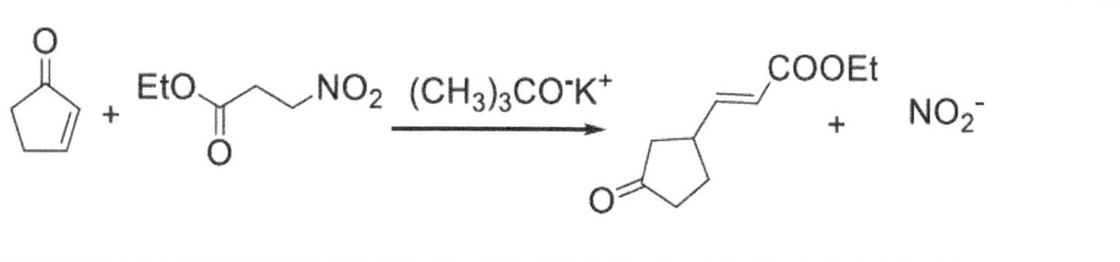$(CH_3)_3CO^-K^+$

255	The compound shown below is stable when isolated as the hydrochloride salt. However, chromatography on silica gel yields the products shown below. Write a mechanism that explains the formation of these products (Hint: silica gel tends to absorb moisture) 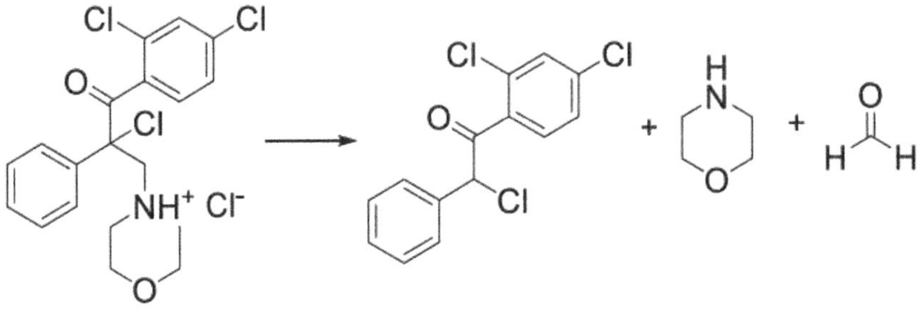
256	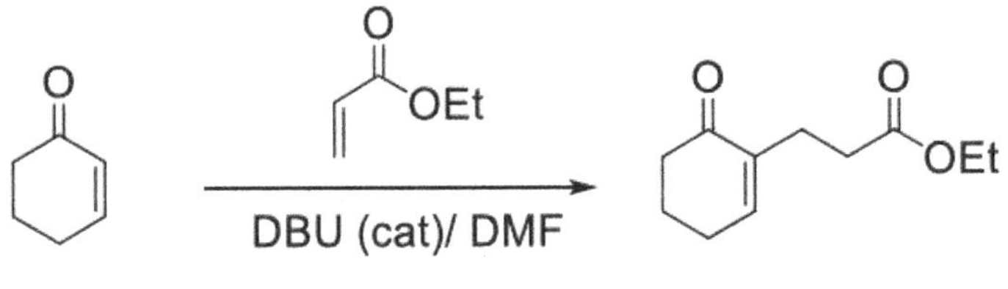
257	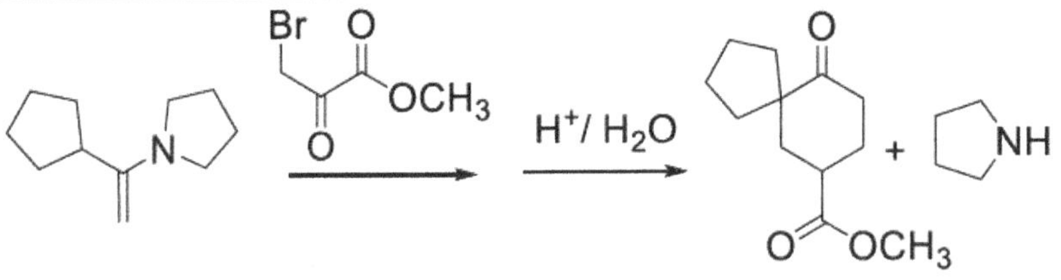
258	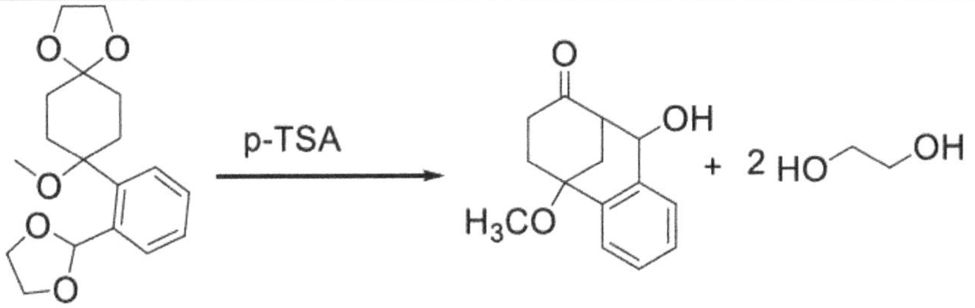
259	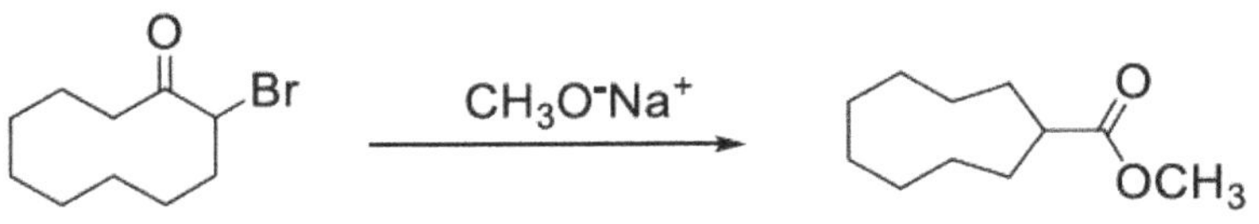

260

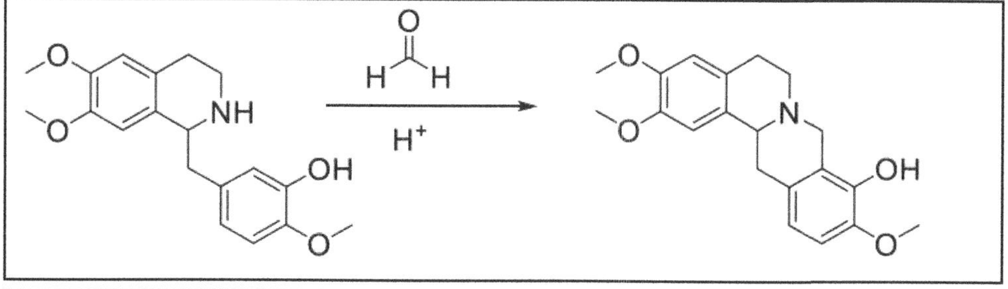

NH
H
H
H+
N
OH
OH

261

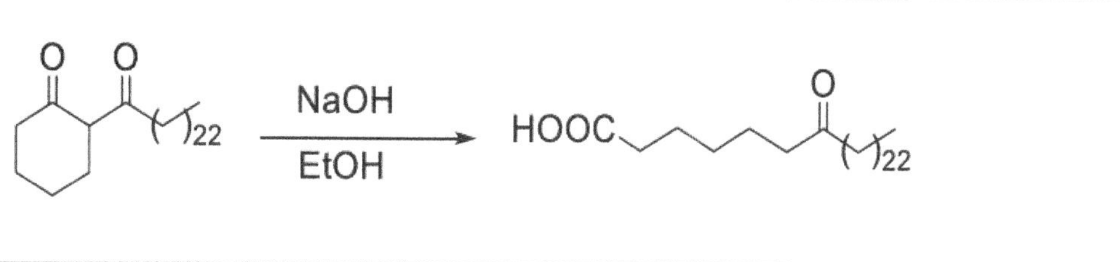

22
NaOH
EtOH
HOOC
22

262

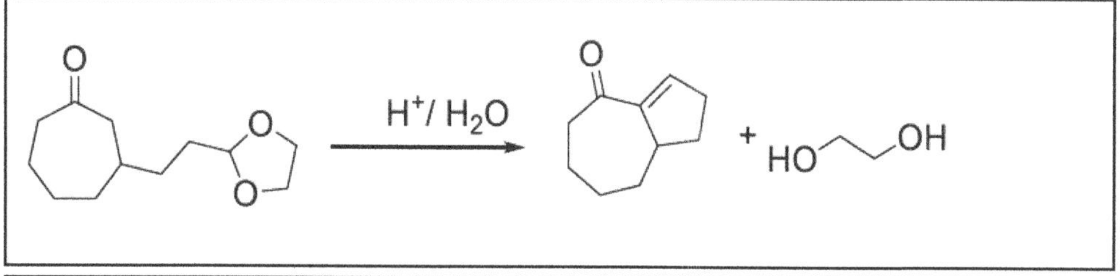

H+/ H2O
+
HO
OH

263

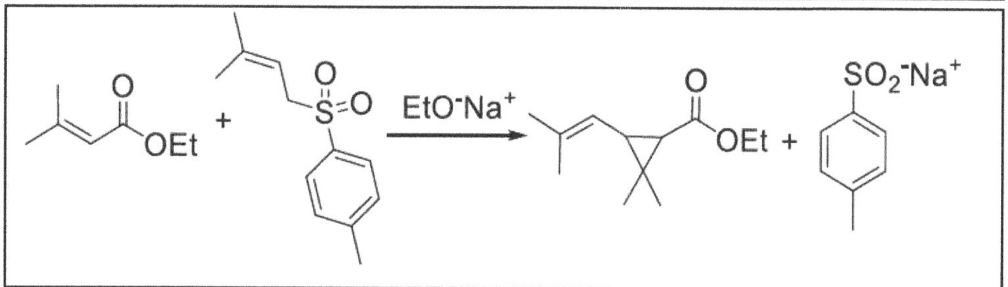

OEt
+
EtO-Na+
OEt
+
SO2-Na+

264

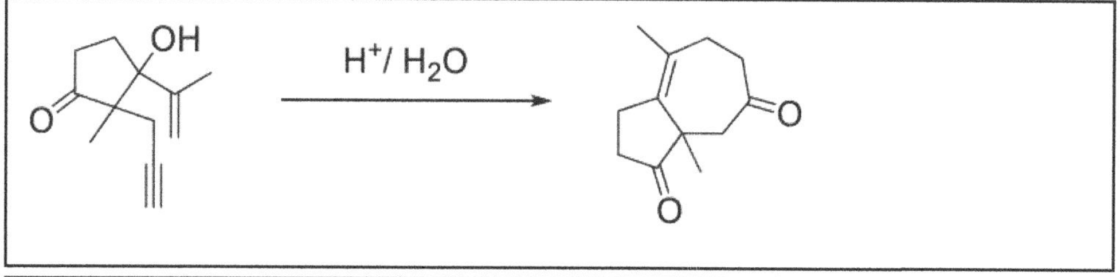

OH
H+/ H2O

265

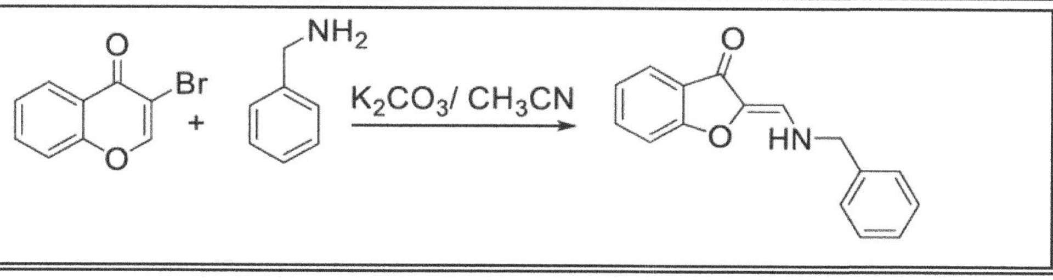

Br
+
NH2
K2CO3/ CH3CN
HN

266

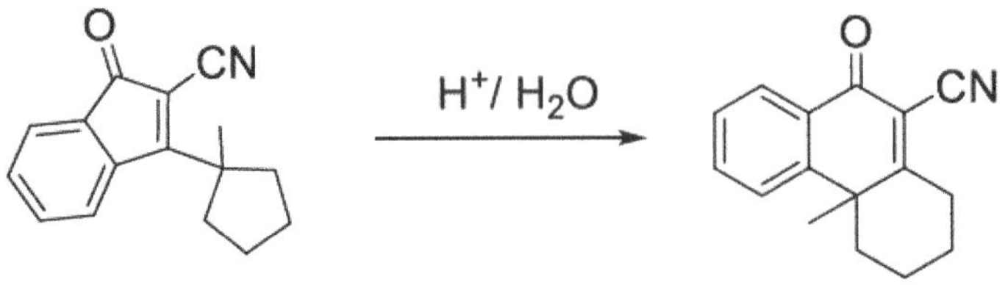

O
CN
H+/ H2O
O
CN

267

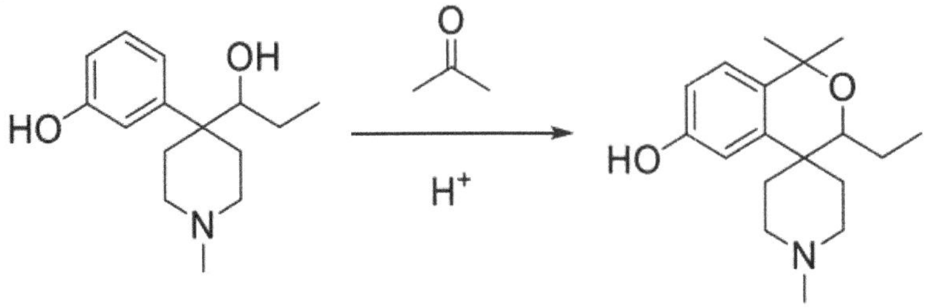

OH
HO
N
O
H+
O
HO
N

268

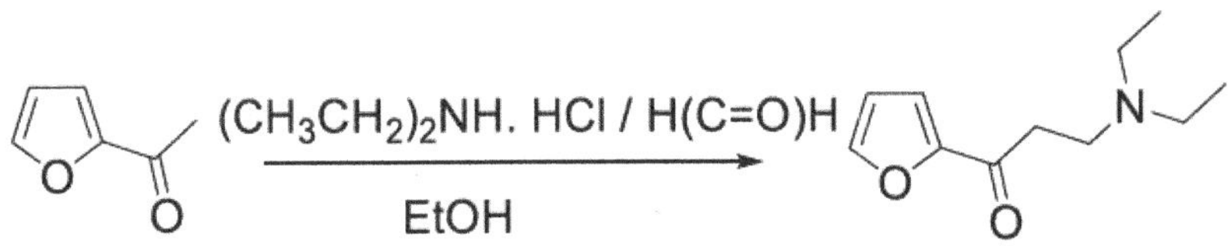

O
O
(CH3CH2)2NH. HCl / H(C=O)H
EtOH
N
O
O

269

Aldehydes can be converted into alpha-chloromethyl ketones by reacting an aldehyde with the reagent shown below, and refluxing the product formed in toluene. Write a mechanism for this transformation.

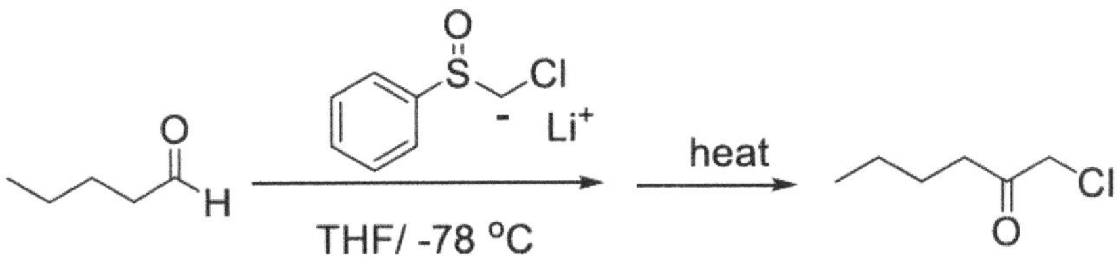

O
H
O
S
Cl
- Li+
THF/ -78 °C
heat
Cl
O

270

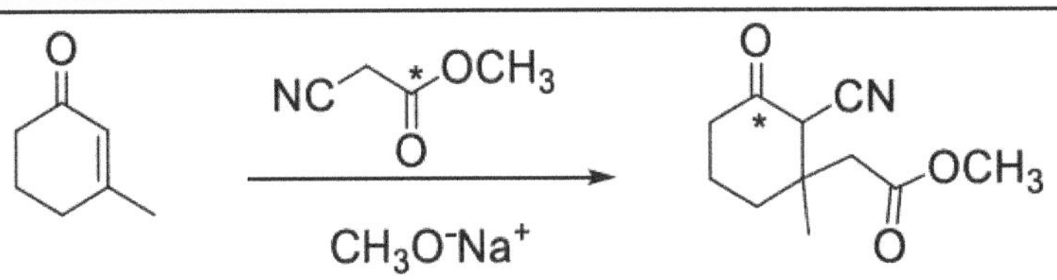

O
NC
*
OCH3
O
CH3O-Na+
O
*
CN
OCH3
O

(mechanism must account for the location of (*) carbon)

271	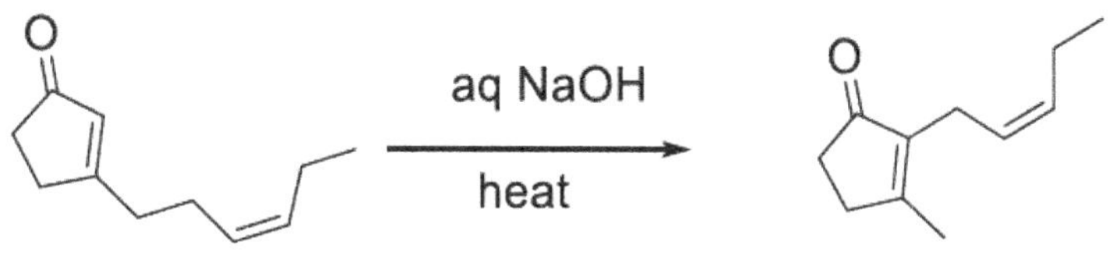aq NaOH heat
272	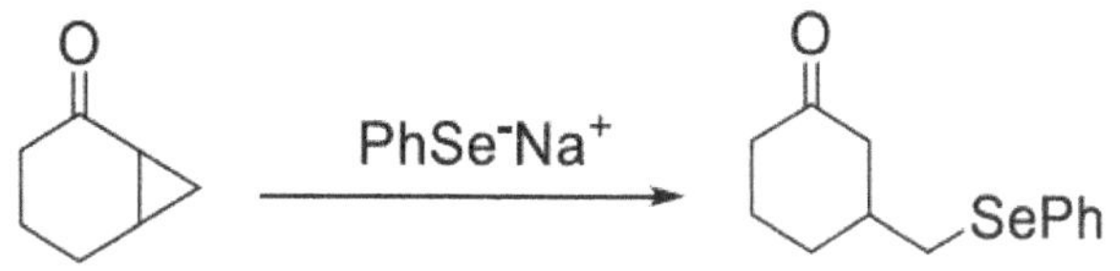PhSe⁻Na⁺ SePh
273	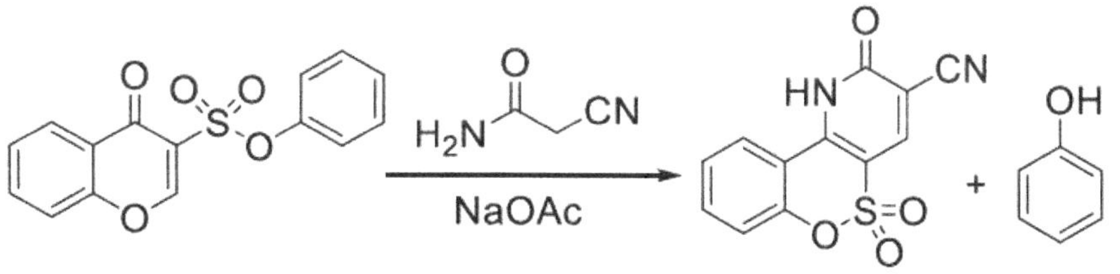NaOAc CN HN OH
274	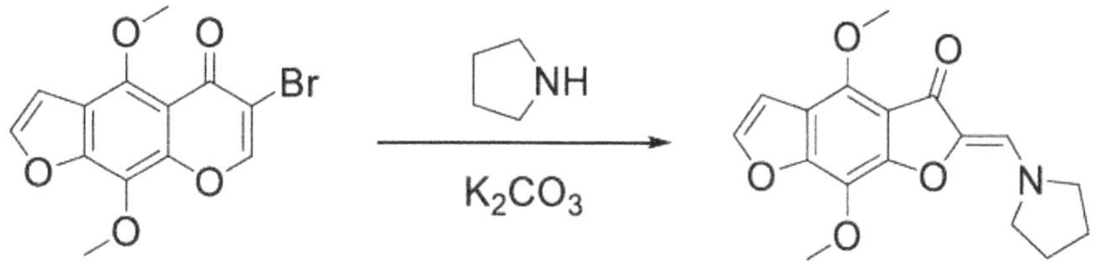Br NH K2CO3 N
275	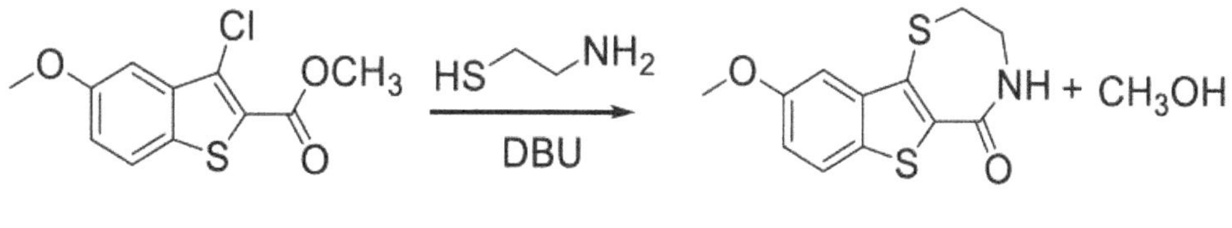Cl OCH3 HS NH2 DBU NH + CH3OH
276	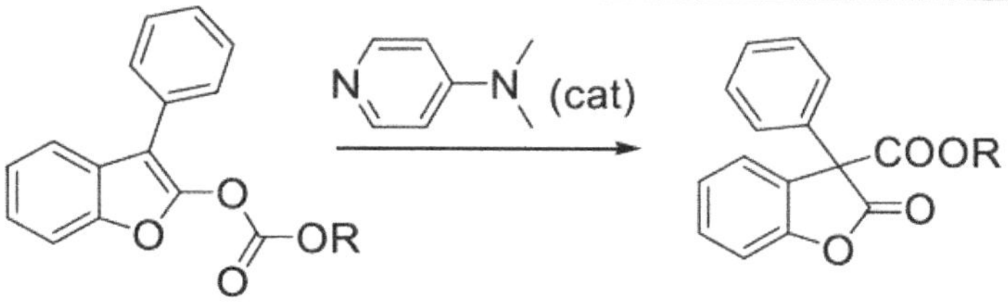OR (cat) COOR

277	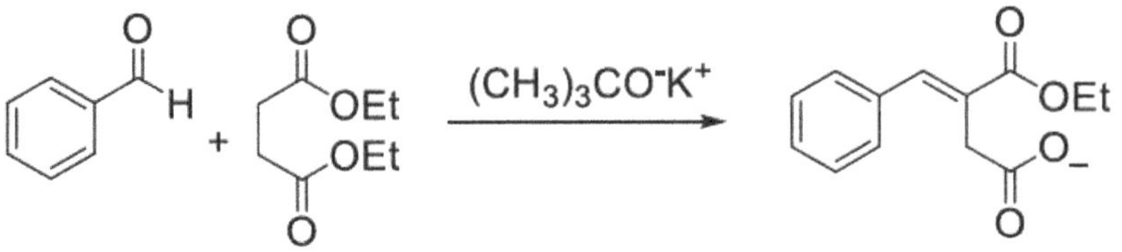(CH3)3CO-K+
278	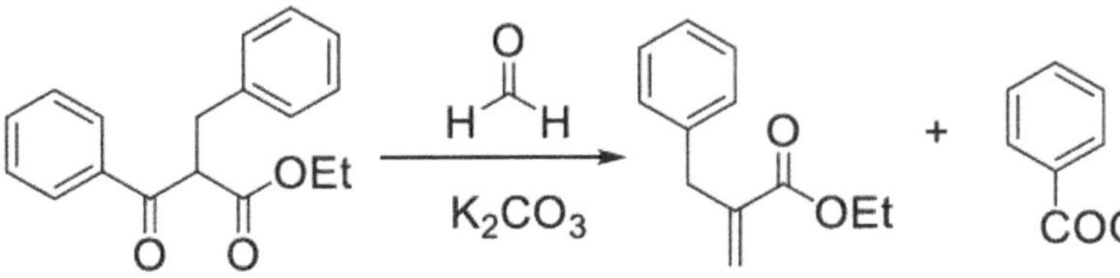K2CO3
279	Primary and secondary alcohols can be readily oxidized by dimethyl sulfoxide/phosphorus pentoxide/triethylamine under mild conditions (an example is shown below). Use your knowledge of Lewis structures and Lewis acid/base reactions to write a reasonable mechanism for this reaction. 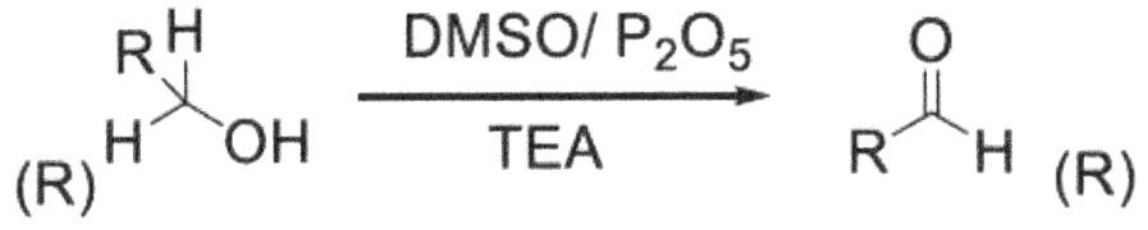DMSO/ P2O5 TEA 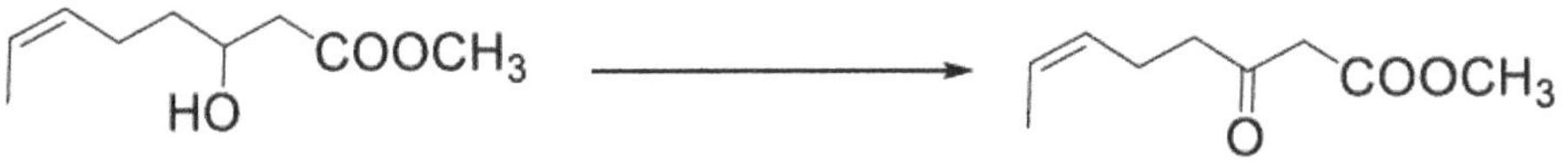
280	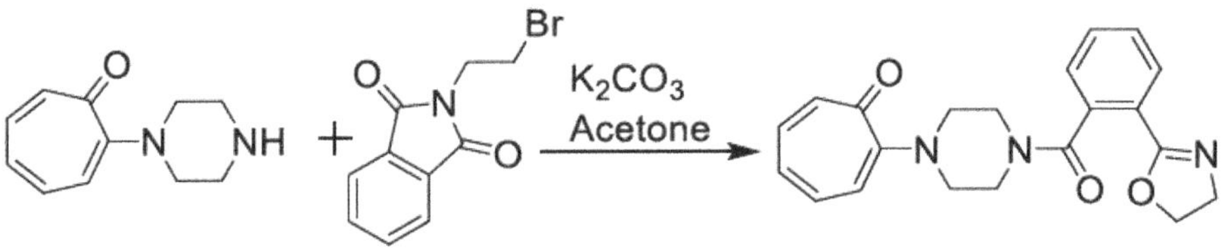K2CO3 Acetone
281	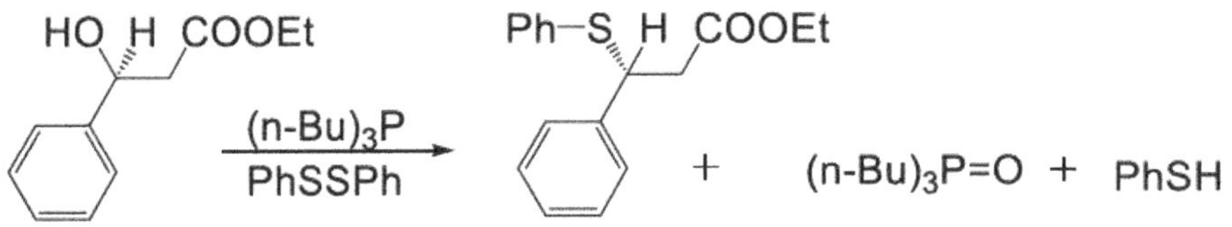(n-Bu)3P PhSSPh (n-Bu)3P=O + PhSH

282	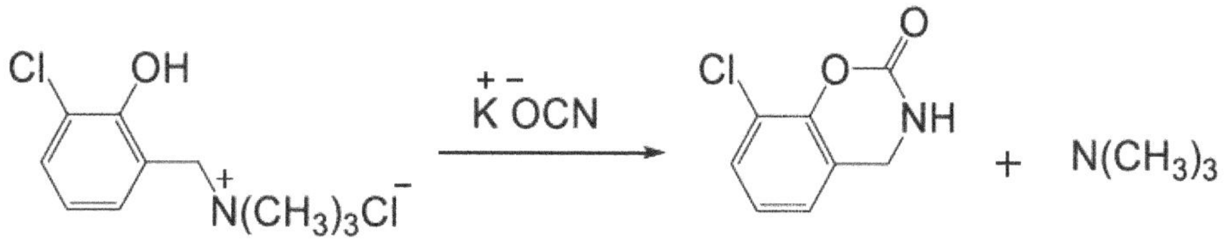 K OCN
283	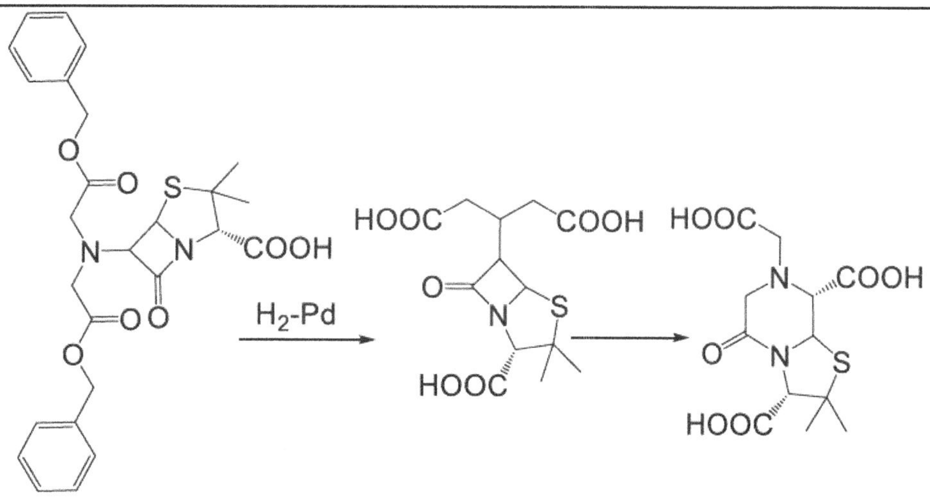 H2-Pd
284	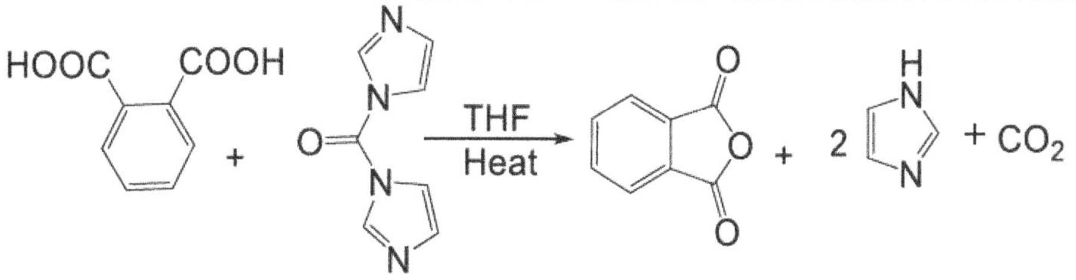 THF Heat
285	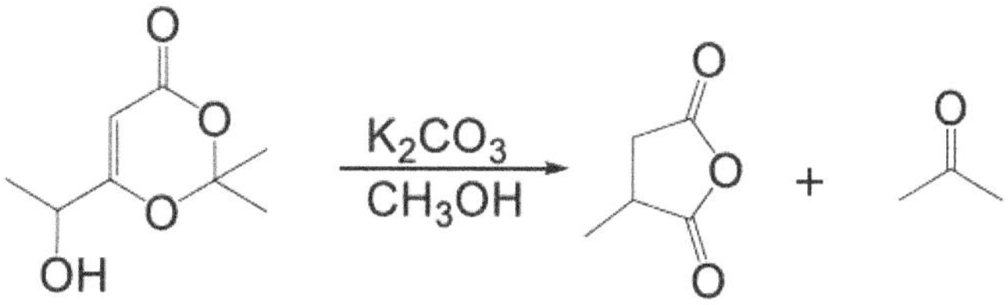 K2CO3 CH3OH
286	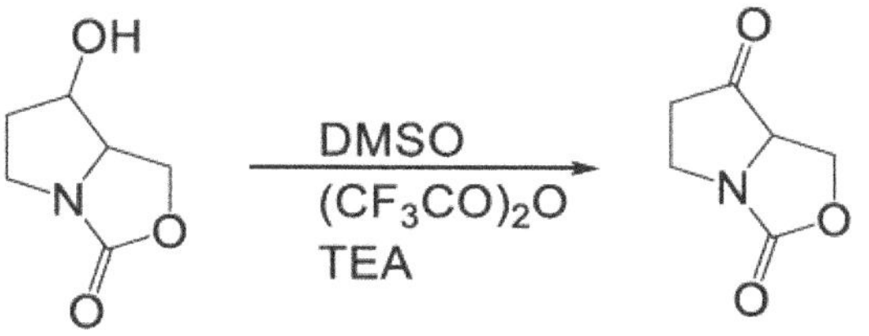 DMSO (CF3CO)2O TEA

287

Alcohols can be readily converted to alkyl chlorides or bromides under mild conditions using PPh_3 /CCl_4 or PPh_3/$CHBr_3$, respectively (a specific example is shown below). Write a mechanism for this transformation

$$R\text{-}CH_2\text{-}OH \xrightarrow[CX_4]{PPh_3} R\text{-}CH_2\text{-}X + O{=}PPh_3 \text{ (X=Cl. Br)}$$

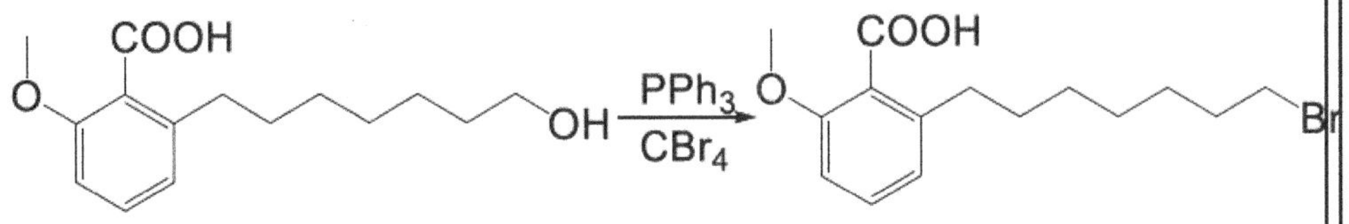
COOH
O
OH
PPh3
CBr4
COOH
O
Br

288

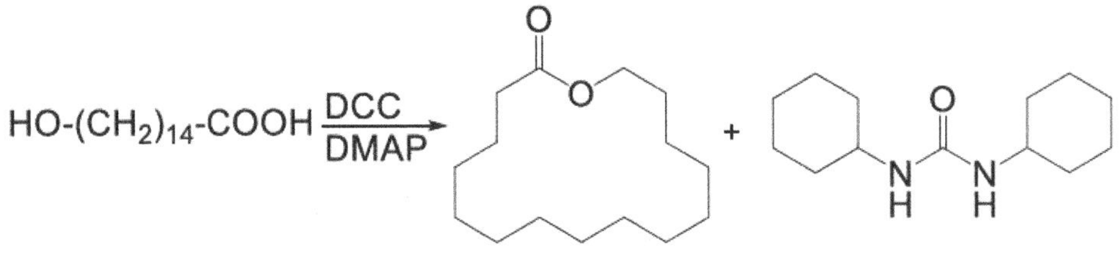
HO-(CH2)14-COOH
DCC
DMAP
O
O
+
O
N
H
N
H

289

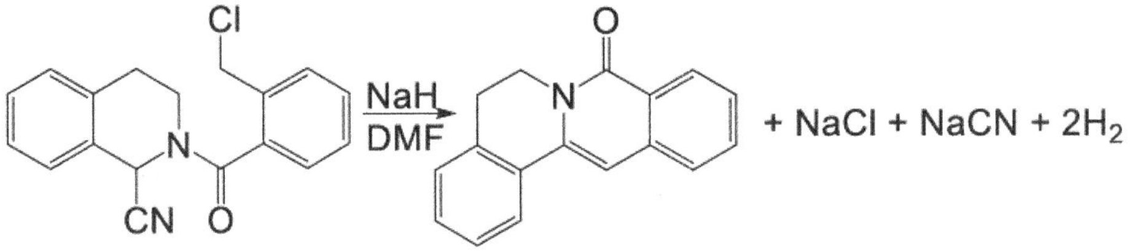
Cl
N
CN
O
NaH
DMF
O
N
+ NaCl + NaCN + 2H2

290

Primary amides can be readily dehydrated under mild conditions using chloromethylene iminium salts called *Vilsmeier reagents* to form the corresponding nitriles. When dimethyl formamide (DMF) is treated with oxalyl chloride it forms a *Vilsmeier* reagent which is then reacted with primary amines in the presence of pyridine to yield the corresponding nitrile as shown below. Write a mechanism for the formation of the Vilsmeier reagent and a second mechanism for the formation of the nitrile.

	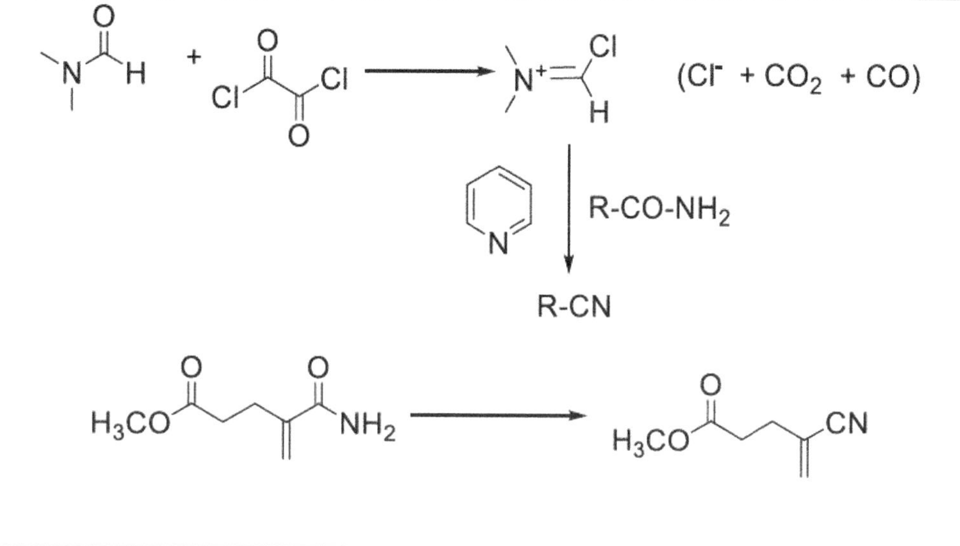
291	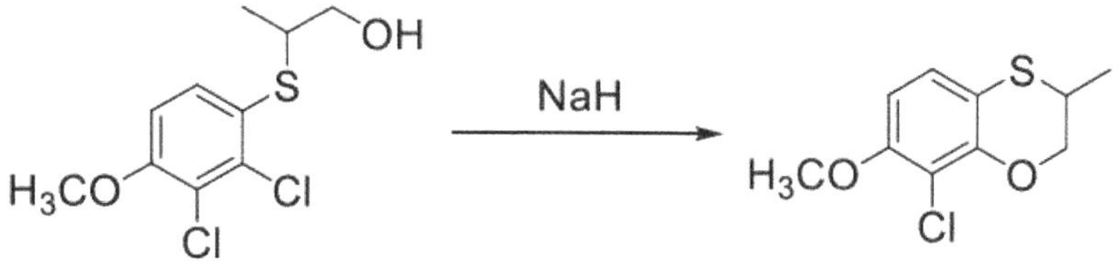
292	
293	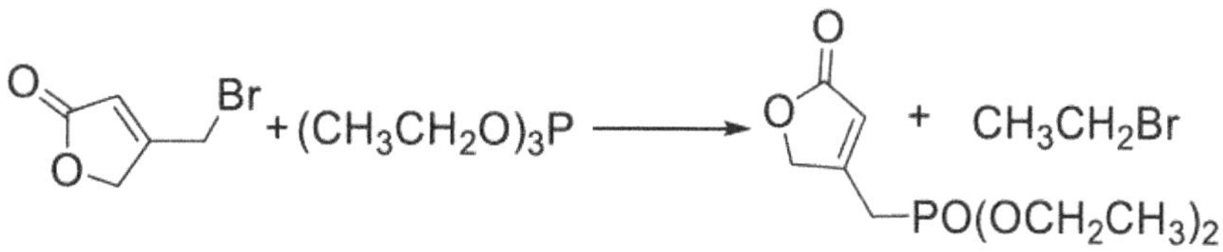
294	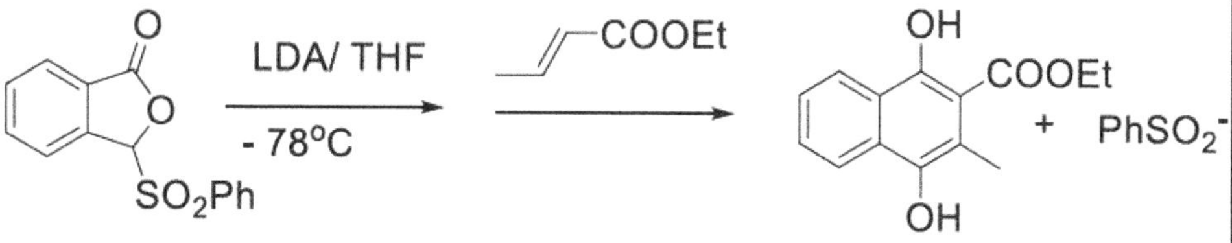

295

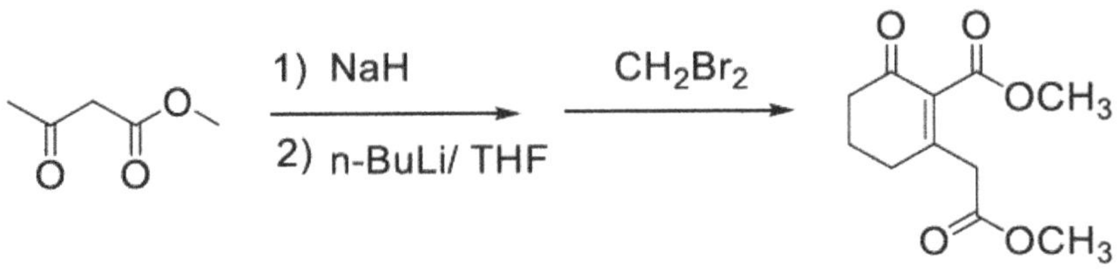
1) NaH
2) n-BuLi/ THF
CH2Br2
OCH3
OCH3

296

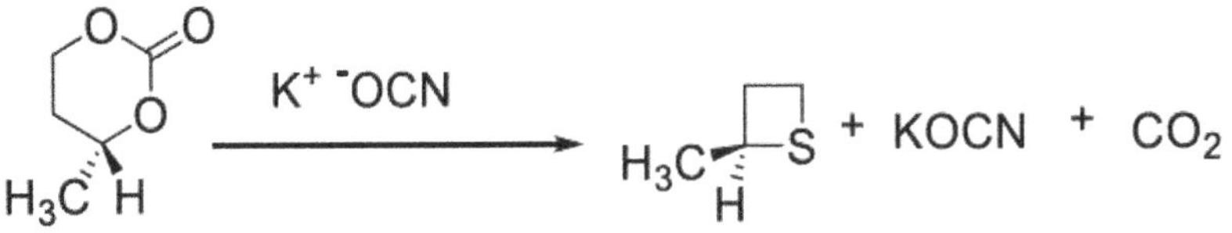
K+ -OCN
H3C H
H3C
H
S
+ KOCN + CO2

297

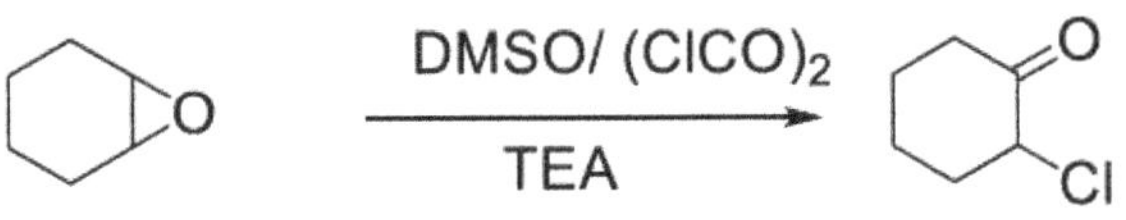
DMSO/ (ClCO)2
TEA
Cl

298

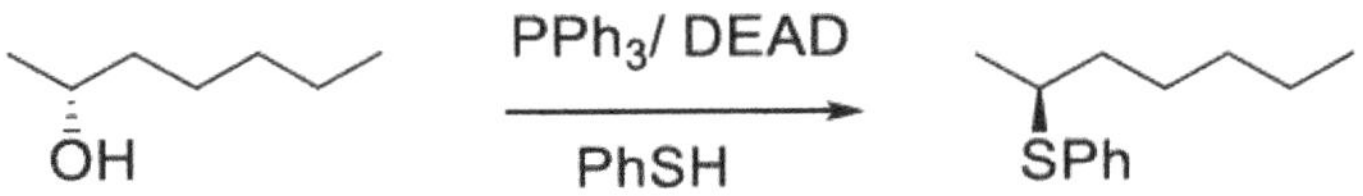
OH
PPh3/ DEAD
PhSH
SPh

299

t-Butyl ethers and esters can be readily made from the corresponding alcohols and carboxylic acids by mixing the reagent shown below with t-butanol in the presence of a catalytic amount of BF_3 etherate. Write a mechanism for these transformations.

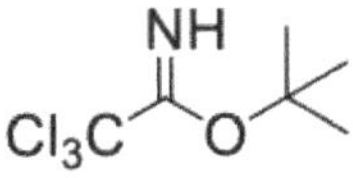
NH
Cl3C
O

300

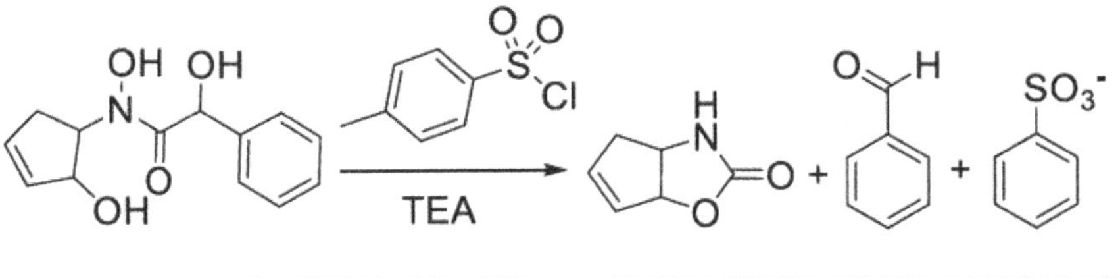
OH OH
N
OH
Cl
TEA
SO3-

301	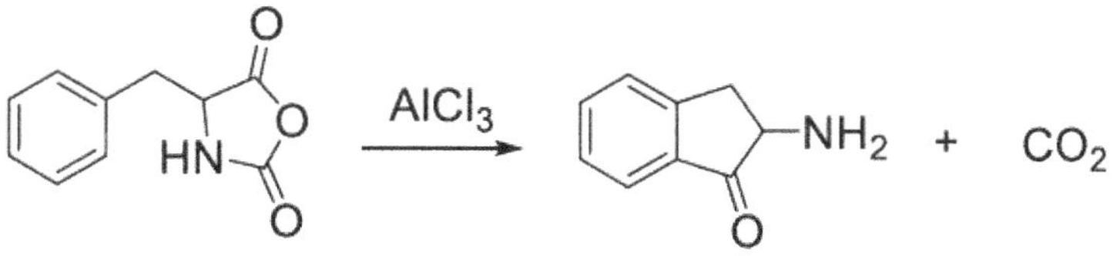AlCl3 HN NH2 + CO2
302	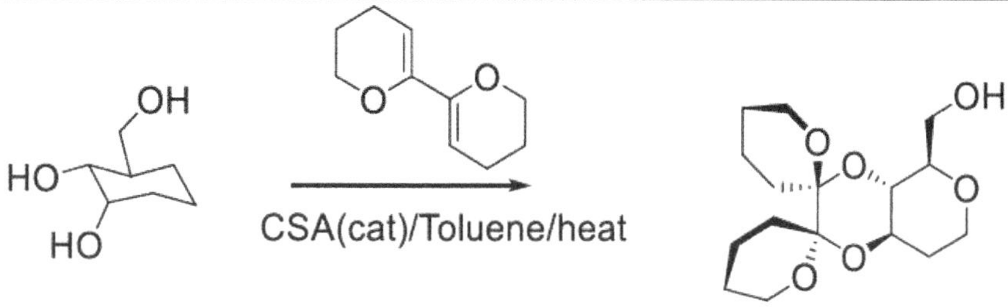OH HO HO CSA(cat)/Toluene/heat OH
303	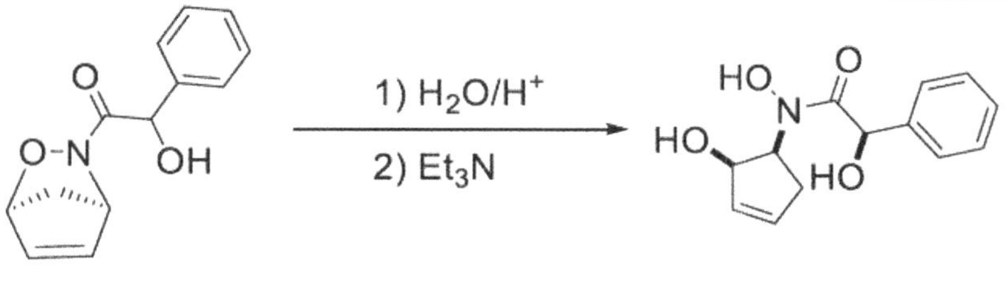O-N OH 1) H2O/H+ 2) Et3N HO HO HO
304	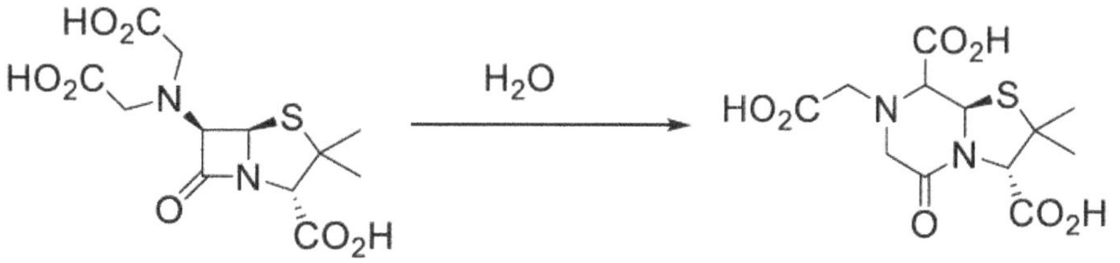HO2C HO2C CO2H H2O CO2H HO2C CO2H
305	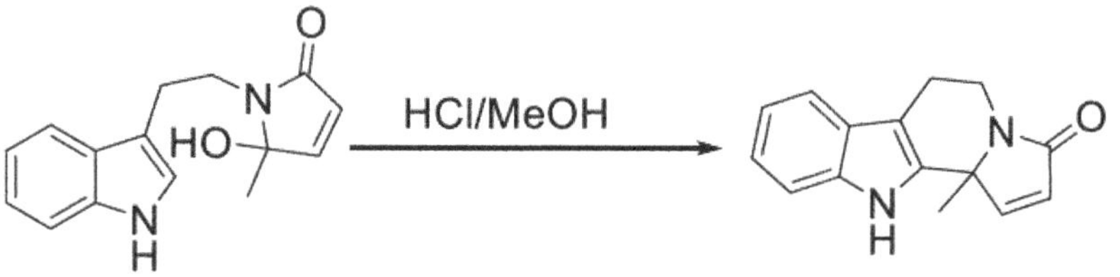HO HCl/MeOH
306	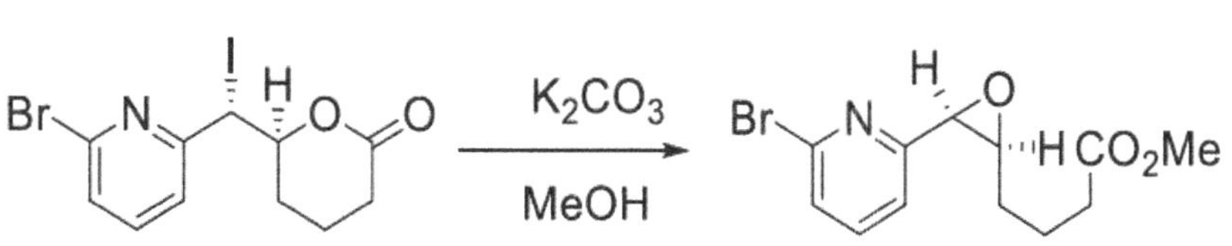Br I H K2CO3 MeOH Br H H CO2Me

307	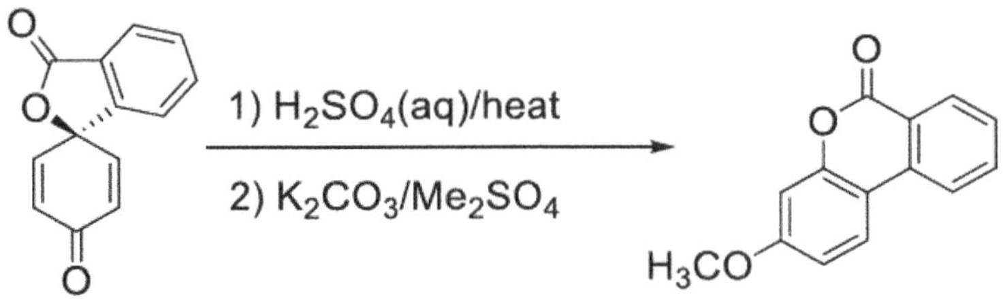1) H_2SO_4(aq)/heat 2) K_2CO_3/Me_2SO_4 H_3CO
308	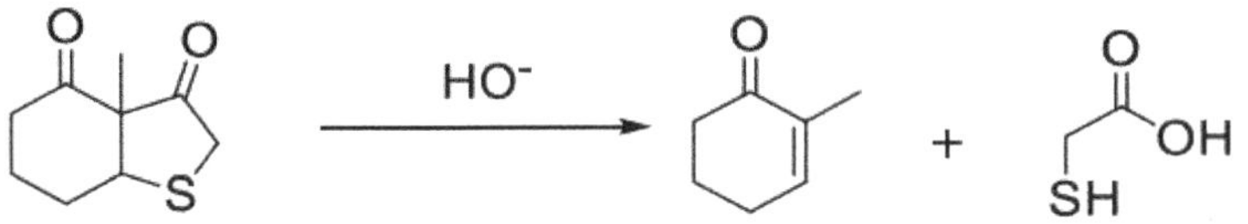HO^- + OH SH
309	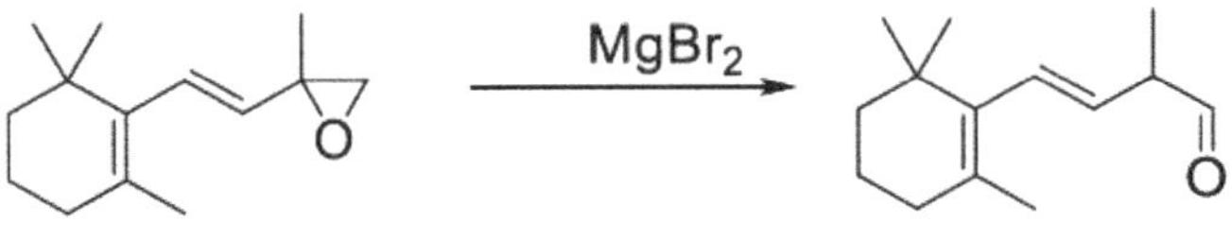$MgBr_2$
310	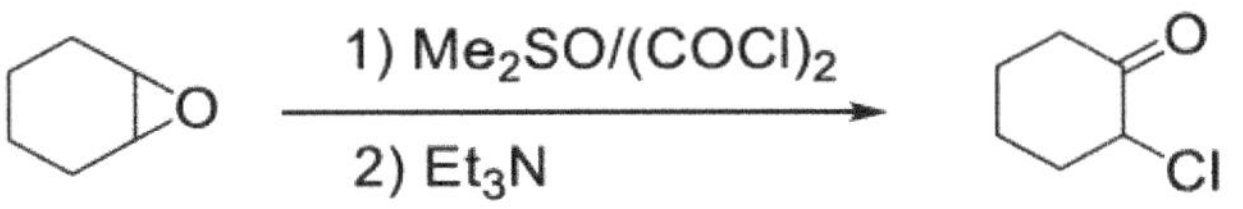1) $Me_2SO/(COCl)_2$ 2) Et_3N Cl
311	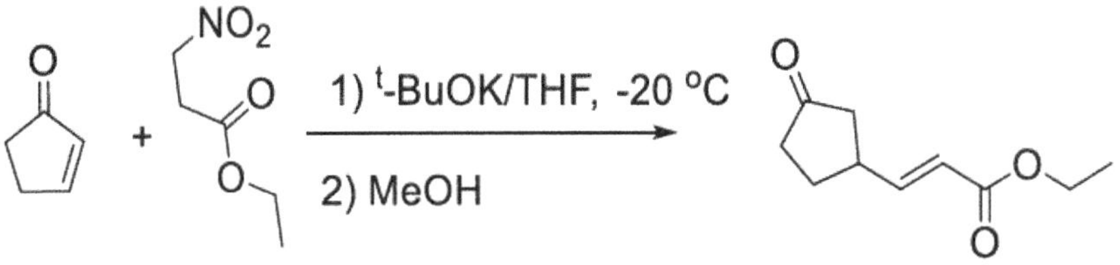NO_2 + 1) ^{t}-BuOK/THF, -20 °C 2) MeOH
312	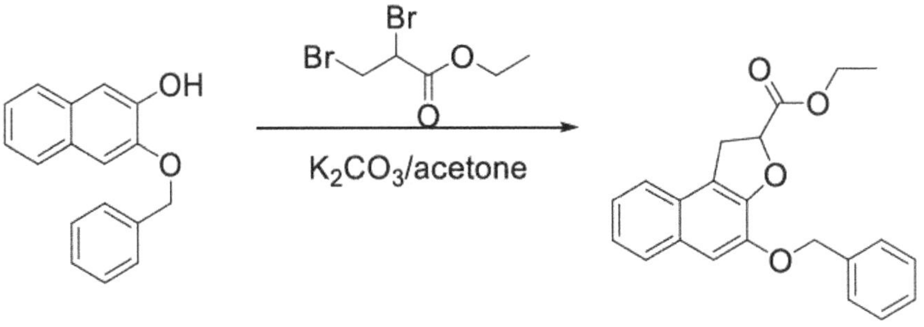OH Br Br K_2CO_3/acetone

313

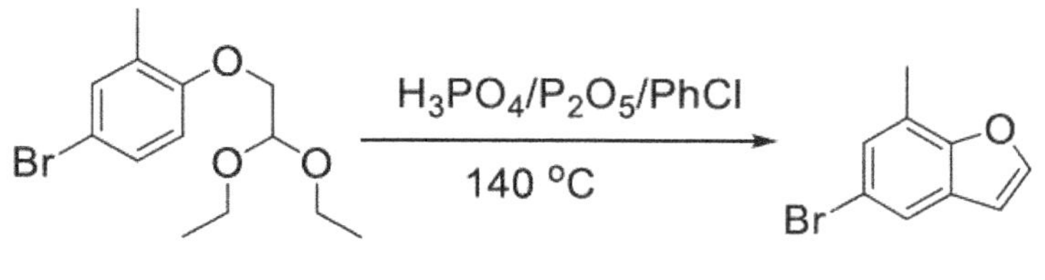
Br
H3PO4/P2O5/PhCl
140 °C
Br

314

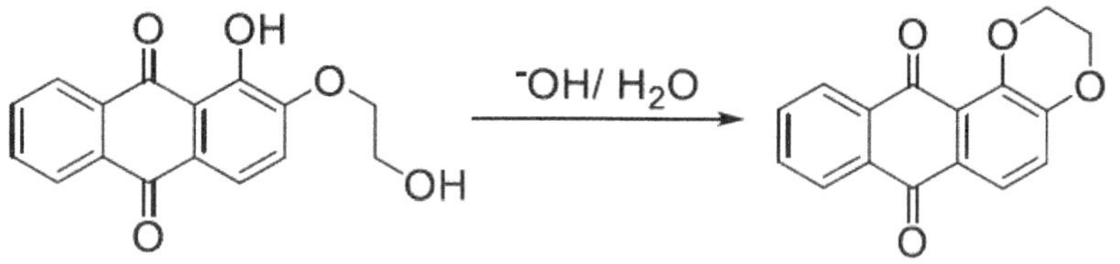
OH
-OH/ H2O

315

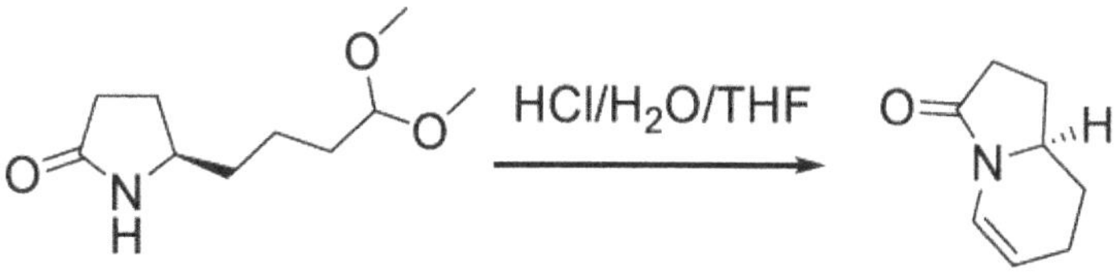
HCl/H2O/THF

316

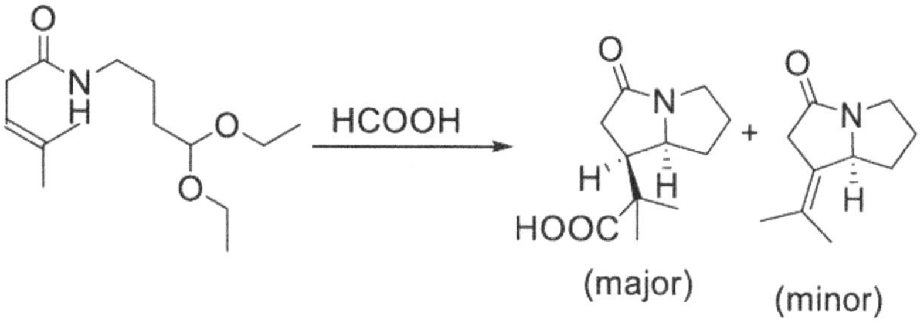
HCOOH
HOOC
(major)
(minor)

317

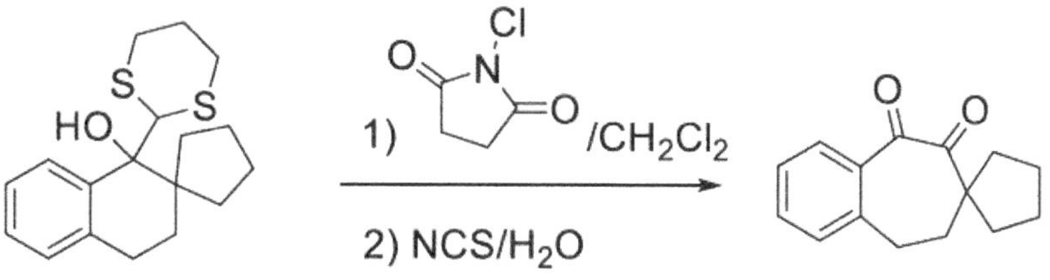
HO
Cl
1)
/CH2Cl2
2) NCS/H2O

318

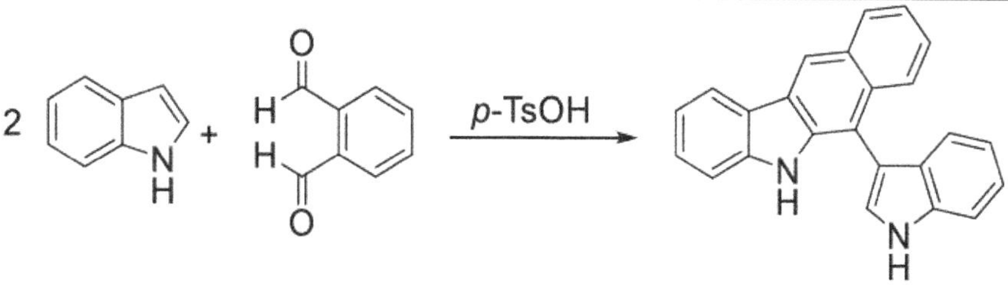
2
p-TsOH

319	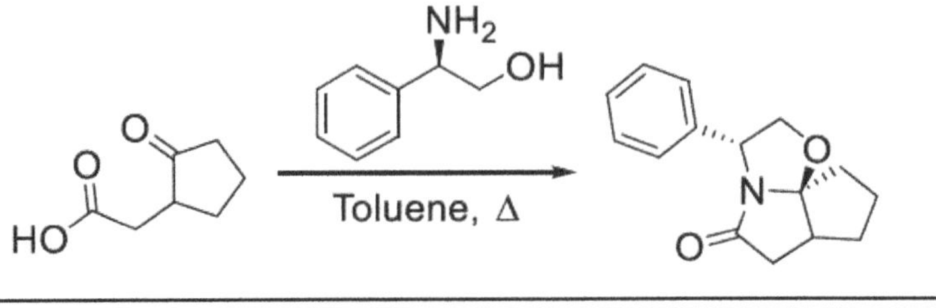Toluene, Δ
320	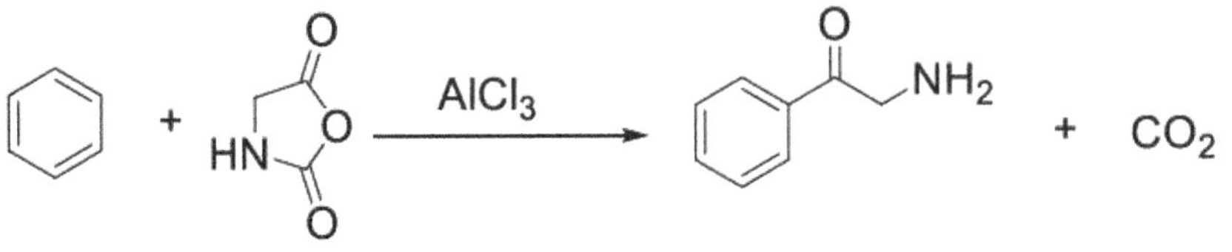AlCl3 + CO2
321	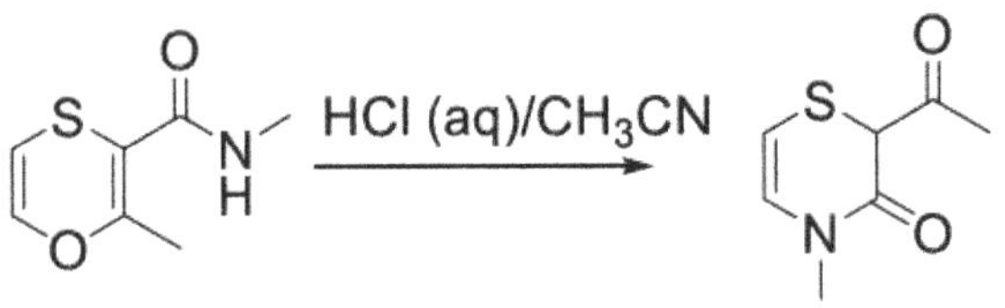HCl (aq)/CH3CN
322	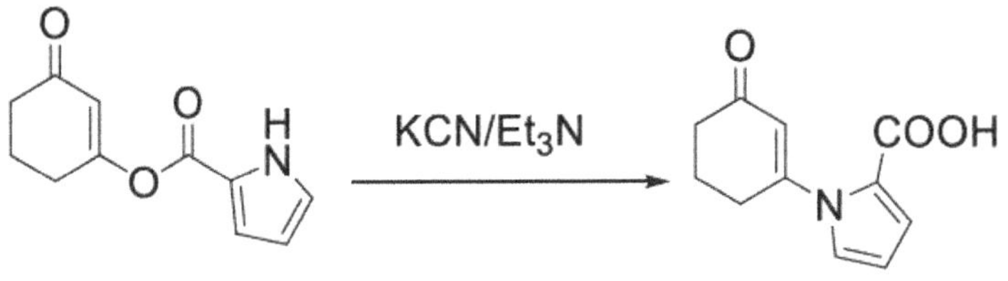KCN/Et3N
323	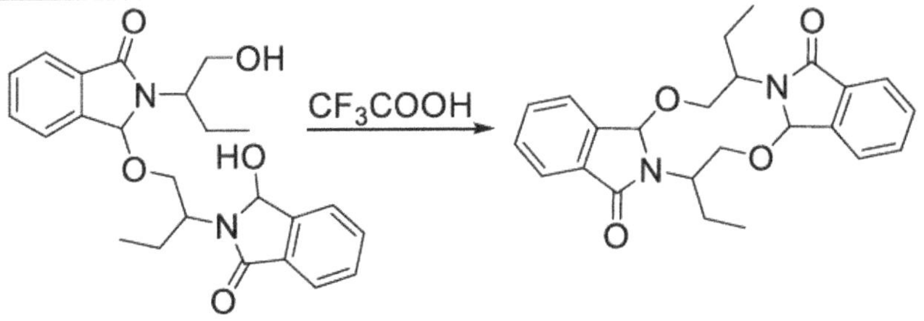CF3COOH
324	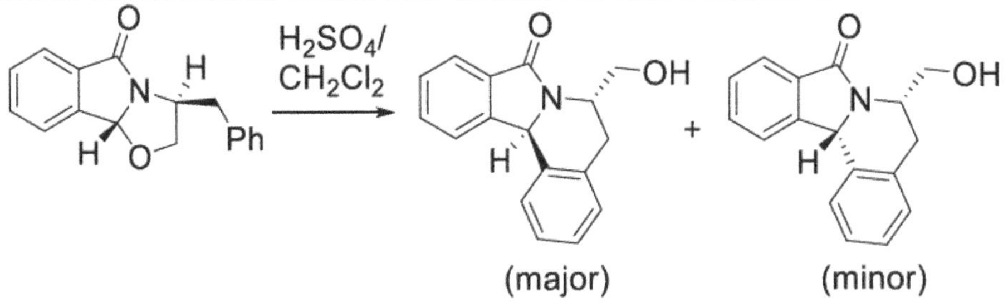H2SO4/ CH2Cl2 (major) (minor)

325	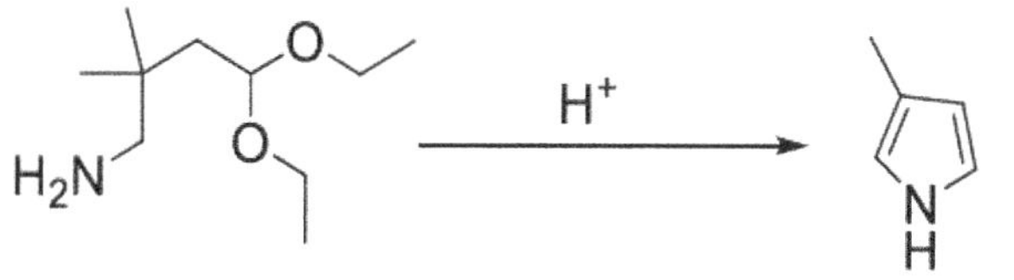H+
326	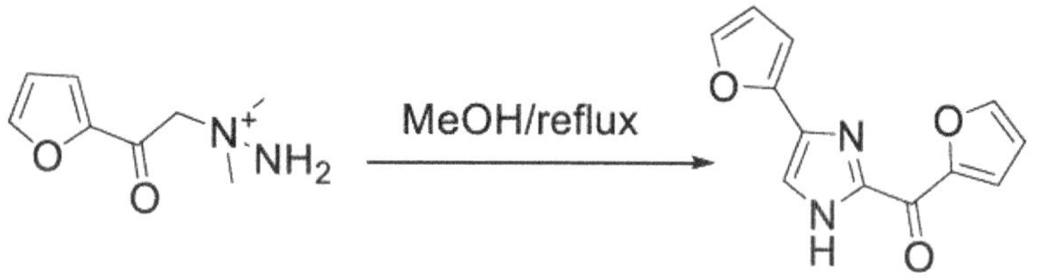MeOH/reflux
327	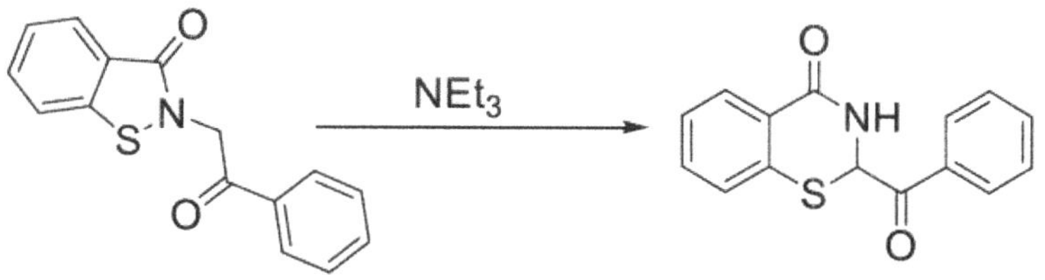NEt3
328	When the antibiotic Emycin F is treated with acid, it rearranges to Emycin E. Write a mechanism for this transformation. 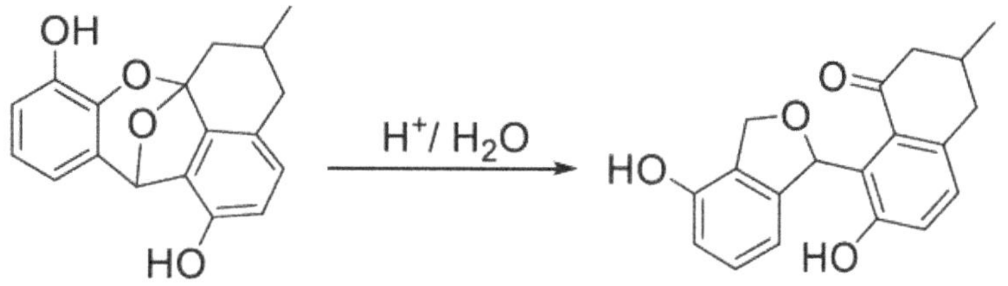H+/ H2O Emycin F Emycin E
329	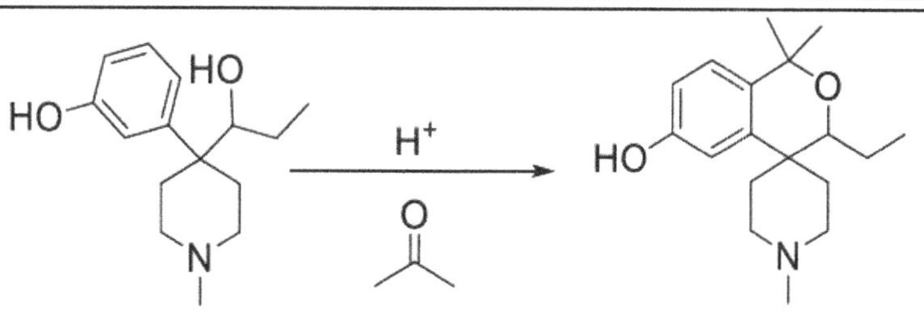H+

330

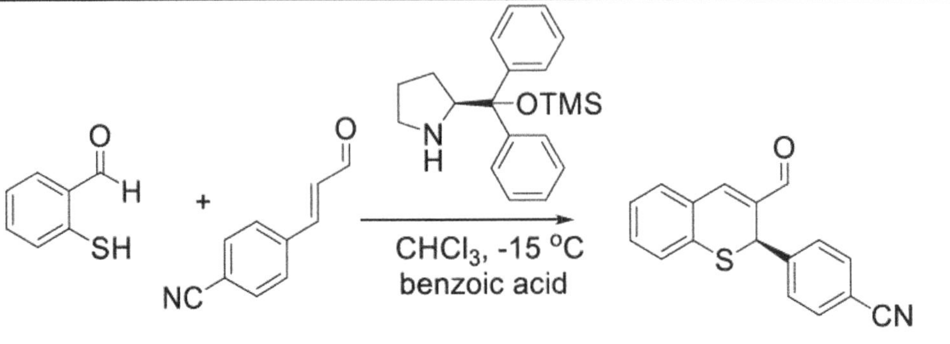
O
H
SH
+
O
NC
N
H
OTMS
CHCl$_3$, -15 °C
benzoic acid
O
S
CN

331

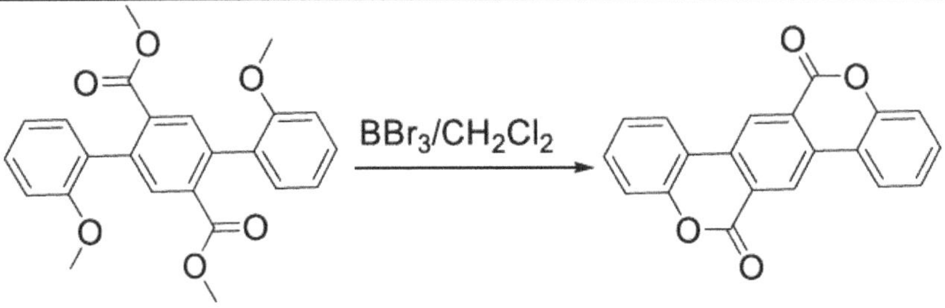
O
O
O
O
O
O
BBr$_3$/CH$_2$Cl$_2$
O
O
O
O

332

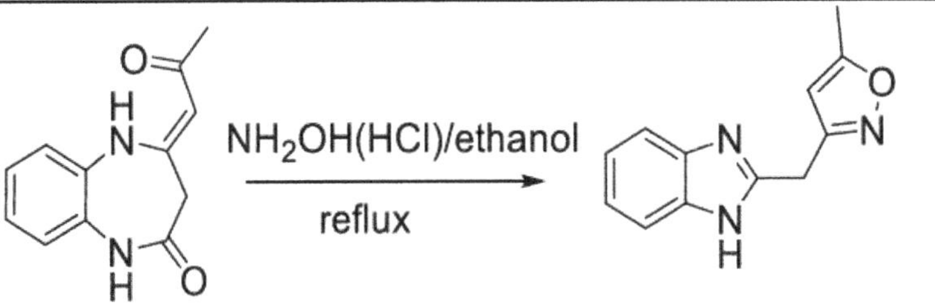
O
H
N
N
H
O
NH$_2$OH(HCl)/ethanol
reflux
O
N
N
N
H

333

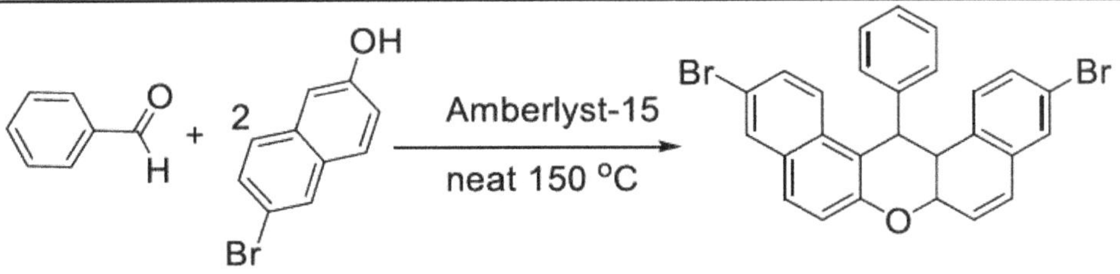
O
H
+
2
OH
Br
Amberlyst-15
neat 150 °C
Br
Br
O

334

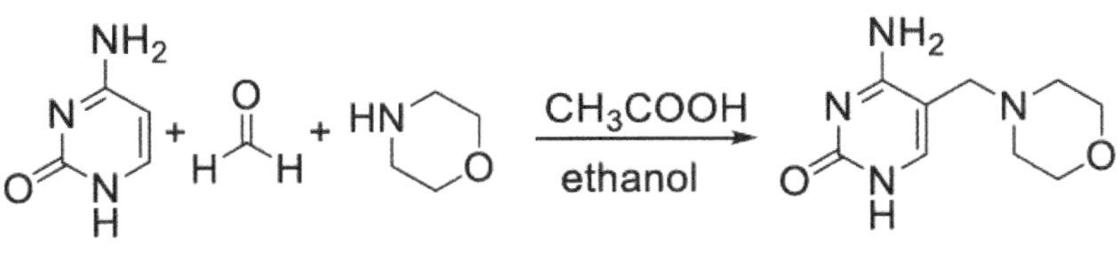
NH$_2$
N
O
N
H
+
O
H
H
+
HN
O
CH$_3$COOH
ethanol
NH$_2$
N
N
O
N
H
O

335

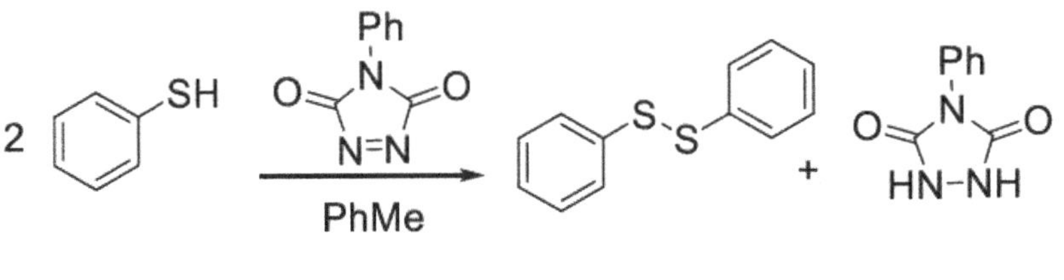
2
SH
Ph
N
O
O
N=N
PhMe
S
S
+
Ph
N
O
O
HN-NH

336	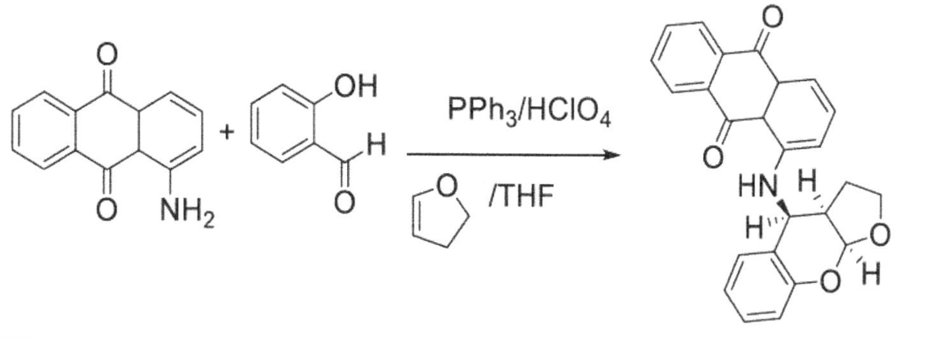PPh3/HClO4 /THF
337	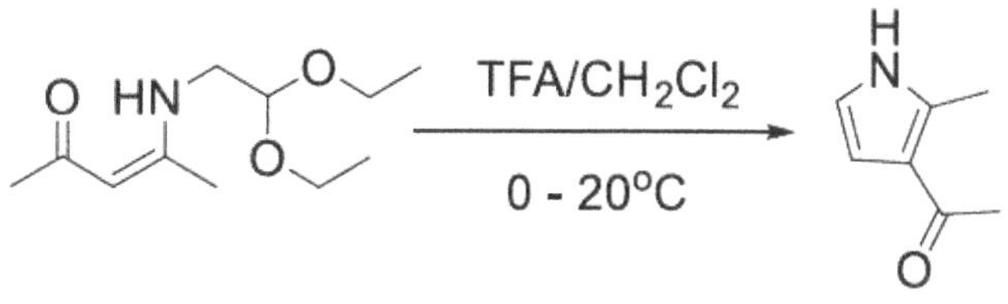TFA/CH2Cl2 0 - 20°C
338	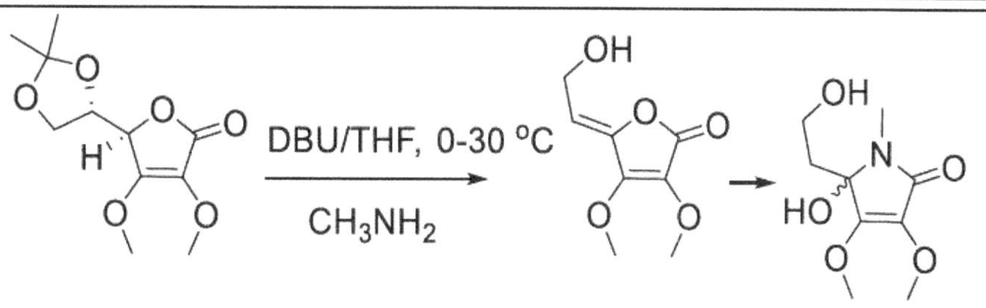DBU/THF, 0-30 °C CH3NH2
339	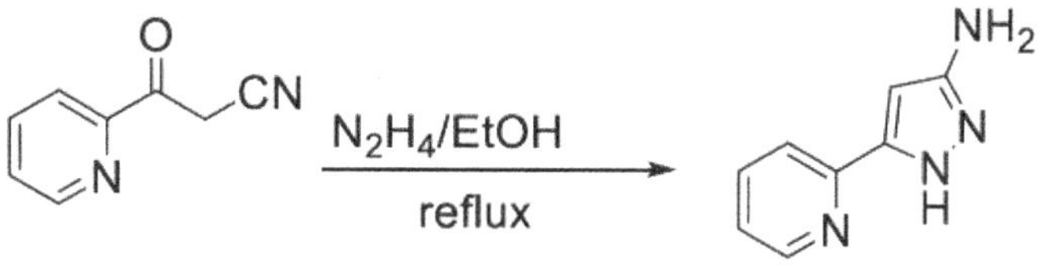N2H4/EtOH reflux
340	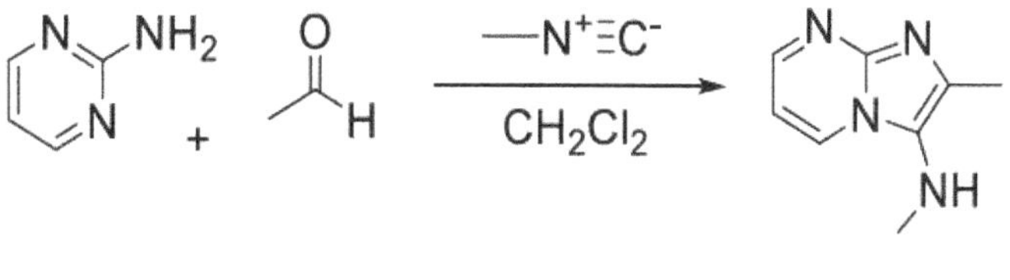CH2Cl2
341	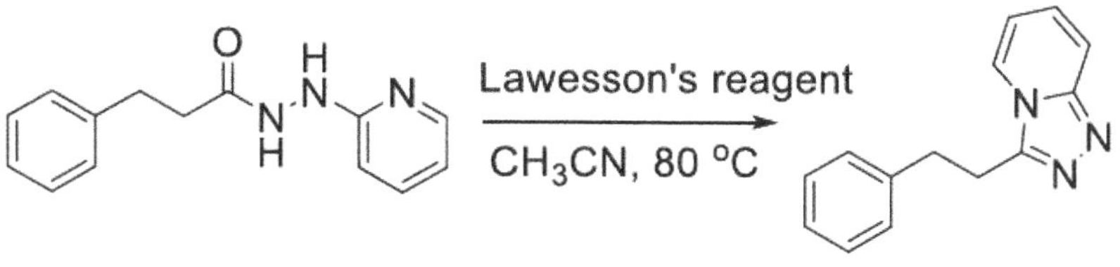Lawesson's reagent CH3CN, 80 °C

342	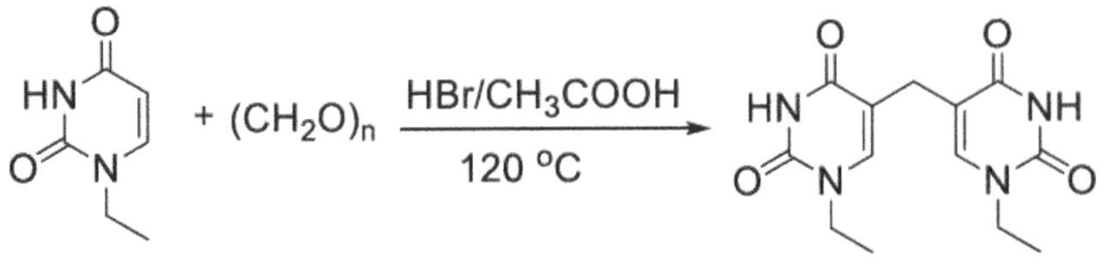HN O N + $(CH_2O)_n$ HBr/CH_3COOH 120 °C NH
343	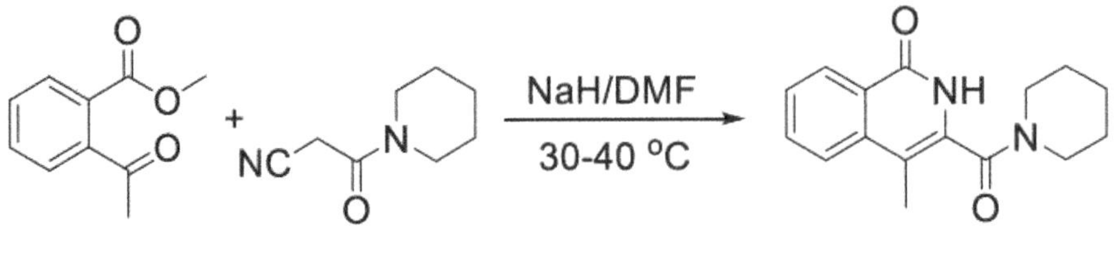O + NC N NaH/DMF 30-40 °C NH
344	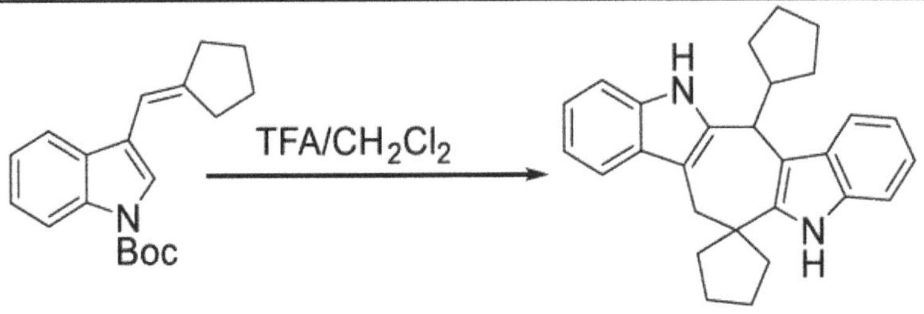N Boc TFA/CH_2Cl_2 H N
345	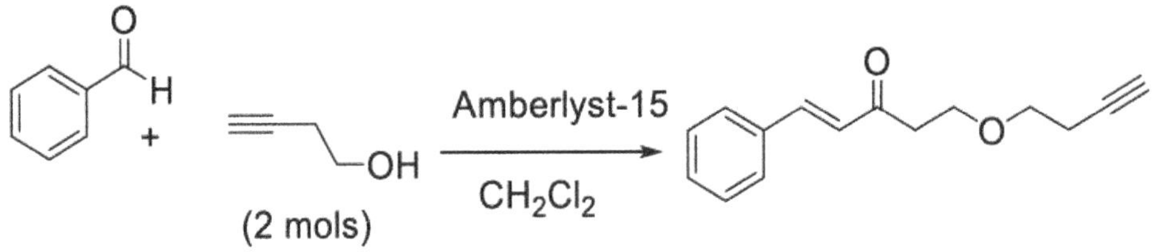O H + OH (2 mols) Amberlyst-15 CH_2Cl_2

346	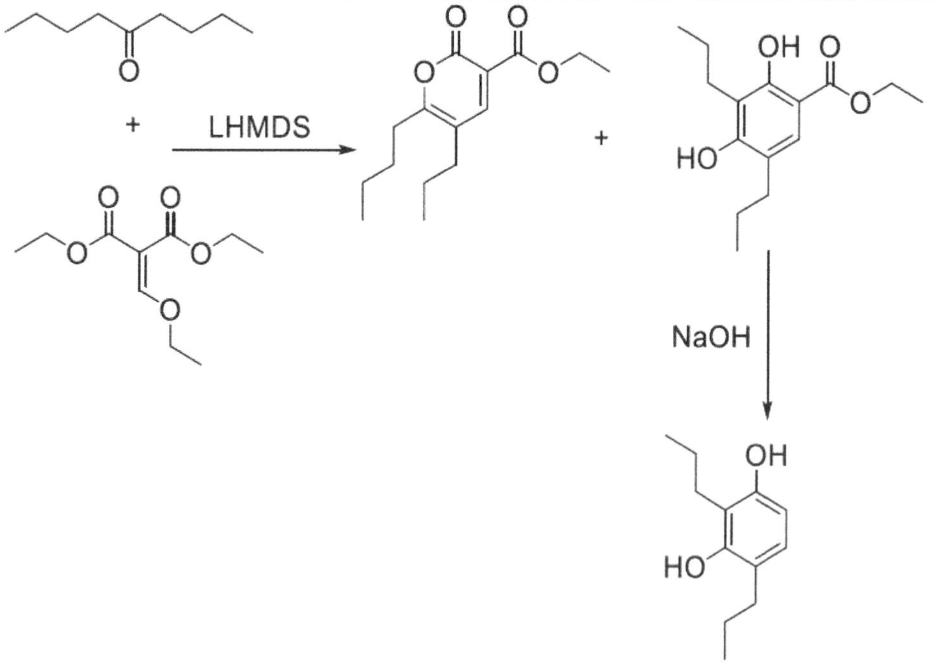 LHMDS NaOH OH HO
347	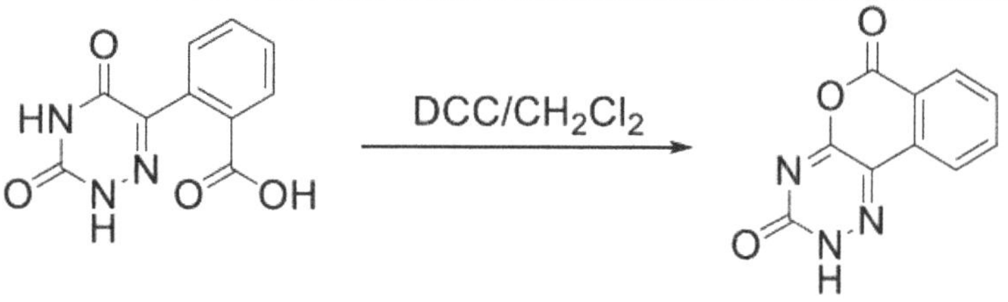 DCC/CH2Cl2 HN OH
348	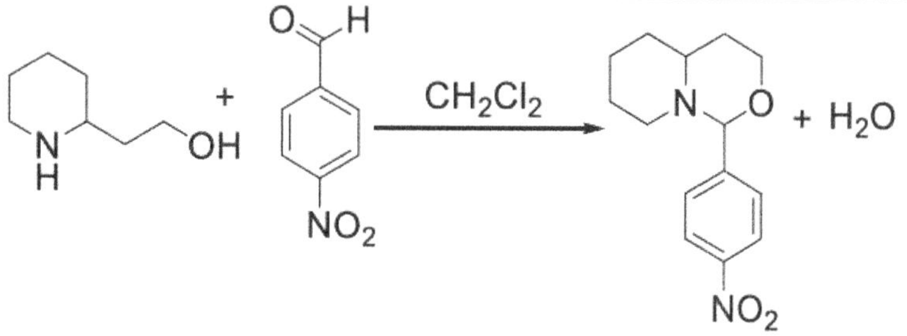 CH2Cl2 NO2 OH + H2O
349	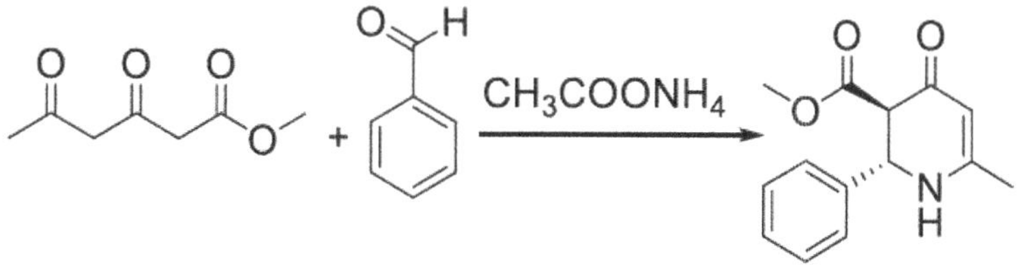 CH3COONH4

350	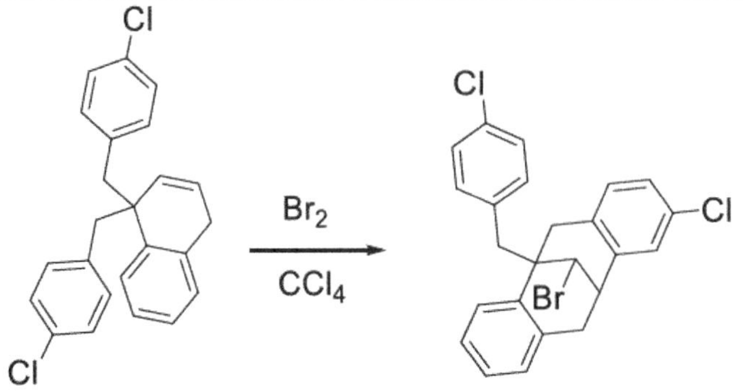 Cl Cl Br2 CCl4 Cl Cl Br
351	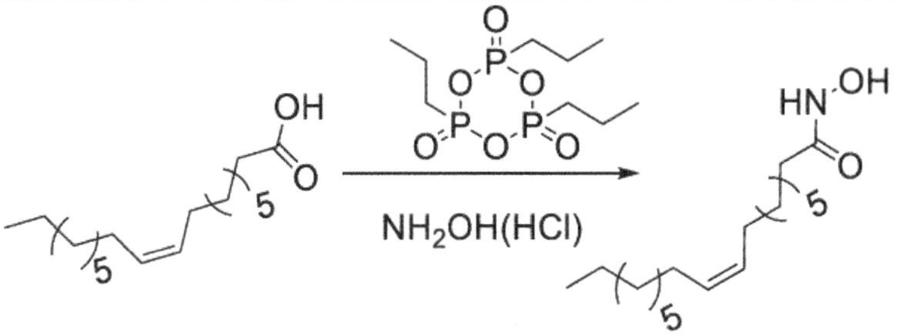 OH O 5 5 O P O O P O P O O O NH2OH(HCl) HN OH O 5 5
352	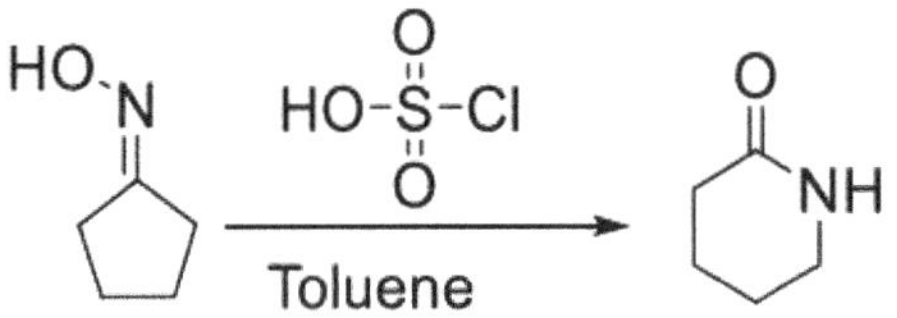 HO N O HO S Cl O Toluene O NH
353	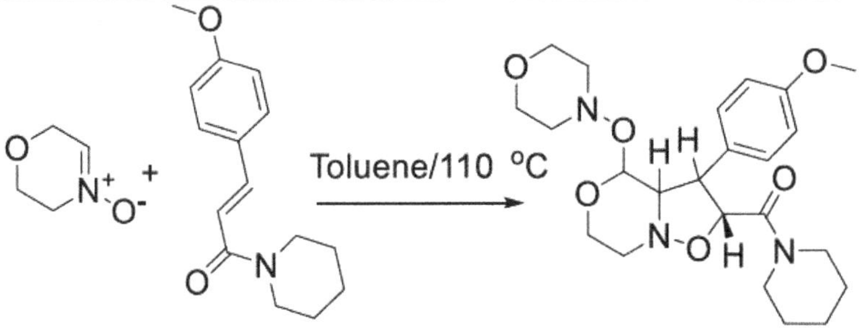 O N+ O- + O O N Toluene/110 °C O N O H H O O N O H N O
354	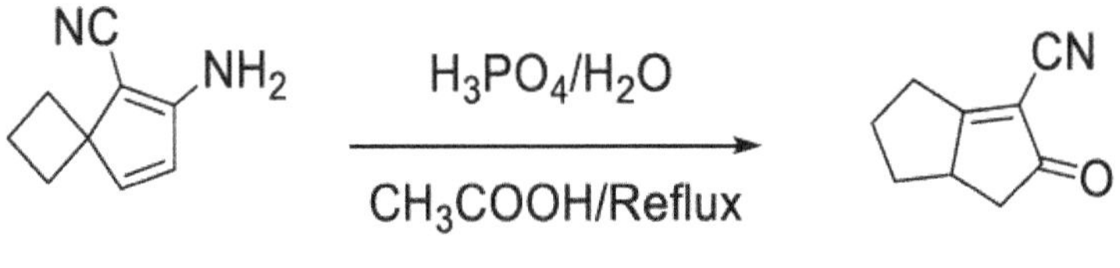 NC NH2 H3PO4/H2O CH3COOH/Reflux CN O

355	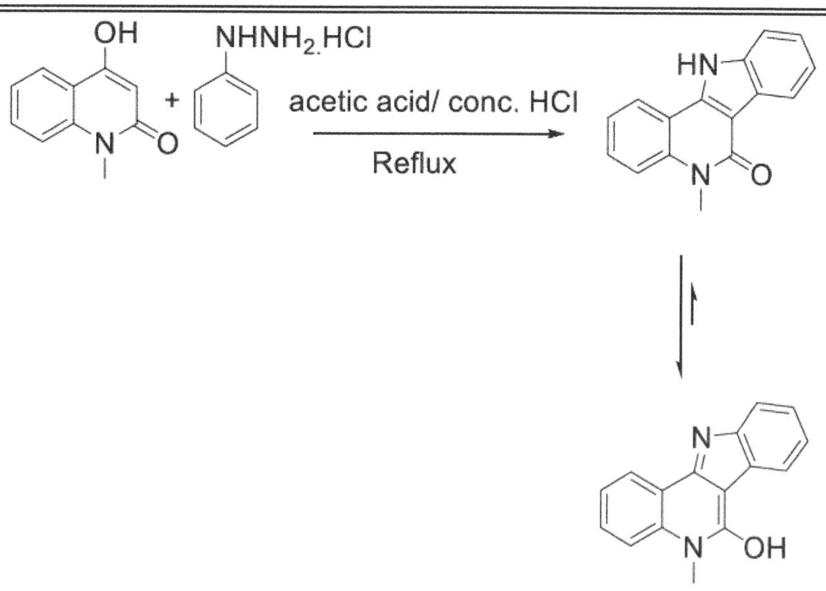 OH NHNH$_2$.HCl acetic acid/ conc. HCl Reflux N O HN N O N N OH
356	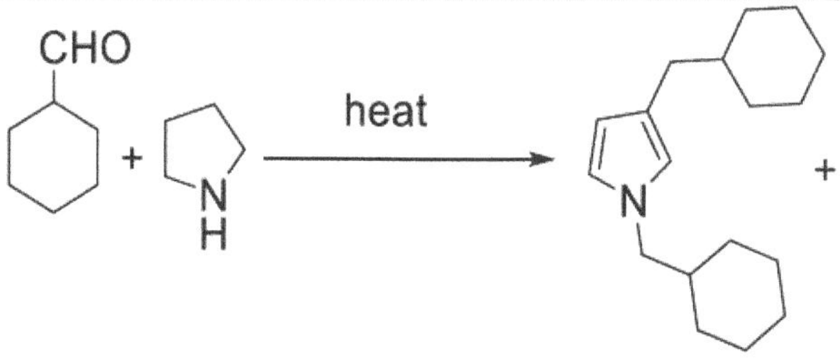 CHO heat N H N N
357	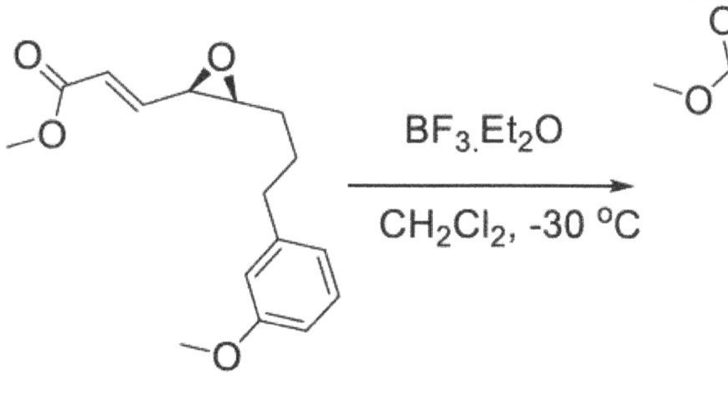 O O O O BF$_3$.Et$_2$O CH$_2$Cl$_2$, -30 °C O O HO O
358	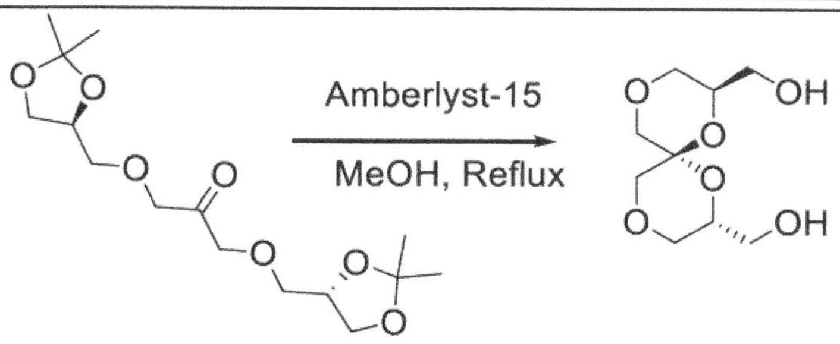 O O O O O O O Amberlyst-15 MeOH, Reflux O O OH O O OH

359	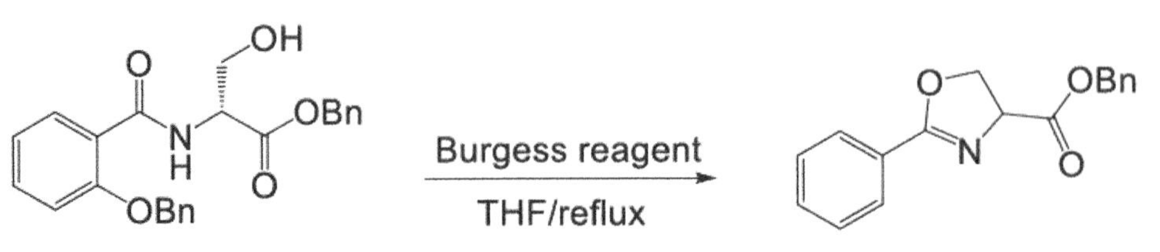OH O N H OBn O OBn Burgess reagent THF/reflux O OBn N O
360	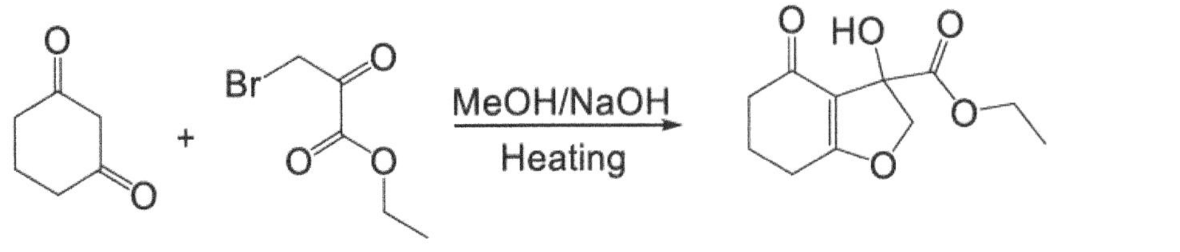O O + Br O O O MeOH/NaOH Heating O HO O O O
361	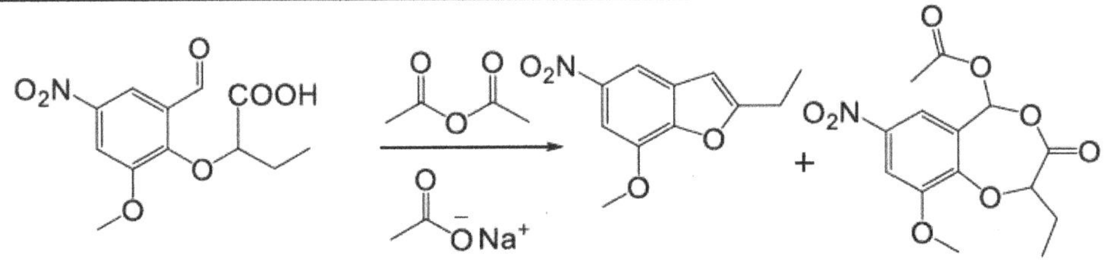O₂N O COOH O O O O O O O O⁻ Na⁺ O₂N O O + O O₂N O O O O O
362	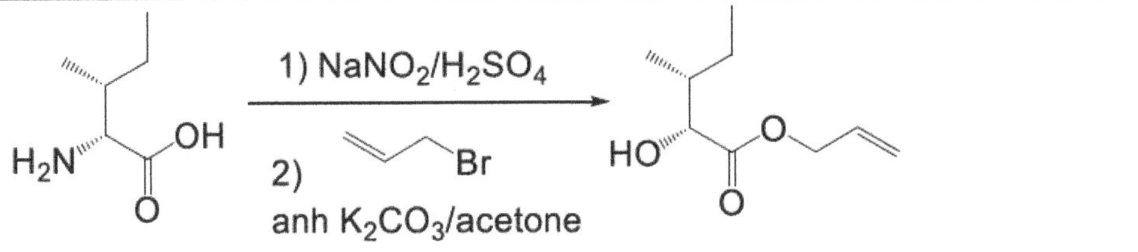H₂N OH O 1) NaNO₂/H₂SO₄ 2) Br anh K₂CO₃/acetone HO O O
363	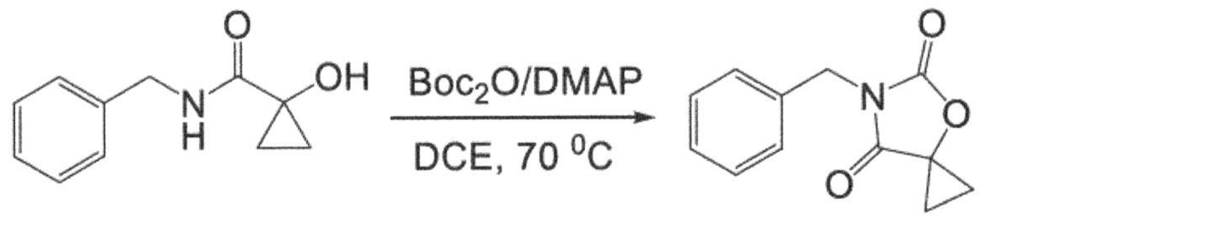O N H OH Boc₂O/DMAP DCE, 70 ⁰C O N O O
364	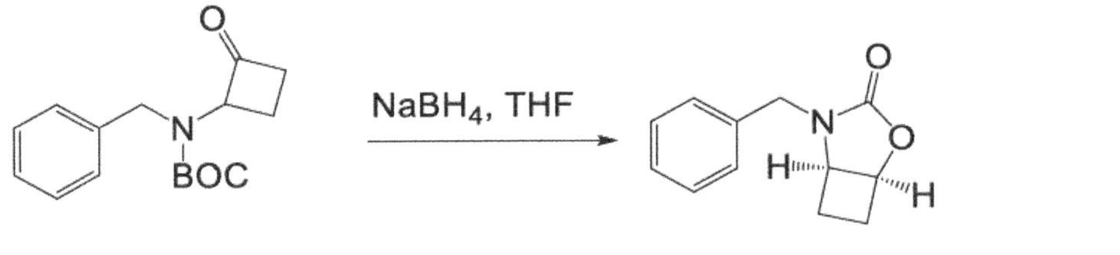O N BOC NaBH₄, THF O N O H H

365	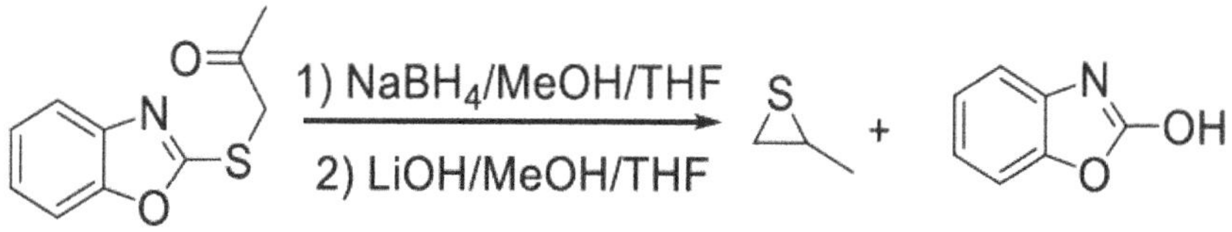1) NaBH4/MeOH/THF 2) LiOH/MeOH/THF OH
366	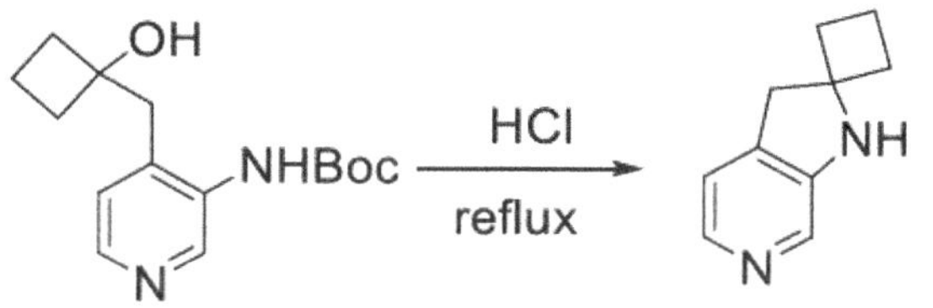OH NHBoc HCl reflux NH
367	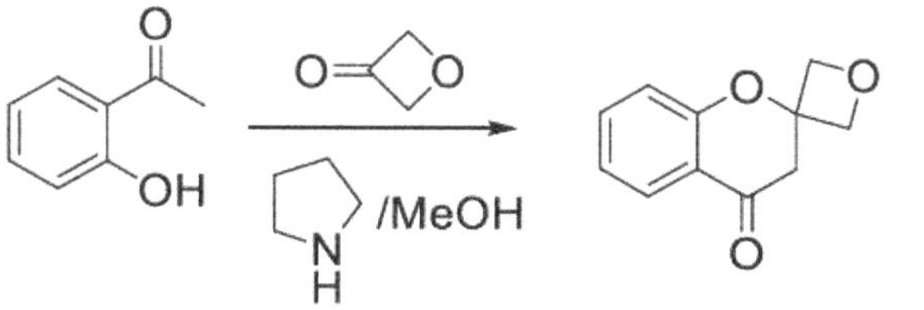OH /MeOH
368	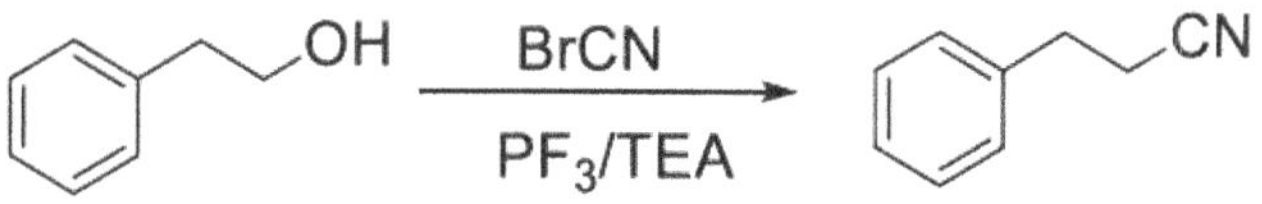OH BrCN PF3/TEA CN
369	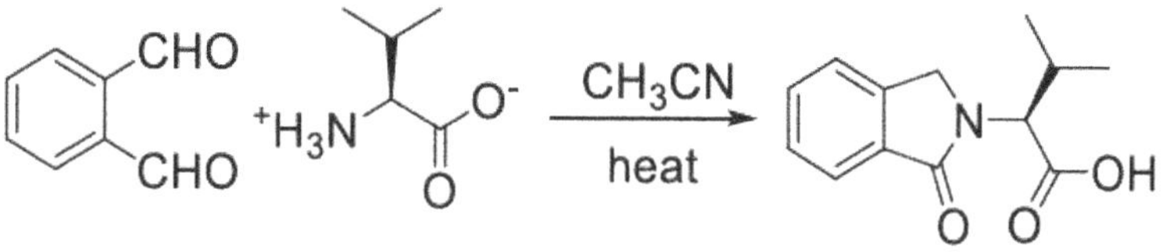CHO CHO +H3N O- CH3CN heat OH
370	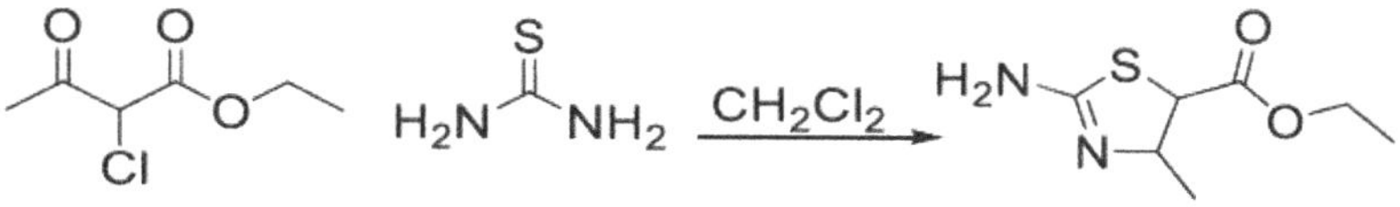Cl H2N NH2 CH2Cl2 H2N

371

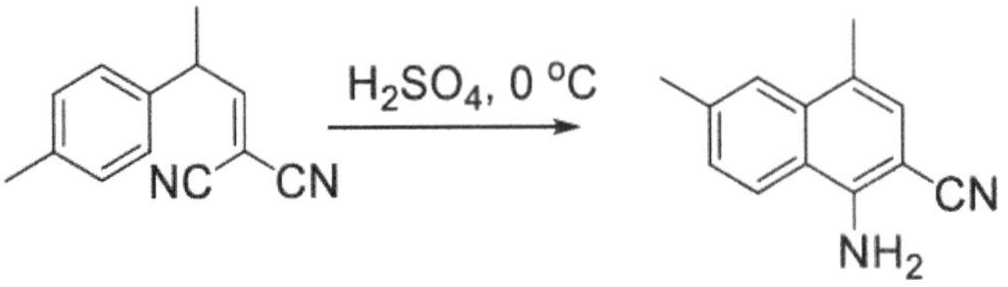
NC
CN
H2SO4, 0 °C
CN
NH2

372

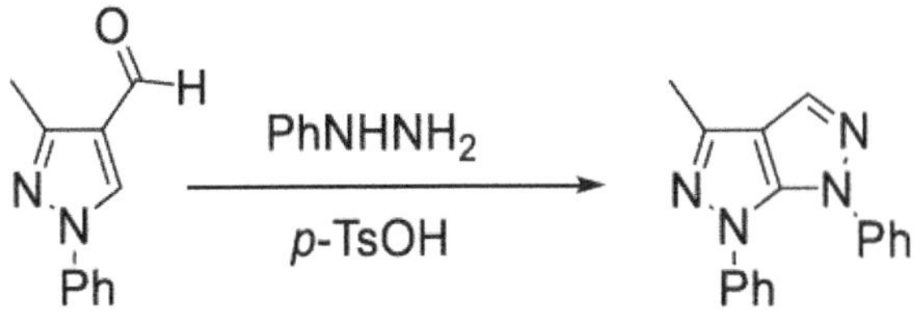
O
H
N
N
Ph
PhNHNH2
p-TsOH
N
N
N
N
Ph
Ph

373

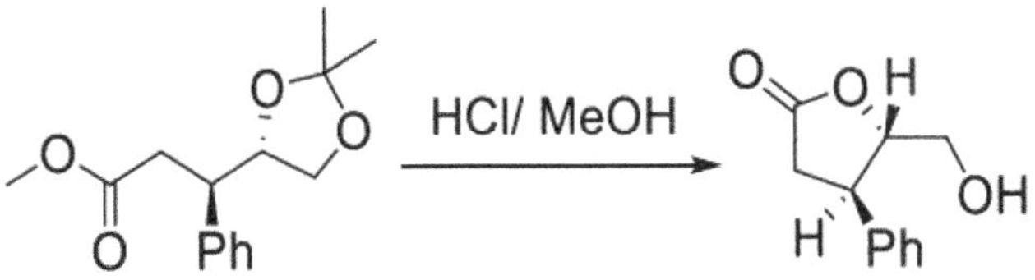
O
O
O
O
Ph
HCl/ MeOH
O
O
H
OH
H
Ph

374

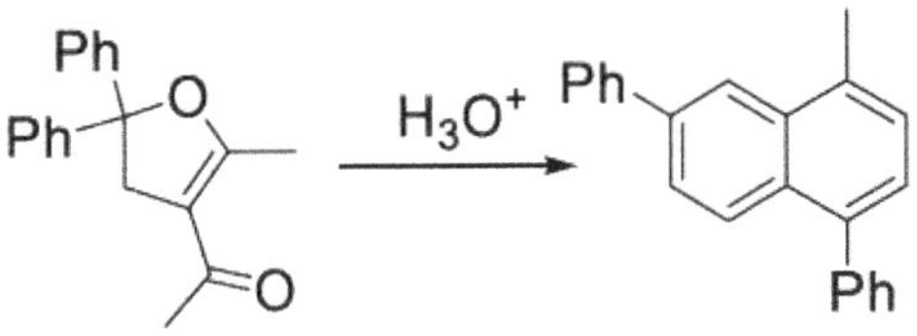
Ph
Ph
O
O
H3O+
Ph
Ph

375

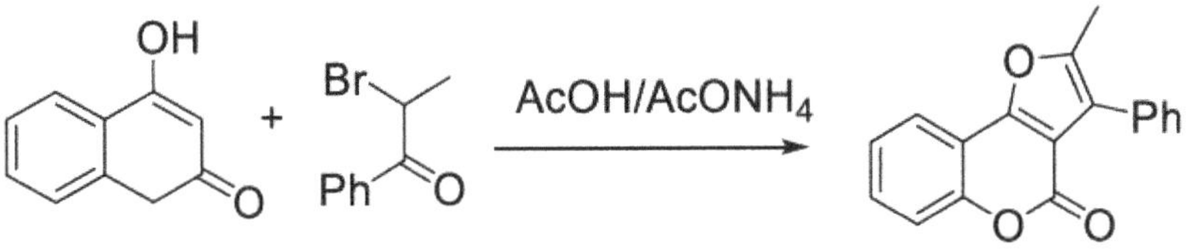
OH
O
+
Br
Ph
O
AcOH/AcONH4
O
Ph
O
O

376

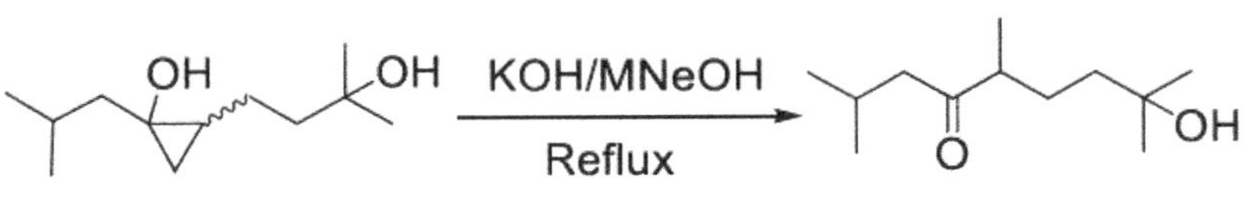
OH
OH
KOH/MNeOH
Reflux
O
OH

377	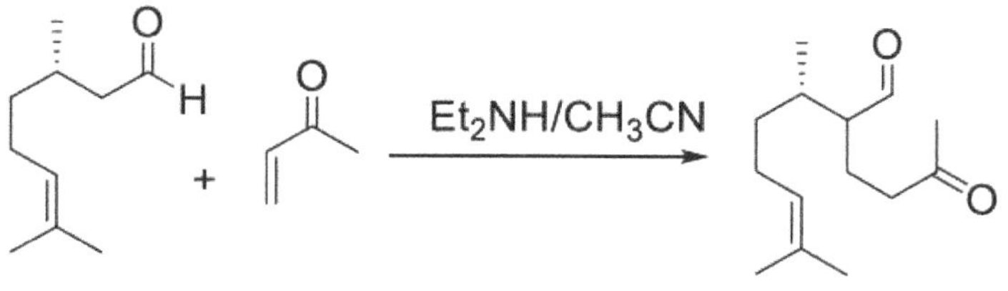 O H + O Et_2NH/CH_3CN O O
378	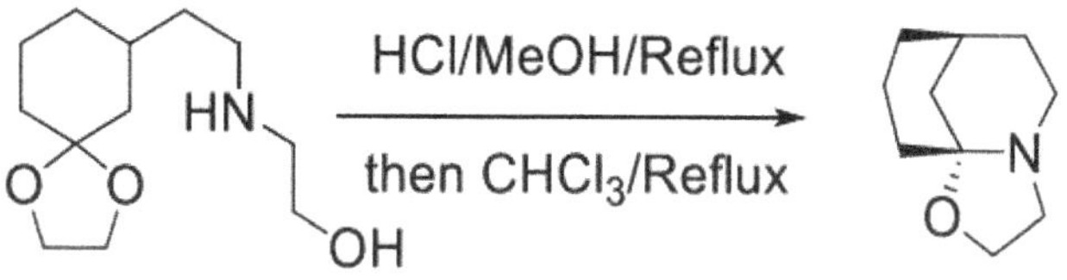 HN O O OH HCl/MeOH/Reflux then $CHCl_3$/Reflux N O
379	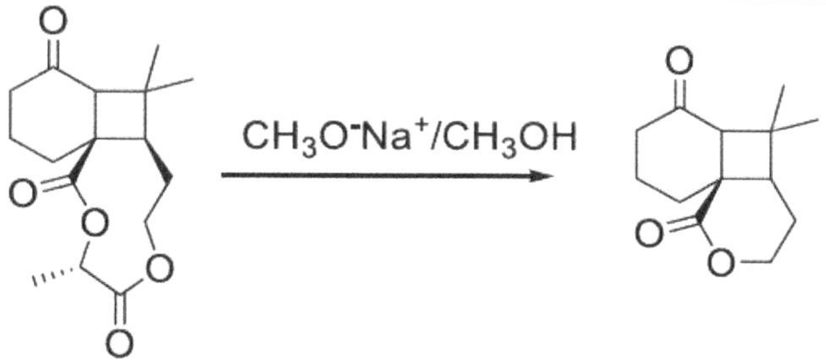 O O O O O $CH_3O^-Na^+/CH_3OH$ O O O
380	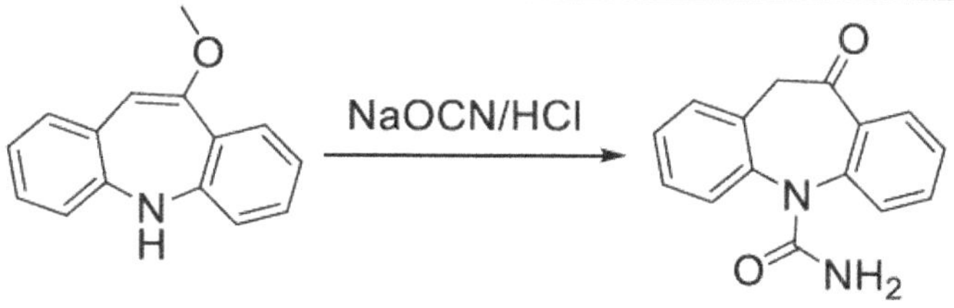 O N H NaOCN/HCl O N O NH_2
381	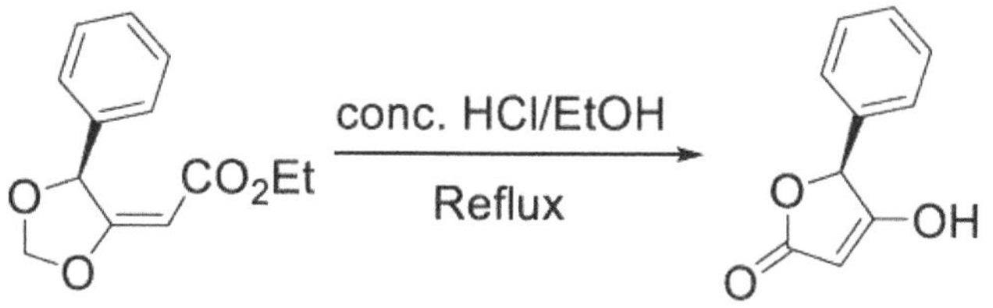 O O CO_2Et conc. HCl/EtOH Reflux O OH O
382	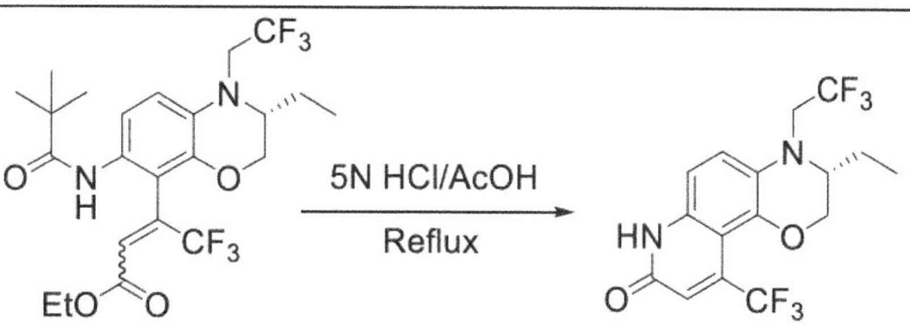 CF_3 N O N H O CF_3 EtO O 5N HCl/AcOH Reflux CF_3 N HN O O CF_3

383

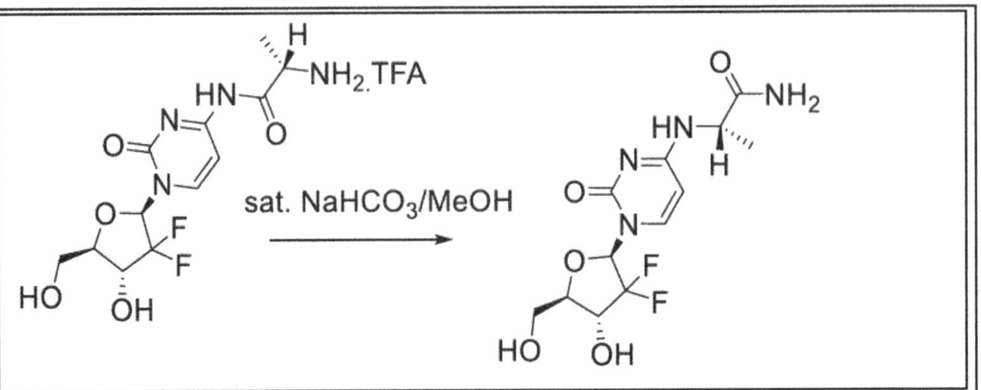
H
NH2.TFA
HN
O
N
O
N
O
F
F
HO
OH
sat. NaHCO3/MeOH
O
NH2
HN
H
N
O
N
O
F
F
HO
OH

384

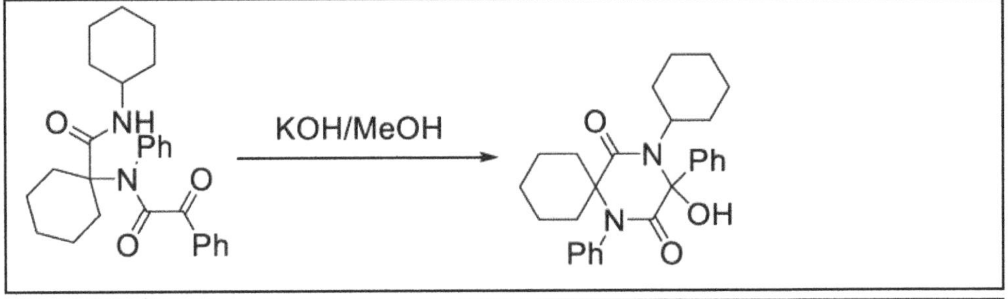
O
NH
Ph
N
O
O
Ph
KOH/MeOH
O
N
Ph
OH
N
Ph
O

385

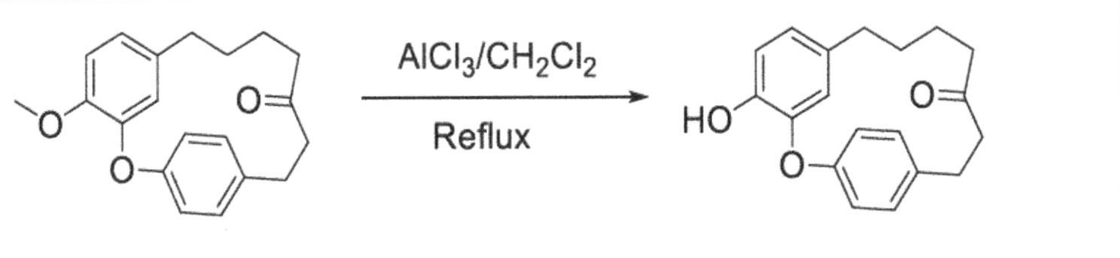
O
O
O
AlCl3/CH2Cl2
Reflux
HO
O
O

386

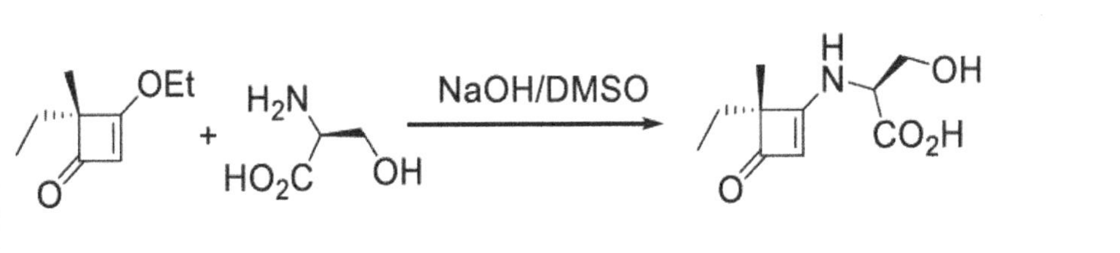
OEt
O
+
H2N
HO2C
OH
NaOH/DMSO
H
N
OH
CO2H
O

387

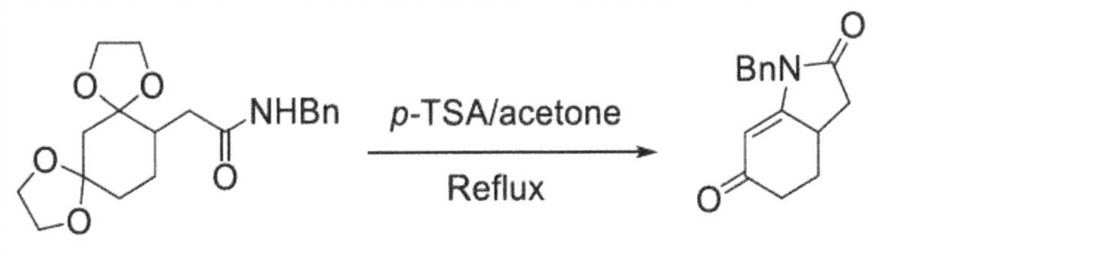
O
O
O
O
NHBn
O
p-TSA/acetone
Reflux
BnN
O
O

388

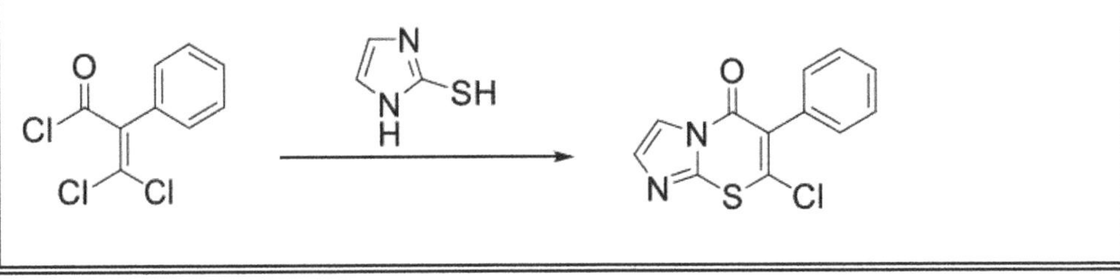
O
Cl
Cl
Cl
N
N
H
SH
O
N
N
S
Cl

389	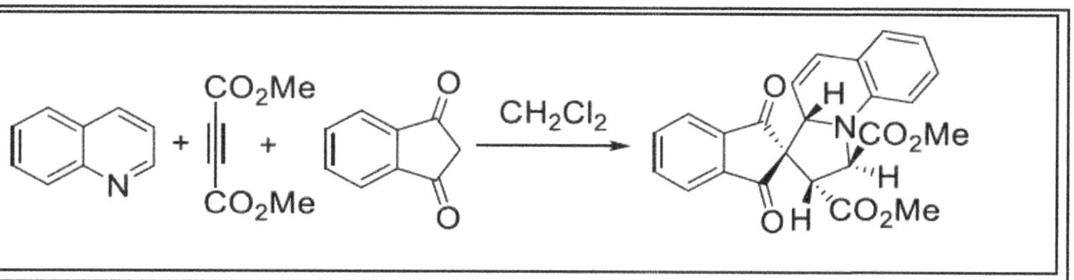CO2Me CO2Me CH2Cl2 N
390	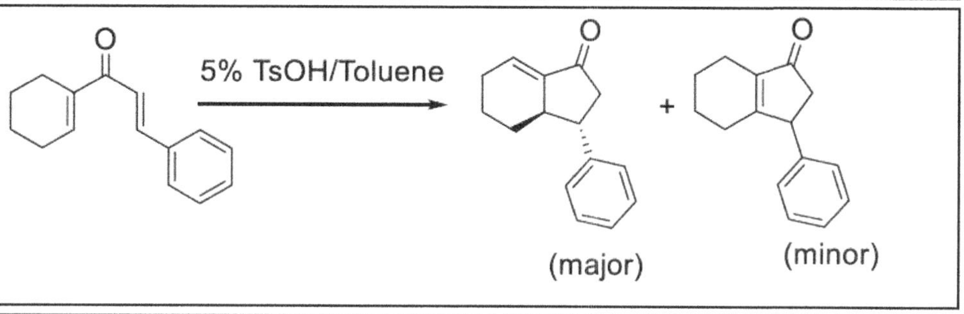5% TsOH/Toluene (major) (minor)
391	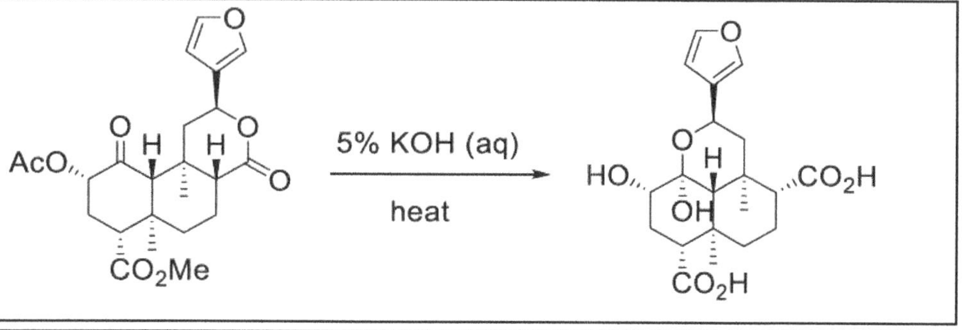AcO CO2Me 5% KOH (aq) heat HO OH CO2H CO2H
392	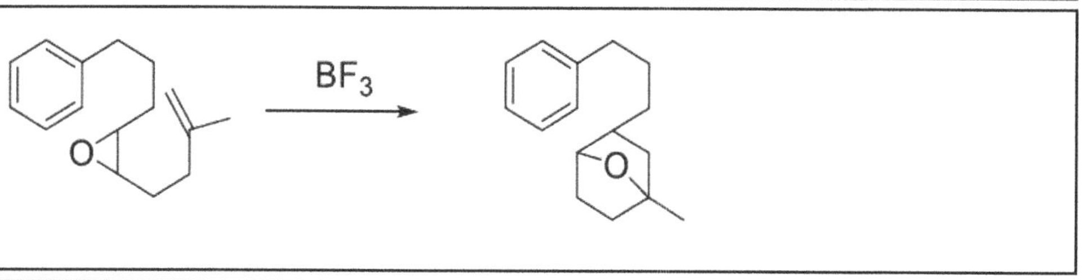BF3
393	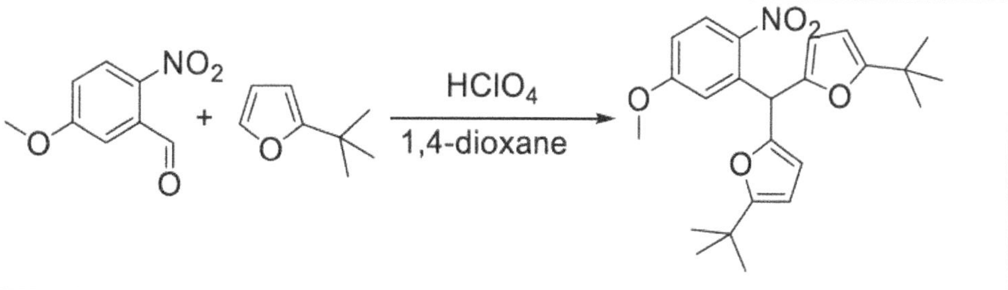NO2 HClO4 1,4-dioxane NO2

394

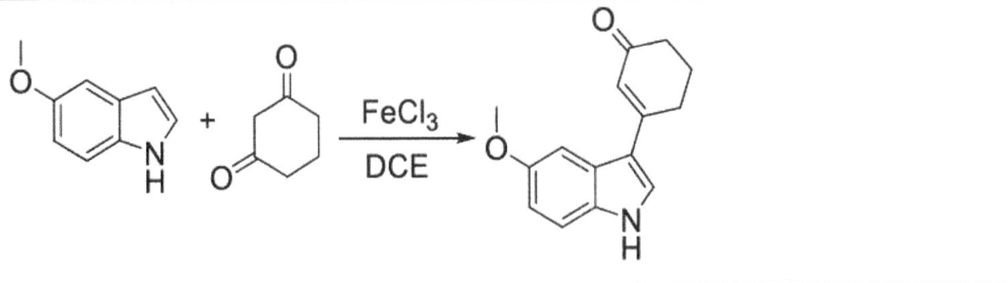

O
+
FeCl3
DCE
N H

395

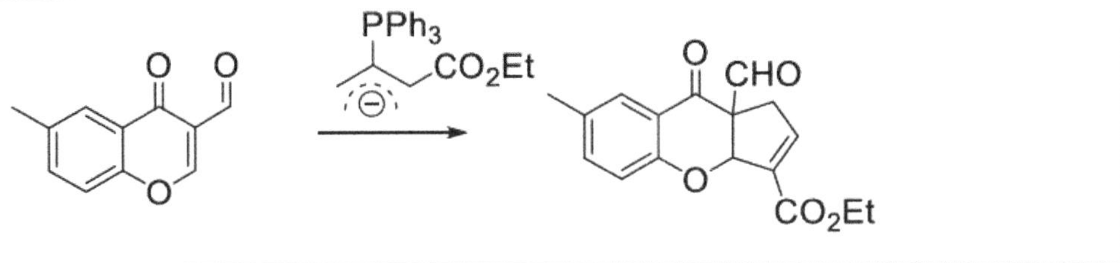

PPh3
CO2Et
CHO
CO2Et

396

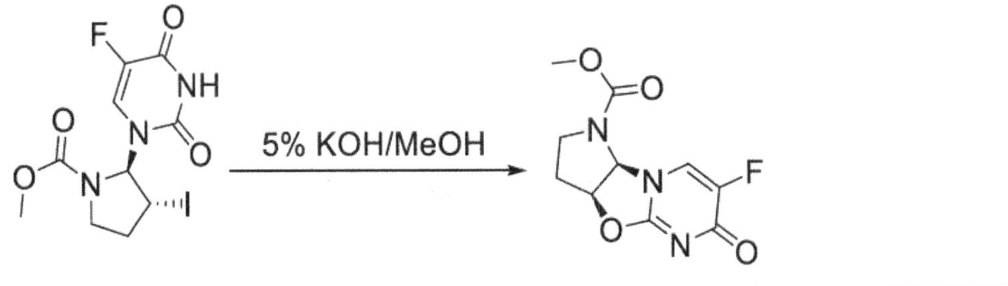

F
NH
5% KOH/MeOH
I

397

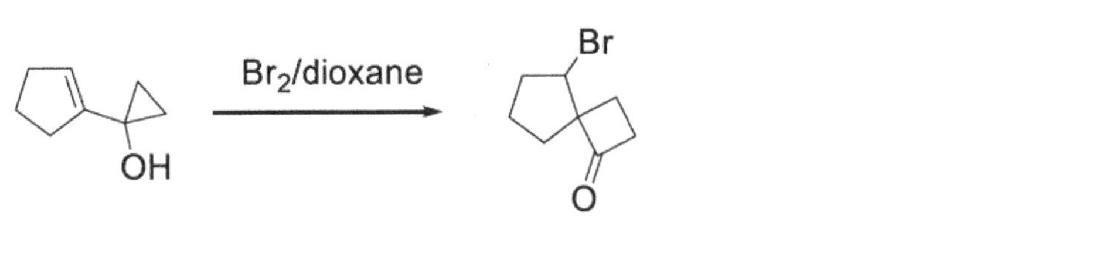

Br2/dioxane
OH
Br

398

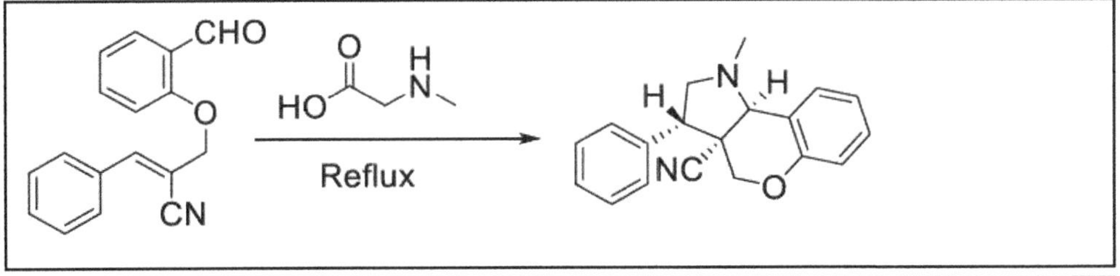

CHO
CN
HO
Reflux
NC

399

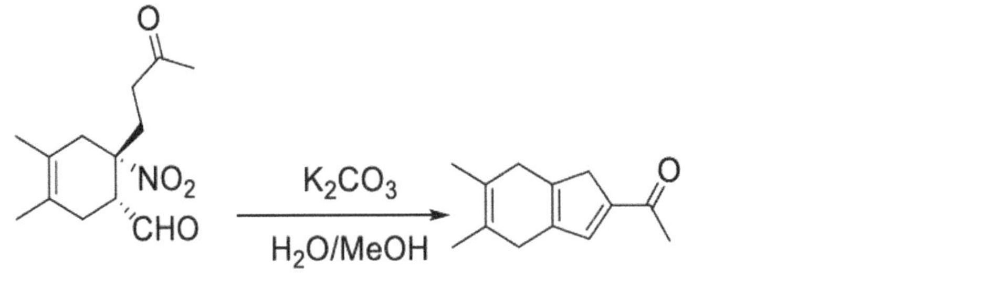

NO2
CHO
K2CO3
H2O/MeOH

400	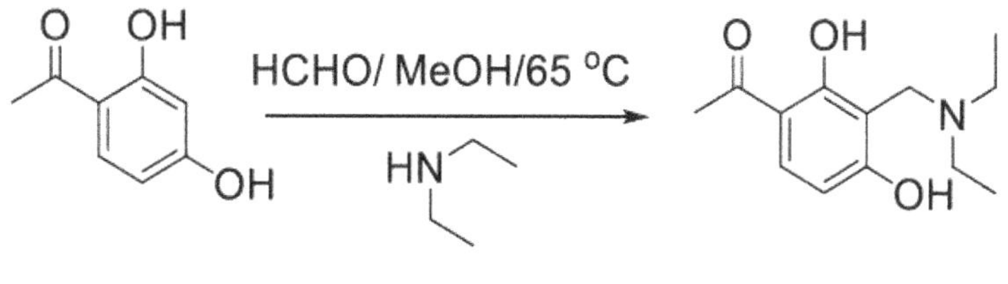O OH OH HCHO/ MeOH/65 °C HN O OH N OH
401	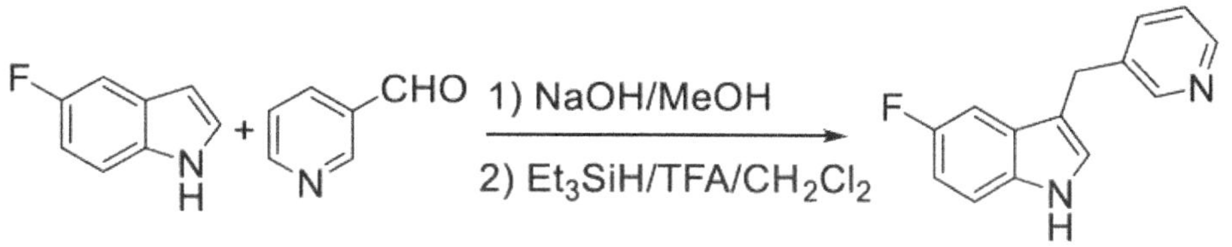F N H + CHO N 1) NaOH/MeOH 2) Et3SiH/TFA/CH2Cl2 F N N H
402	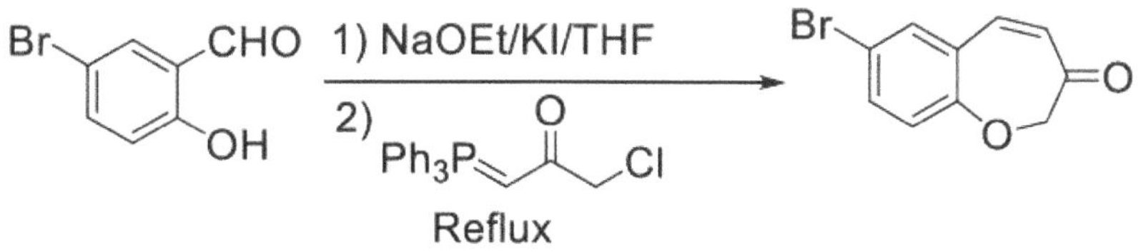Br CHO OH 1) NaOEt/KI/THF 2) Ph3P O Cl Reflux Br O O
403	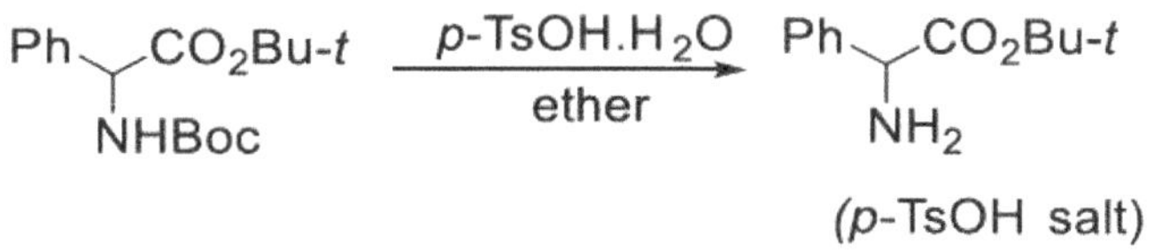Ph CO2Bu-t NHBoc p-TsOH.H2O ether Ph CO2Bu-t NH2 (p-TsOH salt)
404	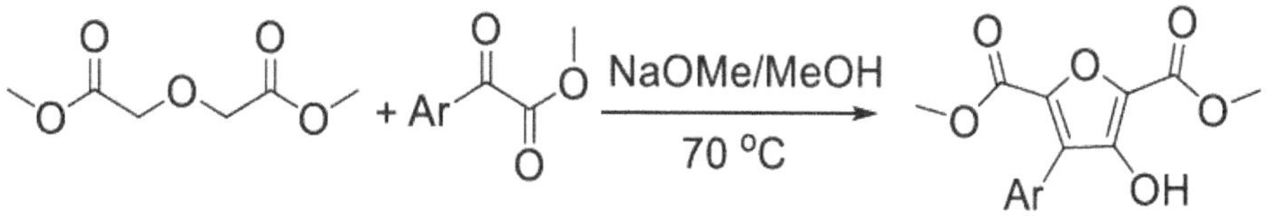O O O O O + Ar O O O NaOMe/MeOH 70 °C O O O O O Ar OH
405	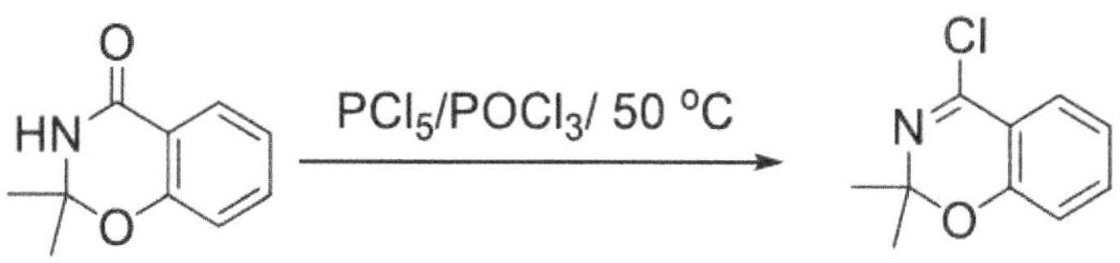O HN O PCl5/POCl3/ 50 °C Cl N O

406

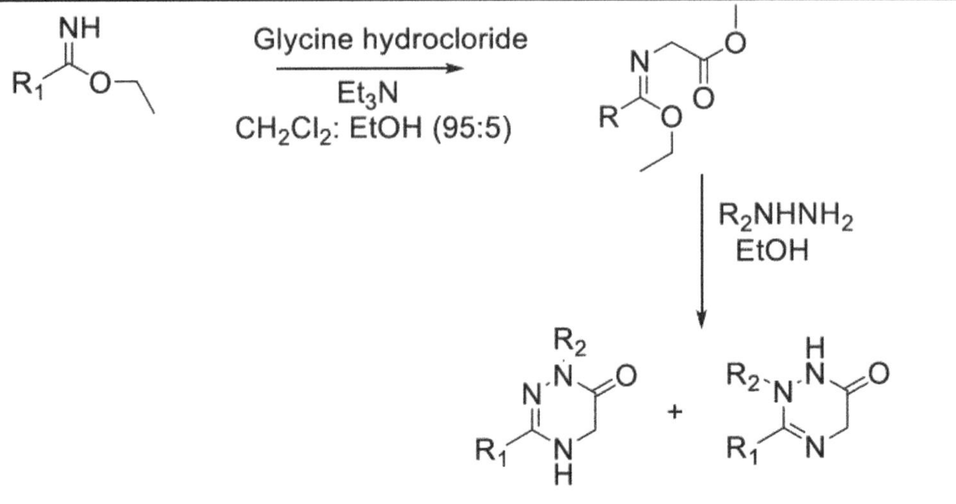
NH
R1
O
Glycine hydrocloride
Et3N
CH2Cl2: EtOH (95:5)
N
O
R
O
O
R2NHNH2
EtOH
R2
N
N
O
R1
N
H
+
H
R2
N
N
O
R1
N

407

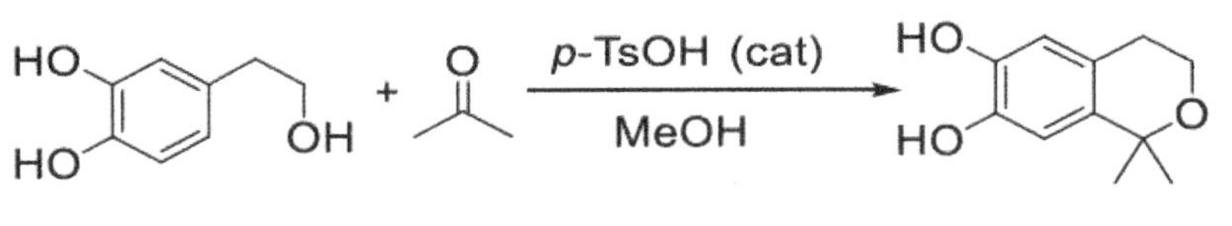
HO
HO
OH
+
O
p-TsOH (cat)
MeOH
HO
HO
O

408

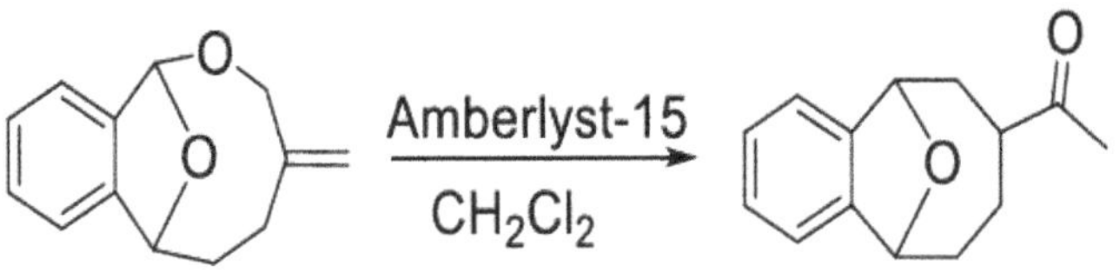
O
O
Amberlyst-15
CH2Cl2
O
O

409

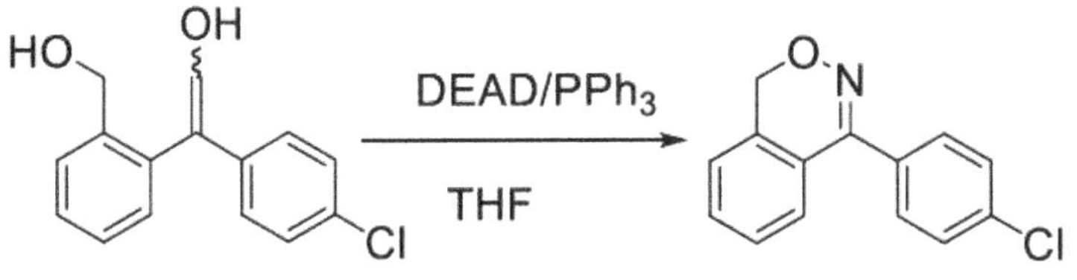
HO
OH
Cl
DEAD/PPh3
THF
O
N
Cl

410

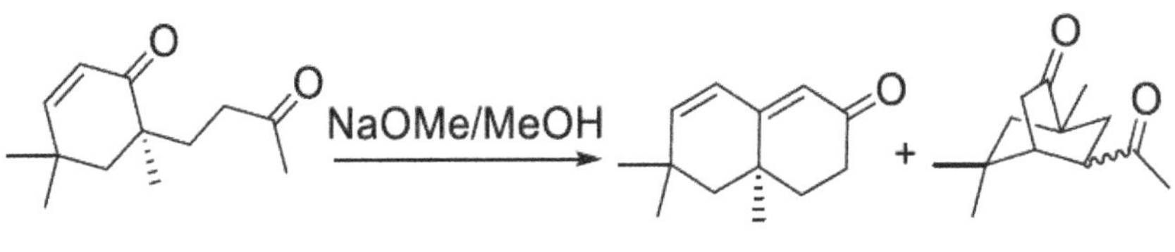
O
O
NaOMe/MeOH
O
+
O
O

411	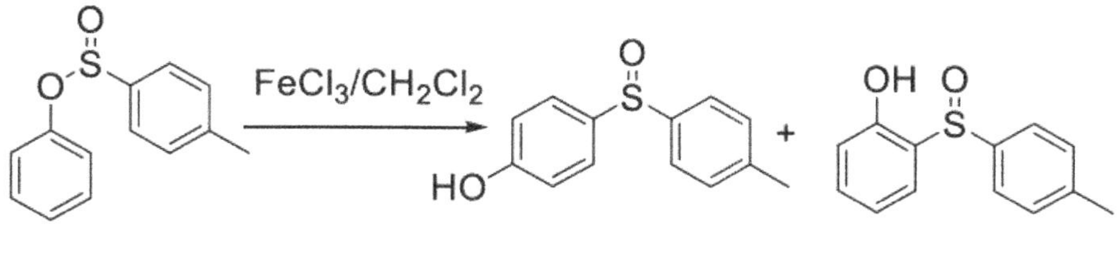FeCl3/CH2Cl2 HO OH +
412	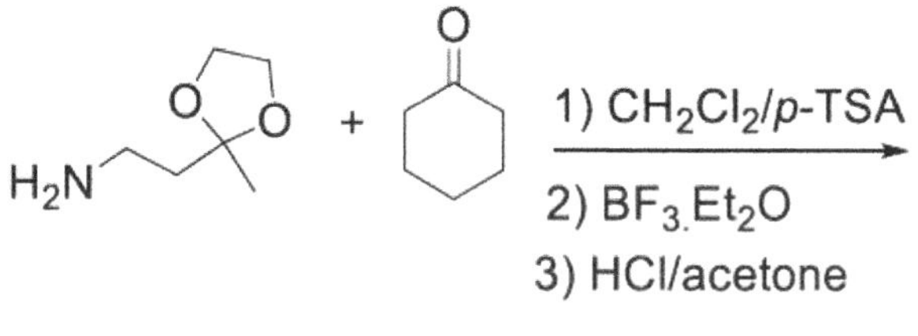H2N + 1) CH2Cl2/p-TSA 2) BF3.Et2O 3) HCl/acetone
413	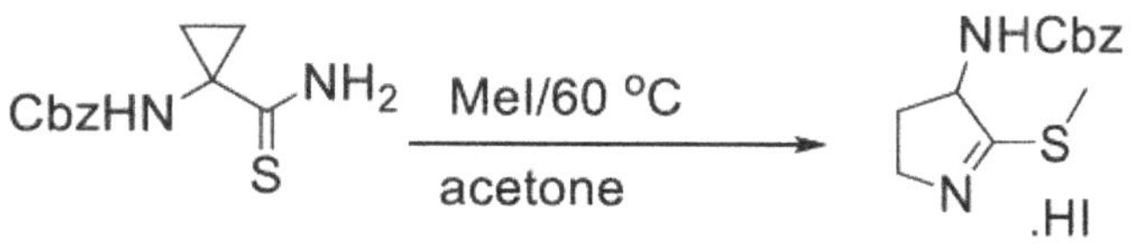CbzHN NH2 MeI/60 °C acetone NHCbz .HI
414	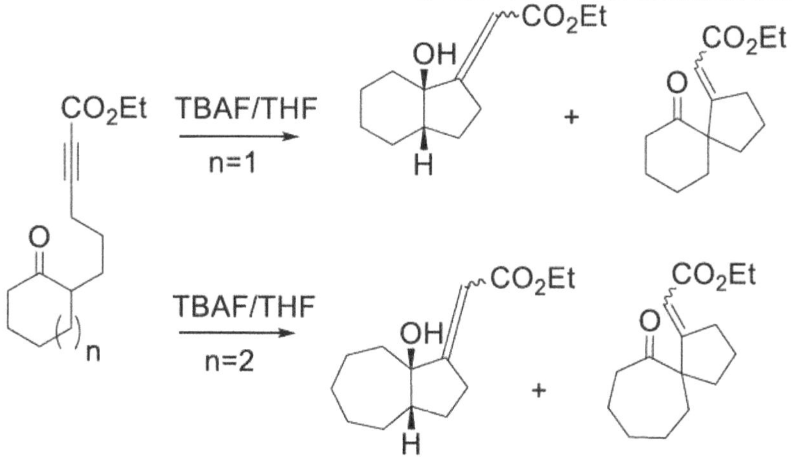CO2Et TBAF/THF n=1 n=2 OH H +
415	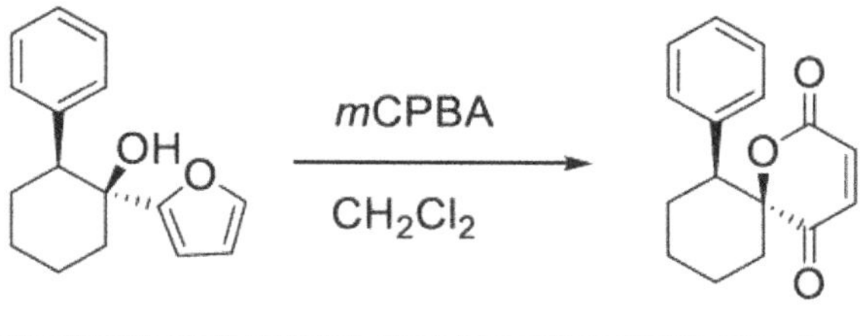OH mCPBA CH2Cl2

416	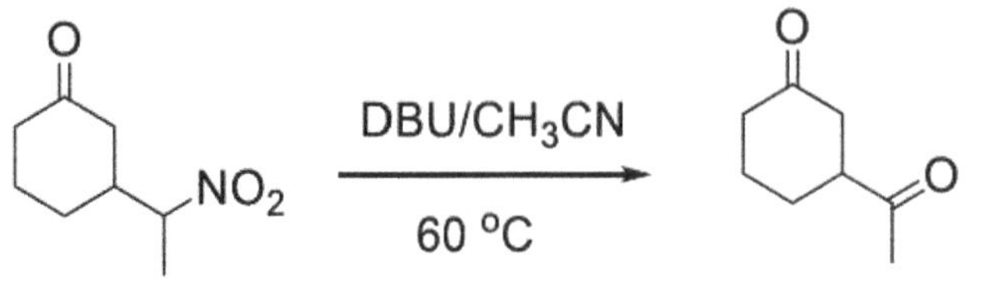NO2 DBU/CH3CN 60 °C
417	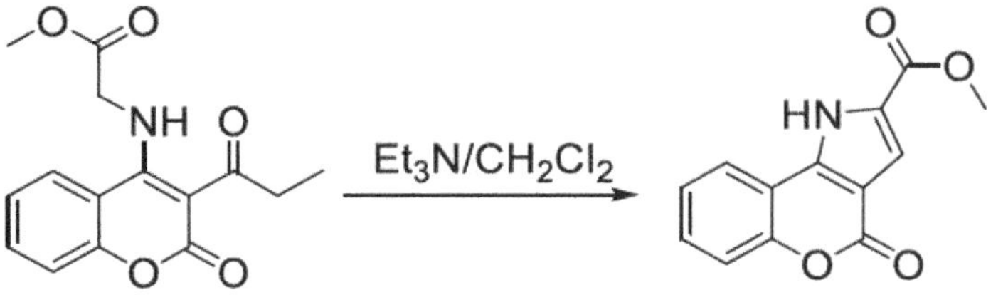NH Et3N/CH2Cl2 HN
418	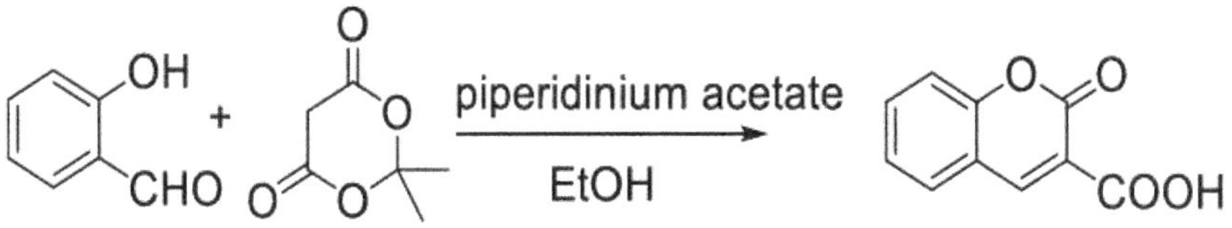OH CHO + piperidinium acetate EtOH COOH
419	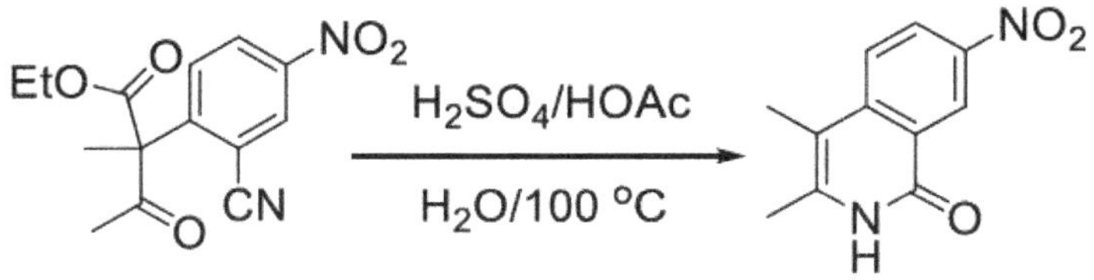EtO NO2 CN H2SO4/HOAc H2O/100 °C NO2 N H
420	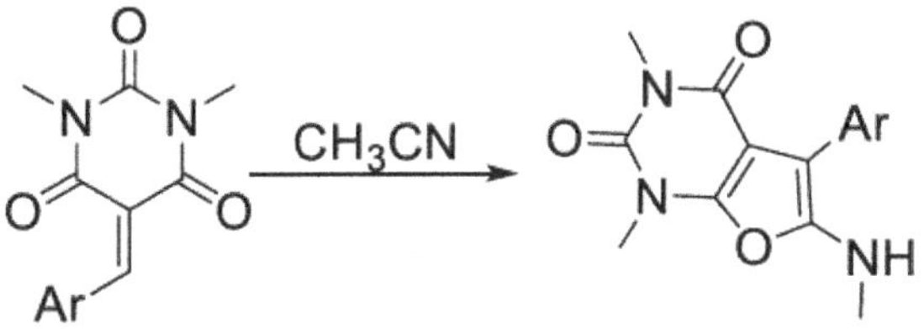Ar CH3CN Ar NH
421	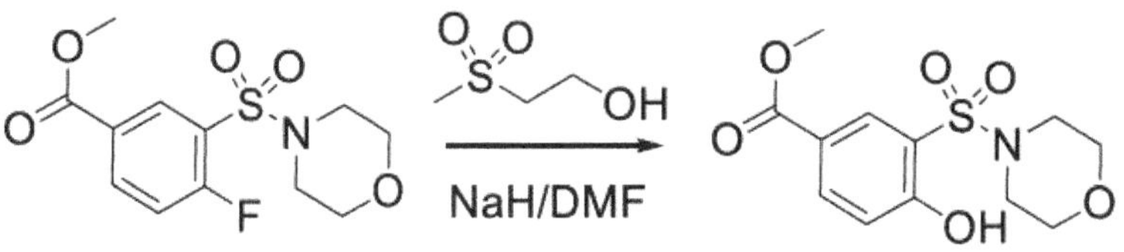F OH NaH/DMF OH

422	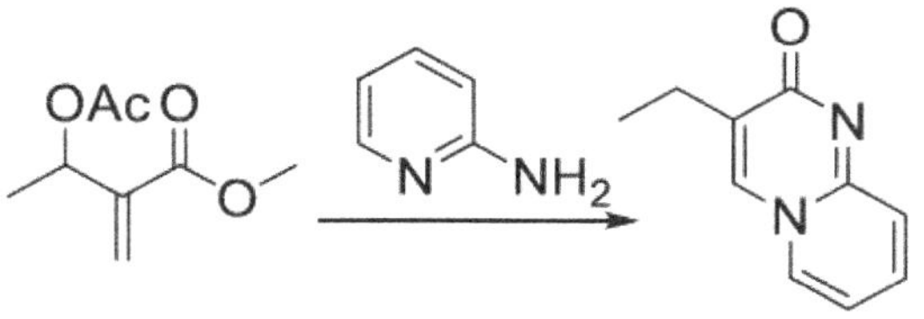OAc O NH2 N O N N
423	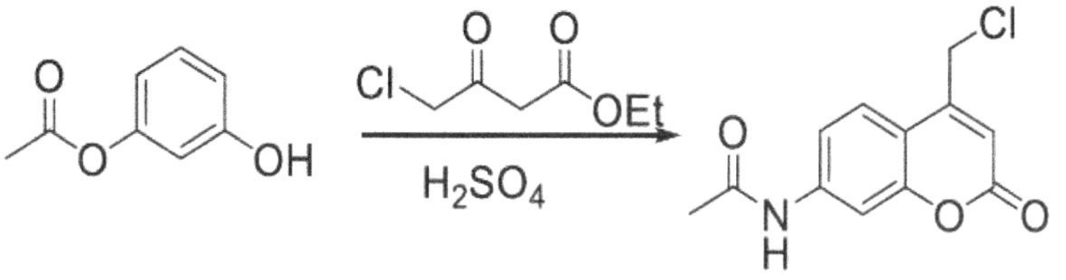Cl OEt H2SO4 OH
424	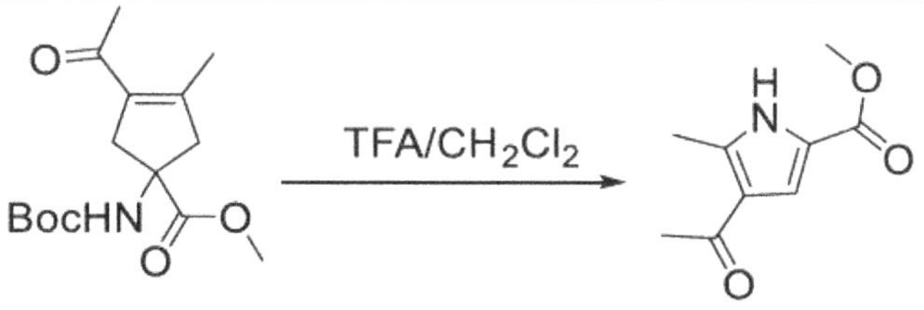BocHN TFA/CH2Cl2
425	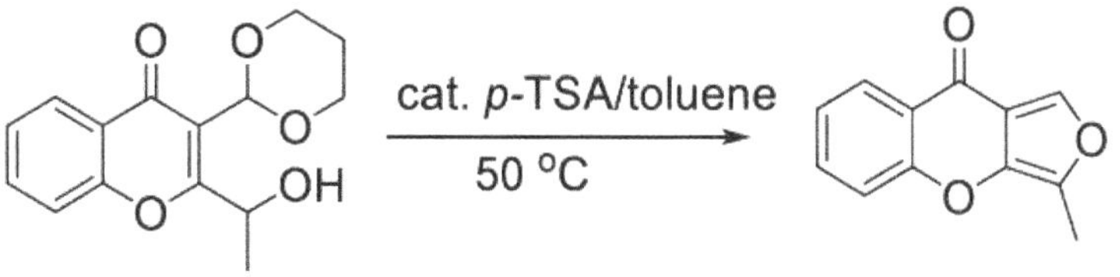cat. p-TSA/toluene 50 °C OH
426	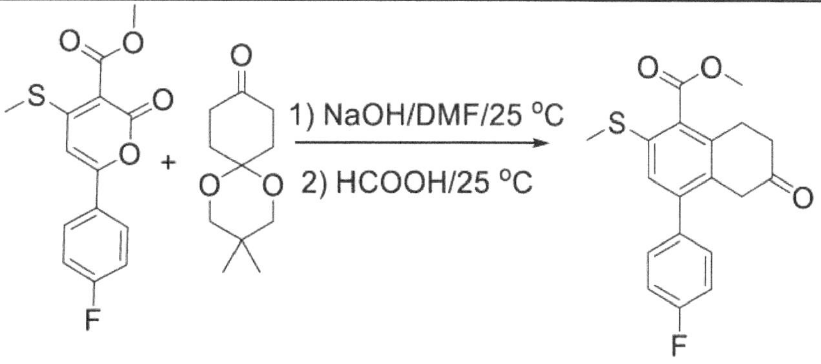1) NaOH/DMF/25 °C 2) HCOOH/25 °C F
427	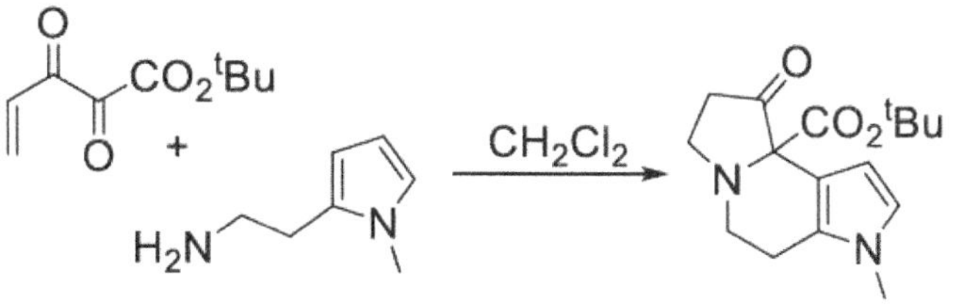CO2tBu H2N CH2Cl2

428	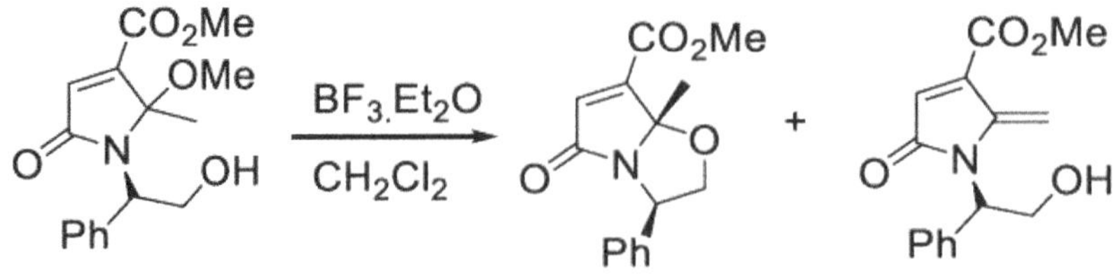BF$_3$.Et$_2$O CH$_2$Cl$_2$
429	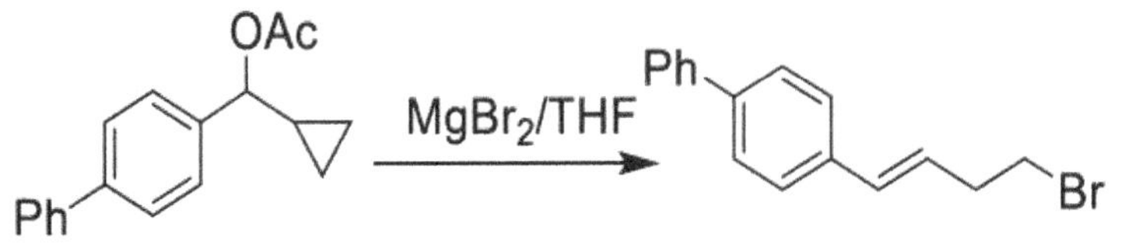MgBr$_2$/THF
430	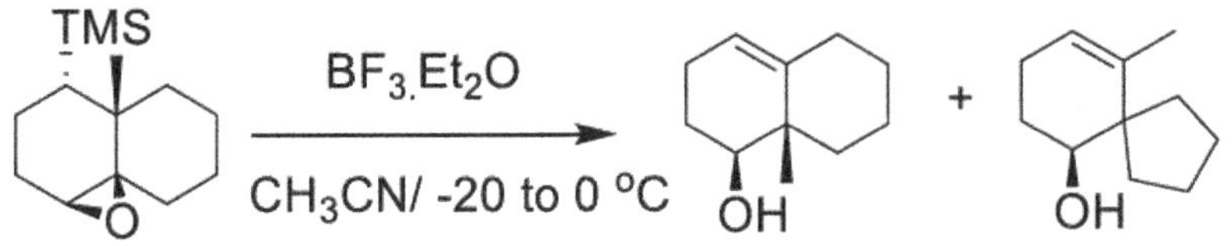BF$_3$.Et$_2$O CH$_3$CN/ -20 to 0 ^{o}C
431	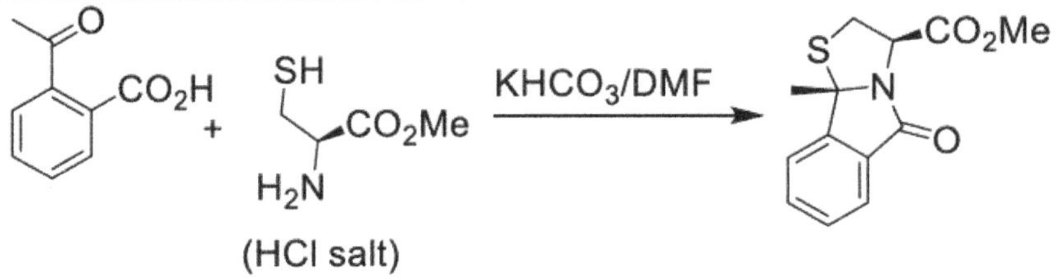KHCO$_3$/DMF (HCl salt)
432	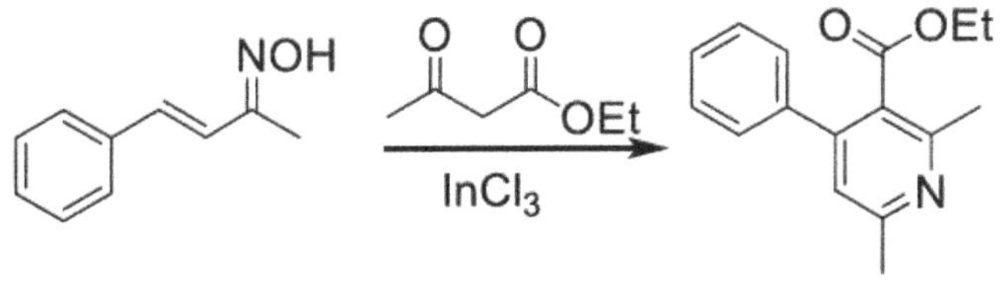InCl$_3$
433	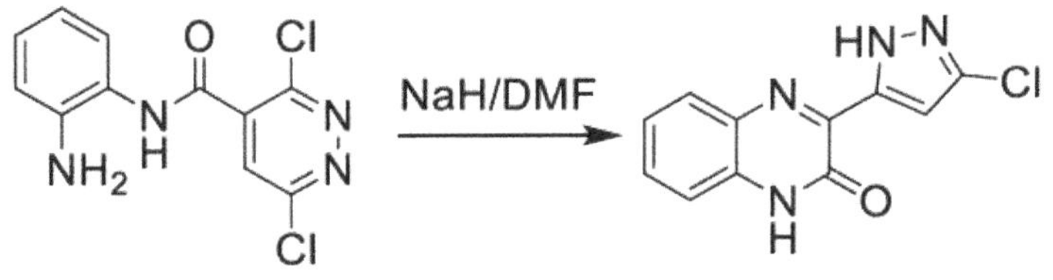NaH/DMF

434

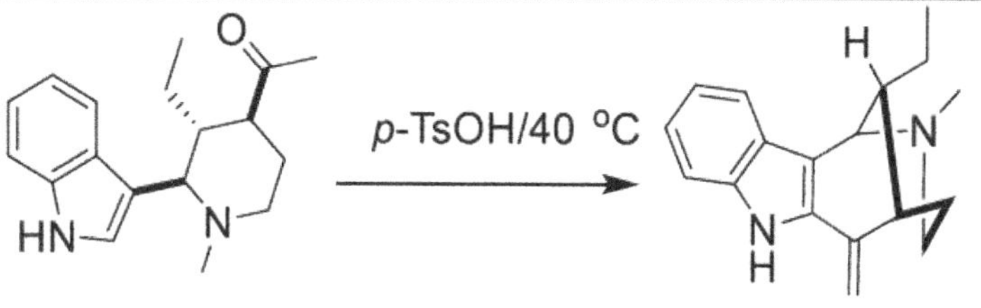
p-TsOH/40 °C
O
HN
N
H
N
N
H

Epiuleine

435

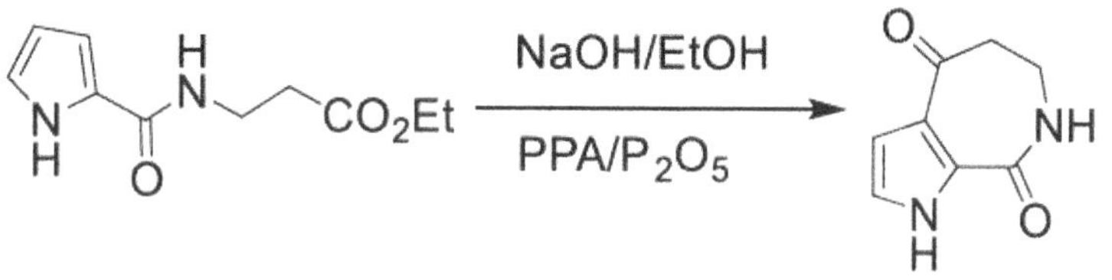
N
H
H
N
O
CO2Et
NaOH/EtOH
PPA/P2O5
O
NH
N
H
O

436

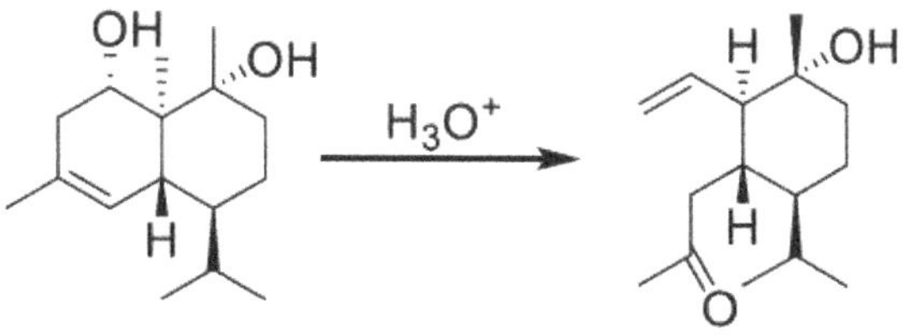
OH
OH
H
H3O+
H
OH
H
O

437

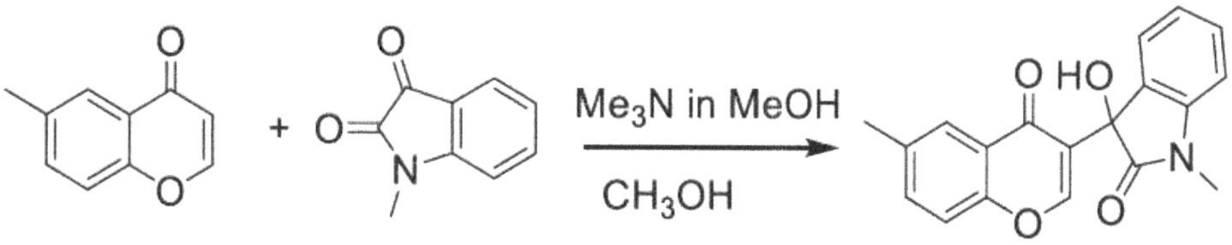
O
O
+
O
O
N
Me3N in MeOH
CH3OH
O HO
O
O
N

438

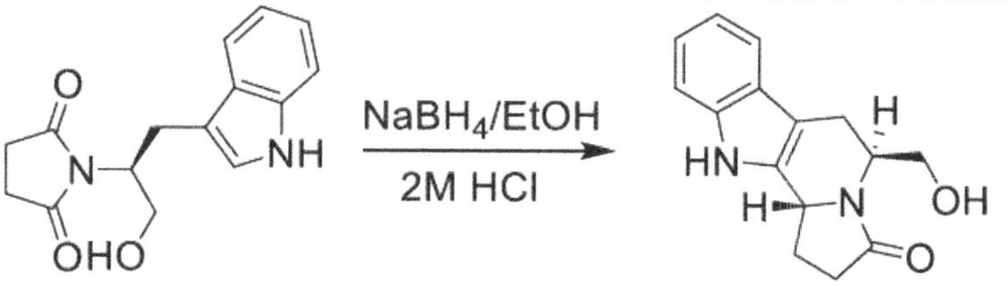
O
N
O HO
NH
NaBH4/EtOH
2M HCl
HN
H
H
N
OH
O

439

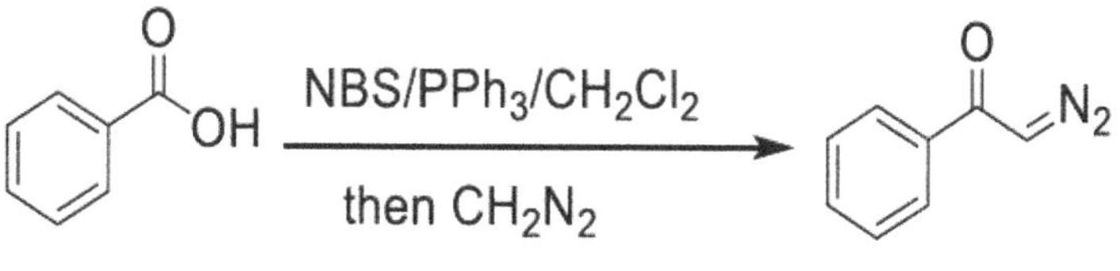
O
OH
NBS/PPh3/CH2Cl2
then CH2N2
O
N2

440

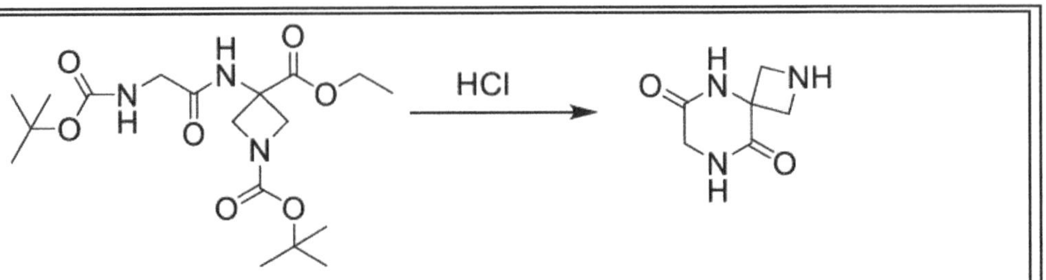
HCl

441

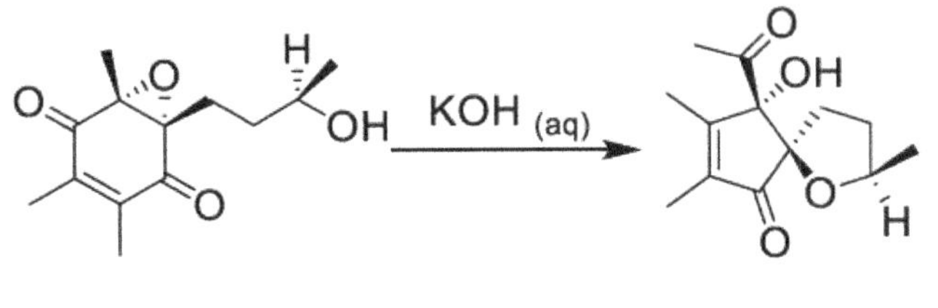
KOH (aq)

442

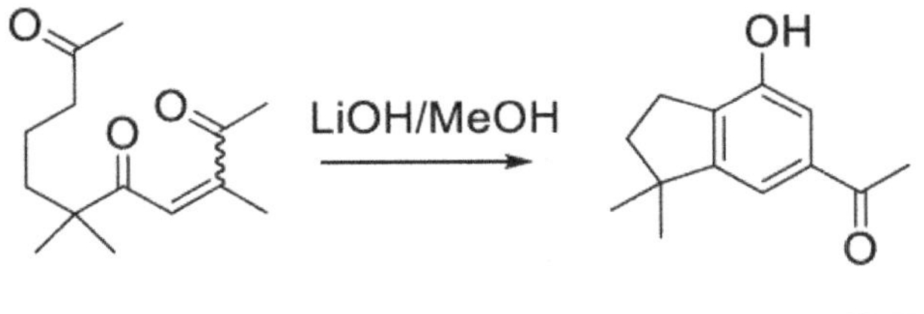
LiOH/MeOH

443

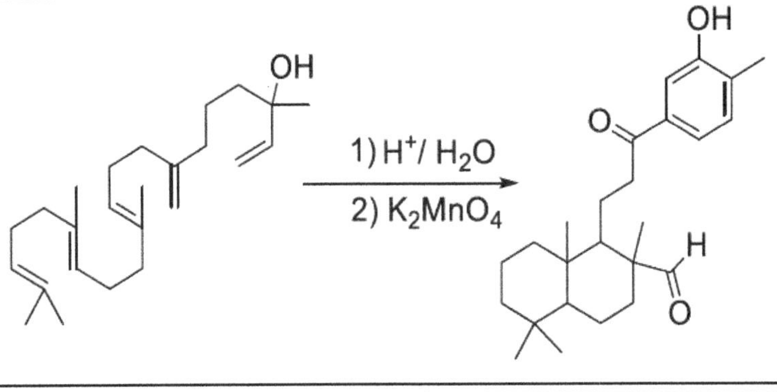
1) H+/ H2O
2) K2MnO4

444

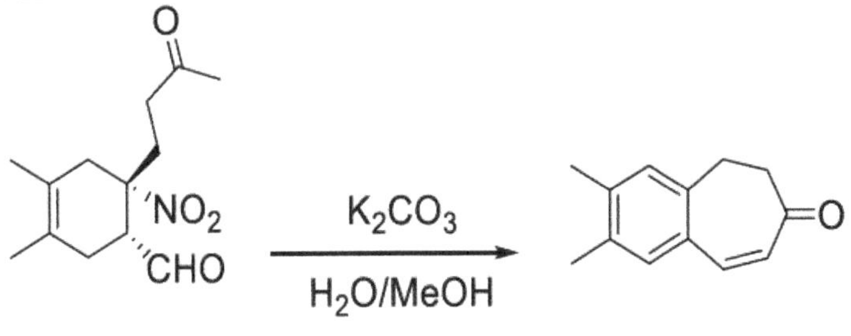
K2CO3
H2O/MeOH

445	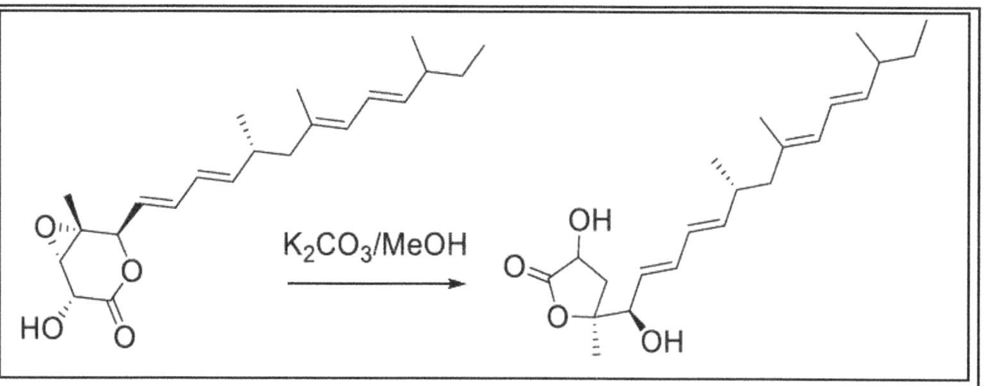 HO O O K2CO3/MeOH OH O O OH
446	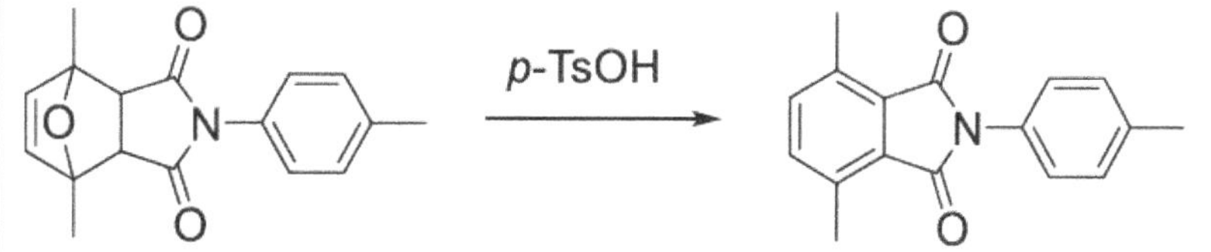 O O N O p-TsOH O N O
447	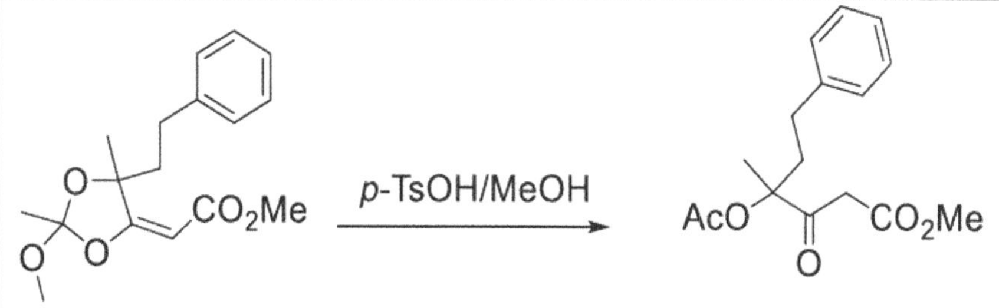 O O O CO2Me p-TsOH/MeOH AcO O CO2Me
448	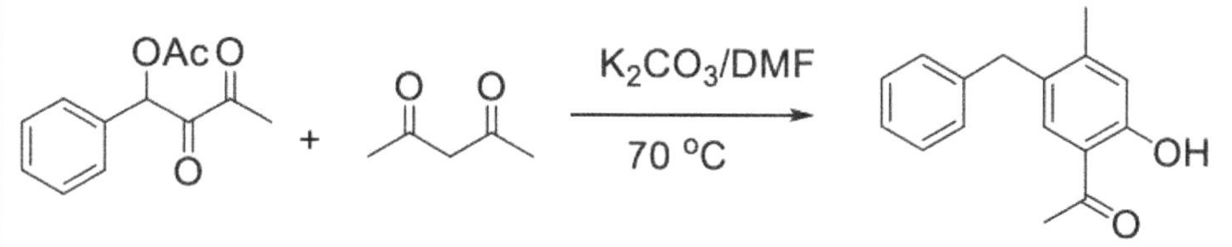 OAc O O + O O K2CO3/DMF 70 °C OH O
449	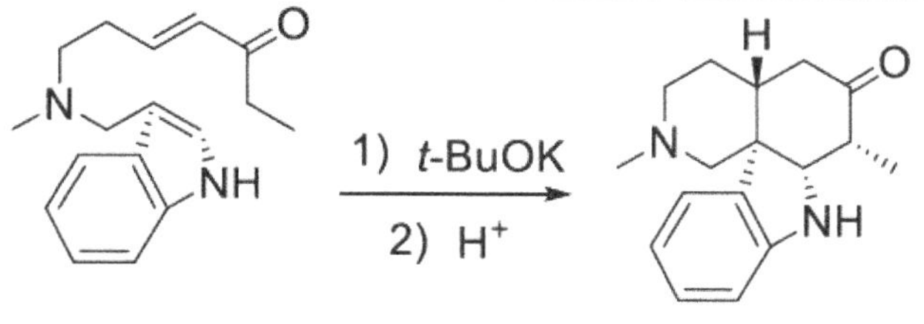 O N NH 1) t-BuOK 2) H+ H O N NH
450	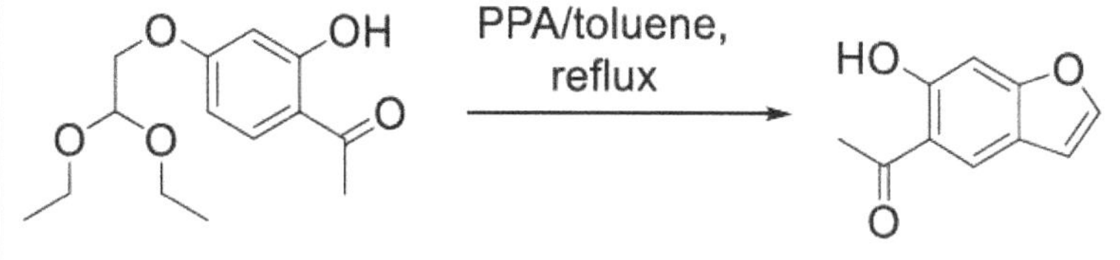 O OH O O O PPA/toluene, reflux HO O O

451

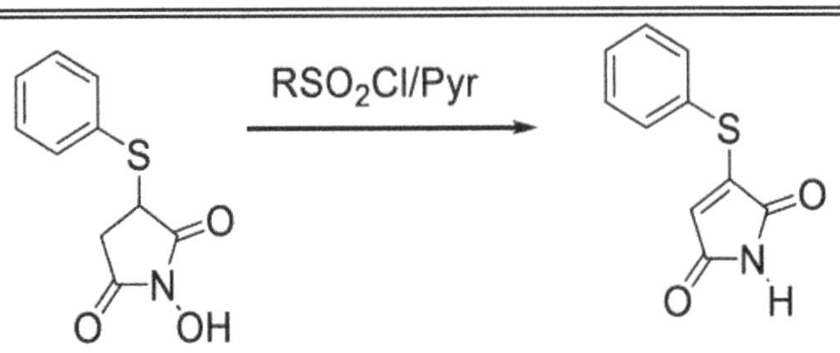
RSO$_2$Cl/Pyr
S
O
N
OH
S
O
N
H
O

452

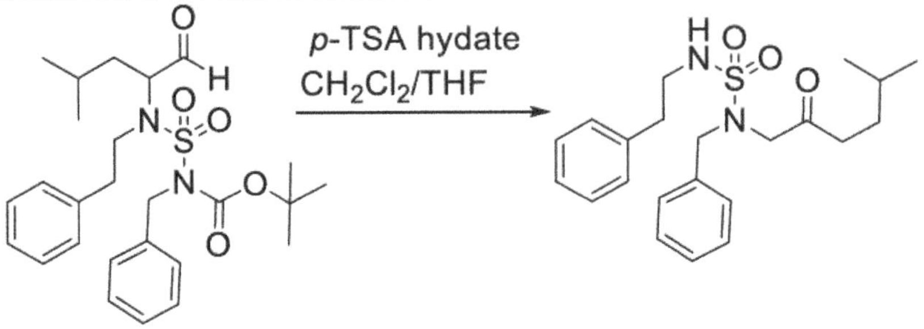
p-TSA hydate
CH$_2$Cl$_2$/THF
O
H
N
S
O
O
N
O
O
H
N
S
O
O
N
O

453

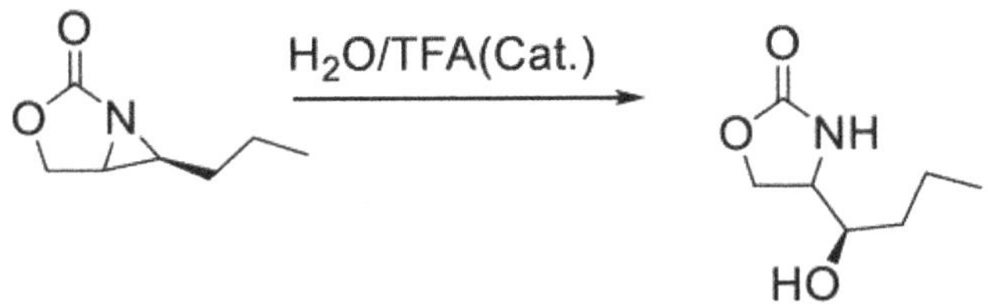
H$_2$O/TFA(Cat.)
O
O
N
O
O
NH
HO

454

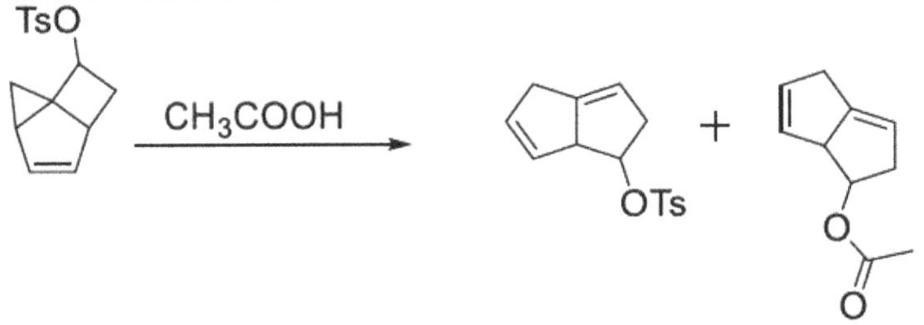
TsO
CH$_3$COOH
OTs
+
O
O

455

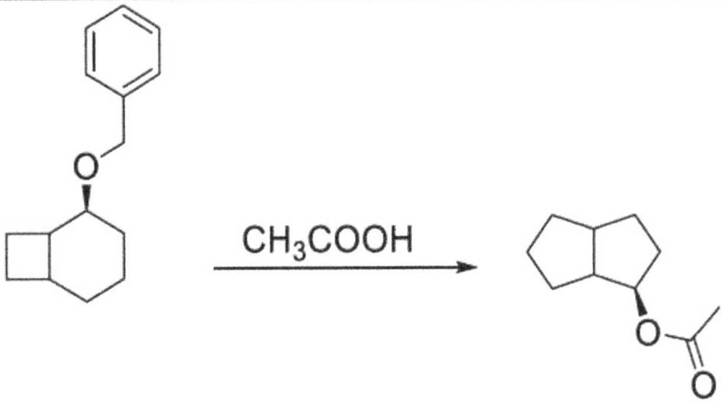
O
CH$_3$COOH
O
O

456	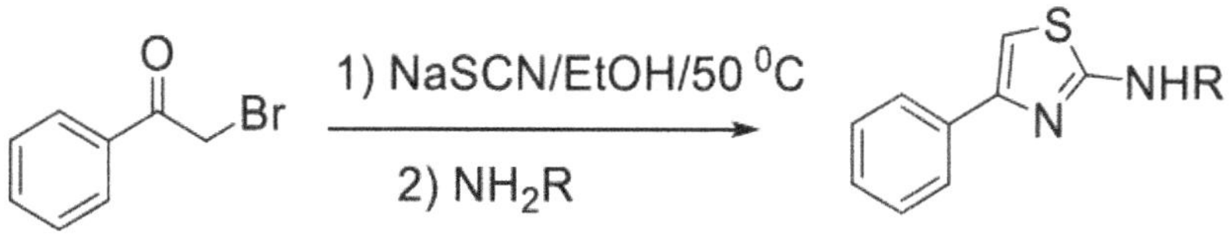Br 1) NaSCN/EtOH/50 ⁰C 2) NH2R S N NHR
457	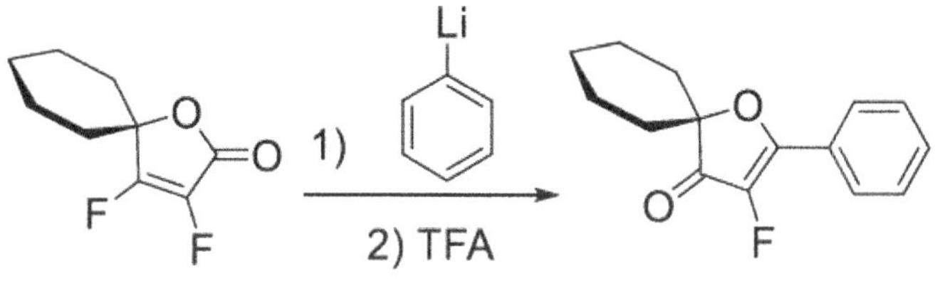O O F F 1) Li 2) TFA O O F
458	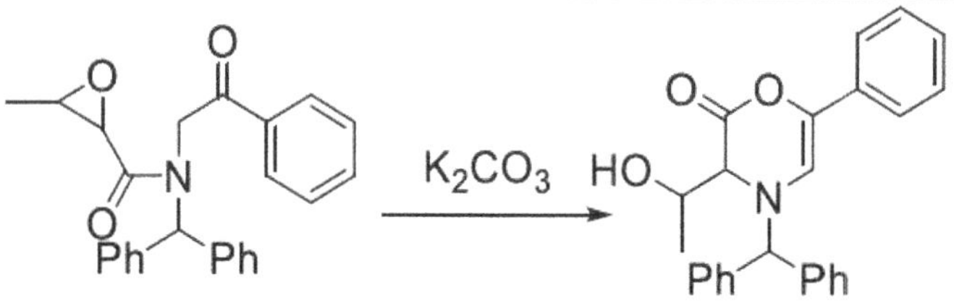O O N O Ph Ph K2CO3 O O HO N Ph Ph
459	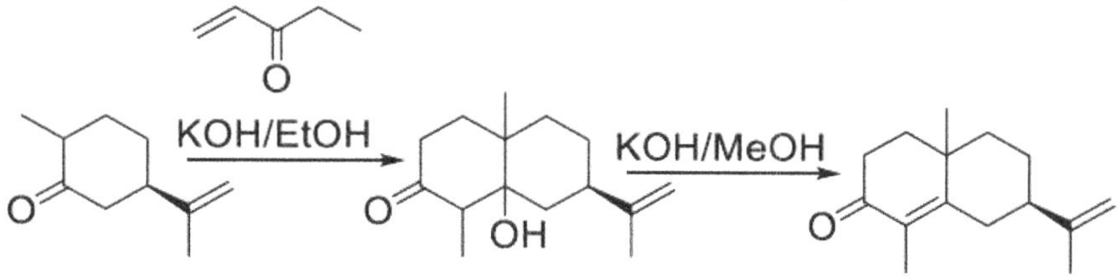O KOH/EtOH O O OH KOH/MeOH O
460	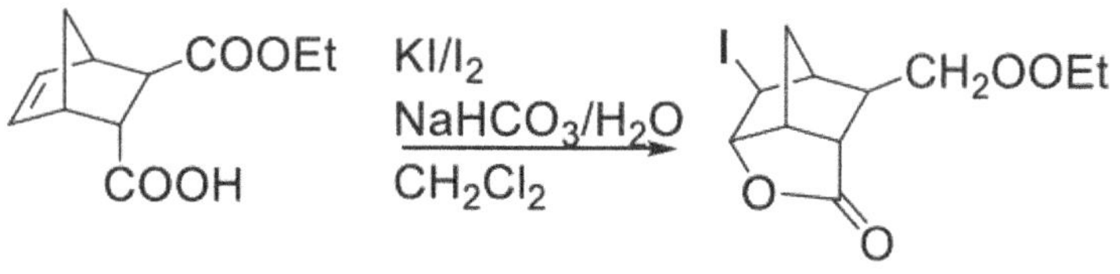COOEt COOH KI/I2 NaHCO3/H2O CH2Cl2 I CH2OOEt O O
461	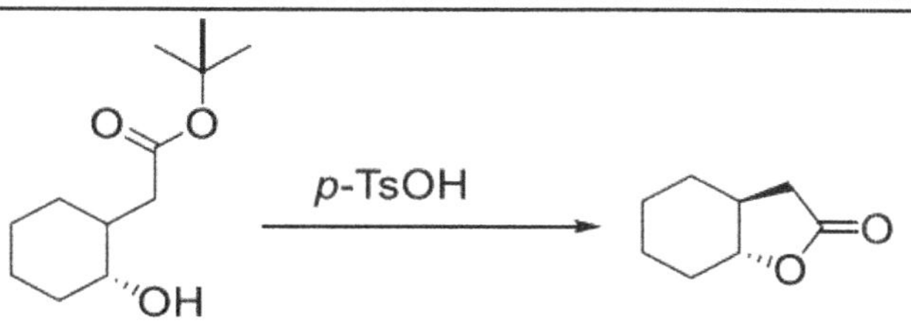O O OH p-TsOH O O

462	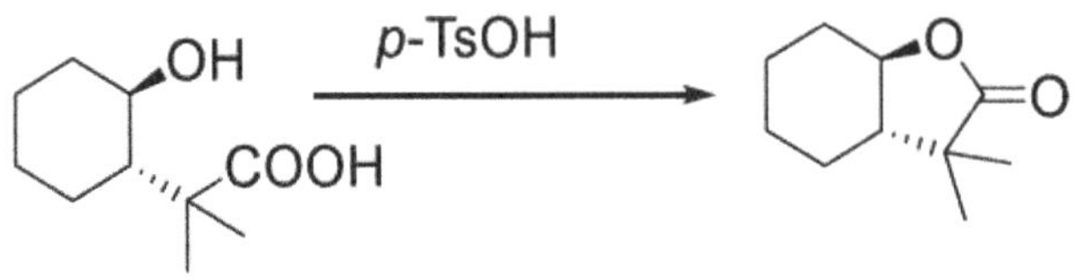 OH COOH p-TsOH O O
463	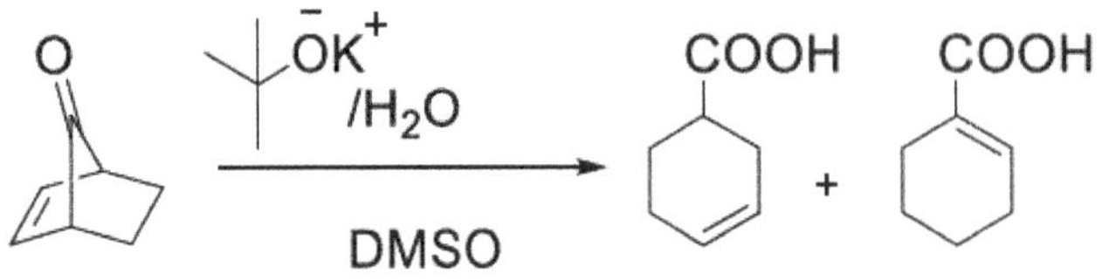 O OK+ /H2O DMSO COOH + COOH
464	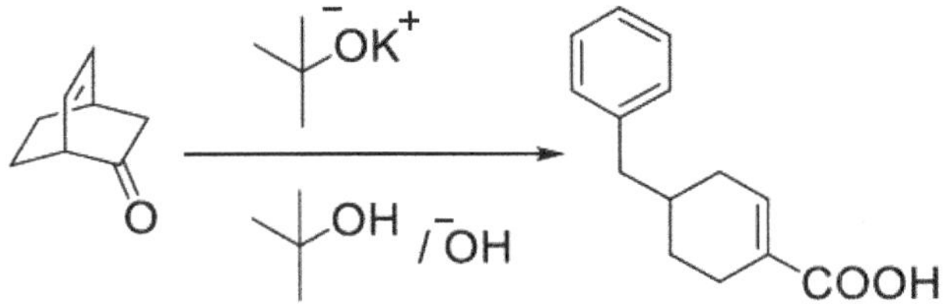 O OK+ OH / OH COOH
465	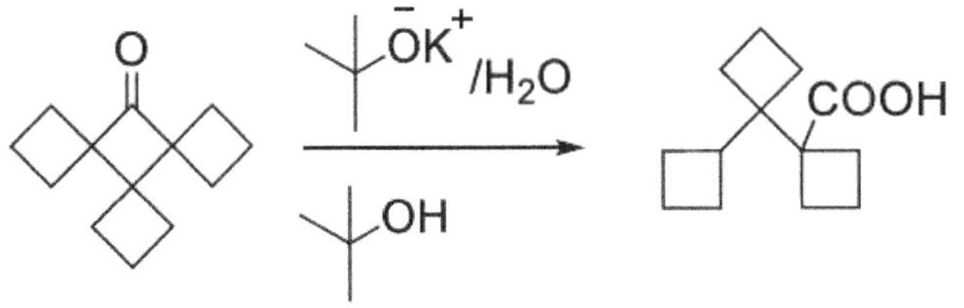 O OK+ /H2O OH COOH
466	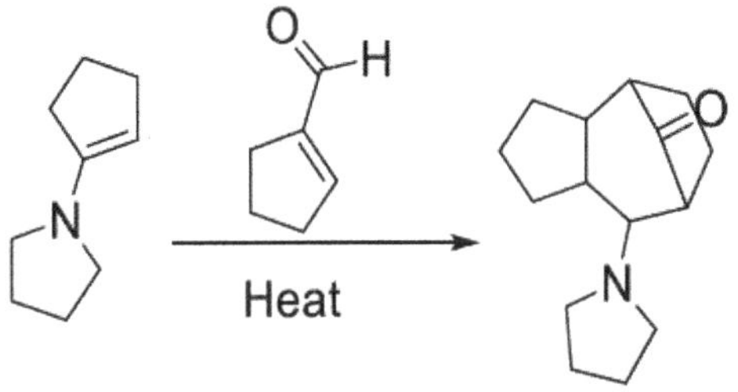 N O H Heat O N
467	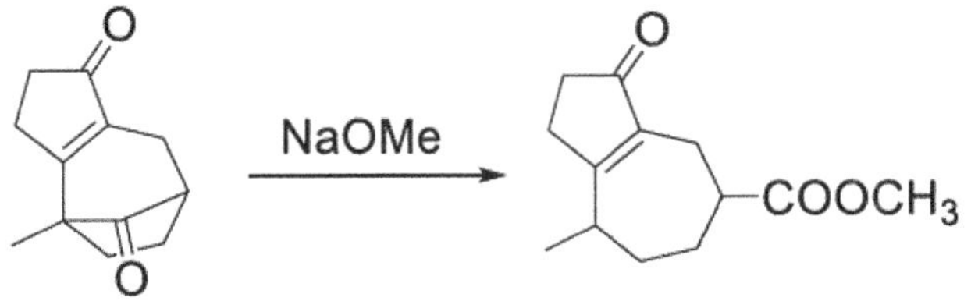 O O NaOMe O COOCH3

468	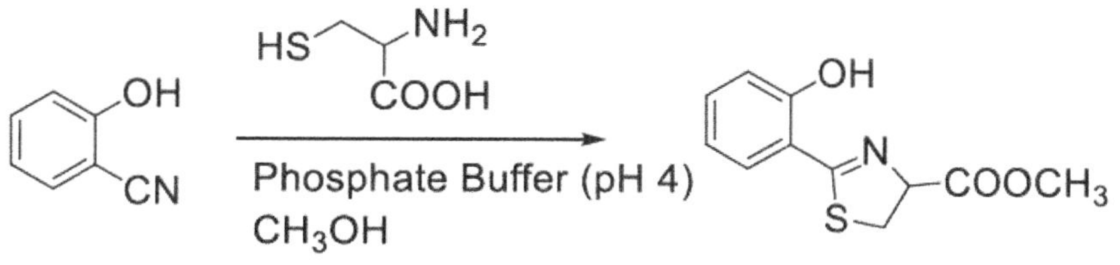 OH CN HS NH2 COOH Phosphate Buffer (pH 4) CH3OH OH N S COOCH3
469	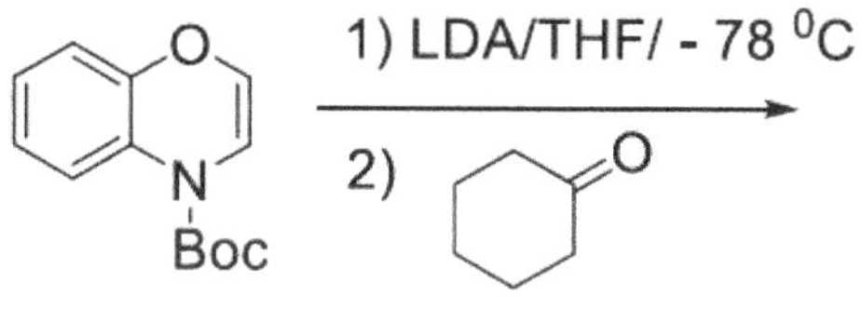 O N Boc 1) LDA/THF/ - 78 0C 2) O O N O O
470	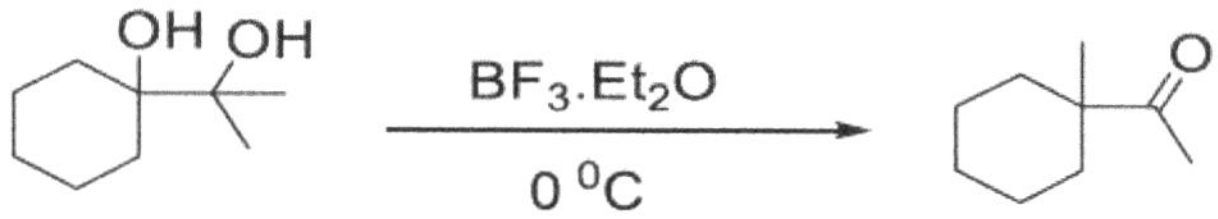 OH OH BF3.Et2O 0 0C O
471	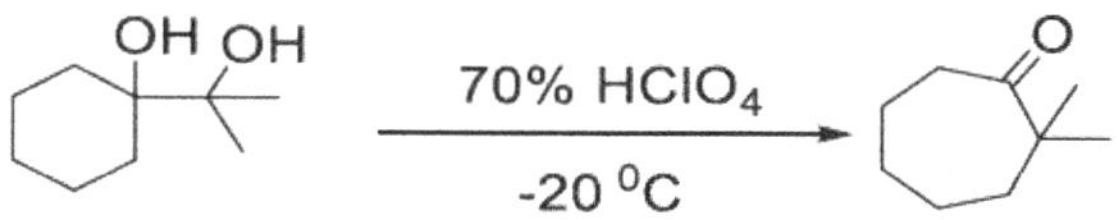 OH OH 70% HClO4 -20 0C O
472	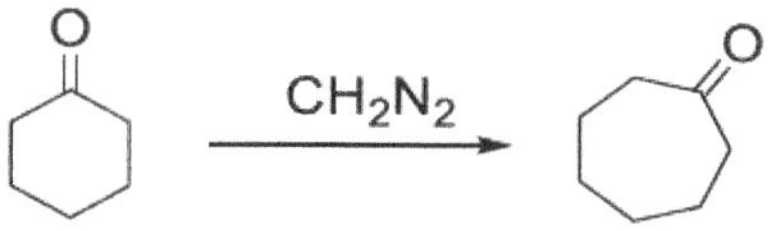 O CH2N2 O
473	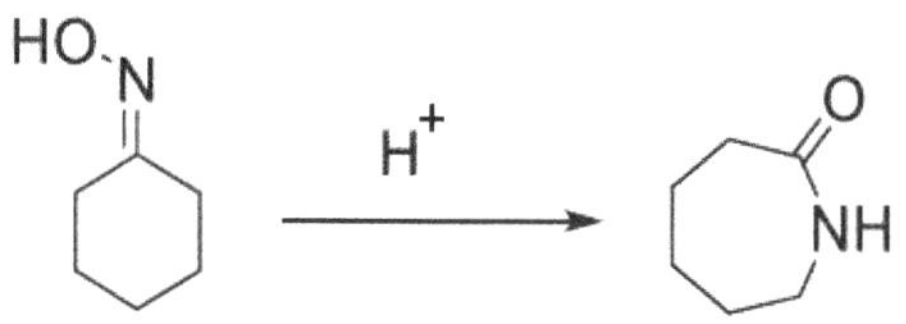 HO N H+ O NH

474	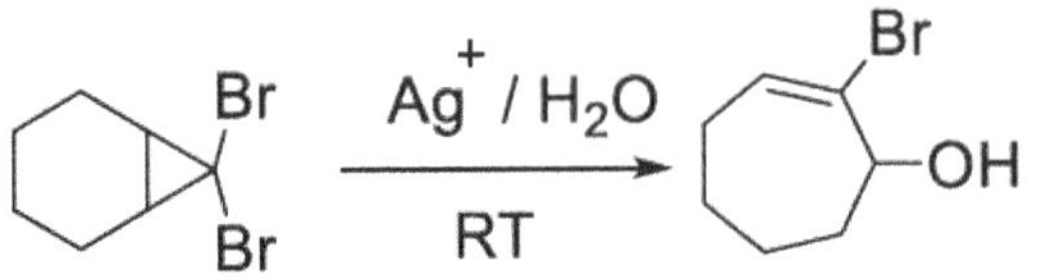 Br Br Ag+ / H2O RT Br OH
475	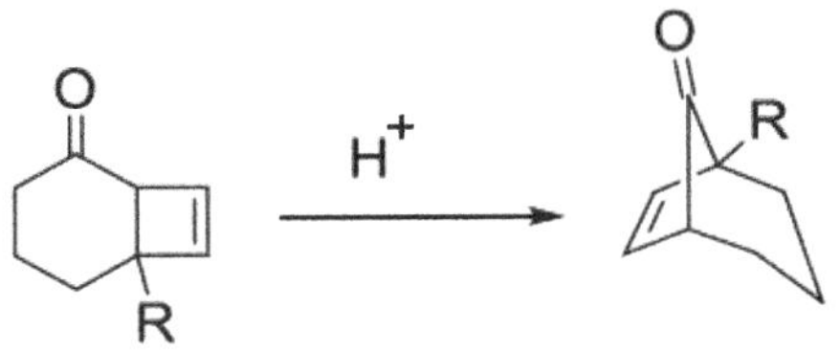 O R H+ O R
476	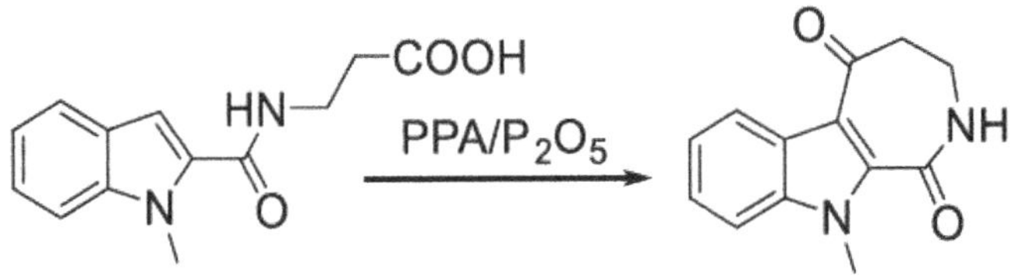 COOH HN N O PPA/P2O5 O NH N O
477	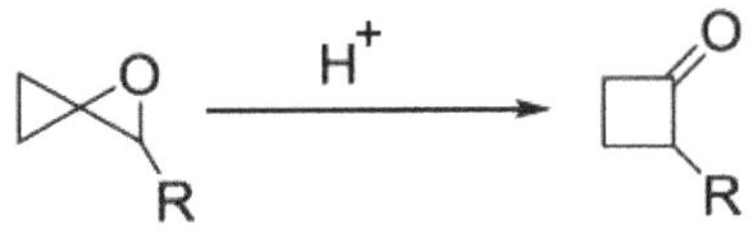 O R H+ O R
478	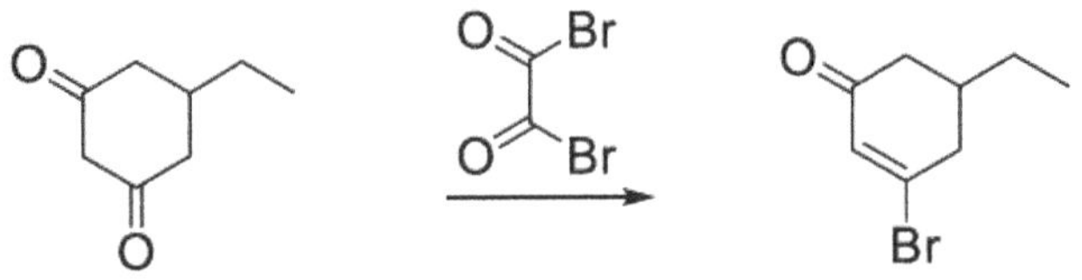 O O O Br O Br O Br
479	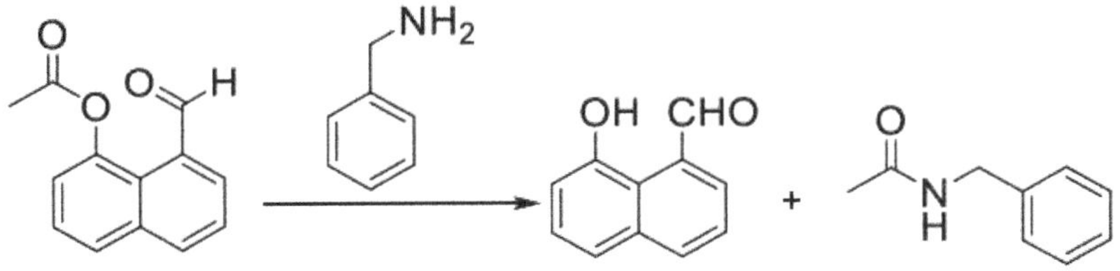 O O O H NH2 OH CHO + O N H

480

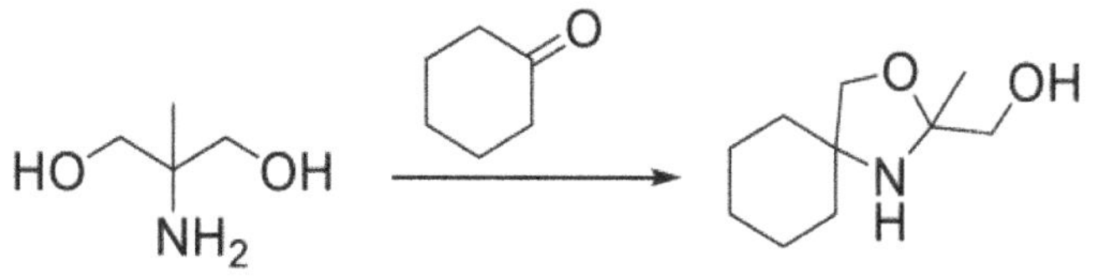
HO
OH
NH2
O
O
OH
N
H

481

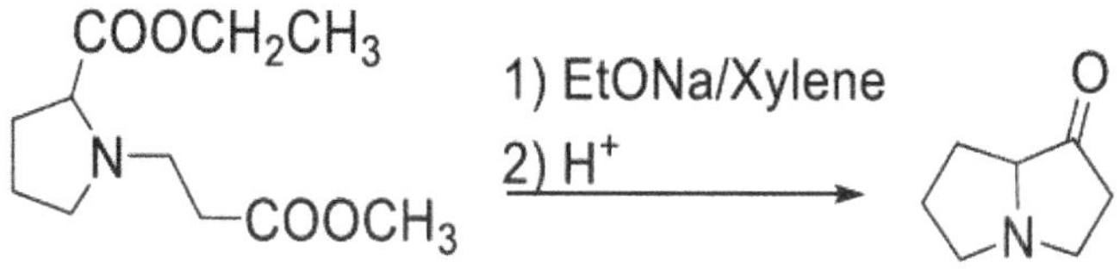
COOCH2CH3
N
COOCH3
1) EtONa/Xylene
2) H+
O
N

482

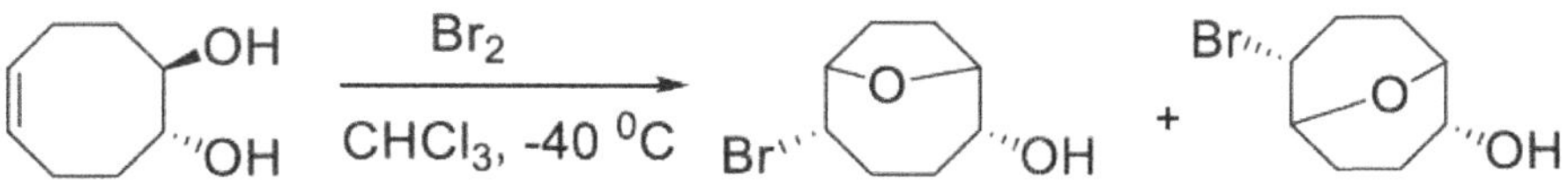
OH
OH
Br2
CHCl3, -40 0C
O
Br
OH
+
Br
O
OH

483

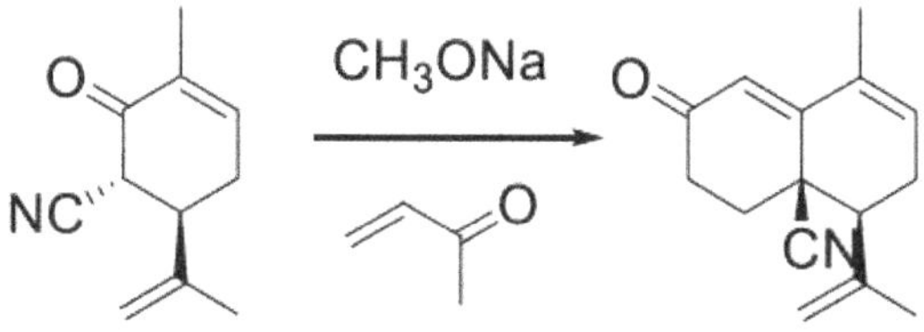
O
NC
CH3ONa
O
O
CN

484

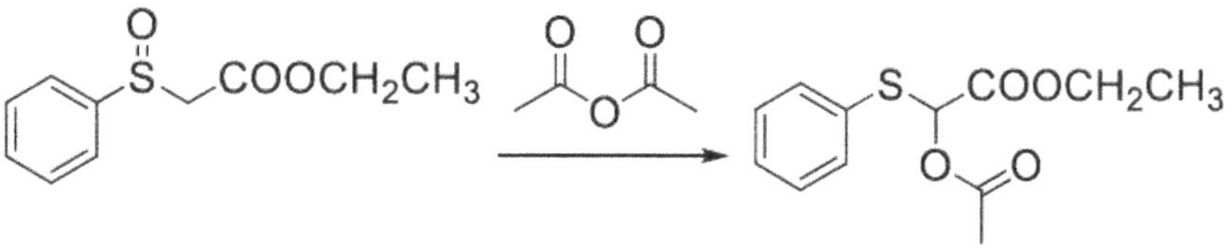
O
S
COOCH2CH3
O
O
O
S
COOCH2CH3
O
O

485

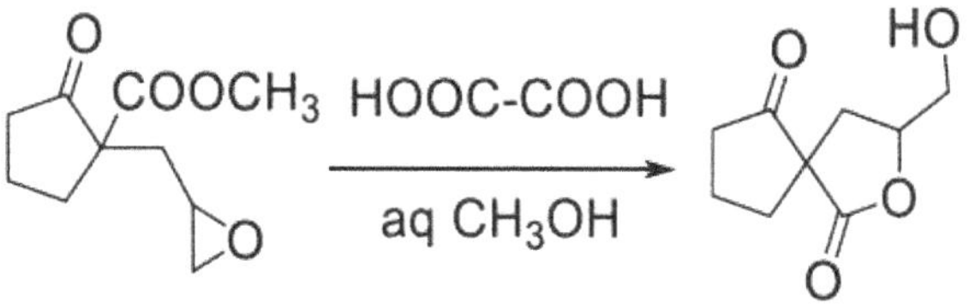
O
COOCH3
O
HOOC-COOH
aq CH3OH
O
HO
O
O

486	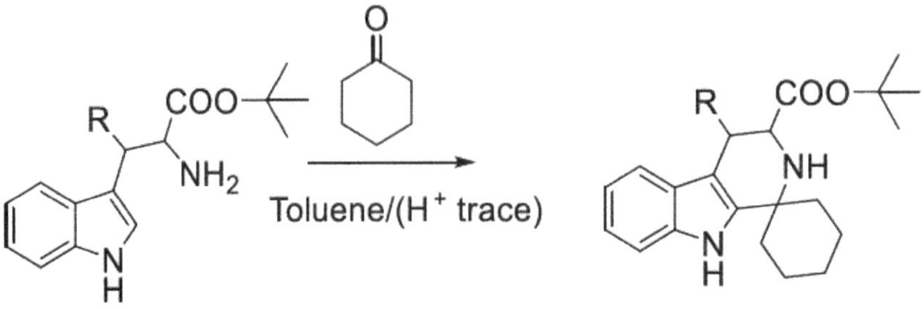R COO NH2 N H O Toluene/(H+ trace) R COO NH N H
487	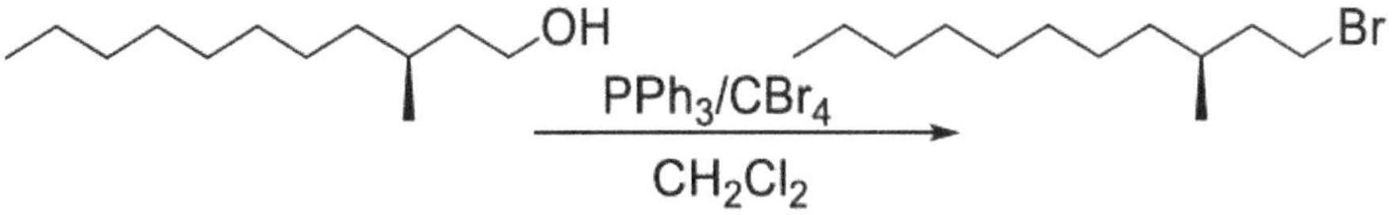OH PPh3/CBr4 CH2Cl2 Br
488	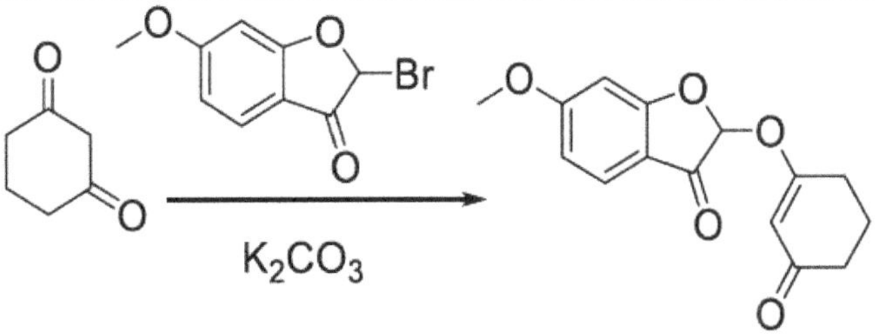O O O O Br O K2CO3 O O O O O
489	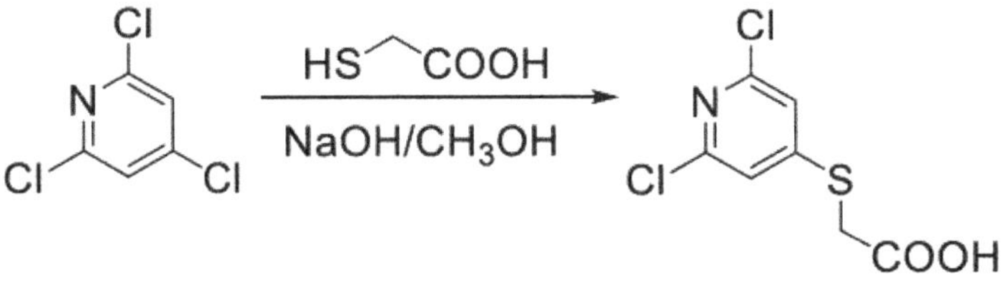Cl N Cl Cl HS COOH NaOH/CH3OH Cl N Cl S COOH
490	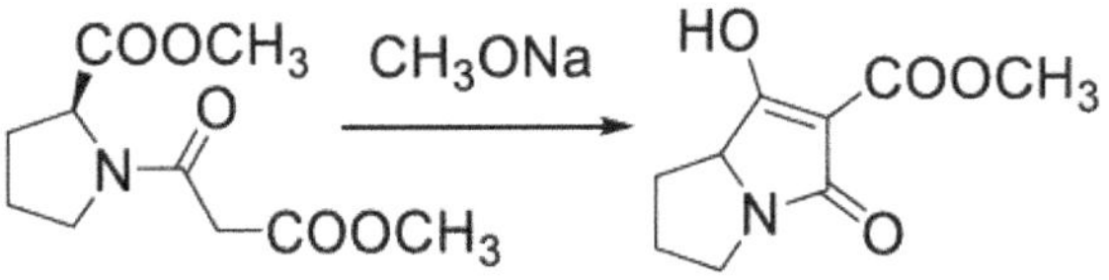COOCH3 O N COOCH3 CH3ONa HO COOCH3 N O

Answers to Questions 201-490

201

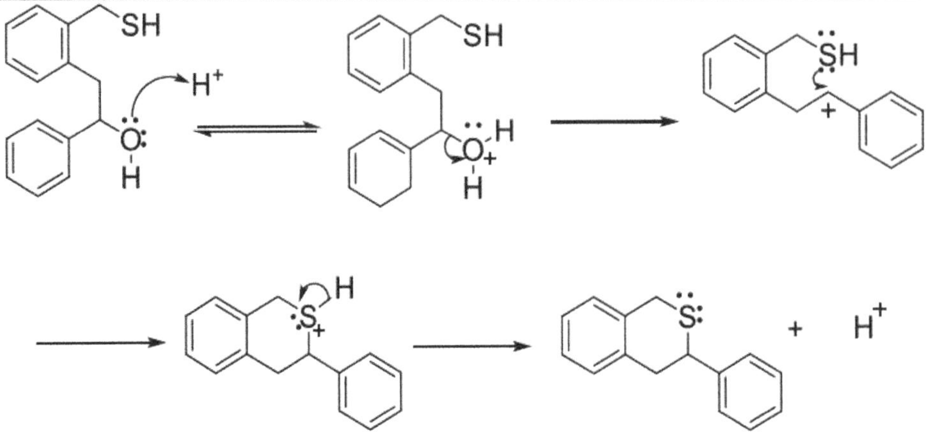
SH
H+
H
SH
H
H
SH
H
H
+
H+

202

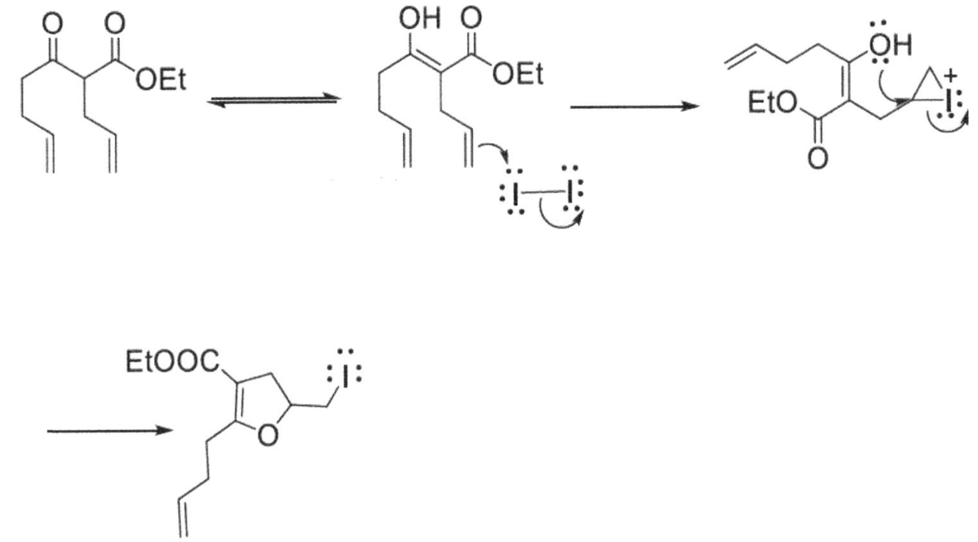
O
O
OEt
OH
O
OEt
I
I
OH
EtO
O
I
EtOOC
I
O

203

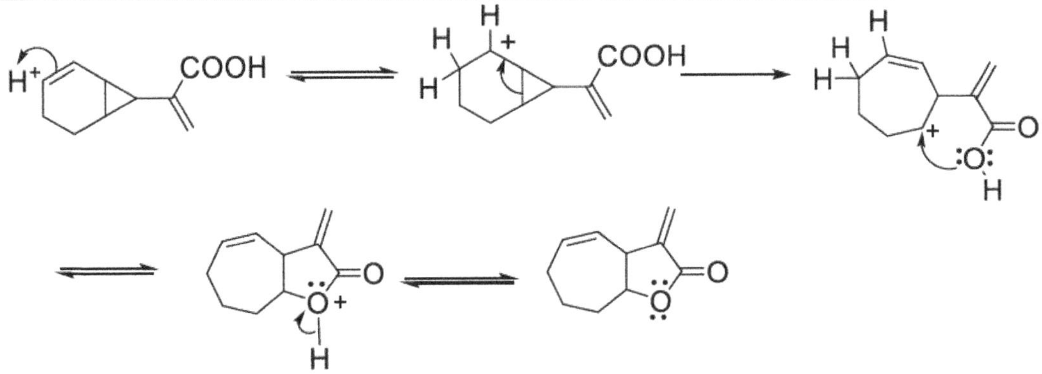
H+
COOH
H
H
H
COOH
H
H
H
O
O
H
O
O
H
O
O

204

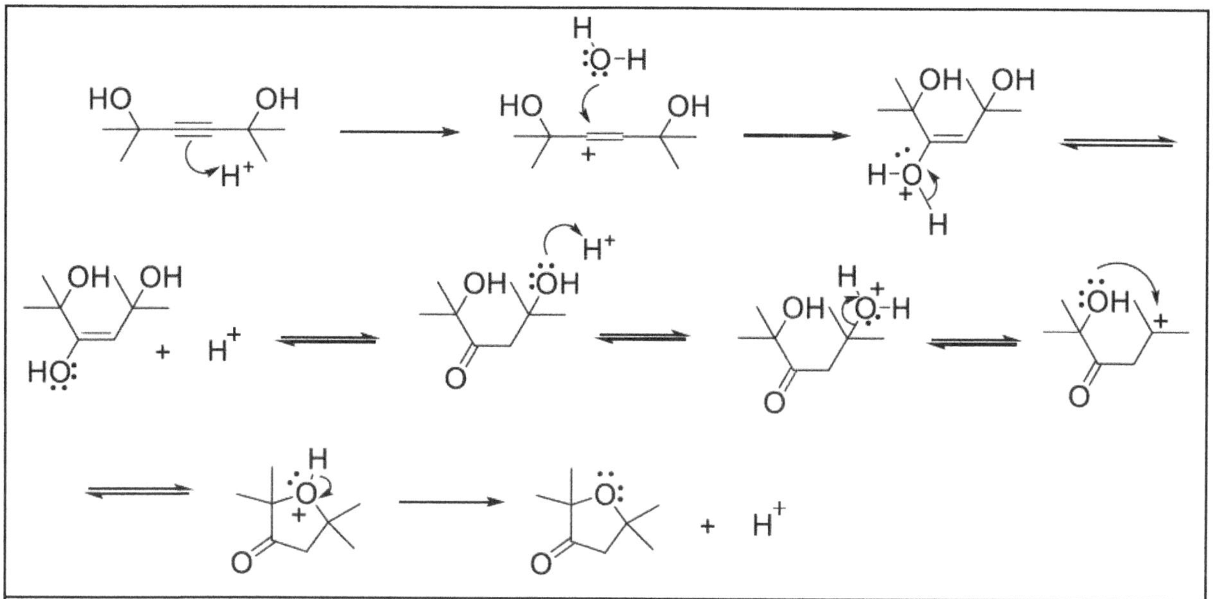

HO
OH
H+
H
O-H
HO
OH
+
OH
OH
H-O
+
H
H
OH
OH
HO
+
H+
H+
OH
OH
O
OH
H
+
O-H
O
OH
+
O
H
O
+
O
O
O
+
H+

205

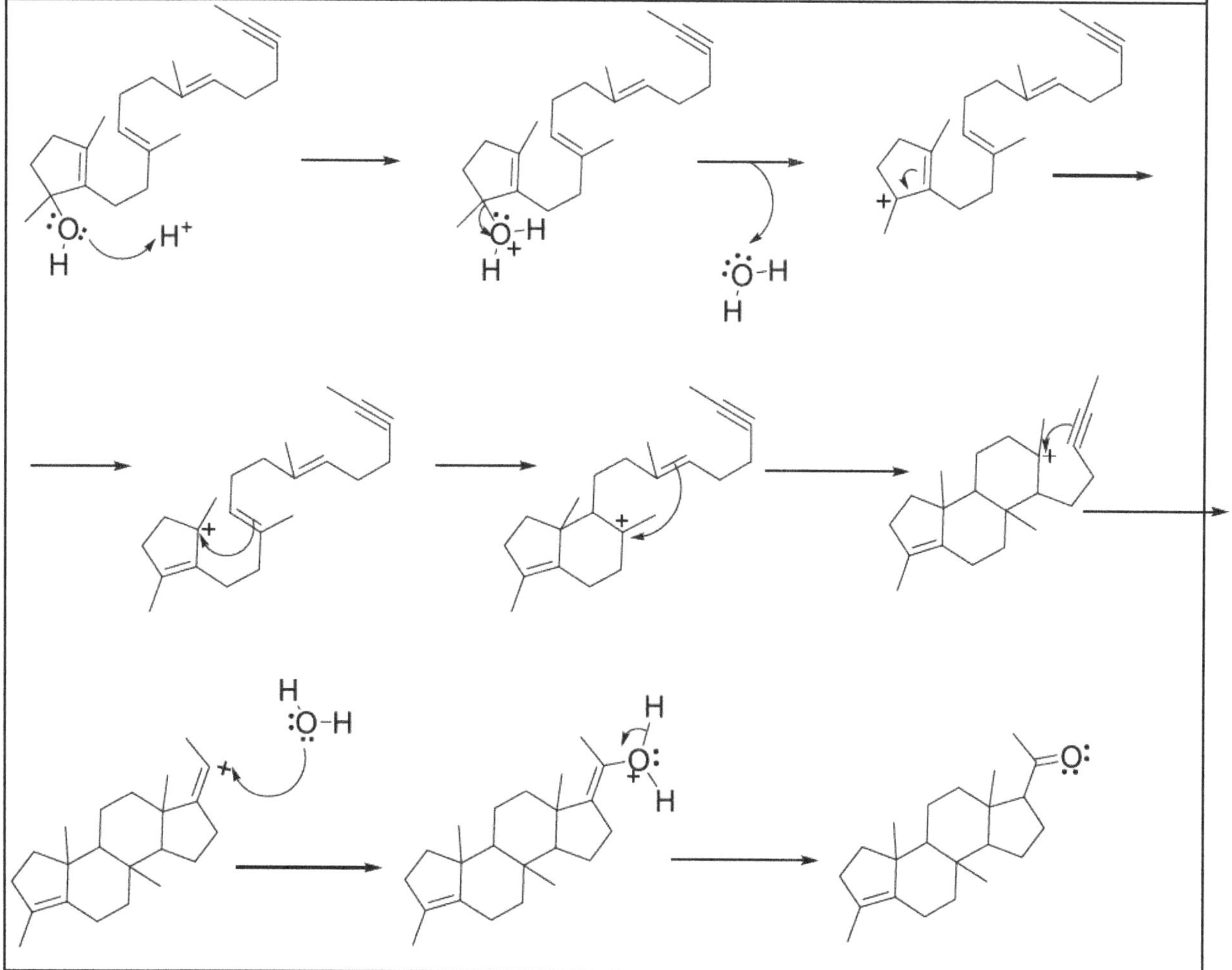

O
H
H+
O-H
H
+
H
O-H
H
+
+
+
+
H
O-H
O
+
H
H
O

206

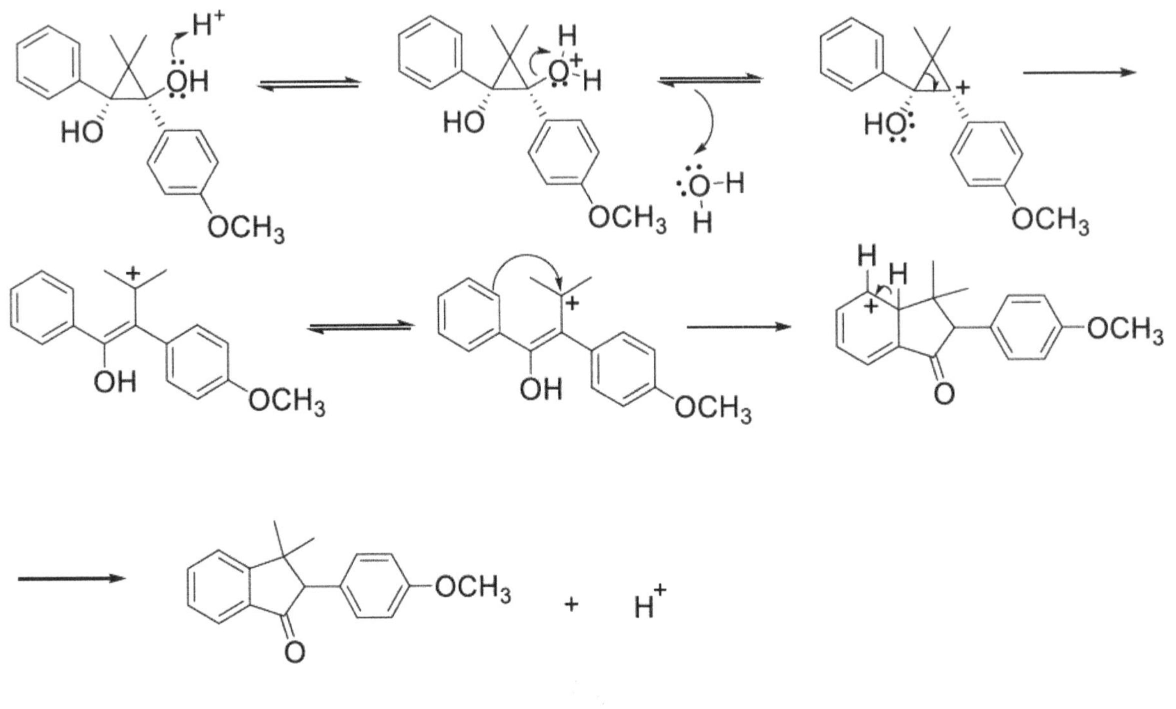
H+
OH
HO
OCH3
H
O+
H
HO
OCH3
O-H
H
+
HO
OCH3
+
OH
OCH3
+
OH
OCH3
H
H
+
OCH3
O
OCH3
O
+
H+

207

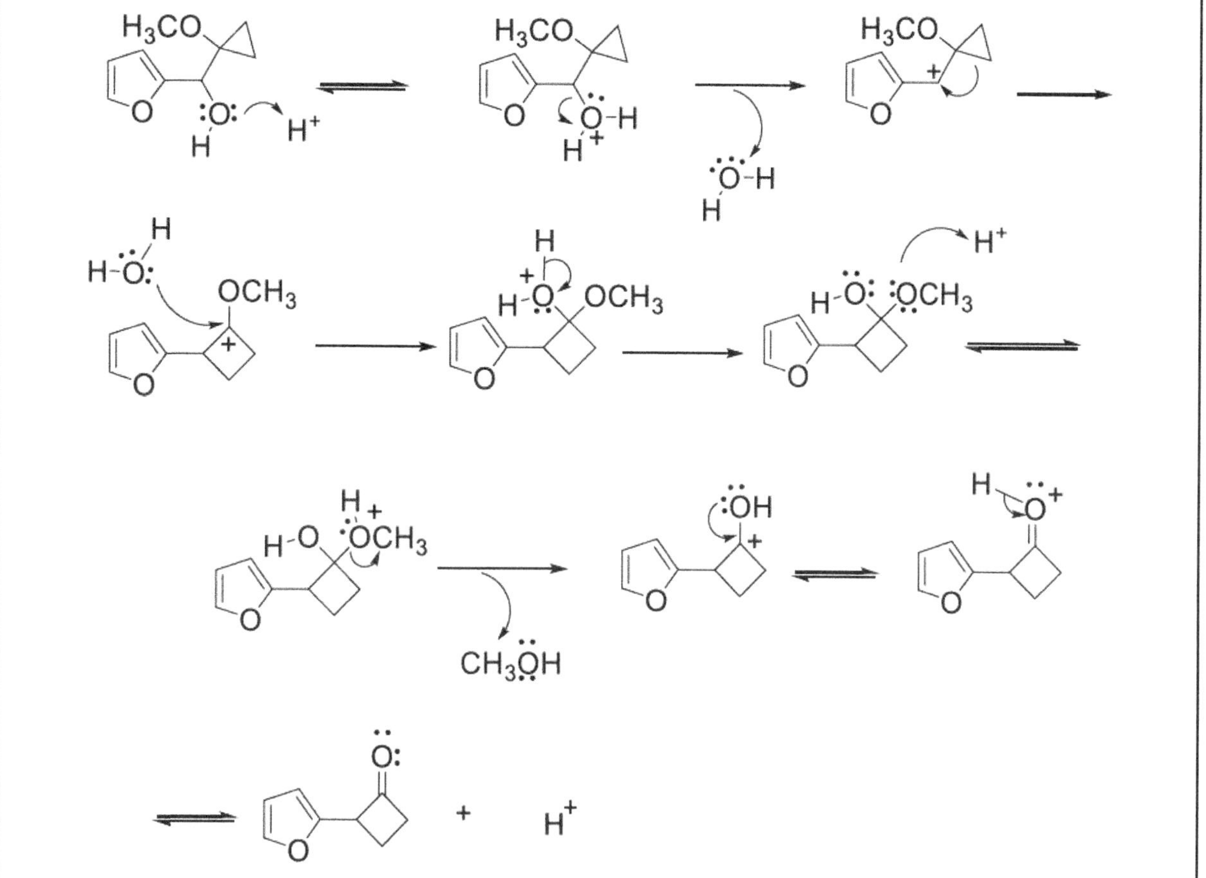

H₃CO
H⁺
H₃CO
H₃CO
OCH₃
OCH₃
OCH₃
H⁺
OCH₃
CH₃OH
OH
H⁺

208

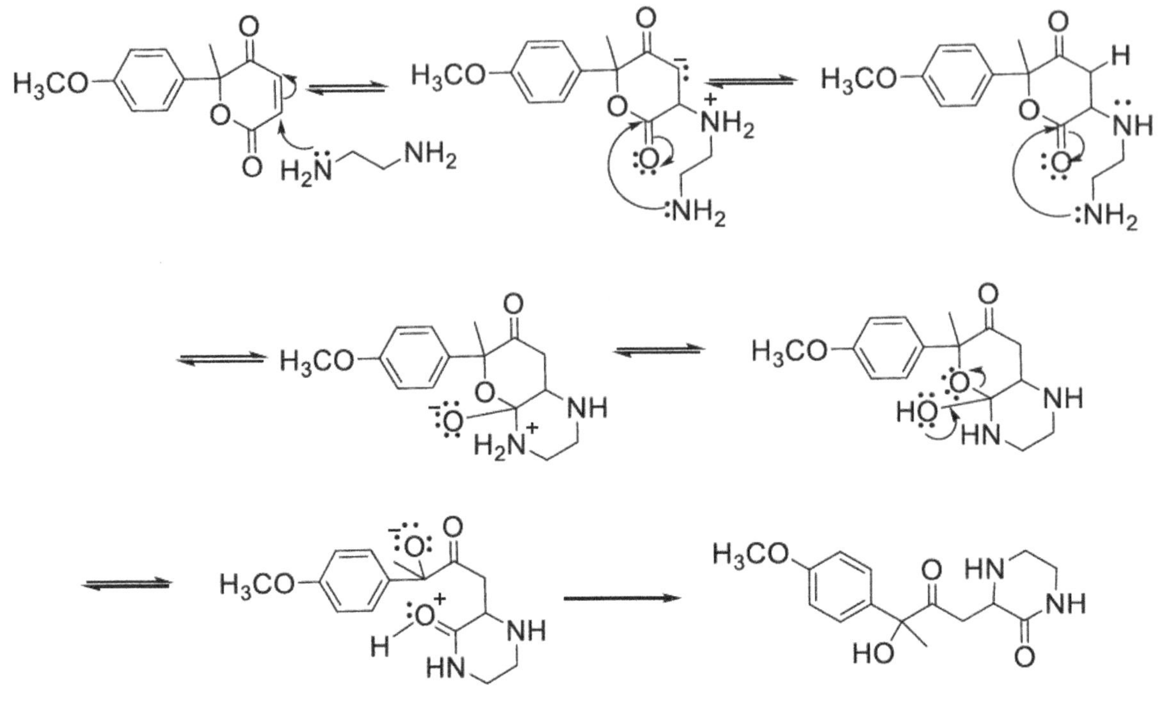

209

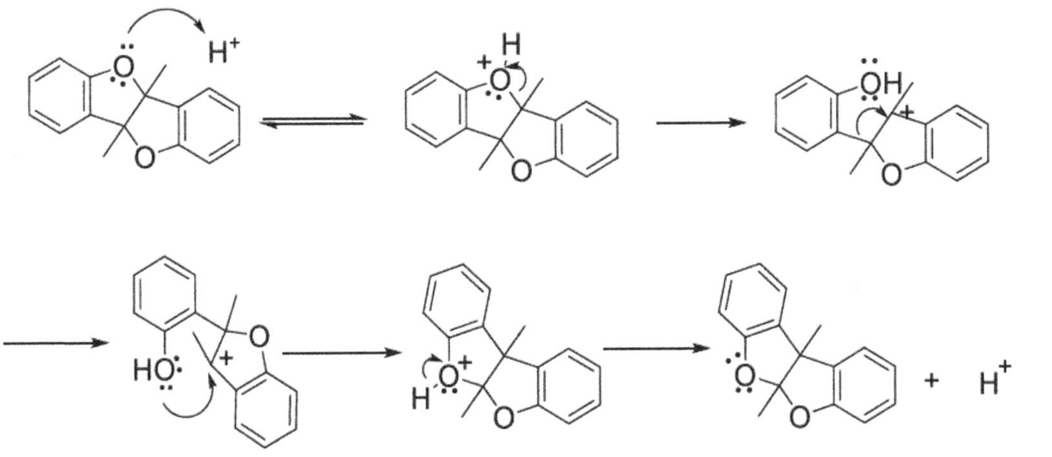

210

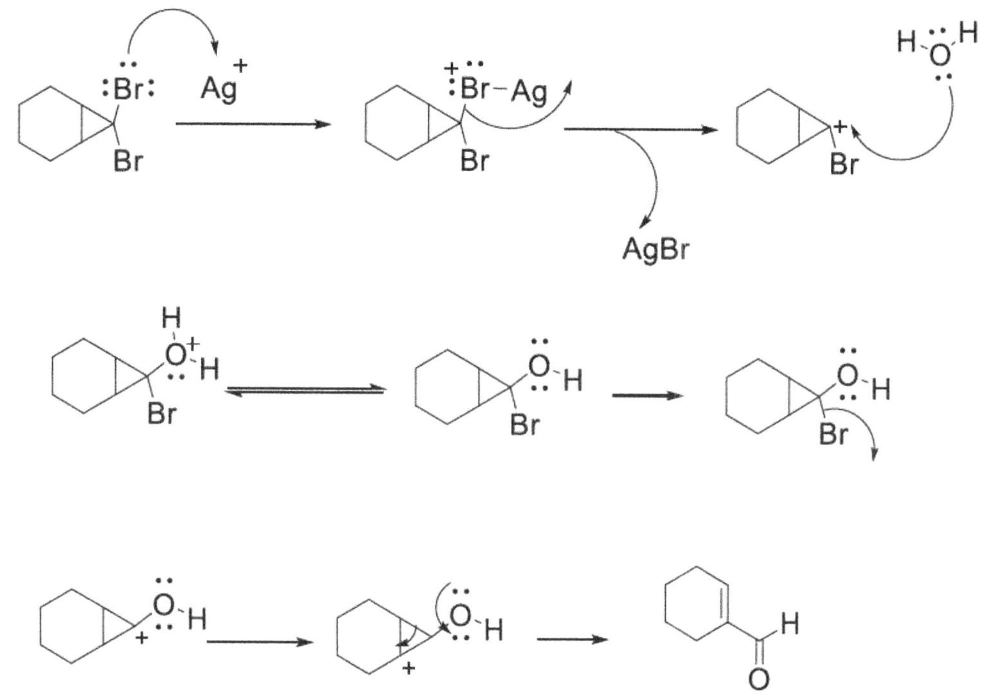
Br
Ag
Br–Ag
Br
AgBr
H–O–H
Br
O–H
H

211

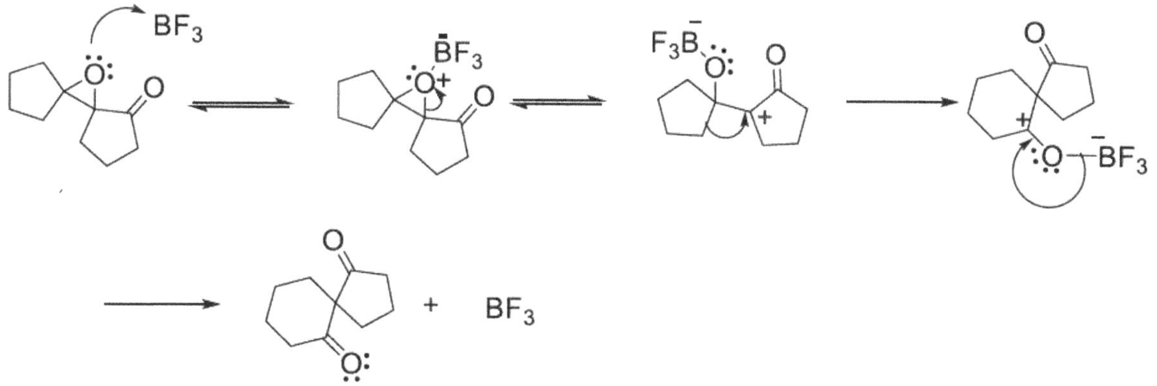
BF3
O
O
BF3
F3B
O
O
O
O–BF3
O
O
+
BF3

212

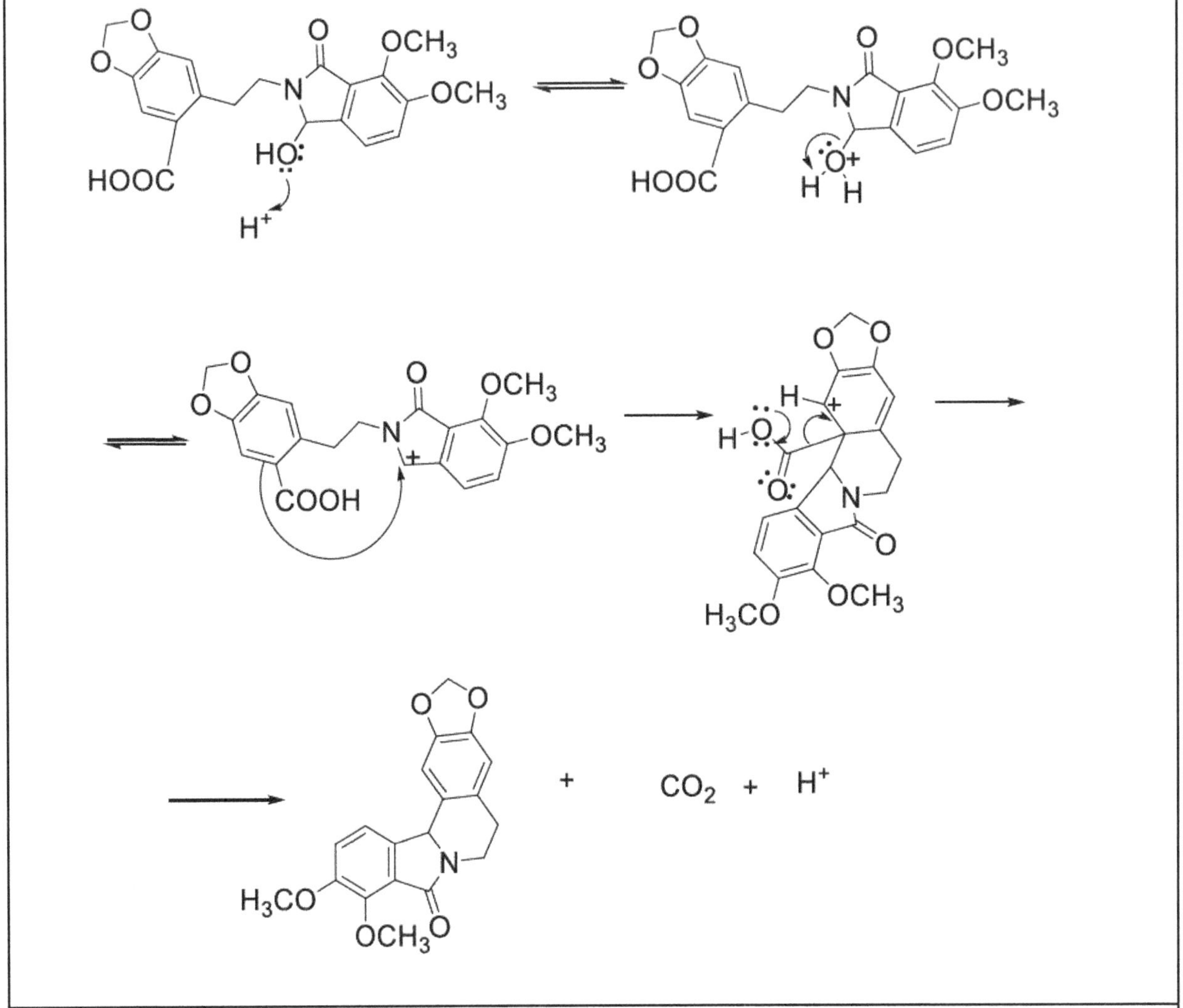

213

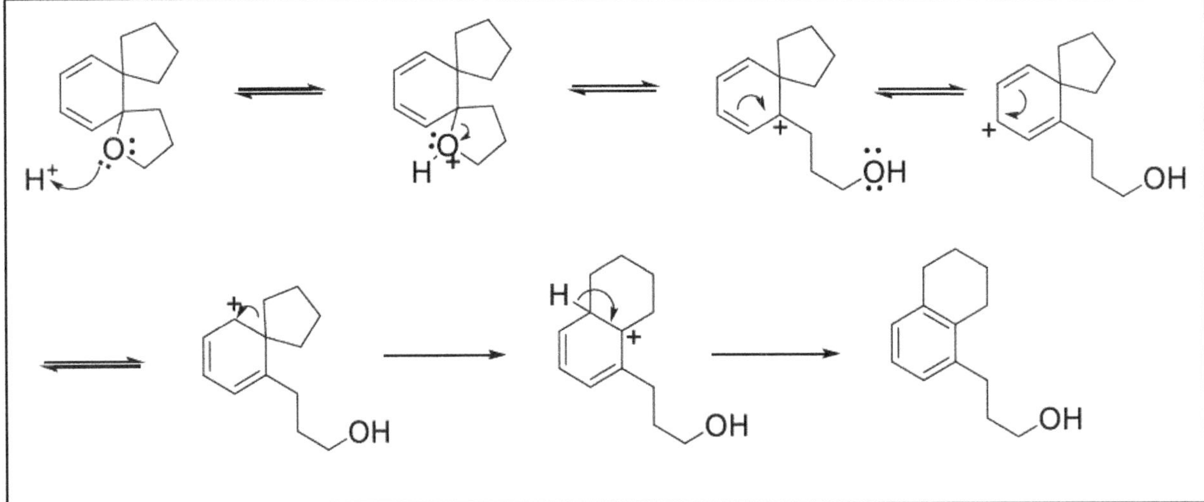

214

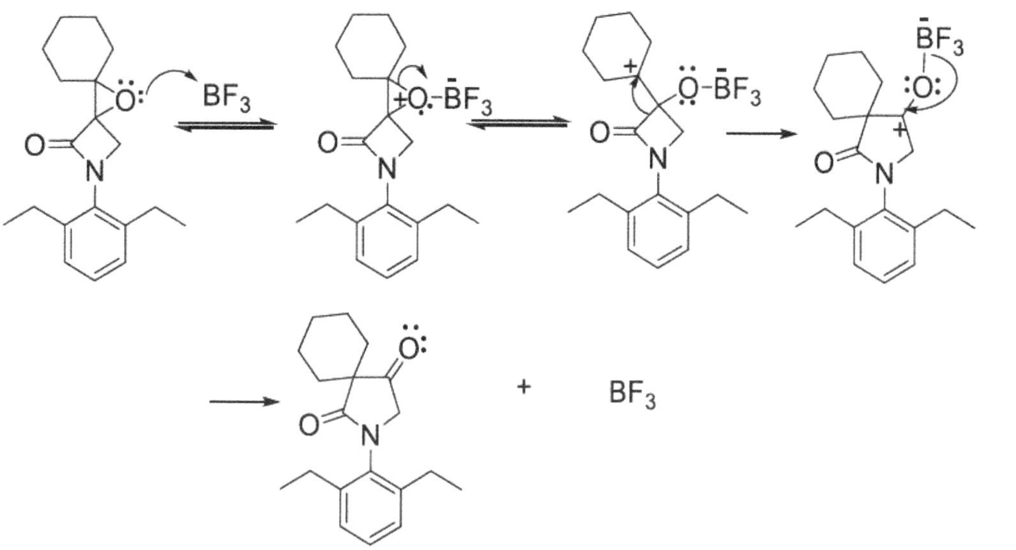
BF3
O
N
BF3

215

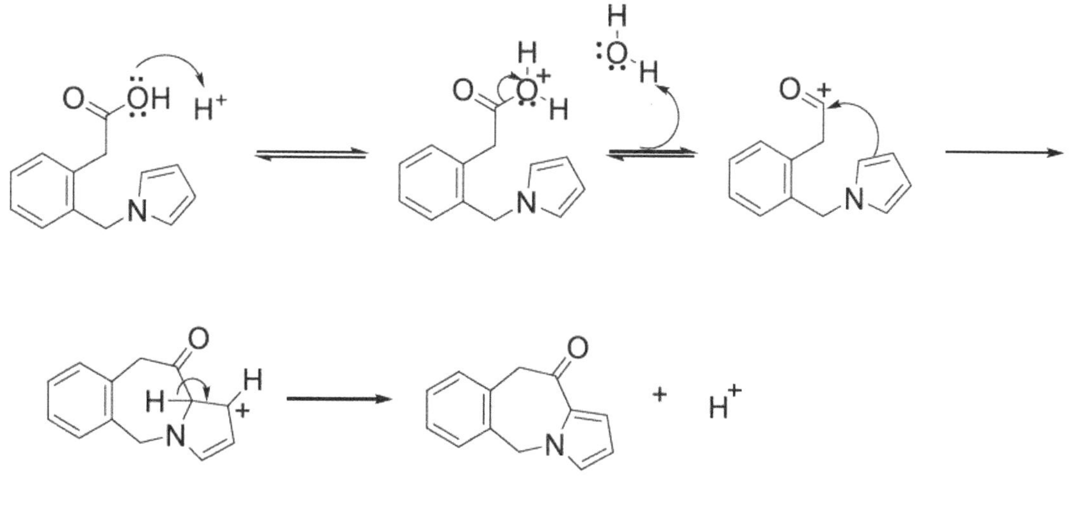
H+
N
H
O
+

216

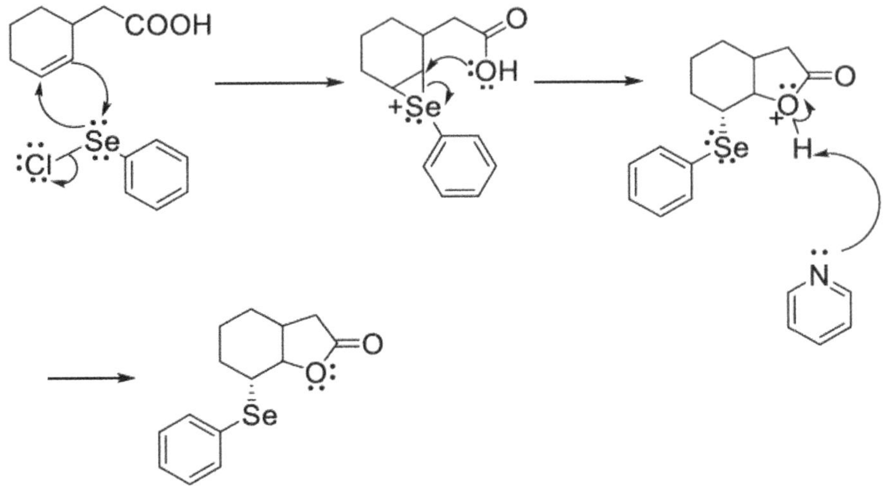
COOH
Cl
Se
OH
+Se
O
O
+
H
Se
N
O
Se

217

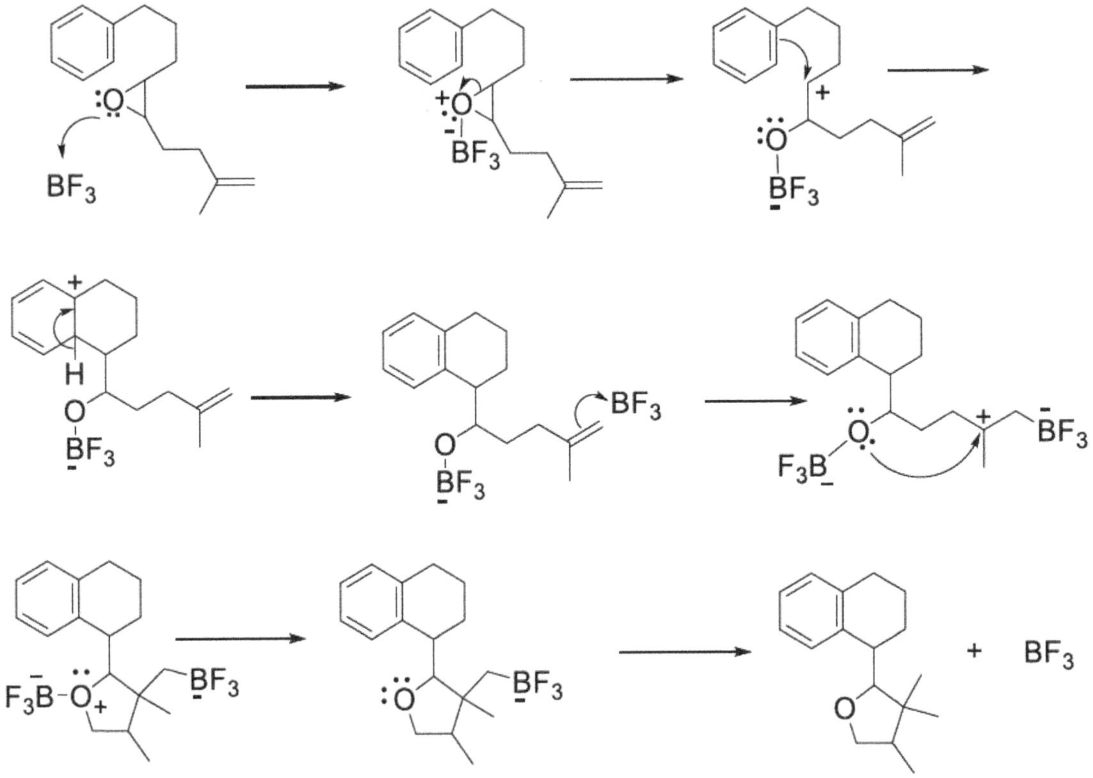
O
BF_3
+
H
F$_3$B

218

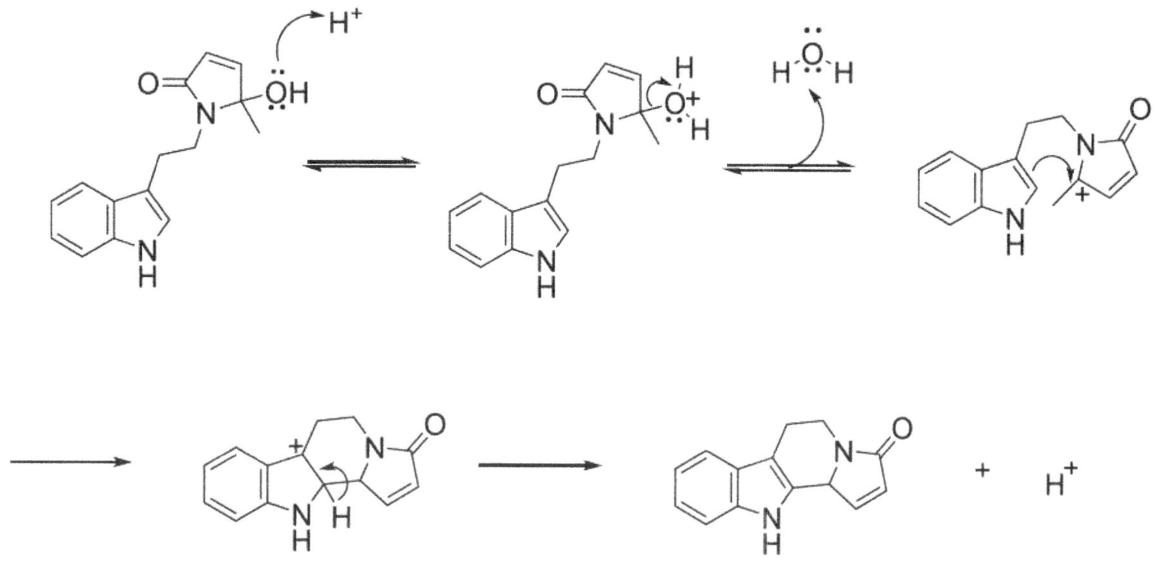

219

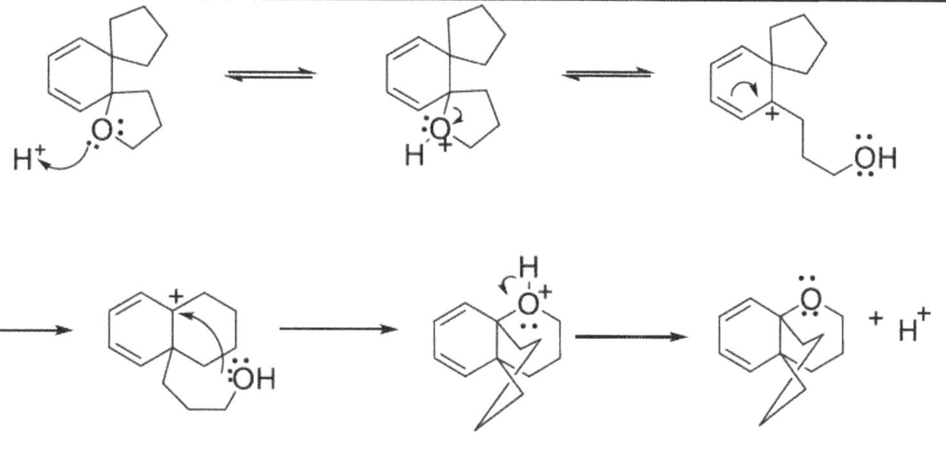

220

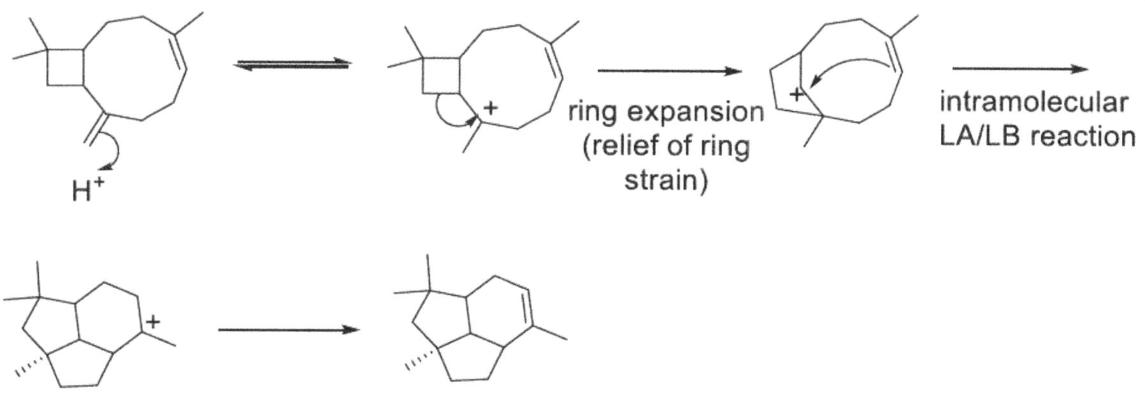
ring expansion
(relief of ring
strain)
intramolecular
LA/LB reaction

221

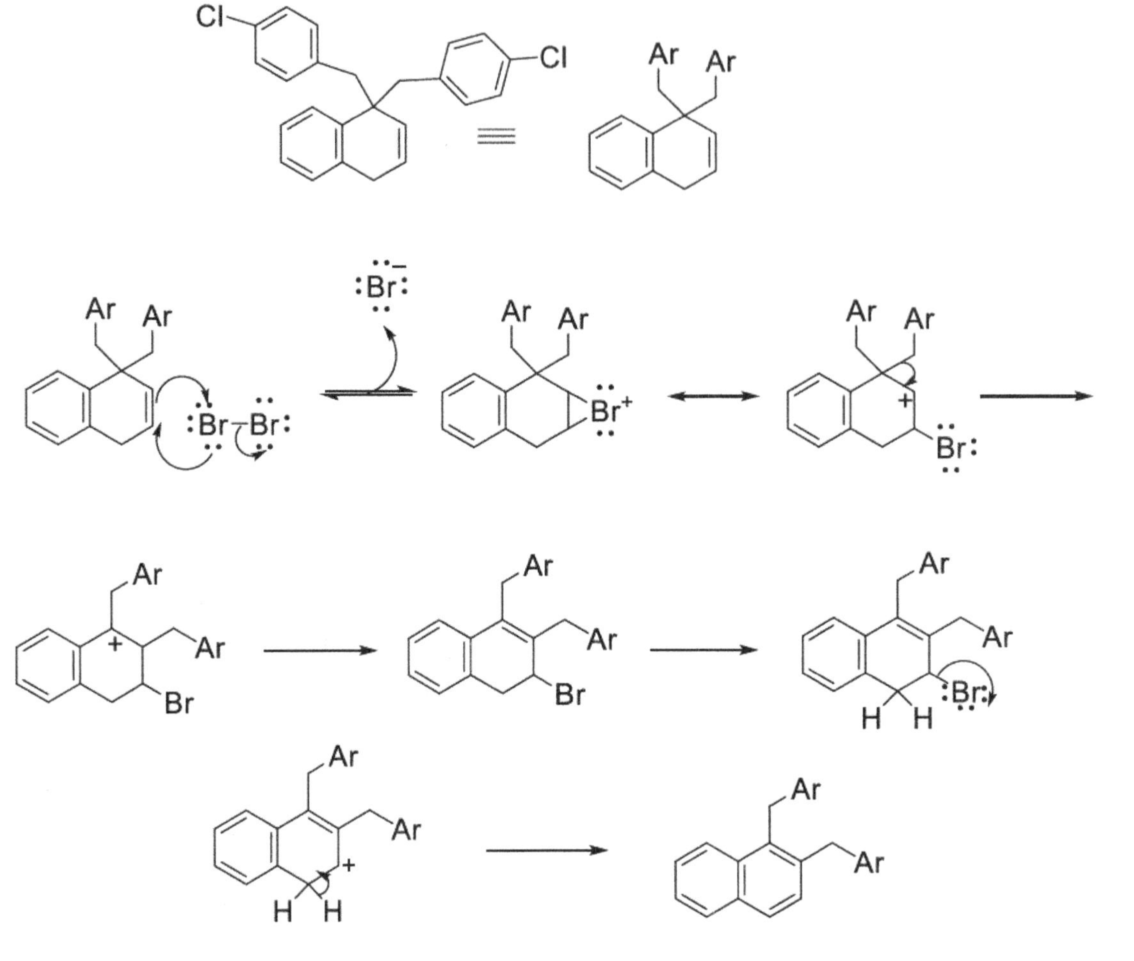

Cl
Cl
Ar
Ar
Br

222

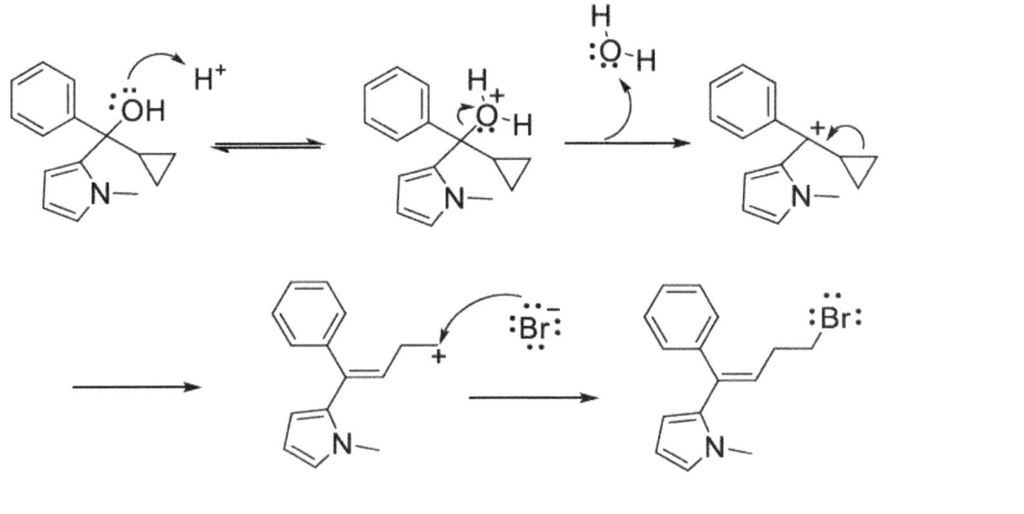

H+
OH
N
Br

223
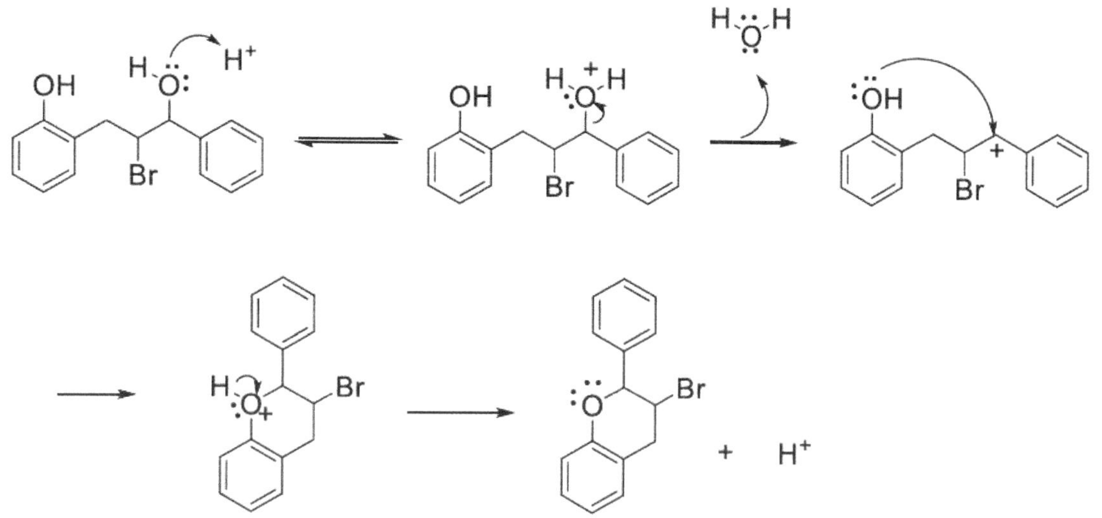
224
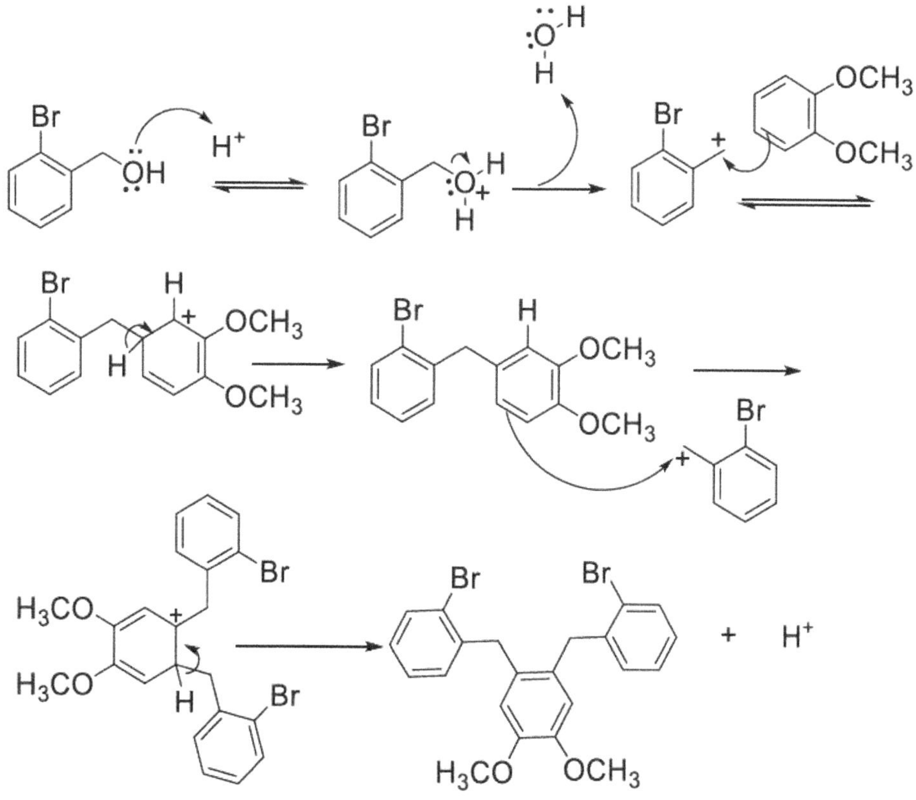
225

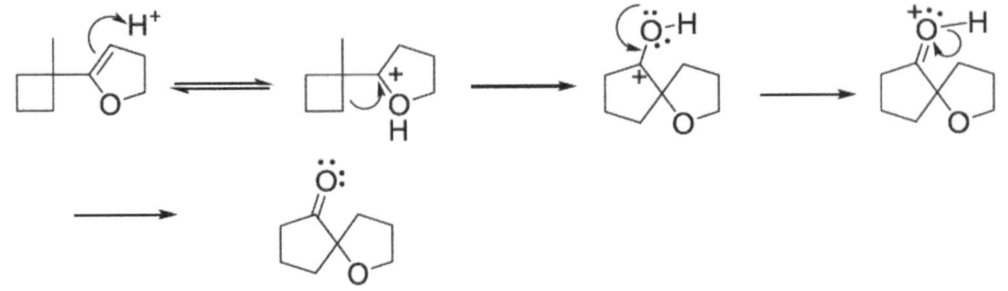

226

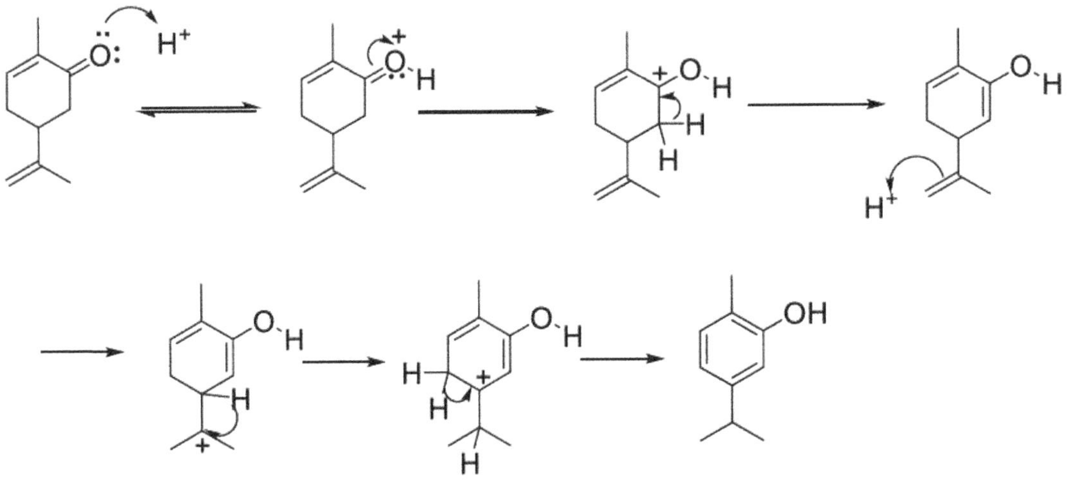

227

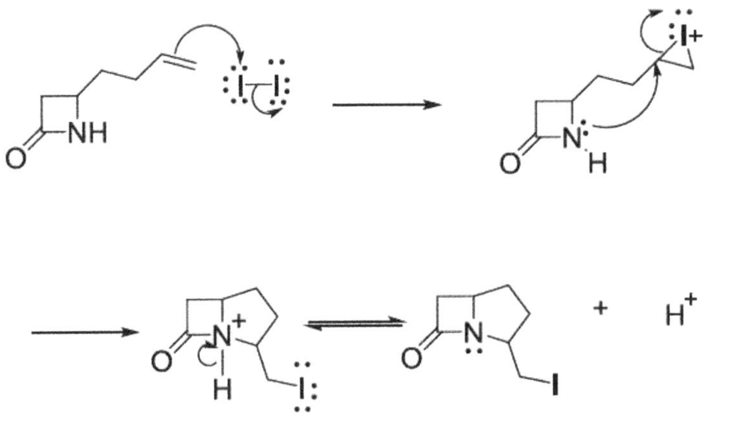

<u>Note</u> The reactivity of the nitrogen is higher than in an acyclic amide because it cannot engage in resonance (ring strain)

228

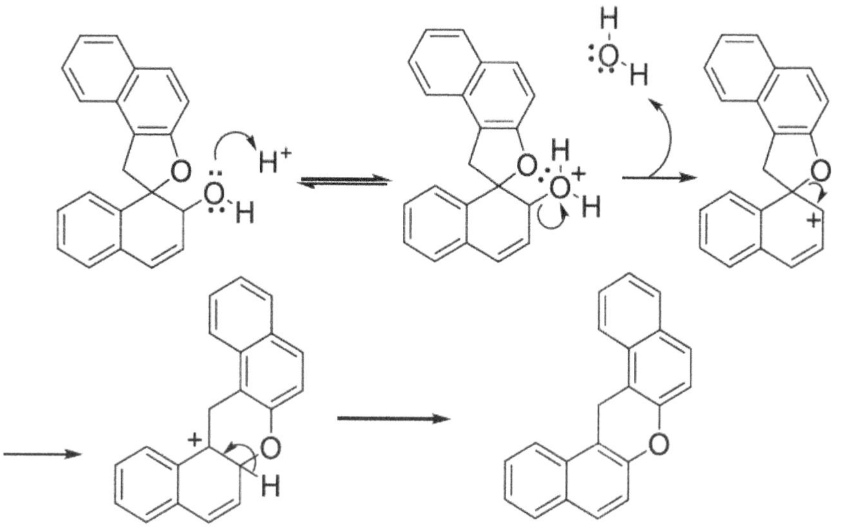

H⁺
O
H

229

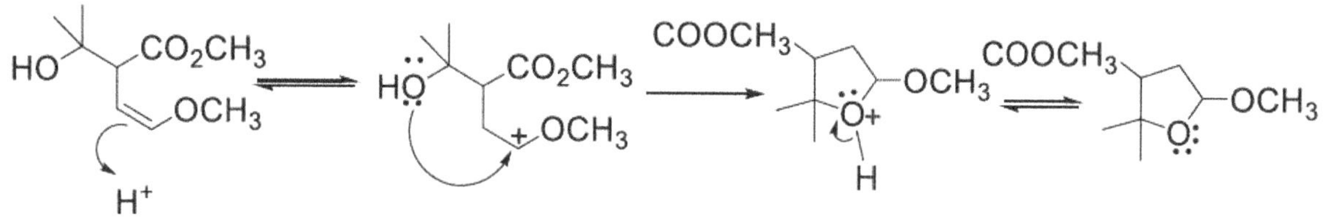

HO
CO2CH3
OCH3
H⁺
COOCH3
OCH3
H

<u>Exercise</u> Why is the carbocation on the carbon next to the oxygen formed?

230

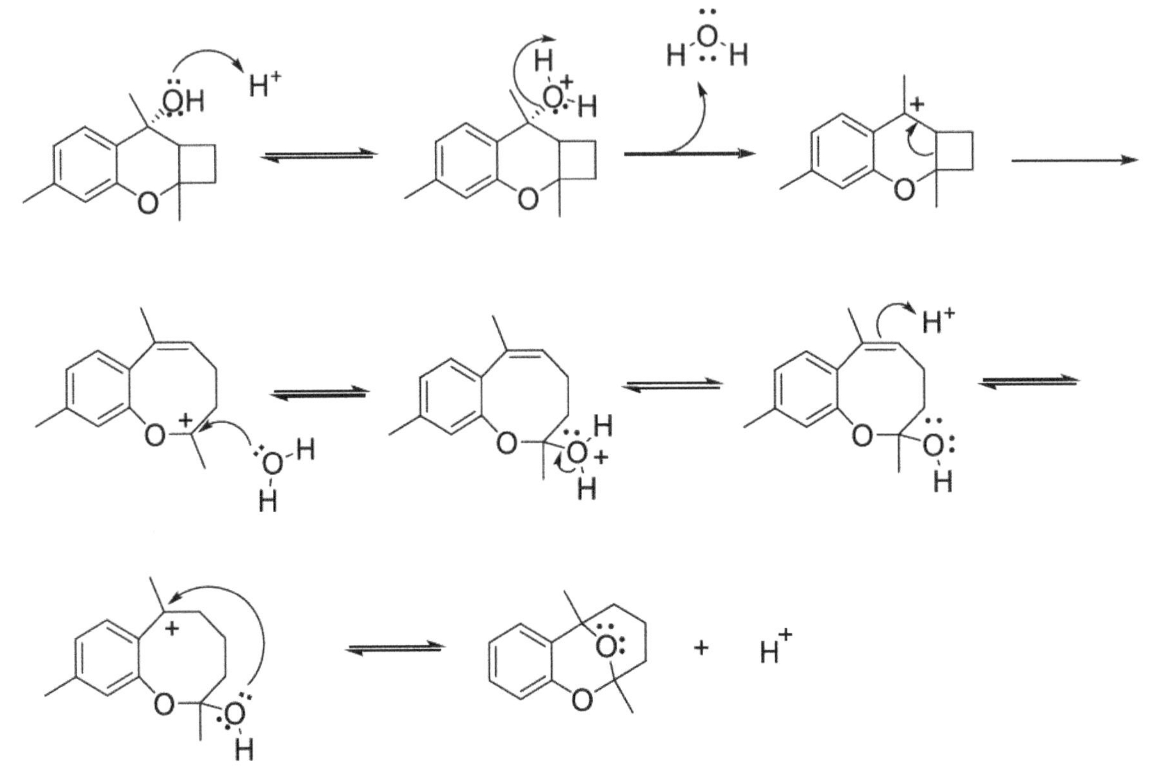

231

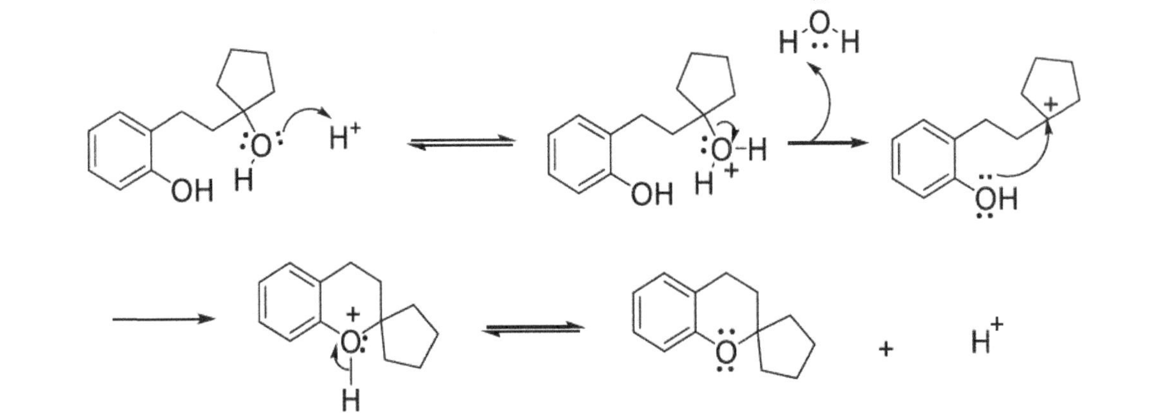

232

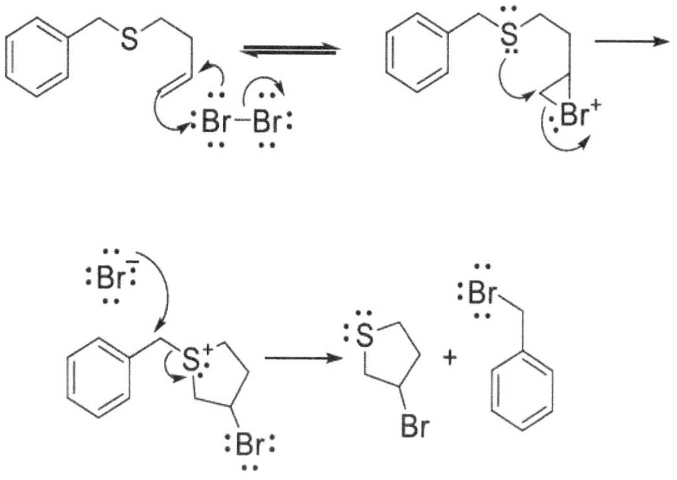

233

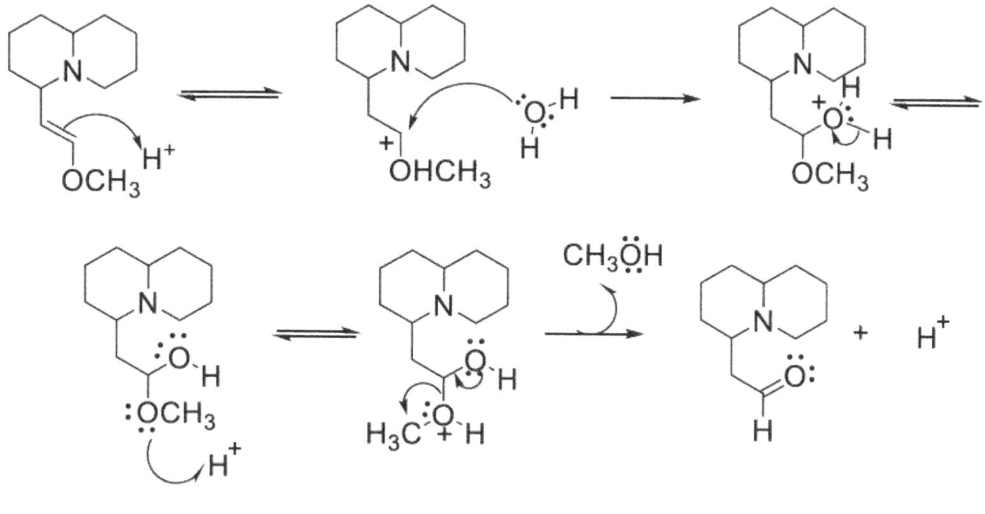

234

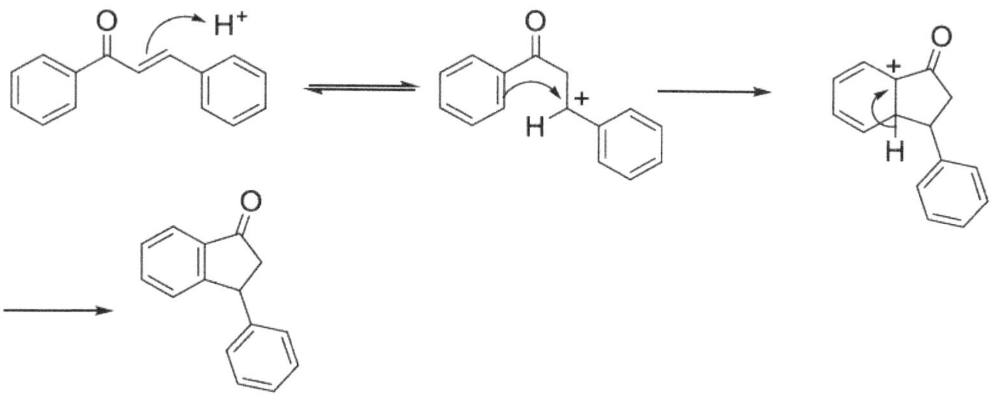

235

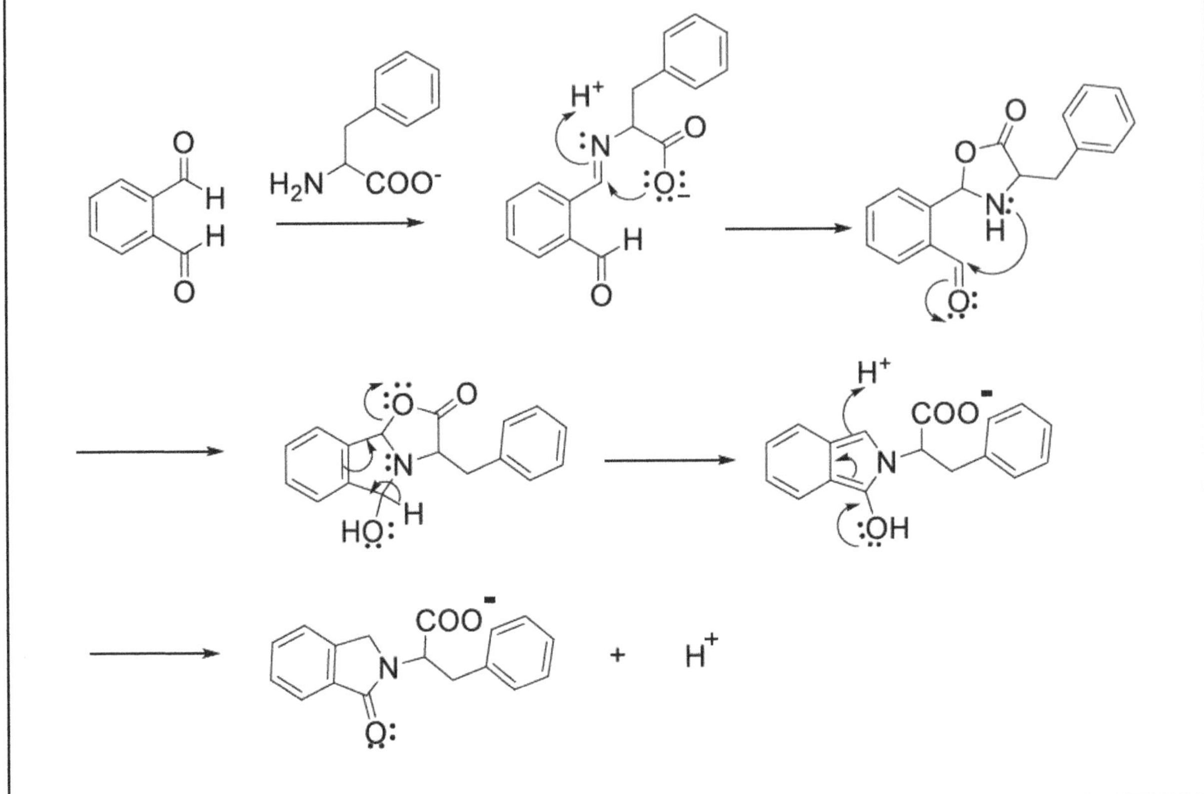

236

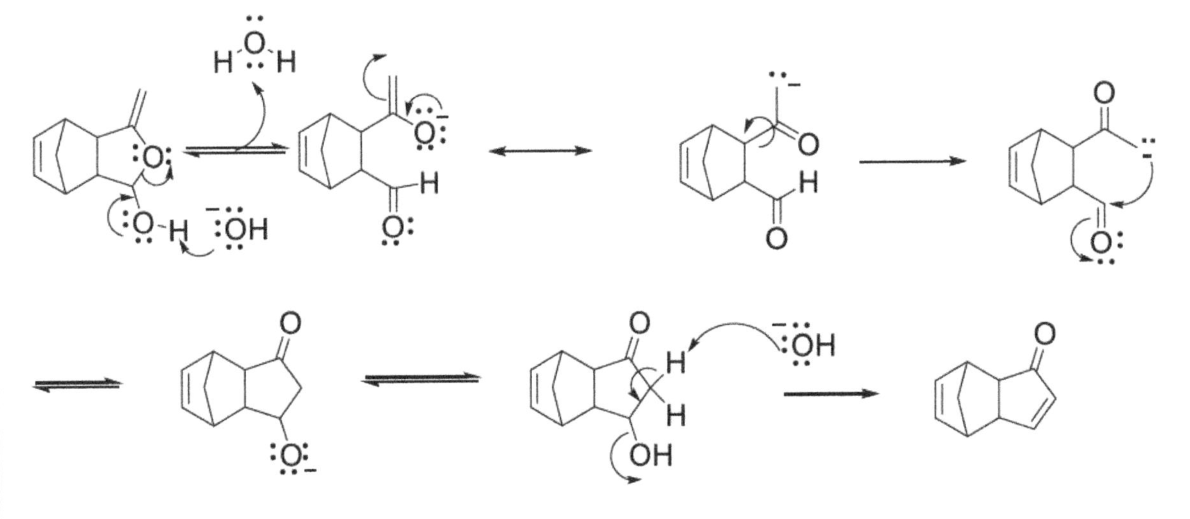

237

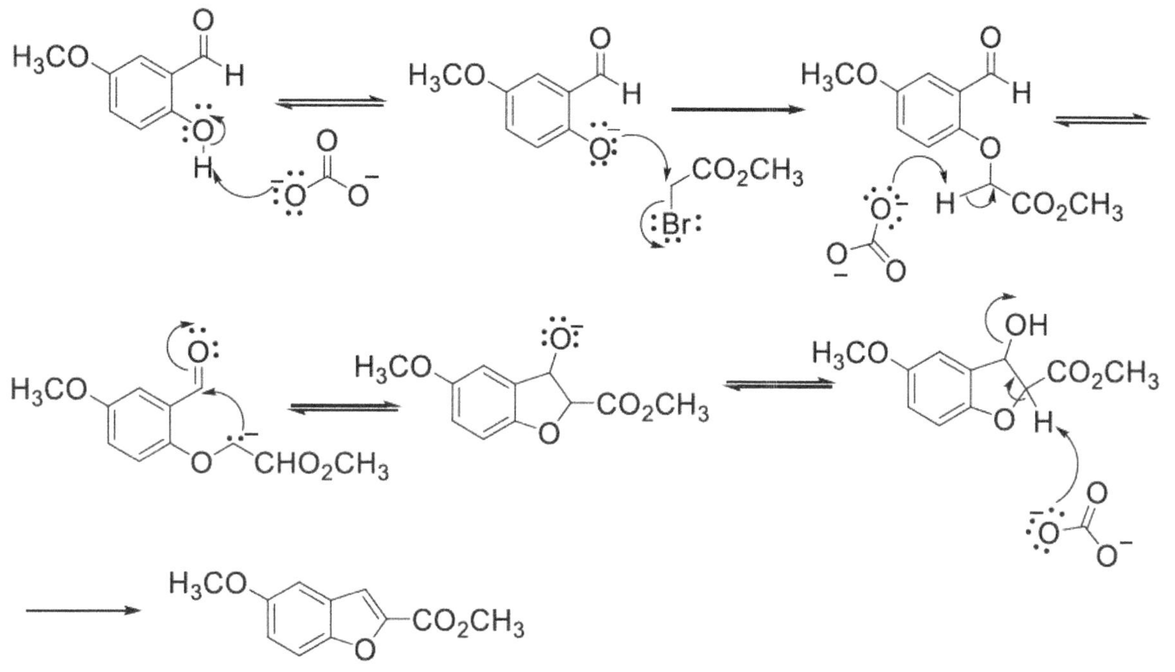

H$_3$CO
H
O
H
CO$_2$CH$_3$
Br
OH
CHO$_2$CH$_3$

238

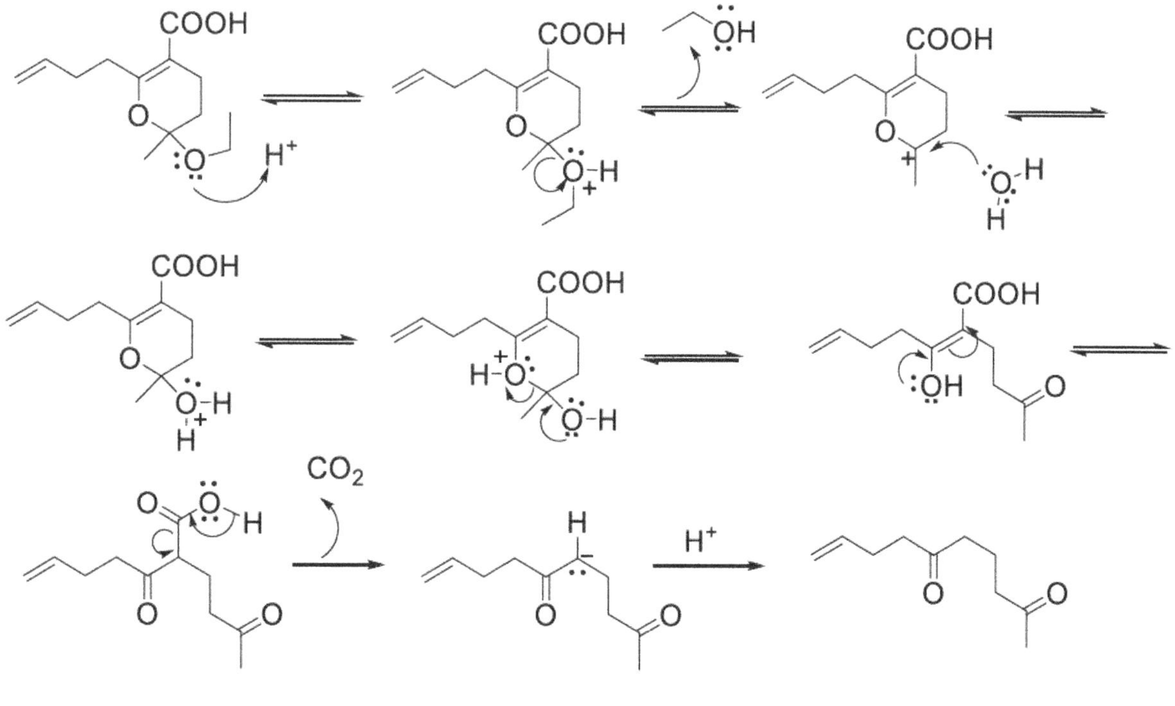

COOH
H$^+$
OH
O
CO$_2$

239

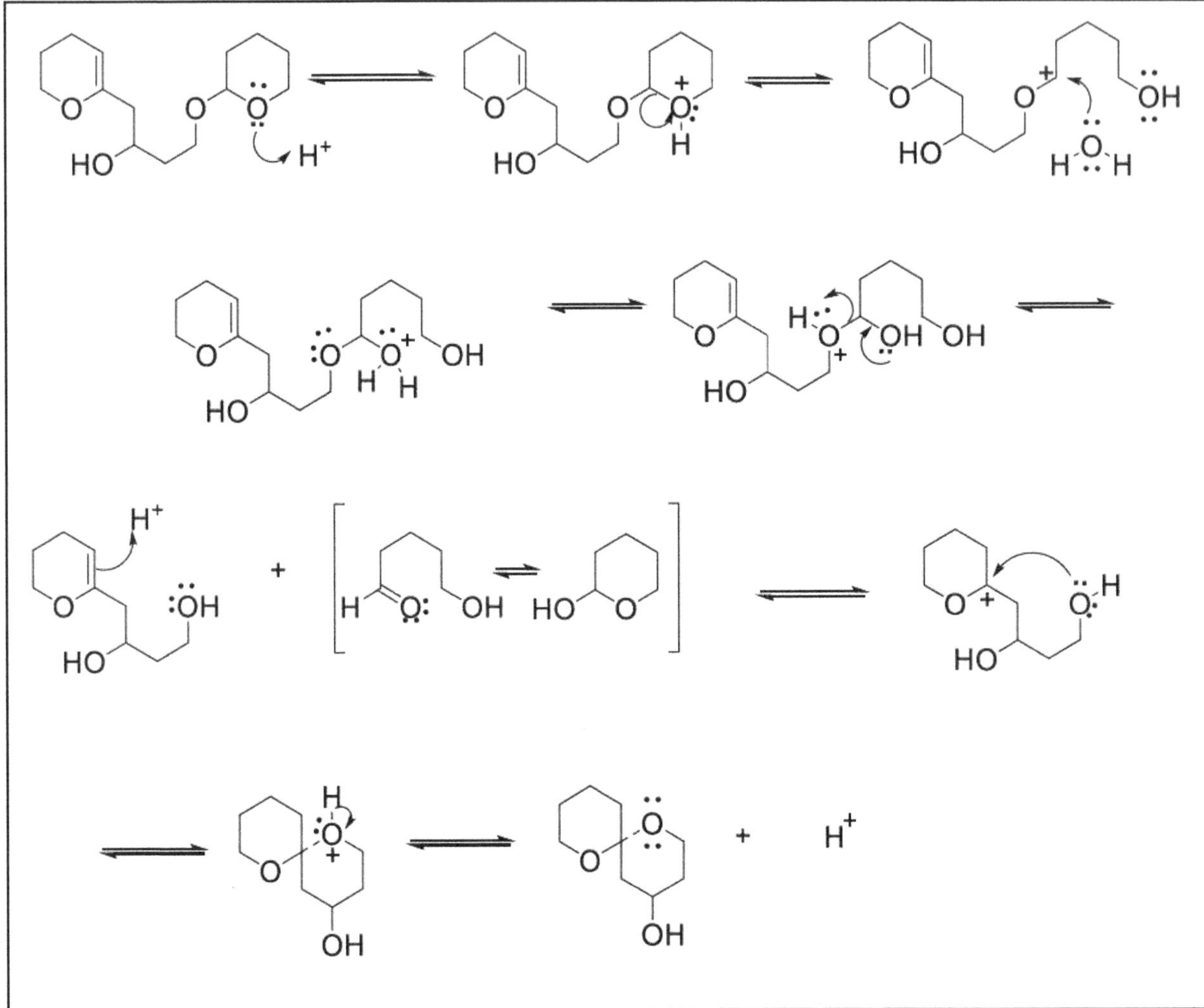

240

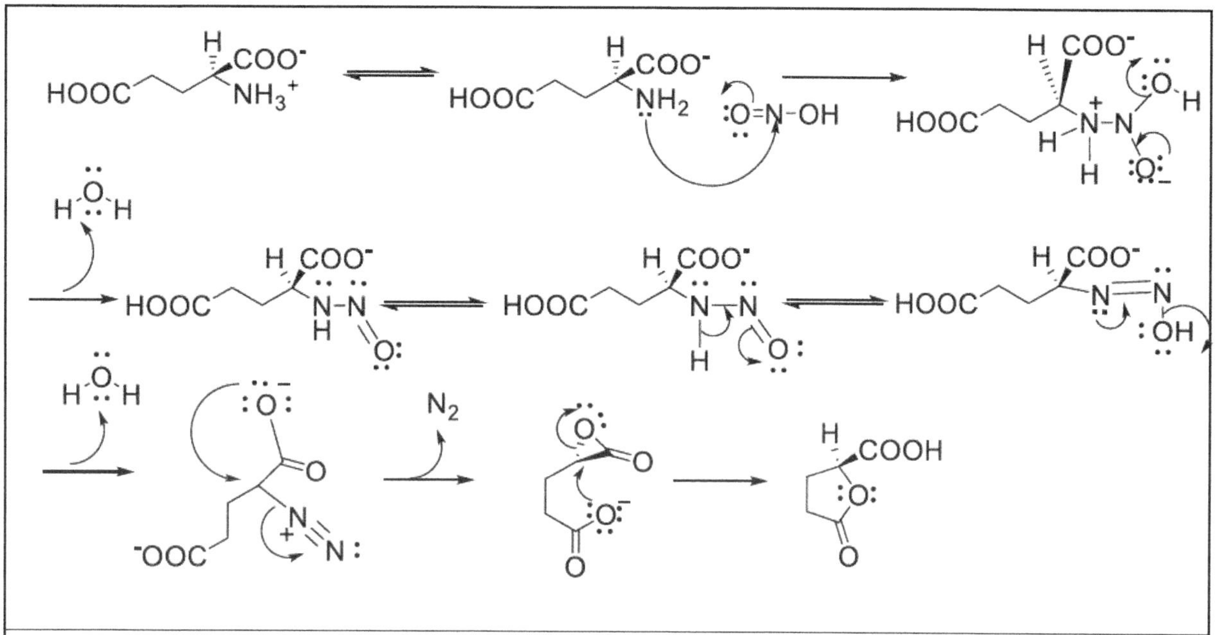
HOOC
COO⁻
NH₃⁺
NH₂
O=N–OH
N₂
COOH
+ H⁺

241

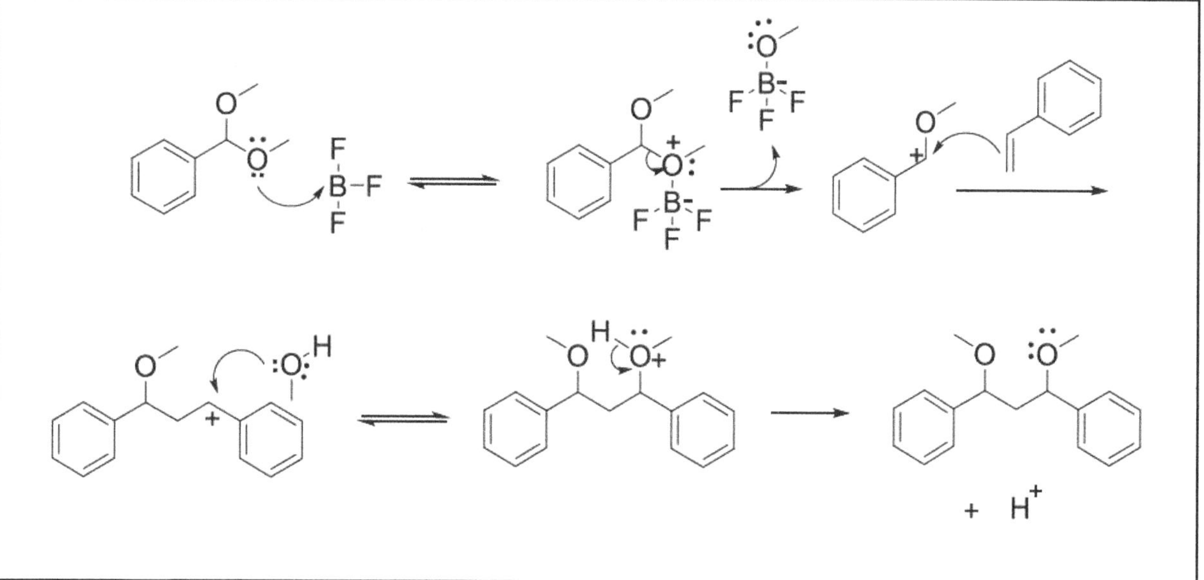
B⁻
F
+ H⁺

242

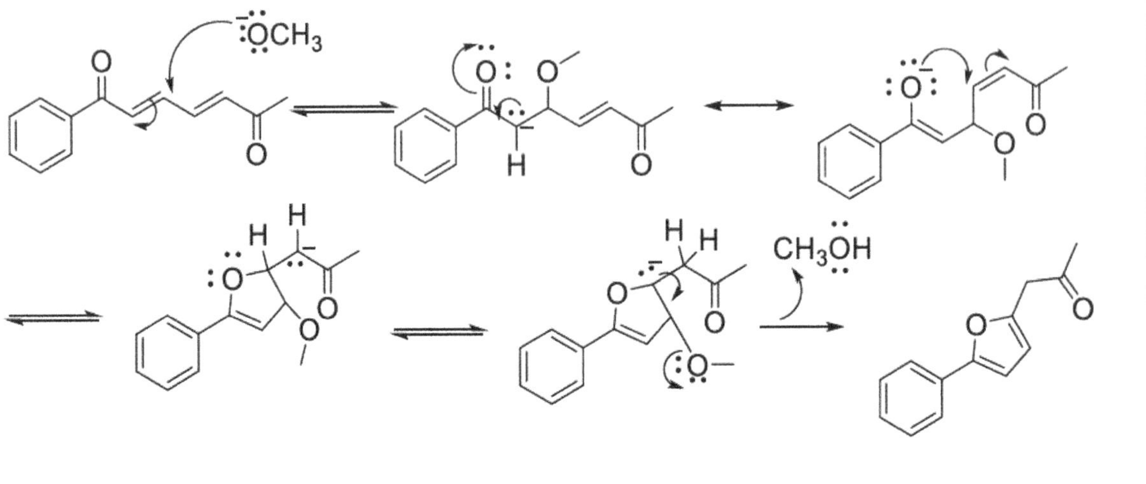

243

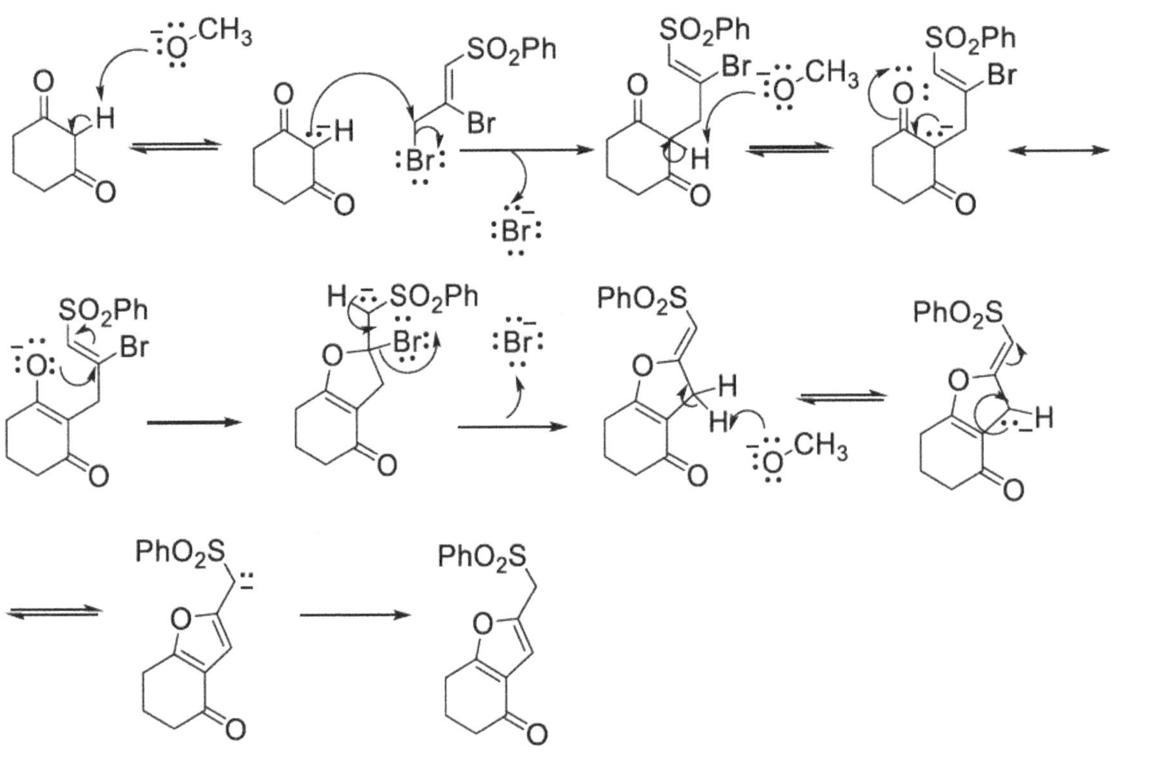

244

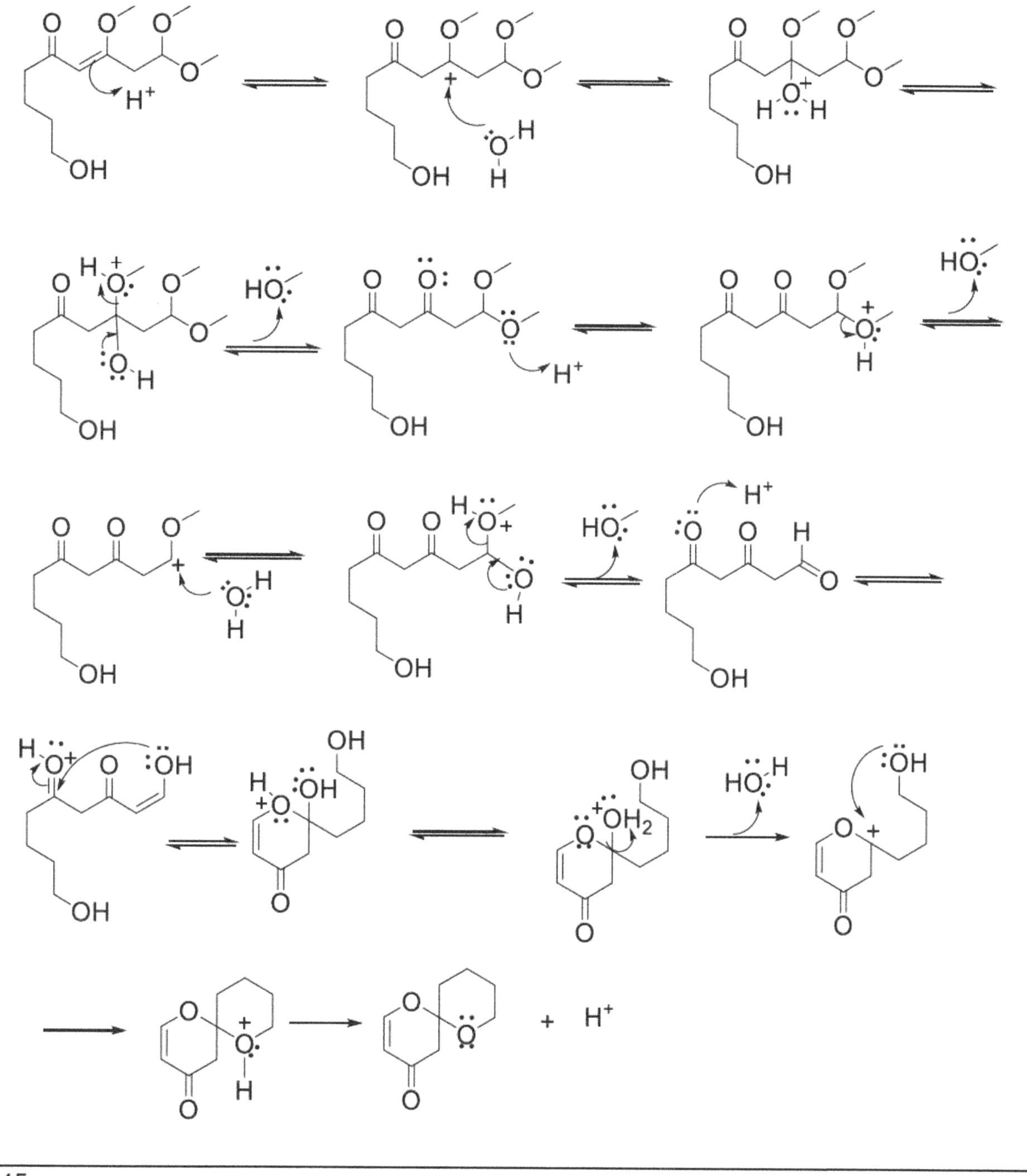

245

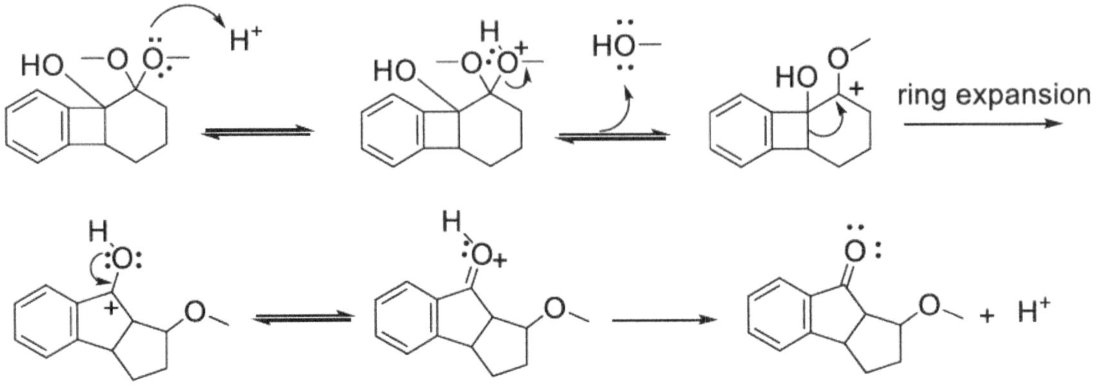
H+
HO
HO
HO
ring expansion
+
H+

246

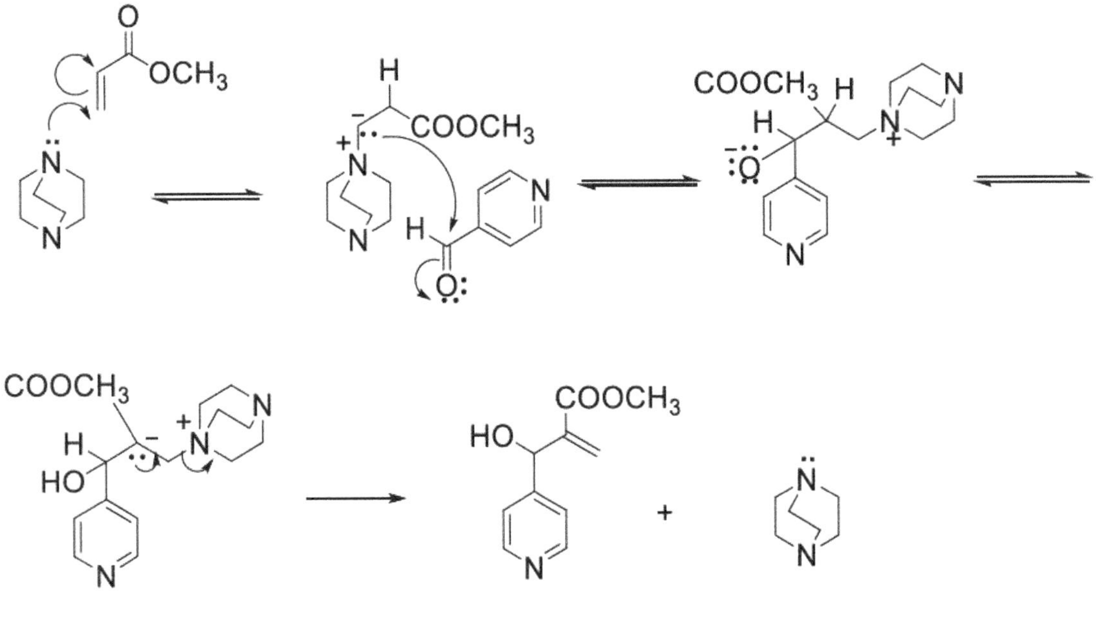
OCH3
COOCH3
COOCH3
COOCH3
COOCH3
HO
N

This is an example of the *Bayliss-Hill reaction*

247

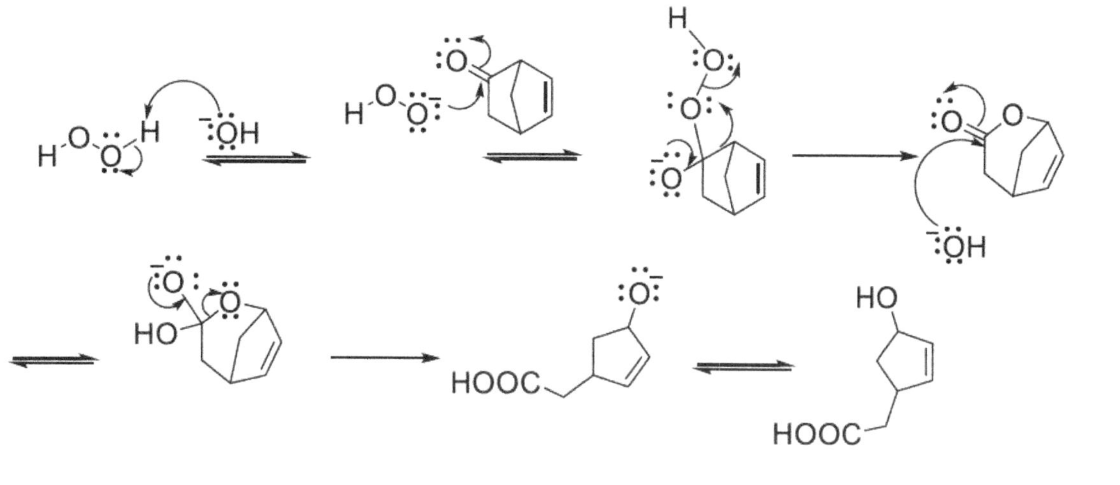

248

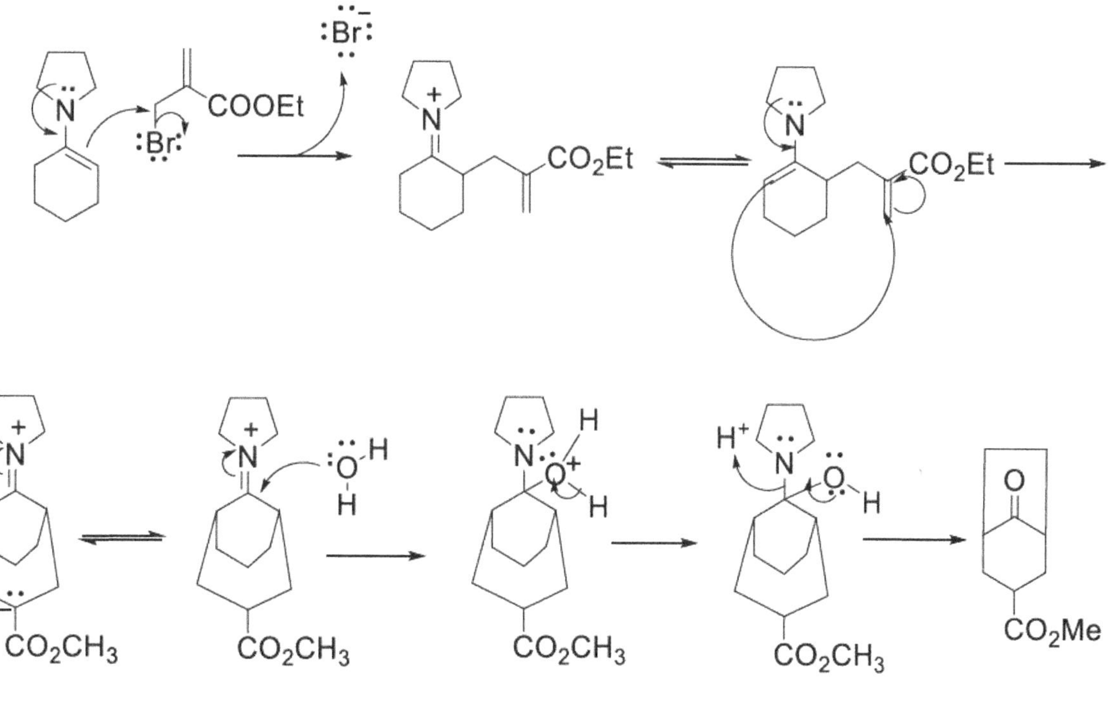

249

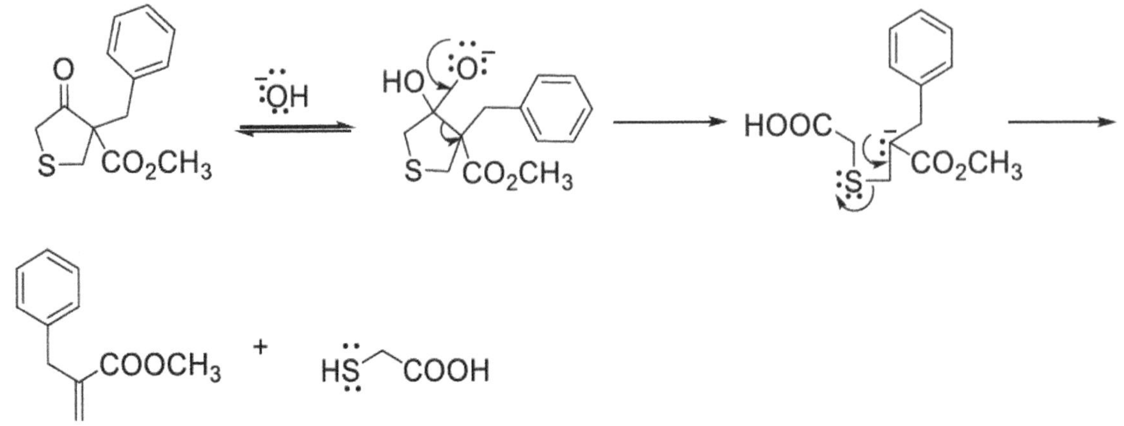

O
:ÖH
HO
:O:
S
CO2CH3
HOOC
COOCH3
+
HS
COOH

250

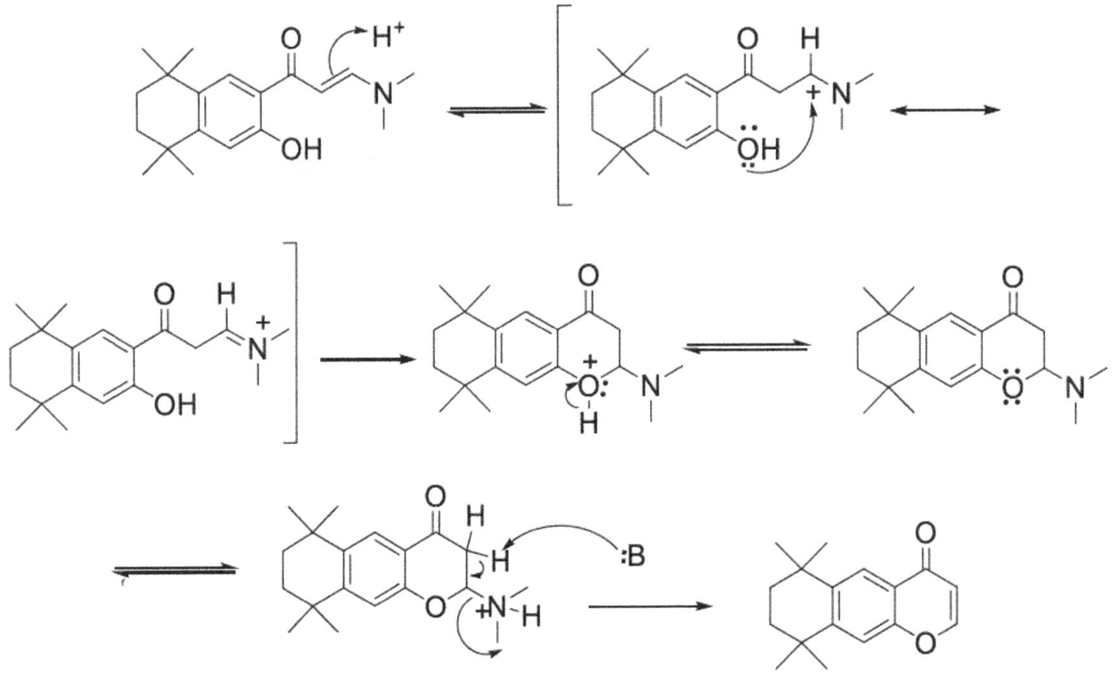

O
H+
N
OH
H
:B

251

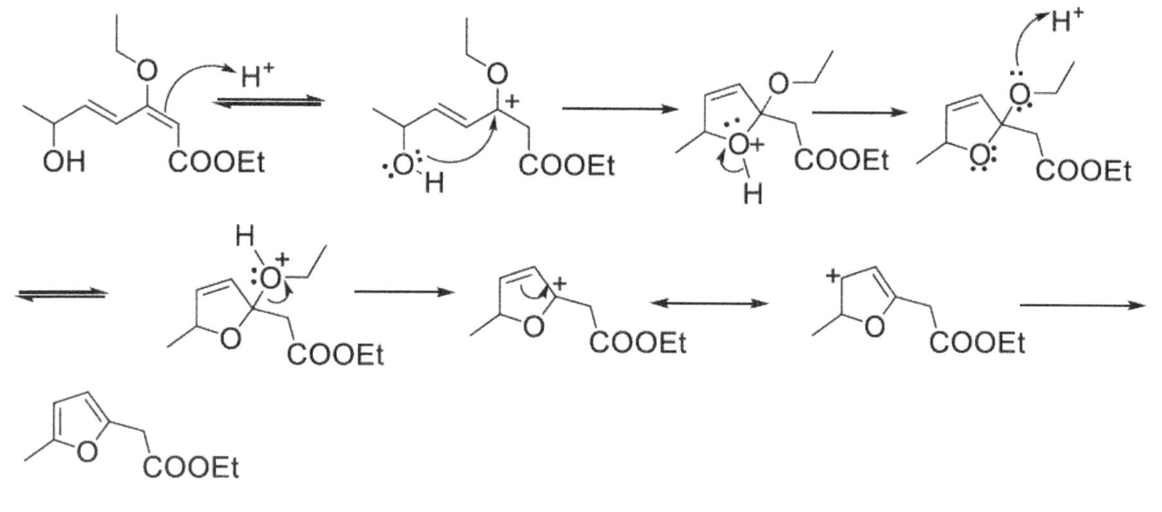

H+
OH
COOEt
O
+
O
H
O+
H

252

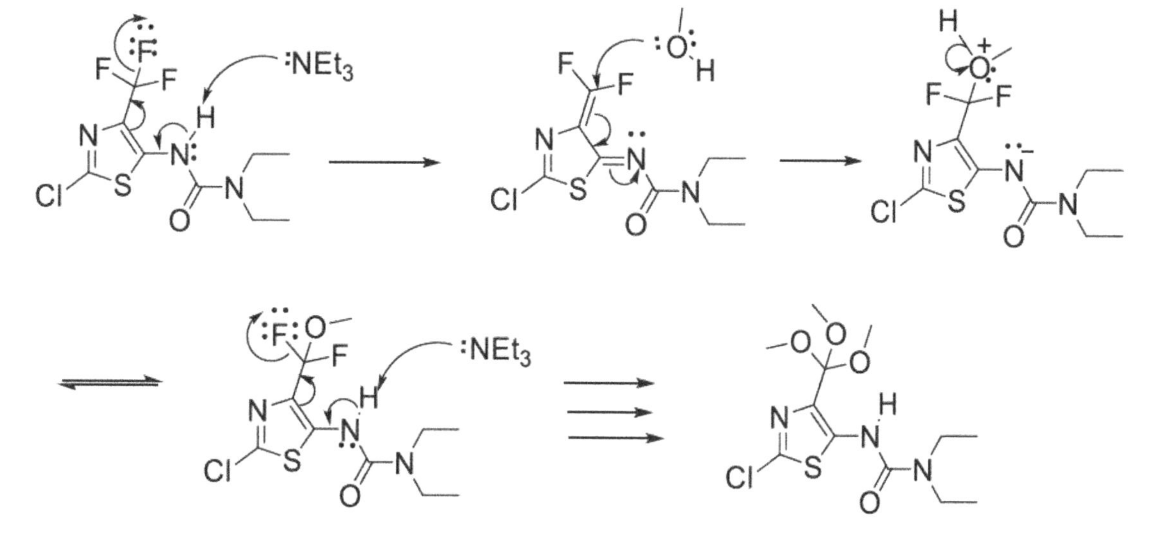

F
NEt3
H
N
Cl
S
O

253

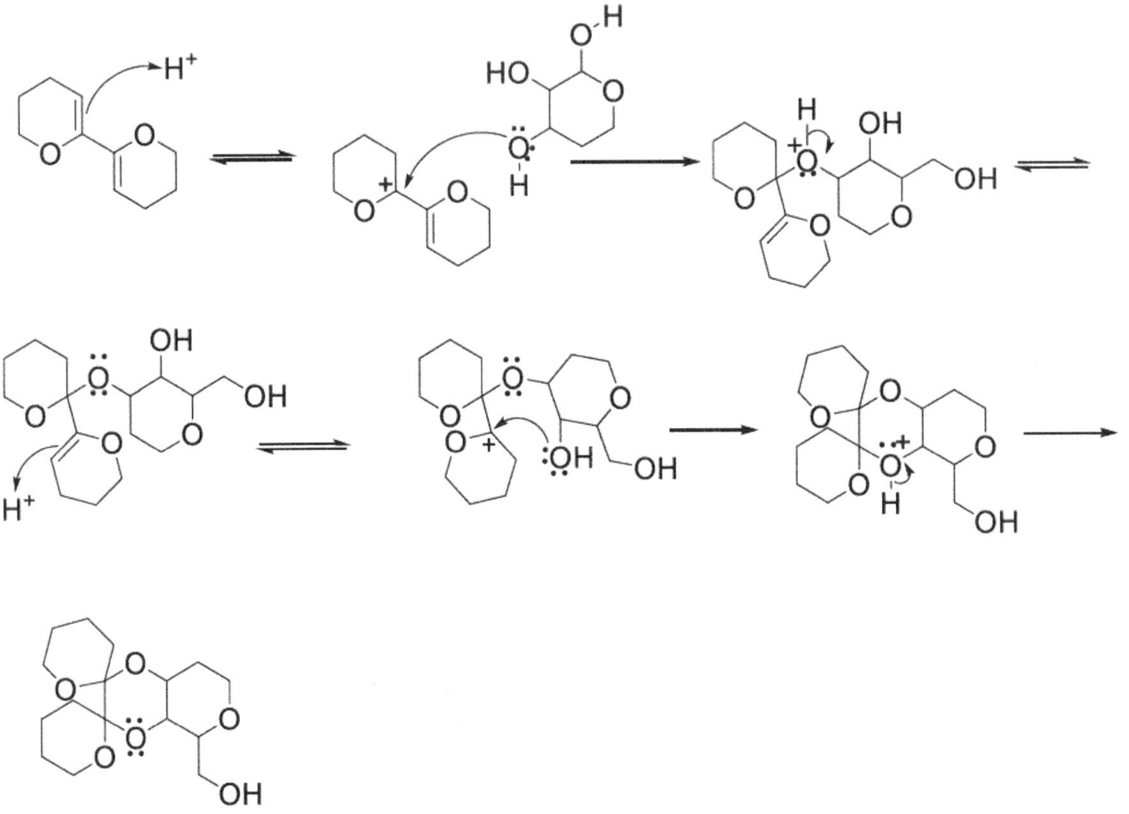

254

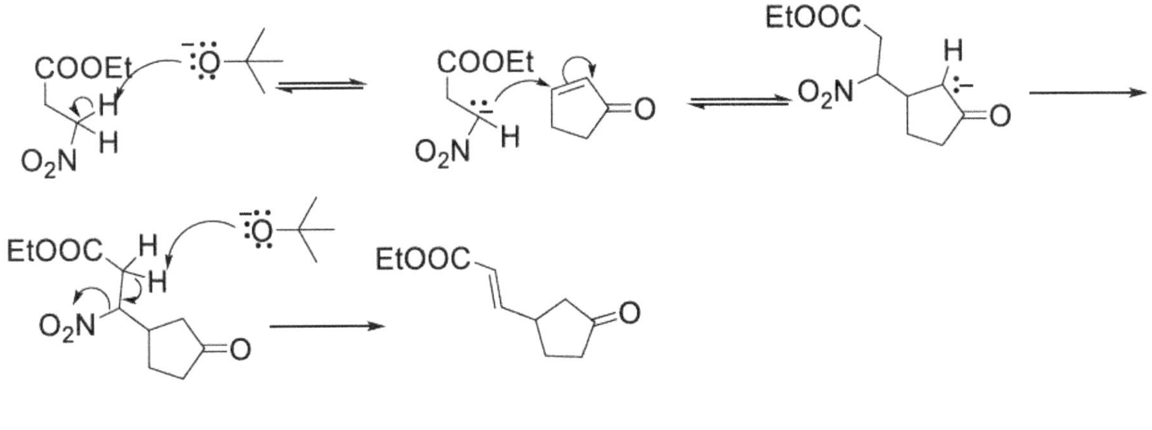

Exercise

a) Why did the initial proton abstraction happen on the carbon next to the nitro group?

b) Why is NO_2^- a good leaving group?

255

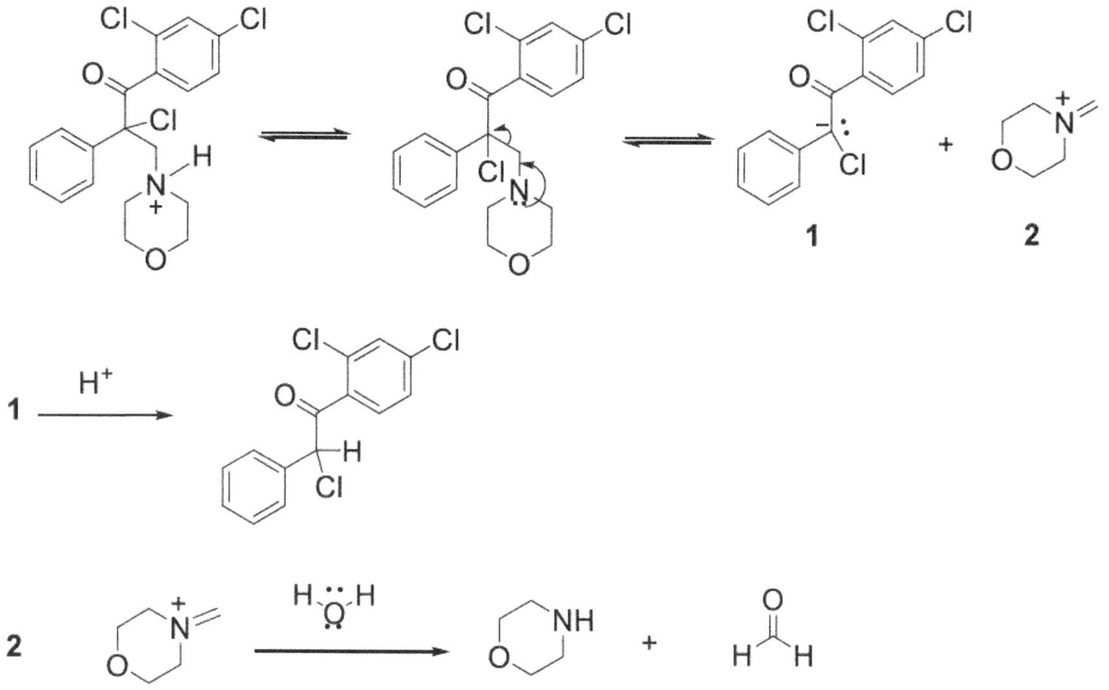

Cl
Cl
O
Cl
N
H
+
O
1
2
H+
H
H
O
NH

<u>Exercise</u> Write a mechanism for step **2**

256

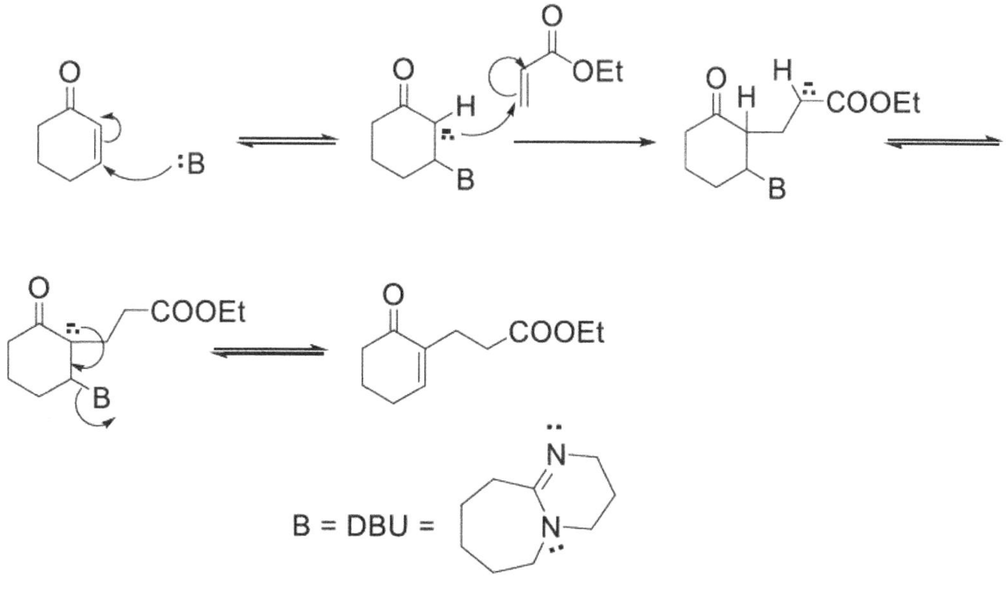

O
:B
H
B
OEt
COOEt
B = DBU =
N

257

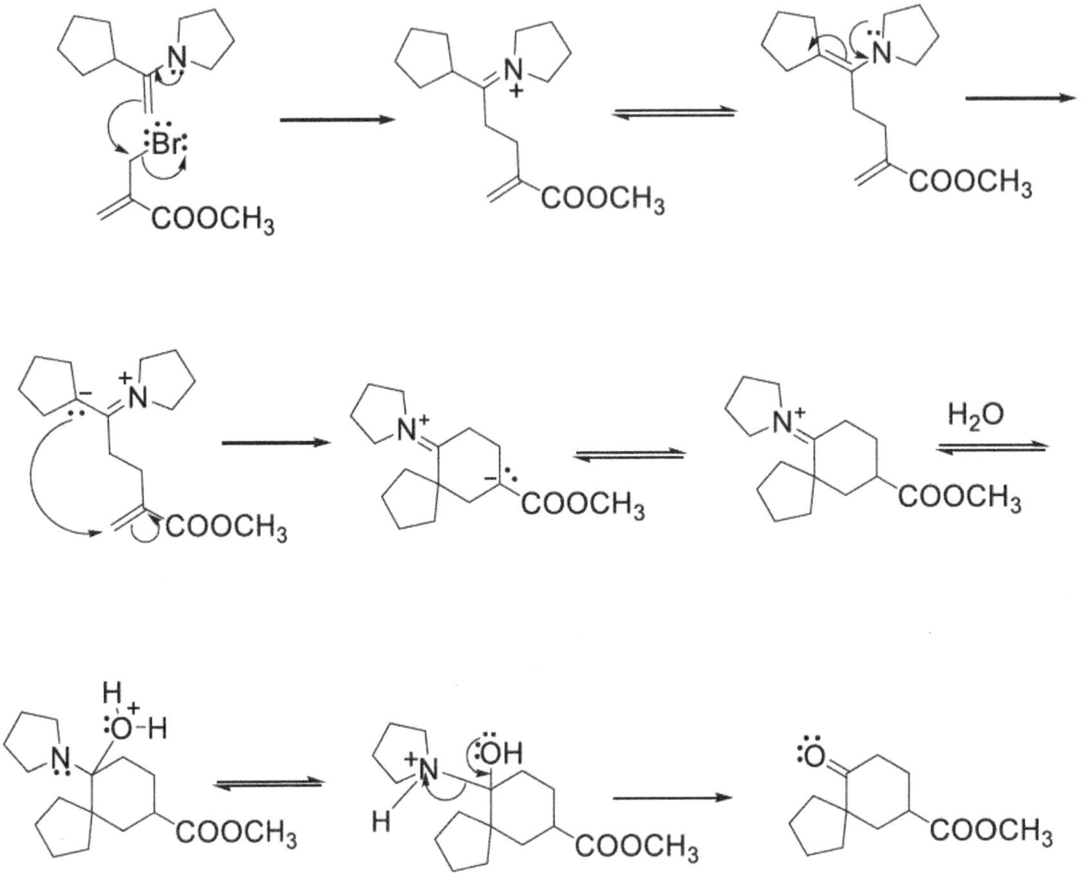

Br
N
COOCH$_3$
H$_2$O
OH
O
H

258

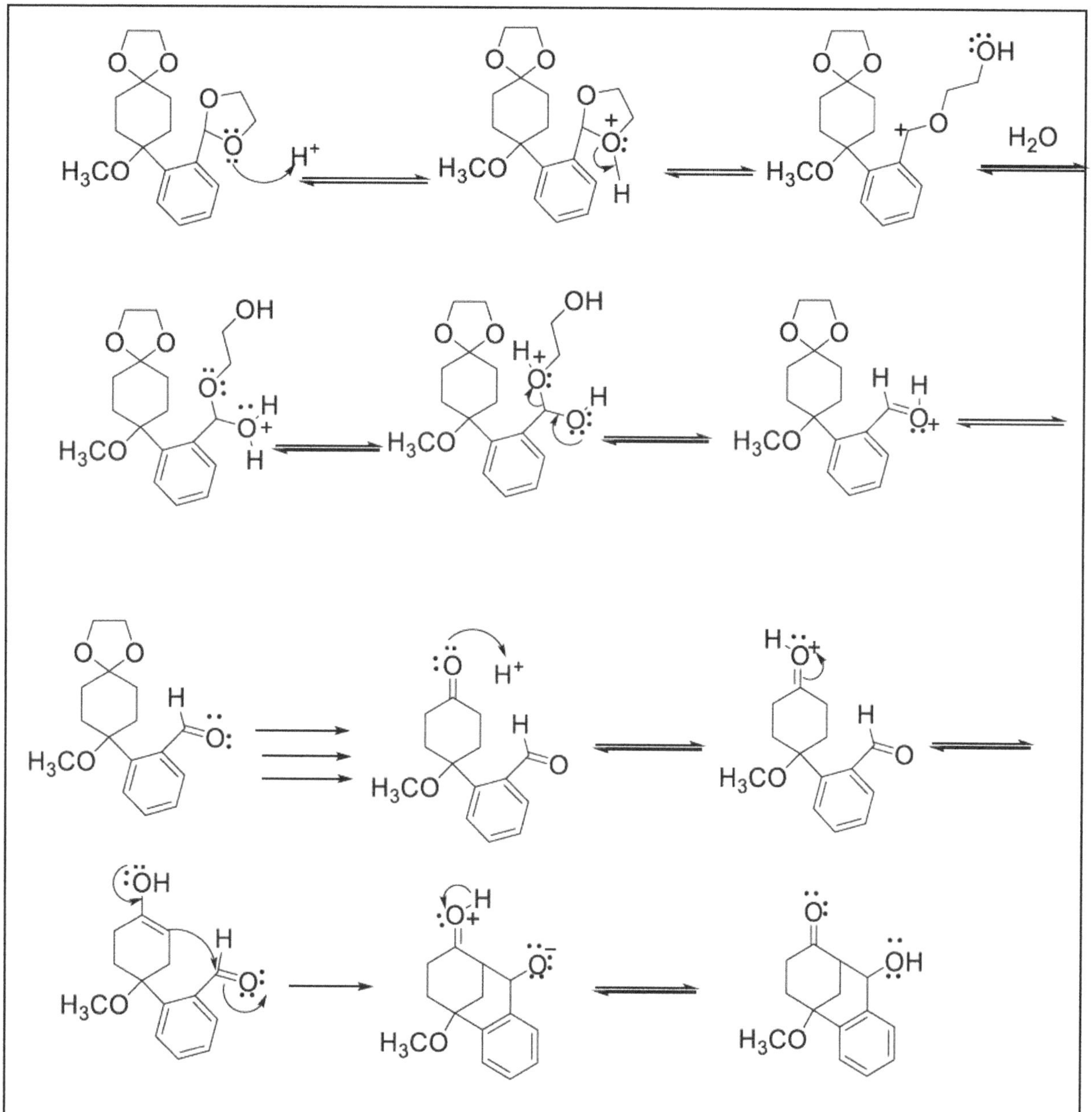
H$_3$CO
H$^+$
H
OH
H$_2$O
O

259

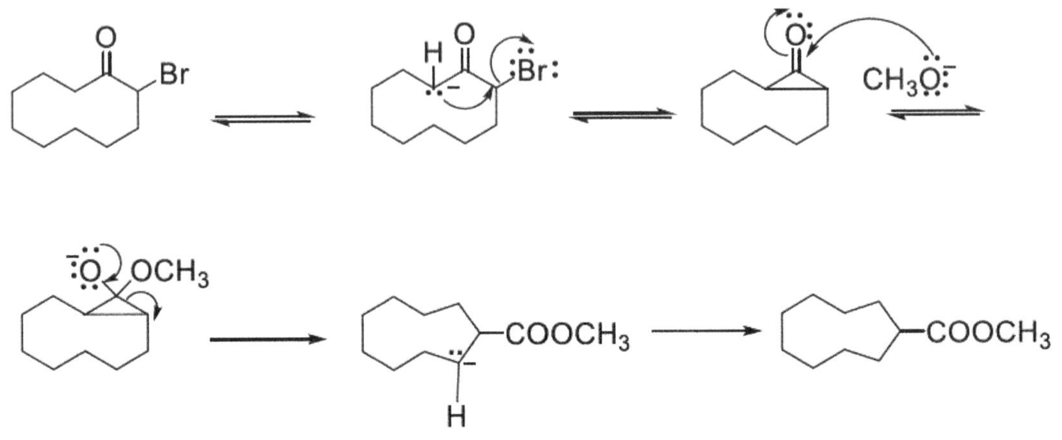

Br
H
Br
CH3O
OCH3
COOCH3
H
COOCH3

This is an example of the *Favorskii rearrangement*

260

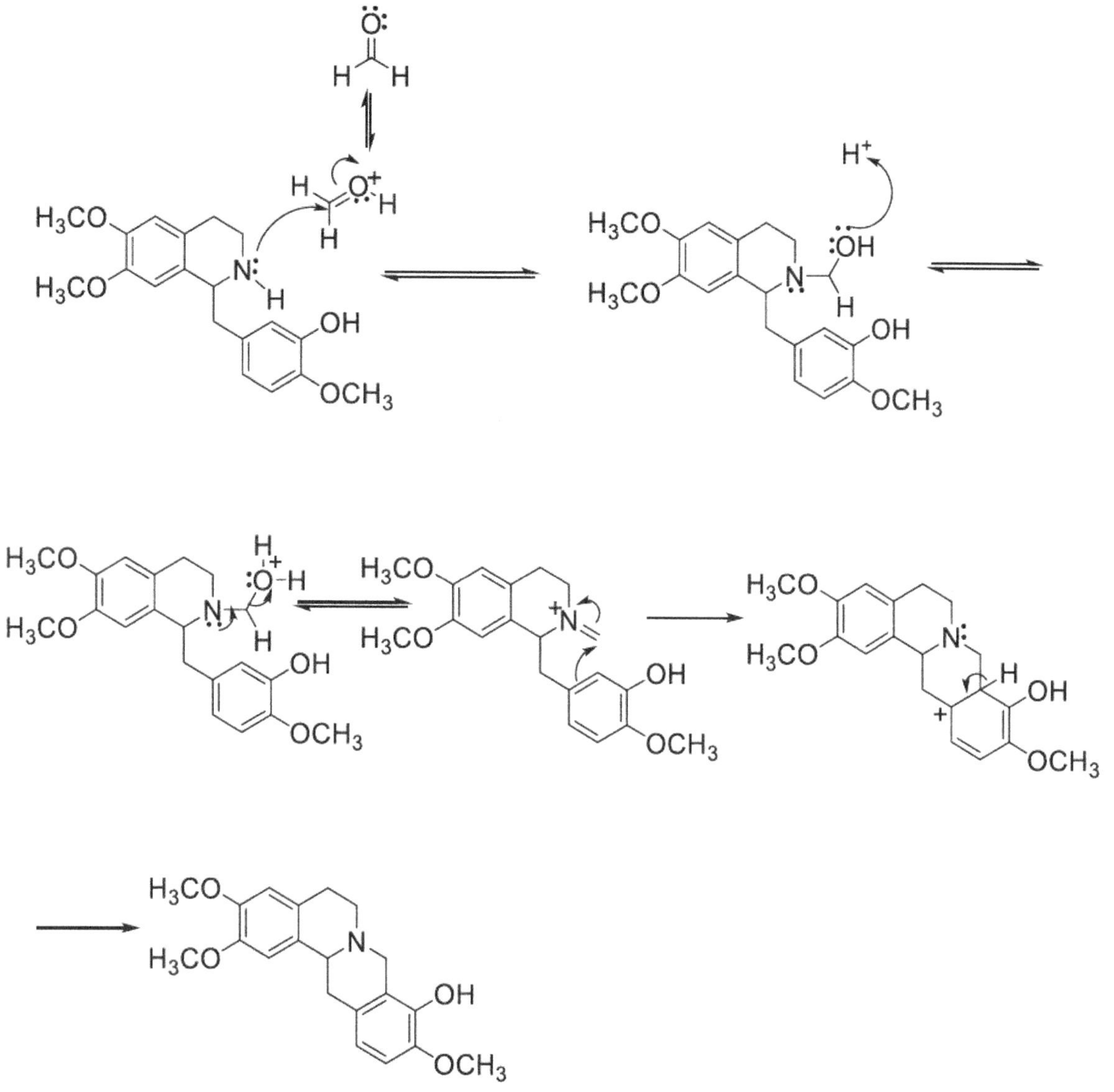

H
H
H
H
H
H3CO
H3CO
N
OH
OCH3
H+
OH
N
H
H3CO
H3CO
OH
OCH3

This is an example of the *Pictet-Spengler reaction*

261

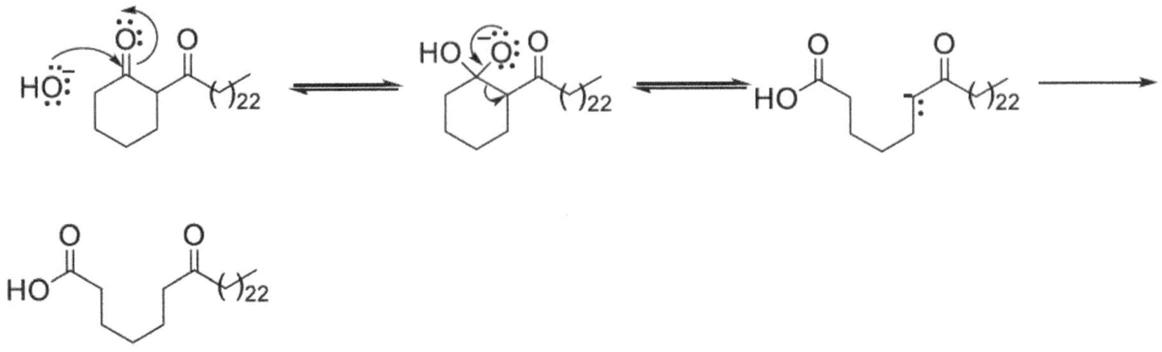

262

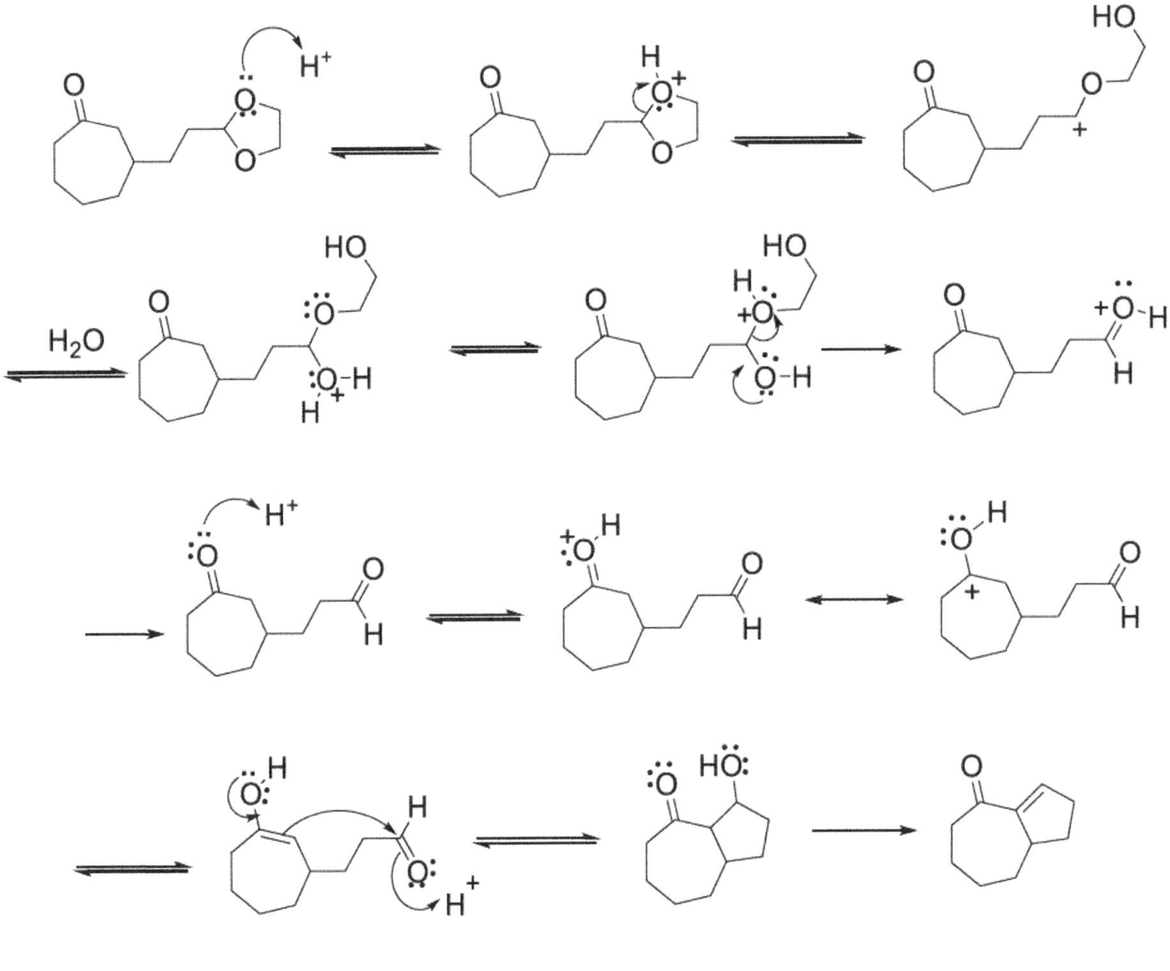

263

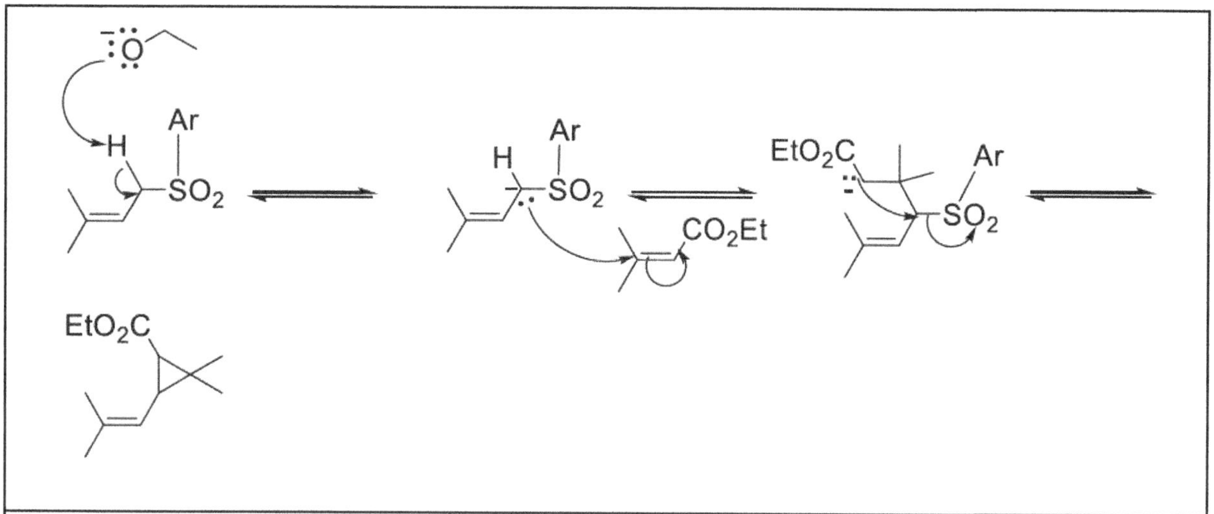
Ar
H
SO_2
EtO_2C
CO_2Et

264

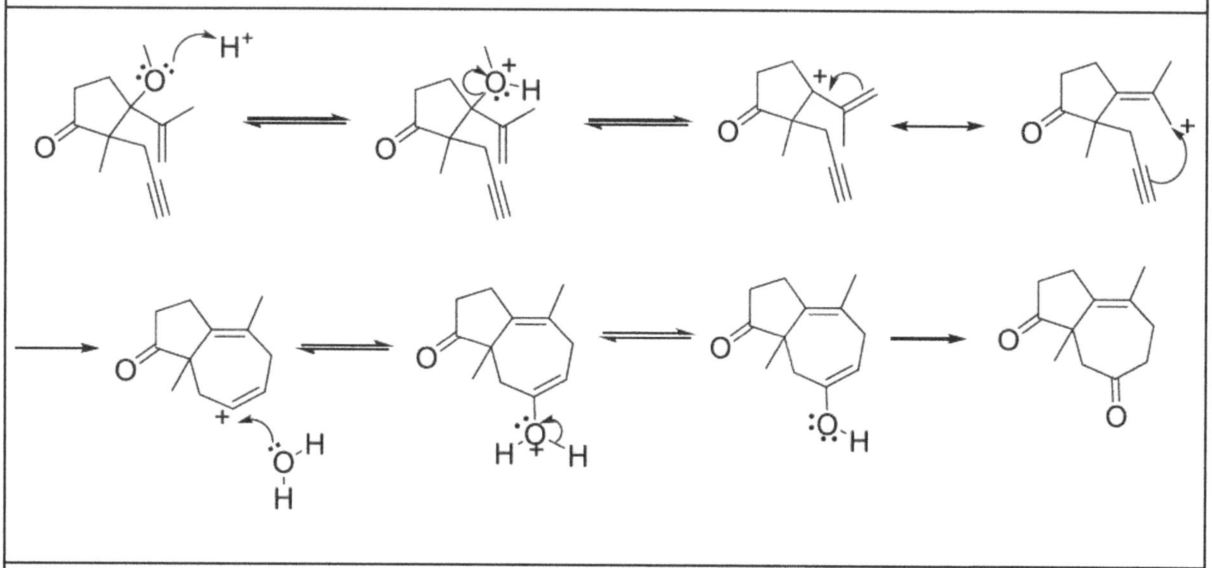
H^+
O
H

265

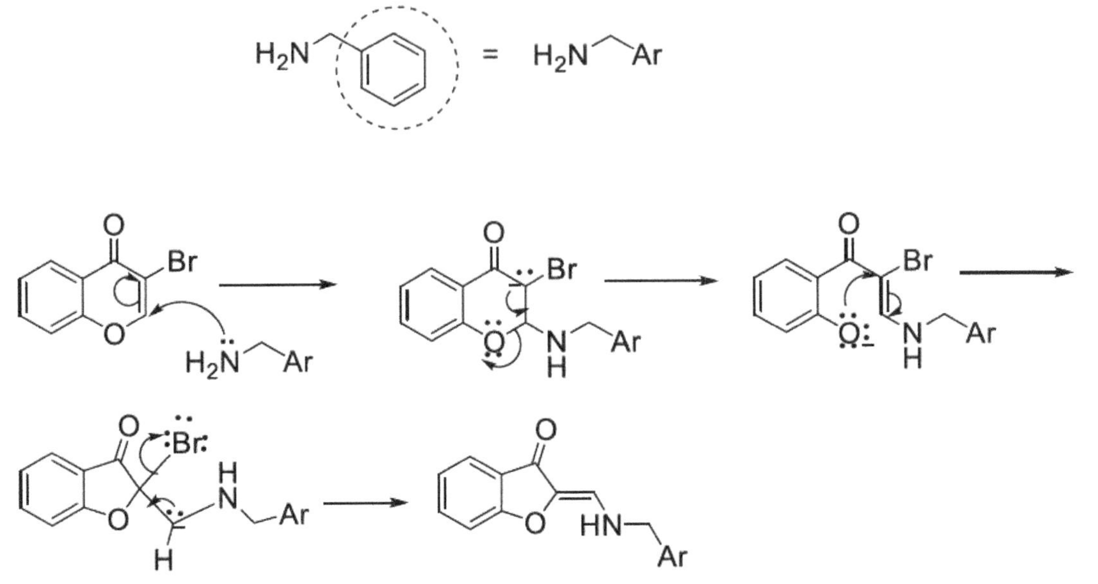
H2N
=
H2N
Ar
O
Br
O
H2N
Ar
O
Br
O
N
H
Ar
O
Br
O
N
H
Ar
O
Br
H
N
Ar
O
H
O
HN
Ar

266

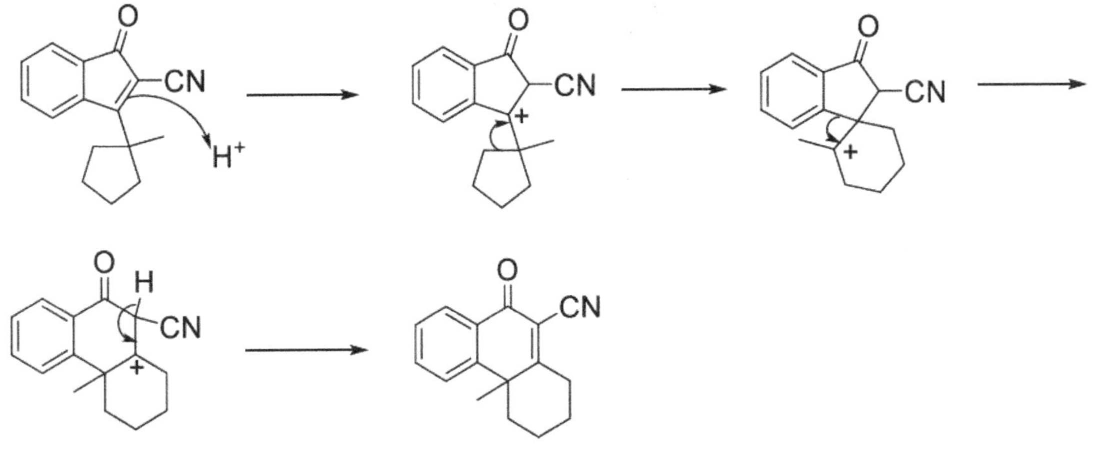
O
CN
H+
O
CN
+
O
CN
+
O
H
CN
+
O
CN

267

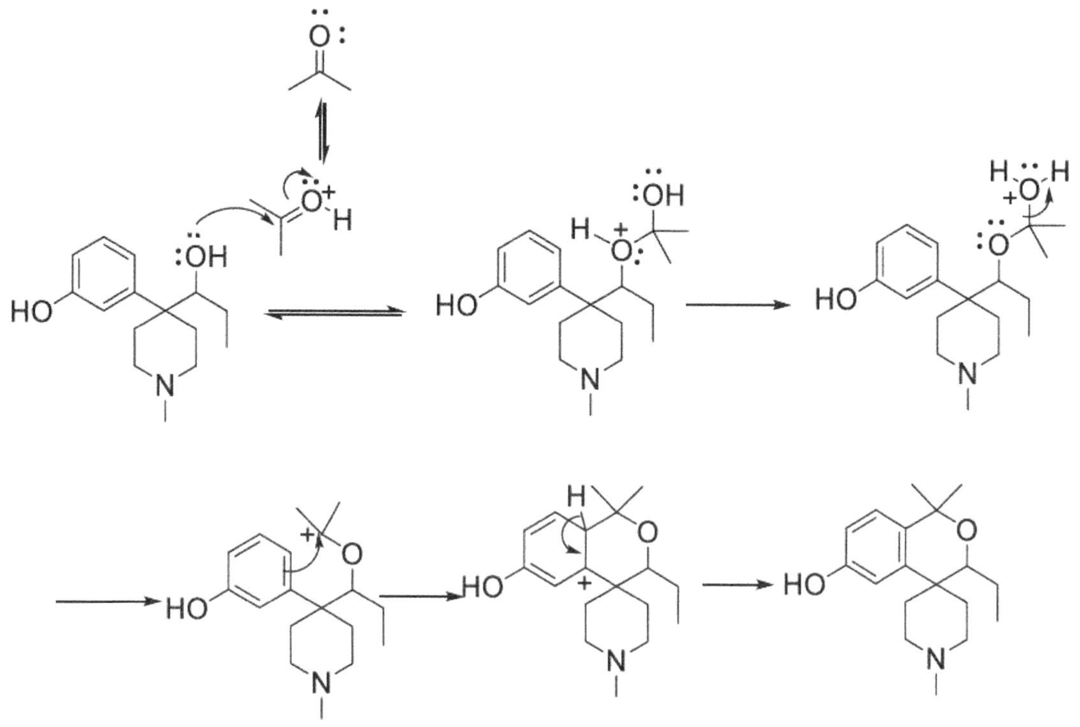

268

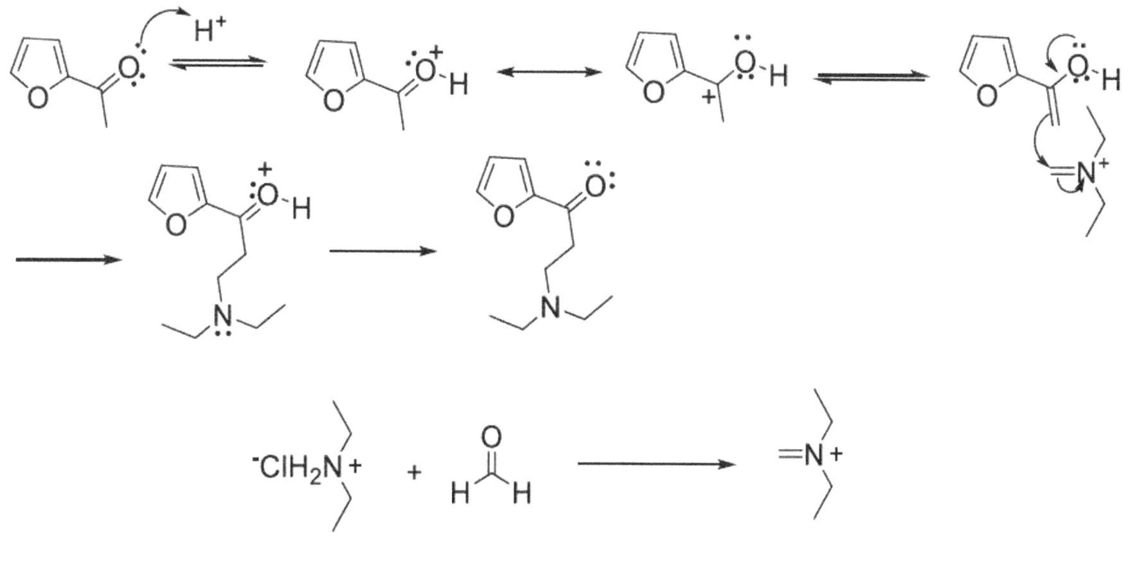

<u>Exercise</u> Write a mechanism for the formation of the imine.

269

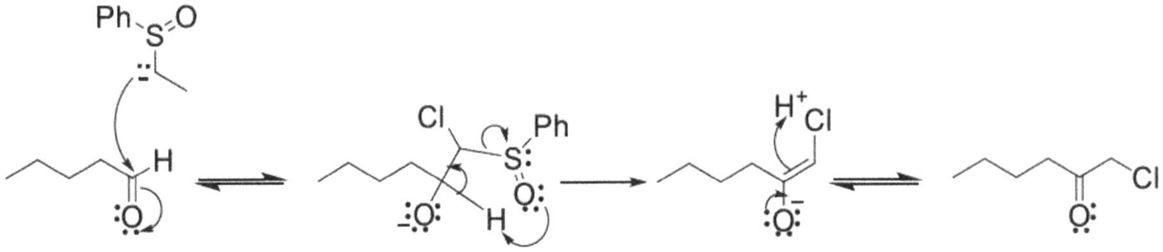

Ph
S
O
Cl
Ph
S
O
H
O
H
H+
Cl
O
O
Cl

This is an example of a sulfoxide *syn elimination reaction*

270

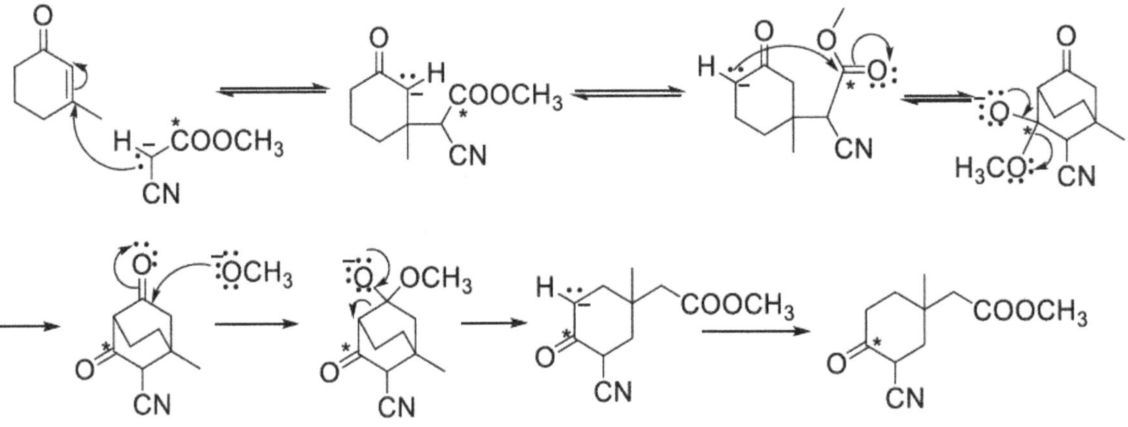

O
H
COOCH3
CN
O
H
COOCH3
CN
H
O
O
O
CN
O
O
H3CO
CN
O
OCH3
O
CN
O
OCH3
O
CN
H
COOCH3
O
CN
COOCH3
O
CN

(prototropic shift refers to the equlibration of two anions with equal stability)

271

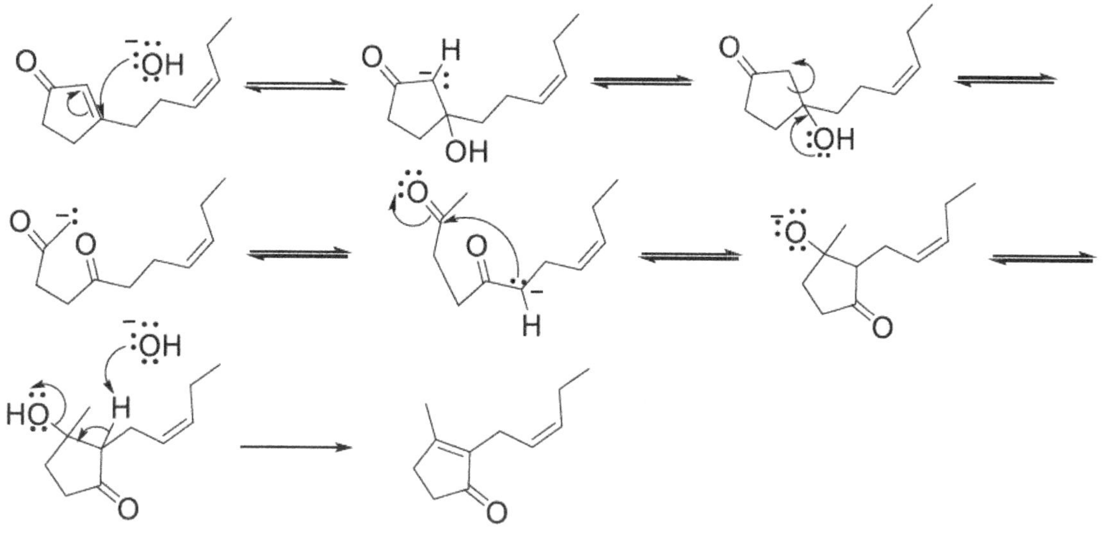

OH
OH
OH
O
O
O

272

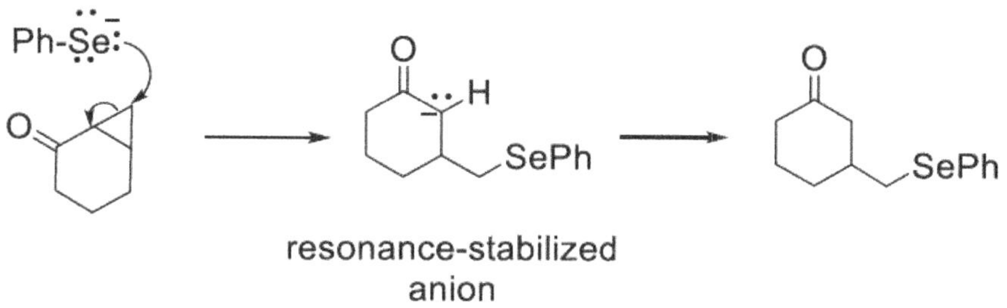

Ph-Se
O
H
SePh
resonance-stabilized
anion

273

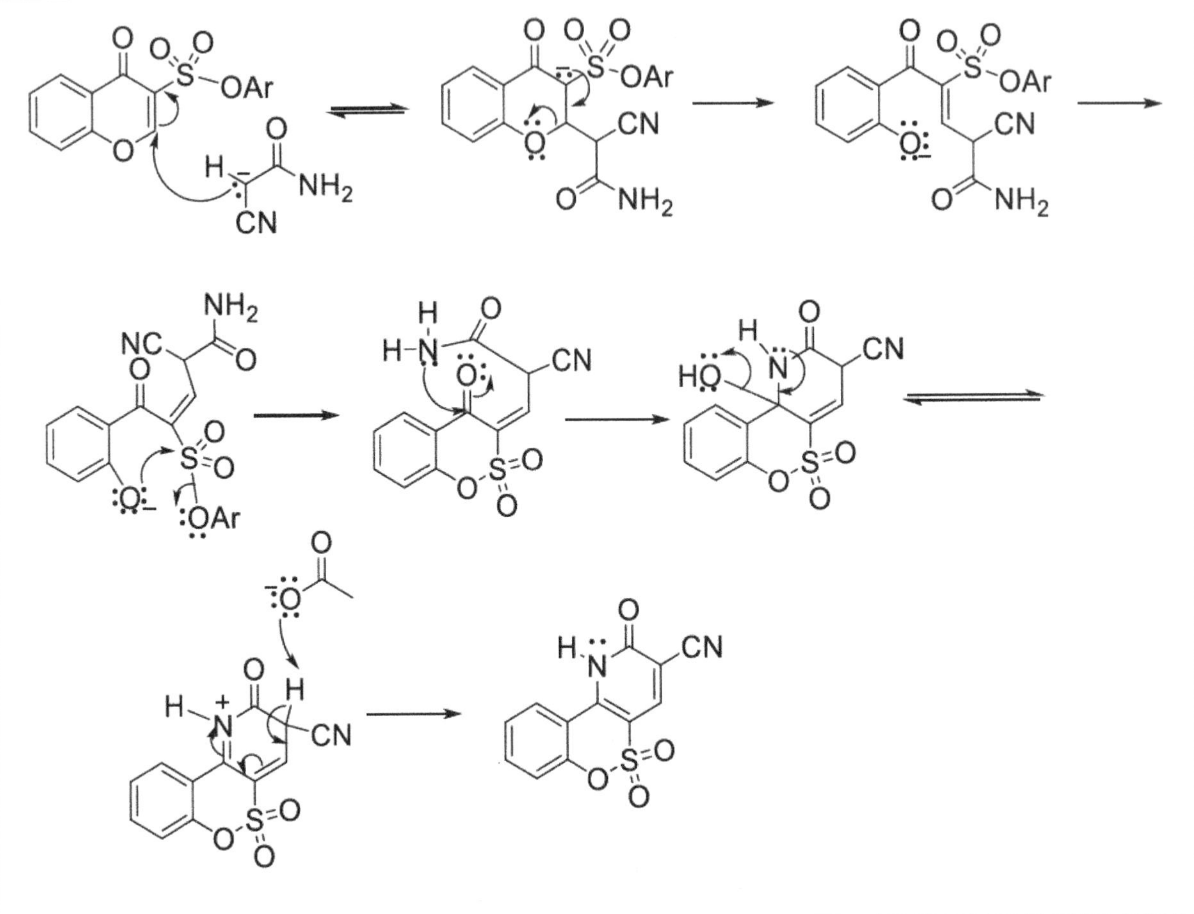
OAr
NH2
CN
OAr
CN
NH2
OAr
CN
NH2
NH2
NC
OAr
CN
CN
H
CN
CN

274

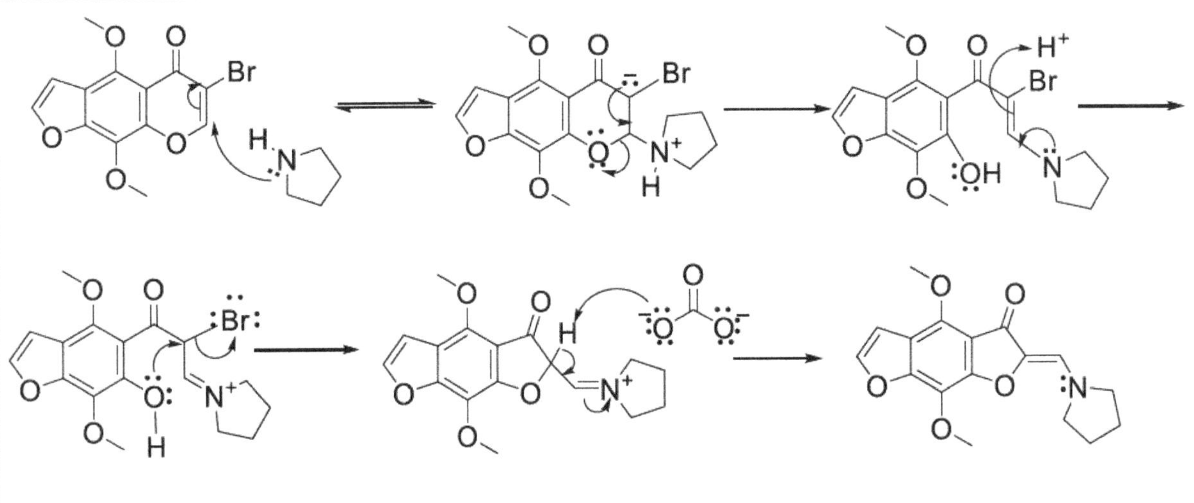
Br
H
N
Br
N+
H
H+
Br
OH
N
Br
N+
H
H
N+
N

275

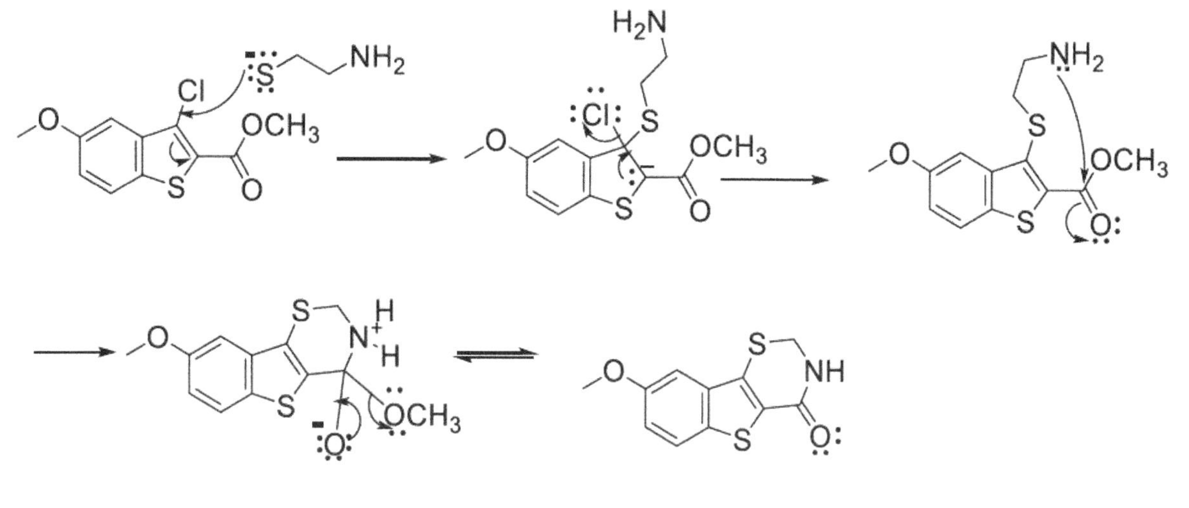

276

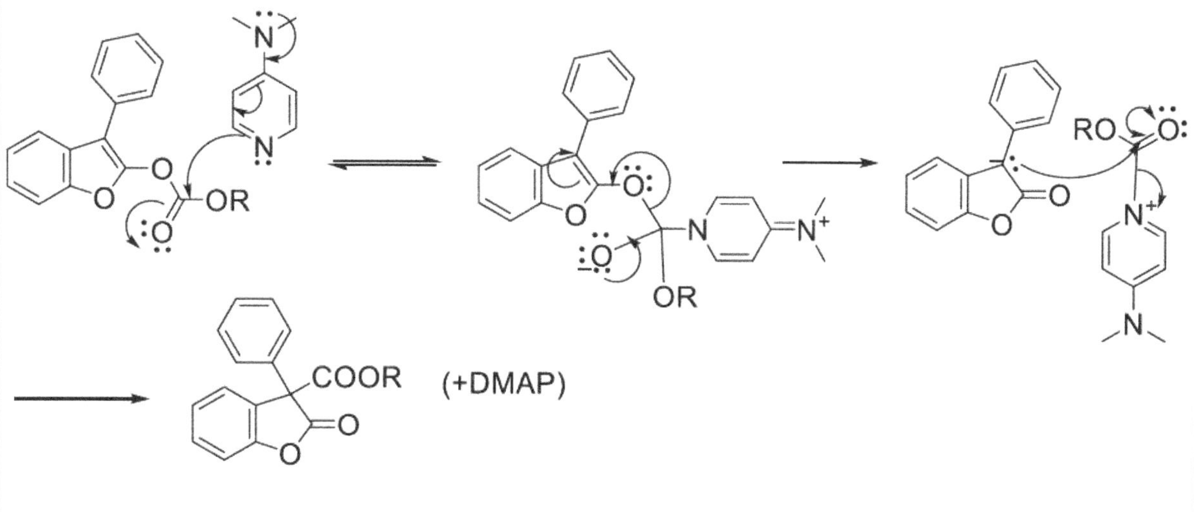
(+DMAP)

277

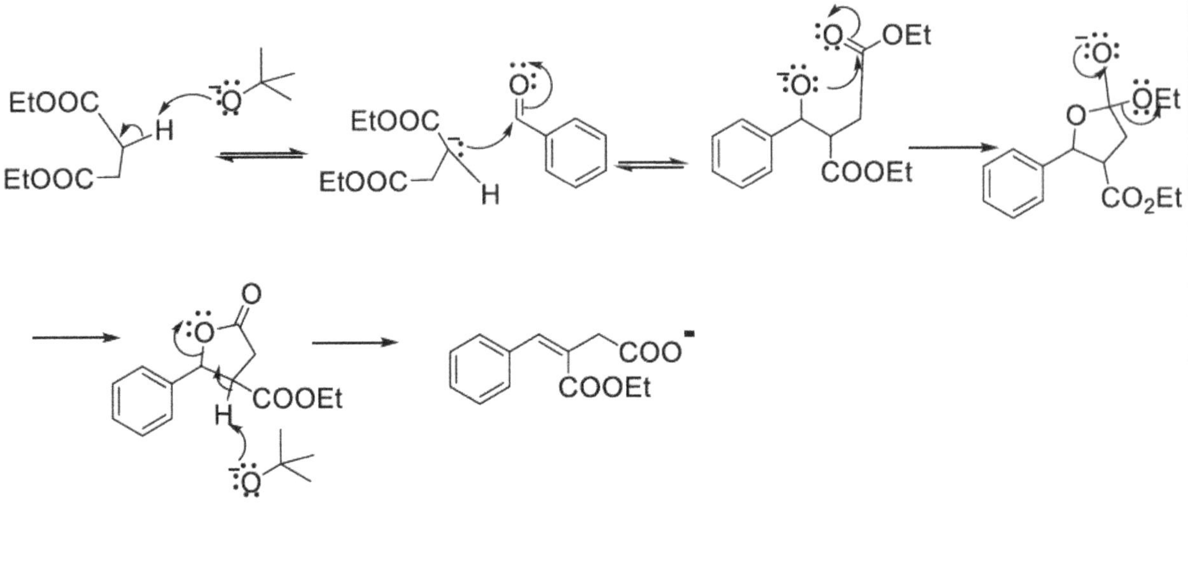

This is an example of the *Stobbe condensation* reaction

278

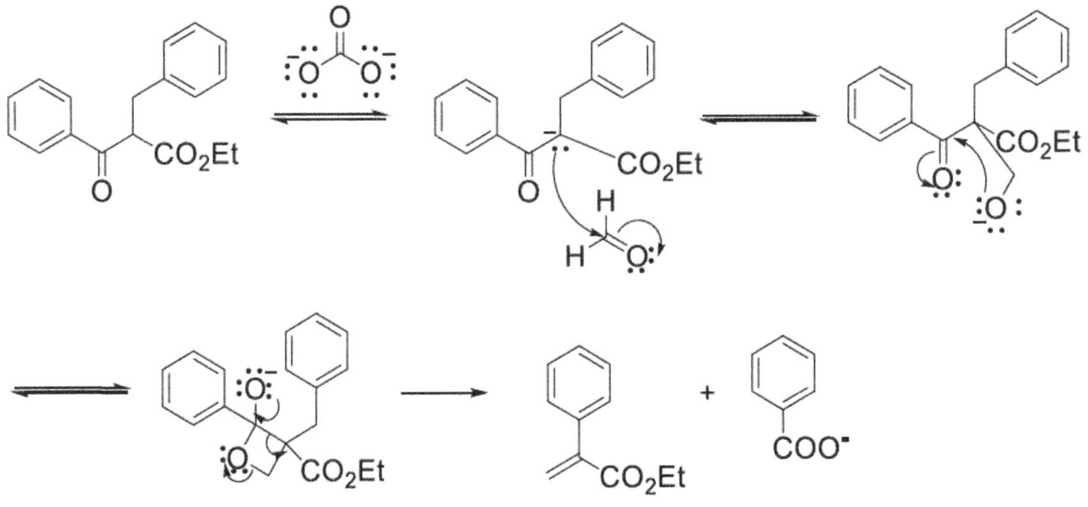

279

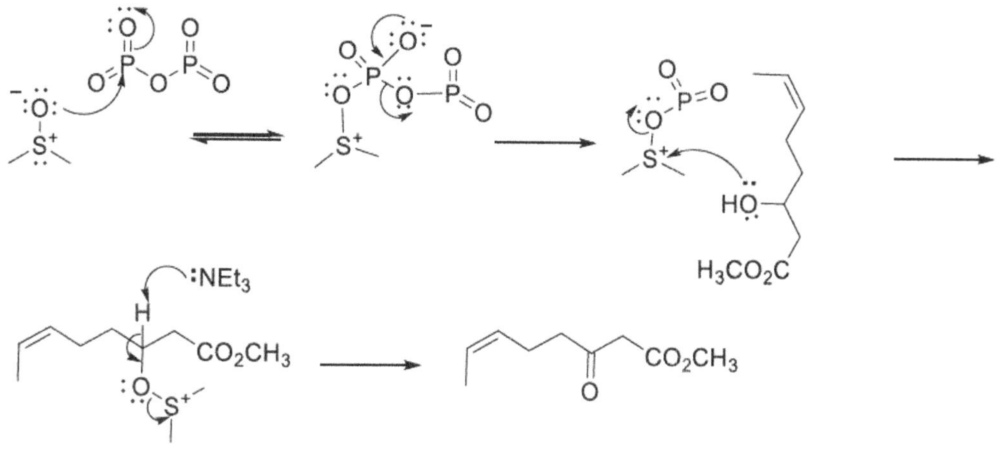

This is a variant of the *Swern oxidation* reaction

280

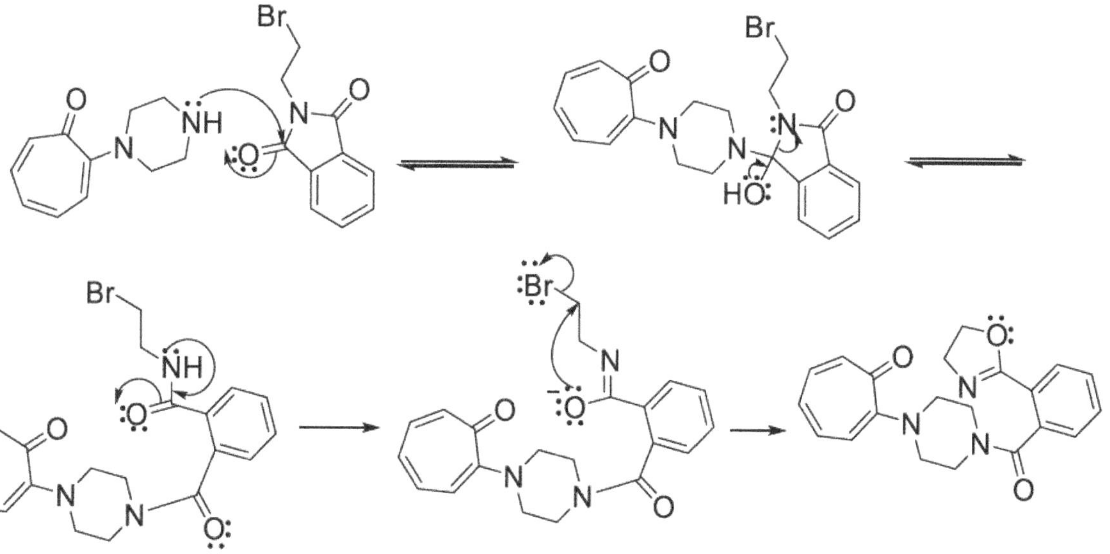

281

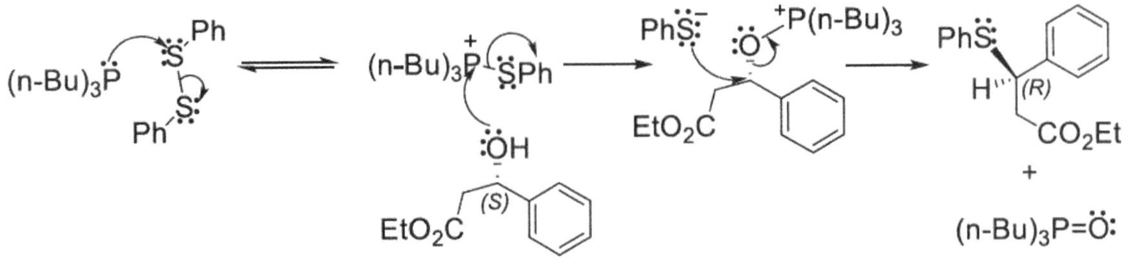

This is an example of the *Mitsunobu* reaction

282

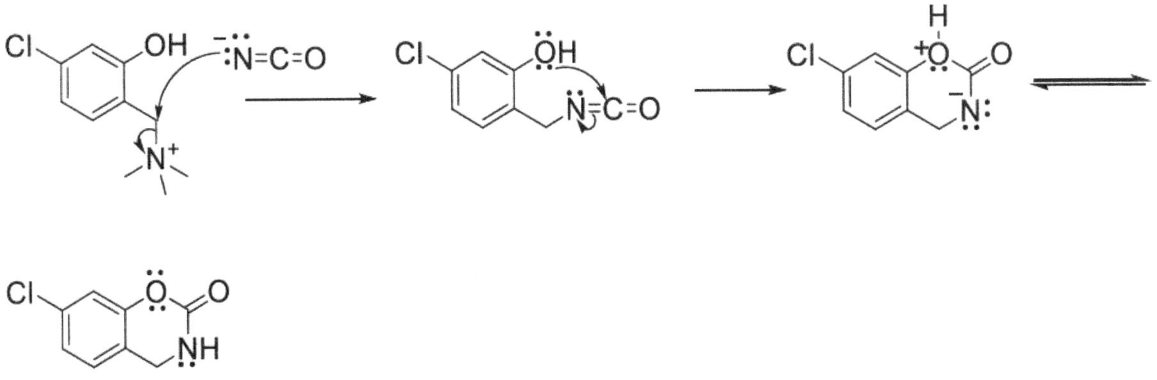

283

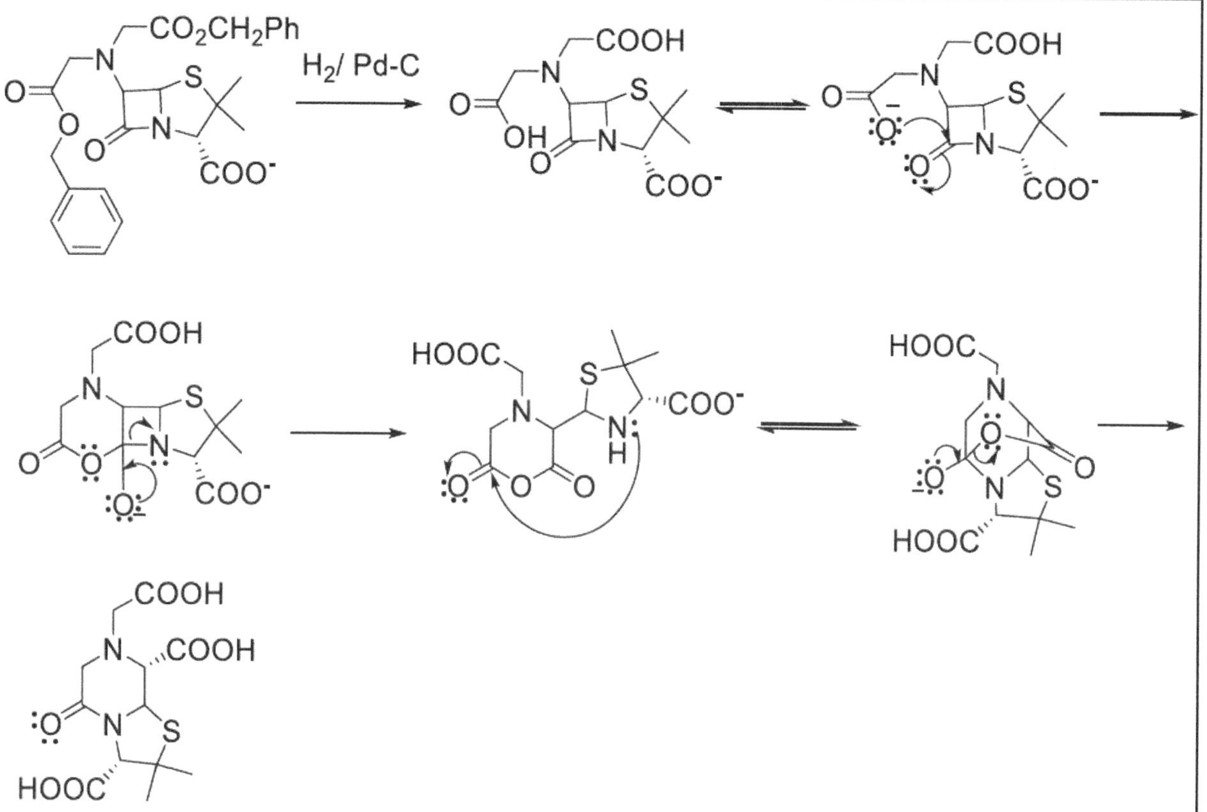

H2/ Pd-C
CO2CH2Ph
COOH
COO-
HOOC

284

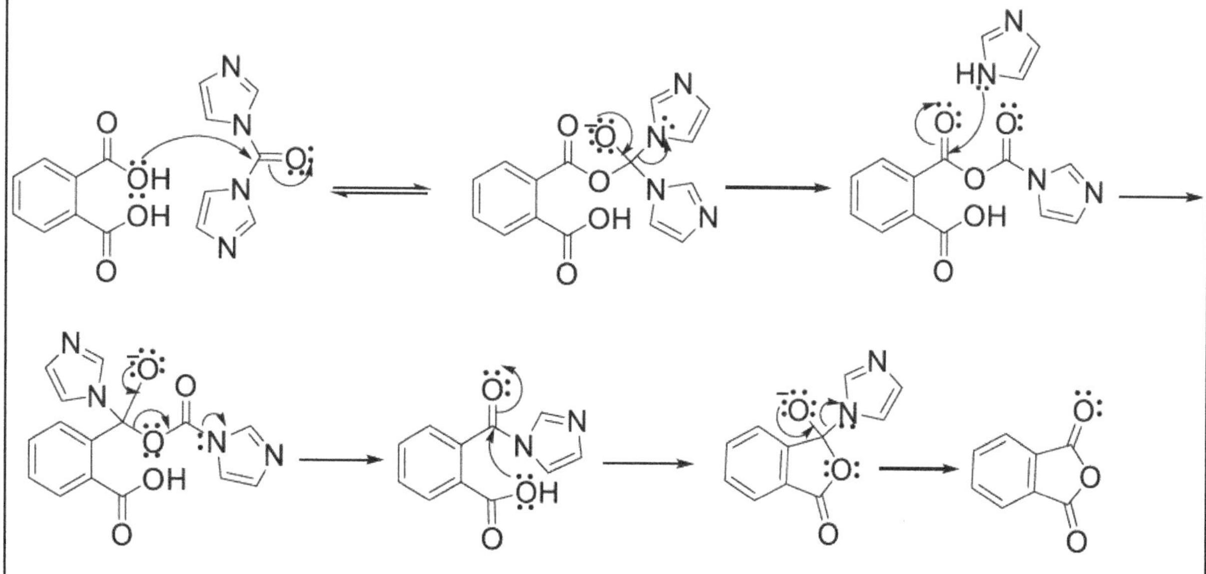

285

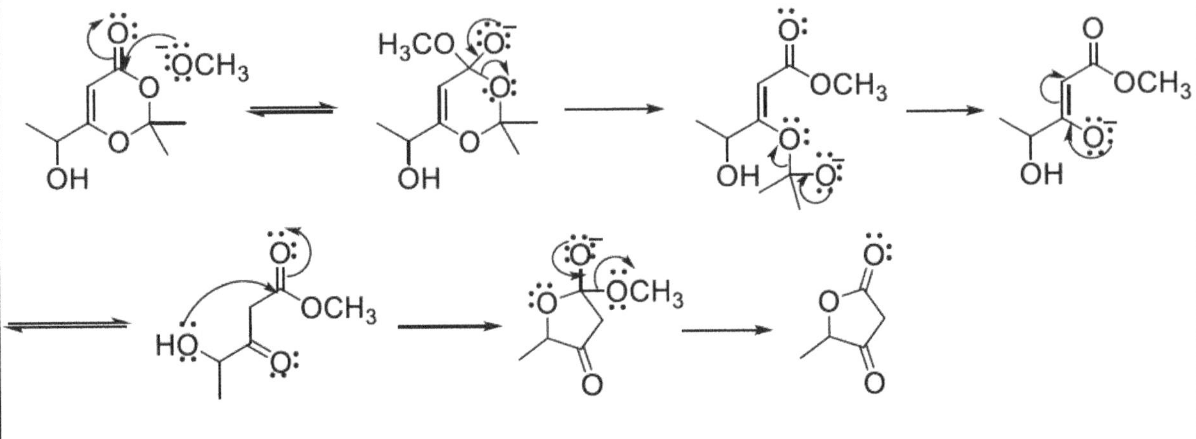

286

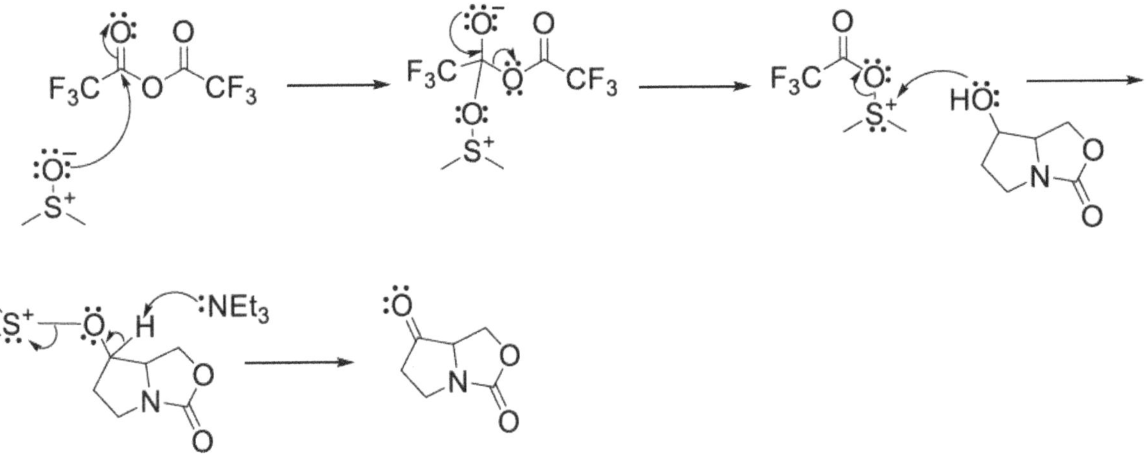

This is a variant of the *Swern oxidation* reaction

287

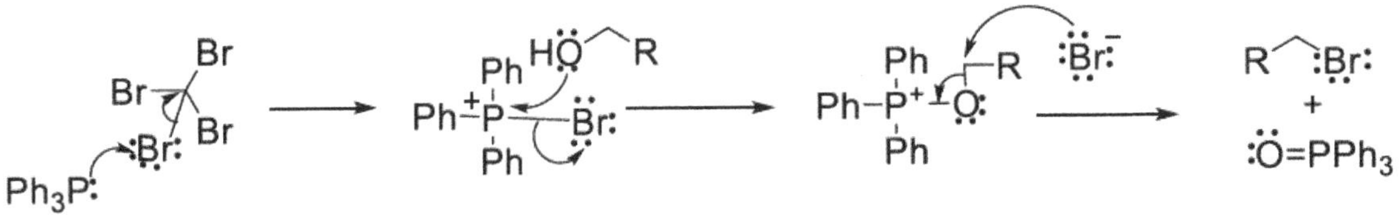

288

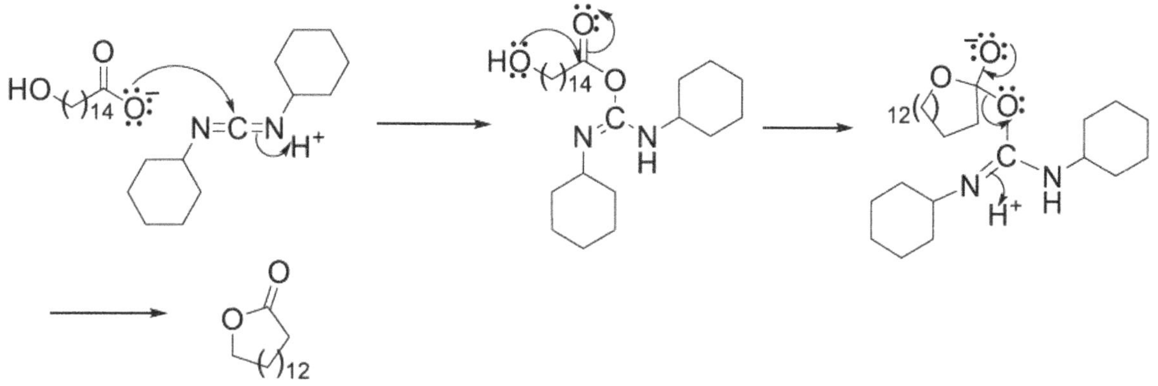

HO
14
O
N=C=N
H+
12

This is an example of a *macrolactonization reaction*

289

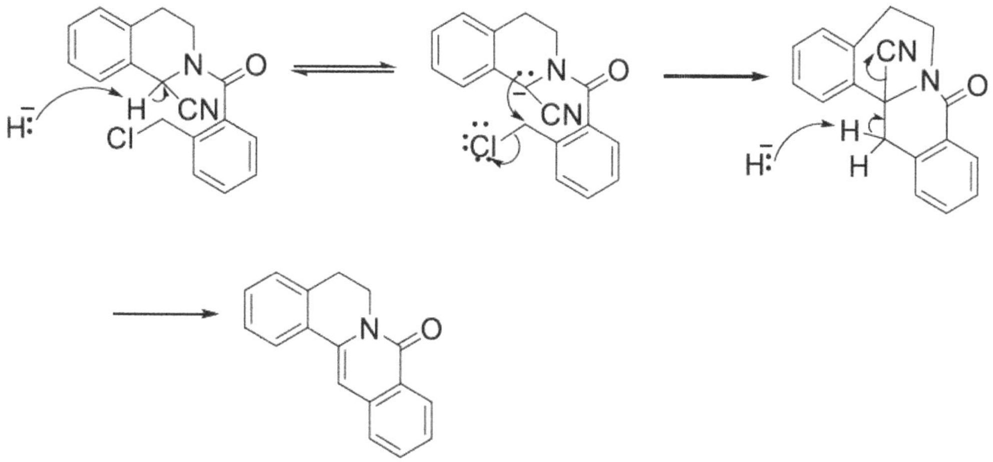

N
O
H
CN
Cl
H:

290

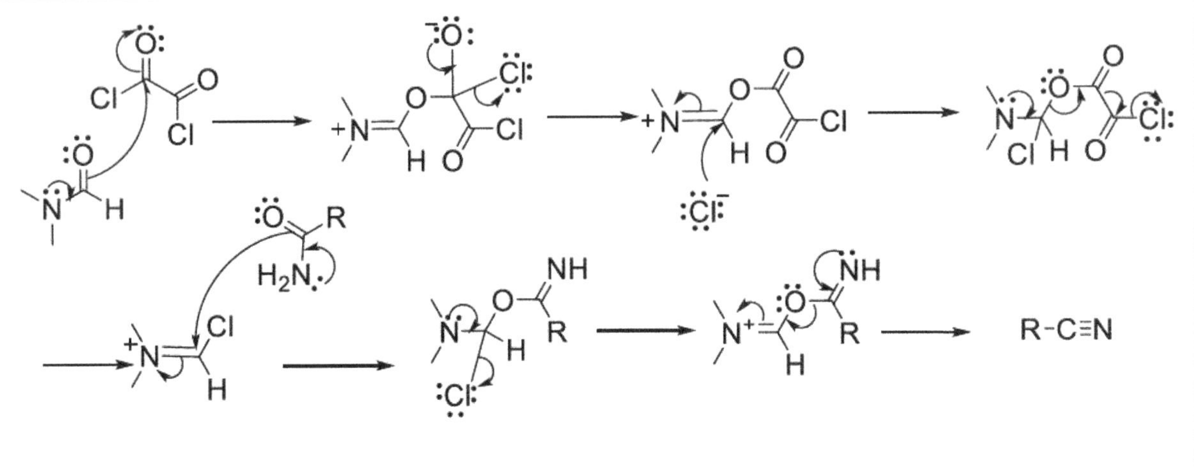

291

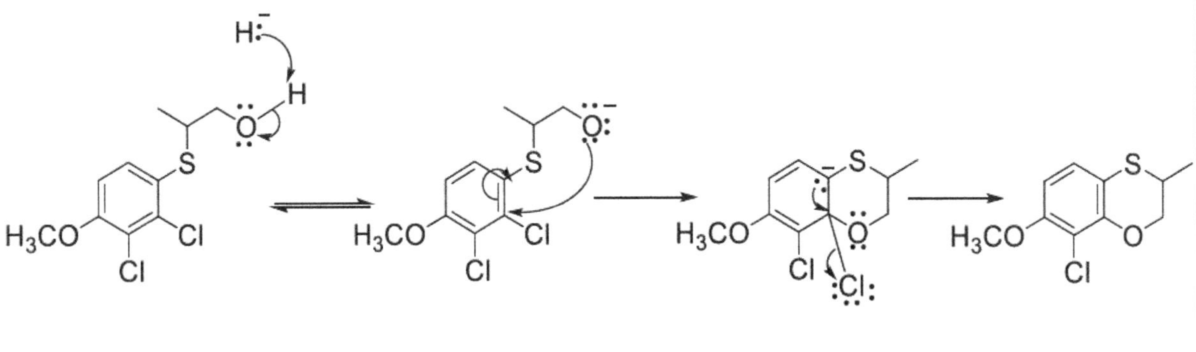

292

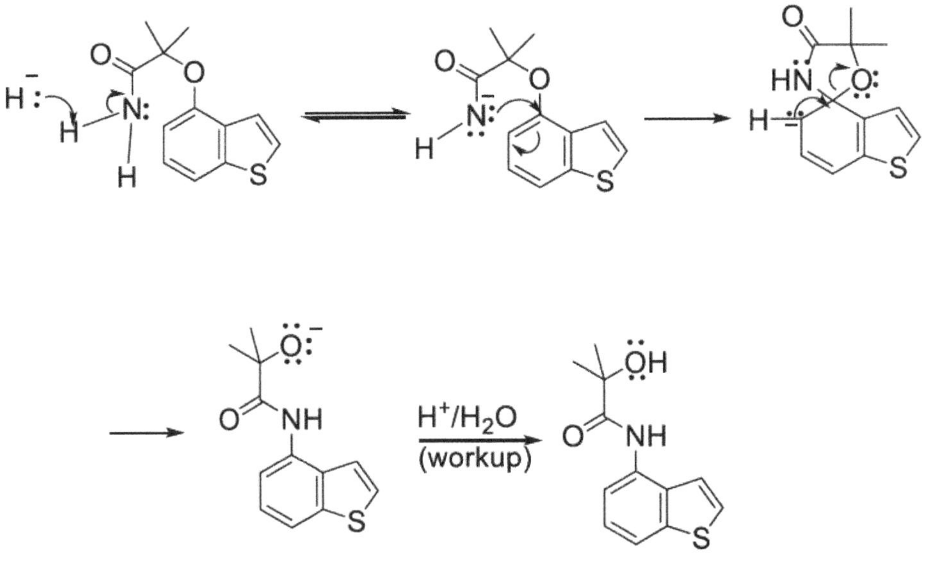

293
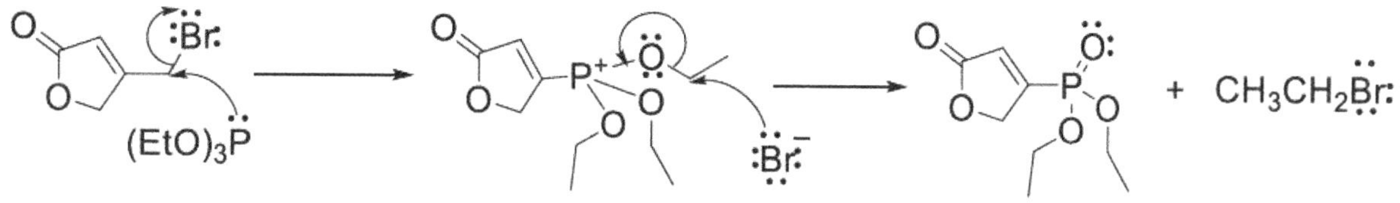:Br: (EtO)$_3$P P$^+$:Br:$^-$ + CH$_3$CH$_2$Br: This is an example of the *Arbusov reaction*
294
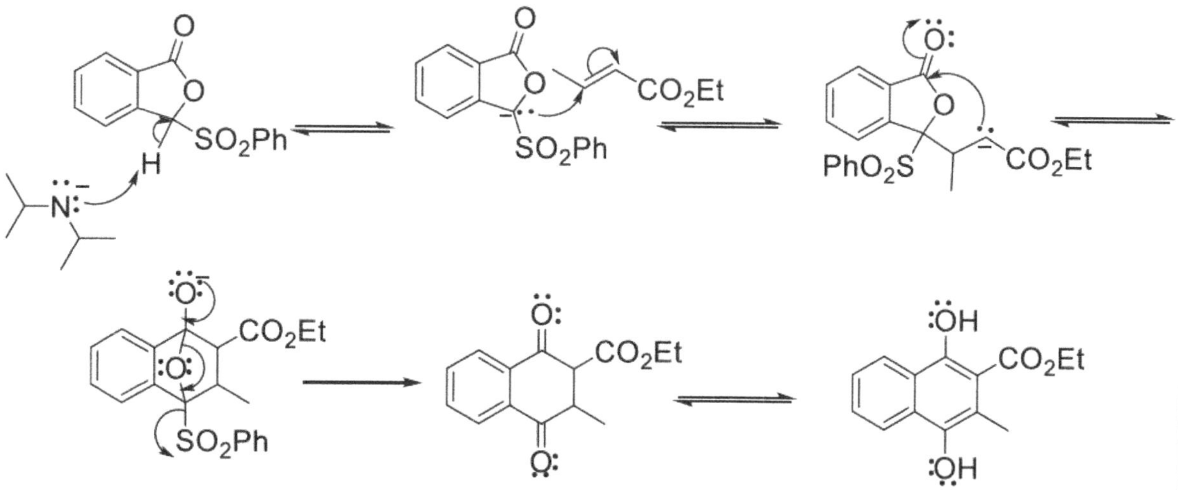SO$_2$Ph H N:$^-$ CO$_2$Et SO$_2$Ph PhO$_2$S CO$_2$Et :O:$^-$ CO$_2$Et :O: SO$_2$Ph CO$_2$Et :OH CO$_2$Et :OH Exercise The keto tautomer is ordinarily more stable than the enol form. In this example, the enol form is favored. Why?
295

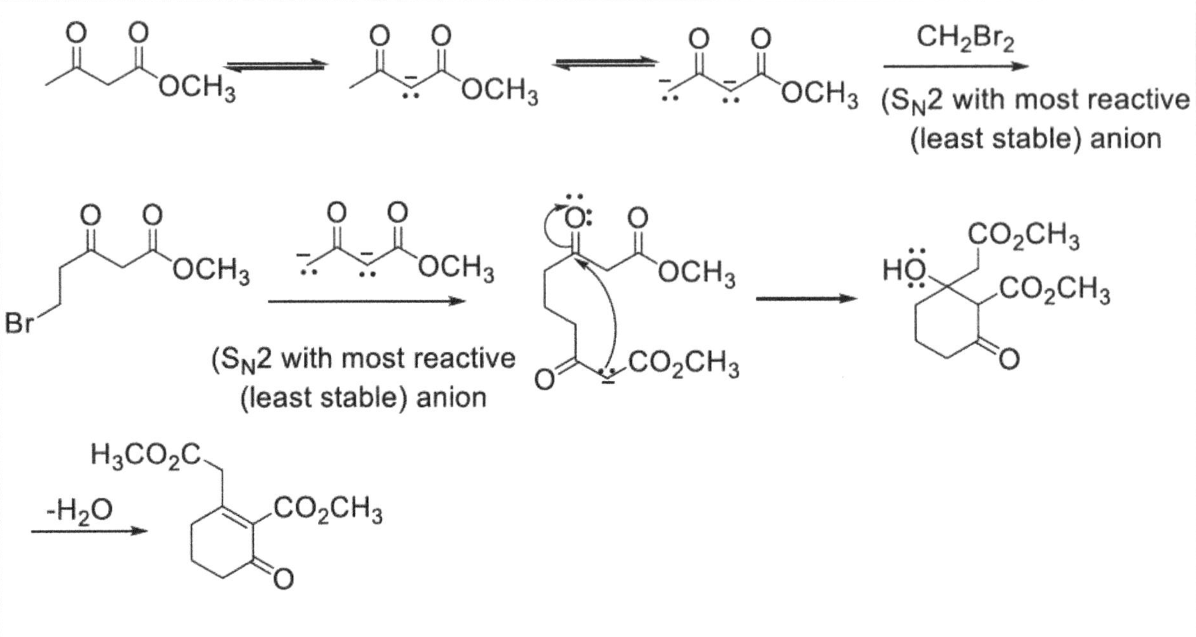
CH2Br2
(SN2 with most reactive
(least stable) anion
(SN2 with most reactive
(least stable) anion
-H2O

296

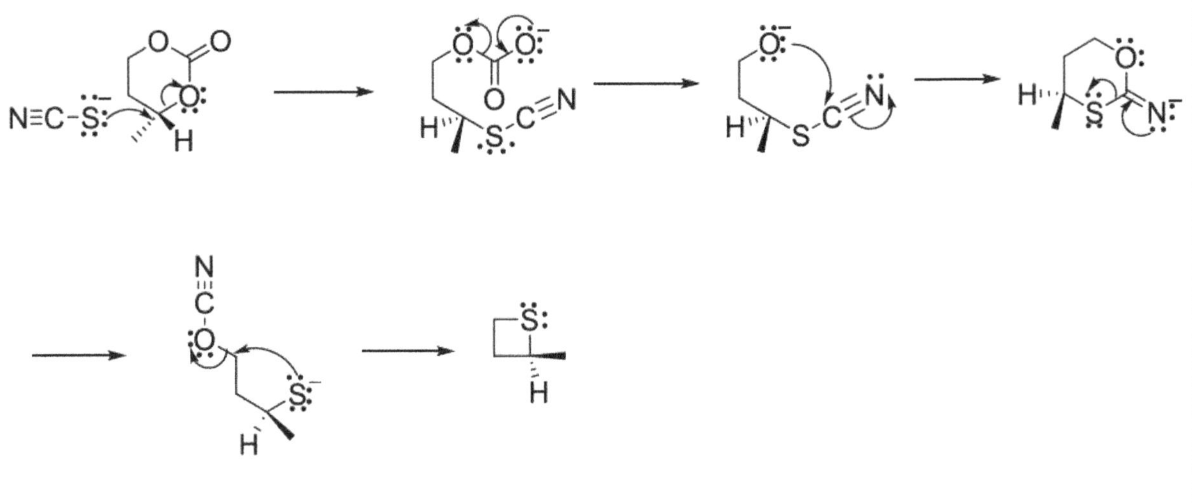

297

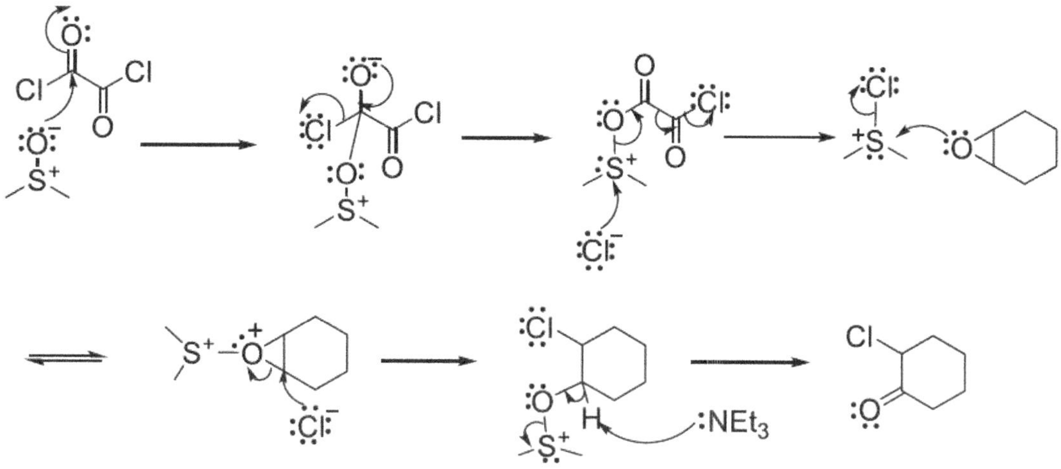
:NEt3

298

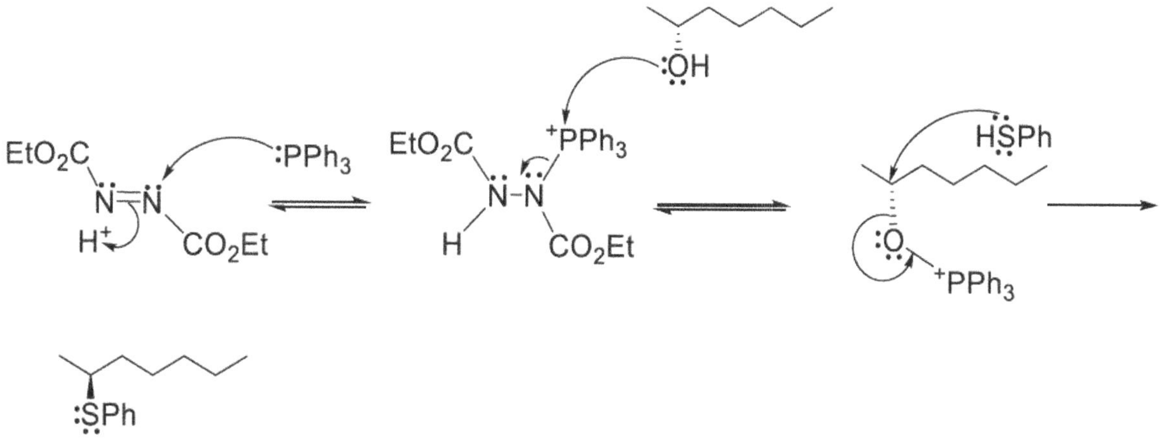
EtO2C
:PPh3
CO2Et
+PPh3
HSPh
:SPh

This is an example of the *Mitsonobu reaction*

299

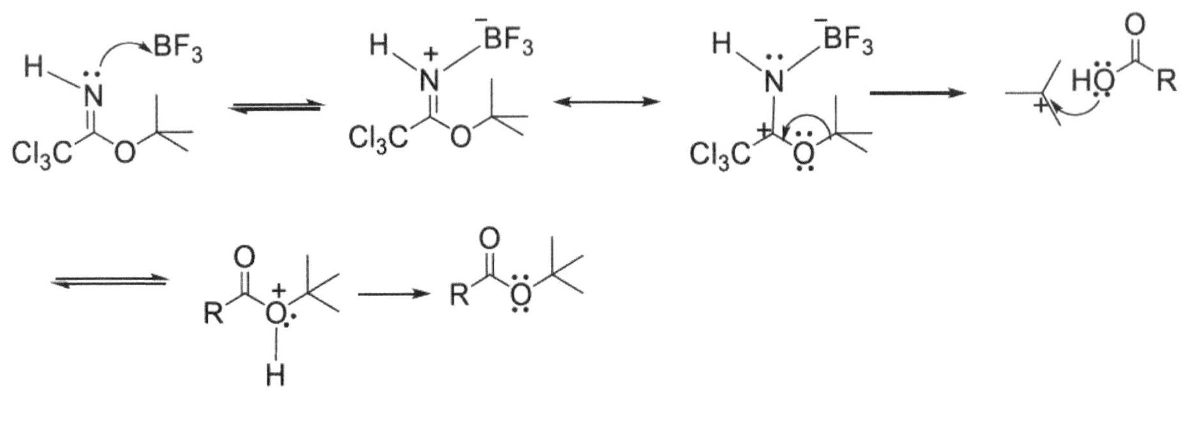
BF3
H
N
Cl3C
O
R
HO

300

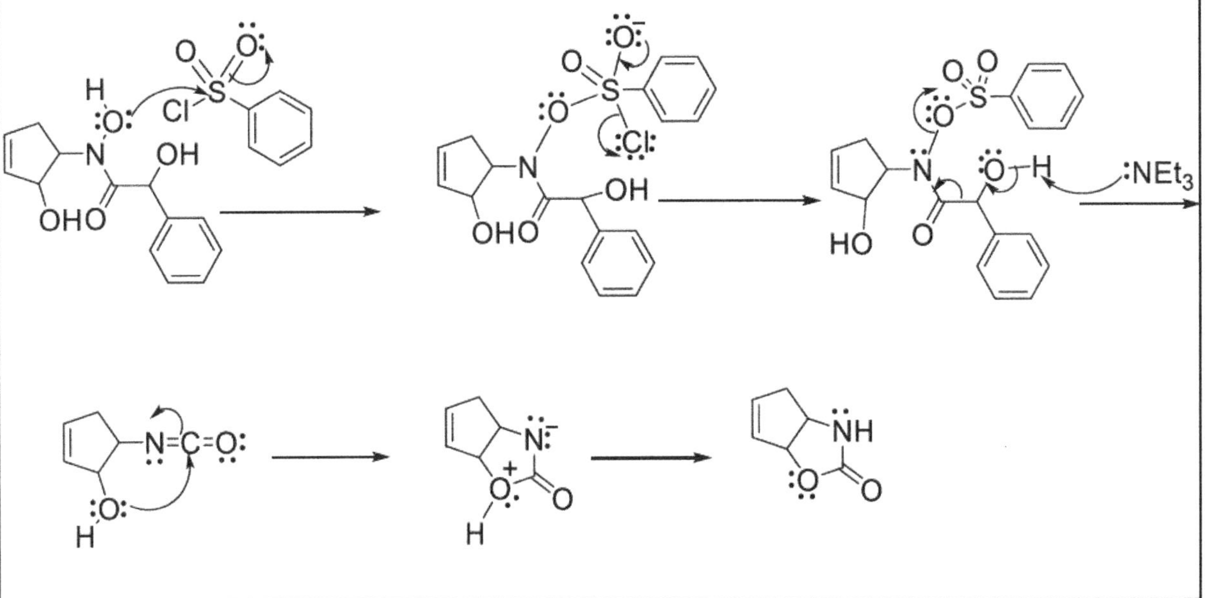
O
S
Cl
H
N
OH
:NEt3
NH

301

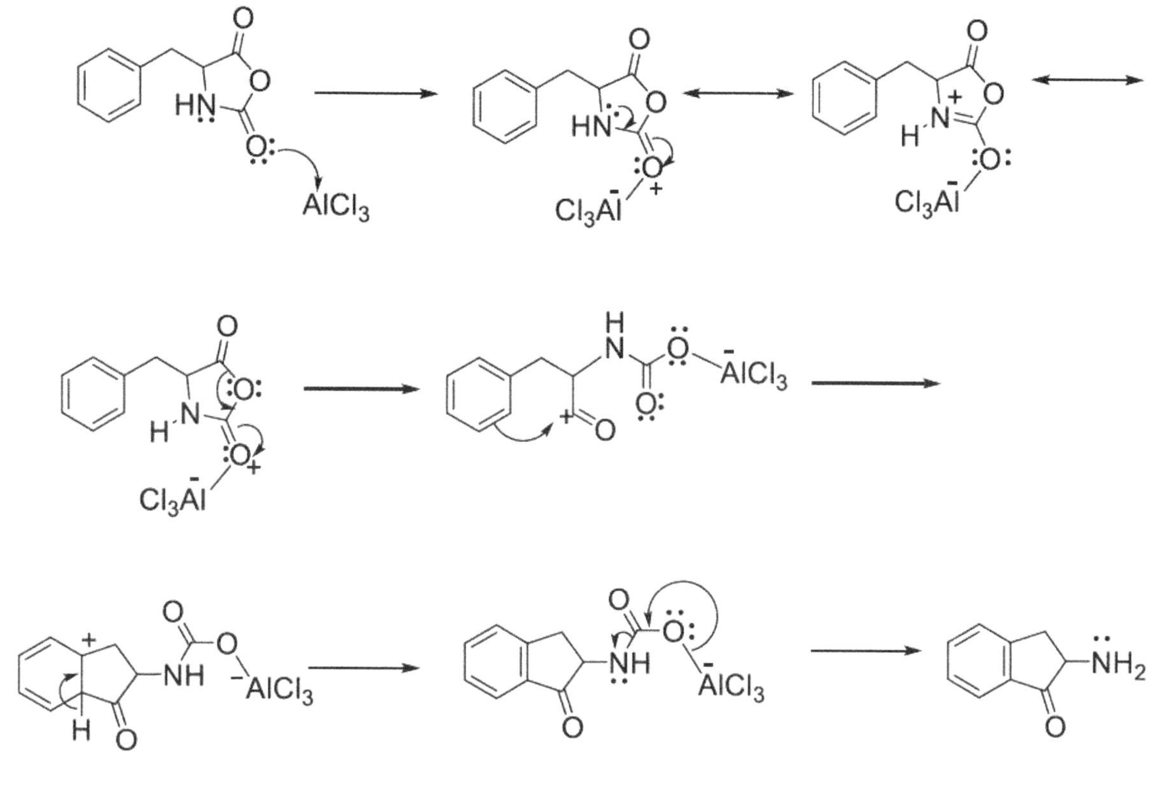

302

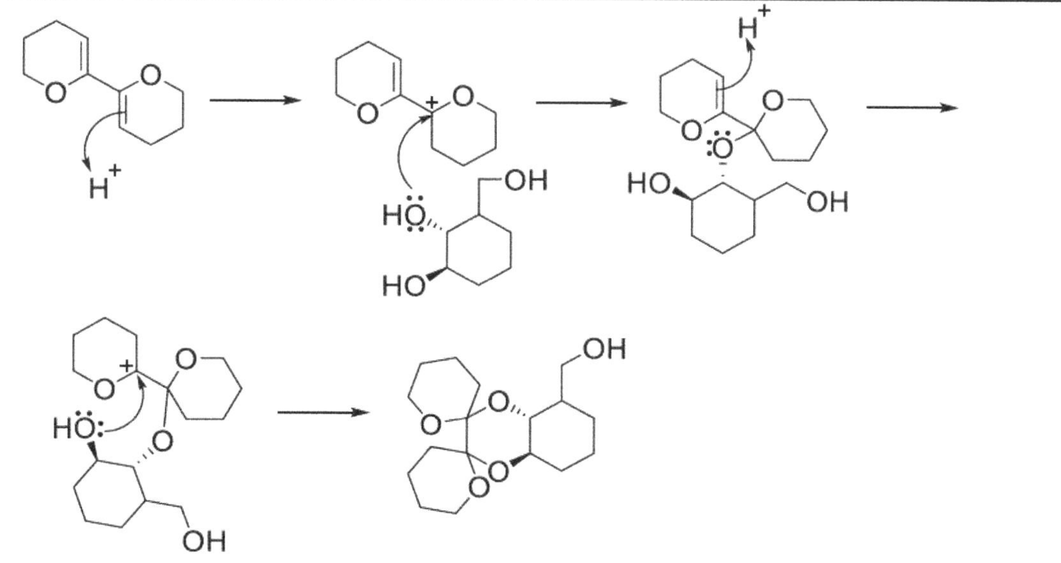

303

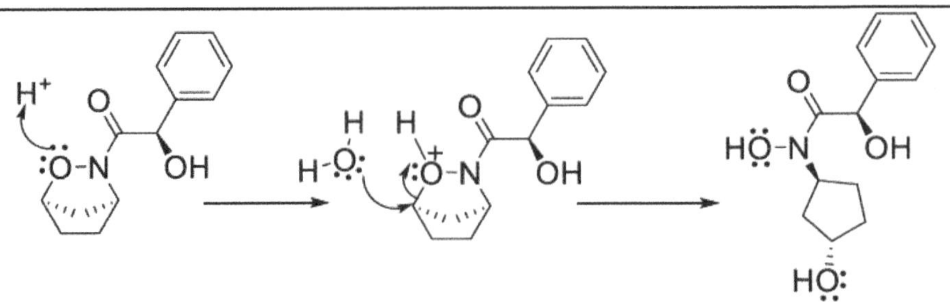

304

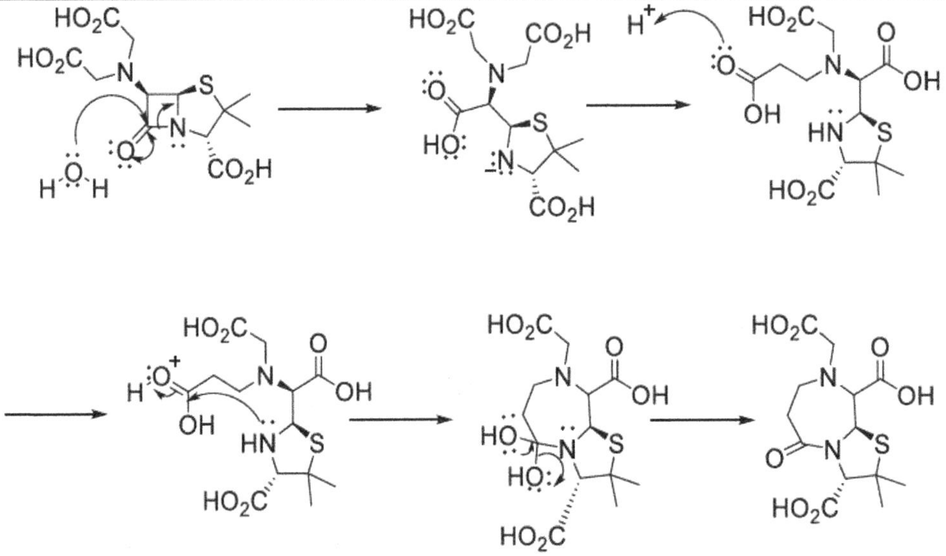

305

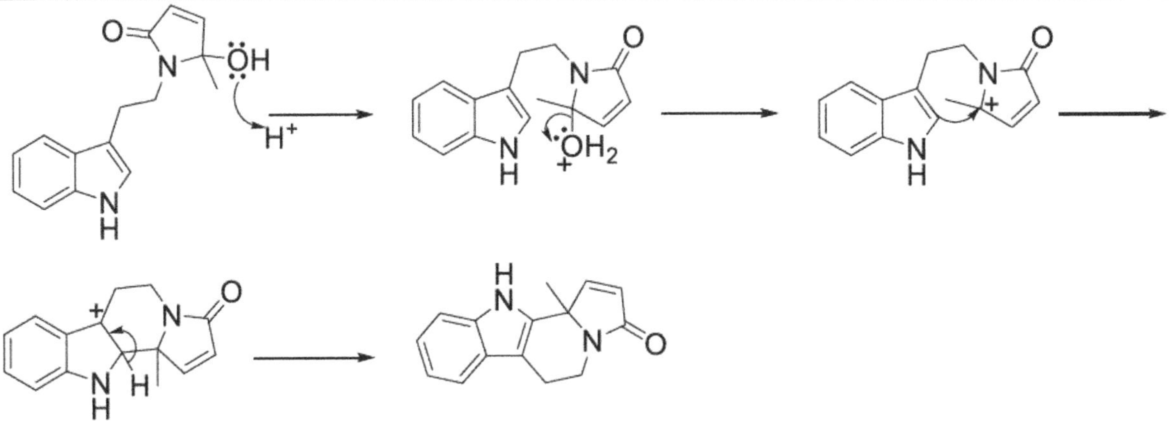

306

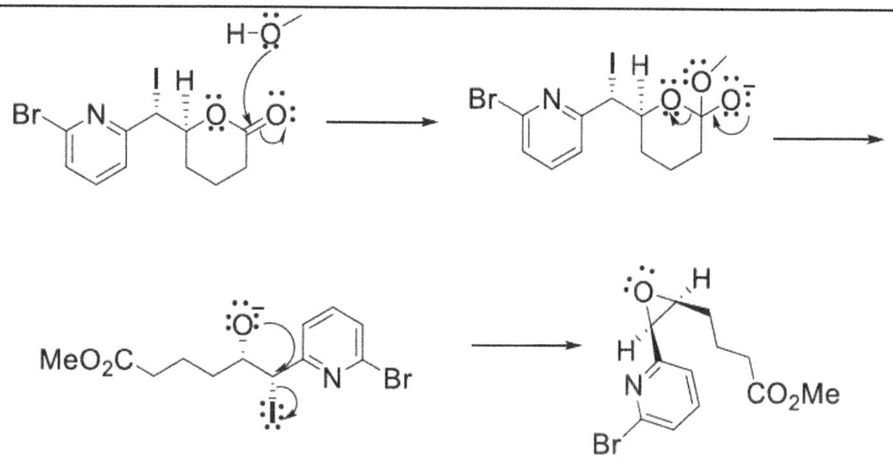
H-O
I
H
Br
N
O
O
Br
N
MeO2C
O
N
Br
H
H
N
Br
CO2Me

307

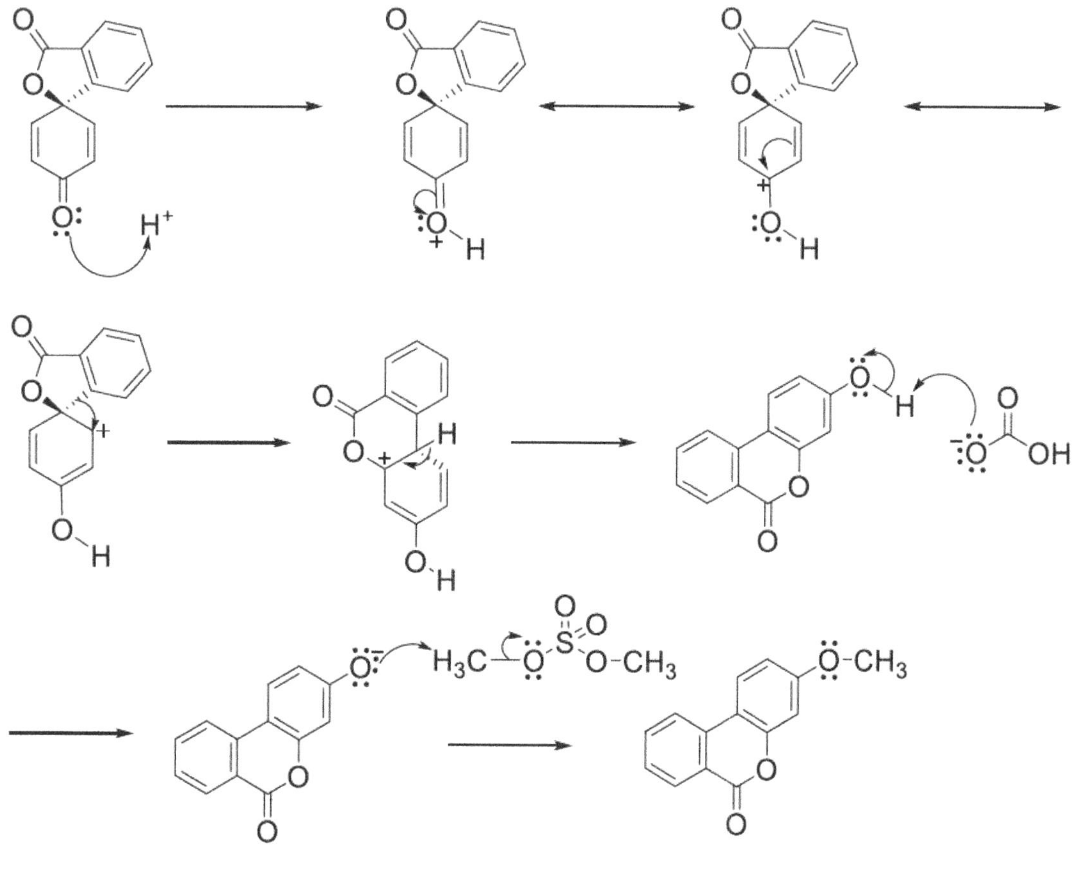
O
O
H+
H
O-
H3C
S
O-CH3
OH

308

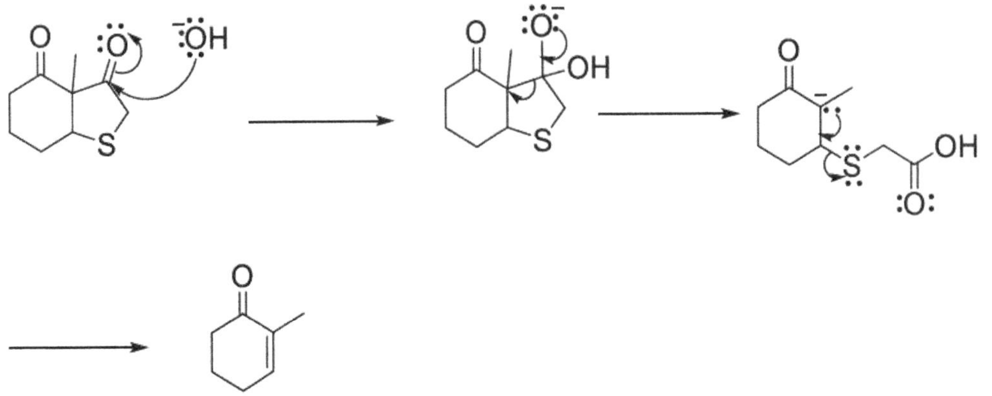

O
OH
S
OH
O
MgBr2
H
O+
H+
Cl
O
S+
N

309

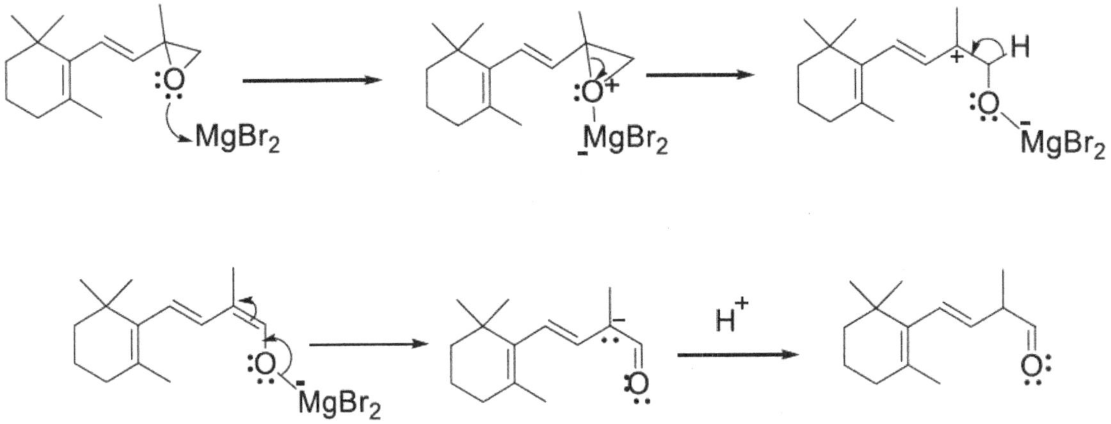

O
MgBr2
O+
H
H+

310

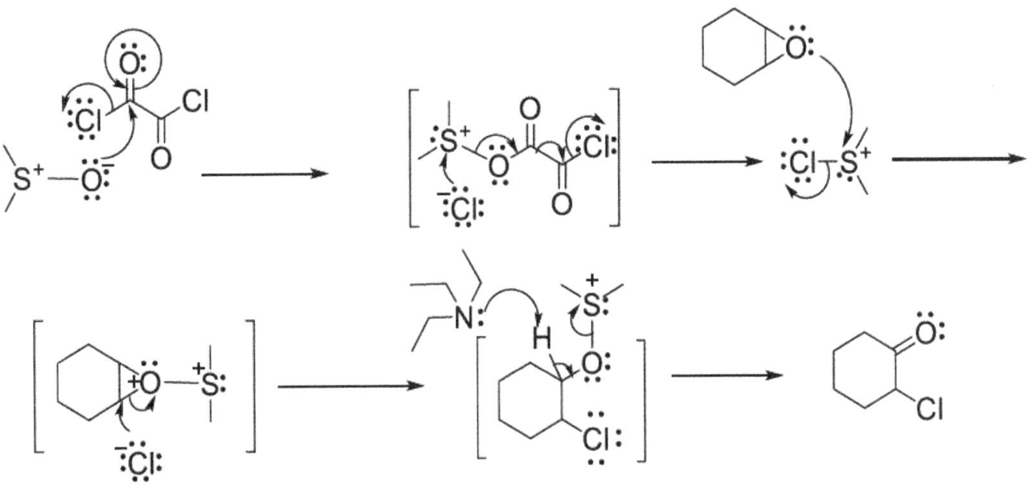

O
Cl
S+
O
N
H

311

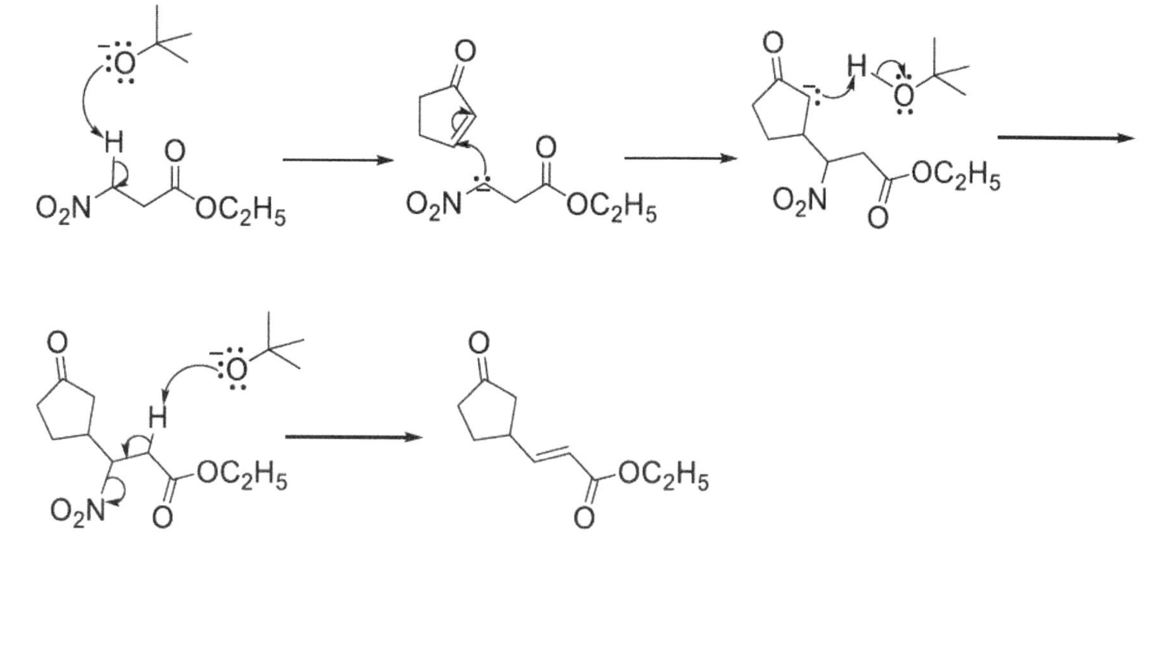
O
H
O
O_2N
OC_2H_5
OC_2H_5
OC_2H_5
OC_2H_5
OC_2H_5
H
O

312

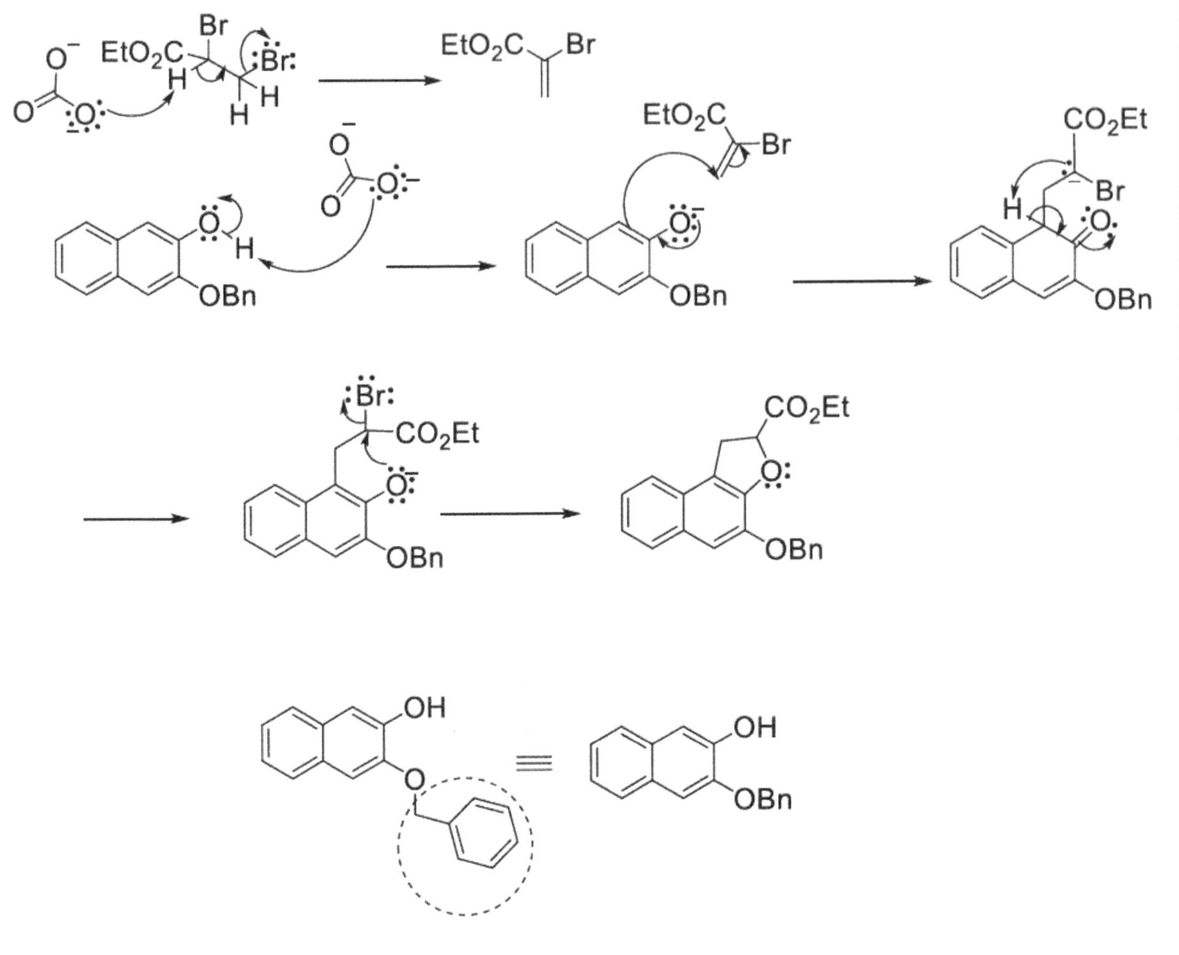

EtO2C
Br
Br
H
H
H
EtO2C
Br
OBn
CO2Et
OBn
CO2Et
OH
O
OH
OBn

313

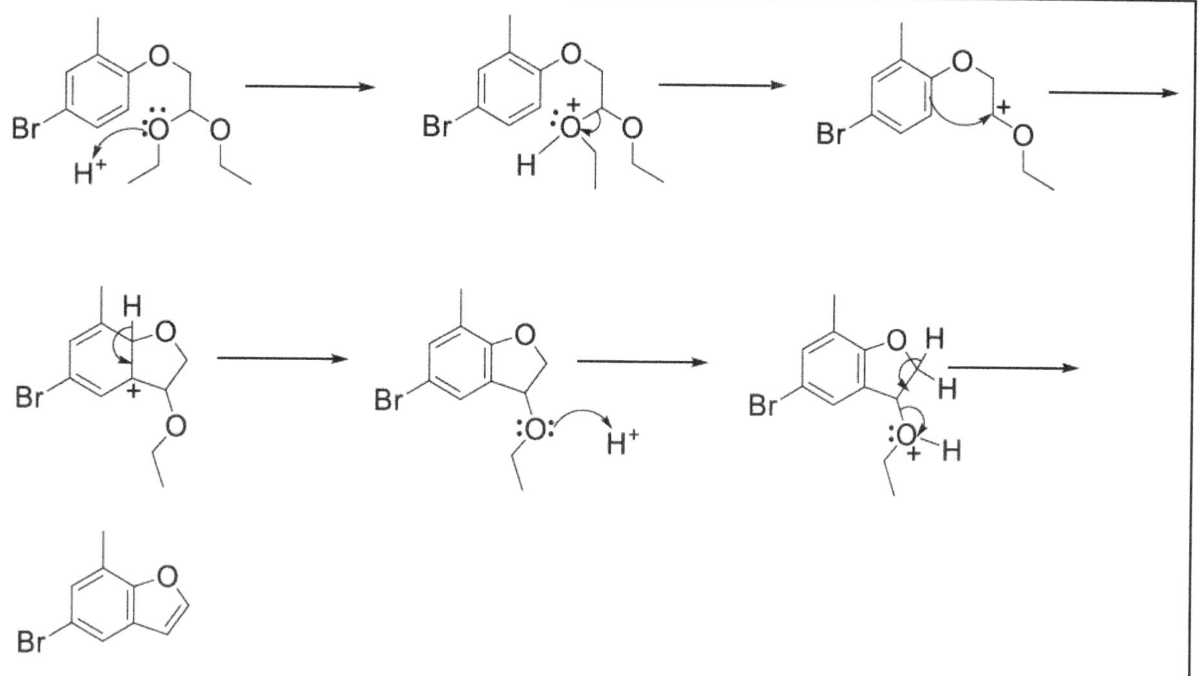

314

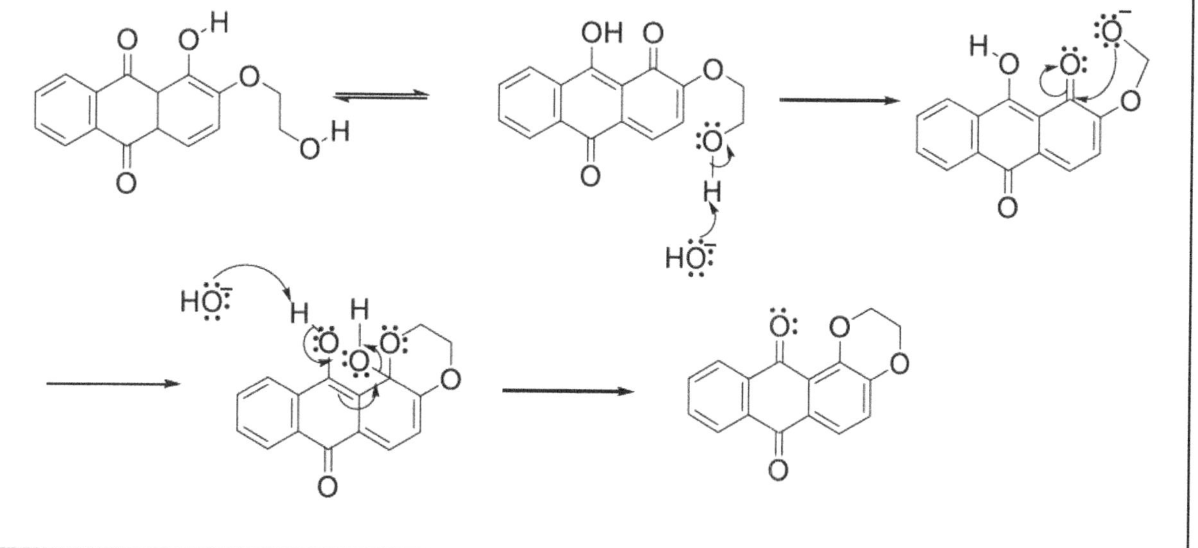

315

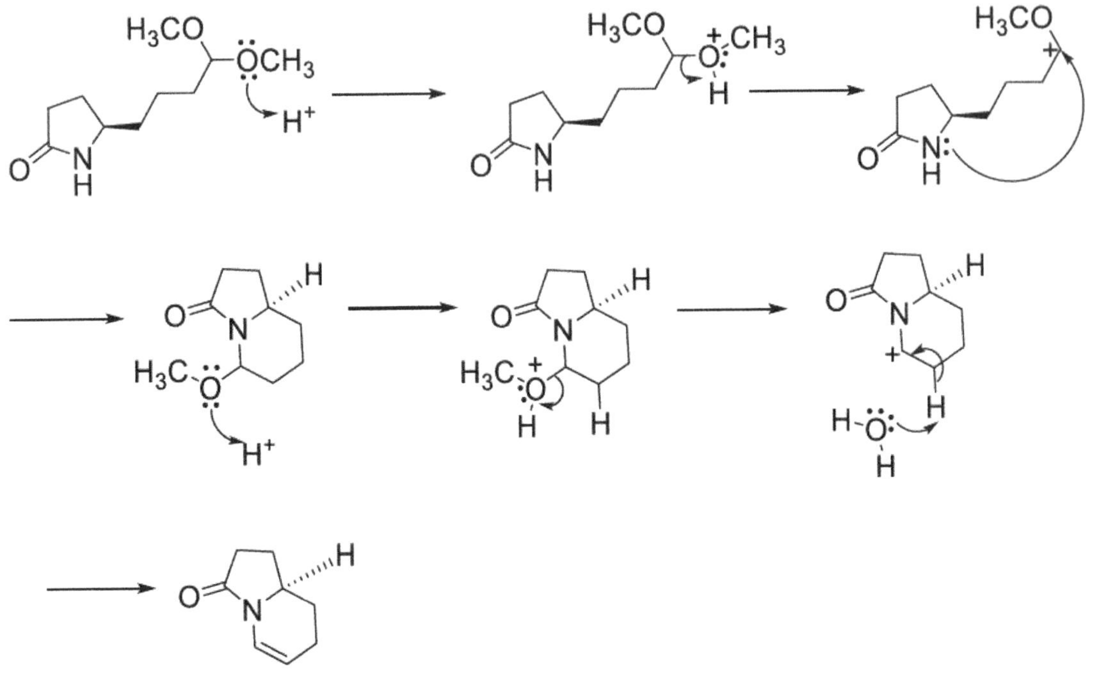

316

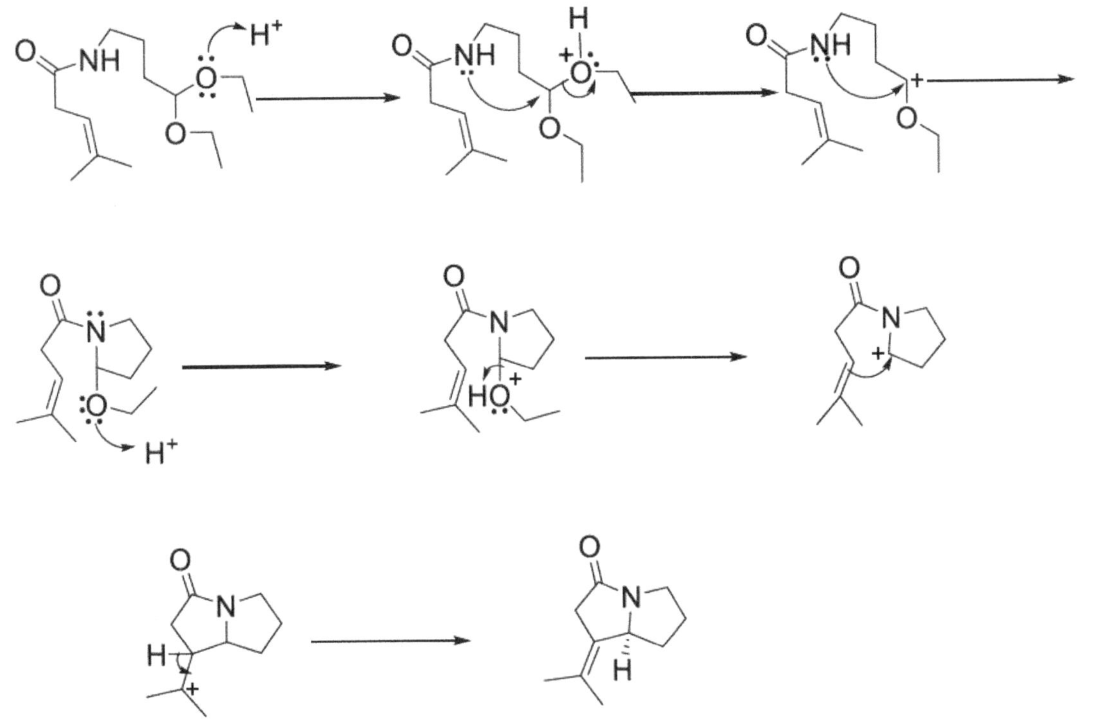

317

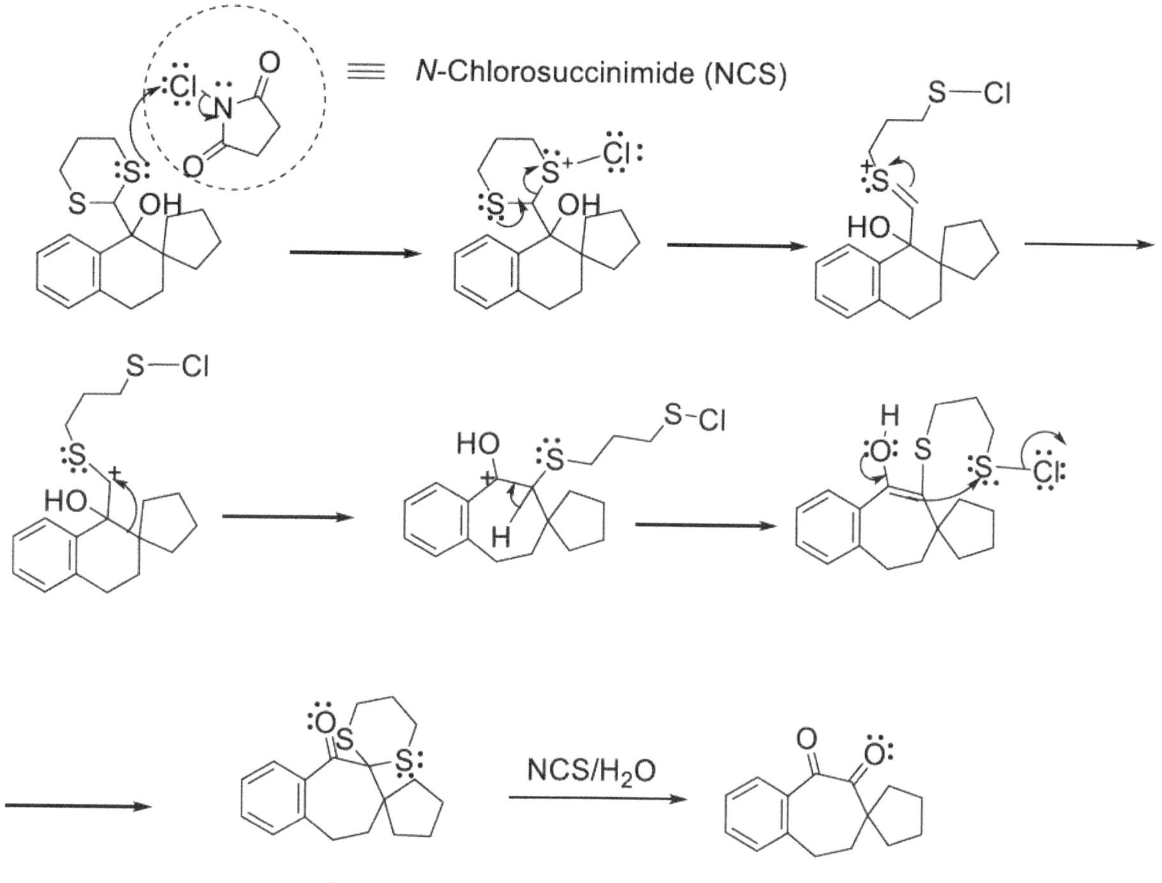
N-Chlorosuccinimide (NCS)
NCS/H_2O

A

Exercise Write the mechanism for the conversion of A to the final product

318

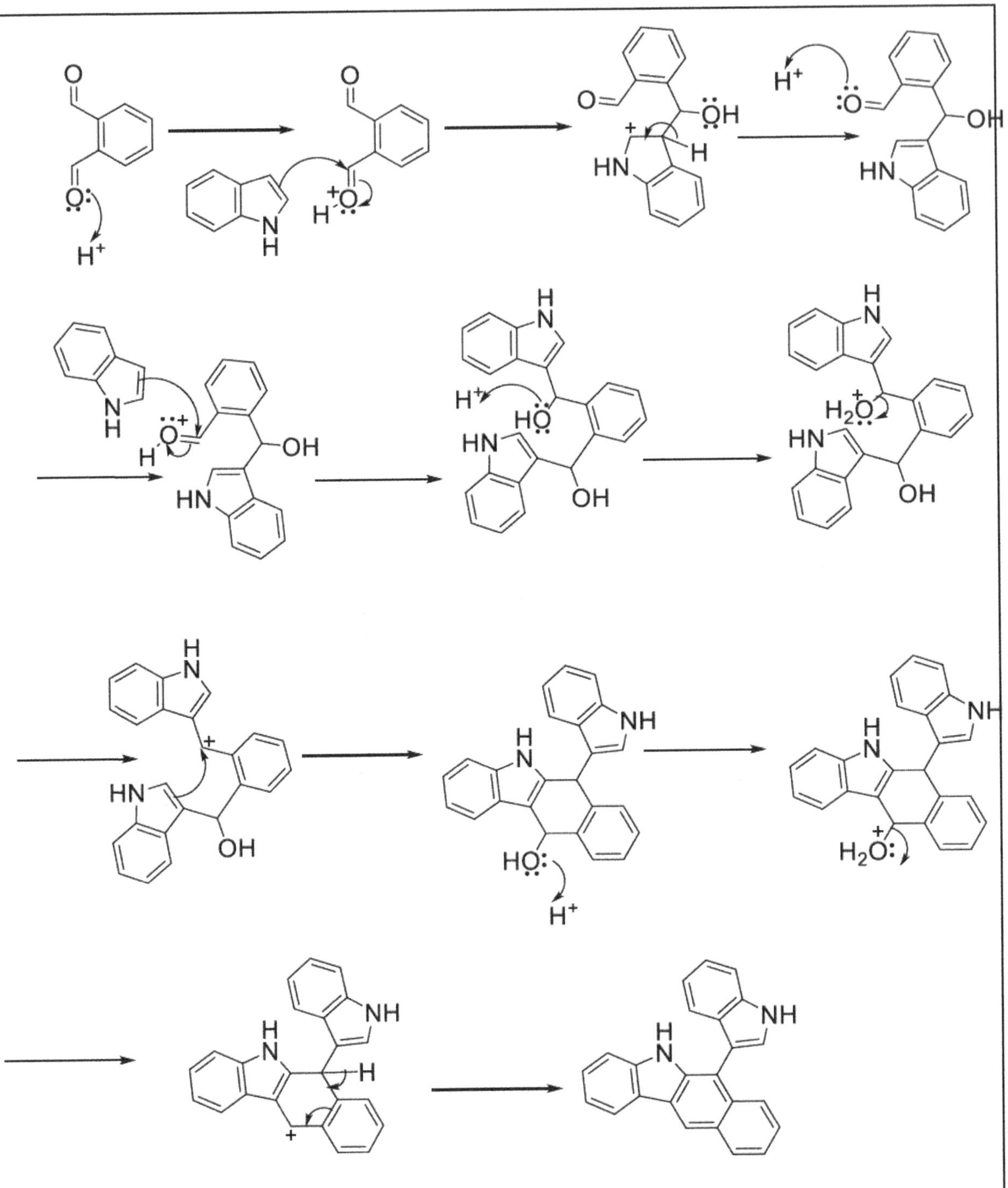

319

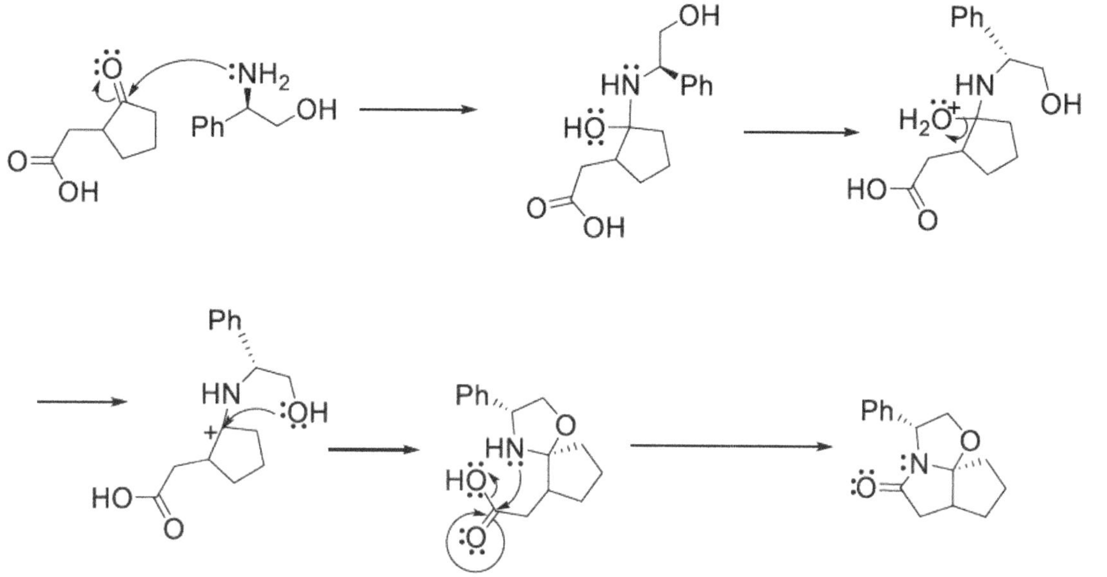

OH
:NH2
Ph
OH
O
OH
HN
Ph
HO
O
OH
Ph
HN
H2O+
OH
HO
O
Ph
HN
OH
+
HO
O
Ph
O
HN
HO
O
Ph
O
N
O

320

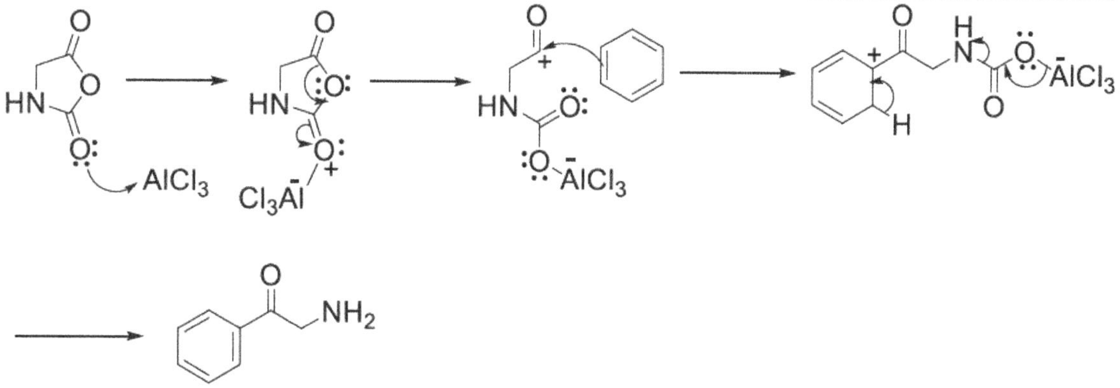

O
O
HN
O
AlCl3
O
HN
O
Cl3Al
O
+
HN
O
AlCl3
O
+
H
N
O
AlCl3
O
H
O
NH2

321

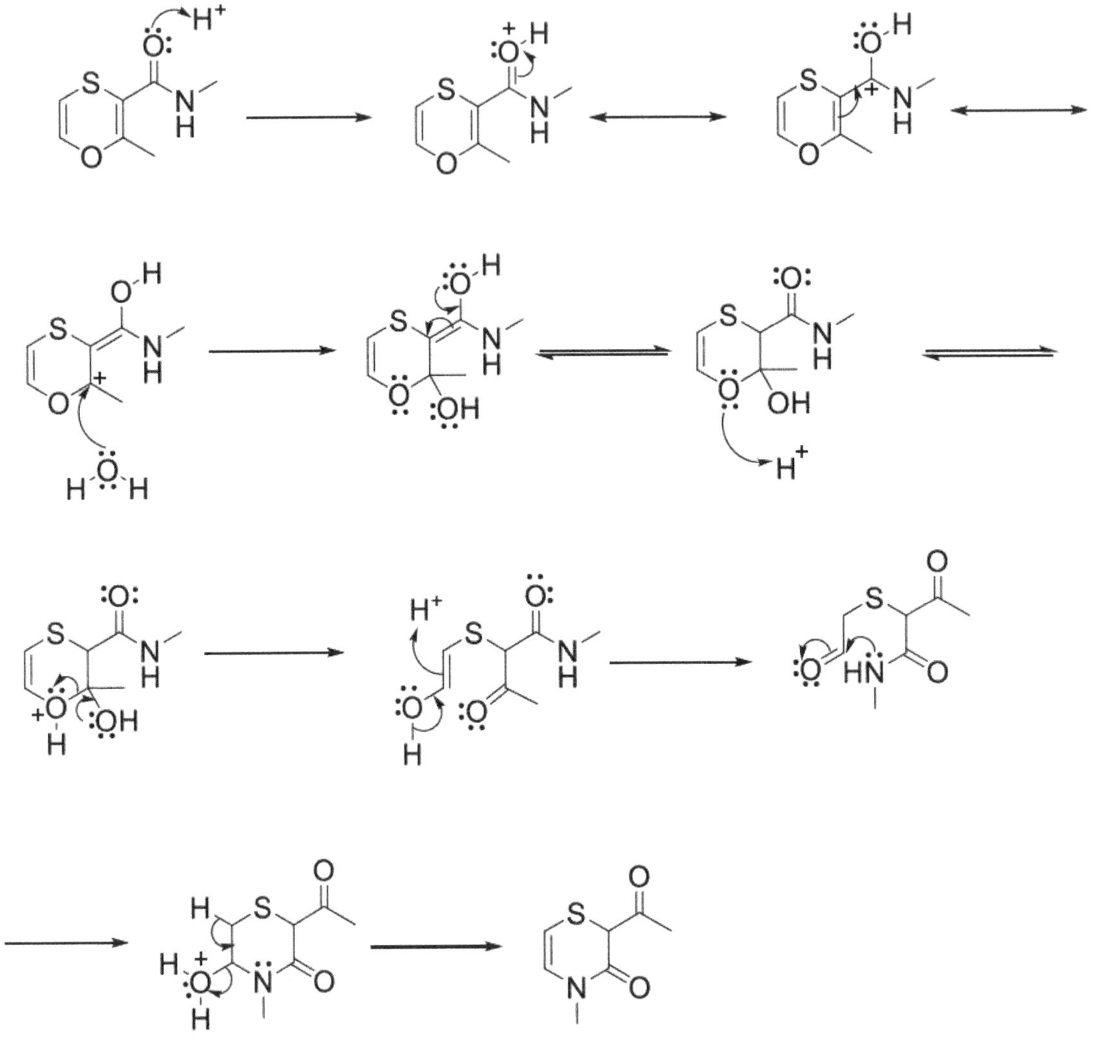

S
O
N
H
H⁺
OH
OH
H

322

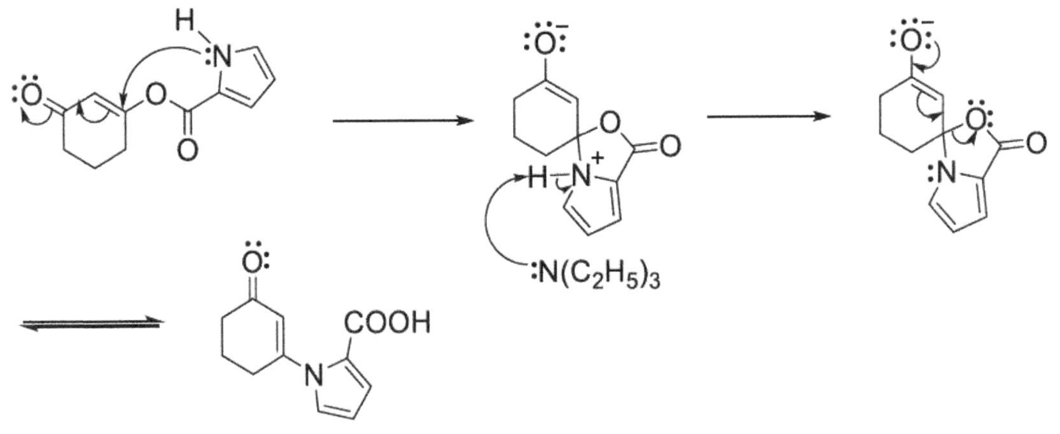

H
N
O
O
O
N(C2H5)3
COOH

323

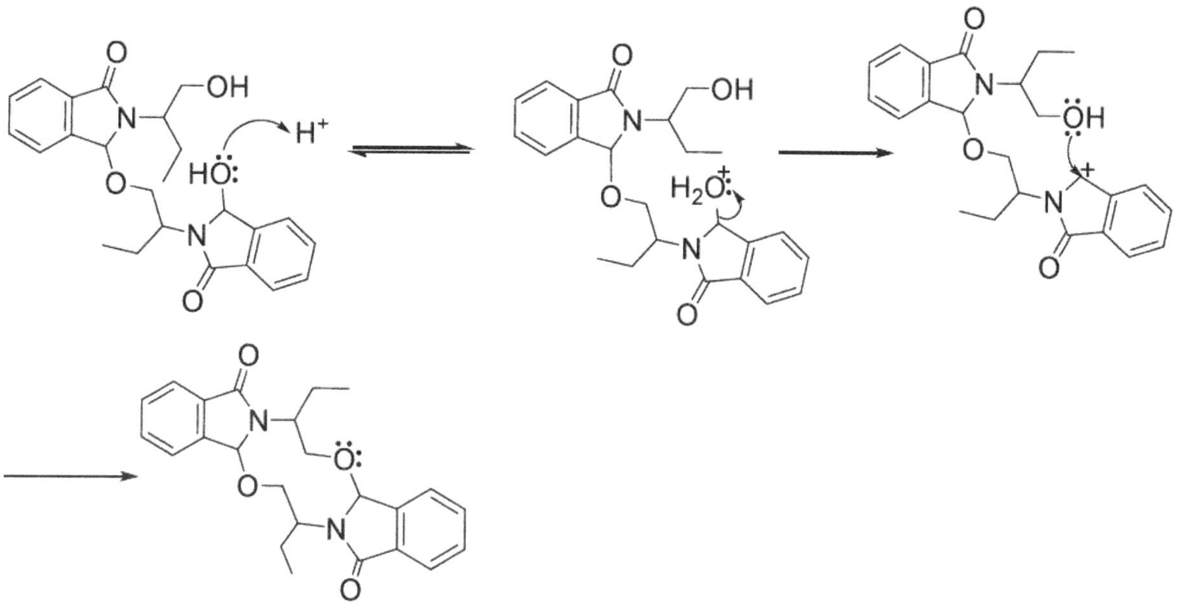

324

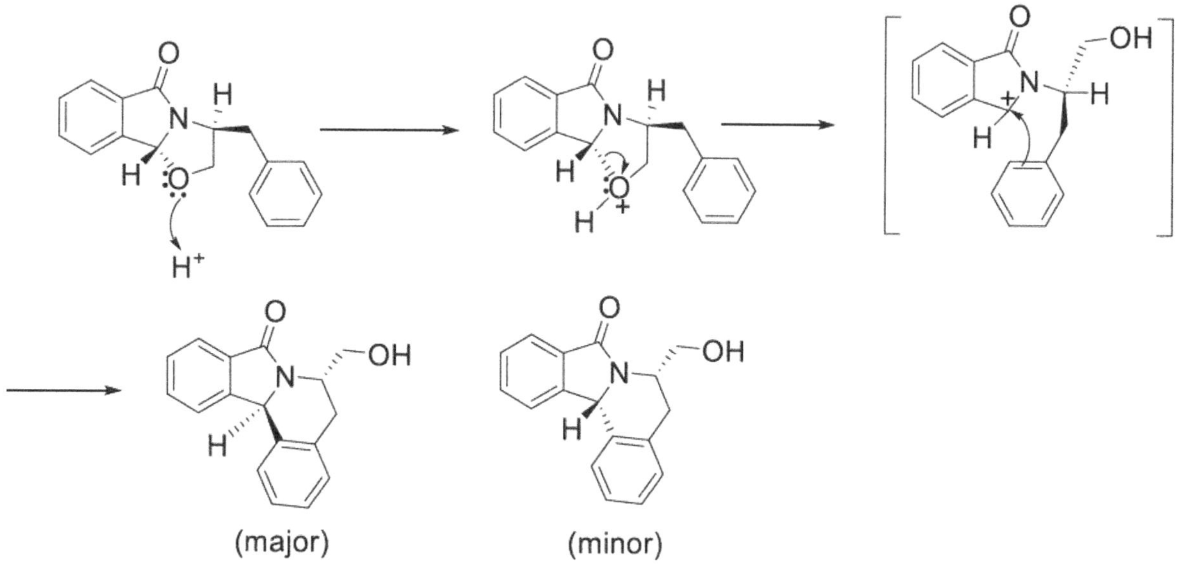
(major)
(minor)

325

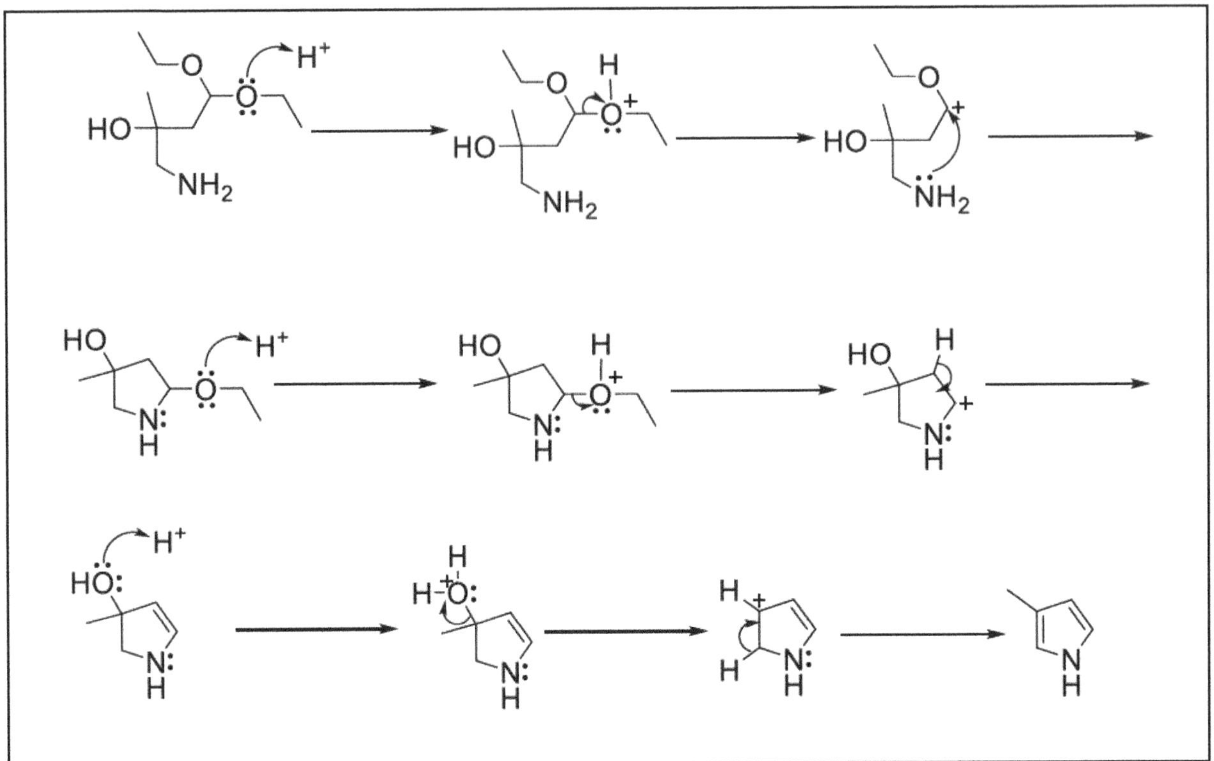

HO
O
H+
NH2
H
+
HO
H
N
+
H

326

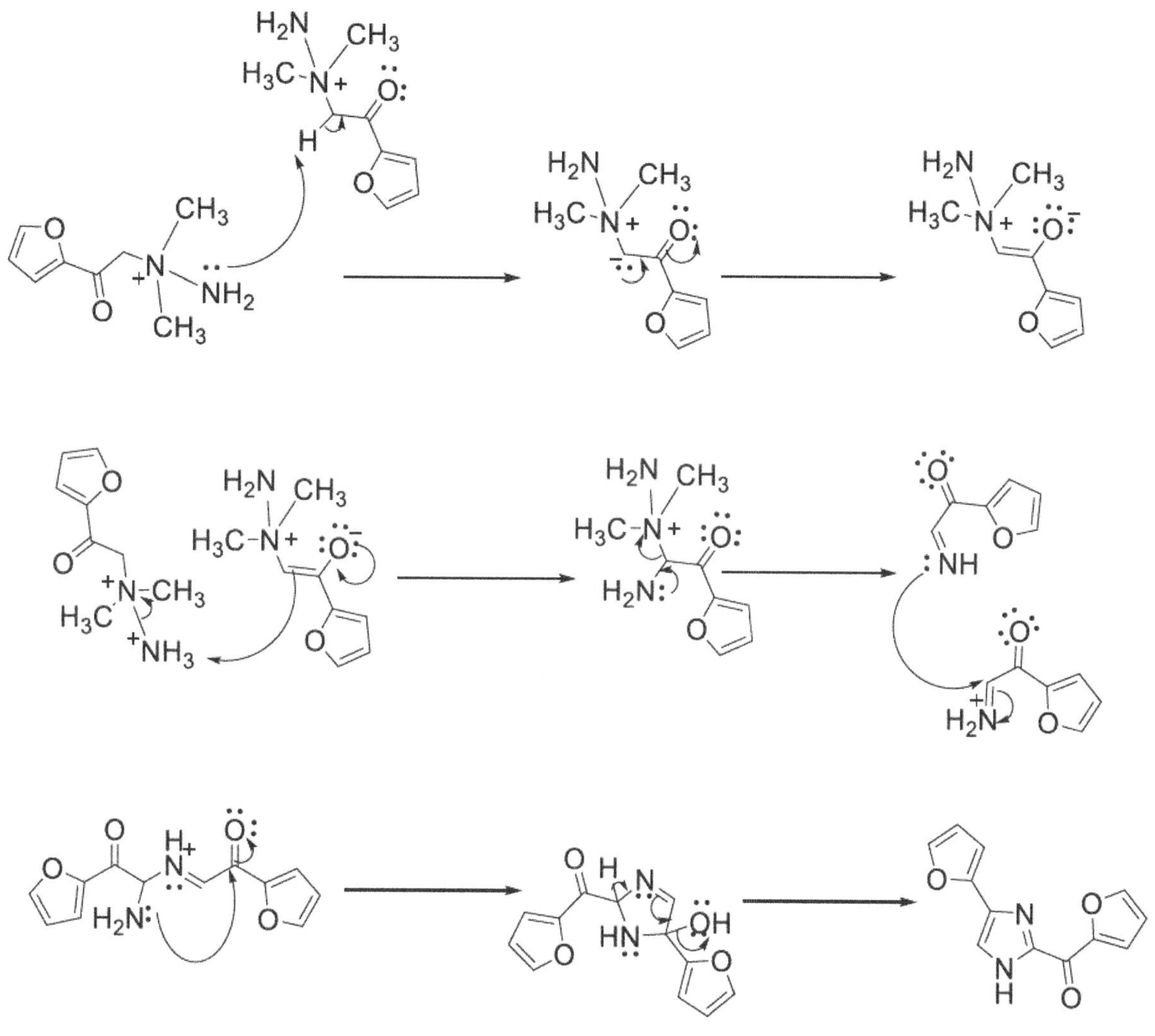

327

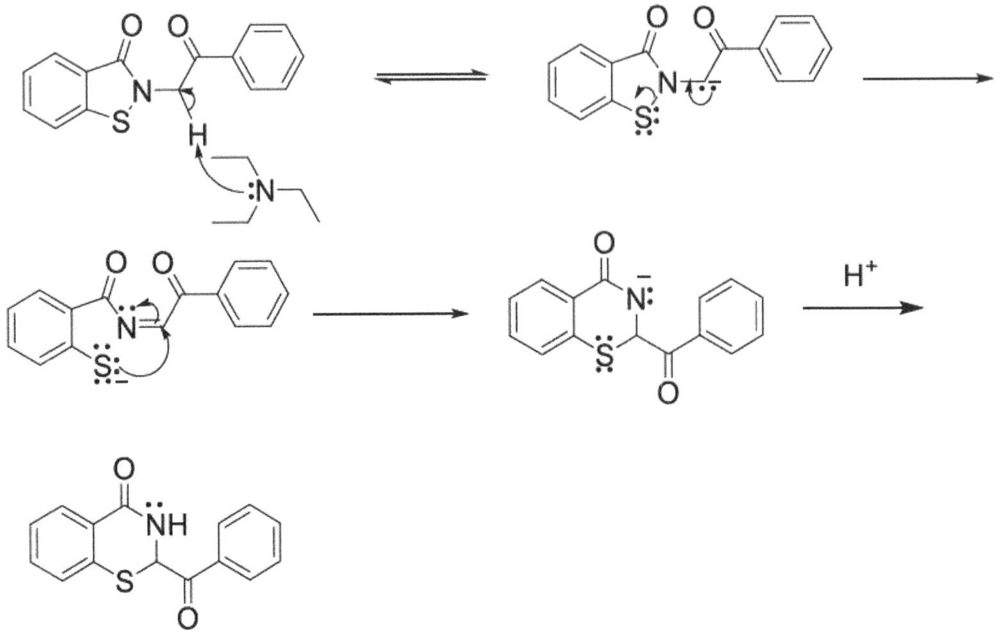
H+

328

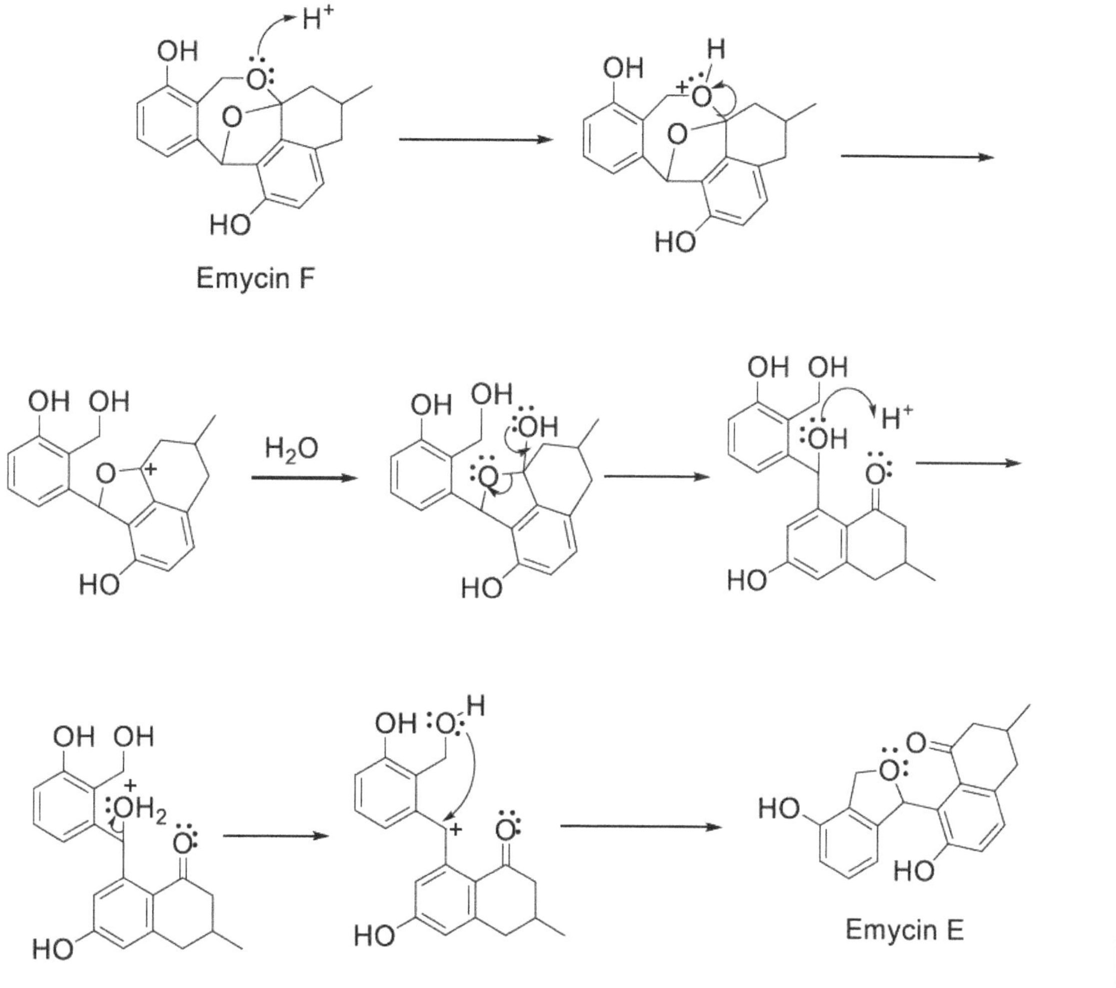

H+
OH
O
HO
Emycin F
H
H2O
Emycin E

329

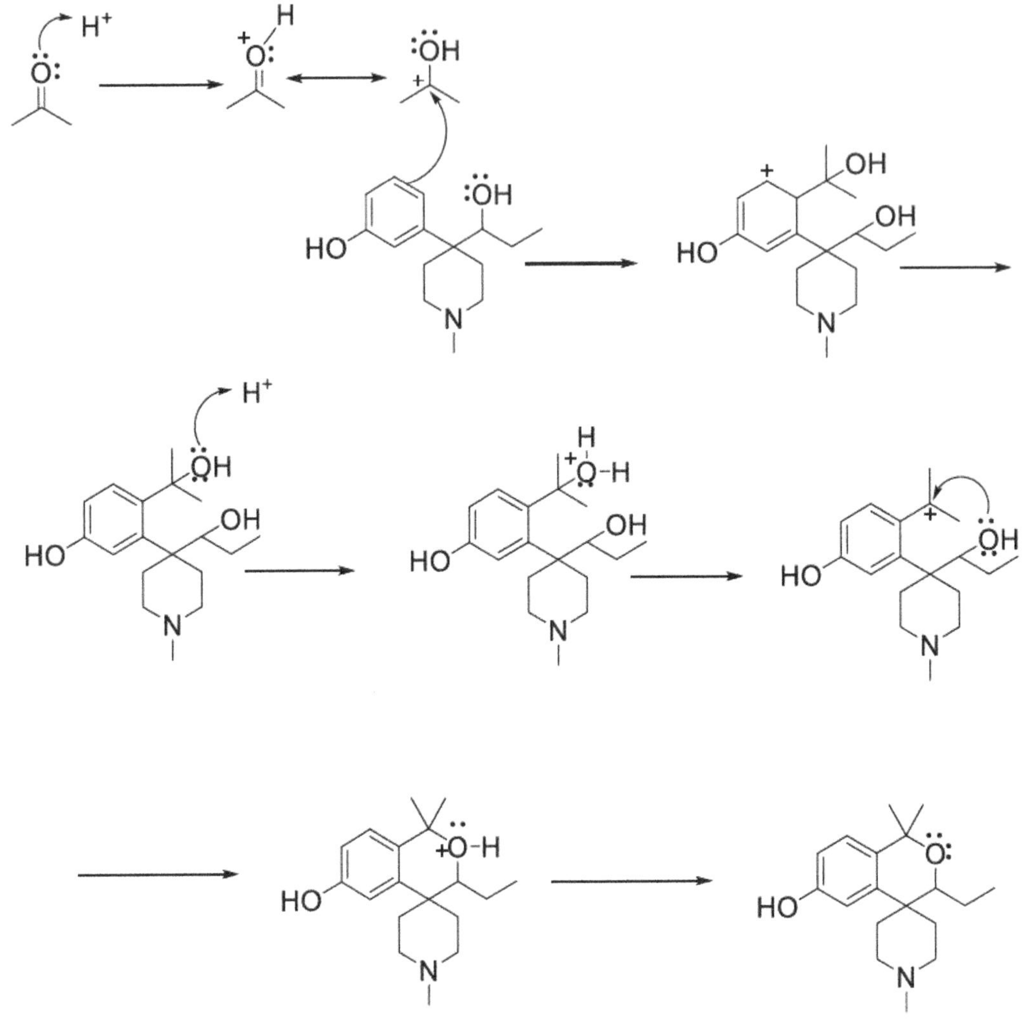
H+
H
OH
HO
OH
OH
HO
OH
H+
OH
OH
HO
H
O-H
OH
HO
OH
HO
O-H
HO
O
HO
N
N
N
N
N
N
N

330

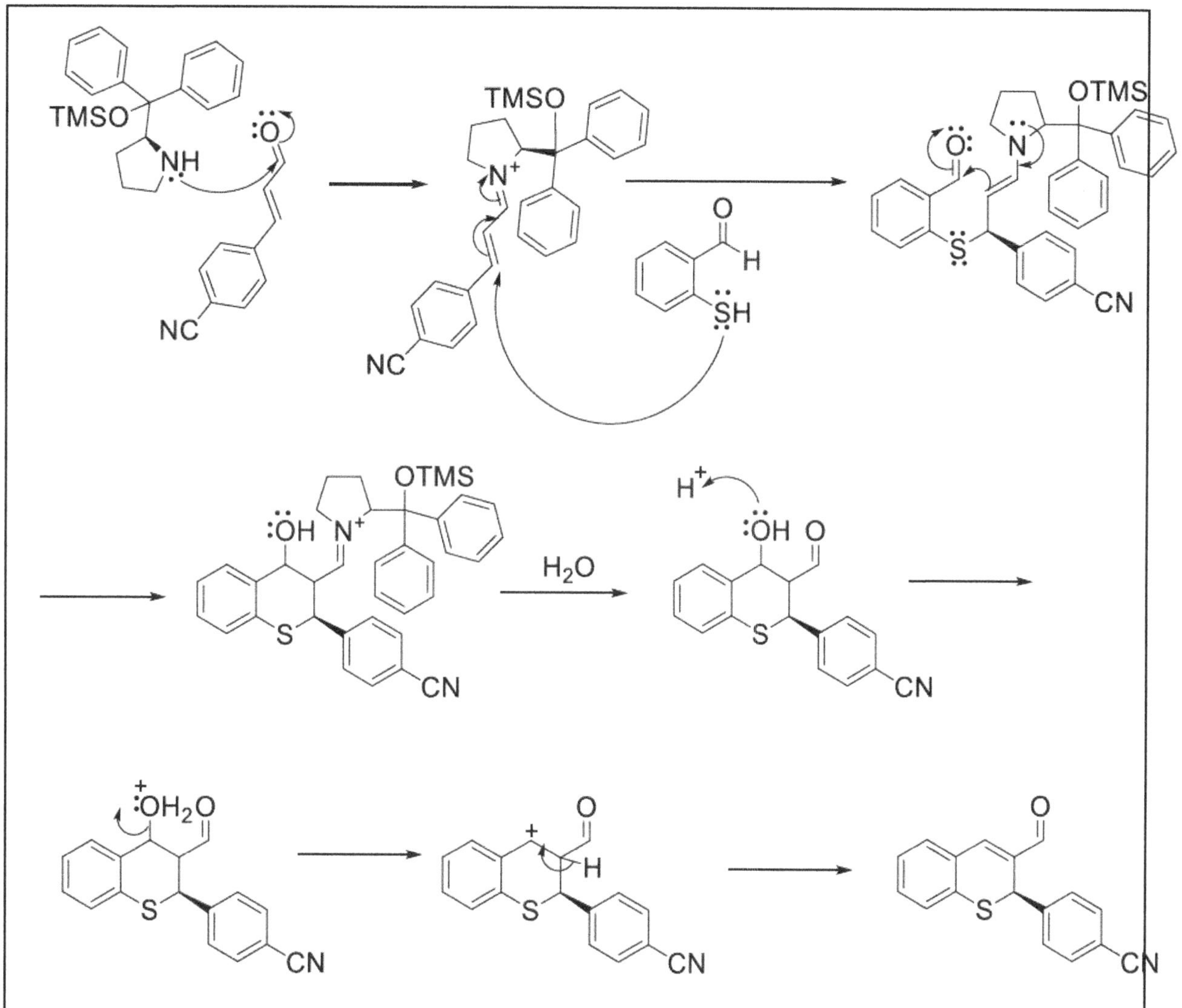
TMSO
NH
O
NC
TMSO
N+
NC
O
H
SH
OTMS
O
N
S
CN
OTMS
OH
N+
S
CN
H_2O
H+
OH
O
S
CN
+
OH2
O
S
CN
+
O
H
S
CN
O
S
CN

331

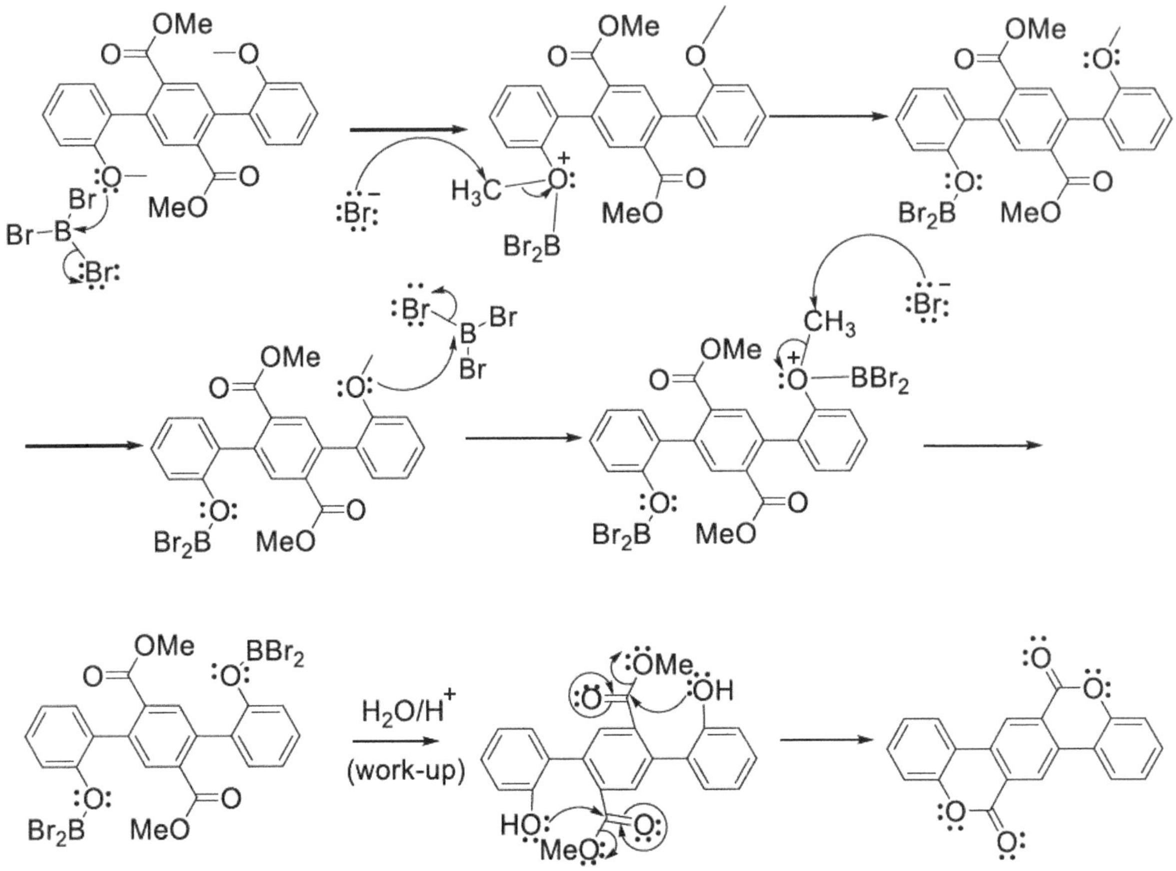
H2O/H+
(work-up)

(two consecutive intramolecular transesterification reactions yield the final product)

332

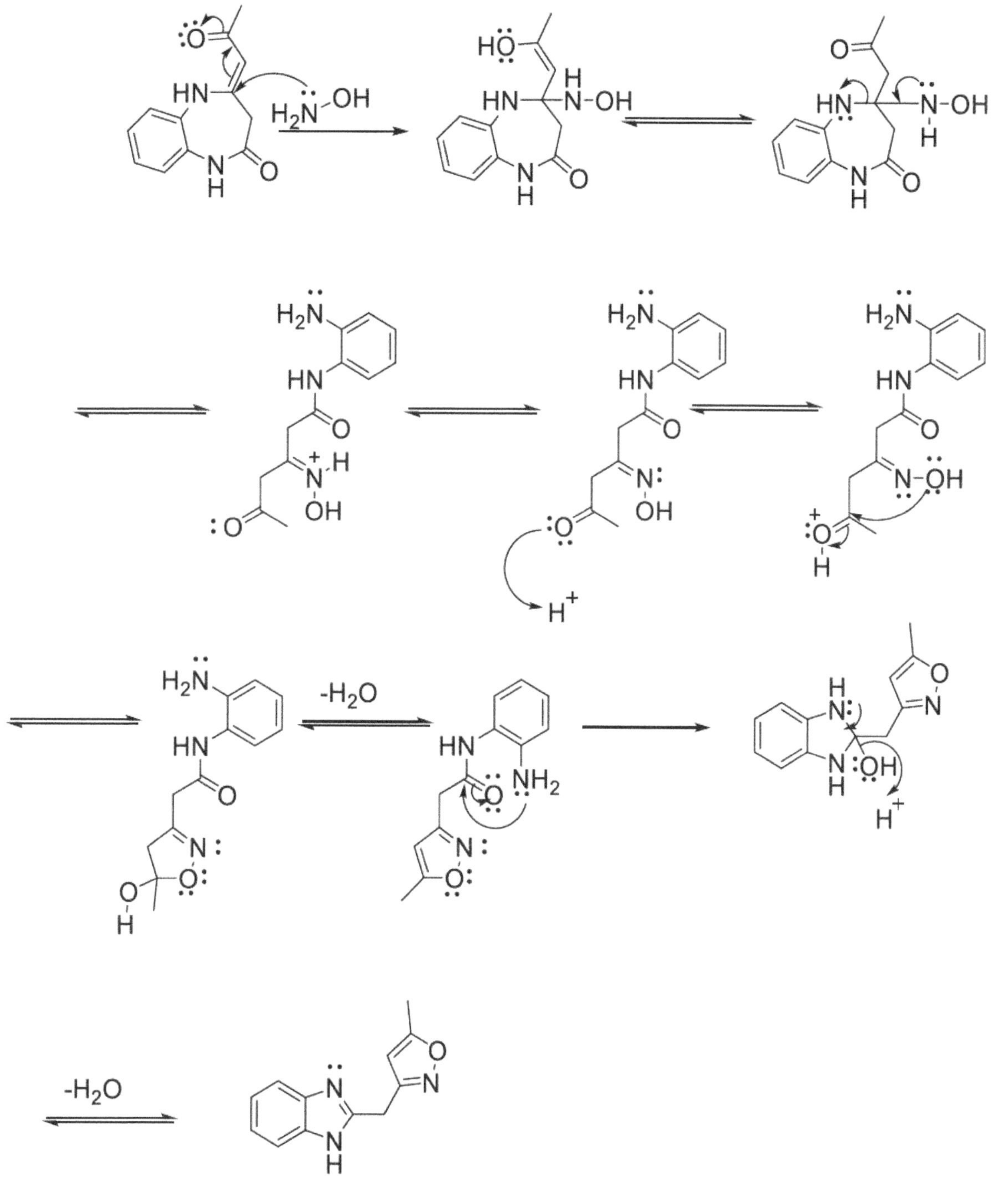

H₂N-OH
HN
N-OH
-H₂O
-H₂O
H⁺

333

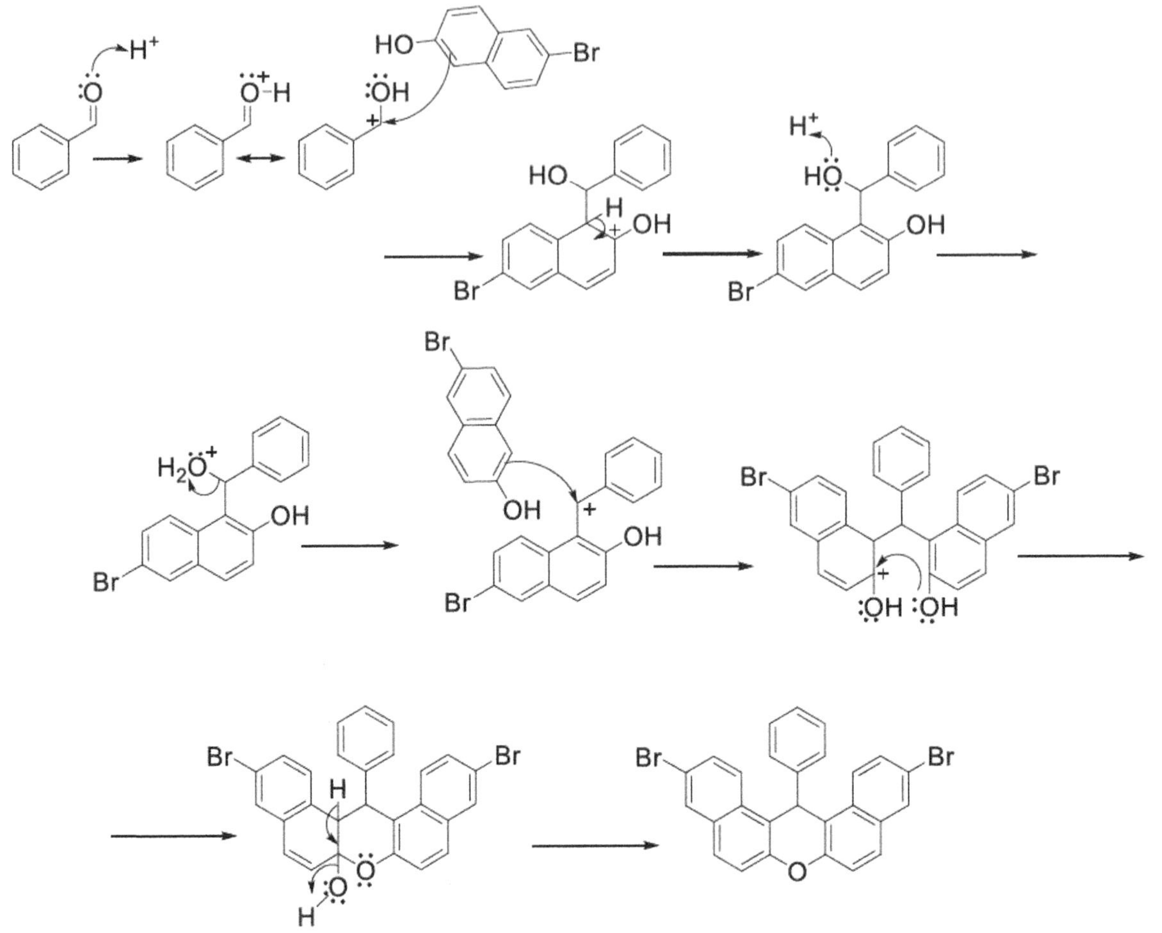

H+
HO
Br
OH
OH
H
Br
H+
HO
OH
Br
H2O+
OH
Br
Br
OH
OH
Br
Br
Br
OH OH
Br
H
Br
O
O
H
Br
Br
O

(Amberlyst-15 is a sulfonic acid polymer resin)

334

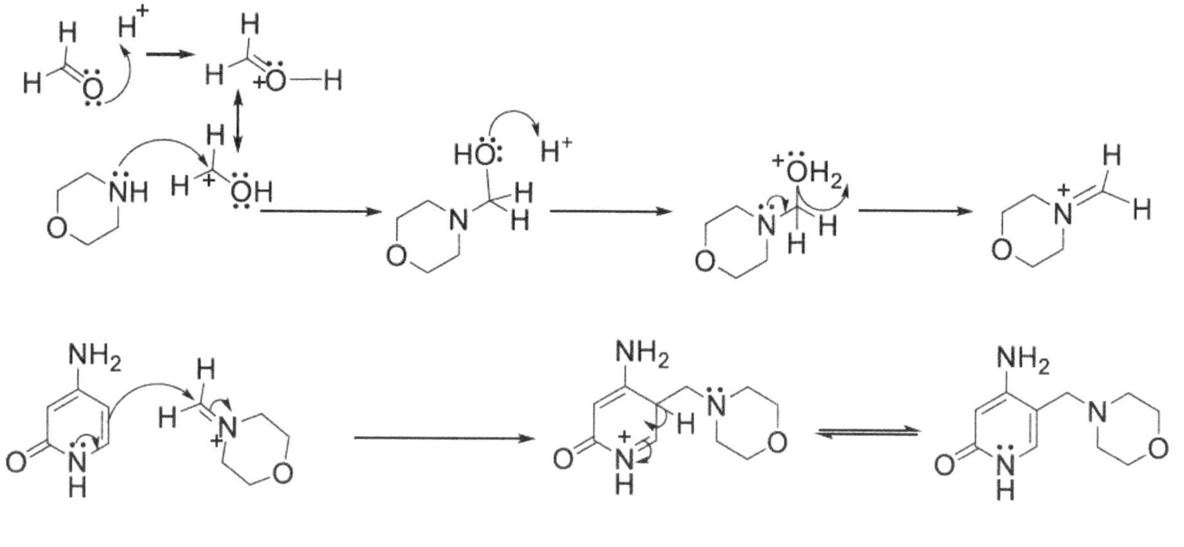

335

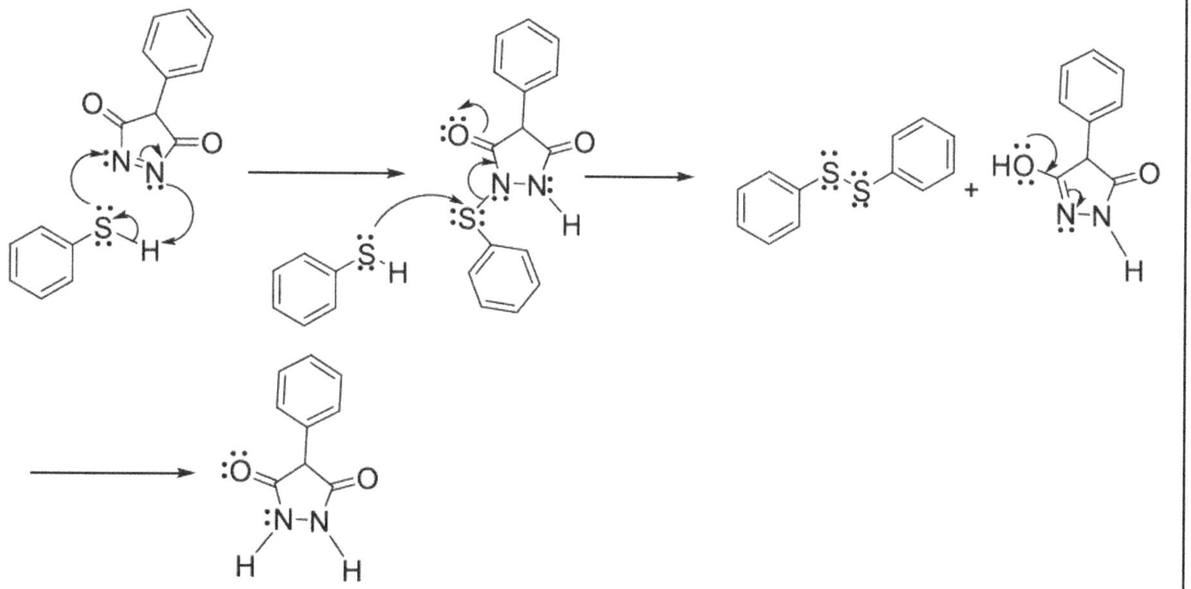

336

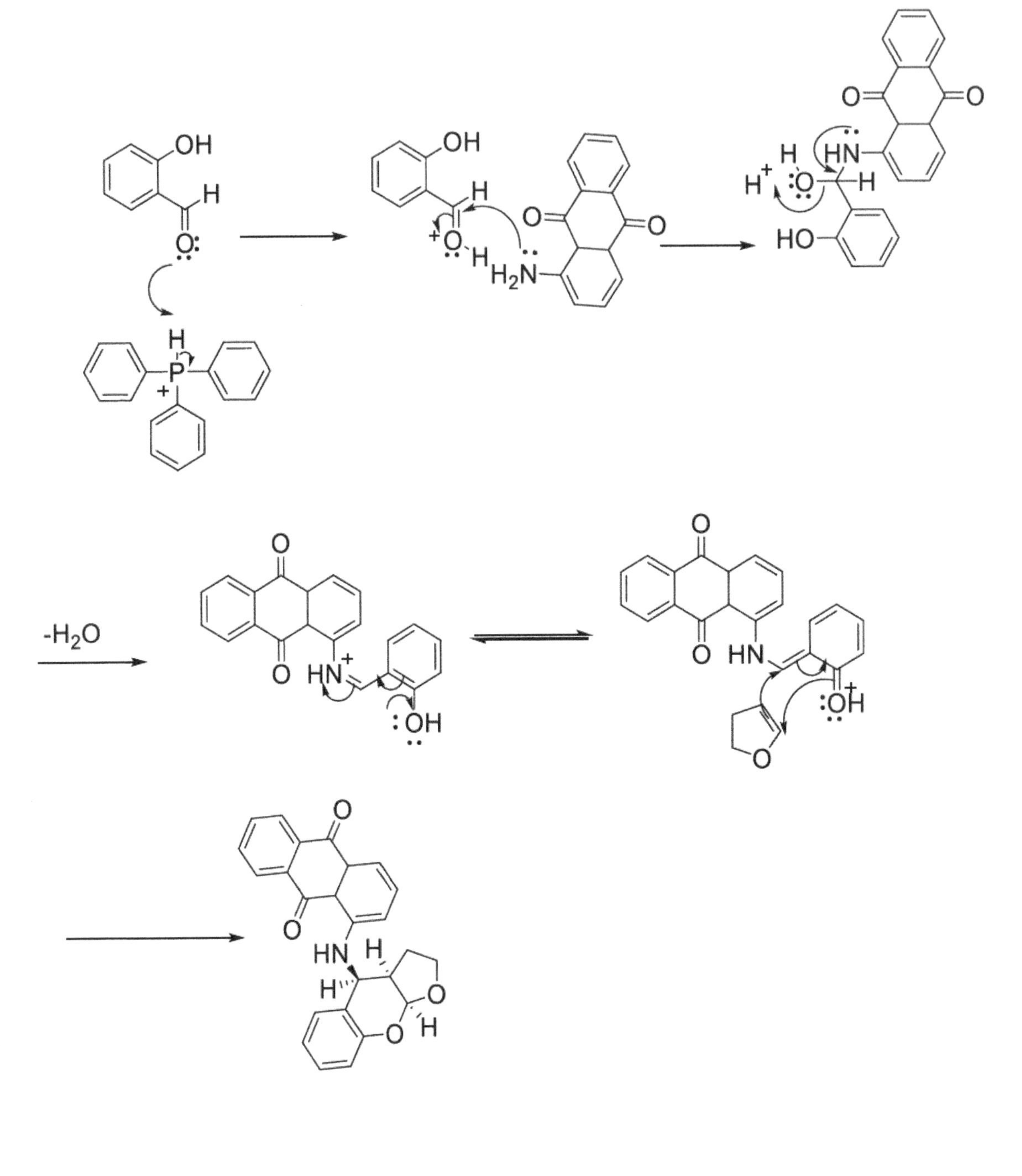

OH
H
O
H
P
+
OH
H
+
O
H
O
O
H2N
O
O
H
H+
:O
HN
H
HO
-H2O
O
O
HN+
:OH
O
O
HN
:OH+
O
O
O
HN
H
H
O
O
H

337

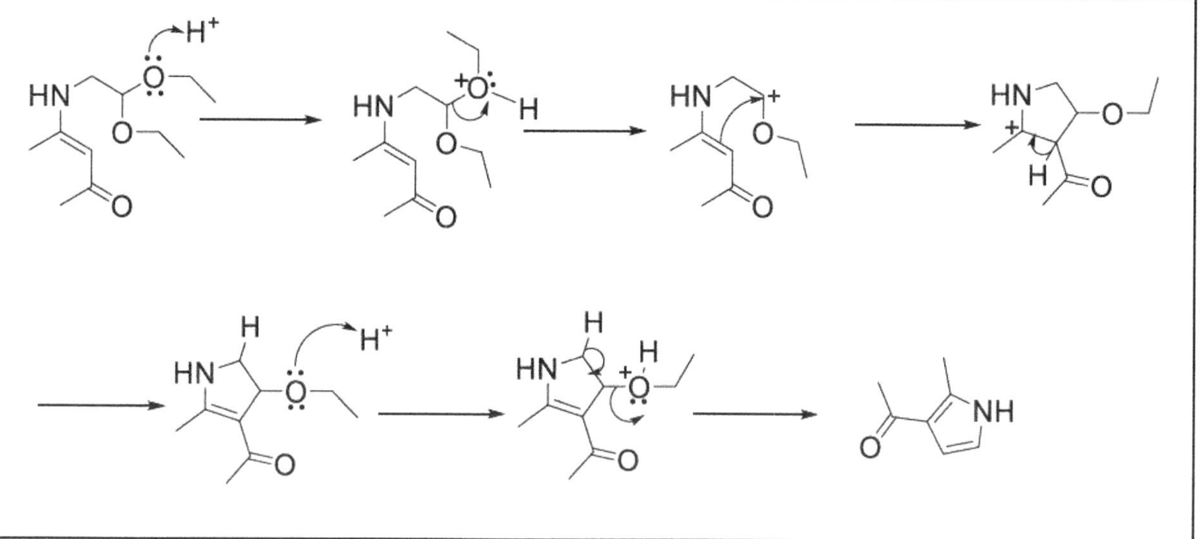

338

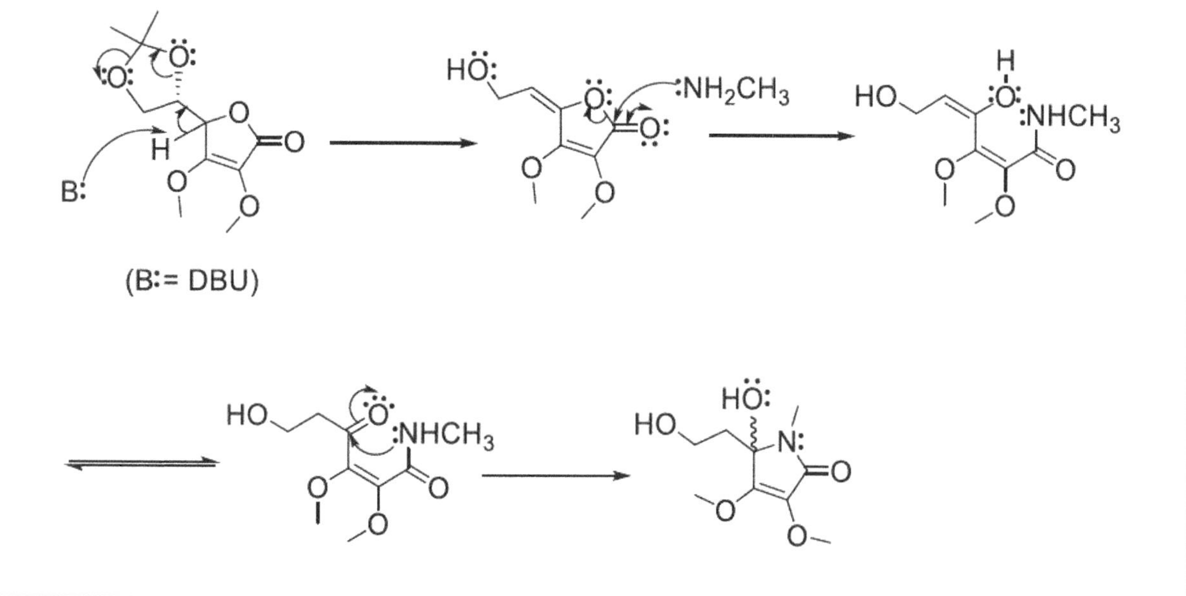
(B:= DBU)

339

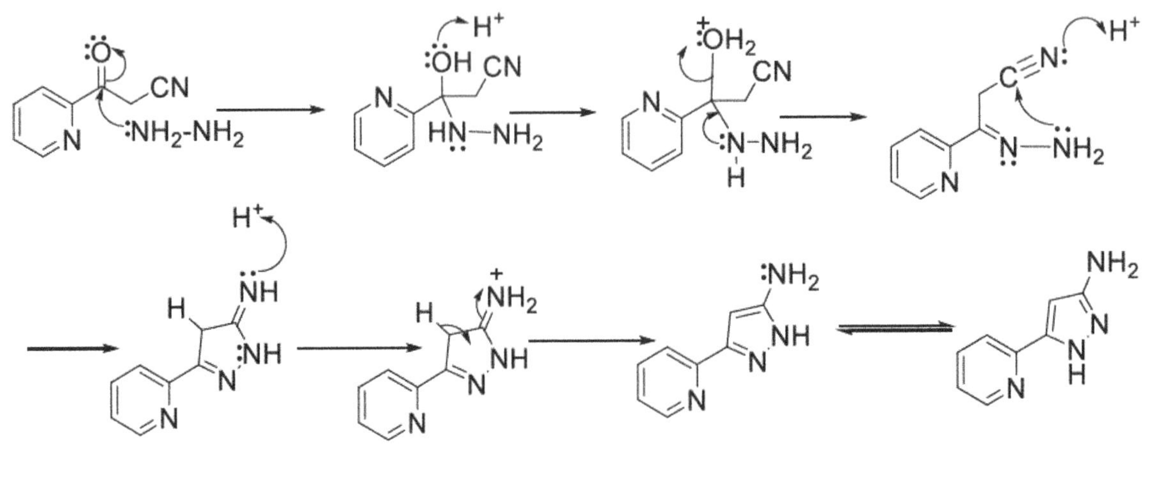

340

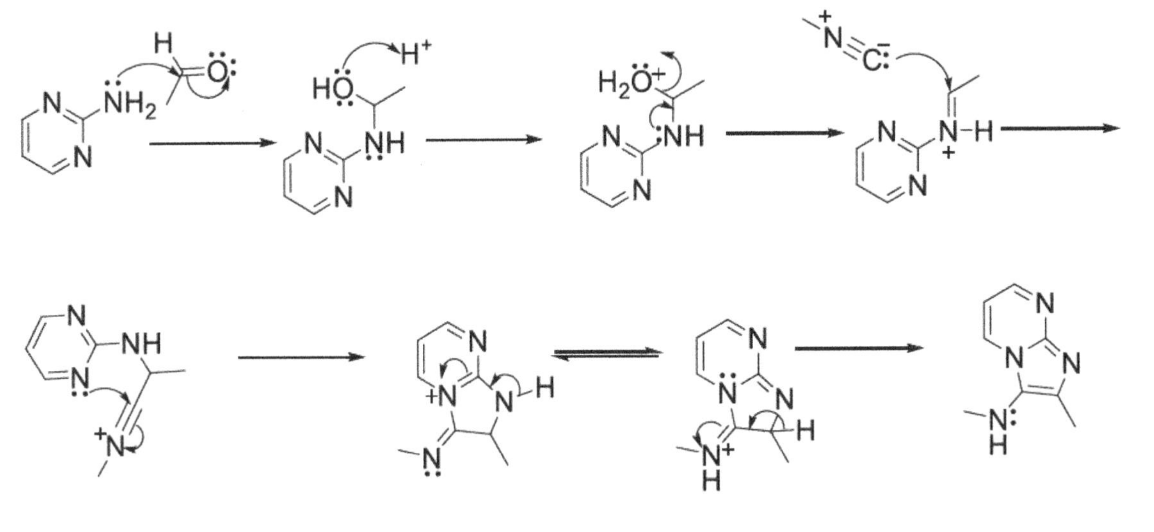

341

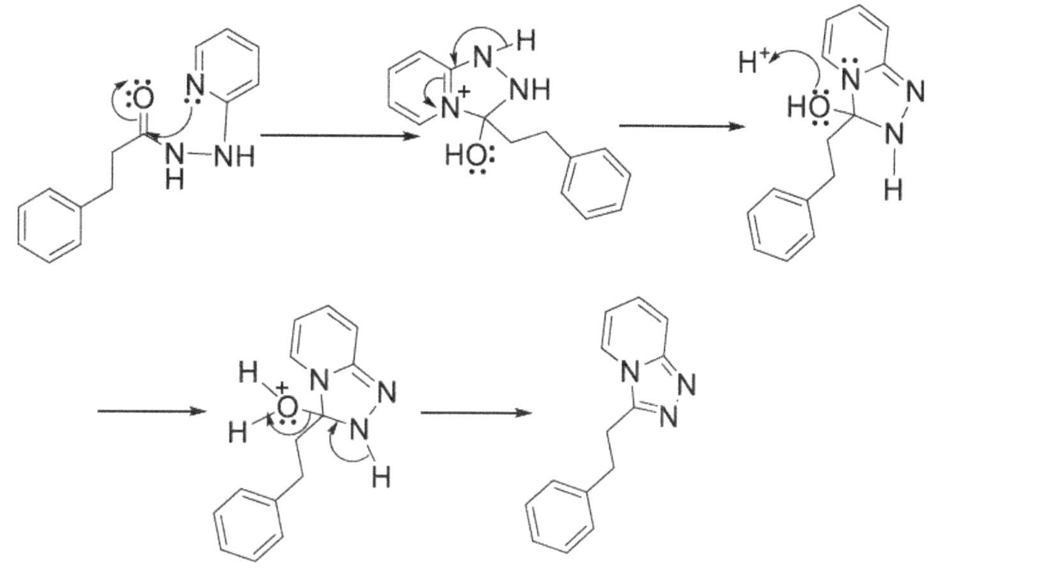

342

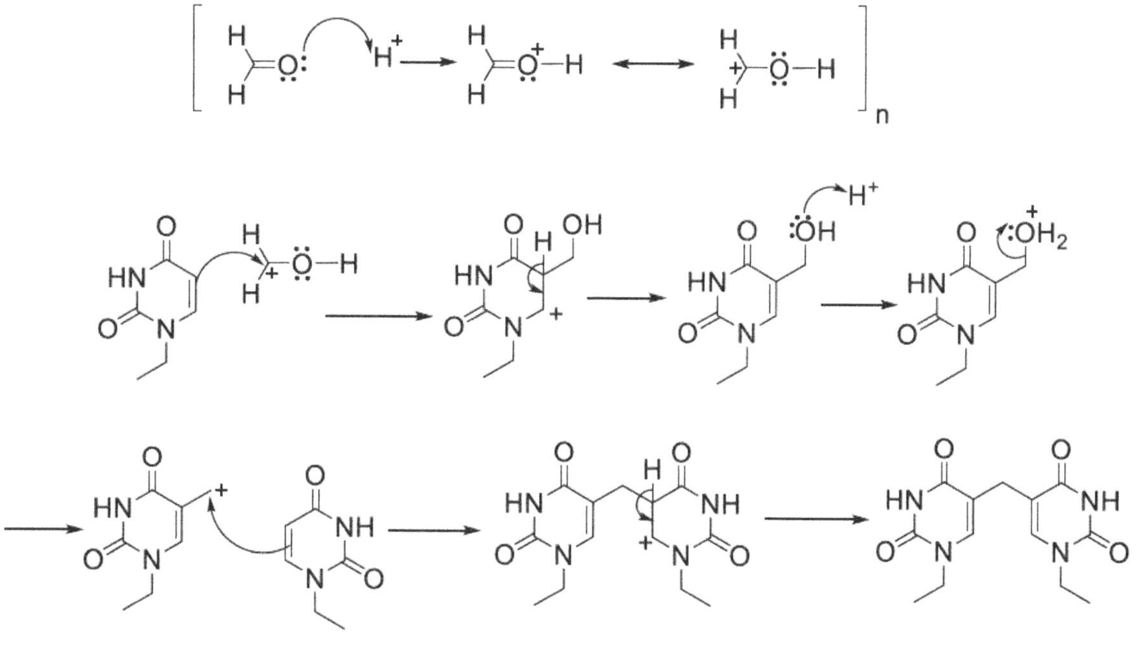

343

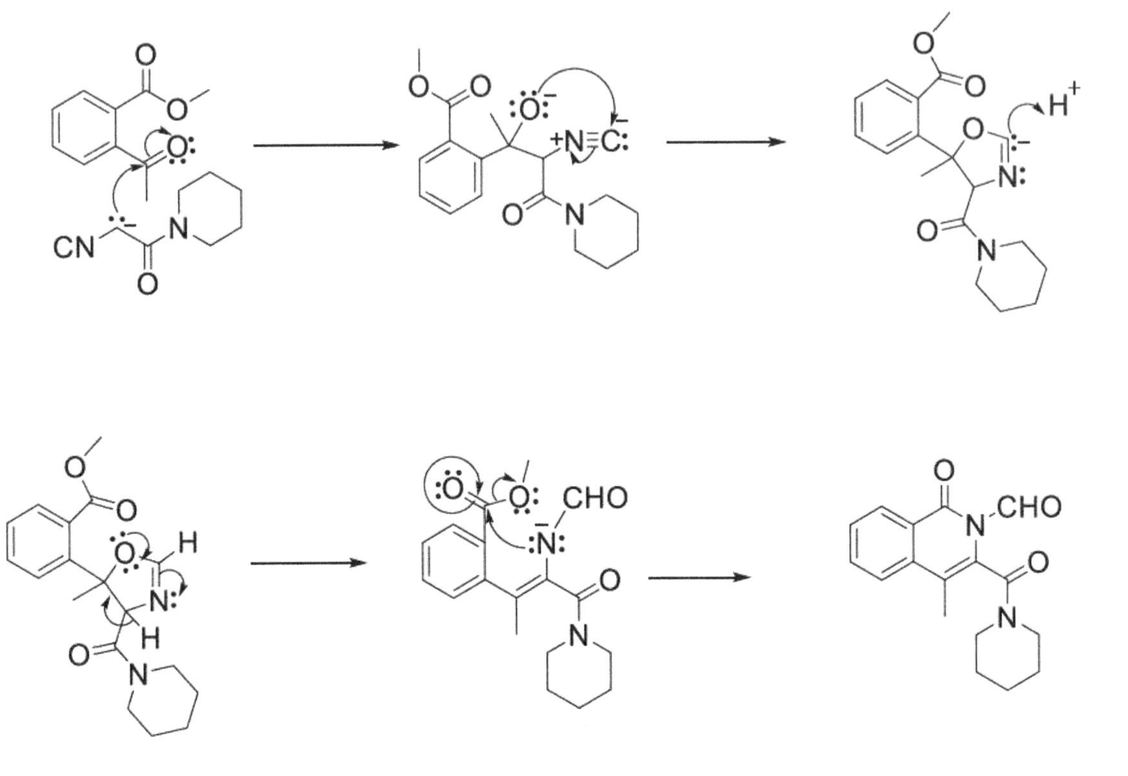

O
O
O:
CN
N
O
O
O
:O:
+N≡C:
O
N
O
O
O
H+
O
N:
O
N
O
O
O
H
N:
H
O
N
:O:
O:
CHO
:N:
O
N
O
N
CHO
O
N

344

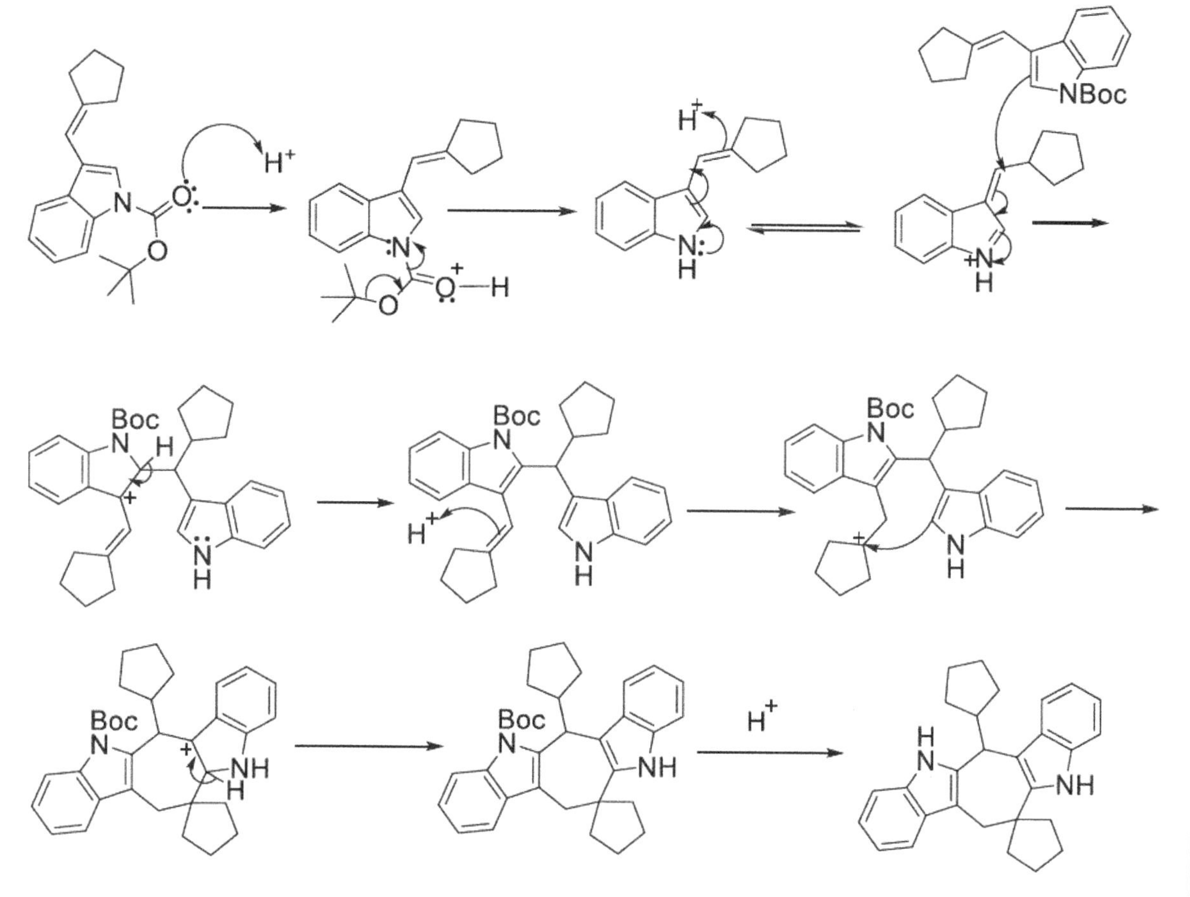
H⁺
NBoc
Boc
NH
H

345

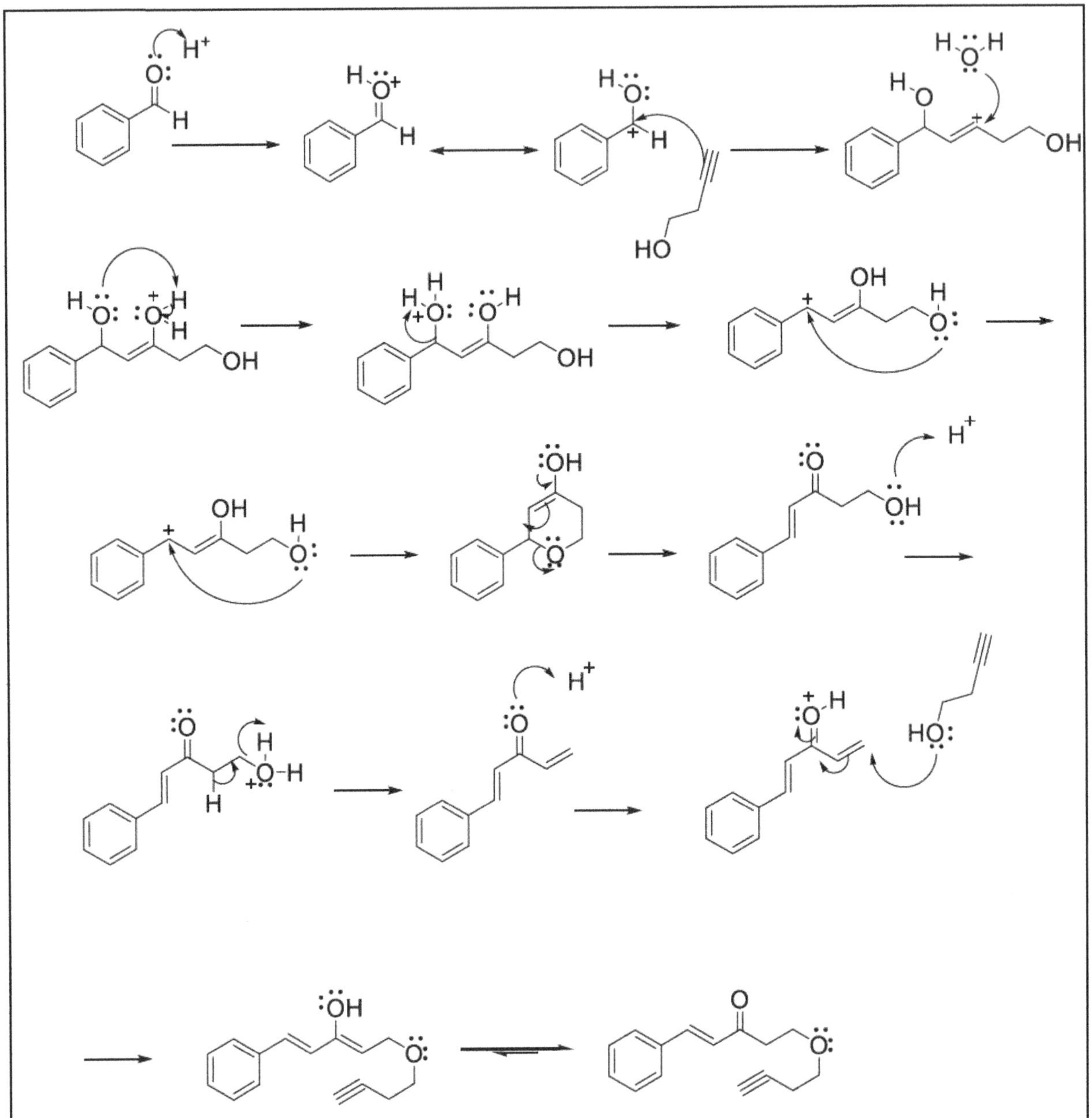

346

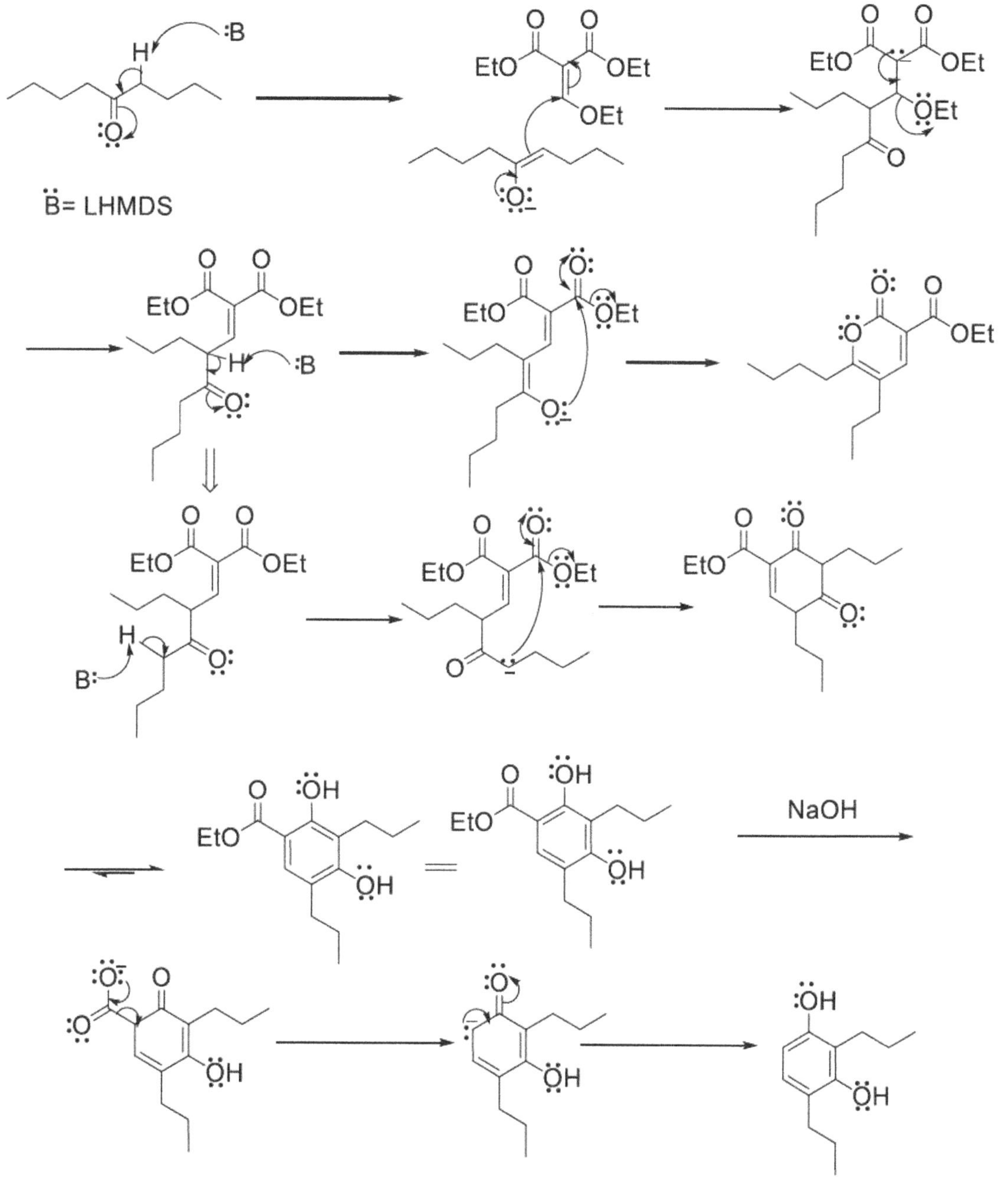
B= LHMDS
NaOH

347

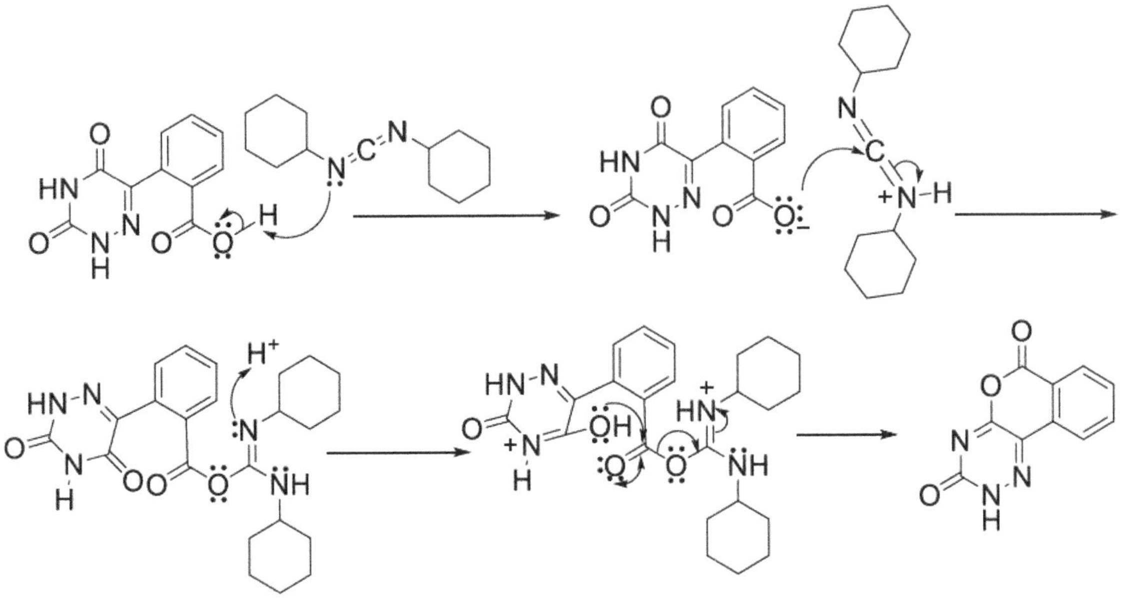

348

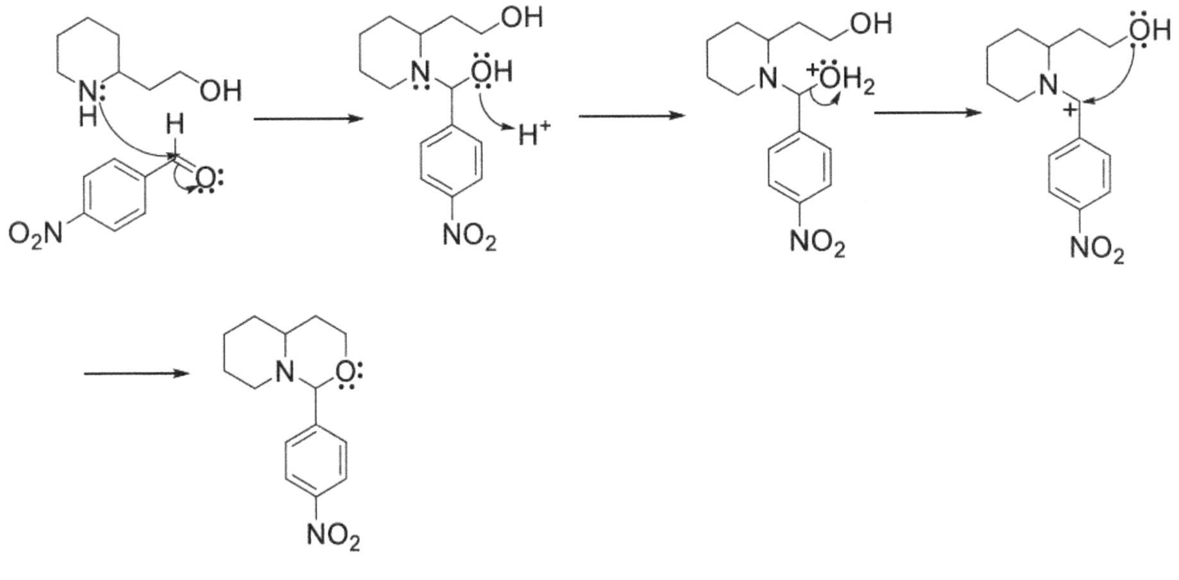

349

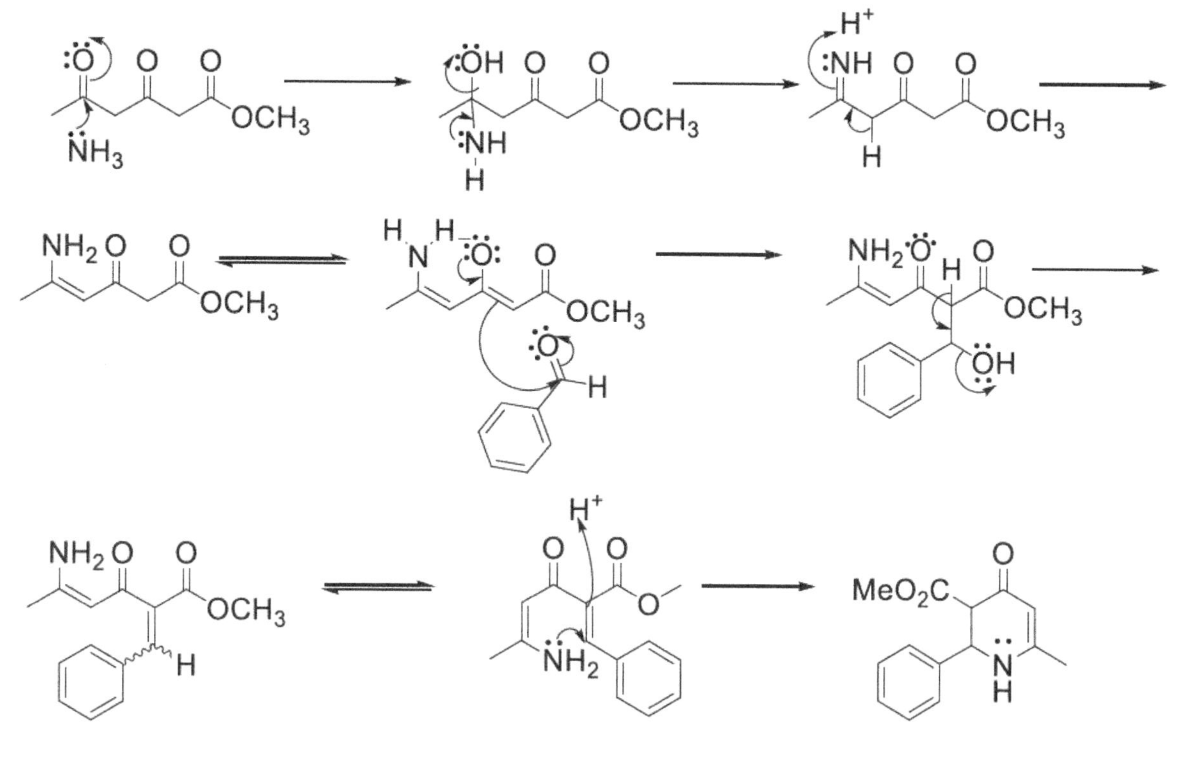

NH3
OCH3
OH
NH
H
H+
NH2
MeO2C

350

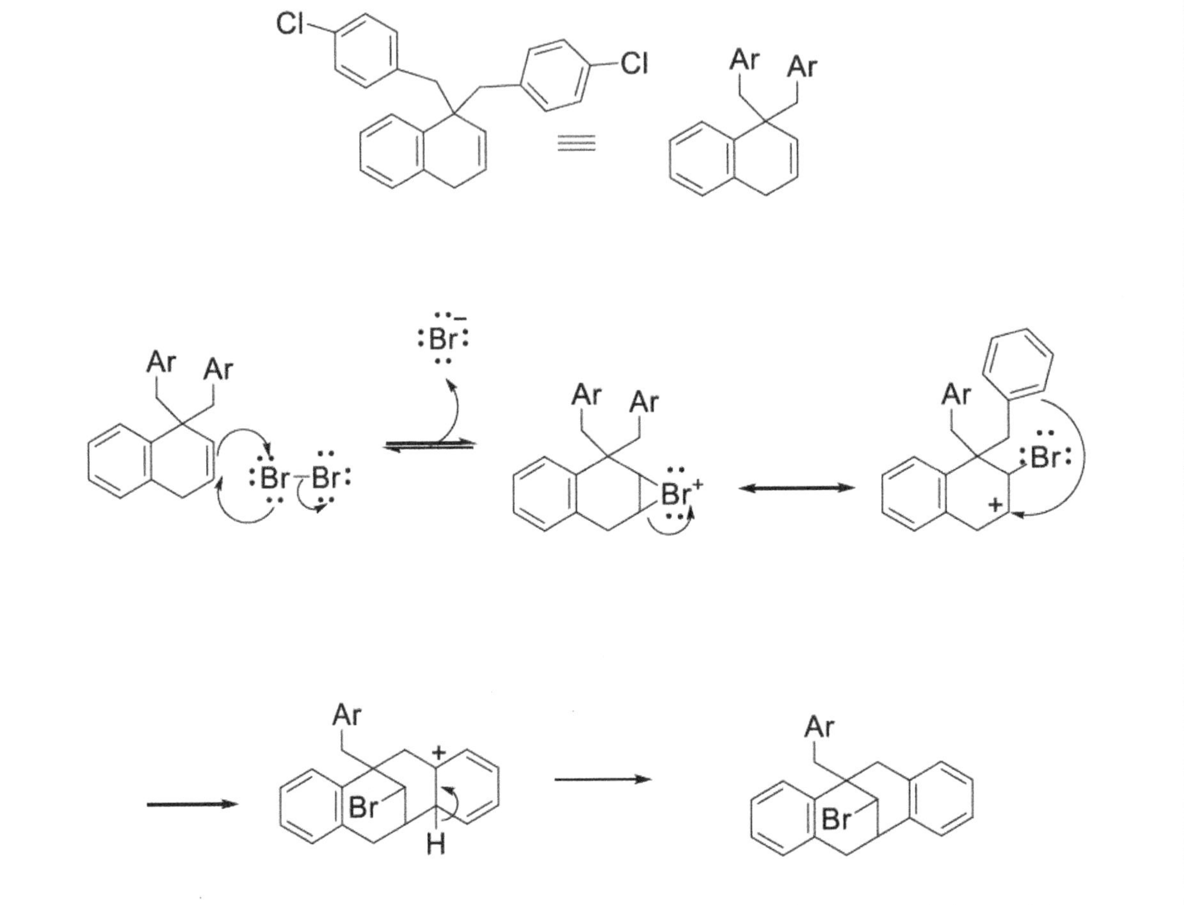
Cl
Cl
Ar
Ar
Br
Br
Br
Br
Br
+
Br
H

351

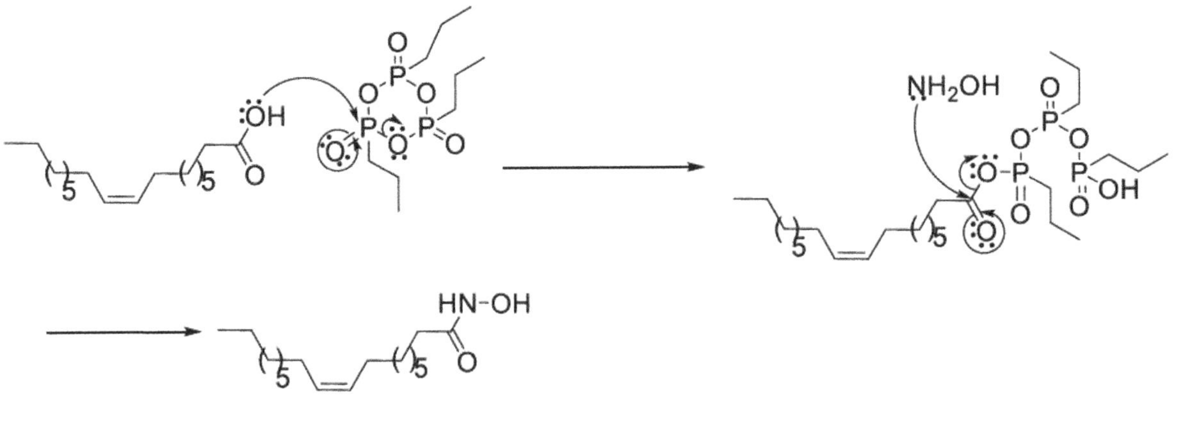
OH
O
O
P
O
P
O
P
O
O
O
5
5
NH_2OH
OH
HN-OH

352

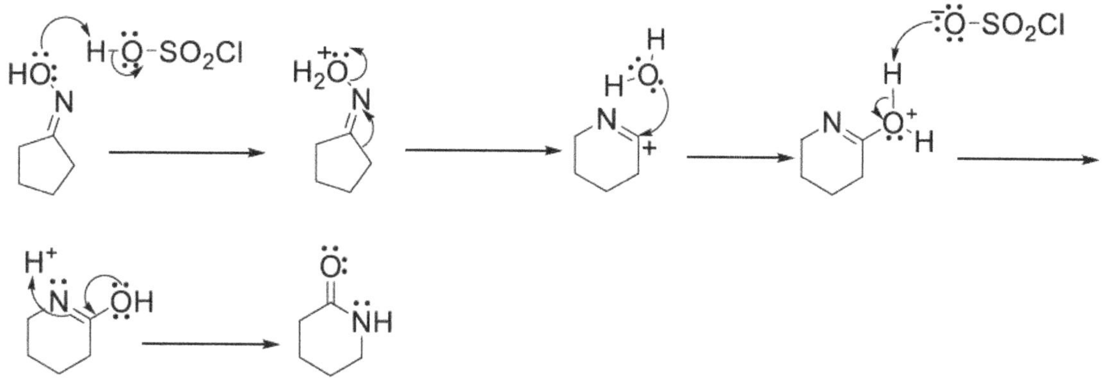

HO
N
H
O
SO_2Cl
H_2O
H
N
+
O
H

This is an example of the *Beckman rearrangement*

353

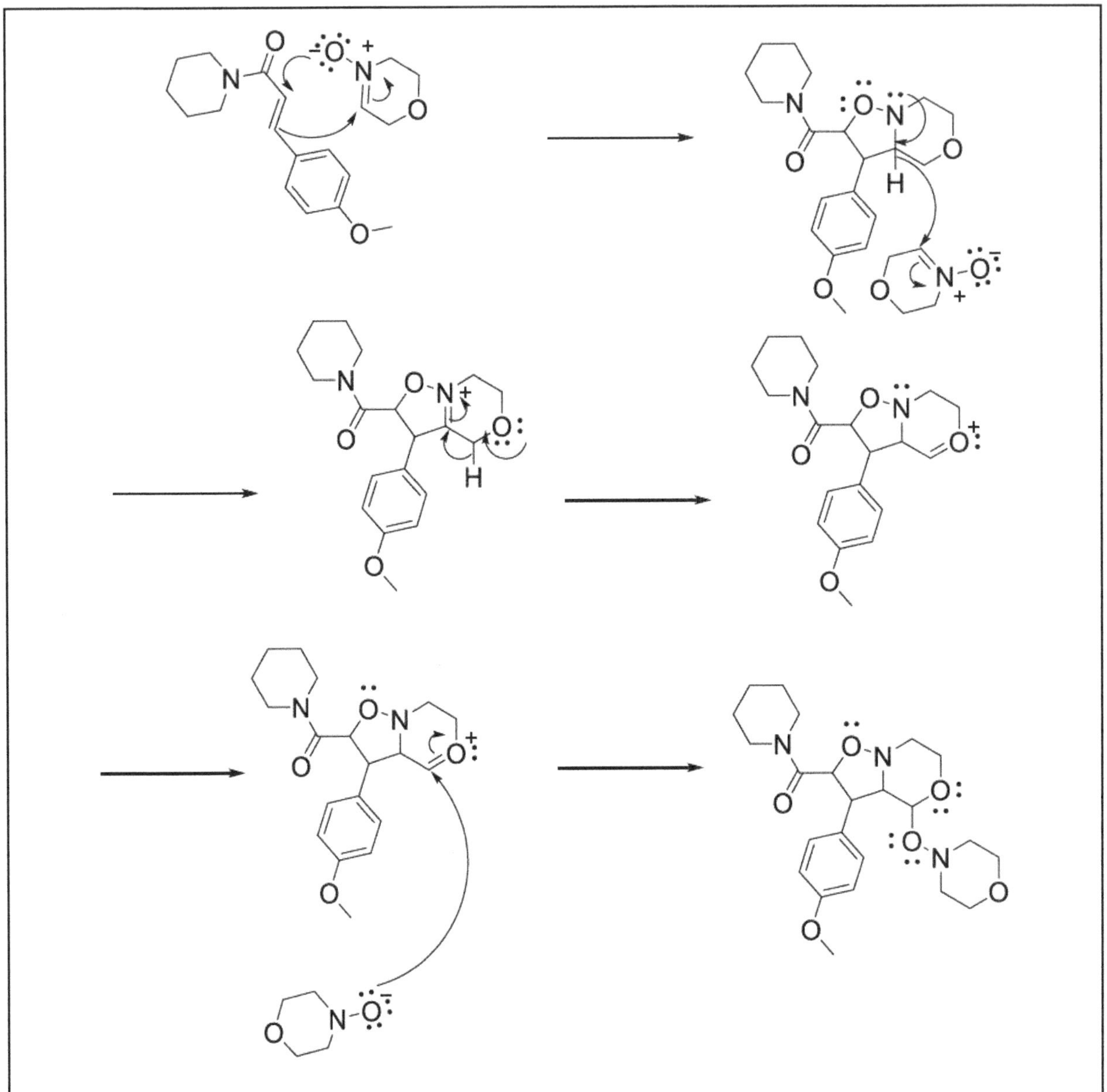
O
N
O
N
O
H
O
N
O
N
O
O
N
O
H
O
N
O
O
N
O
O
N
O
O
N
O
O
N
O
O
N
O
O
N
O

354

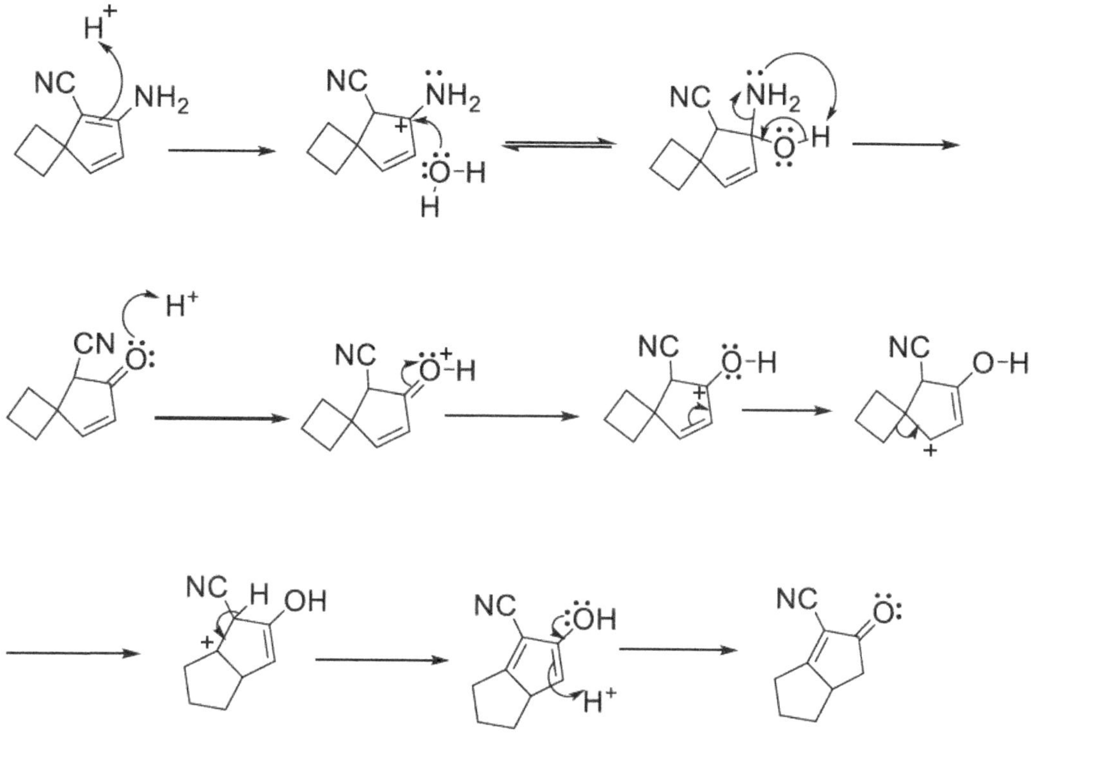
H+
NC
NH2
NC
NH2
+
O-H
H
NC
NH2
O
H
H+
CN
O
NC
O+
H
NC
O-H
+
NC
O-H
+
NC
H
OH
+
NC
OH
H+
NC
O

355

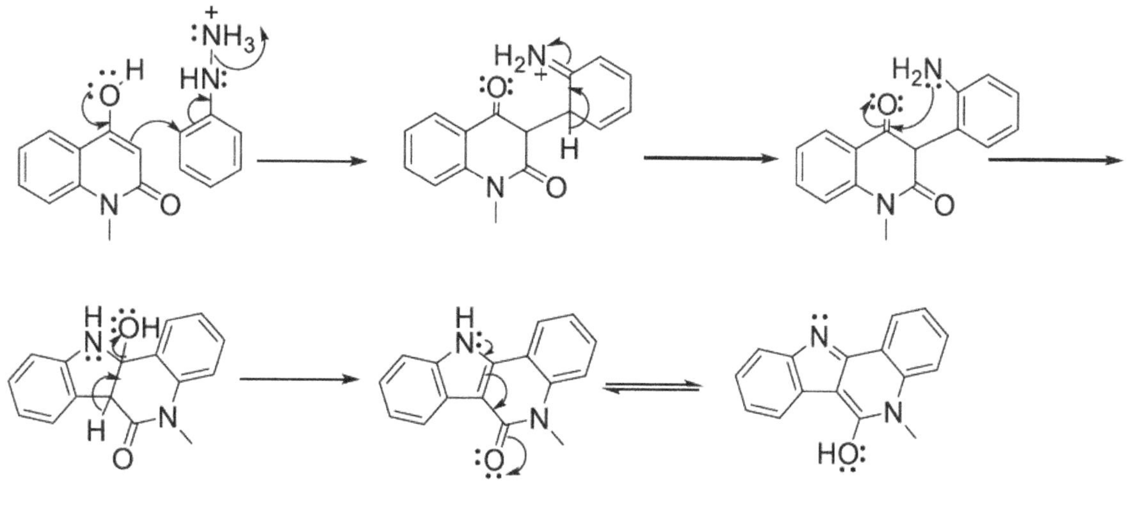
NH3
+
HN
O
H
N
O
H2N
+
O
H
N
O
H2N
O
N
O
H
N
OH
H
N
O
H
N
N
O
N
N
HO

356

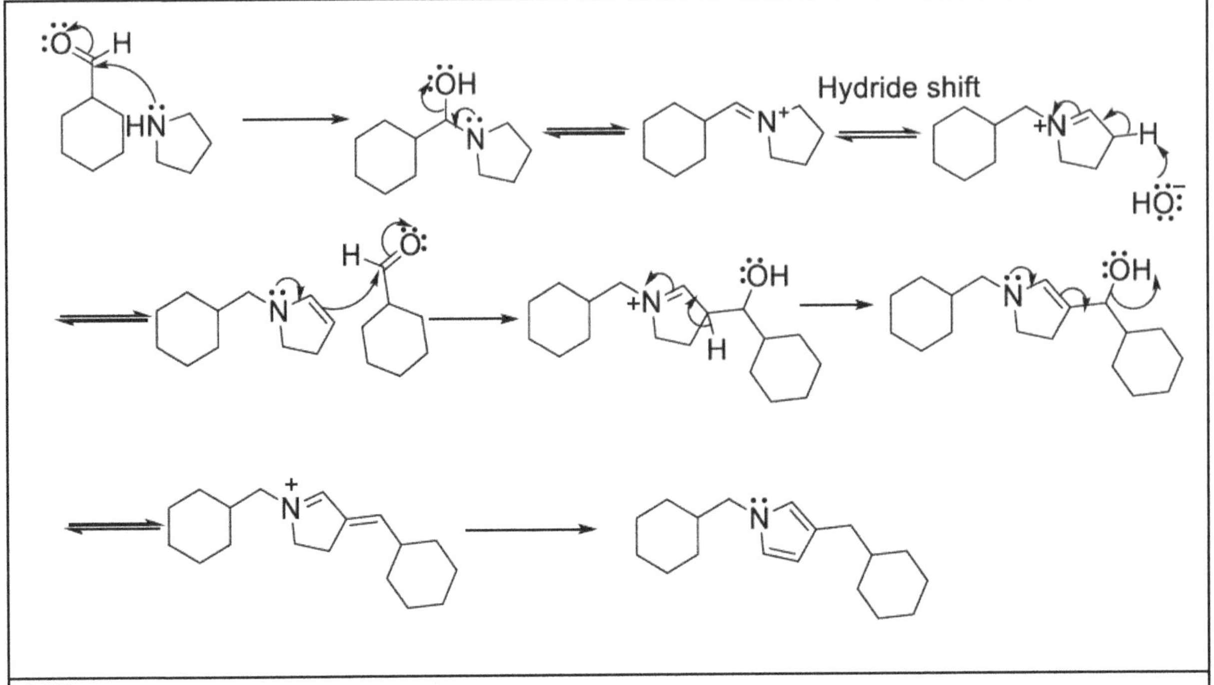
Hydride shift

357

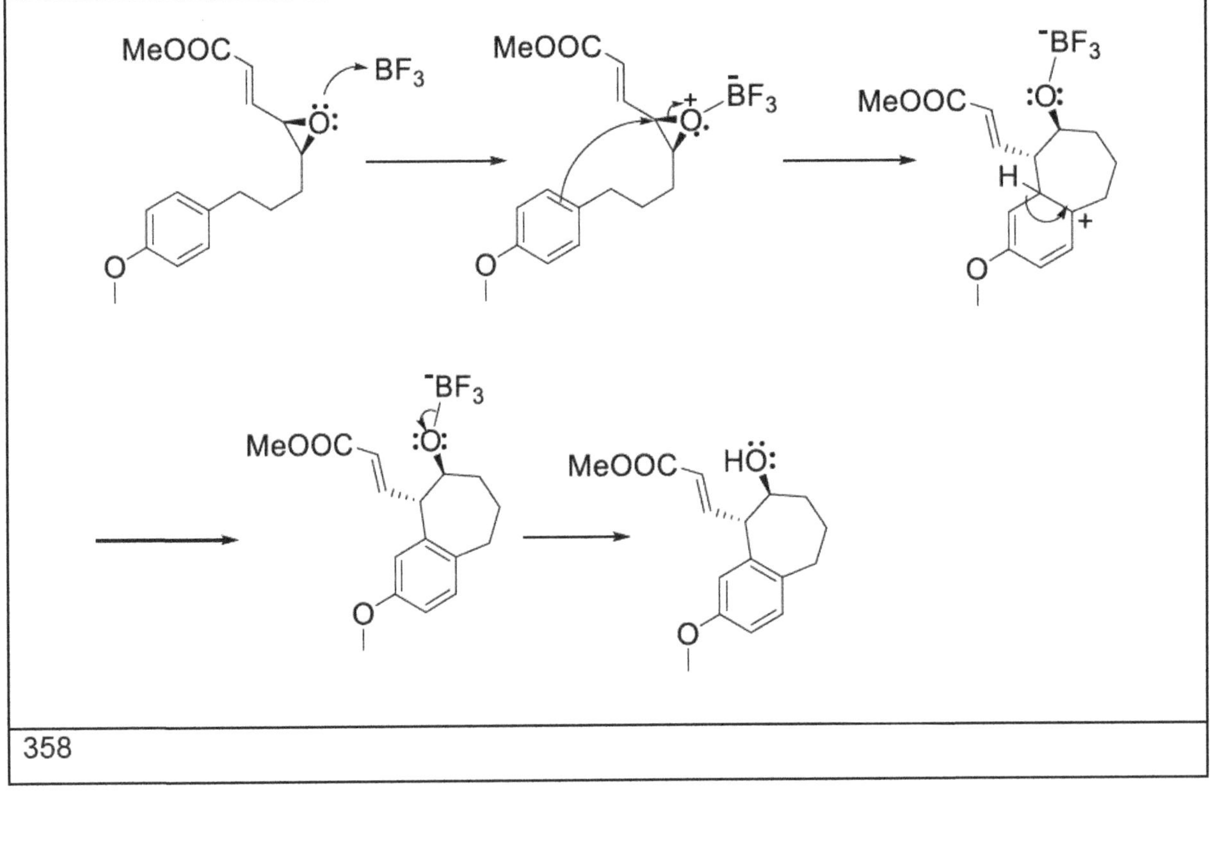
MeOOC
BF3
MeOOC
BF3
MeOOC
BF3
H
MeOOC
BF3
MeOOC
HO

358

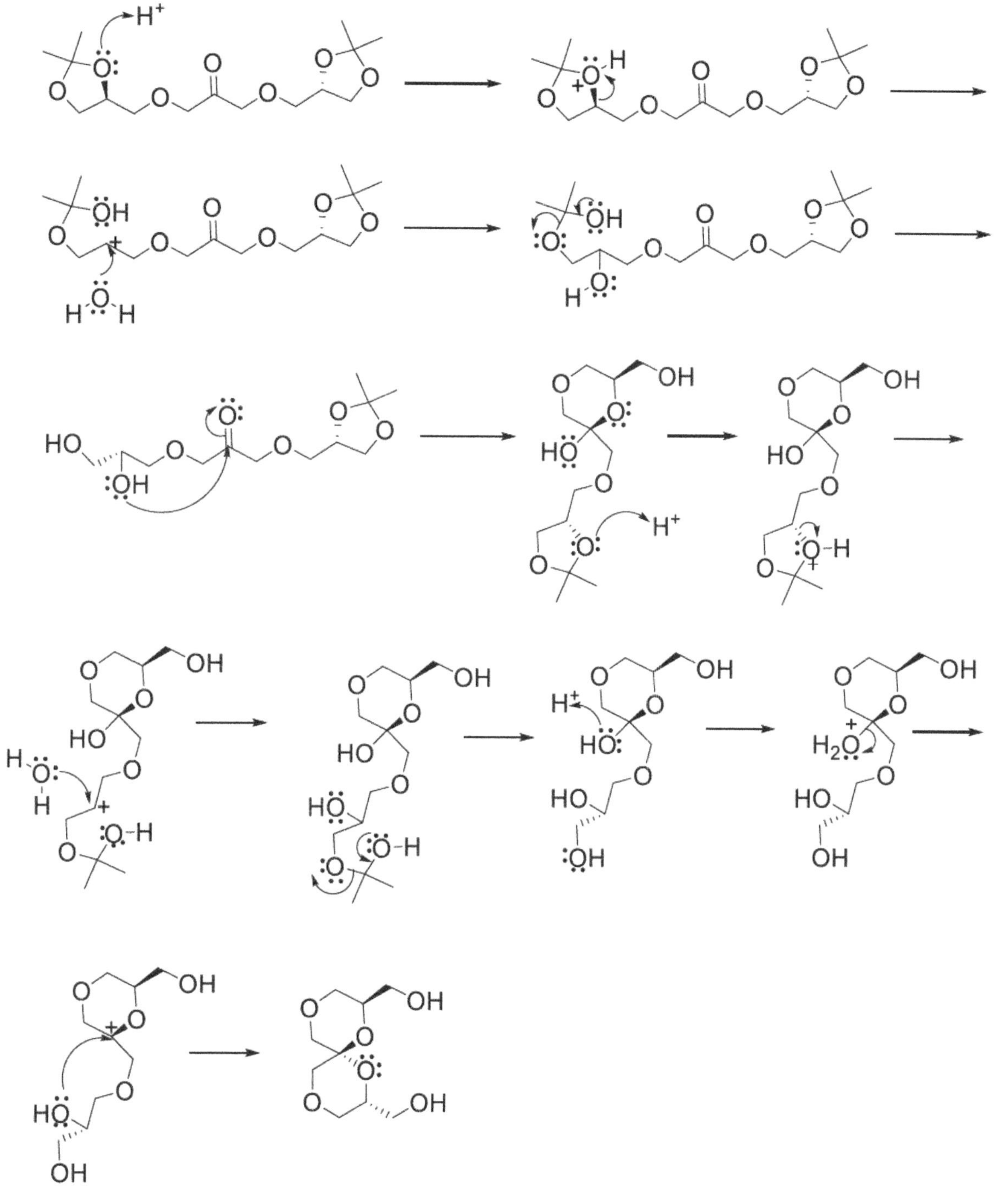

359

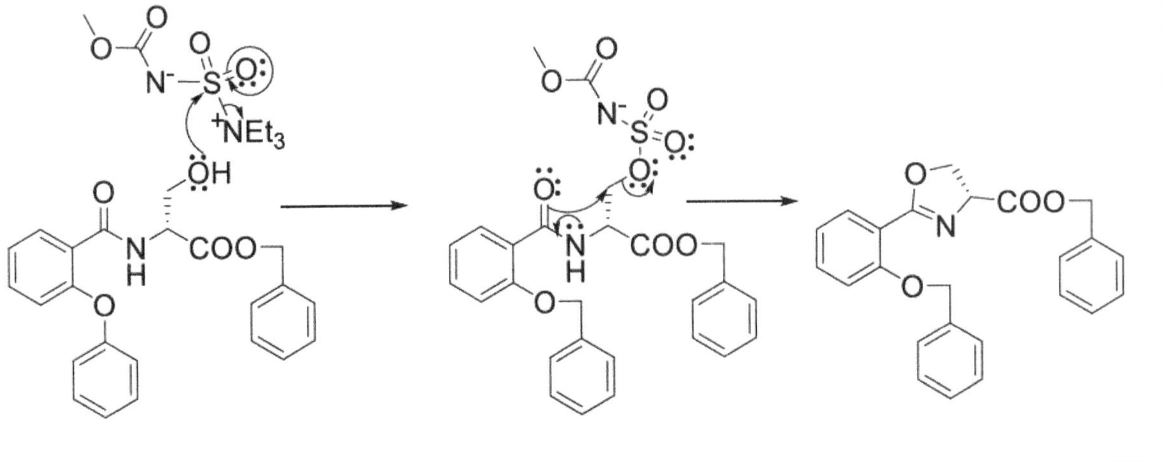
NEt3
OH
N
H
COO
O
COOEt
OEt
Br
HO
OH

360

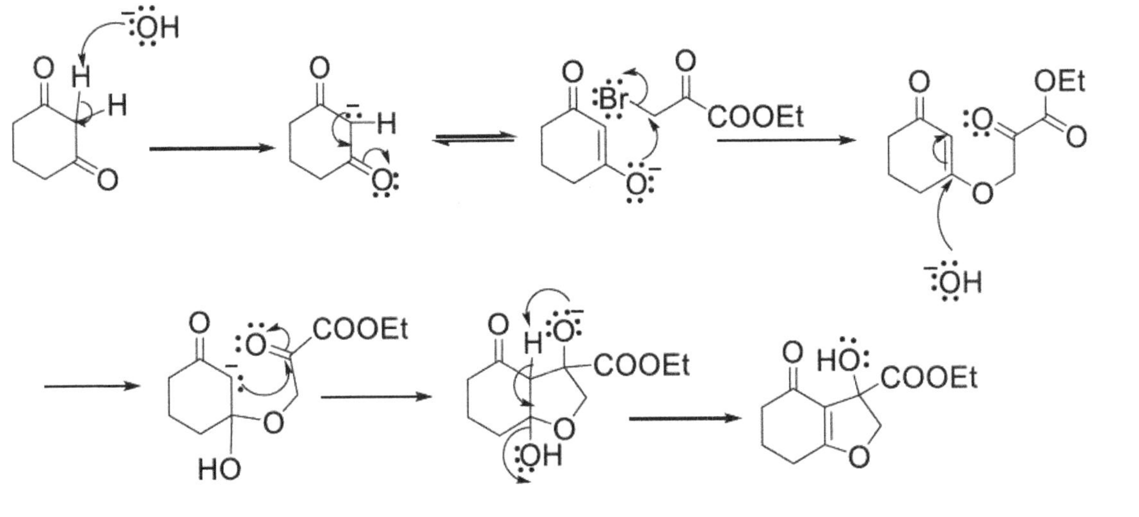

361

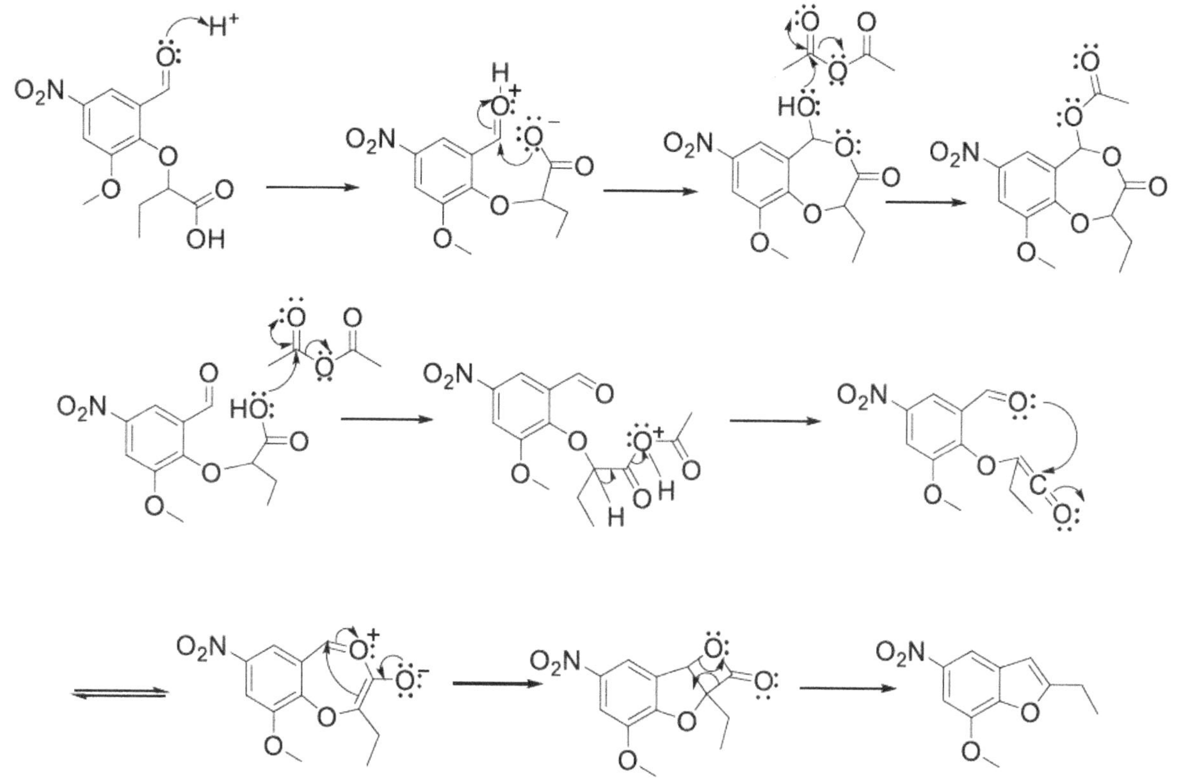

H⁺
O₂N
HO
OH

362

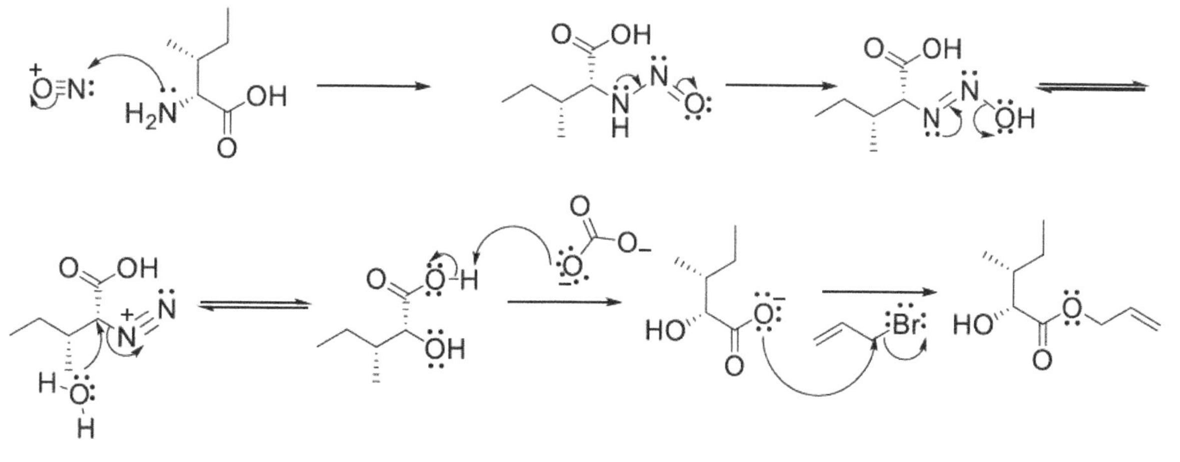

H₂N
OH
NH
HO
Br

363

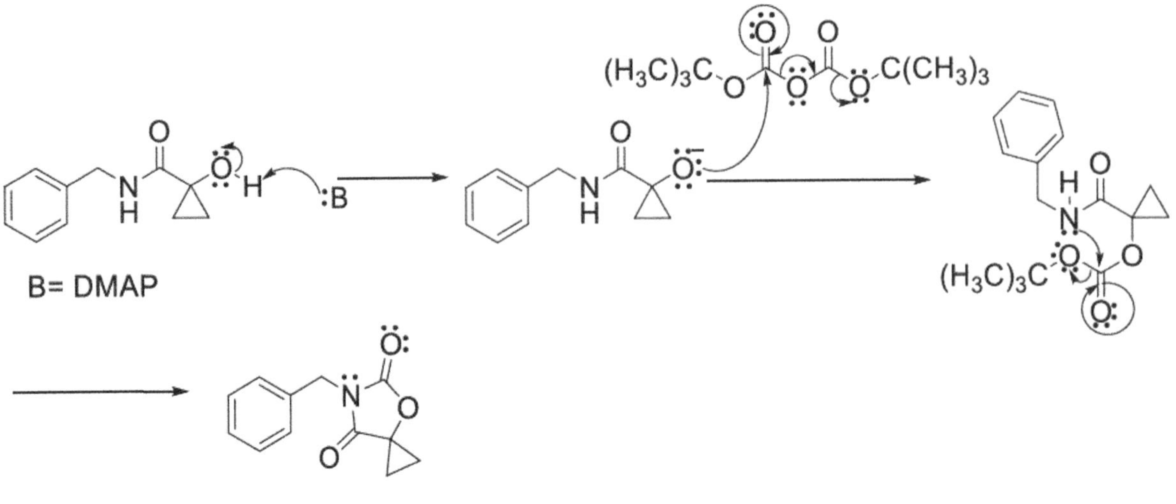
B= DMAP

364

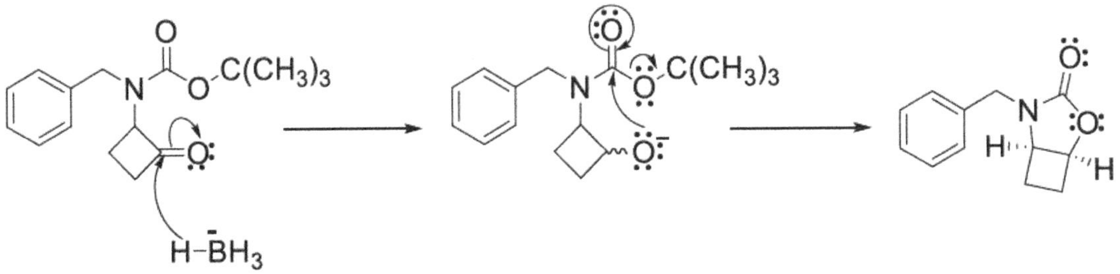

365

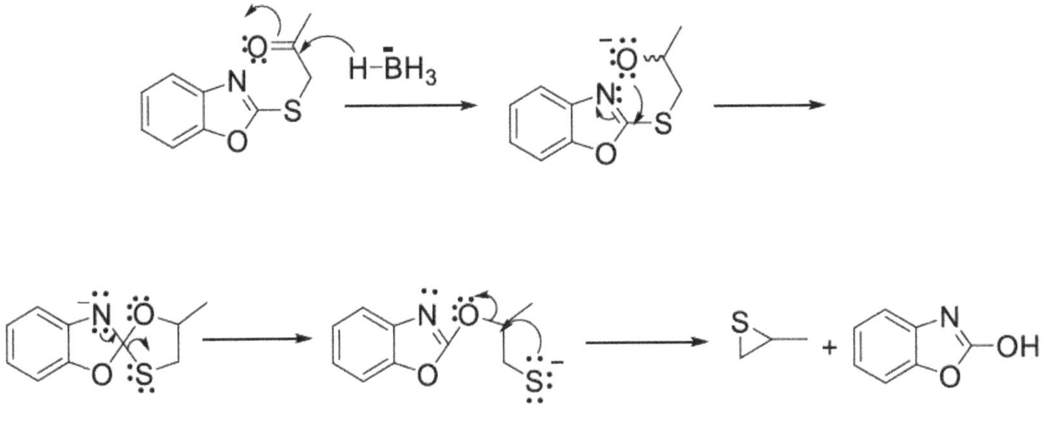

366

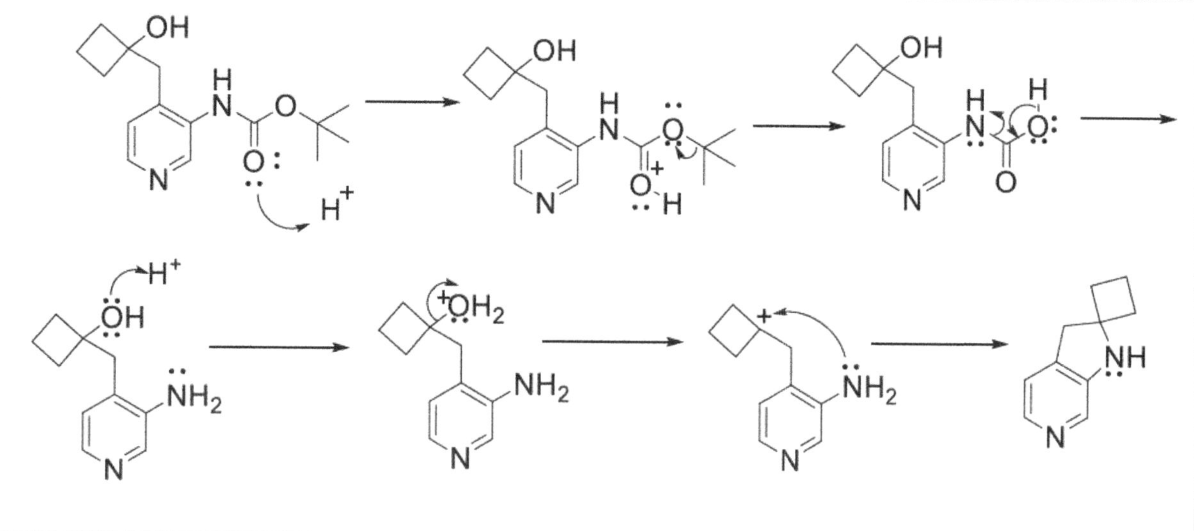

OH
H
N
O
O
N
H+
OH
H
N
O
O+
N
H
OH
H
N
H
O
O
N
H+
OH
NH2
N
+OH2
NH2
N
+
NH2
N
NH
N

367

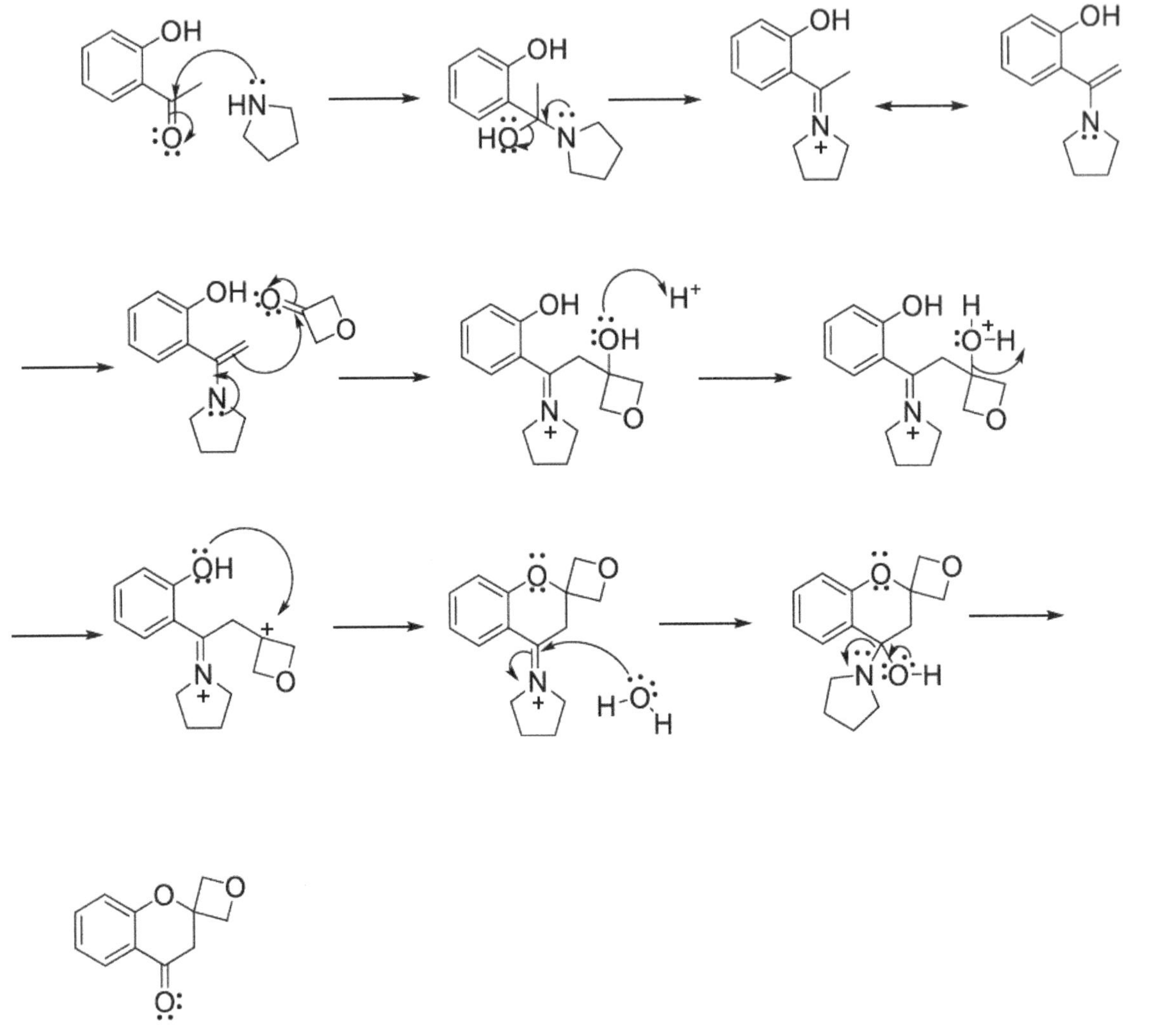

368

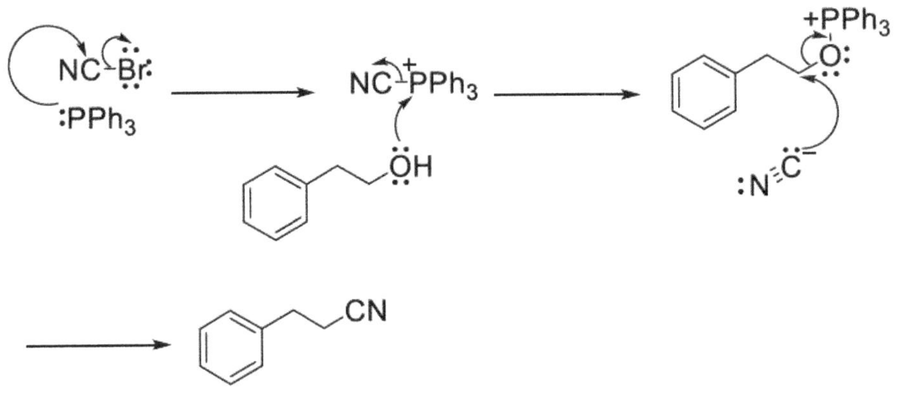

369

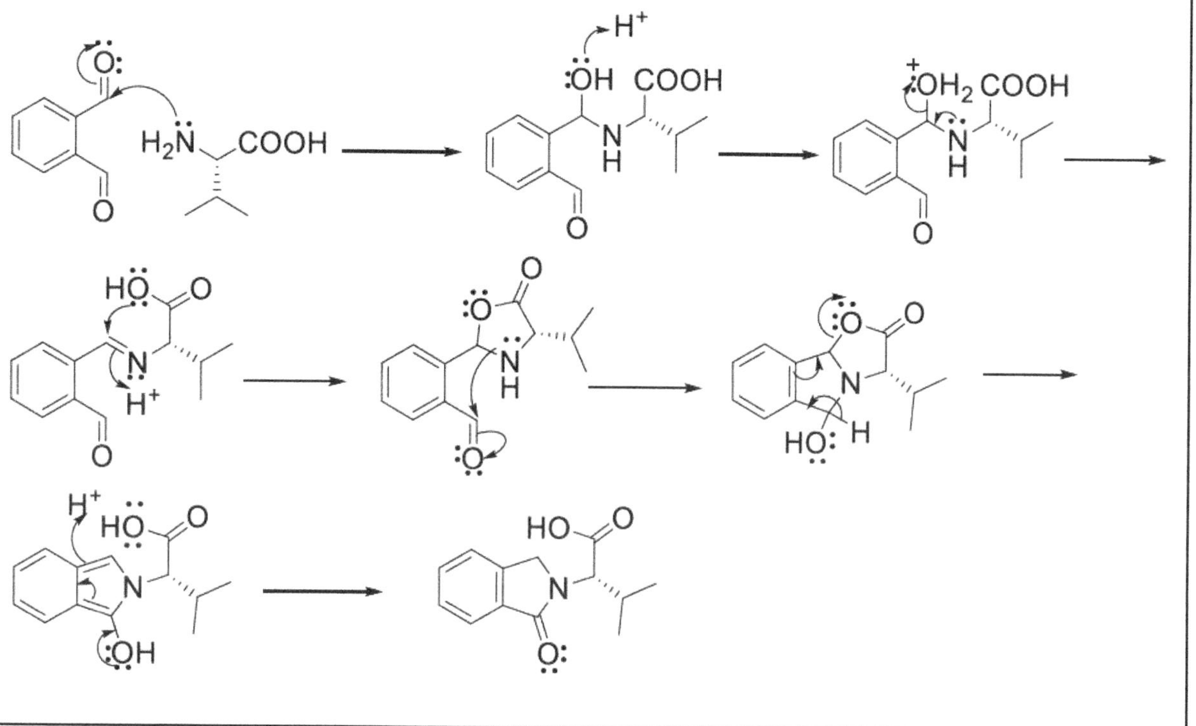
H+
OH
COOH
H2N
COOH
NH
OH2
COOH
N
H+
HO
O
HO
H
H+
OH

370

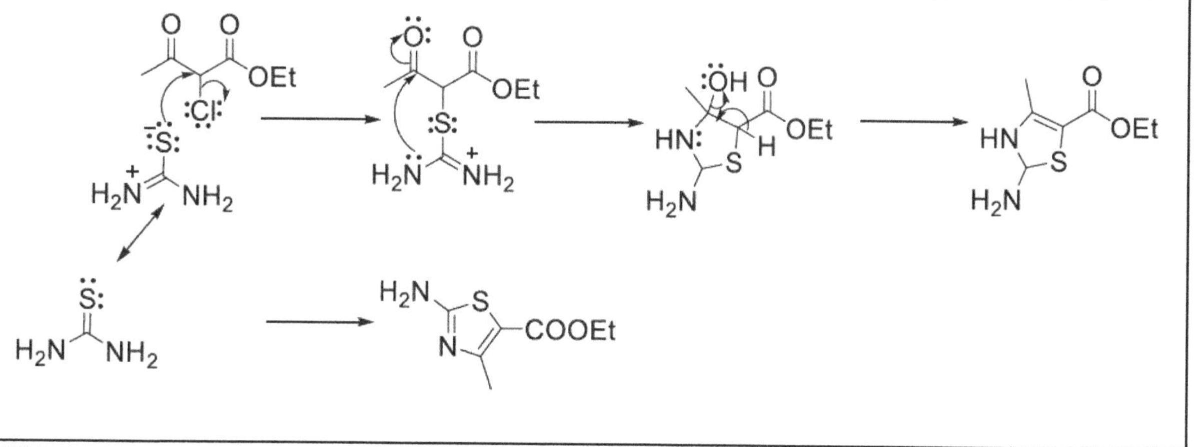
OEt
Cl
S
H2N
NH2
H2N
NH2
OH
HN
H
S
OEt
H2N
HN
S
OEt
H2N
S
H2N
NH2
H2N
S
N
COOEt

371

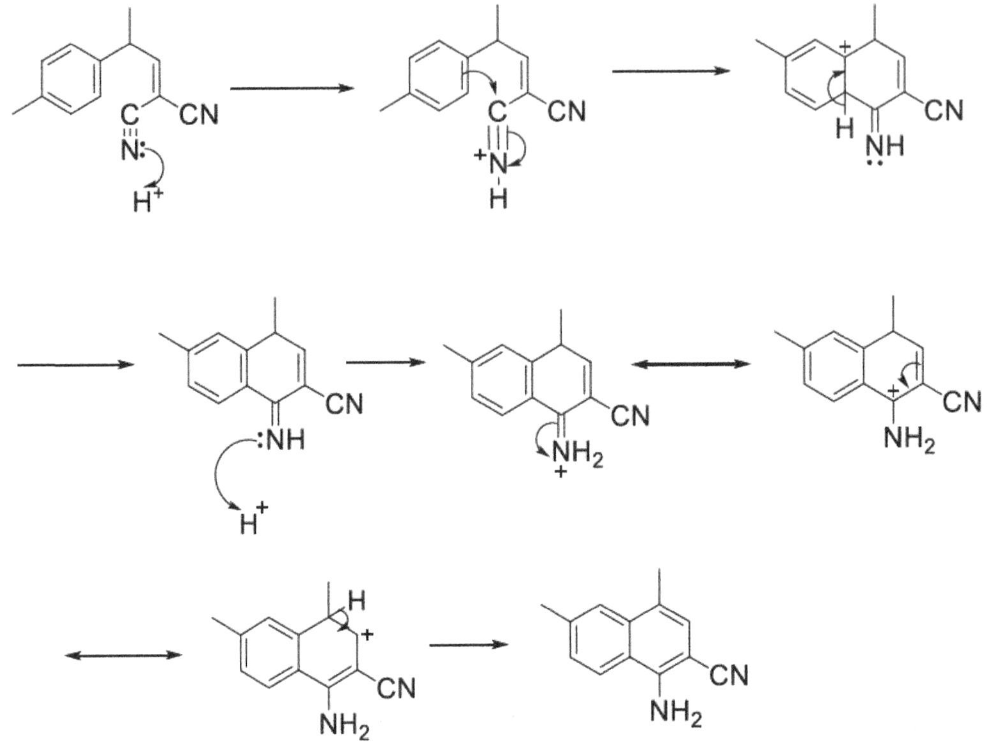

CN
H⁺
H
NH
NH₂

372

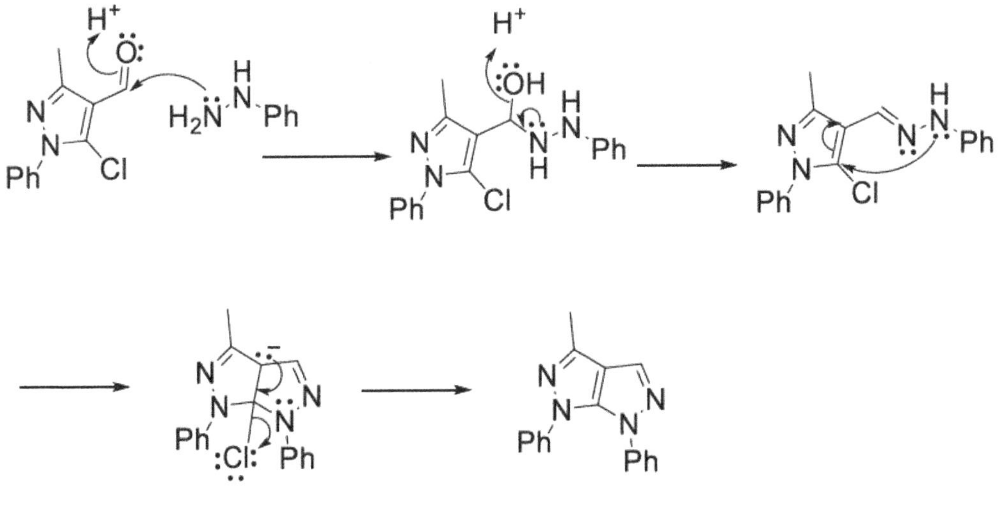

H⁺
O
H₂N
N
H
Ph
N
Cl
OH
Ph

373

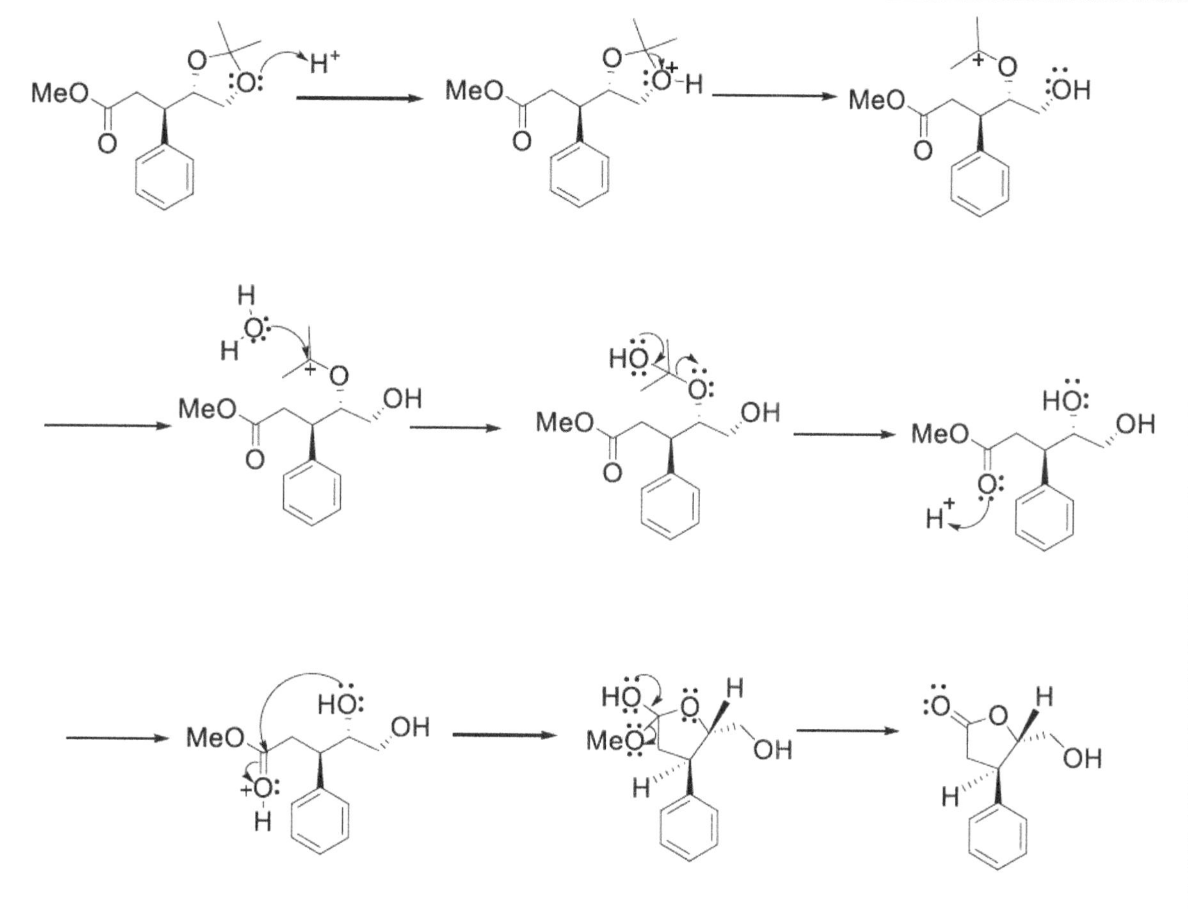
MeO
H+
O
OH
H
HO
OH
H

374

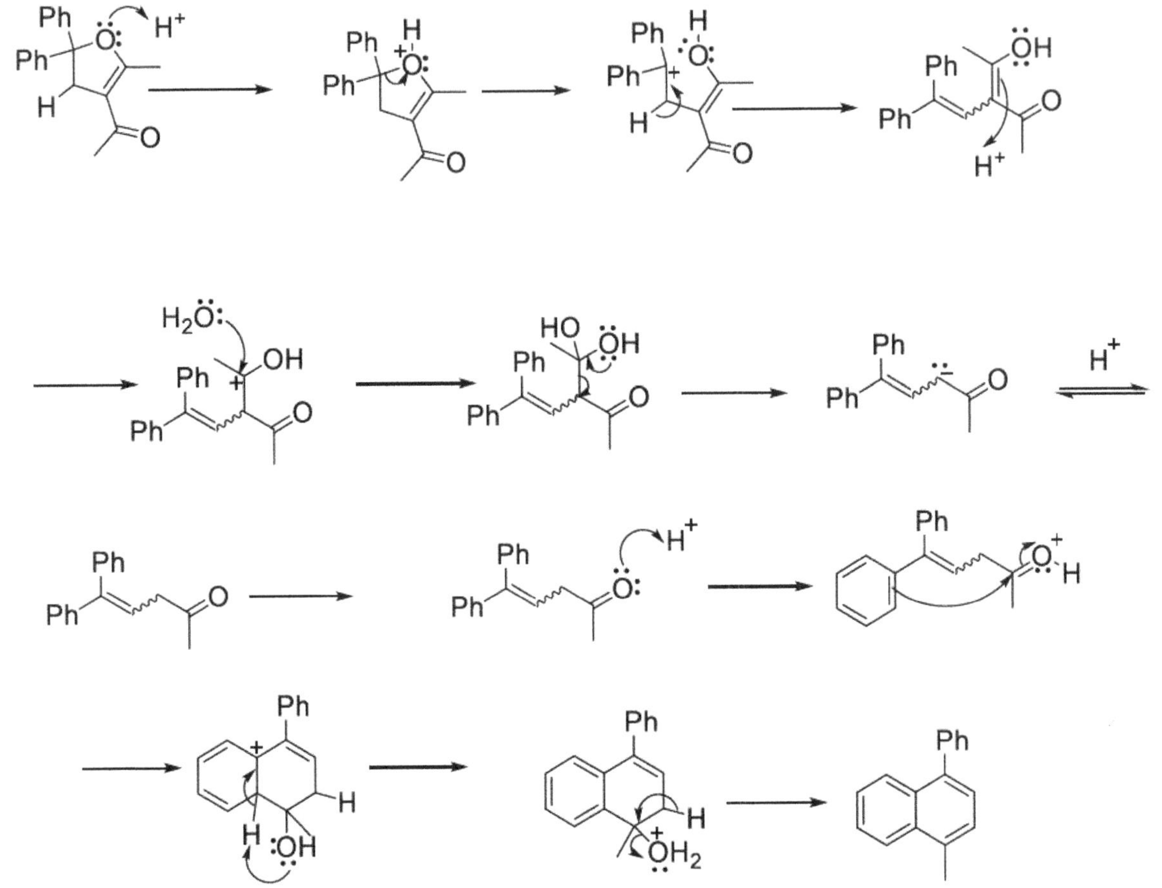

375

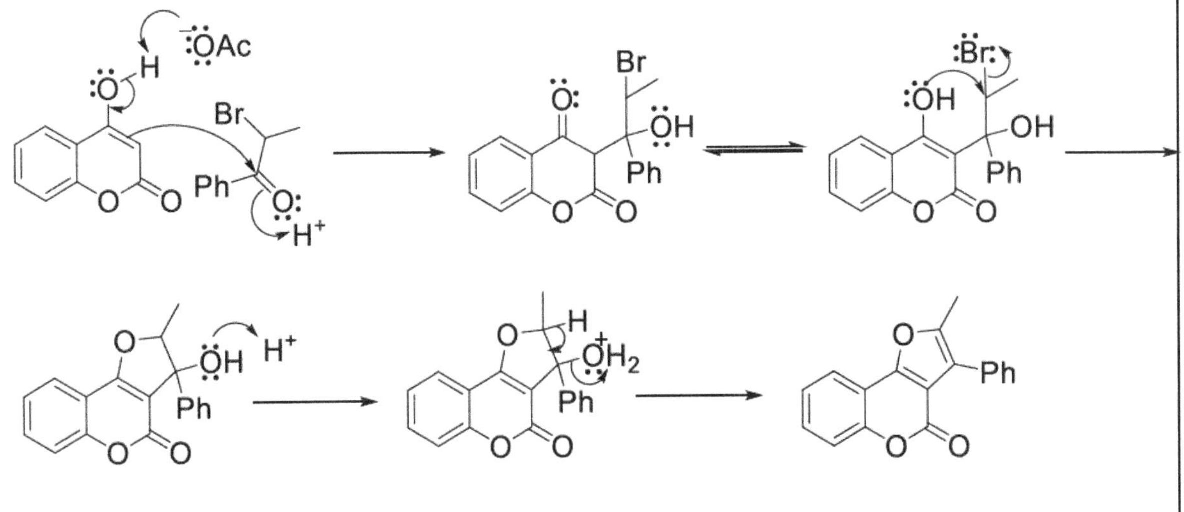

376

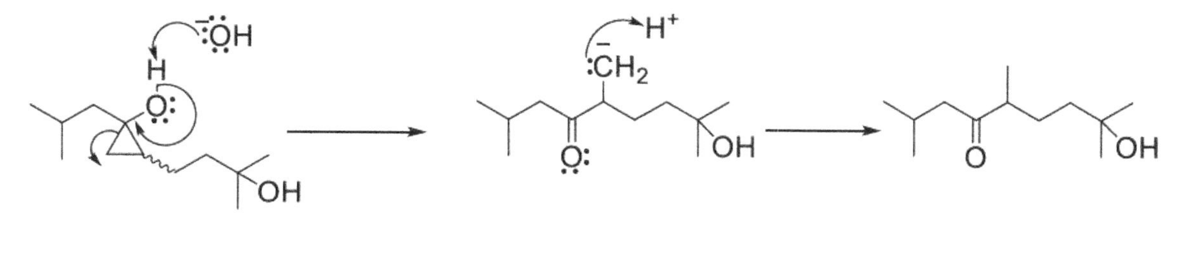

:ÖH
H
O:
OH
H+
:CH2
O:
OH
O
OH

377

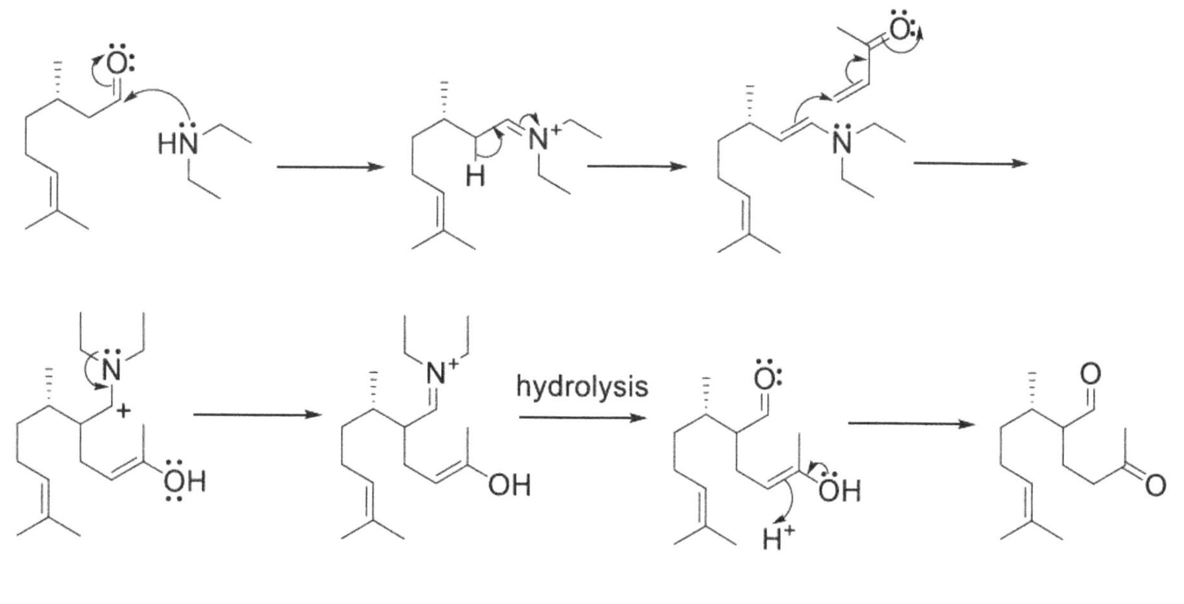

O:
HN
N+
H
O:
N
N
+
OH
N+
OH
hydrolysis
O:
OH
H+
O
O

378

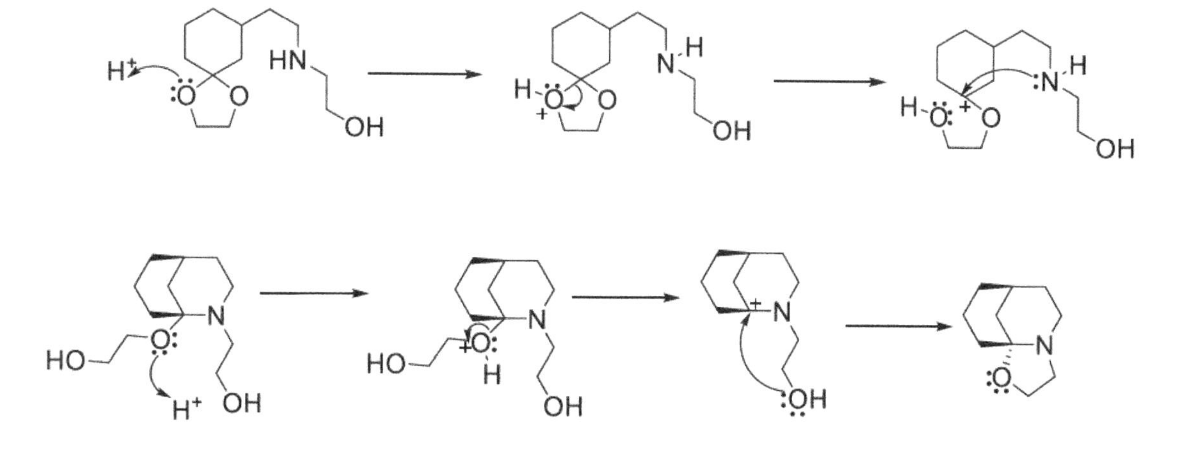

H+
O
O
HN
OH
H
O
O
N
H
OH
H
O
+
O
N
H
OH
N
HO
O:
H+
OH
N
HO
O
H
OH
+
N
OH
N
O

379

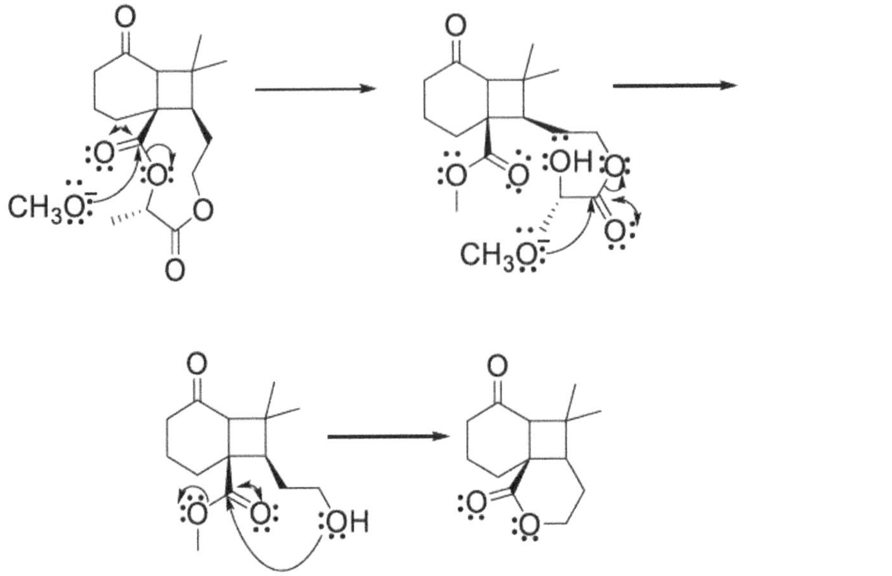

Three consecutive transesterification reactions give the final product

380

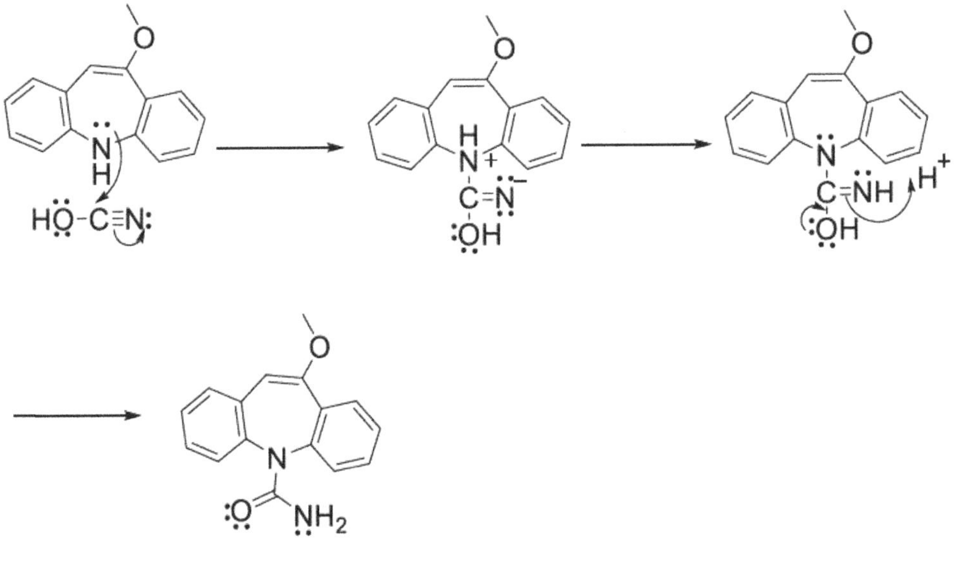

381

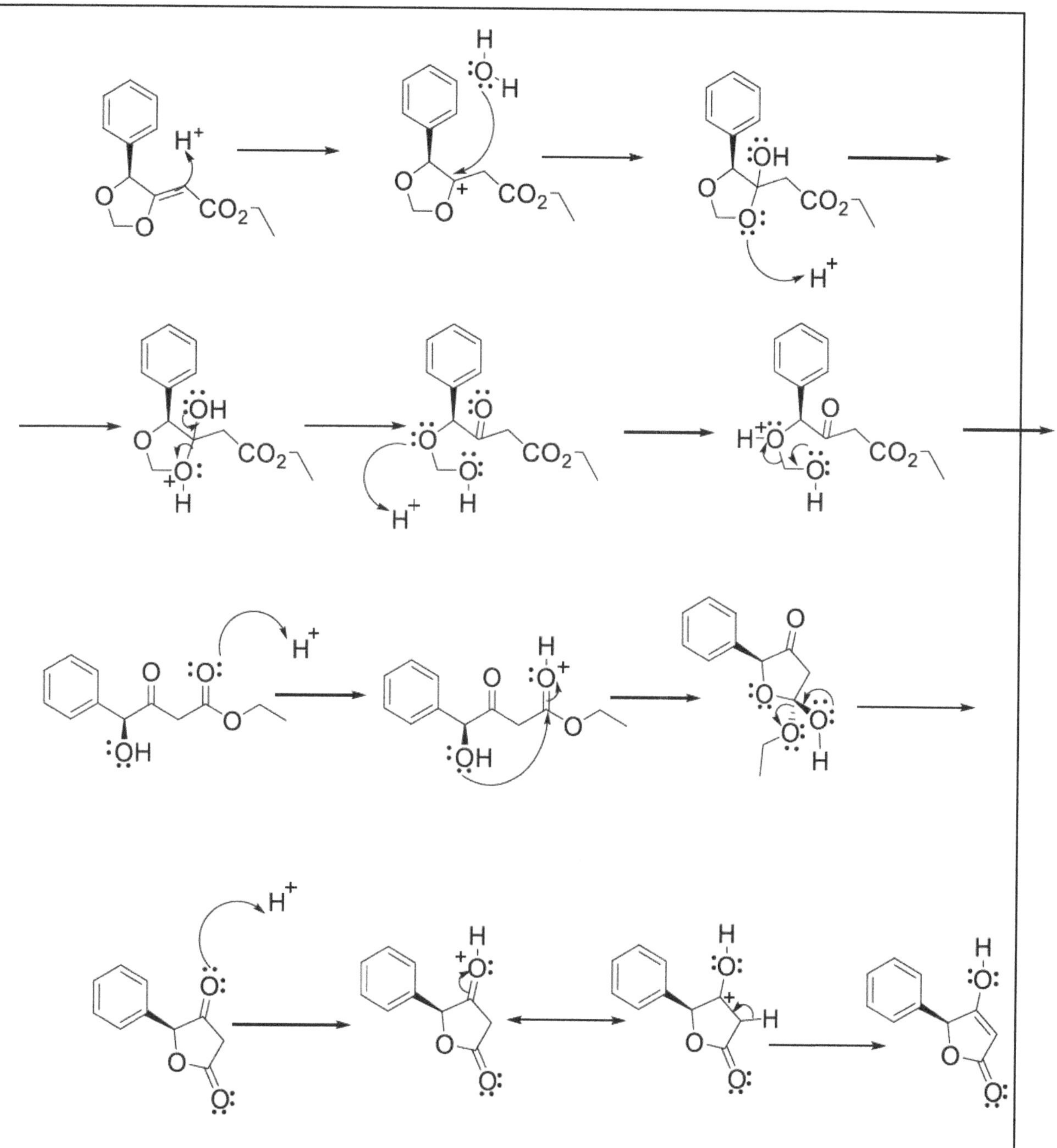

382

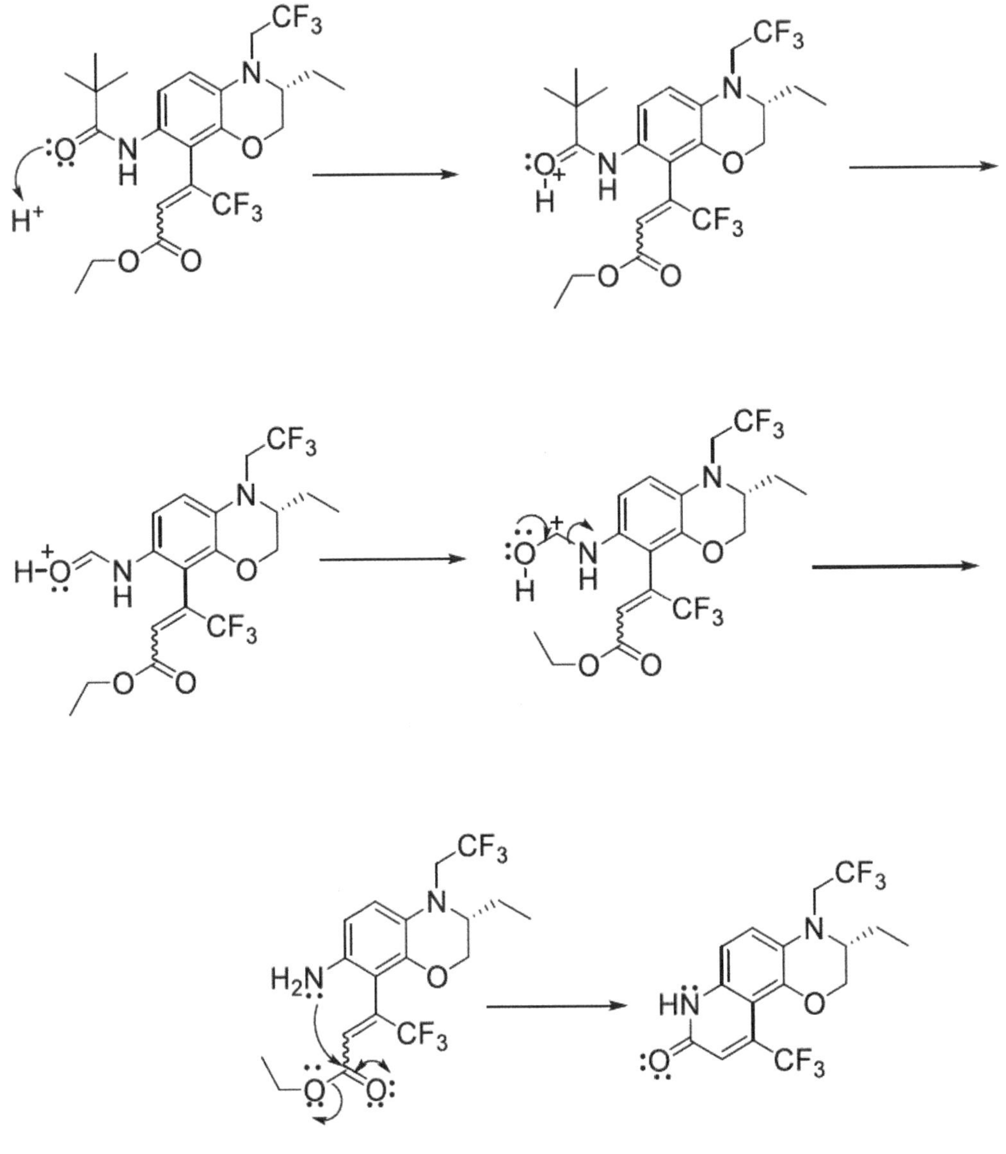

CF3
N
O
NH
H+
CF3
O
O

383

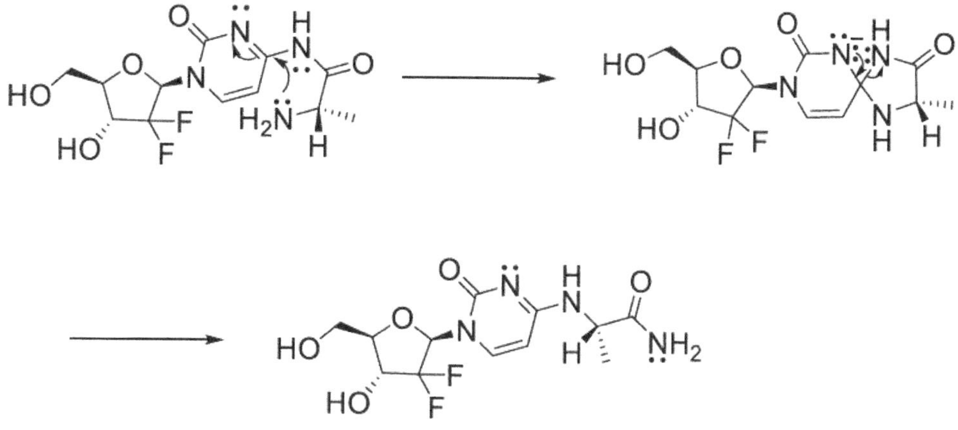

384

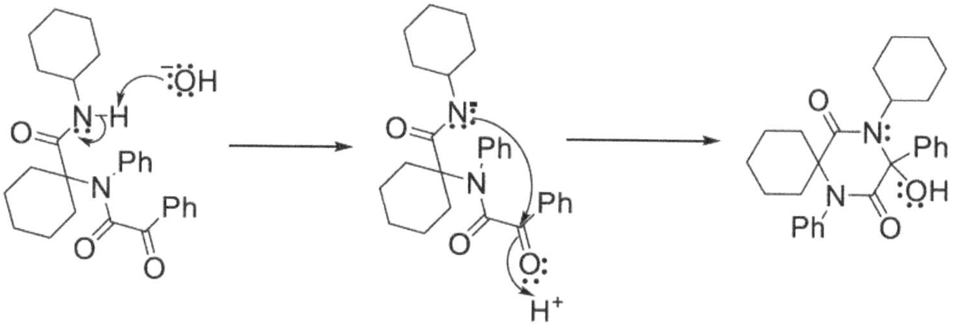

385

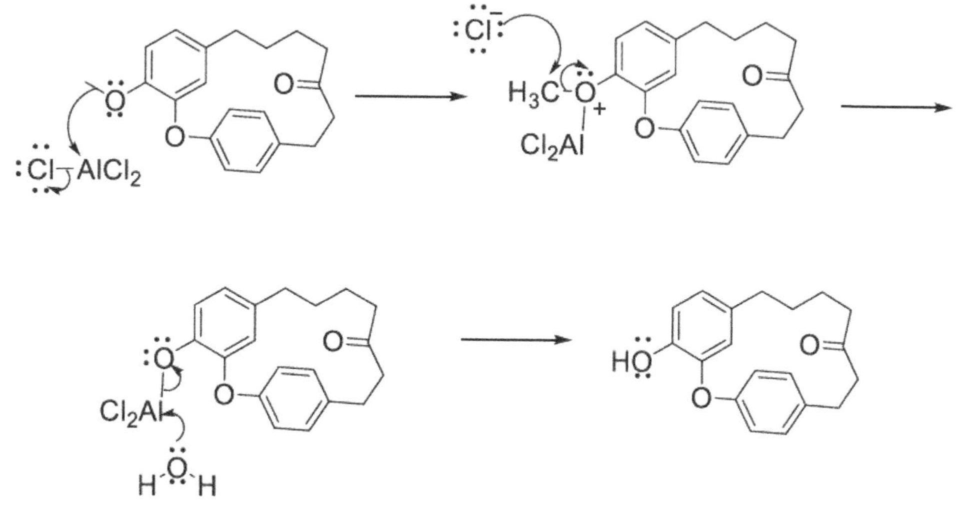

386

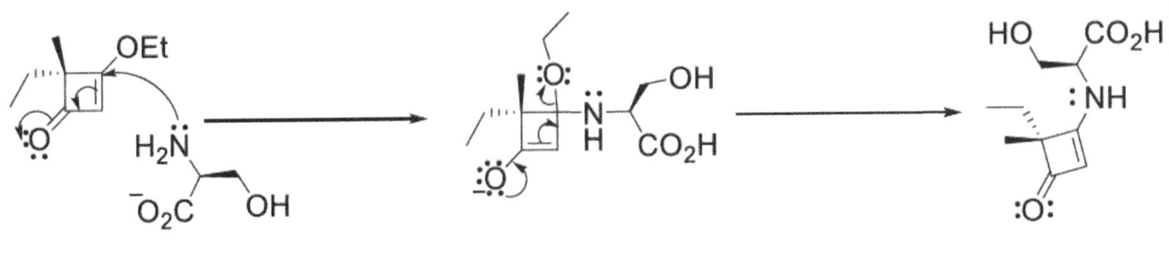

387

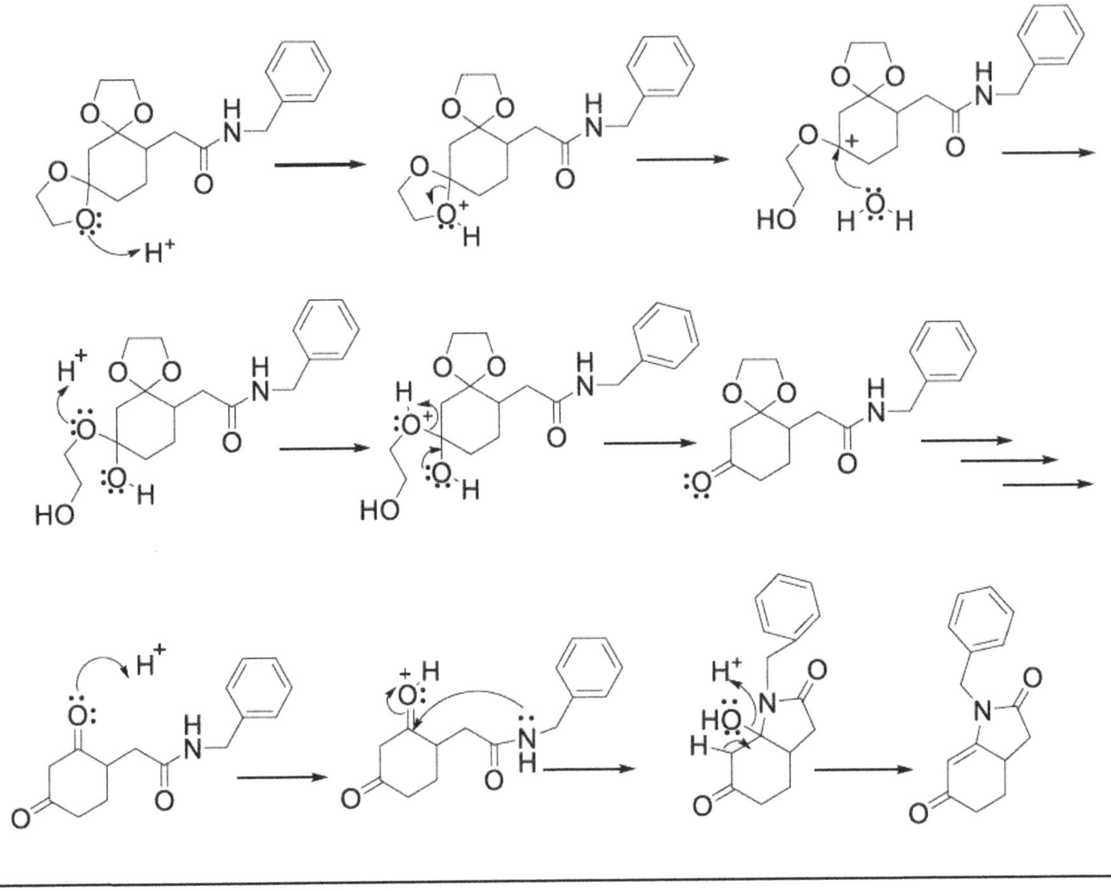

388

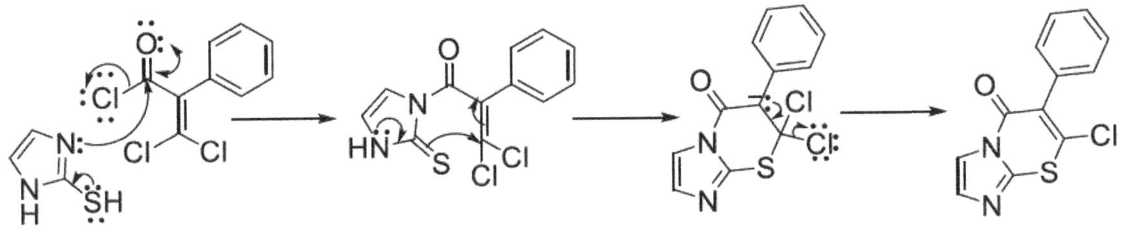

389

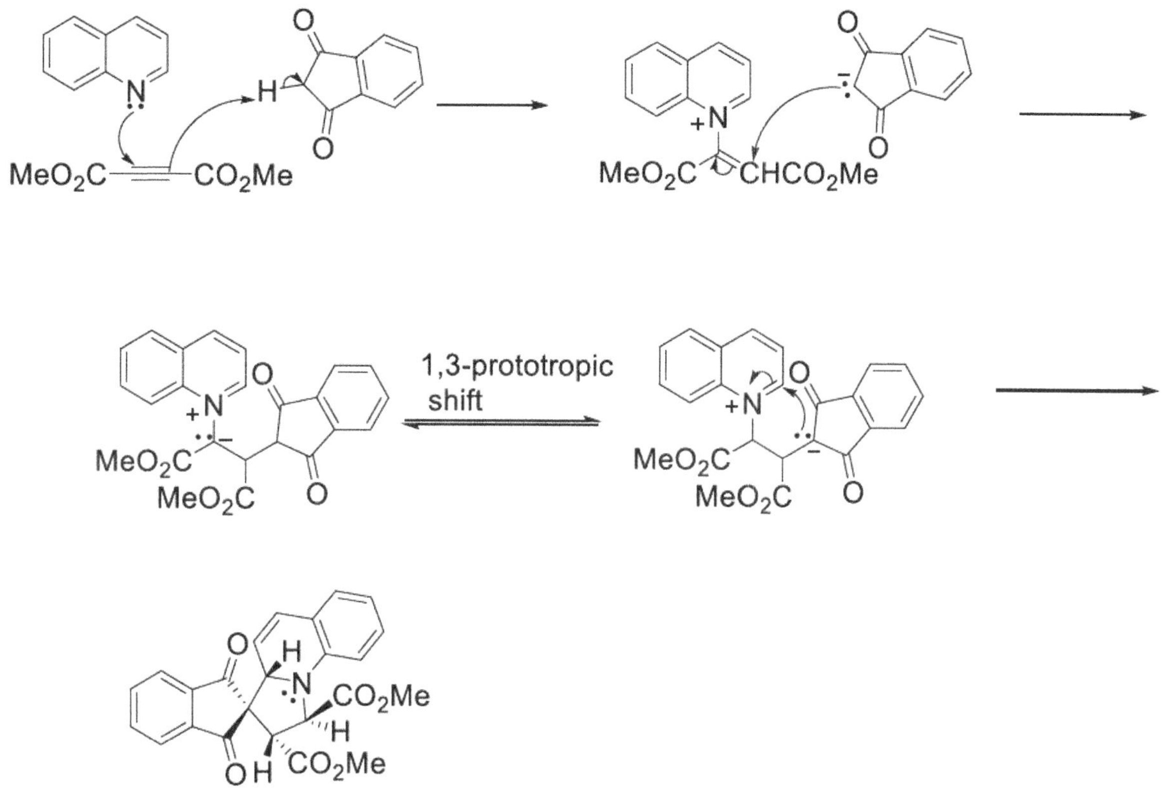
1,3-prototropic shift

390

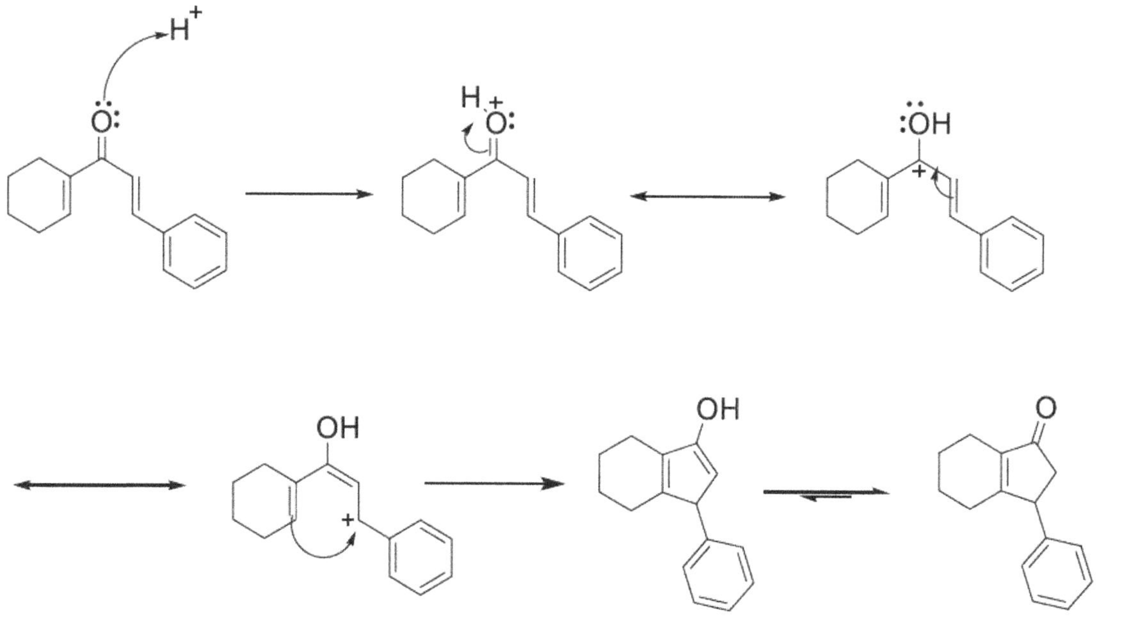

391

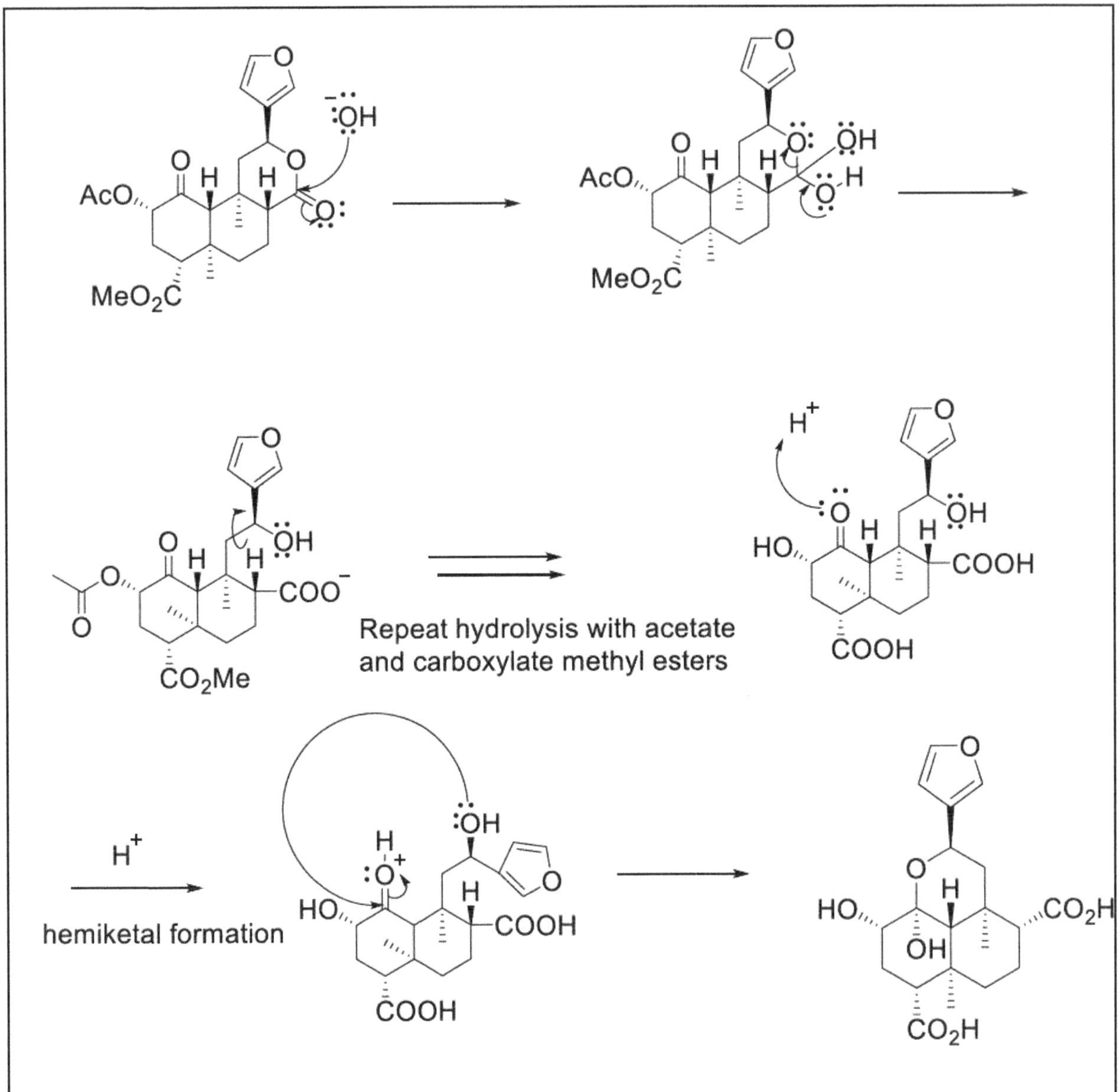
Repeat hydrolysis with acetate and carboxylate methyl esters
hemiketal formation

392

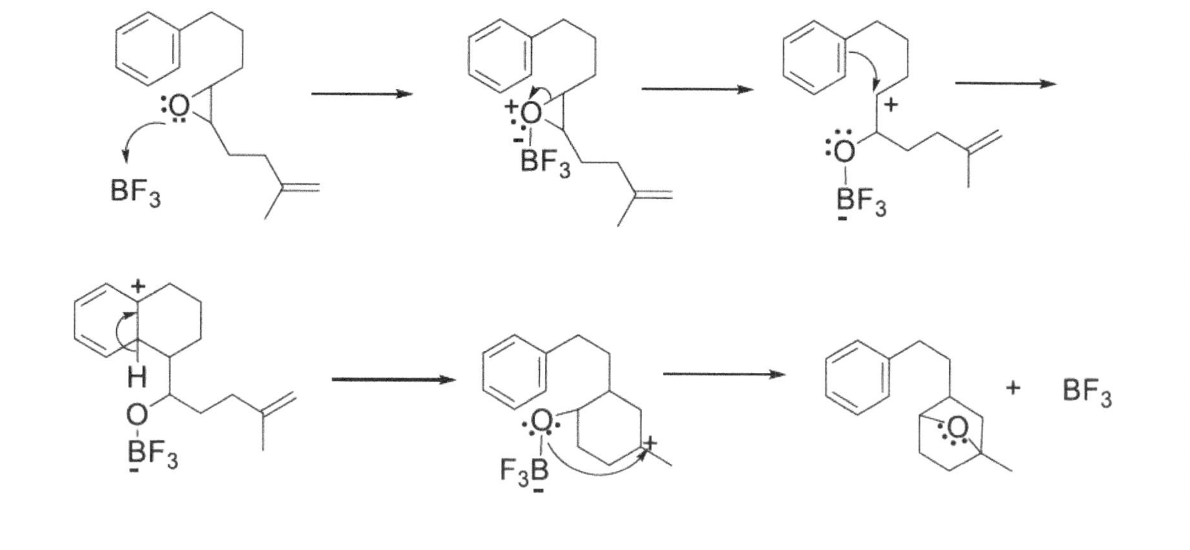
BF3
BF3
BF3
BF3
BF3
F3B
BF3

393

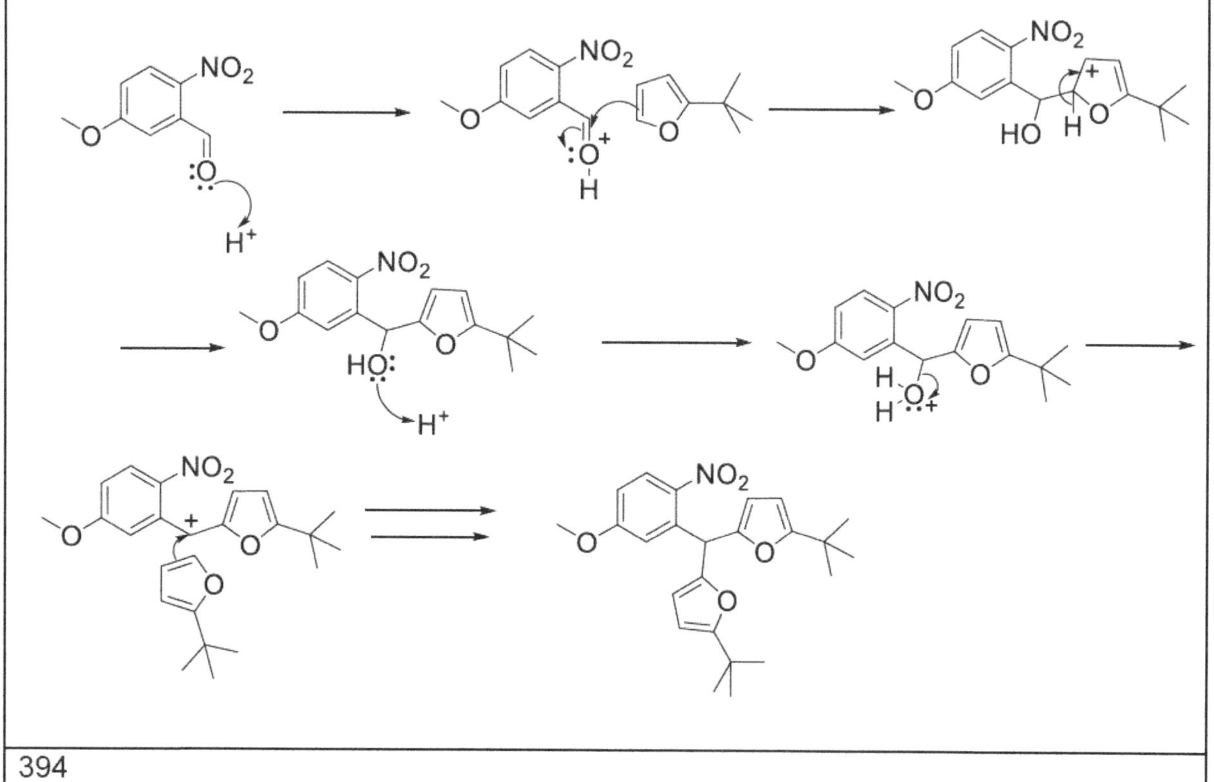
NO2
H+
HO
H
HO
H+
H
H

394

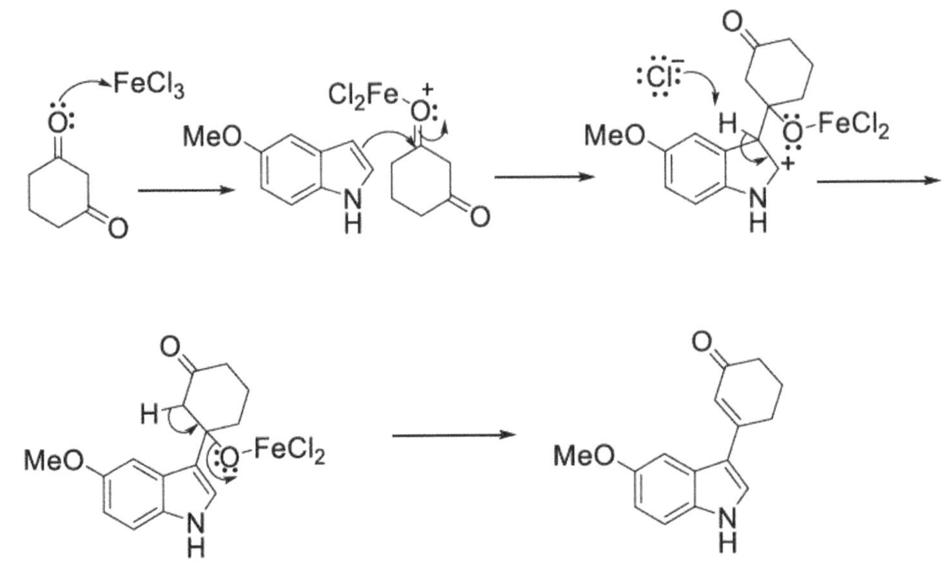
FeCl$_3$
MeO
N
H
Cl$_2$Fe
O
Cl
O-FeCl$_2$
H
O-FeCl$_2$

395

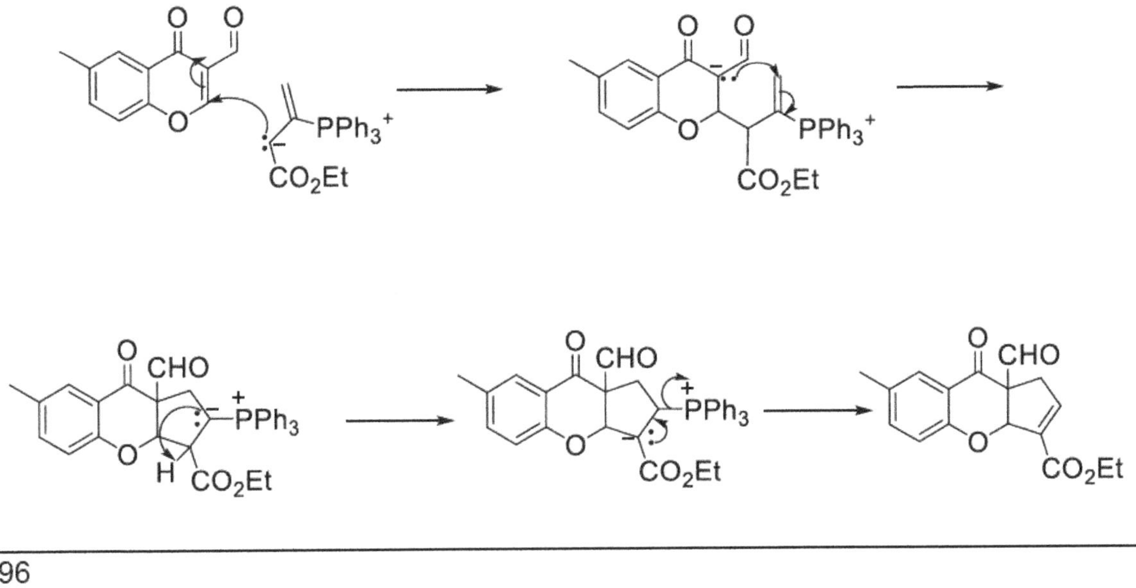
O
PPh$_3^+$
CO$_2$Et
CHO
PPh$_3$
H

396

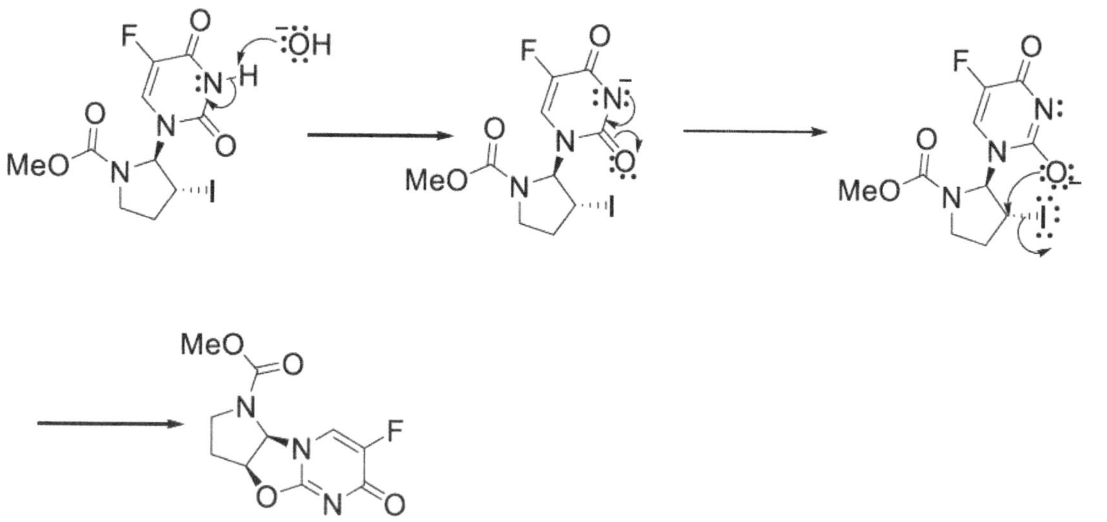

397

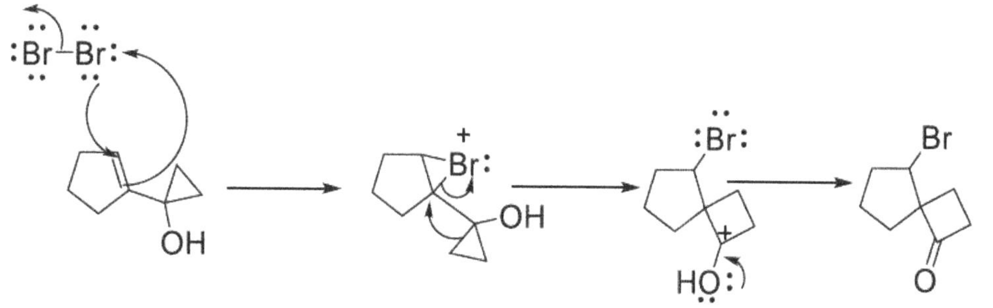

398

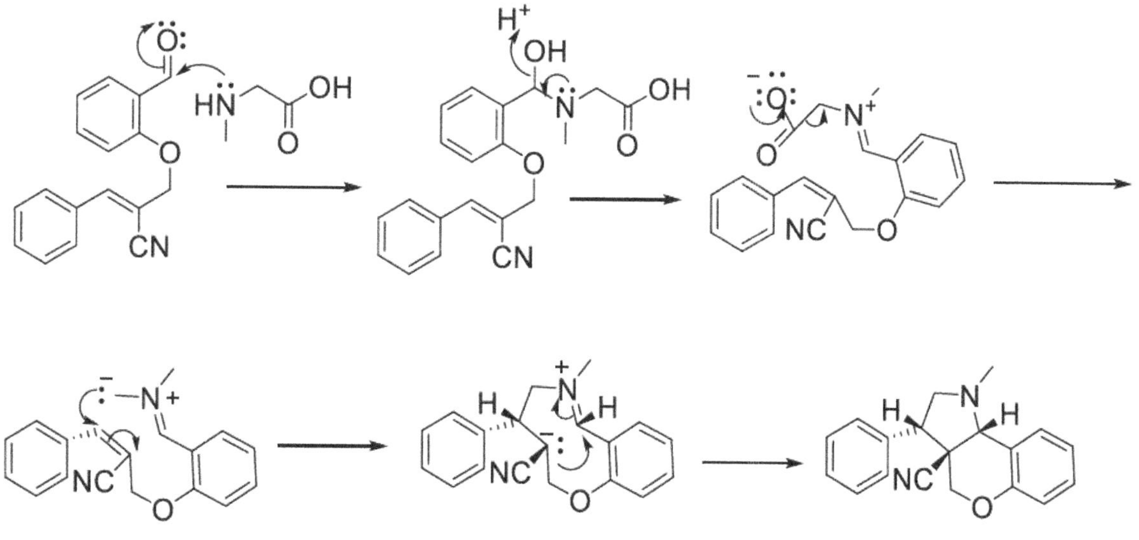

399

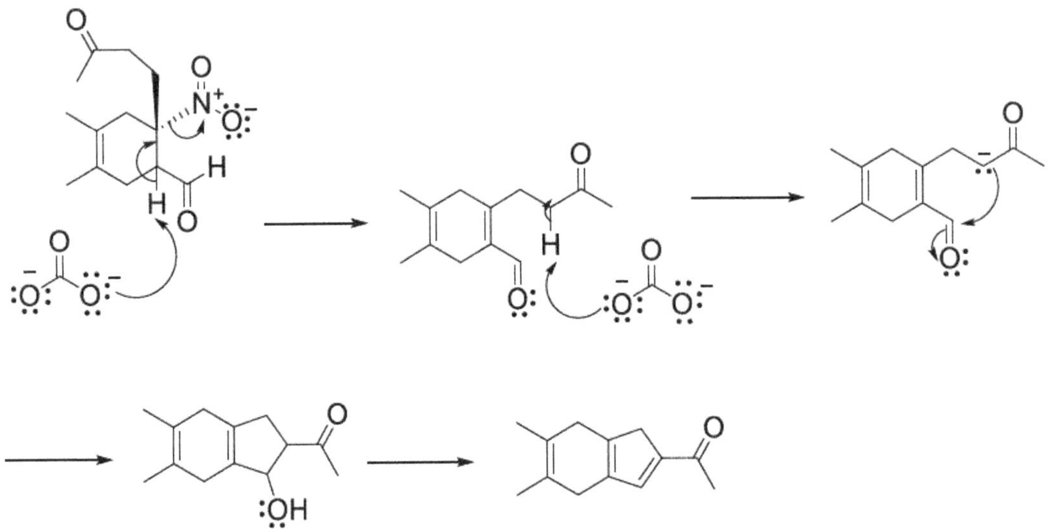

400

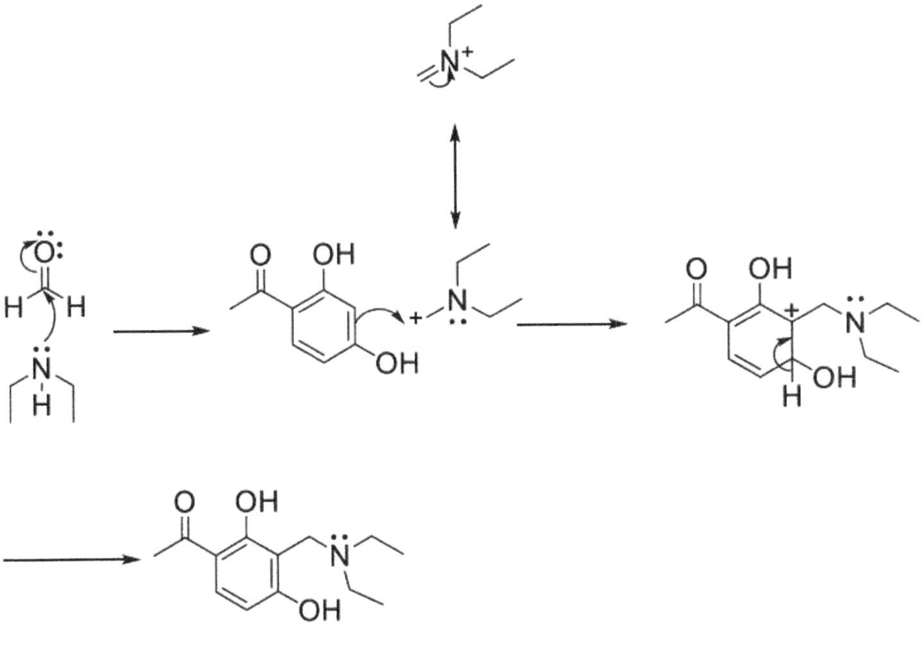

This is an example of the *Mannich reaction* (an aminoalkyl fragment introduced directly onto an aromatic ring)

401

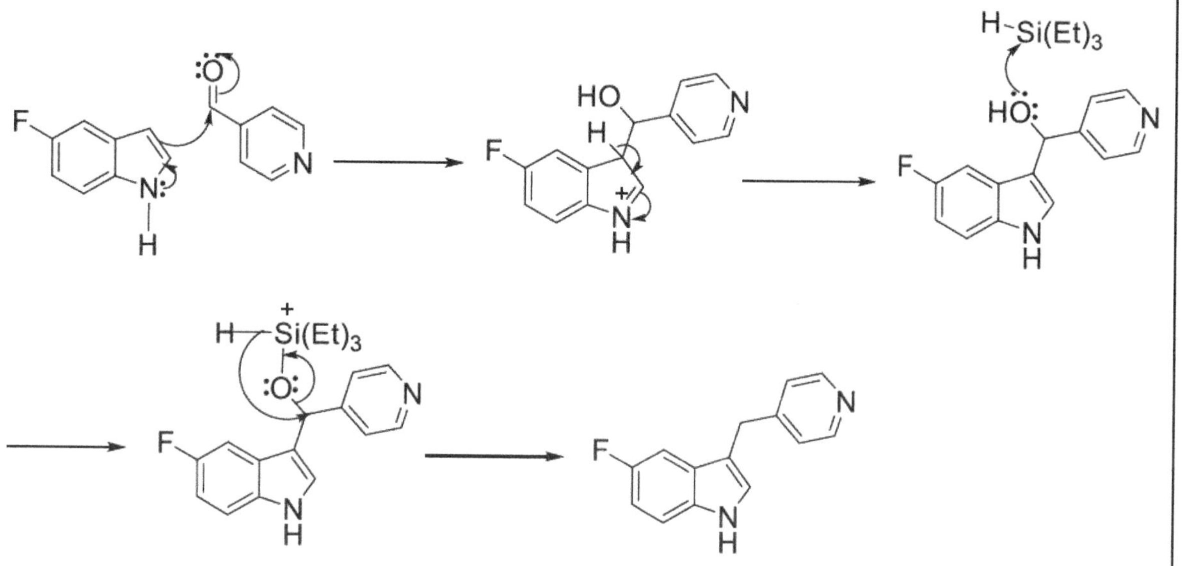
F
N
H
O
N
HO
H
Si(Et)3

402

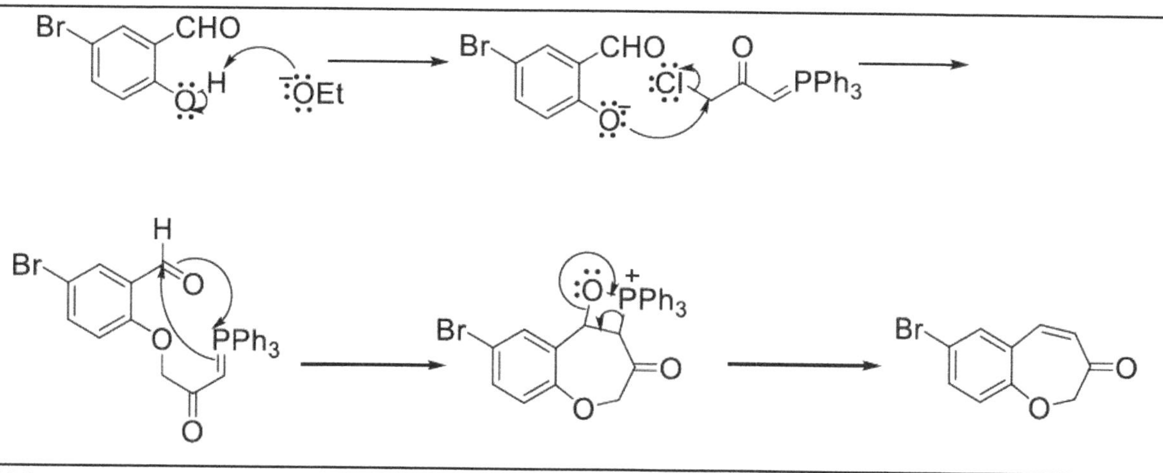
Br
CHO
H
OEt
Cl
O
PPh3
H
O
PPh3

403

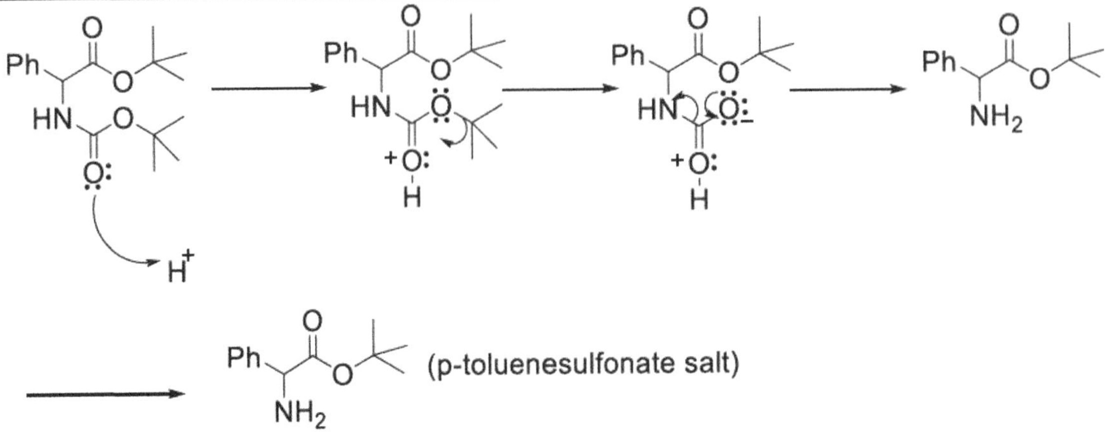
(p-toluenesulfonate salt)

<u>Exercise</u> Provide an explanation for the observed selectivity

404

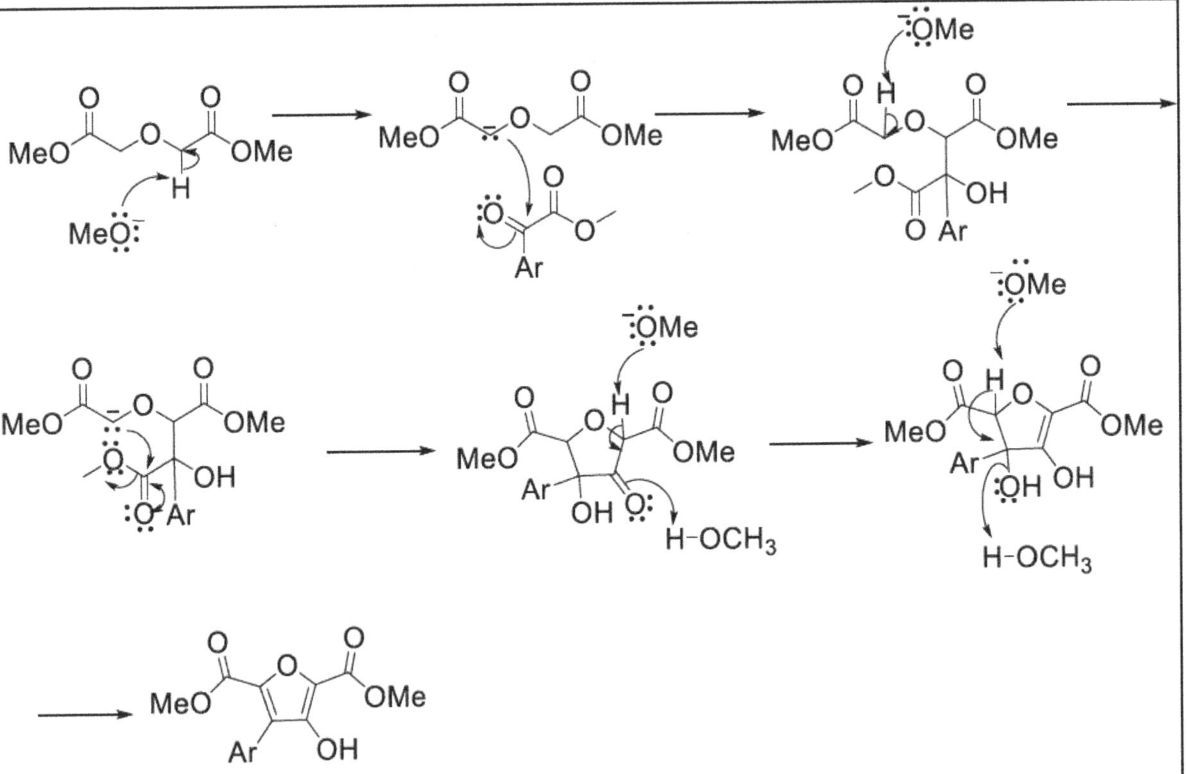

405

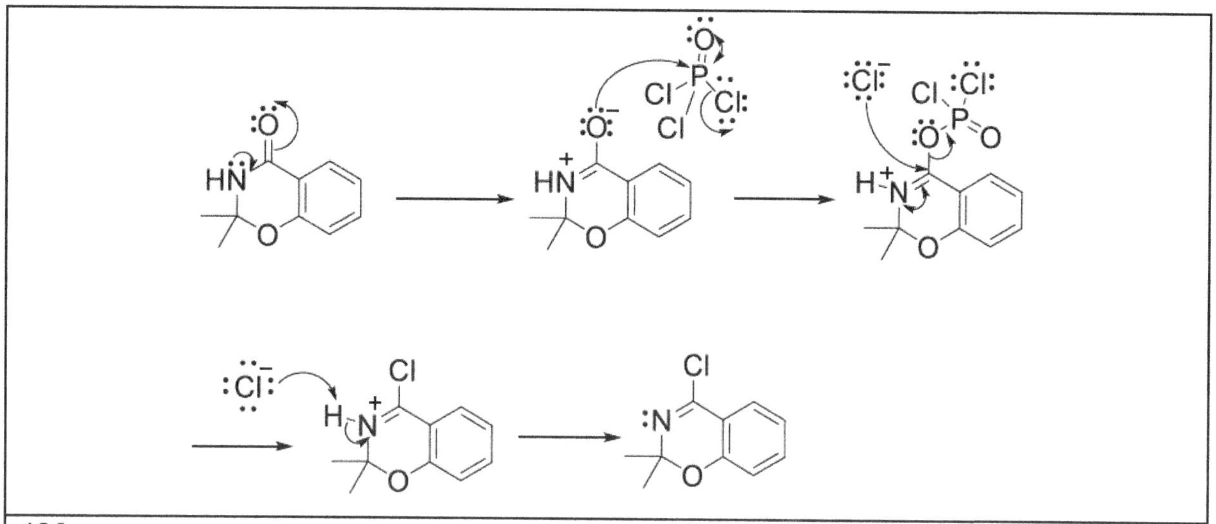

406

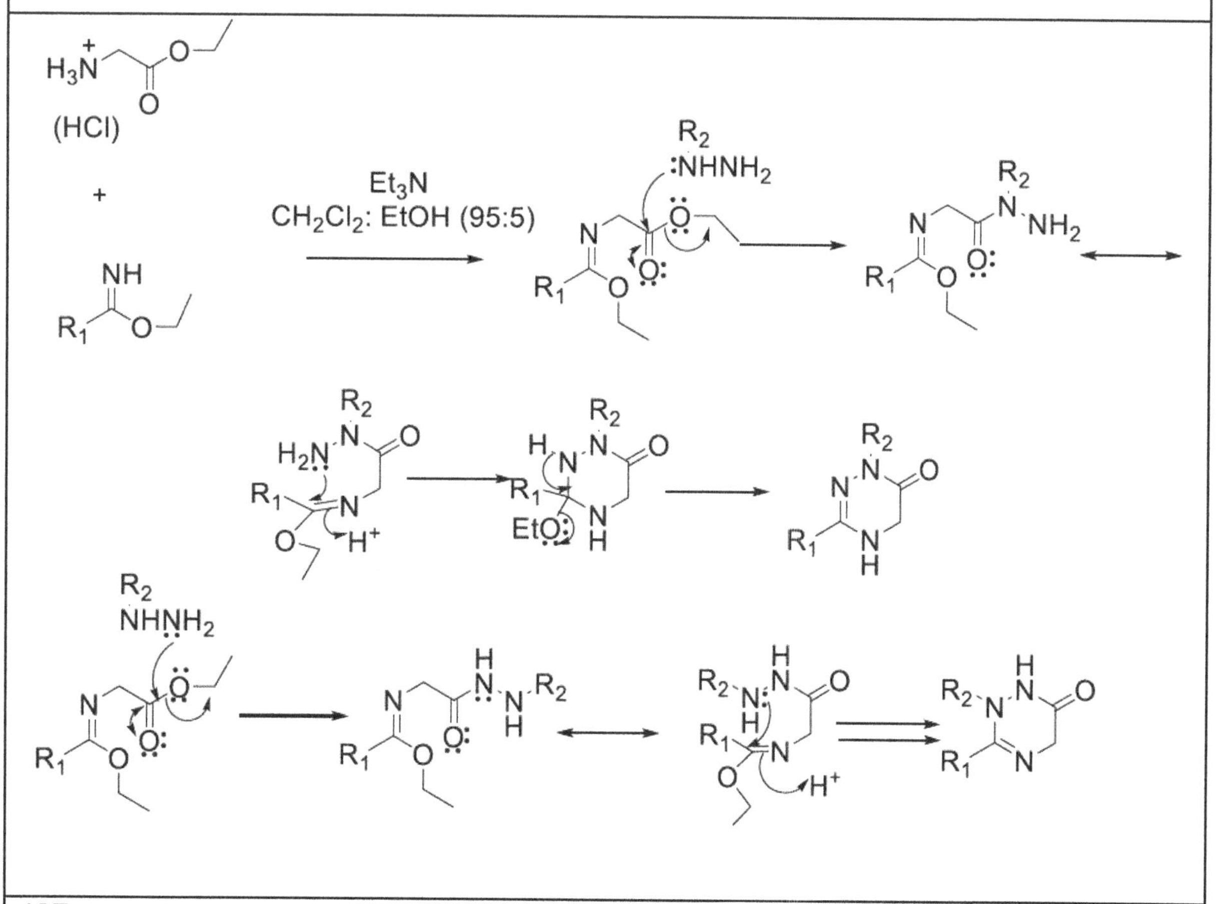
(HCl)
Et$_3$N
CH$_2$Cl$_2$: EtOH (95:5)

407

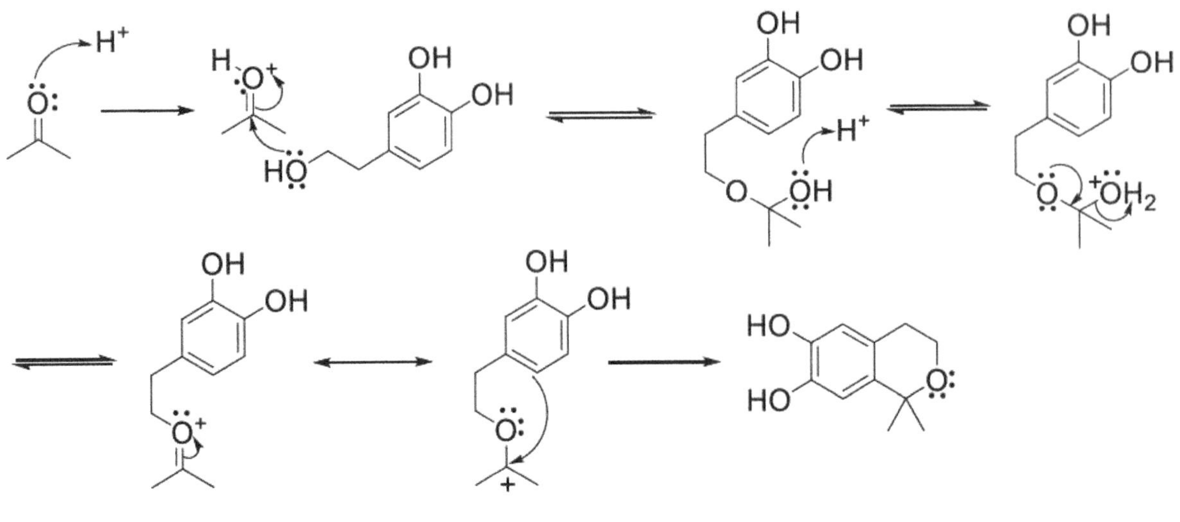

408

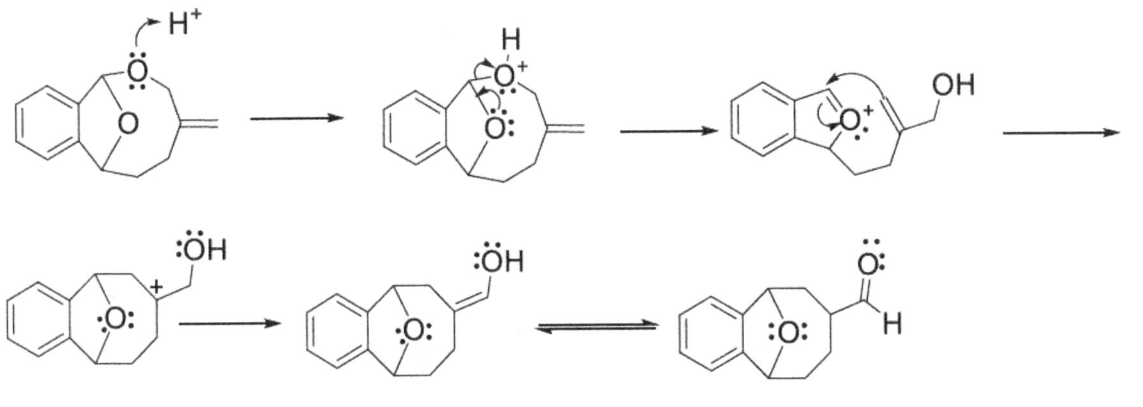

409

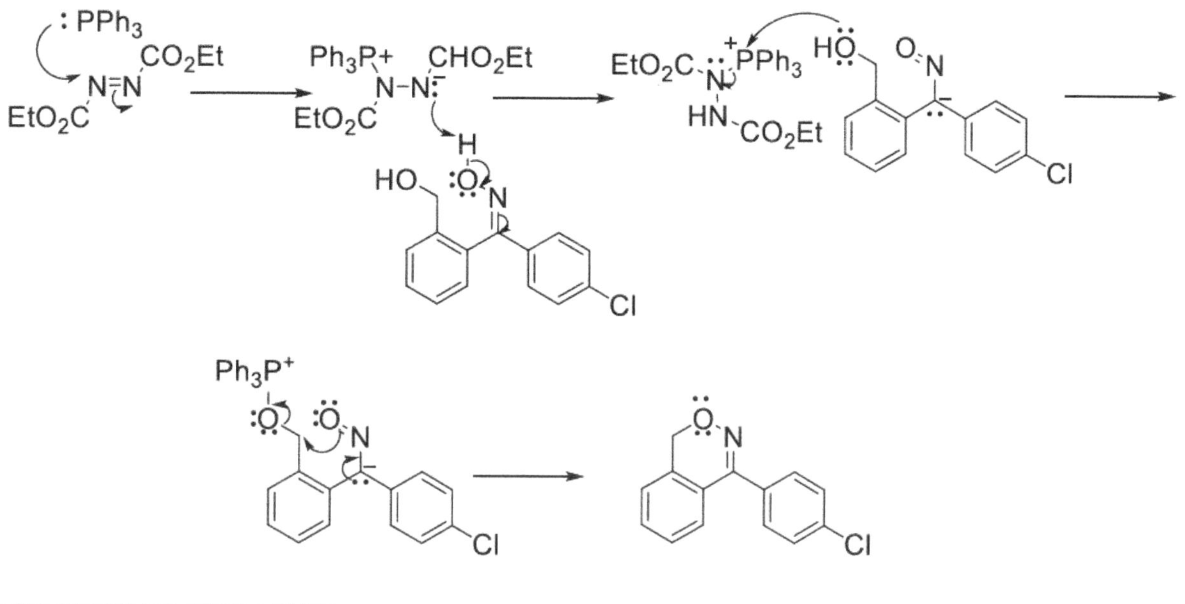

410

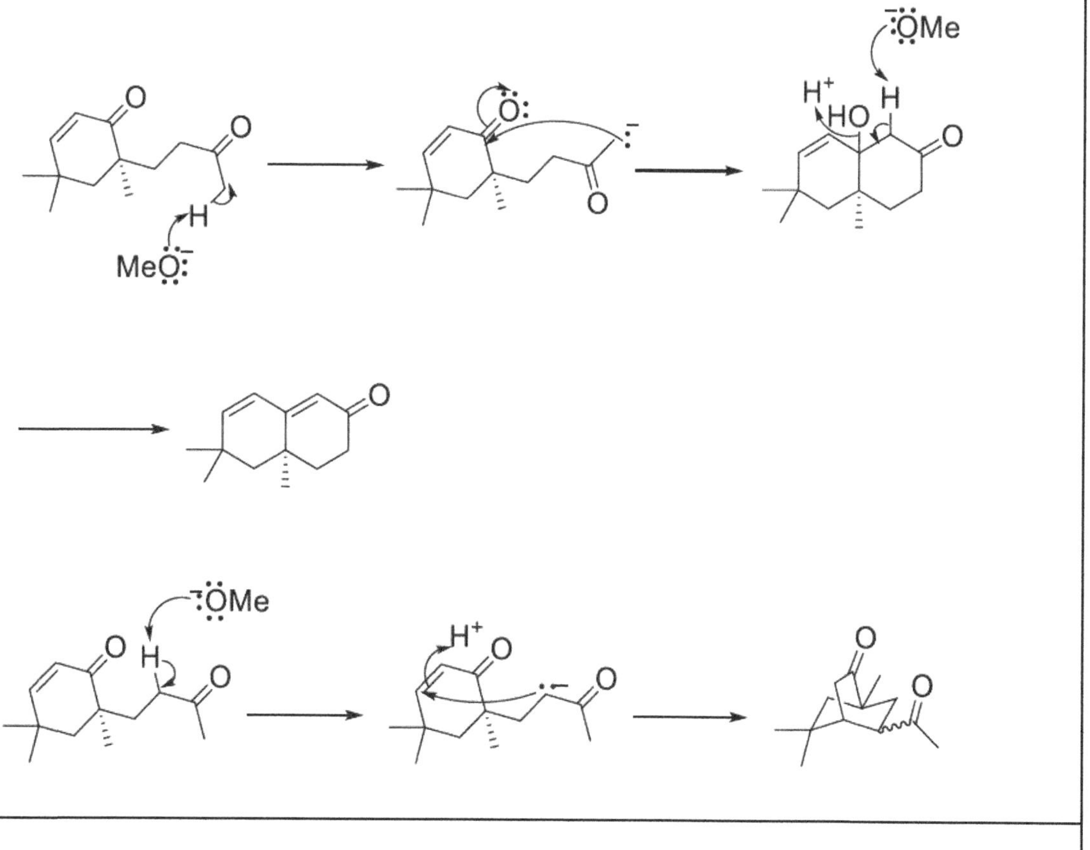

411

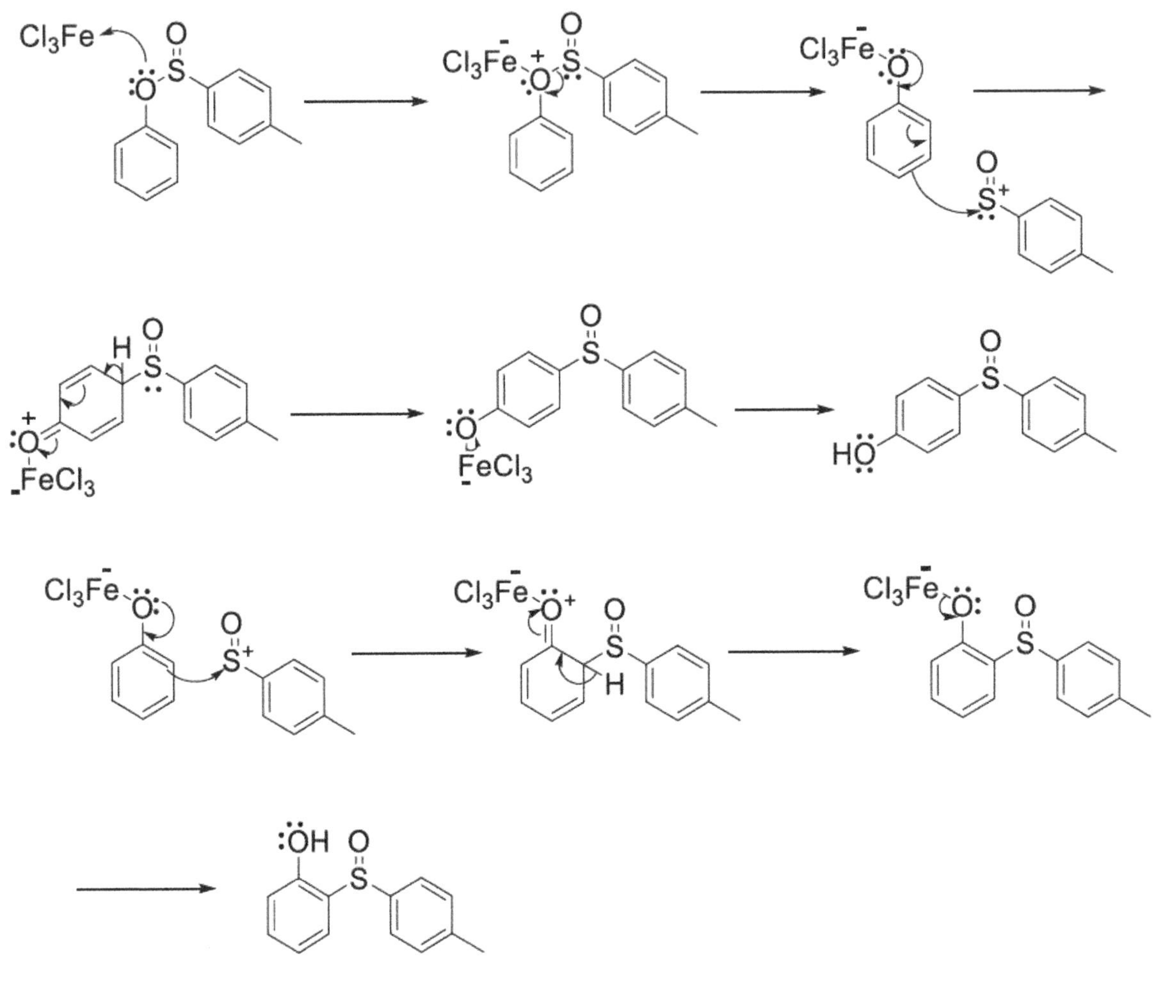
Cl3Fe
Cl3Fe
Cl3Fe
FeCl3
FeCl3
HO
Cl3Fe
Cl3Fe
Cl3Fe
OH

412

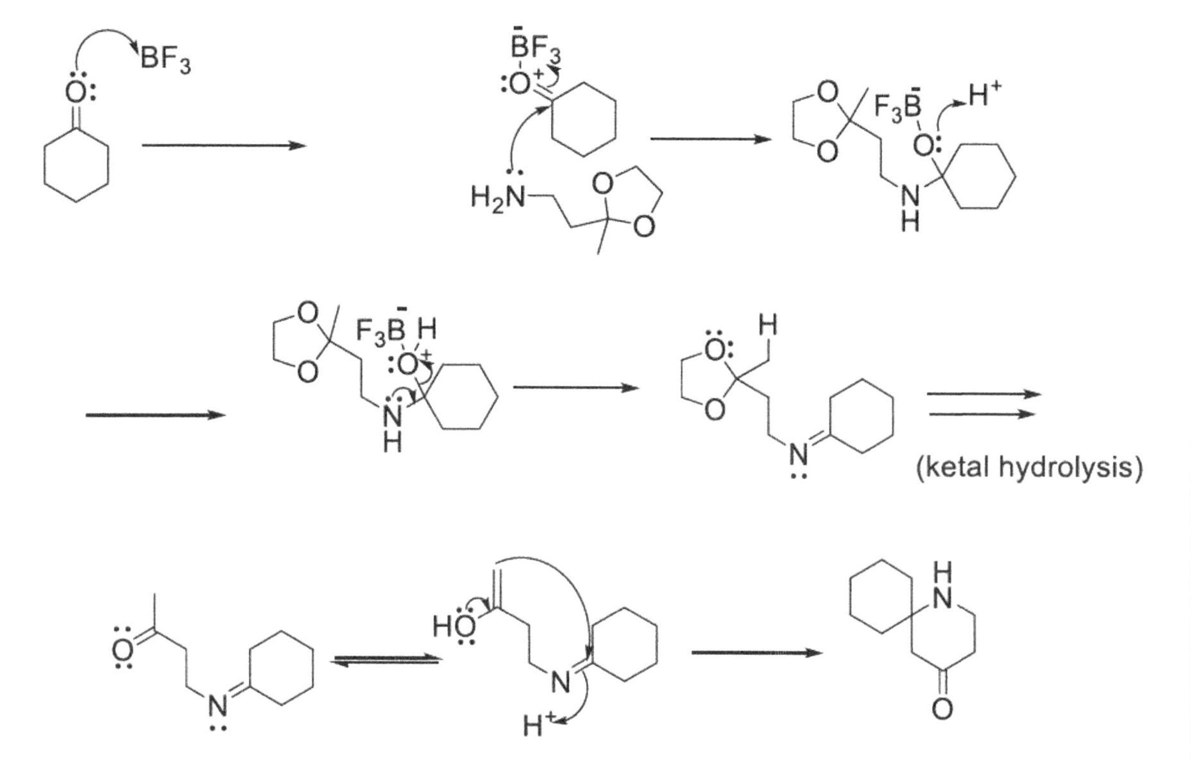
BF3
(ketal hydrolysis)
NHCbz

413

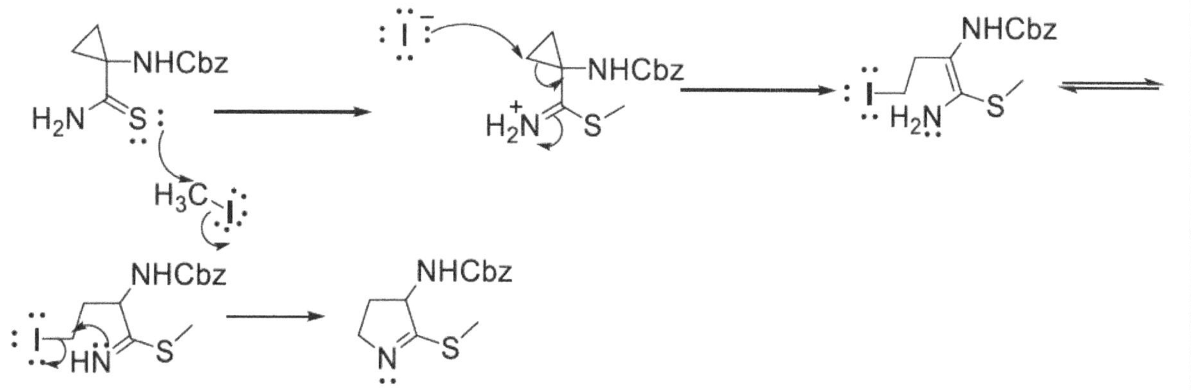
NHCbz

414

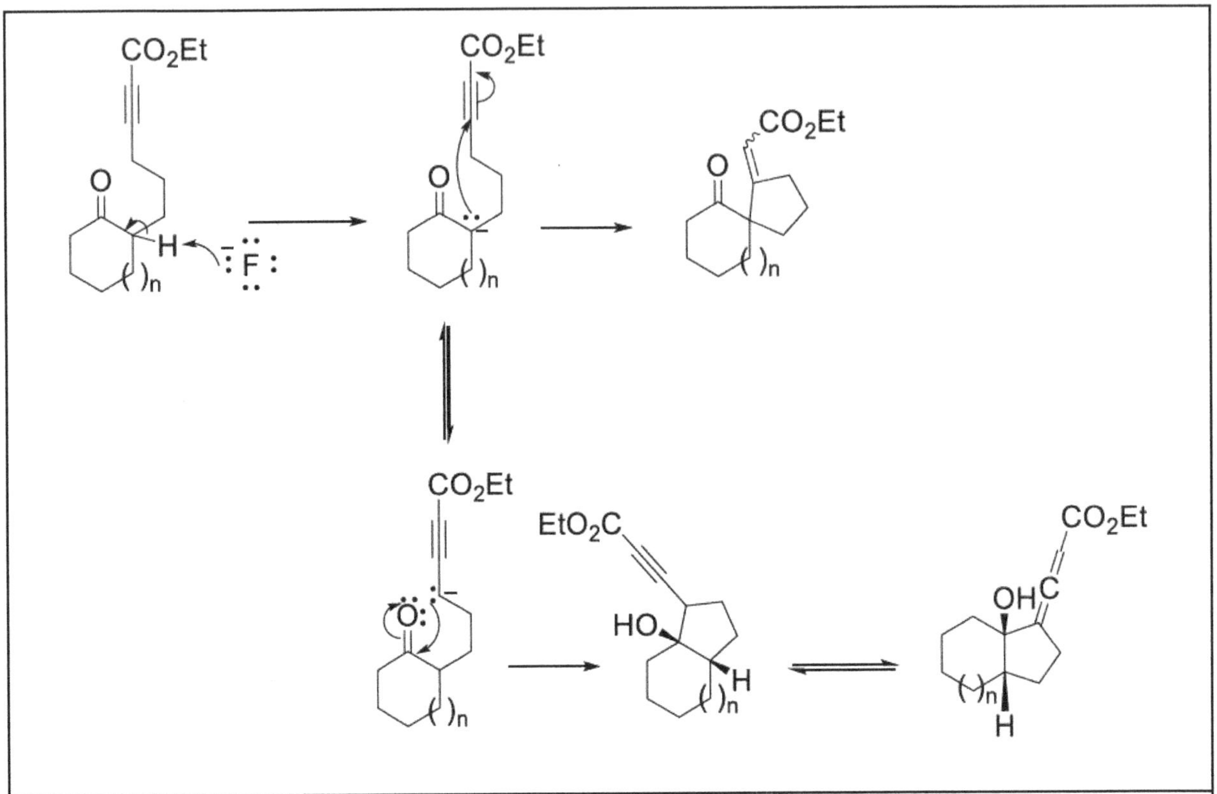
CO2Et
O
H
F
CO2Et
O
( )n
CO2Et
O
( )n
CO2Et
O
( )n
EtO2C
HO
H
( )n
CO2Et
OH
C
( )n
H

415

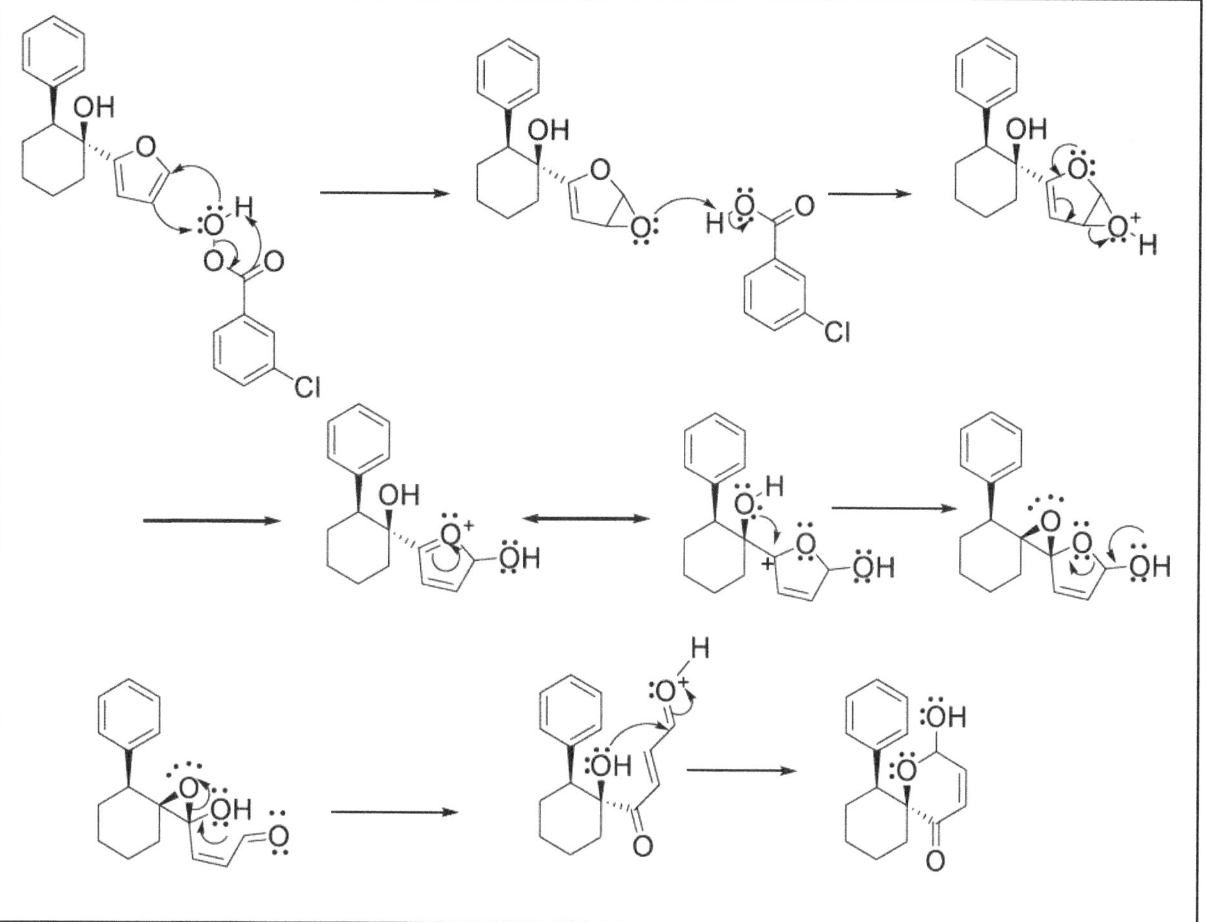

OH
O
H
O
O
Cl
OH
O
O
H
O
O
Cl
OH
O
O
H
OH
O
OH
O
H
O
OH
O
O
OH
O
OH
O
H
O
OH
O
OH
O
O

416

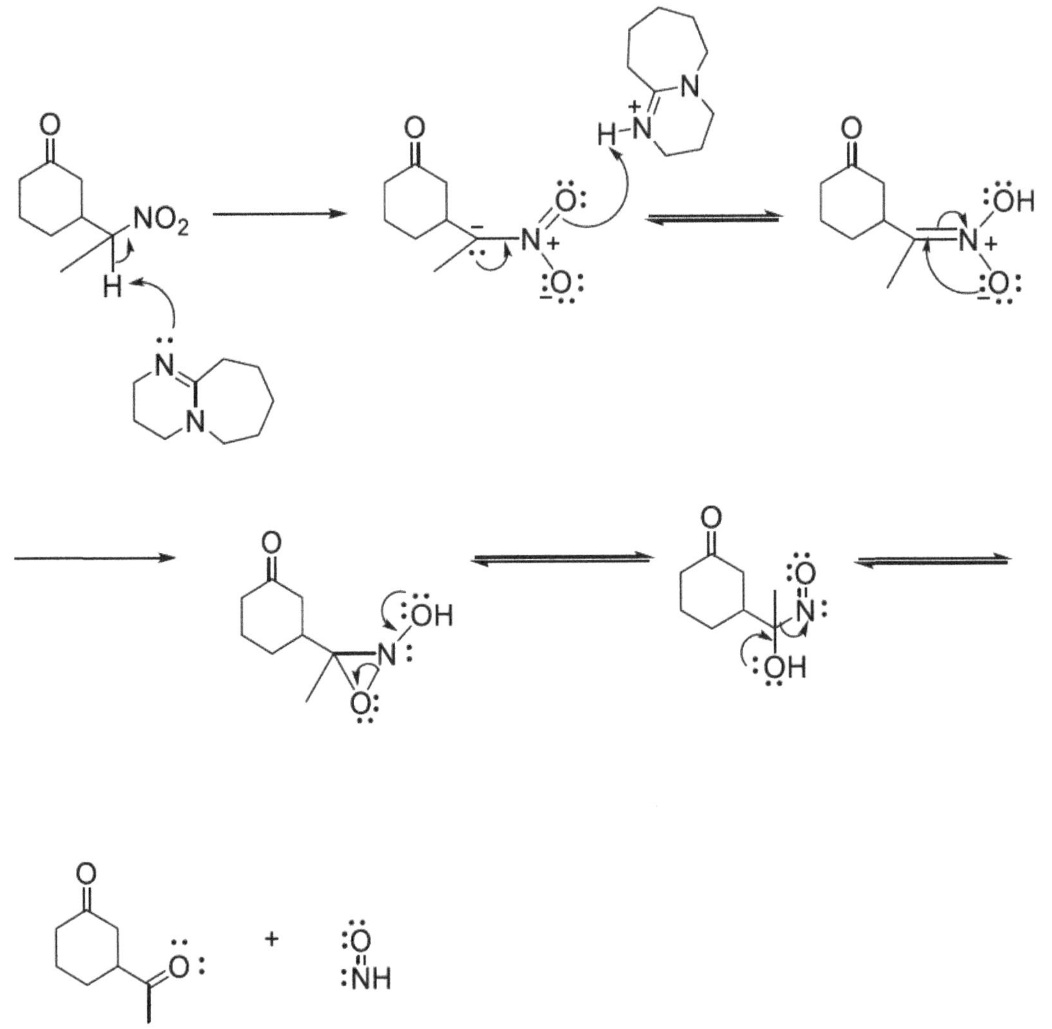
O
NO2
H
N
N
O
O
N+
O
N
N
+
H
O
OH
N+
O
OH
N
O
O
N
OH
O
O
+
O
NH

417

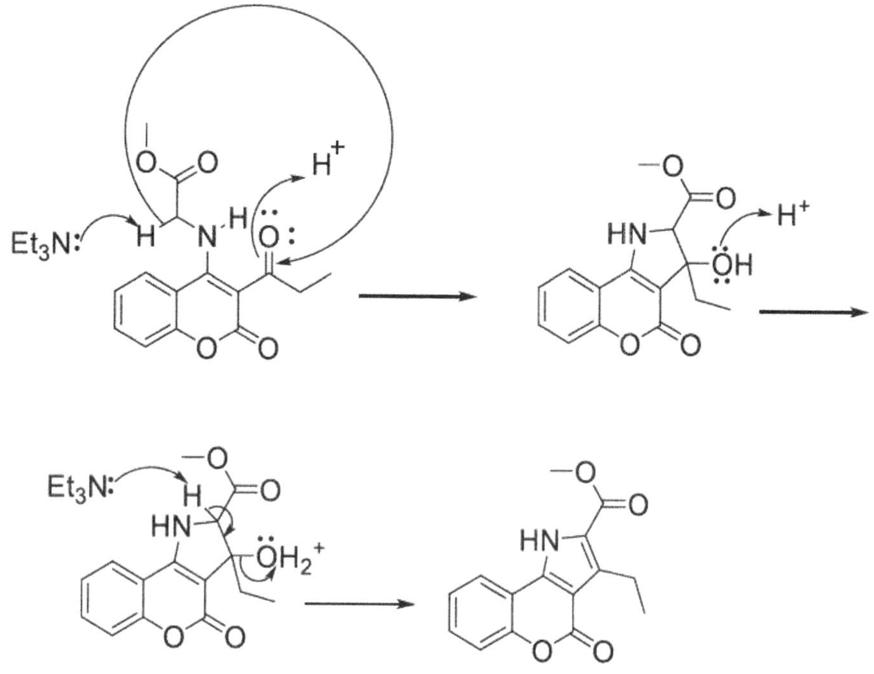

H+
Et3N:
HN
OH
OH2+

418

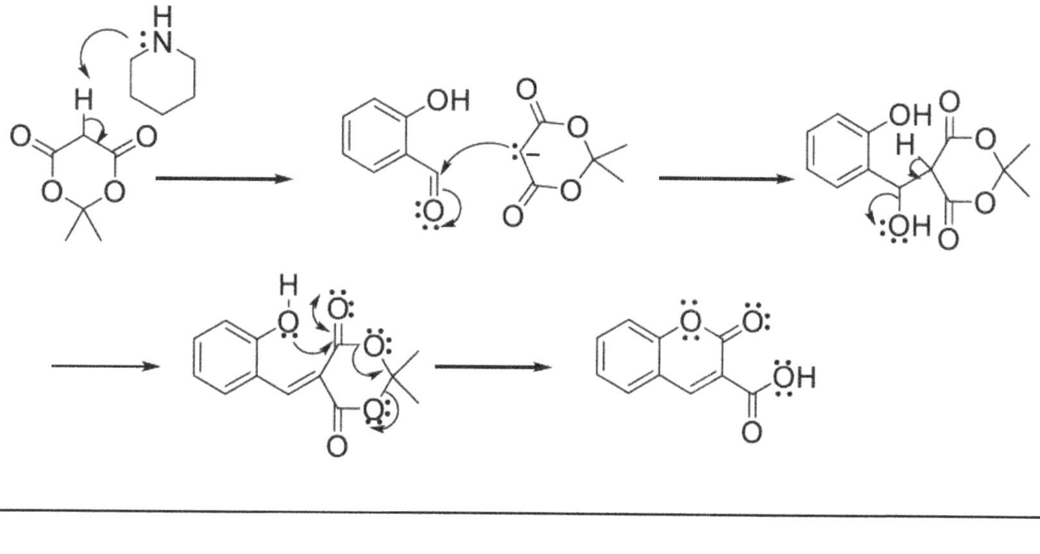

OH
H
OH

419

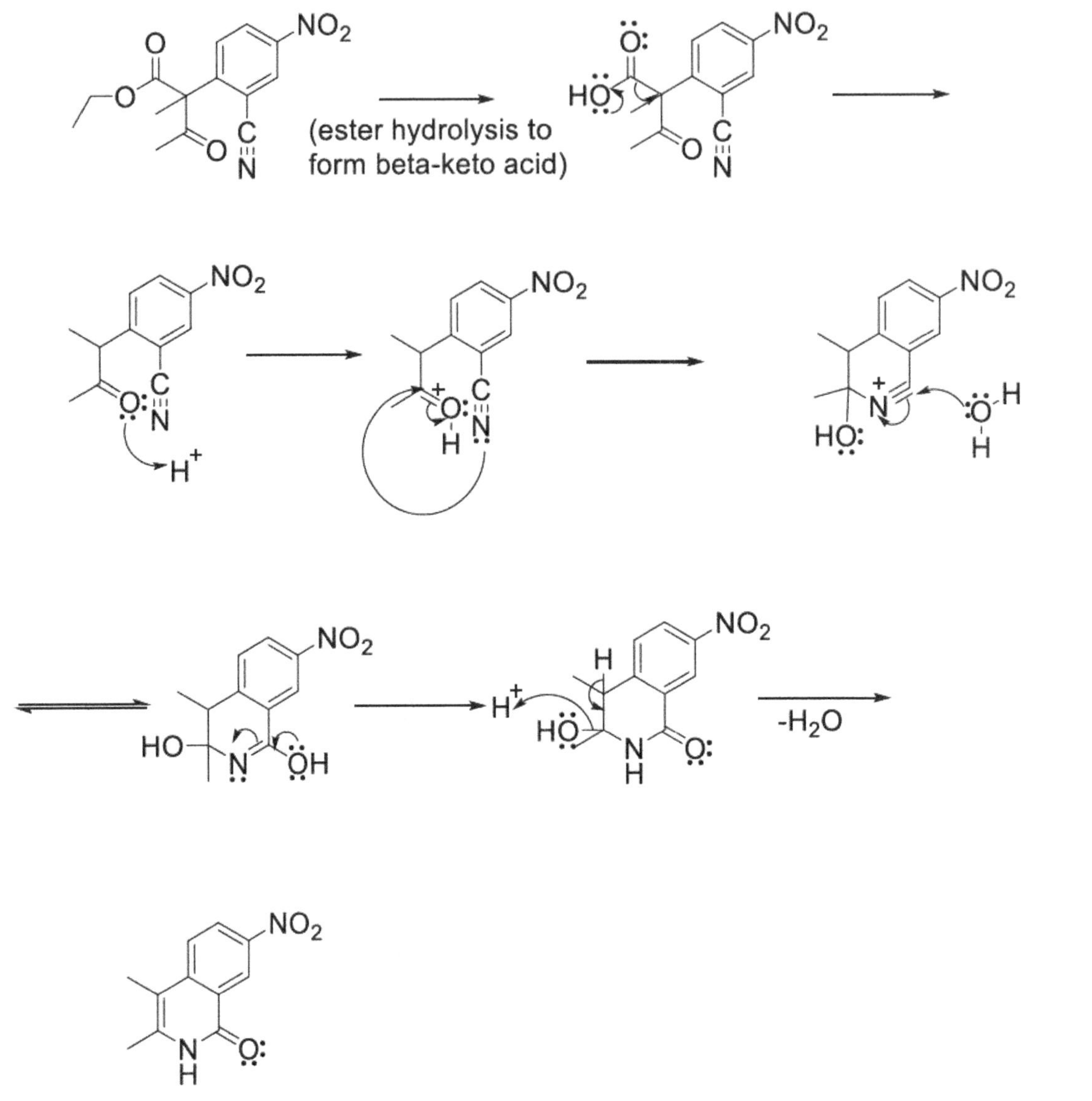

NO2
(ester hydrolysis to
form beta-keto acid)
HO
-H2O

420

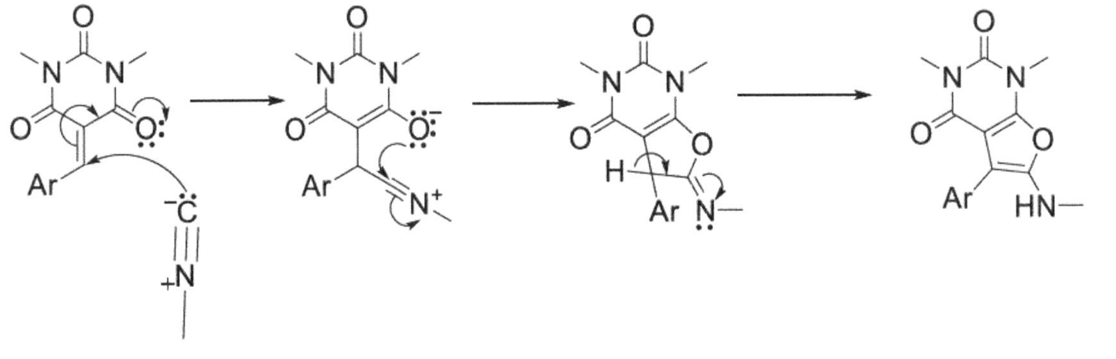

Ar
HN

421

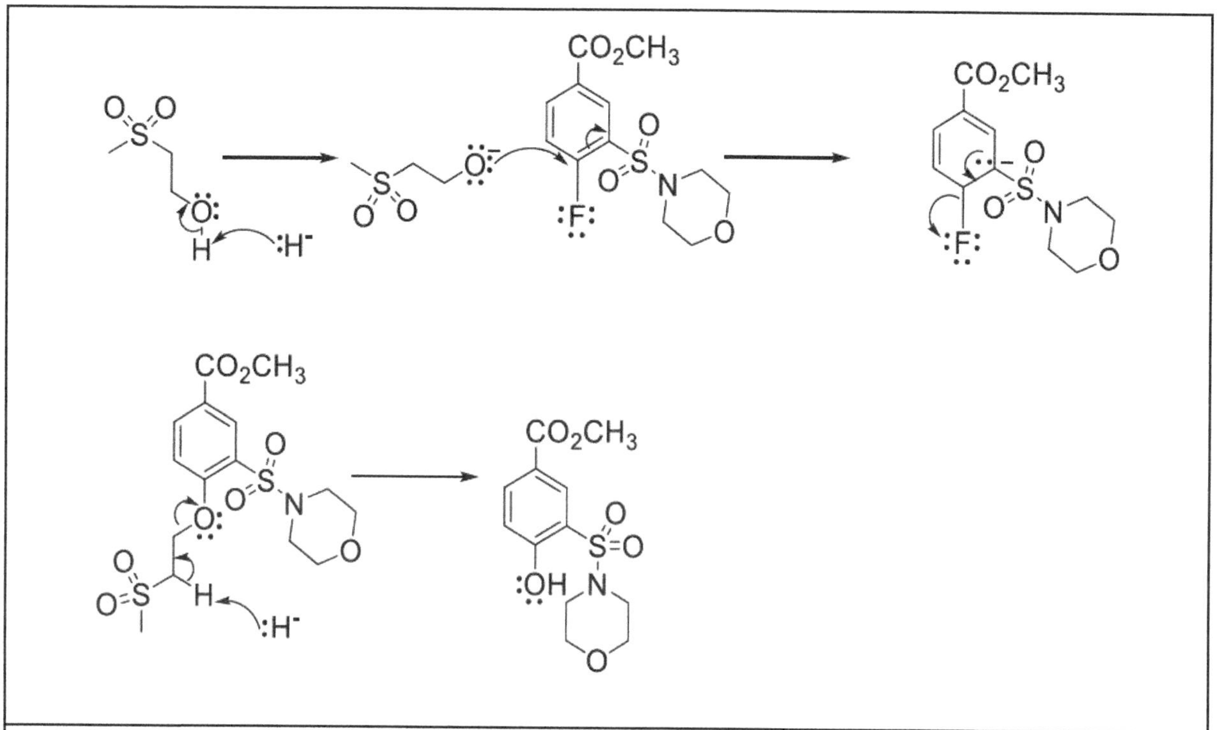
CO2CH3
:H-
CO2CH3
CO2CH3
CO2CH3
:H-
CO2CH3
:OH

422

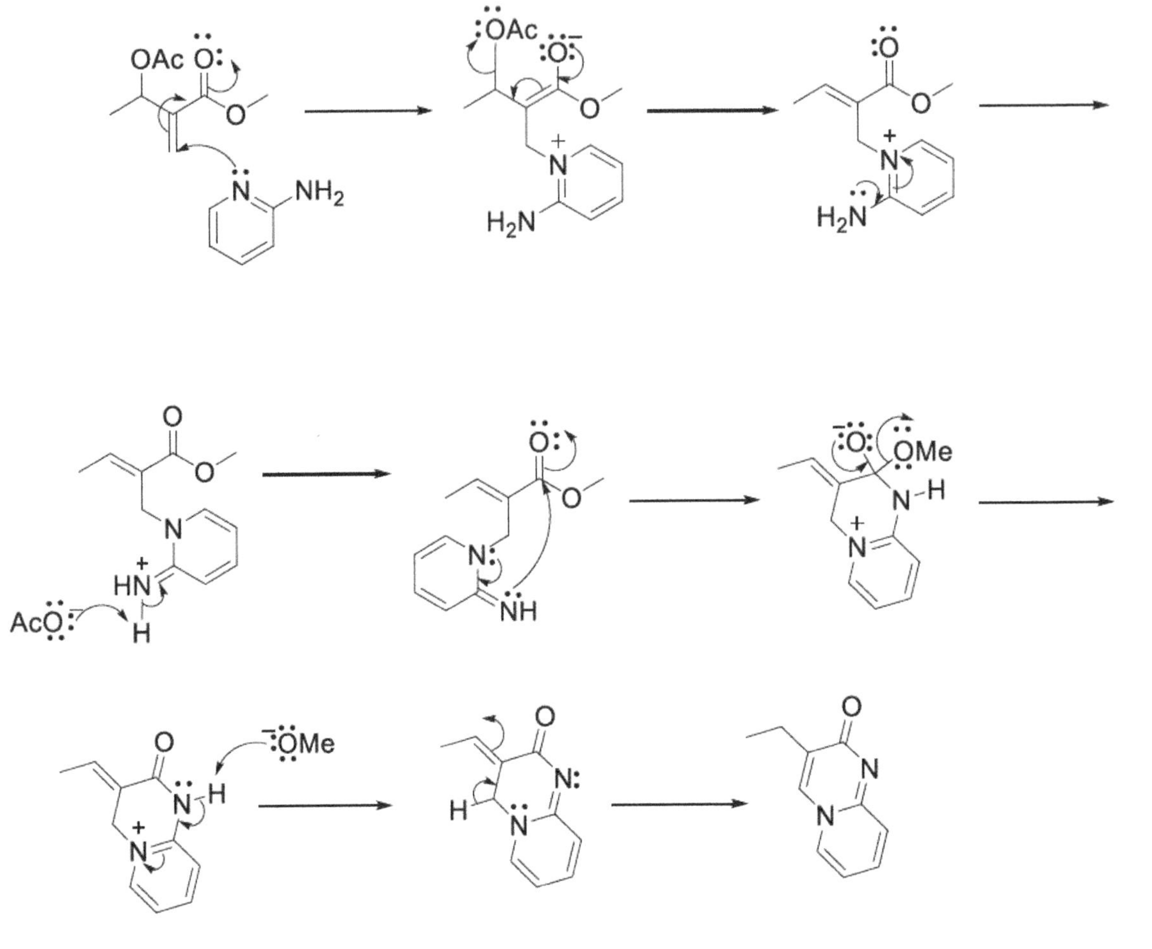

OAc
O
O
N
NH2
OAc
O
O
N
H2N
O
O
N
H2N
O
O
N
HN
AcO
H
O
O
N
NH
O
OMe
N-H
N
O
N-H
N
OMe
O
N
H
N
O
N
N

423

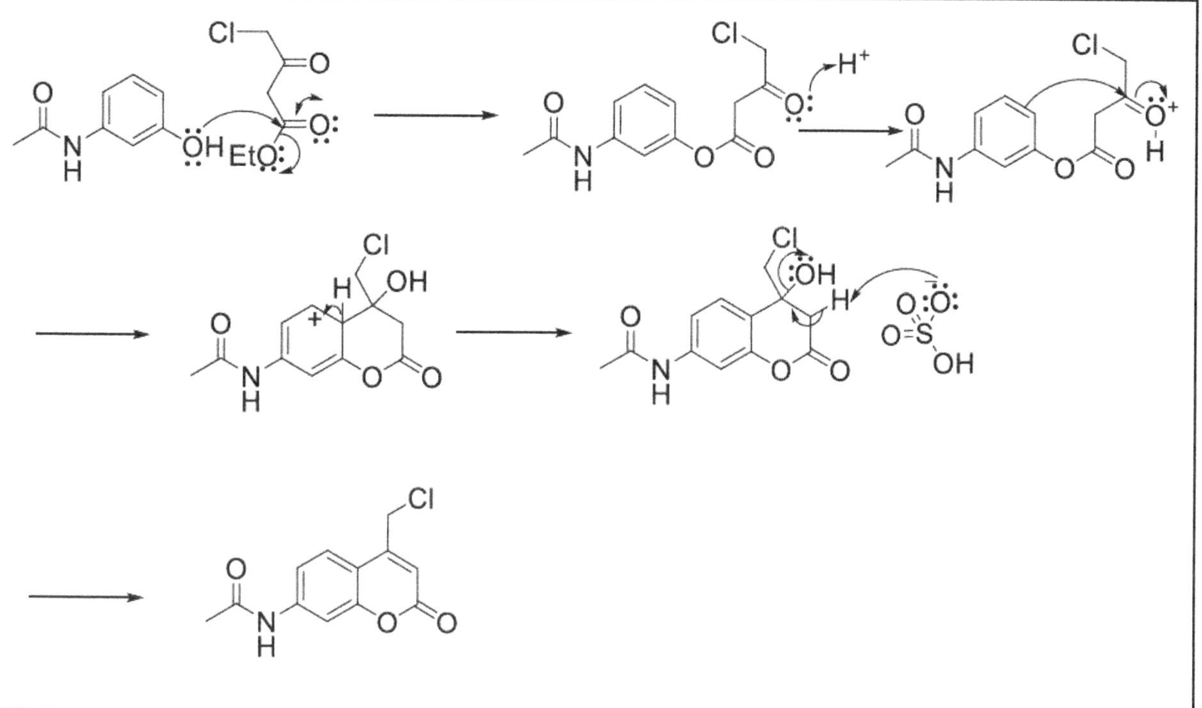

424

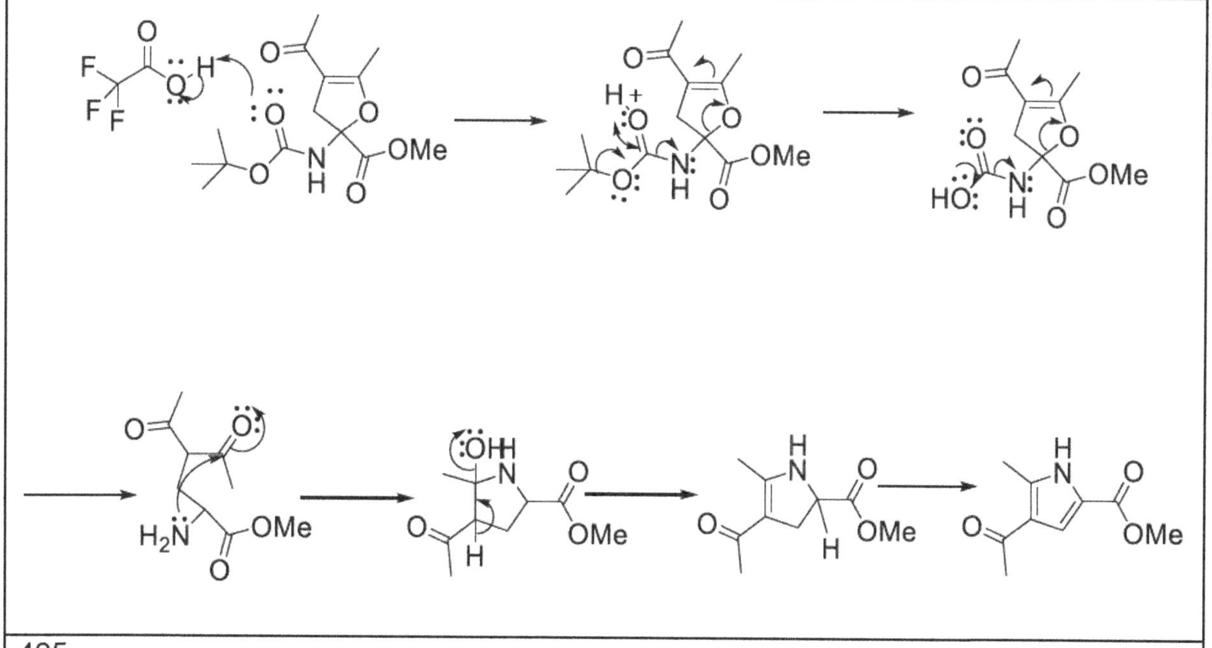

425

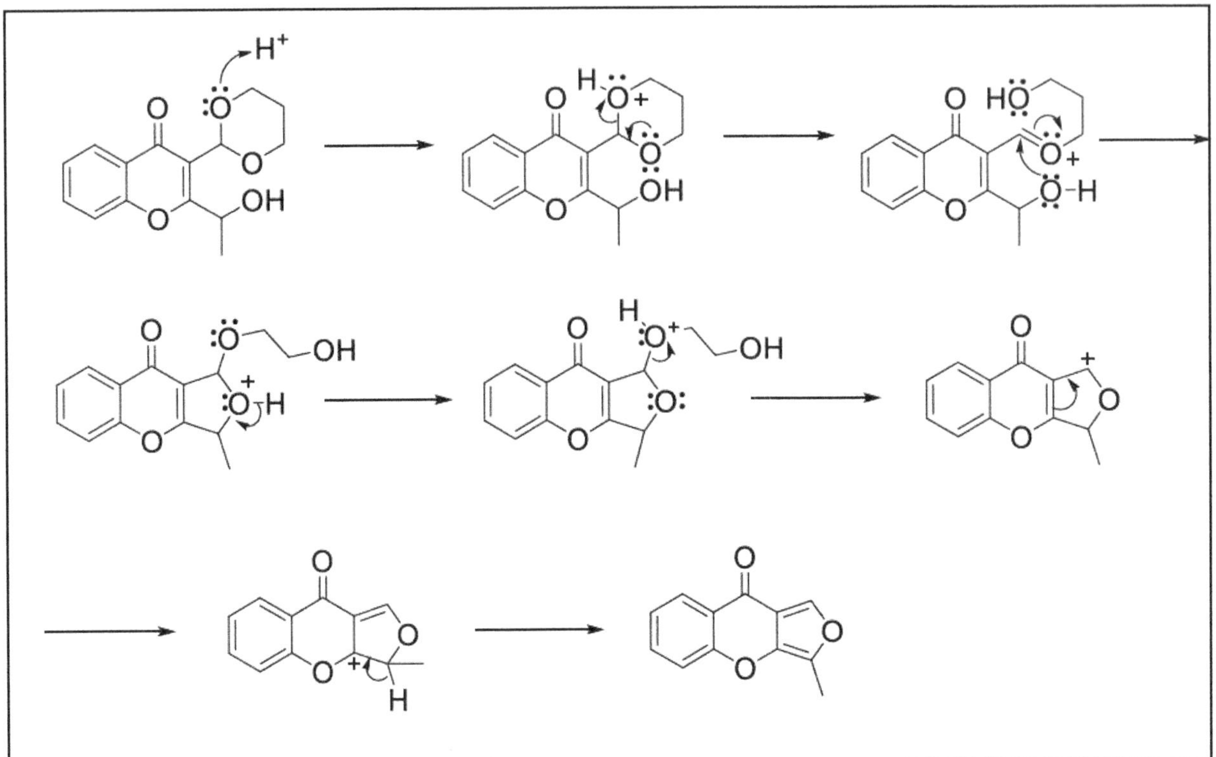

426

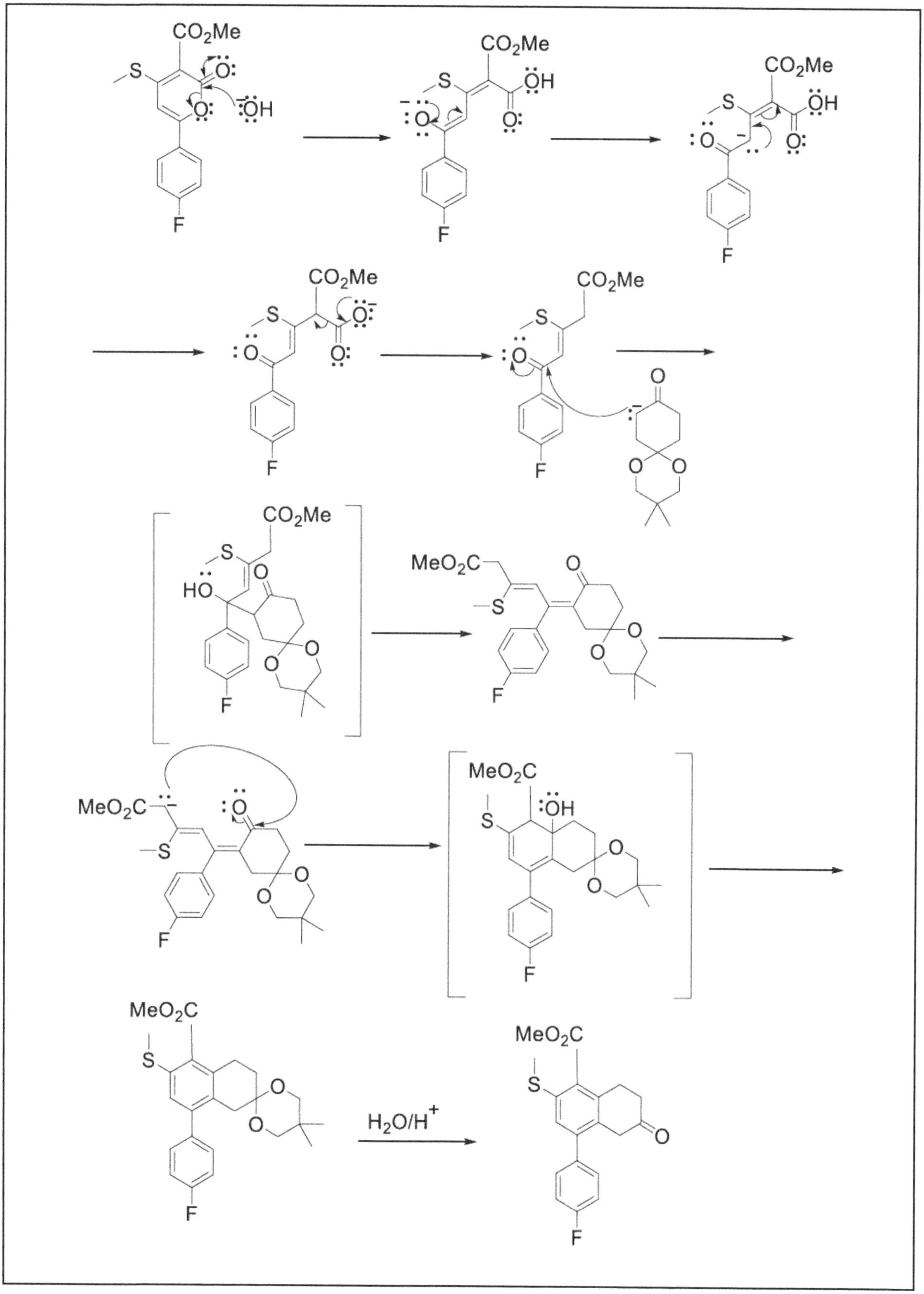

CO2Me
OH
F
H2O/H+
MeO2C
O
S

427

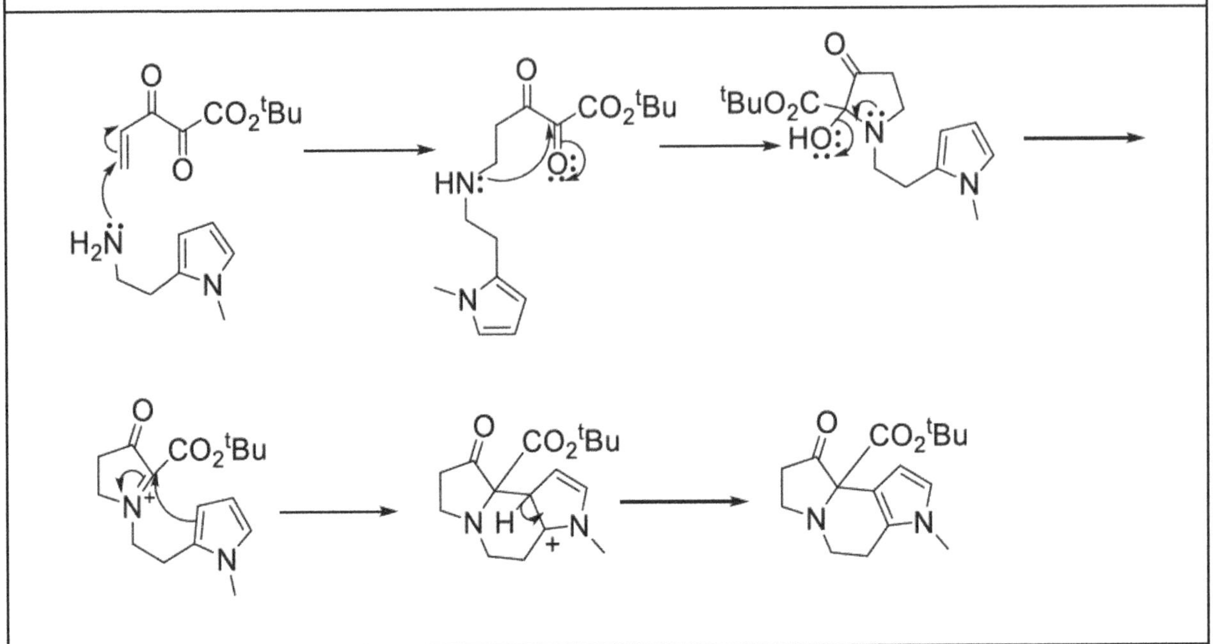

428

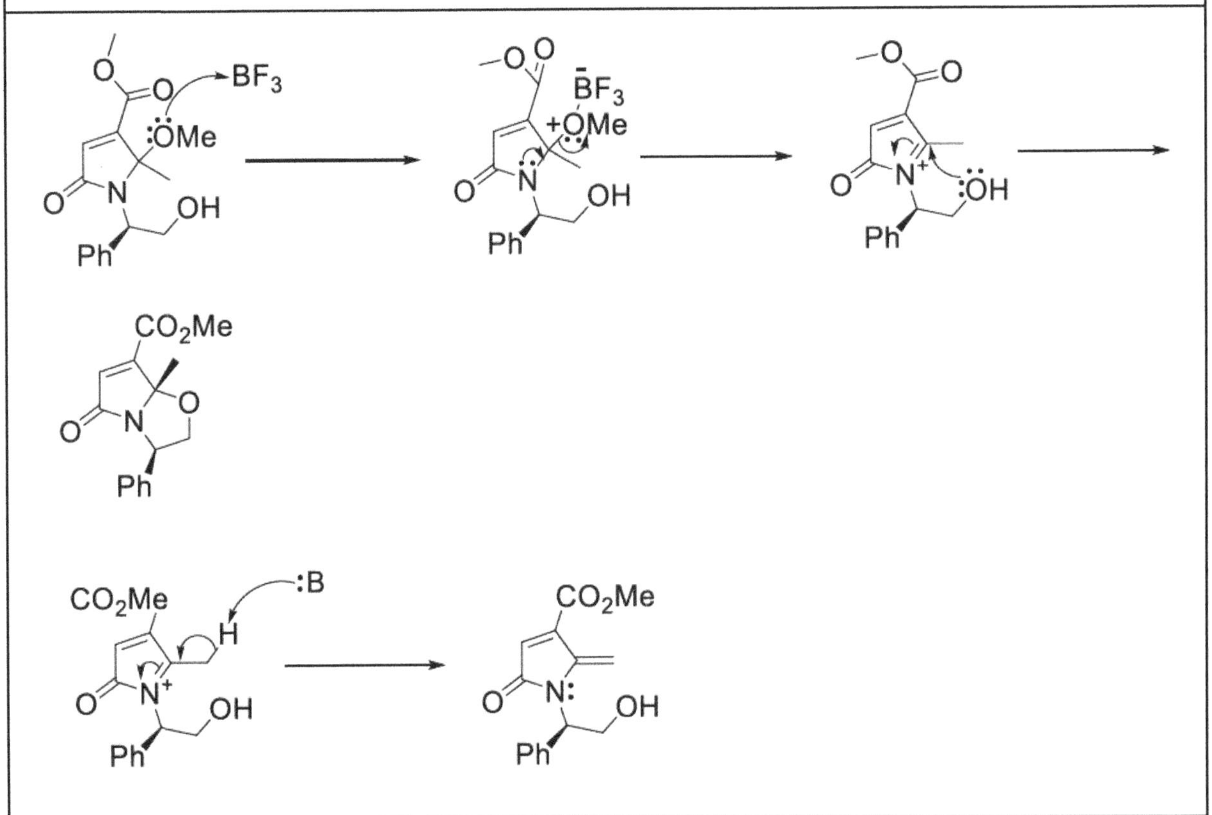

429

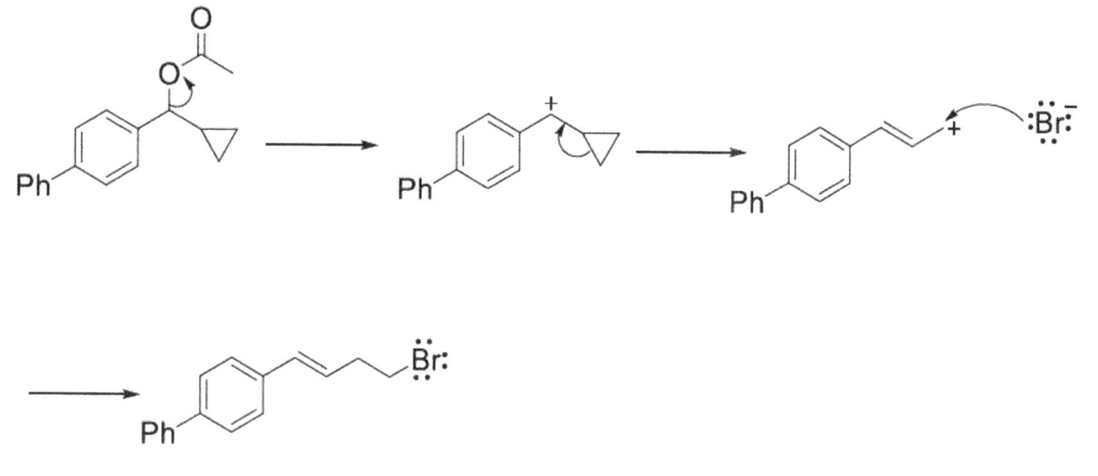
Ph
O
O
Ph
+
Ph
+
:Br:
−
Ph
Br:

430

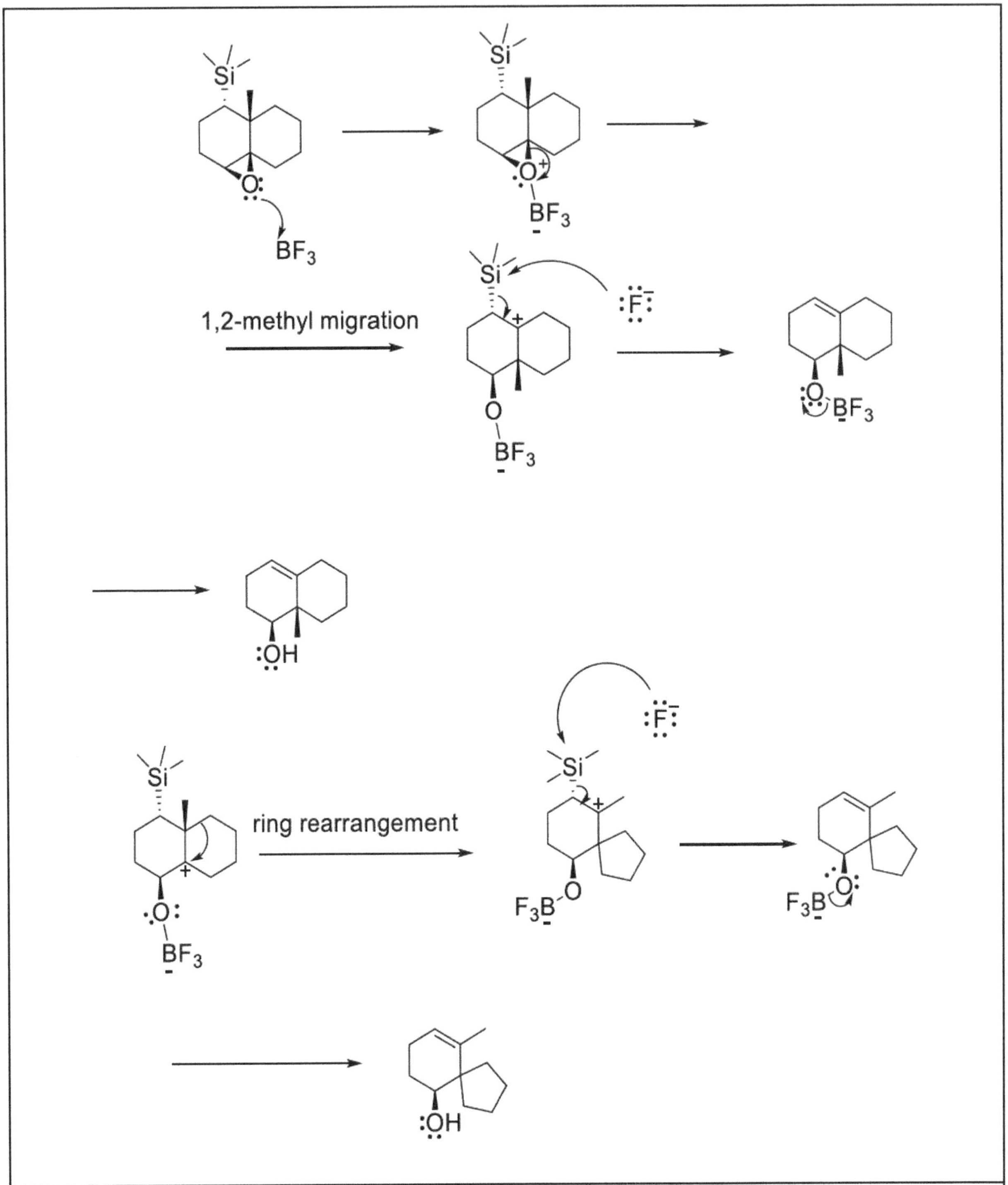
1,2-methyl migration
ring rearrangement

431

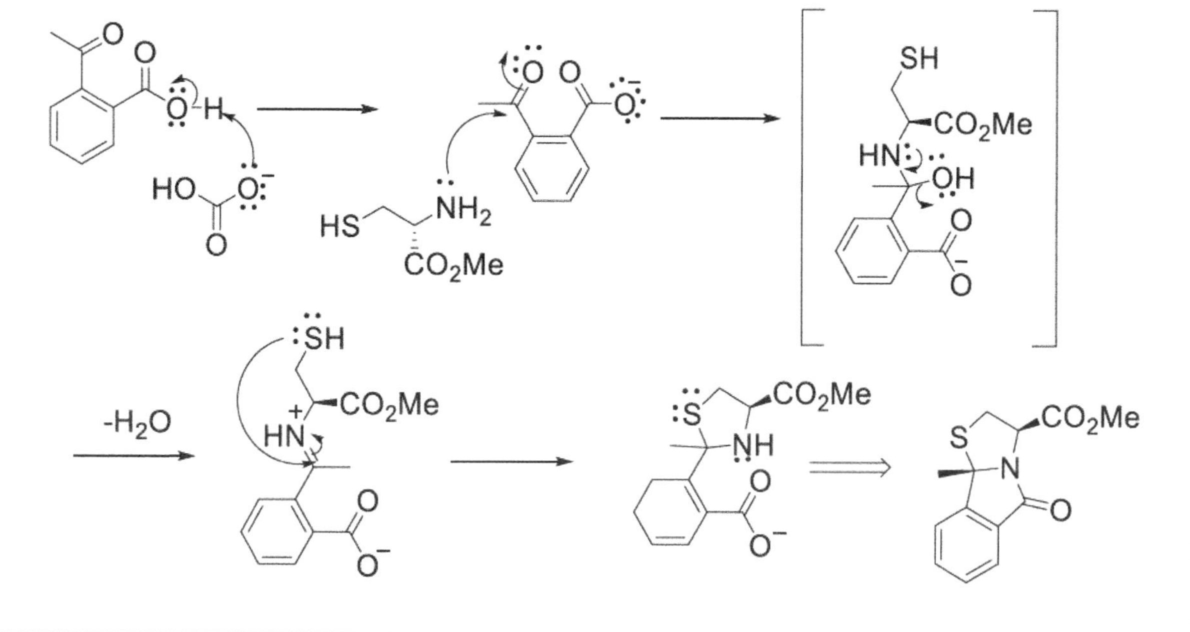
HO
O
HS
NH2
CO2Me
SH
CO2Me
HN
OH
-H2O
SH
CO2Me
HN
S
CO2Me
NH
S
CO2Me
N
O

432

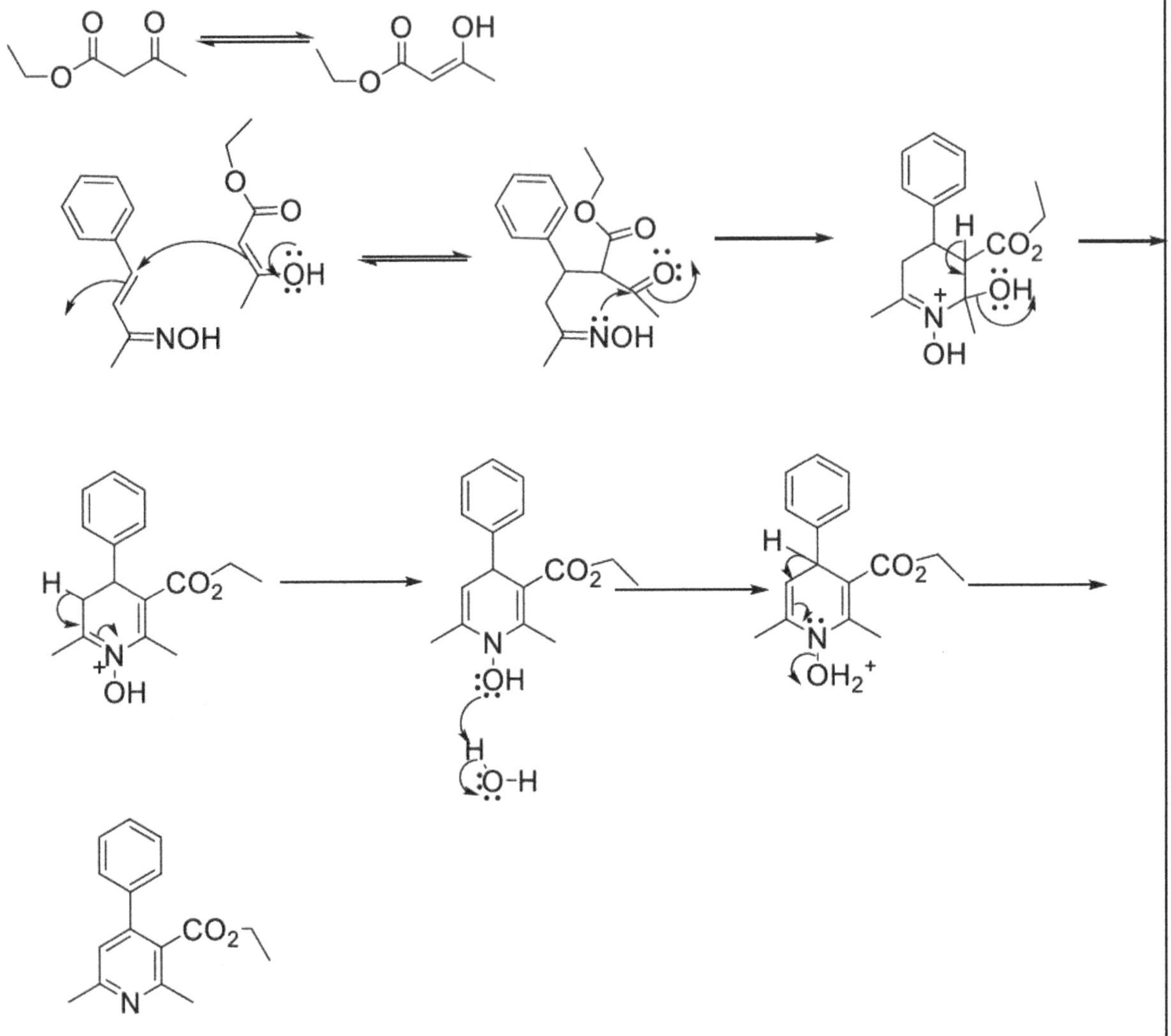
OH
NOH
CO2
OH
N
H
OH2+
O-H
N

433

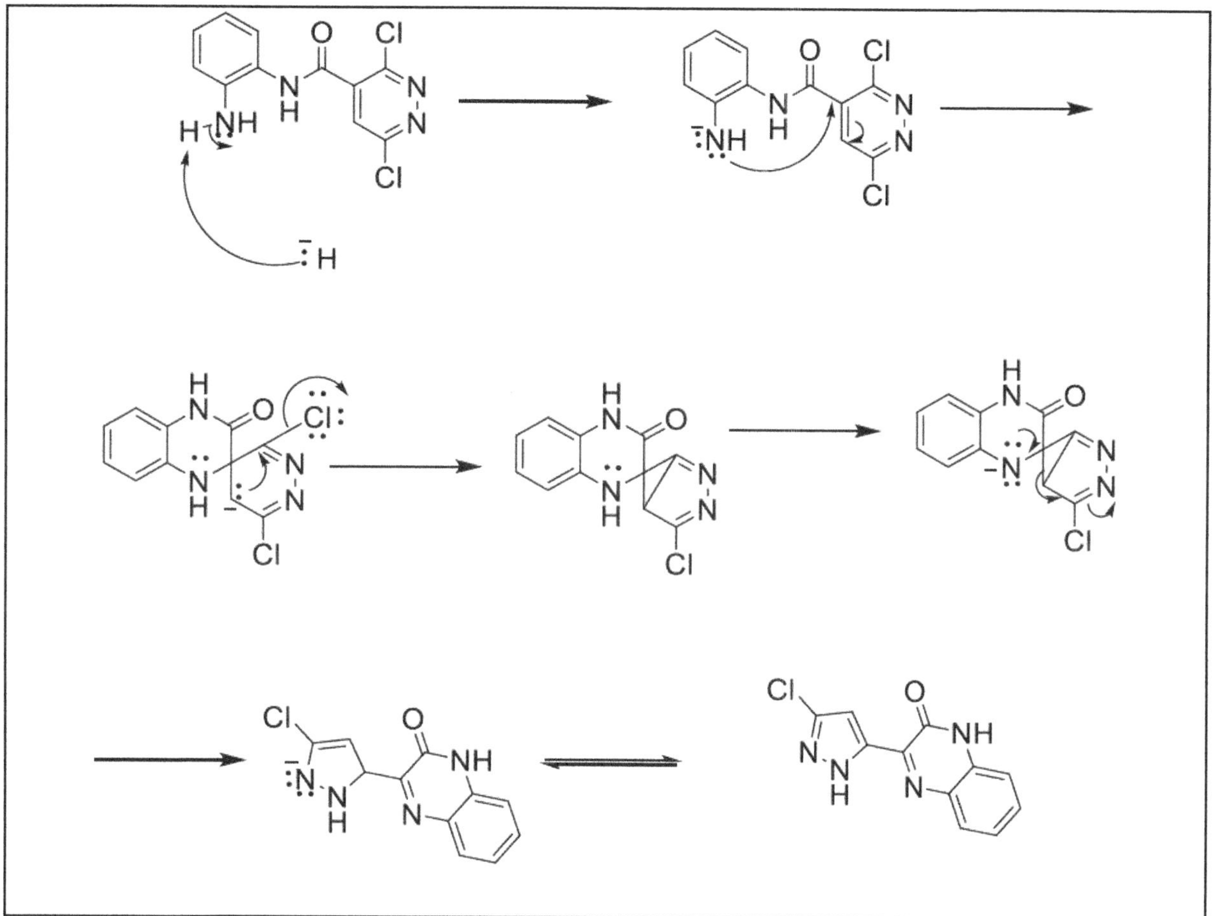

O
Cl
N
H
N
N
H
NH
Cl
H
Cl
N
H
O
NH
H
N
N
O
Cl
N
H
N
Cl

434

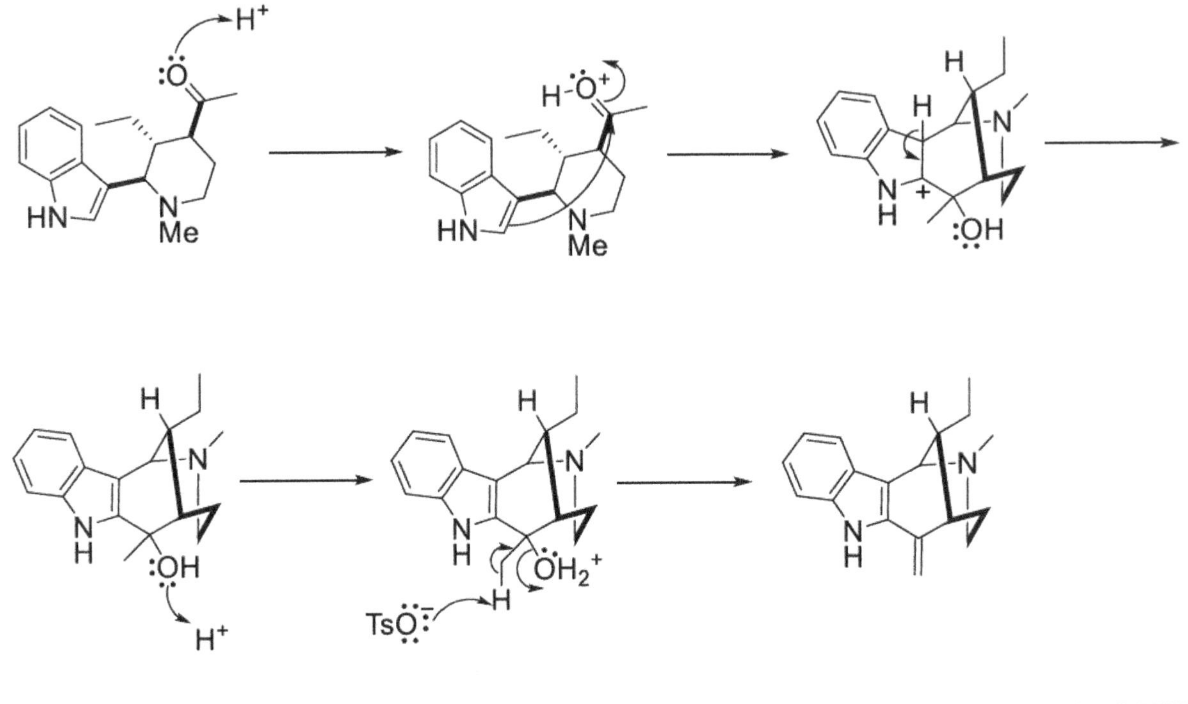
H+
:O:
H-O+
HN
N
Me
H
H
N
+
:OH
H
N
H
N
H
:OH
H+
OH2+
TsO:-

435

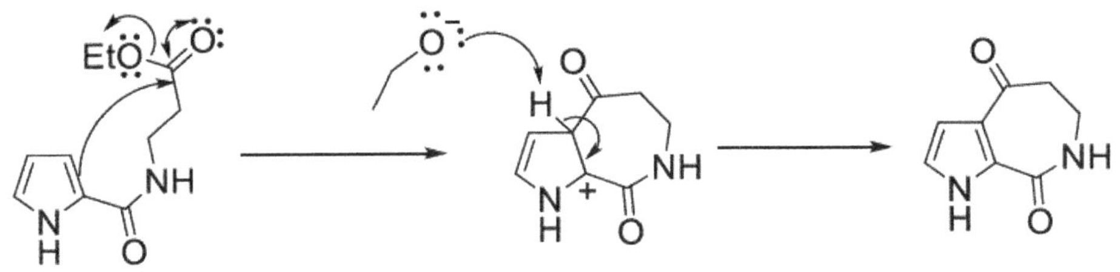
EtO
O:
NH
N
H
O
H
O
NH
N
H
+
O
NH
N
H
O

436

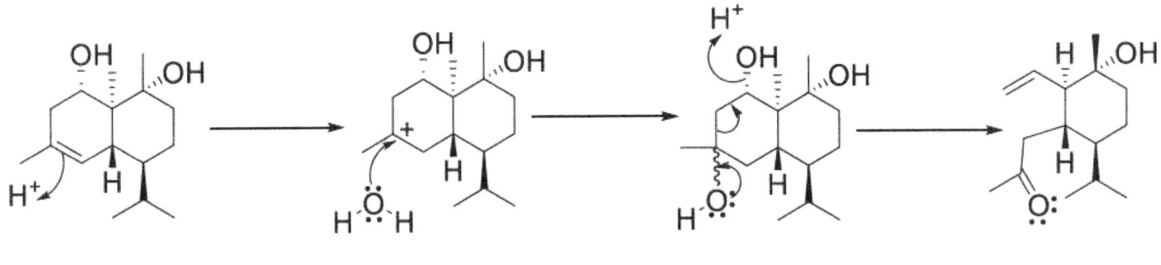
OH
OH
H+
H
+
H
H
O
H
H+
OH
OH
H
H
O
H
OH
H
H
O:

437

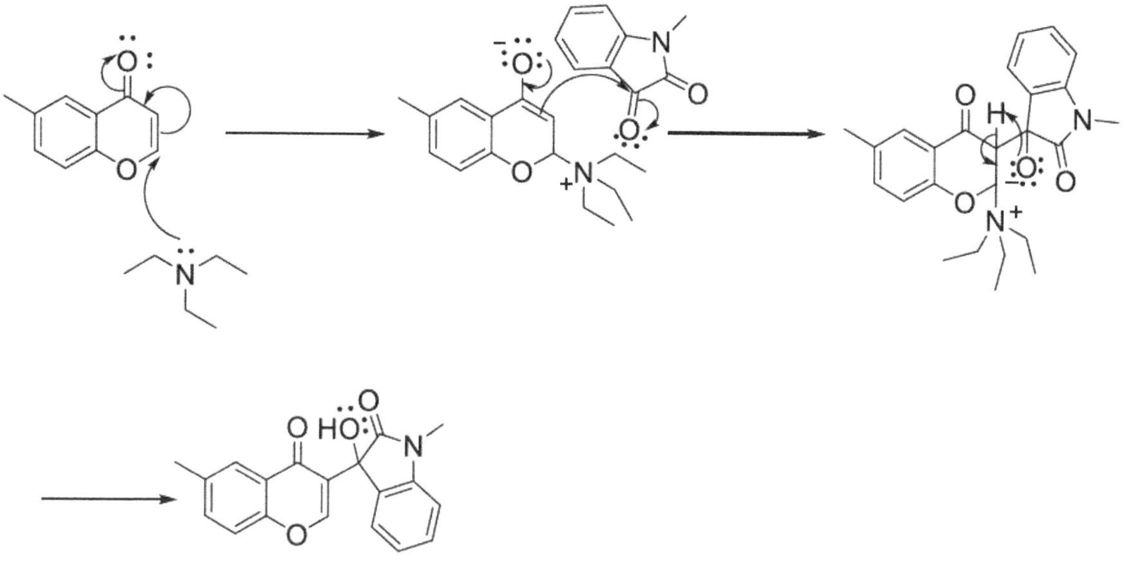

438

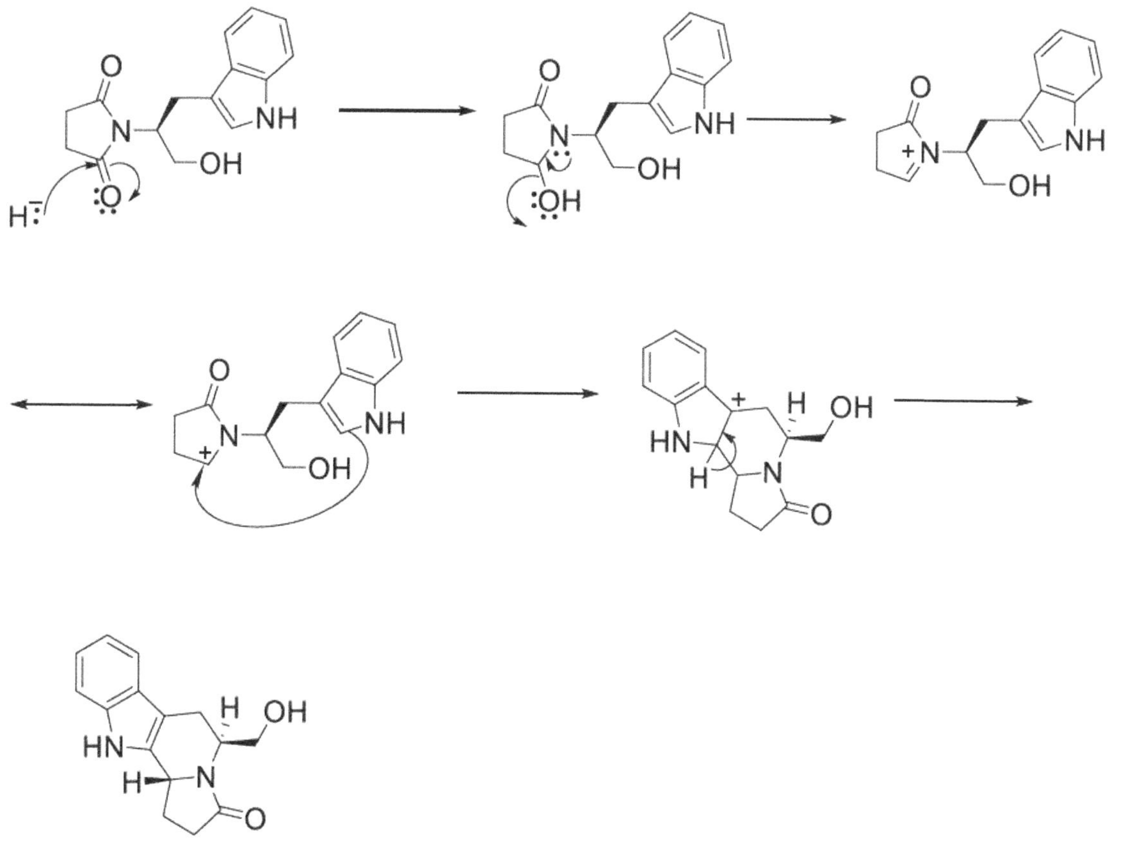

439

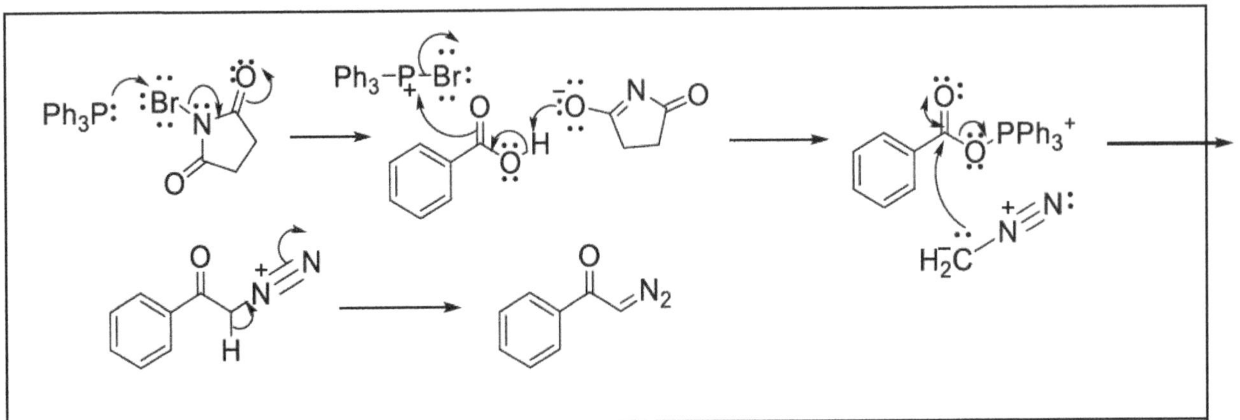

440

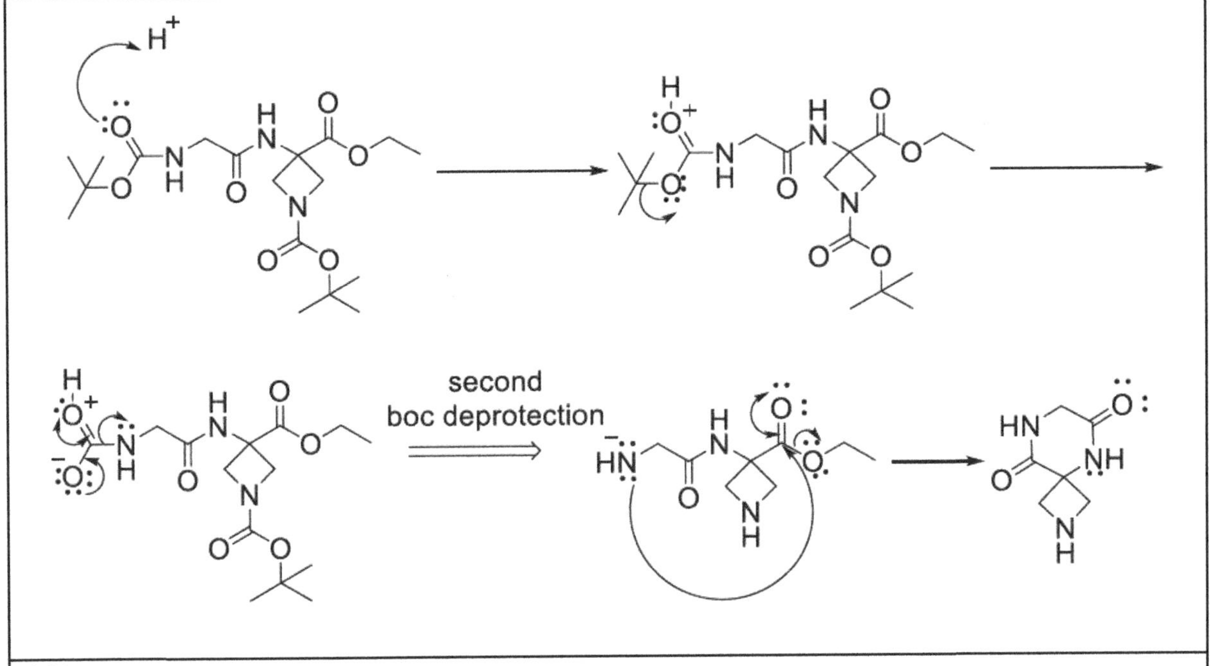
second
boc deprotection

441

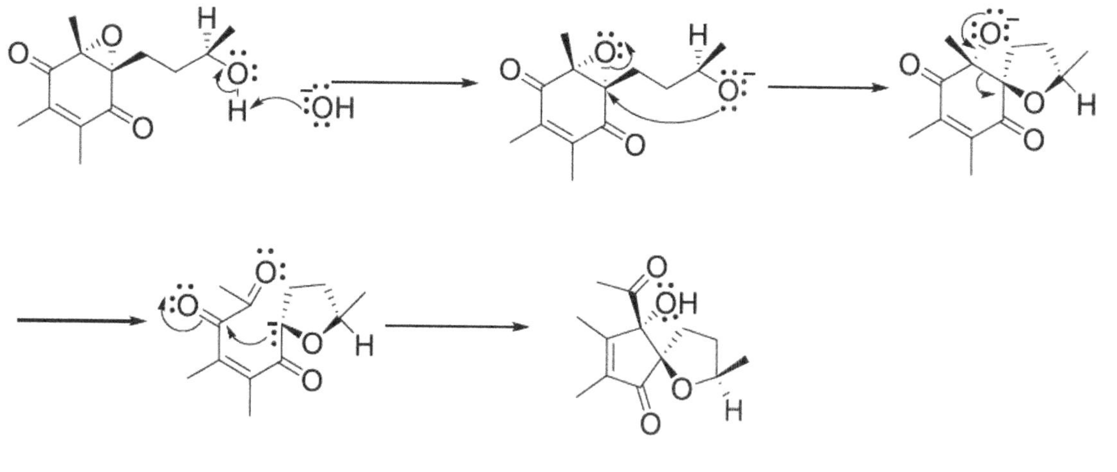

O
OH
H
O
H
OH
O
H
O
O
OH
H

442

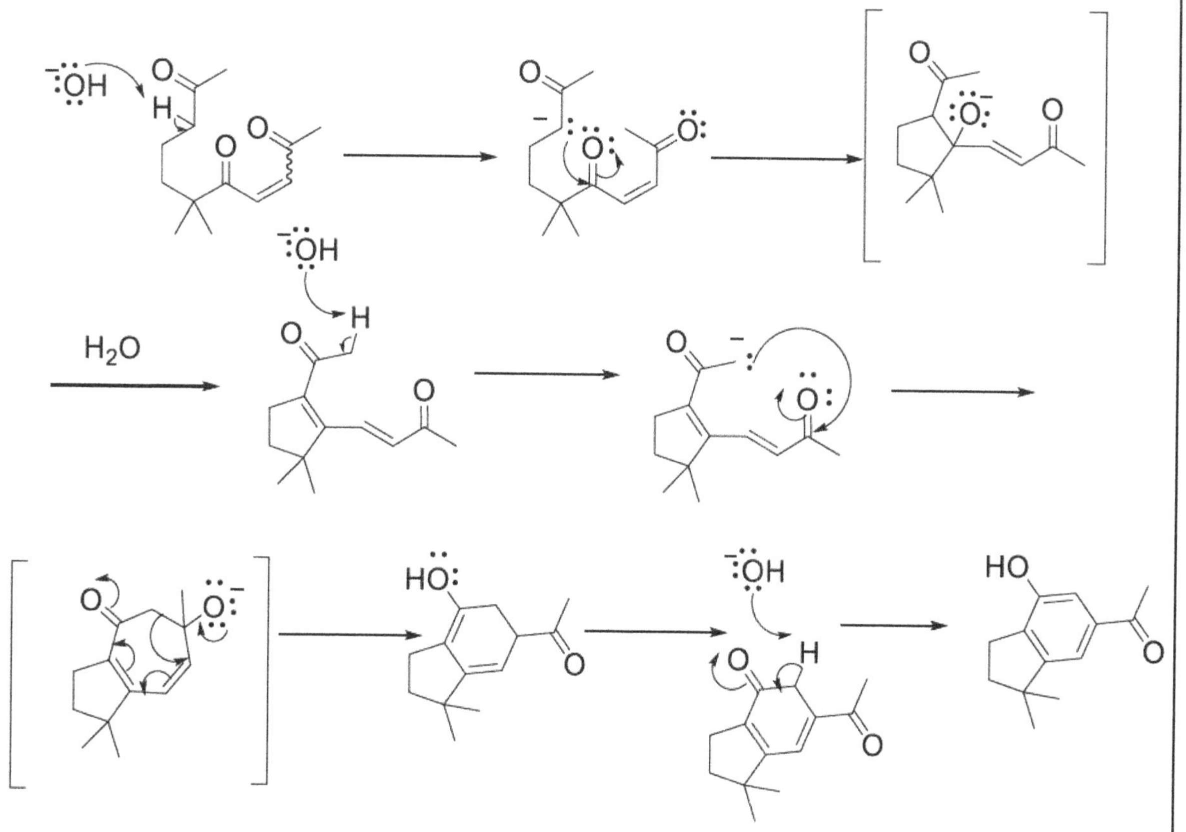

OH
H
O
O
H₂O
H
HO
HO

443

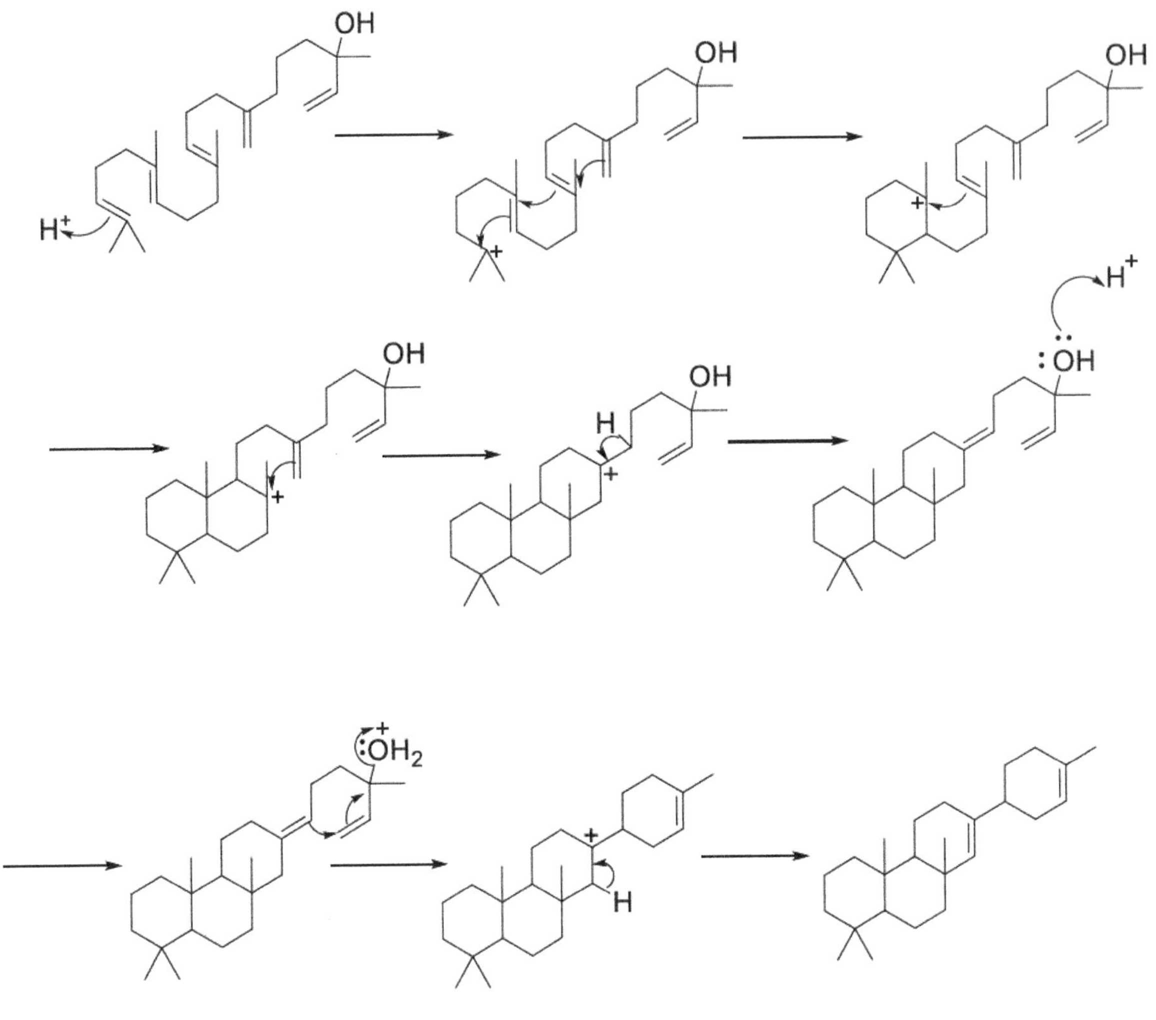
OH
H+
OH
+
OH
+
OH
+
OH
H
+
H+
:OH
+
:OH2
+
H

444

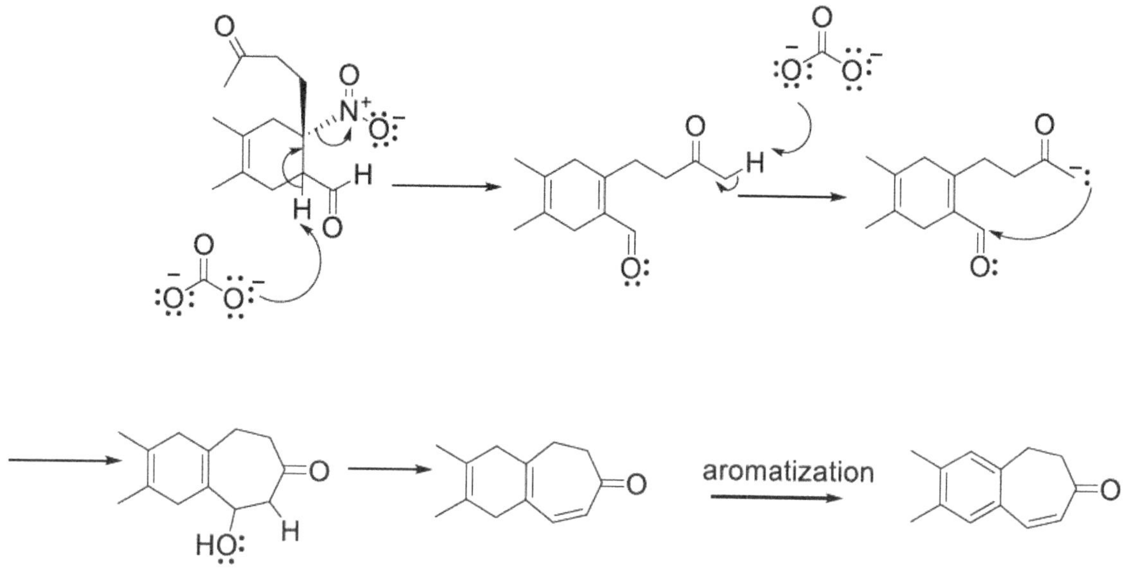
aromatization

445

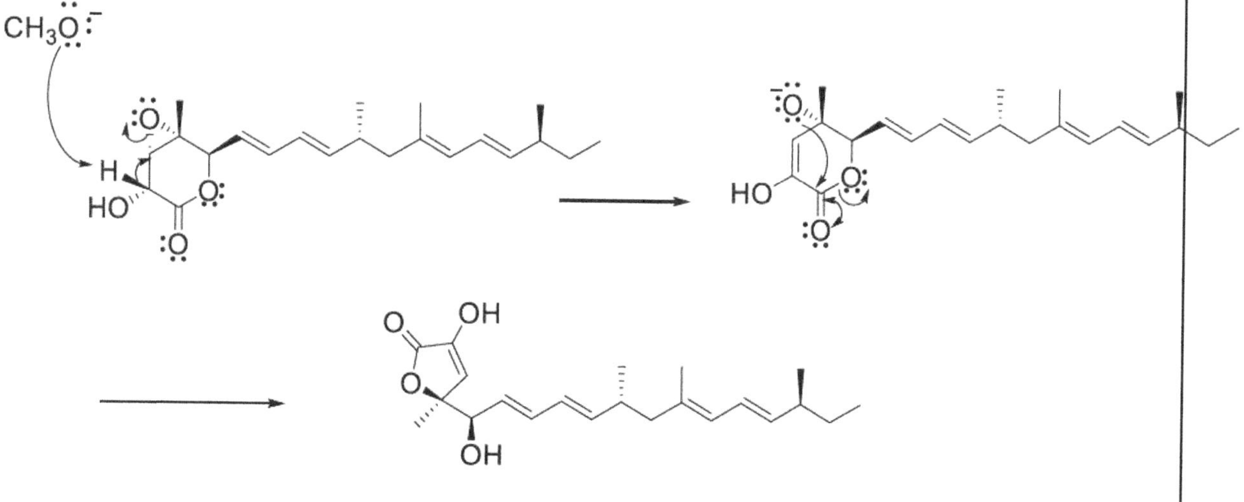

446

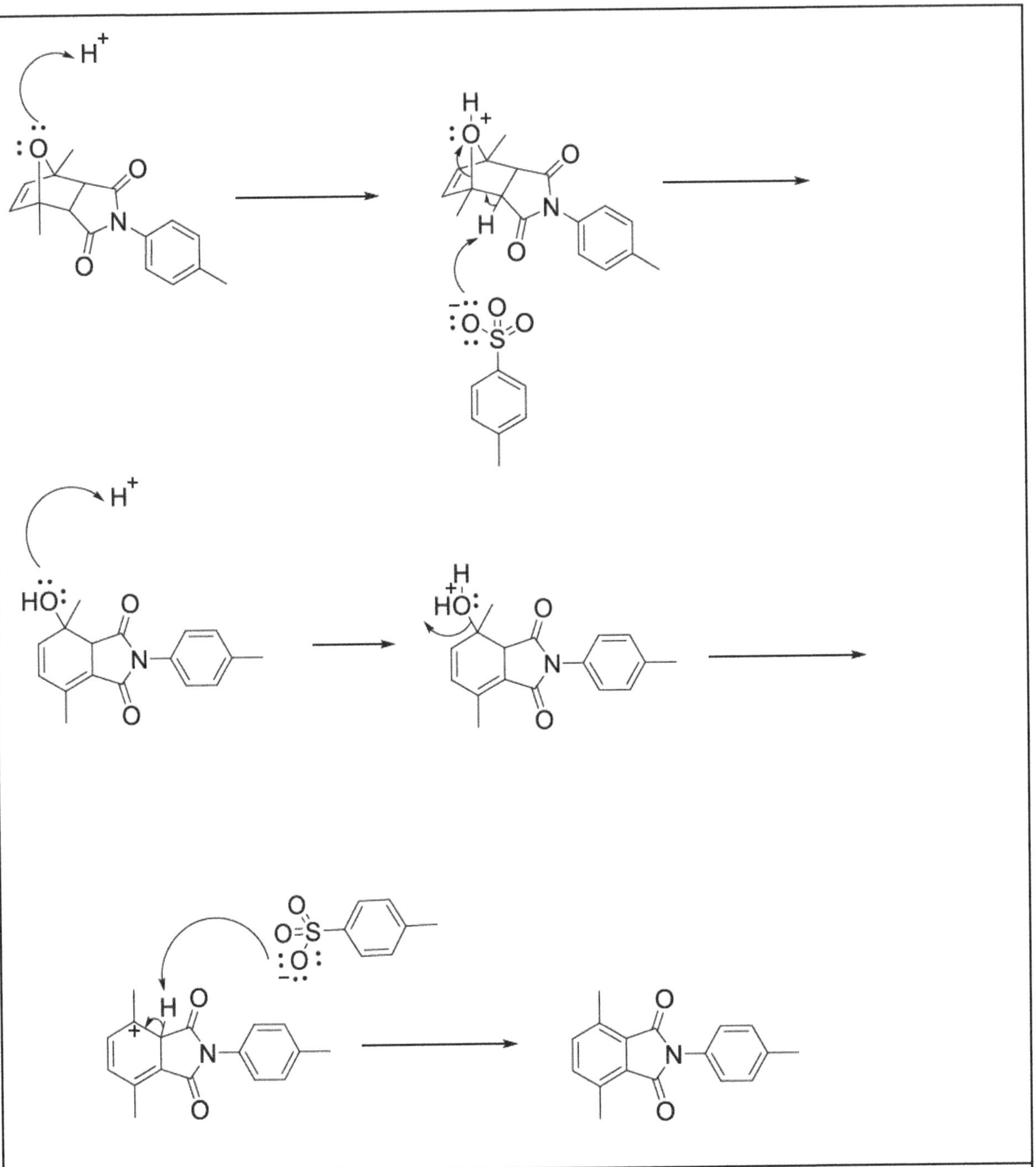
H+
O
O
N
O
H
O+
O
N
O
H
O
O
S
O
H+
HO
O
N
O
H
HO+
O
N
O
O
O
S
O
H
+
O
N
O
O
N
O

447

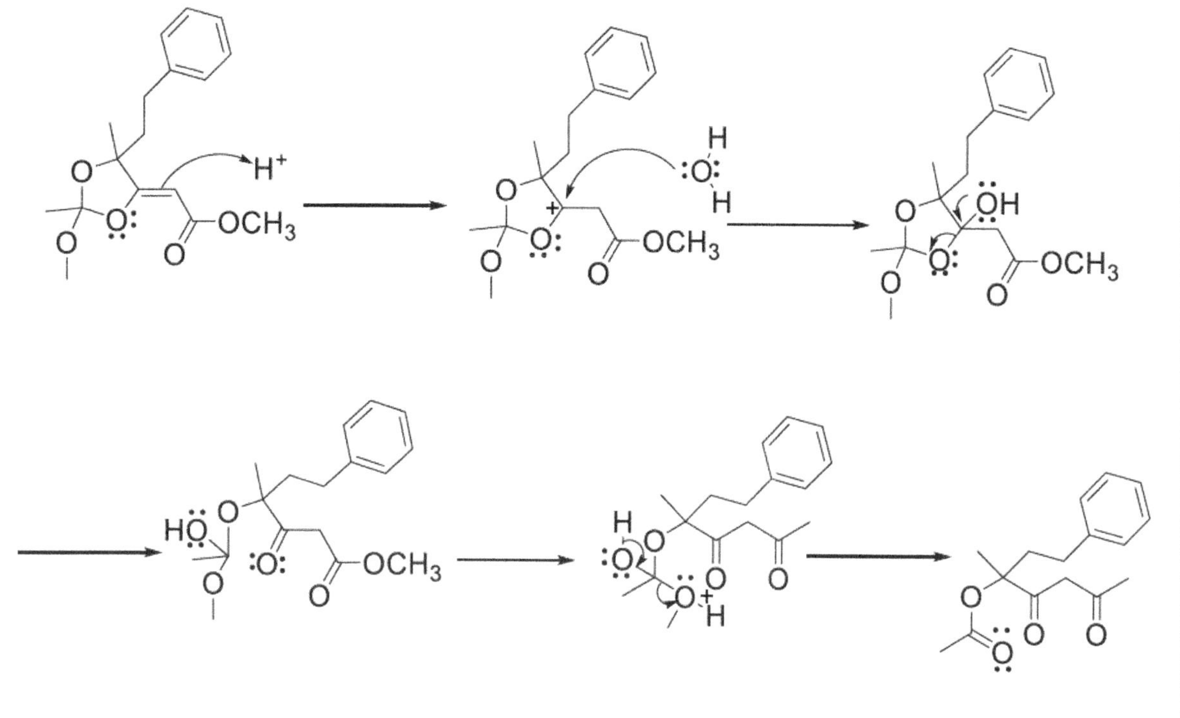
H+
:O:
H
H
OH
OCH3
O

448

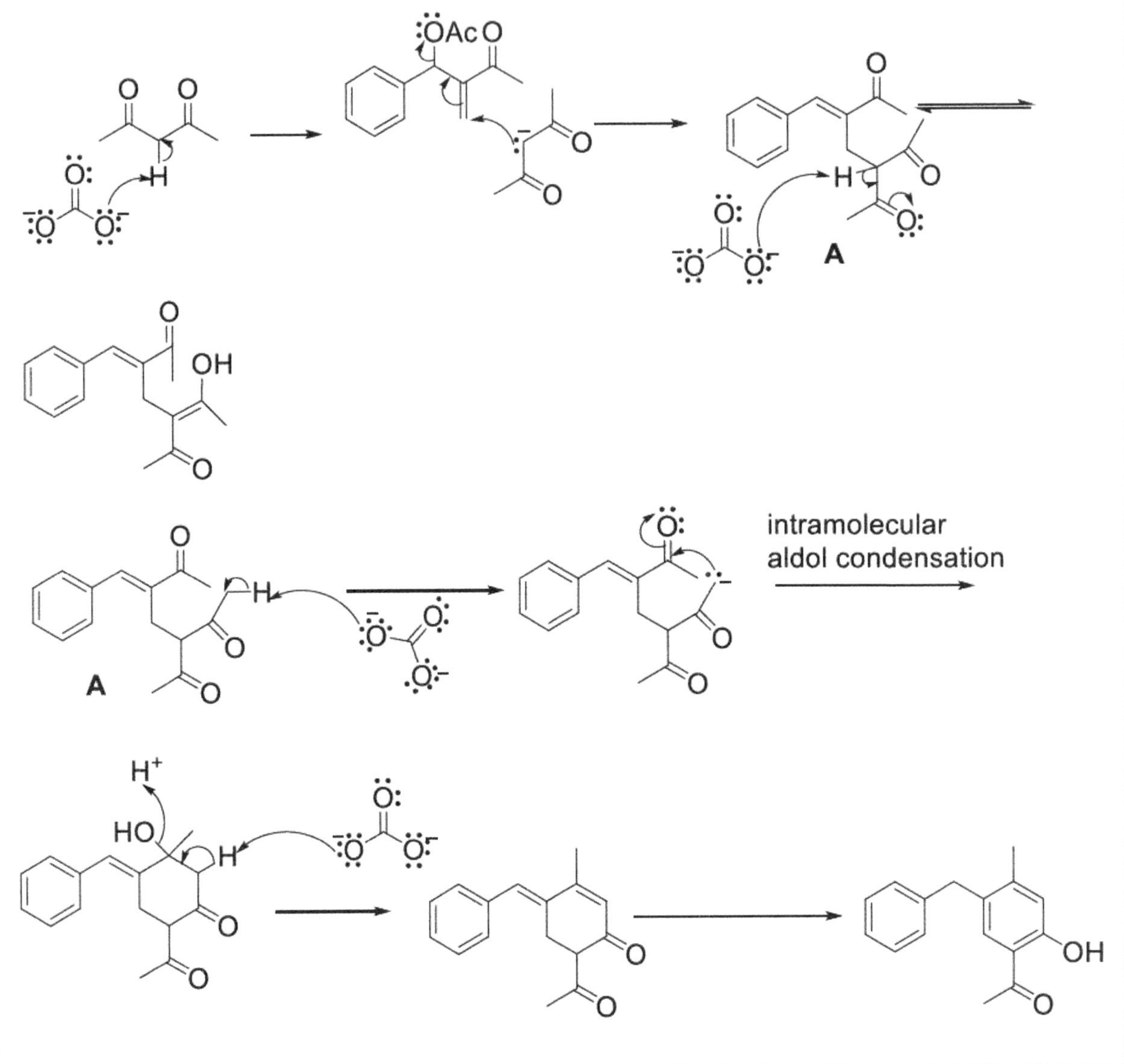
OAc
A
intramolecular
aldol condensation
A
H+
HO
OH

449

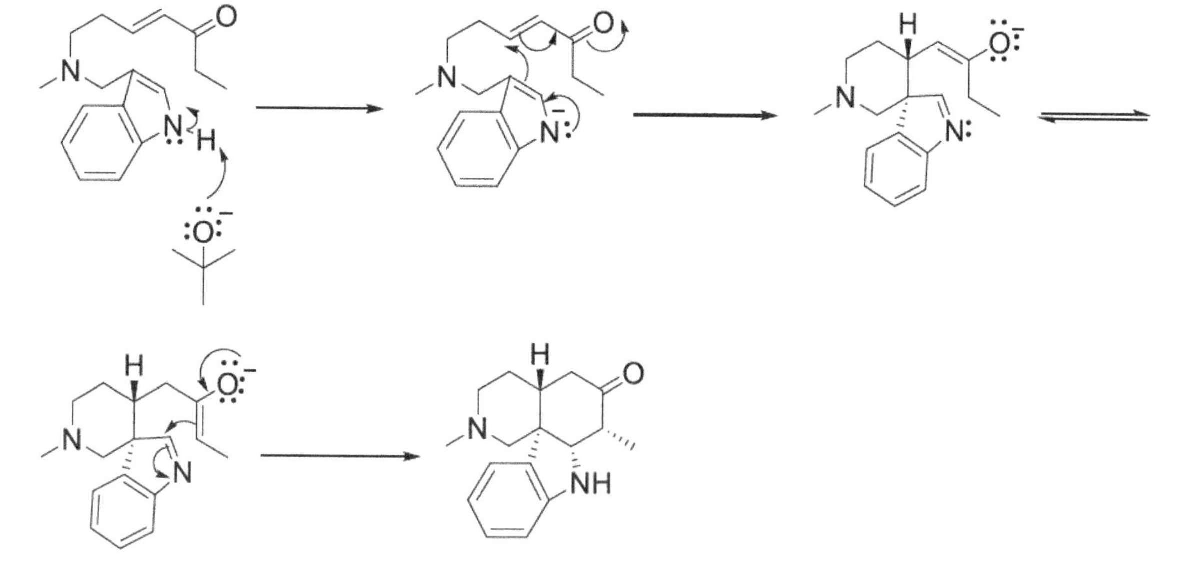

450

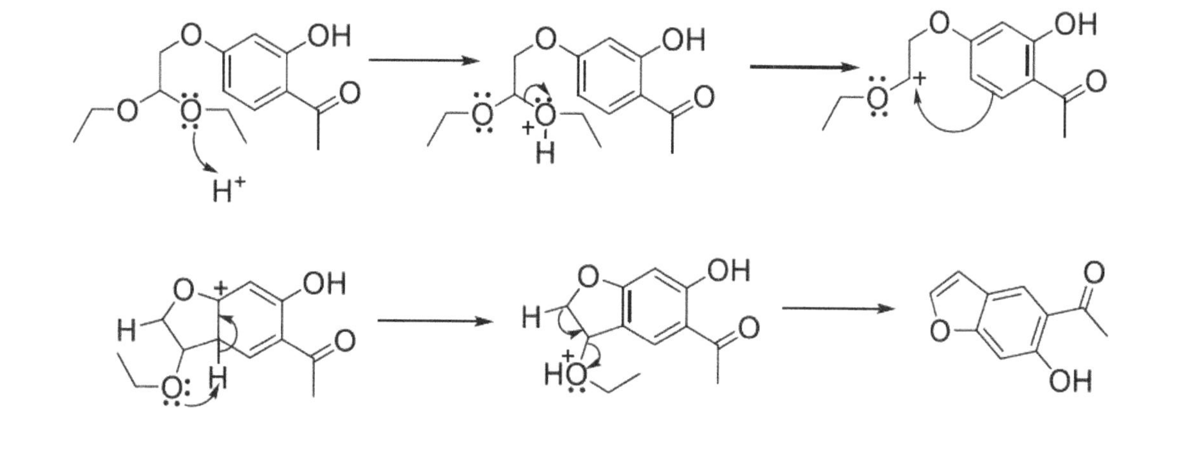

451

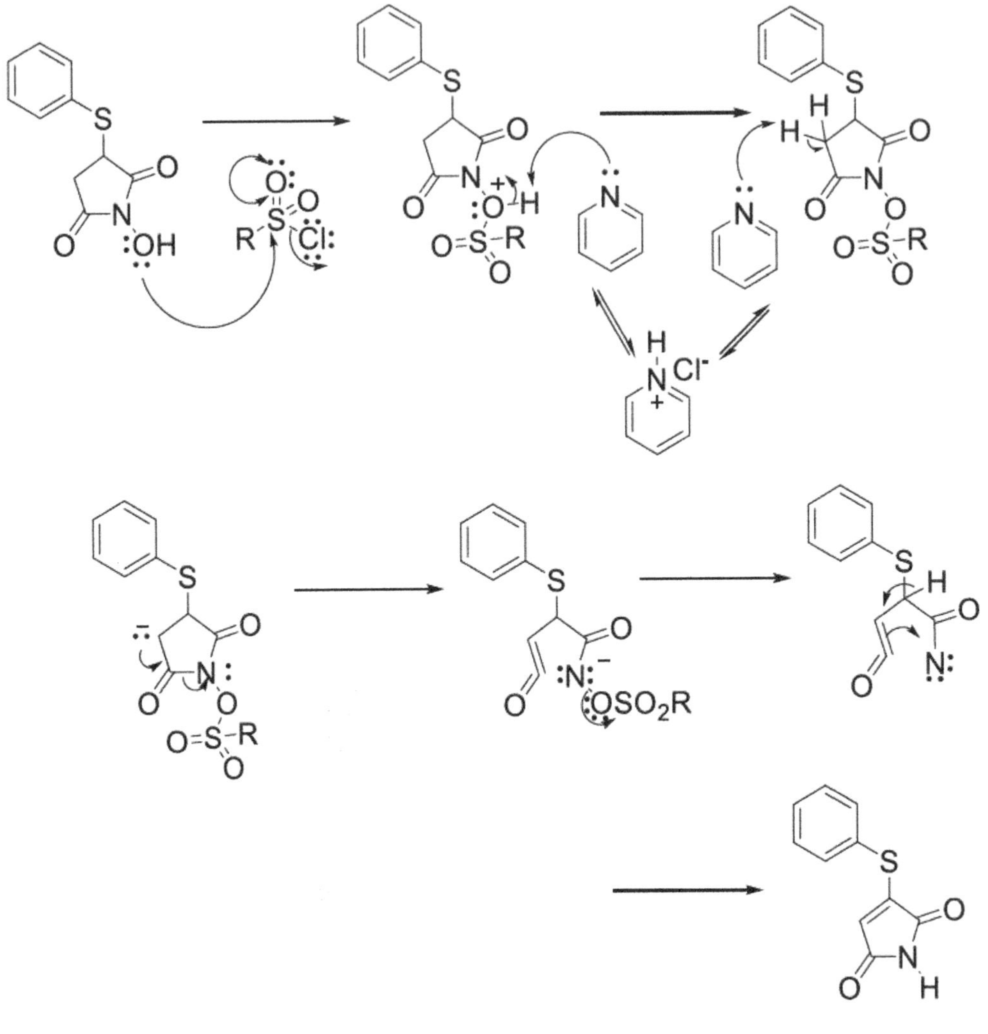

452

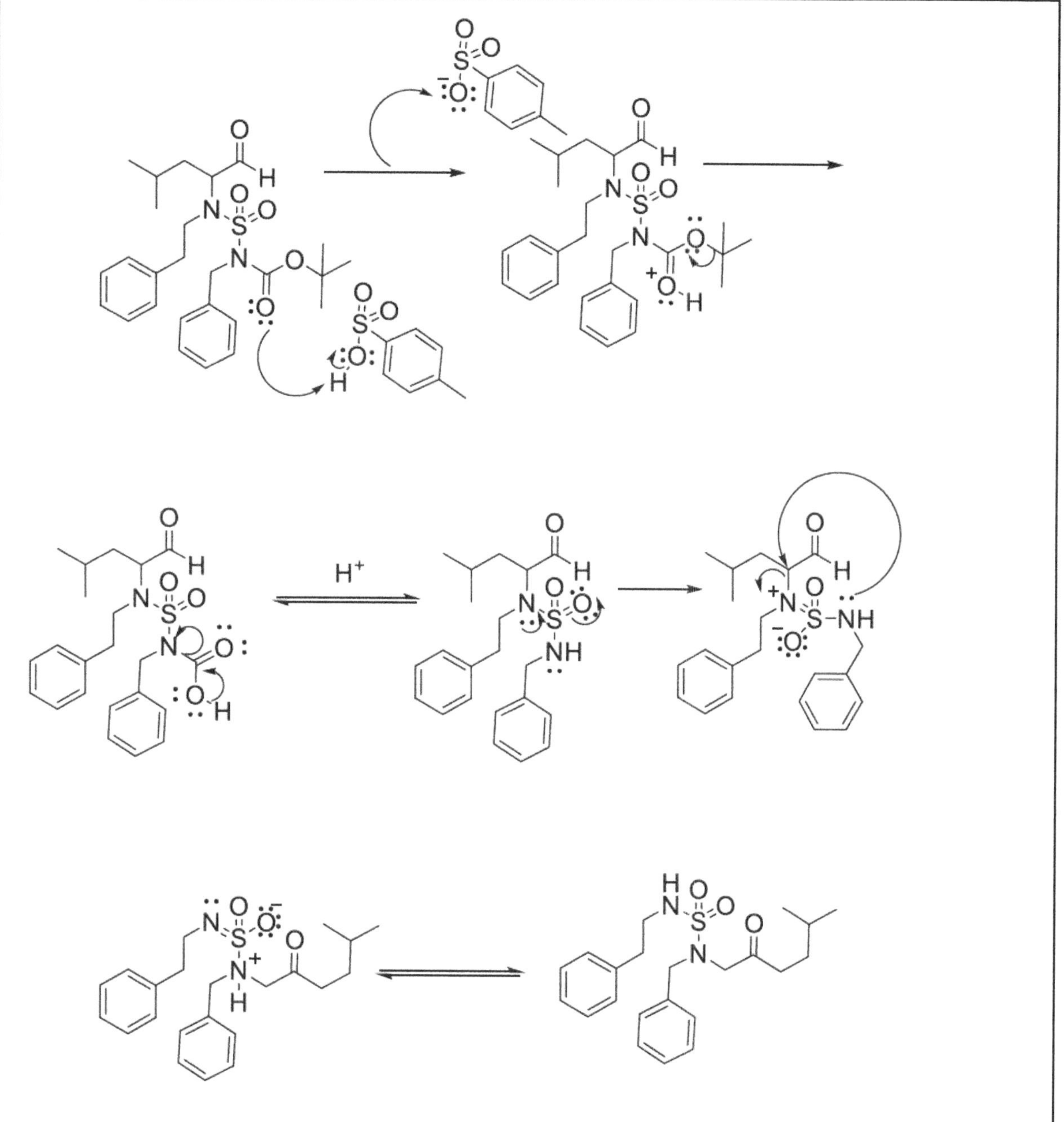

453

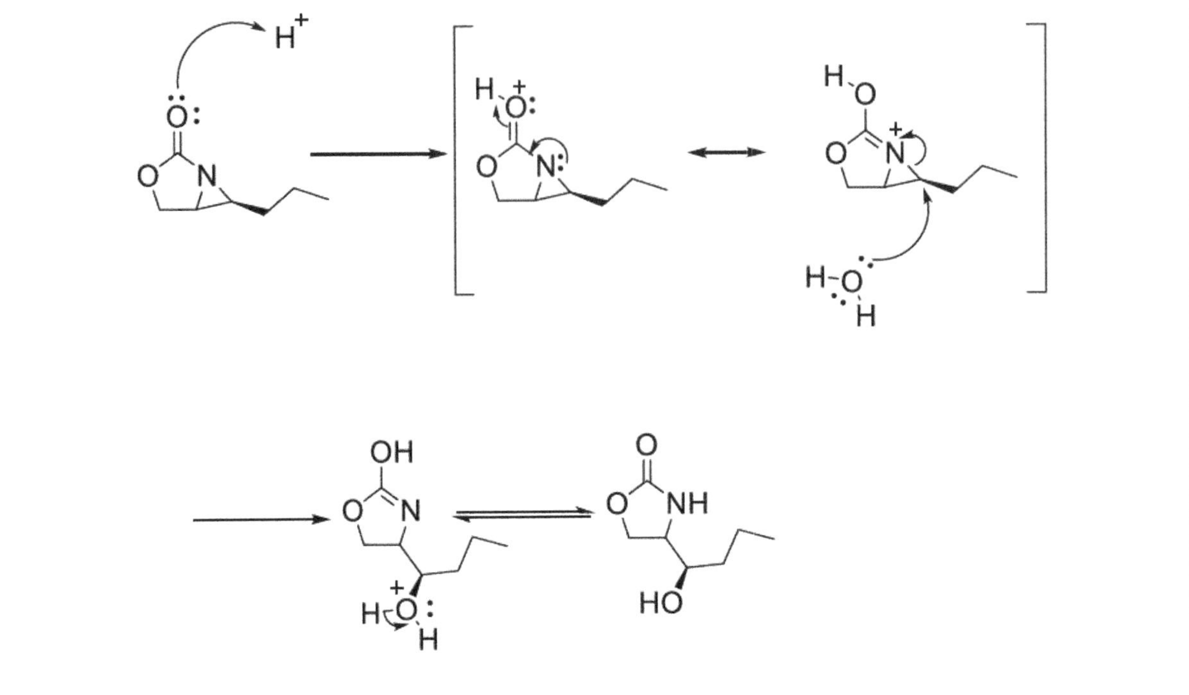
H+
H2O
OH
O
NH
HO

454

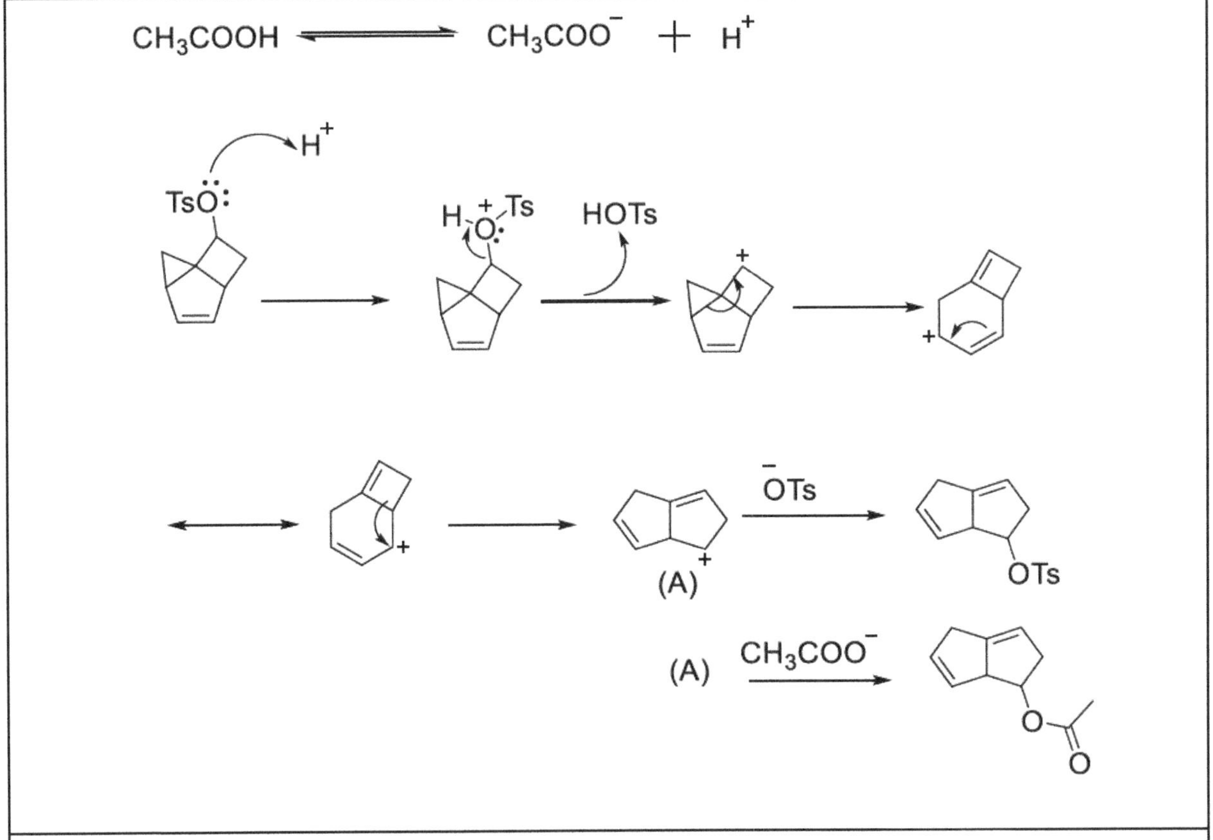
CH3COOH
CH3COO−
H+
TsO
Ts
HOTs
−OTs
(A)
OTs
CH3COO−

455

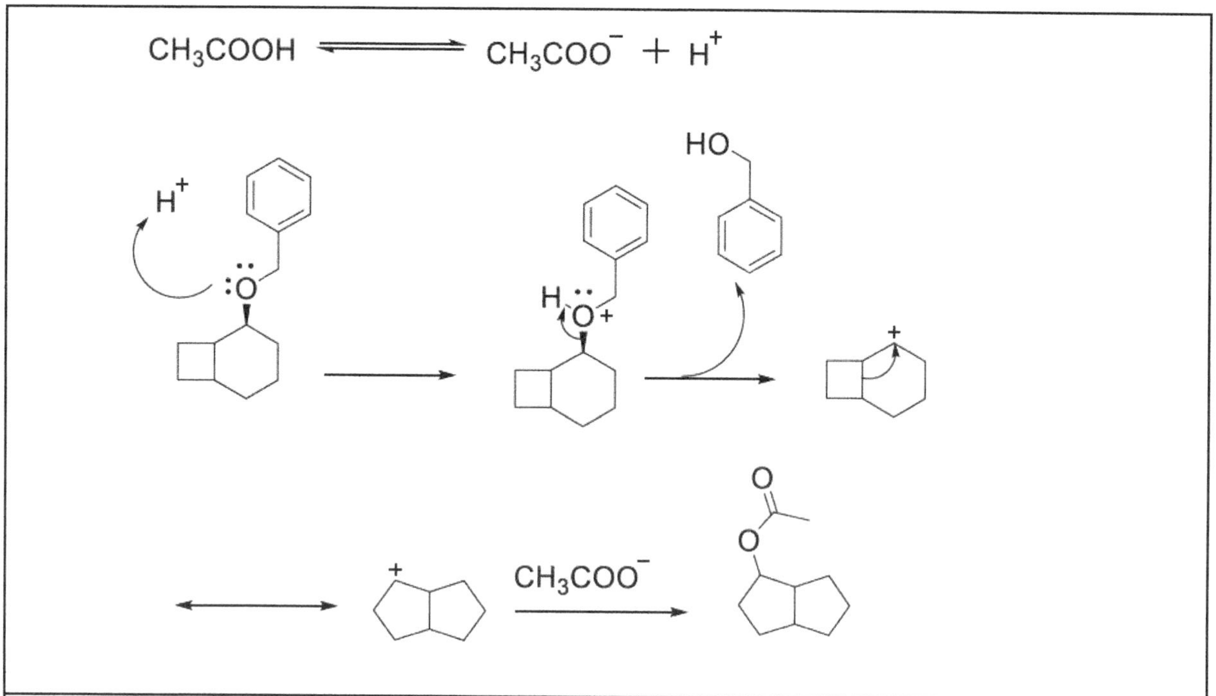
CH3COOH
CH3COO− + H+
H+
O
HO
H
O+
+
CH3COO−
O
O

456

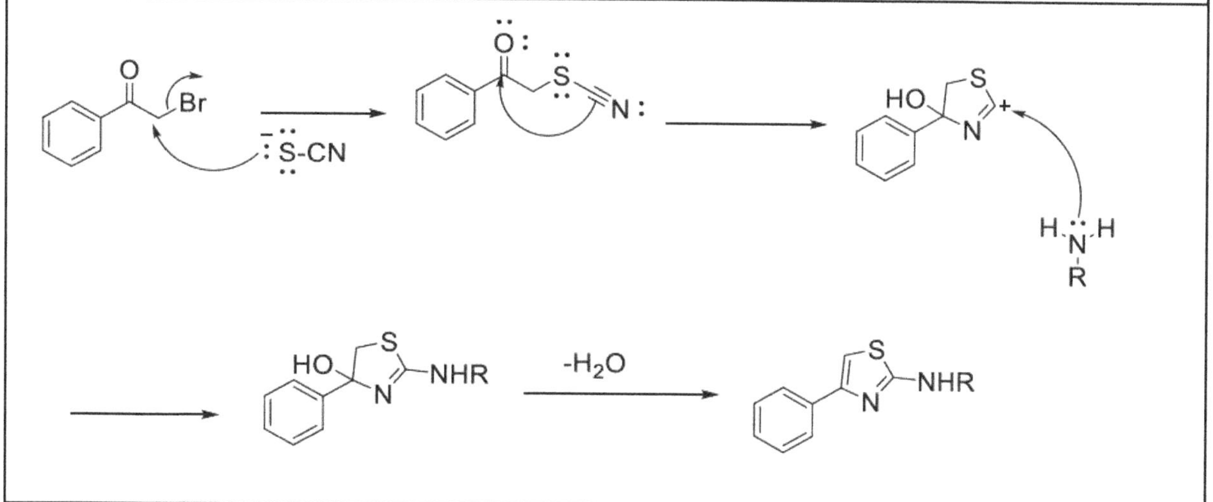
O
Br
S-CN
O
S
N
HO
S
N
+
H
N
H
R
HO
S
NHR
N
-H2O
S
NHR
N

457

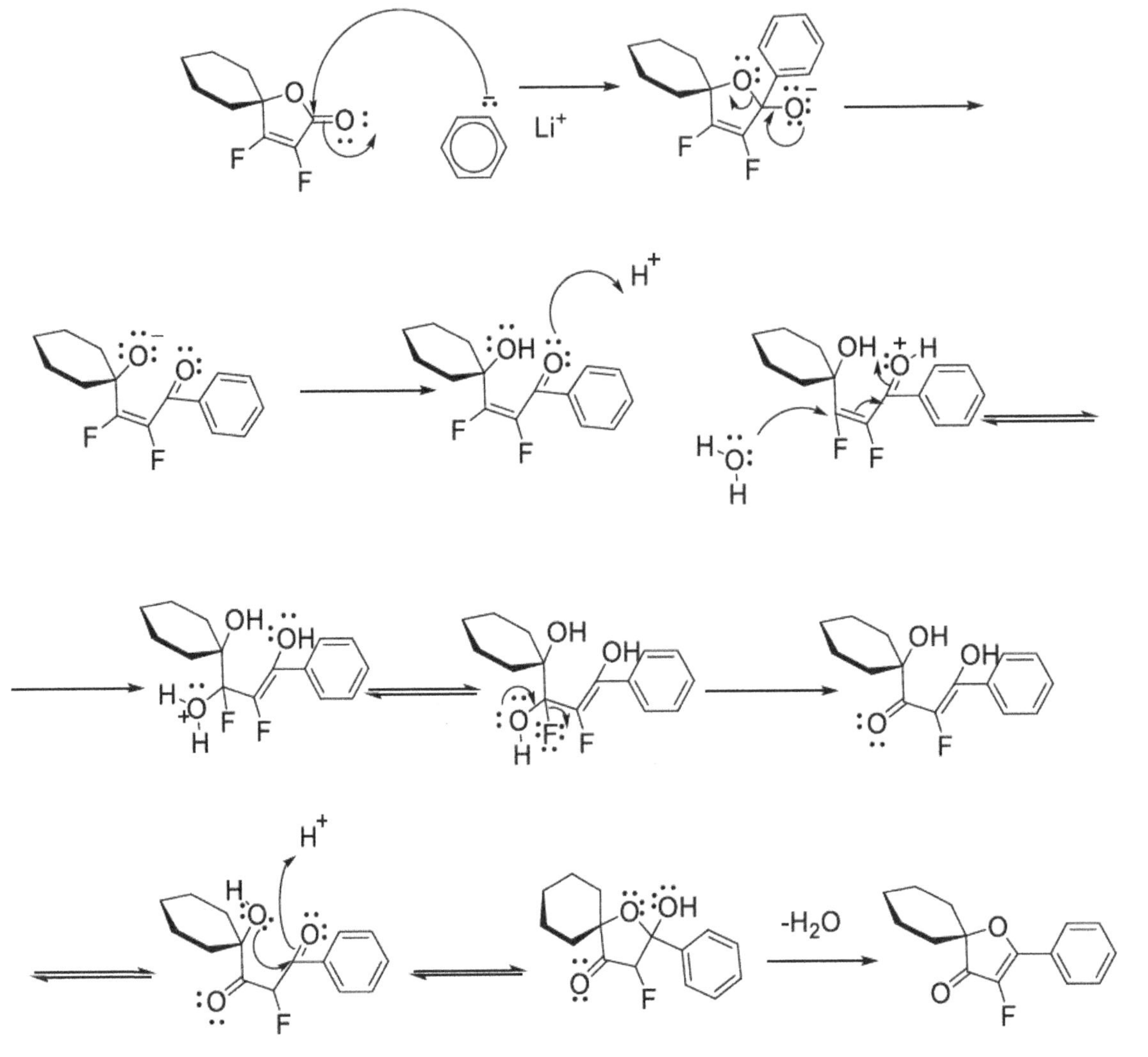

Li+
H+
-H2O

458

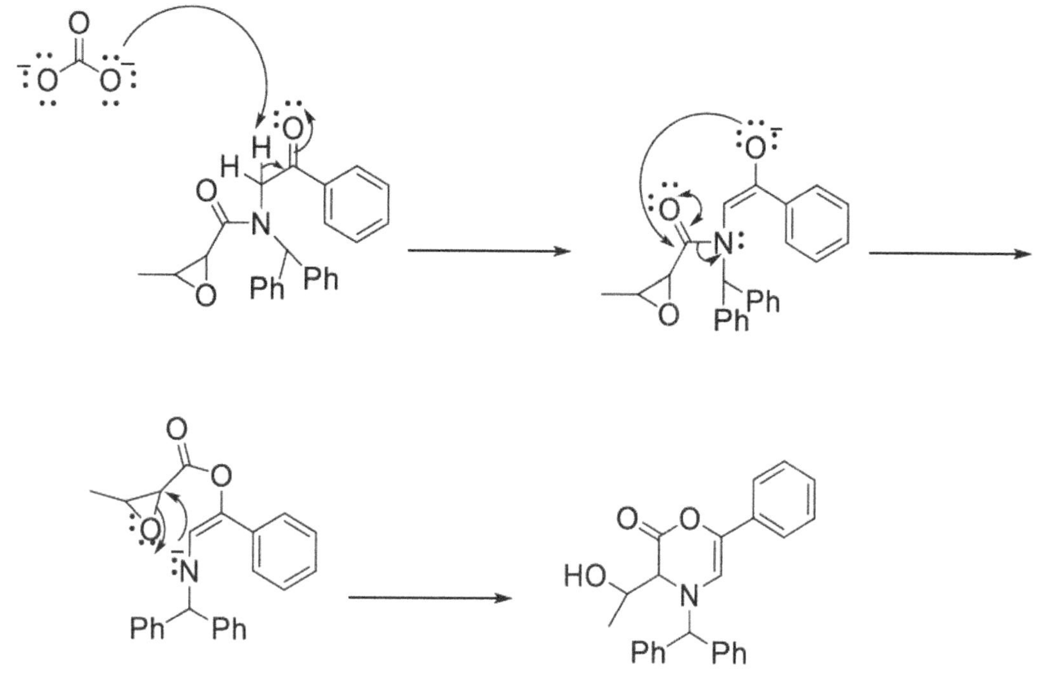

O
H
H
O
N
Ph
Ph
Ph
Ph
HO

459

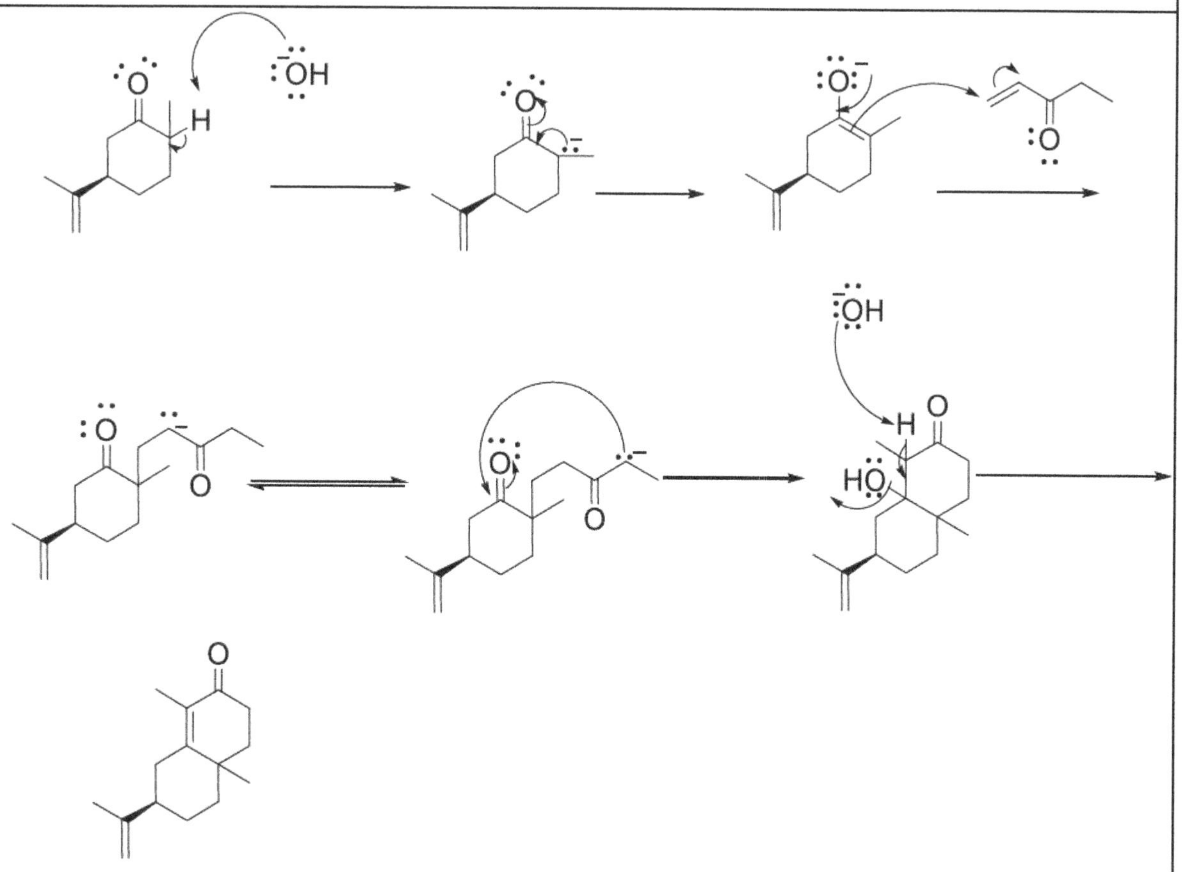

OH
O
H

460
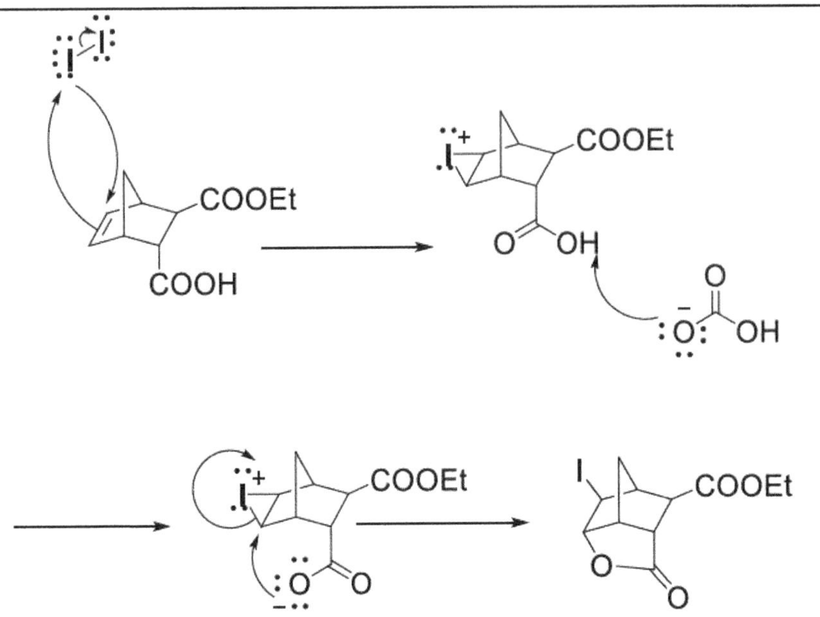
461
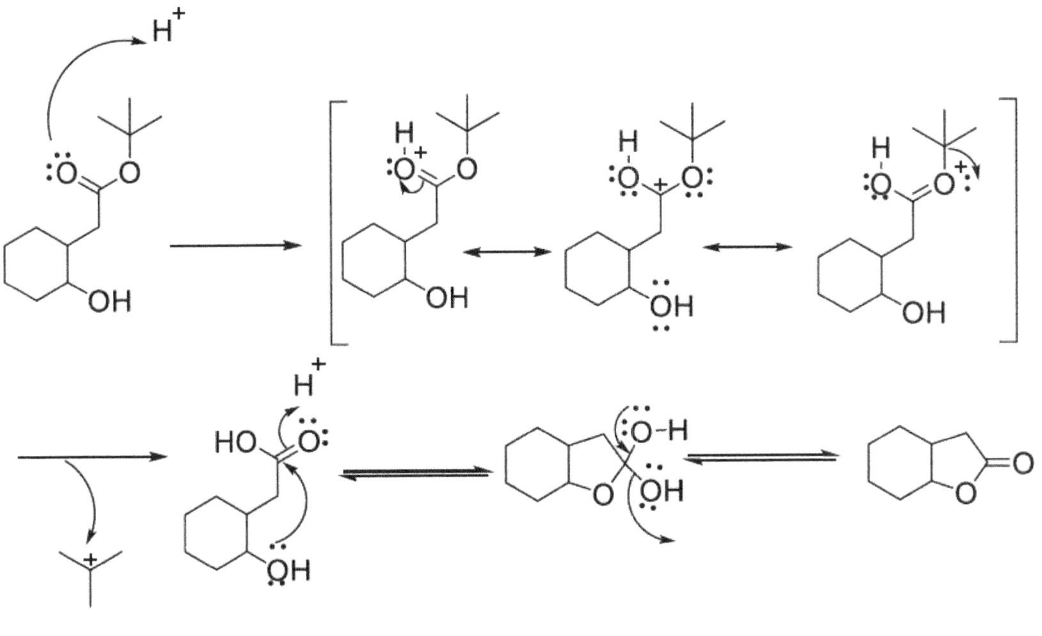
462

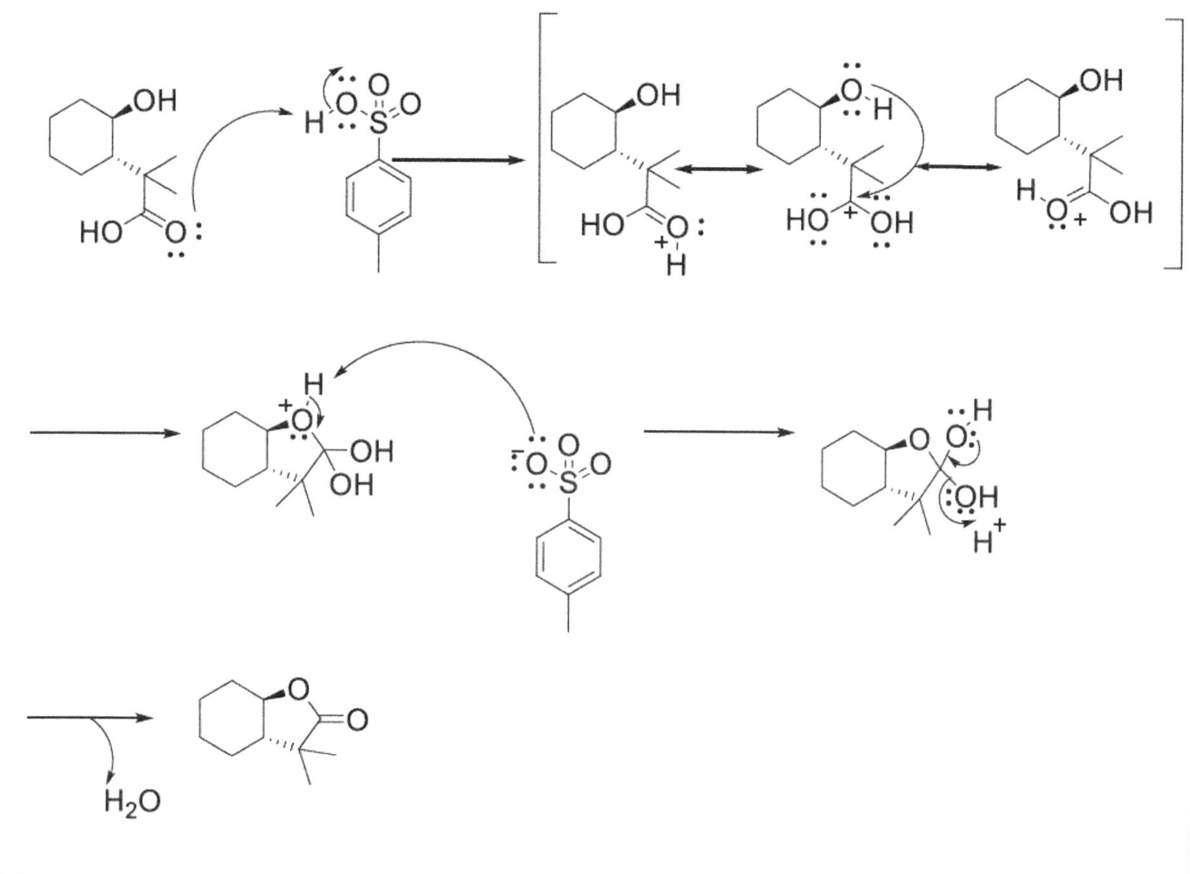

OH
H
O
S
O
O
HO
O
OH
HO
O
H
OH
O
H
HO
+
OH
OH
H
O
+
OH
H
+
O
OH
OH
O
S
O
O
O
O
H
OH
H+
O
O
H_2O

463

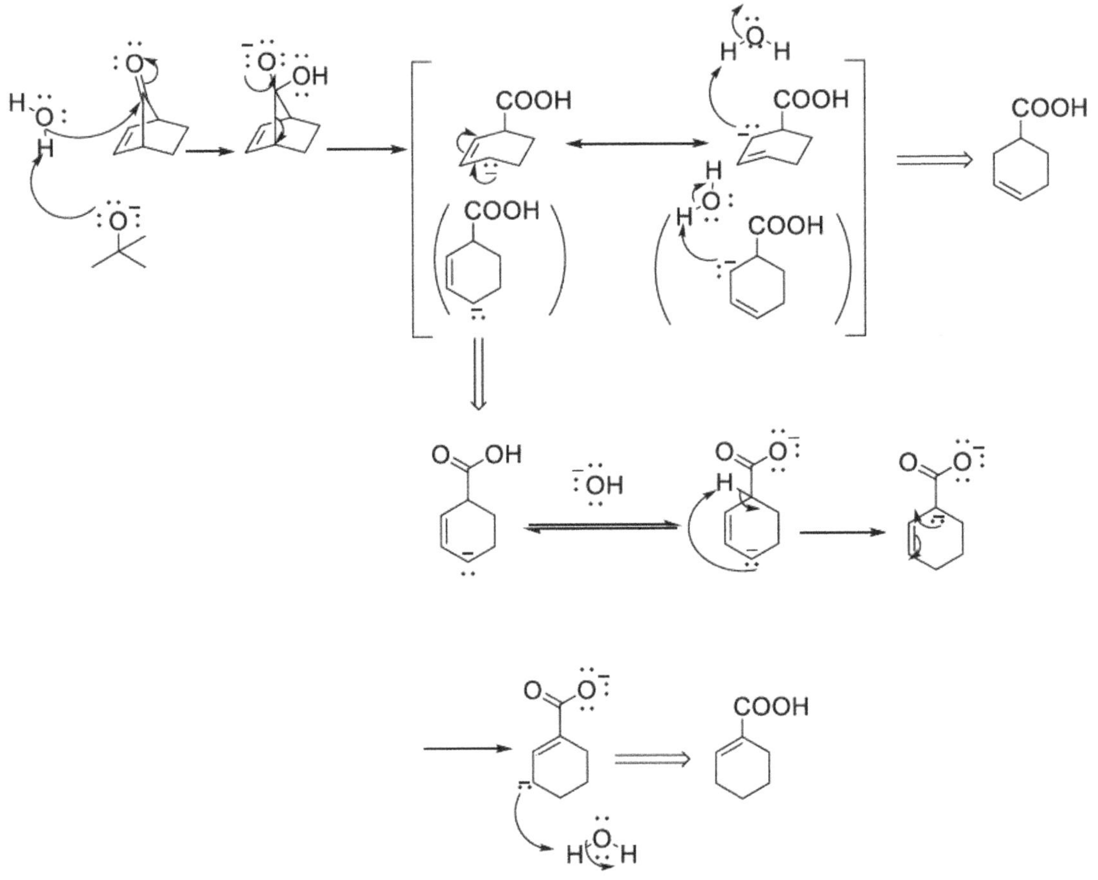
COOH
O
OH

464

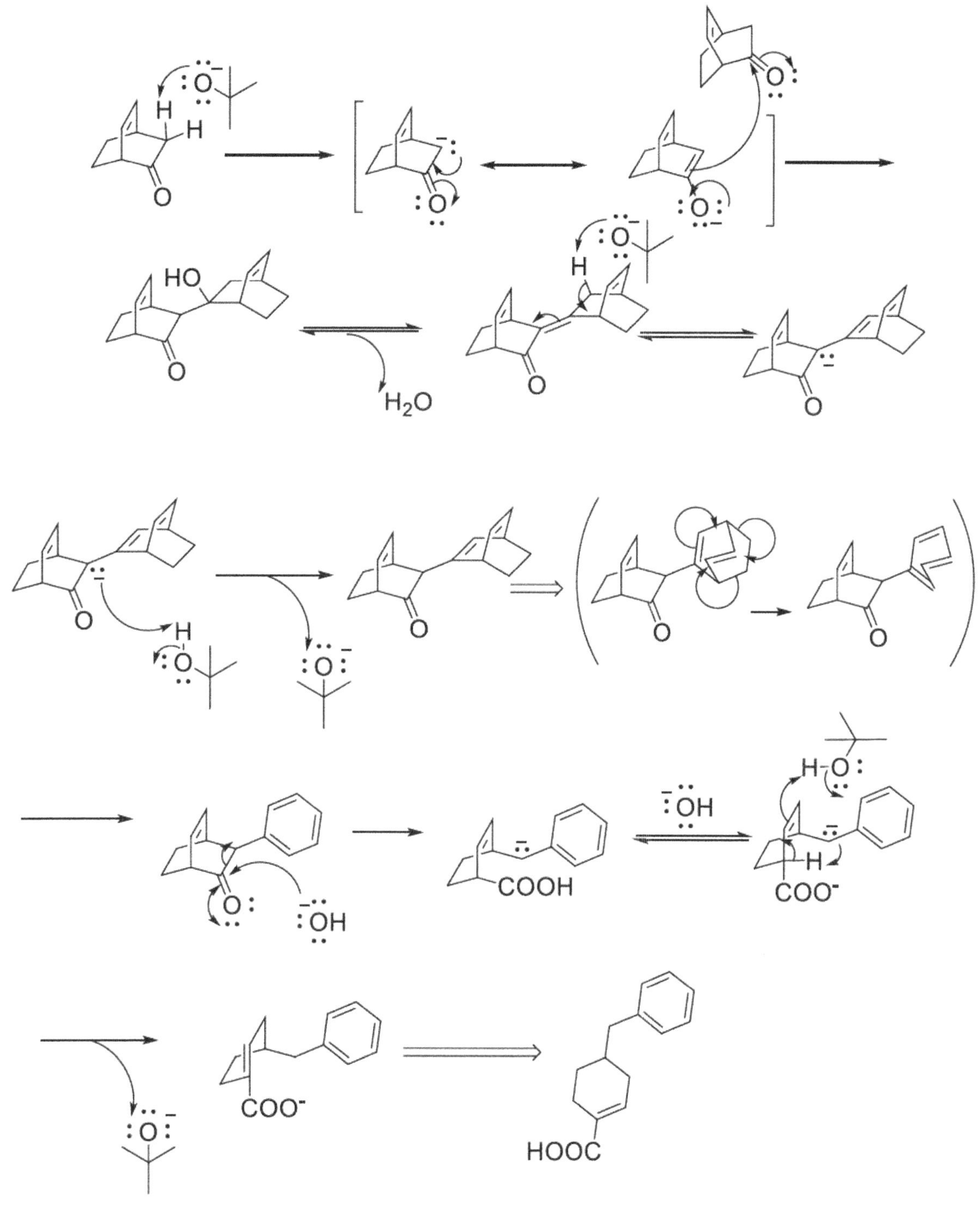
H
H
O
HO
H_2O
COOH
OH
COO-
HOOC

465

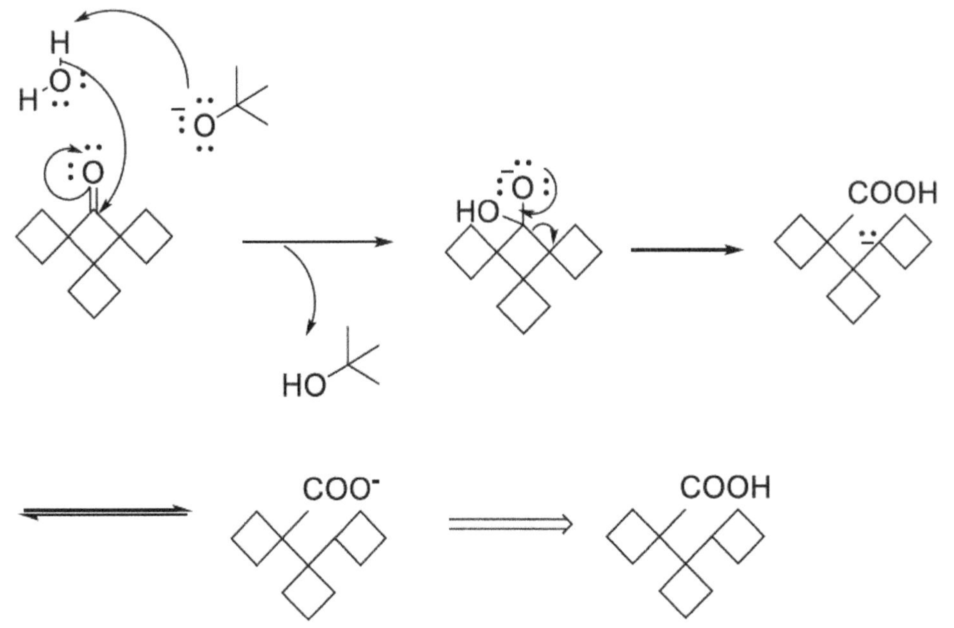
H
H
O
O
O
HO
O
HO
COOH
COO-
COOH

466

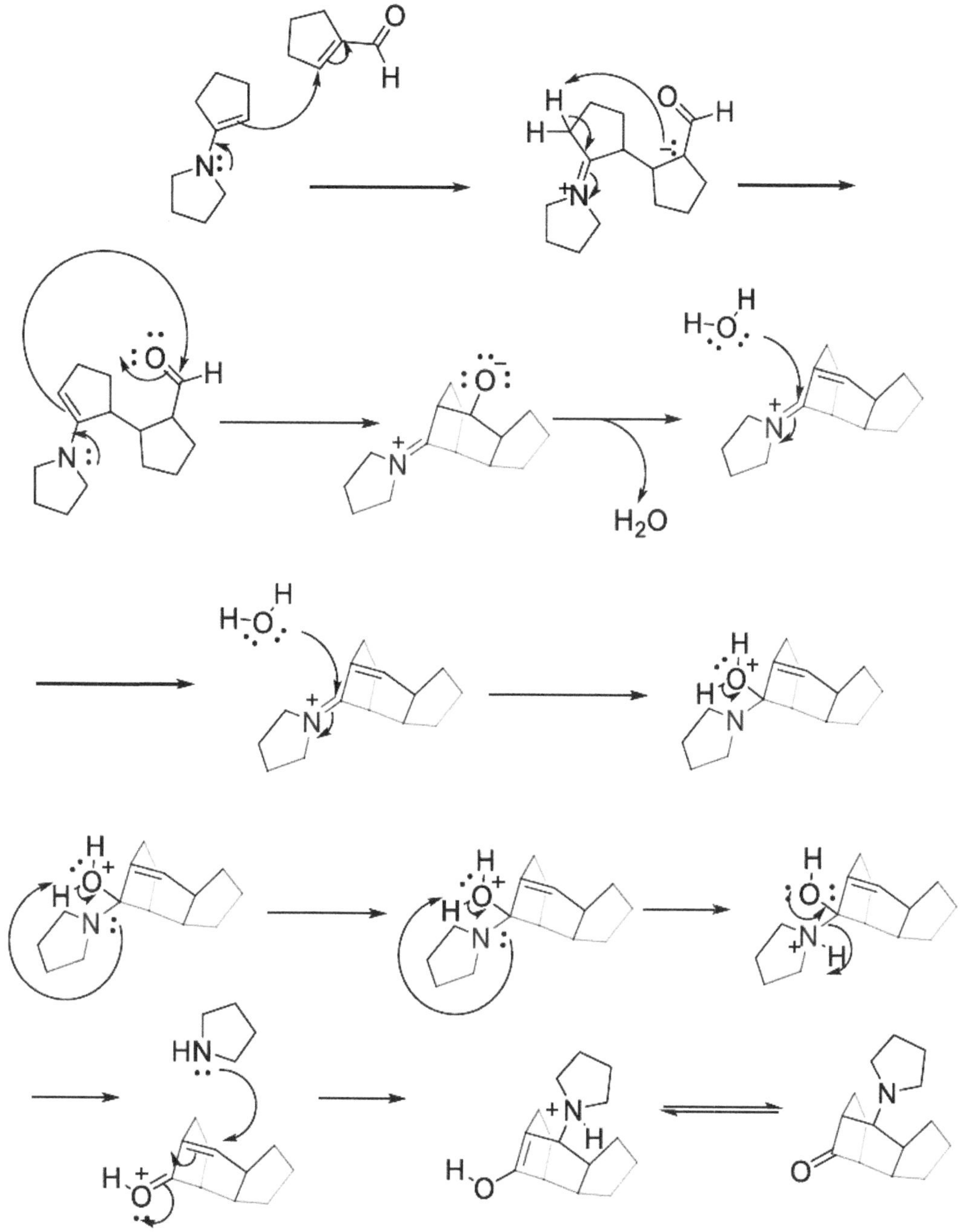
H
N
O
H
H
H
N
H
O
N
H
O
N
H
O
H
H
N
H
O
H
H
N
H
O
H
H
N
H
O
H
H
N
H
O
H
H
N
H
O
H
H
N
H
N
H
O
H
N
H
N
O

467

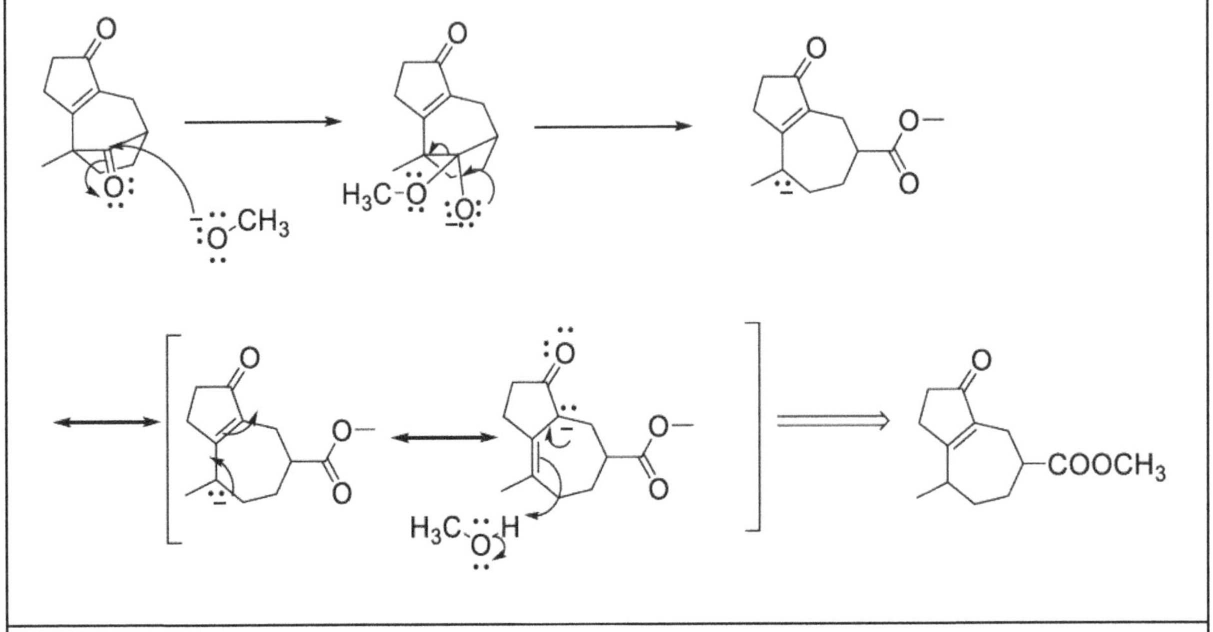

468

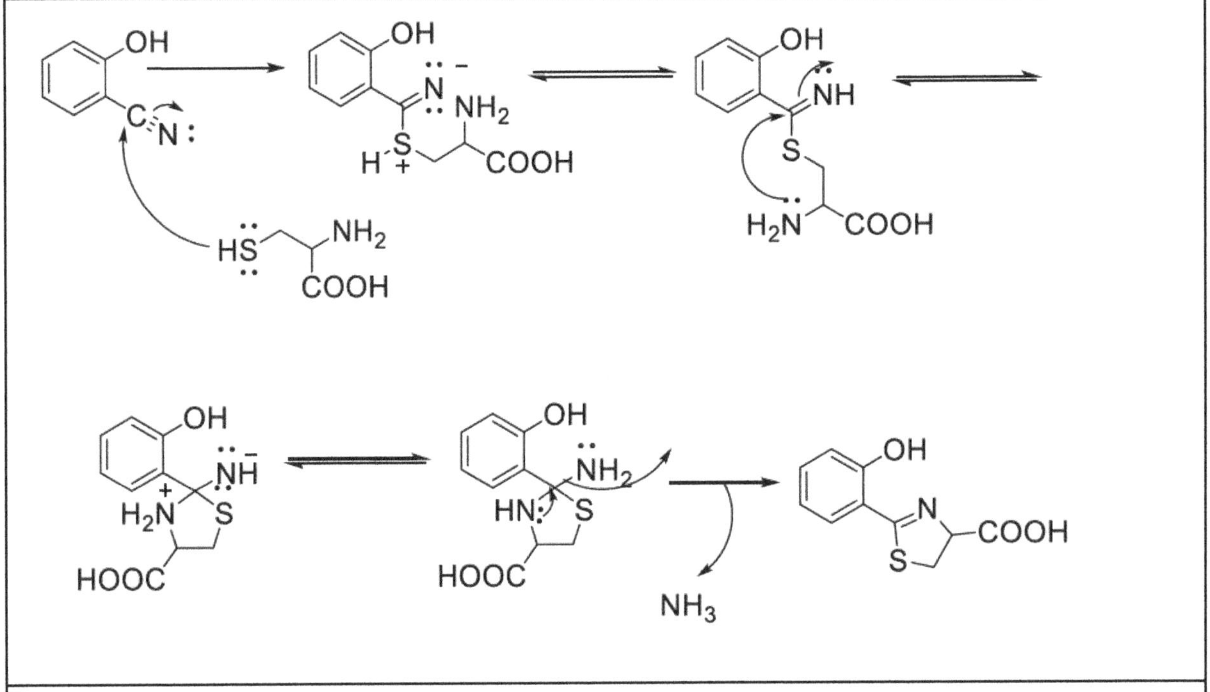

469

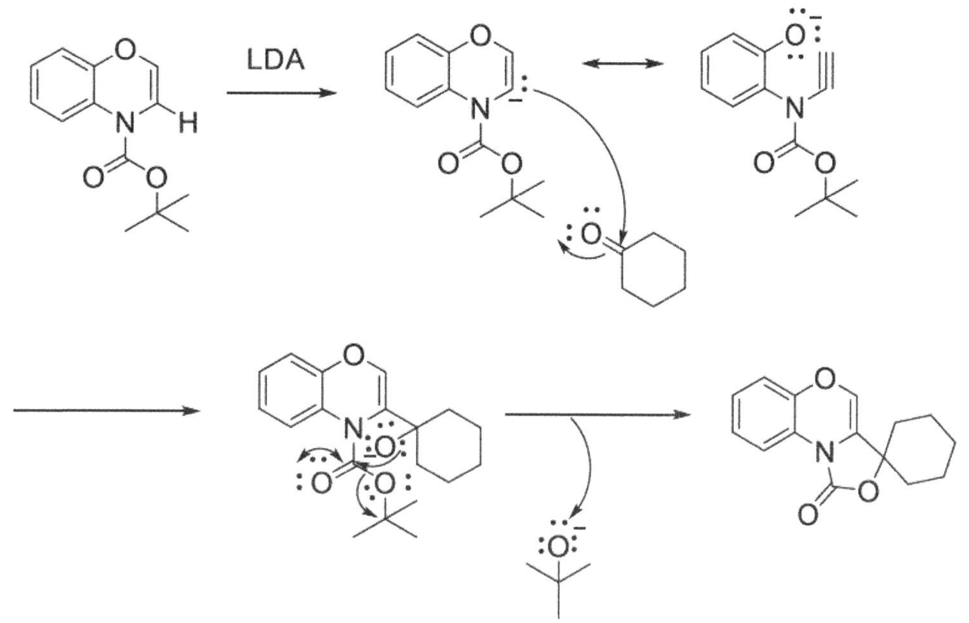

LDA

470

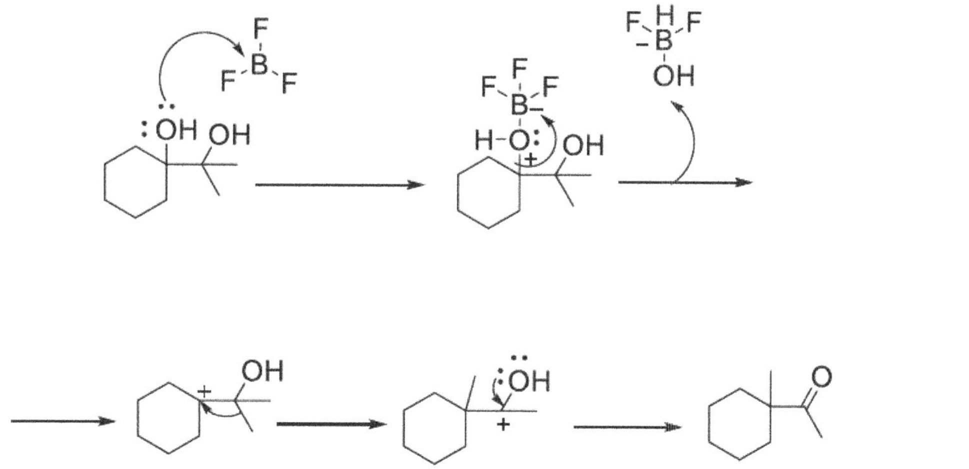

471

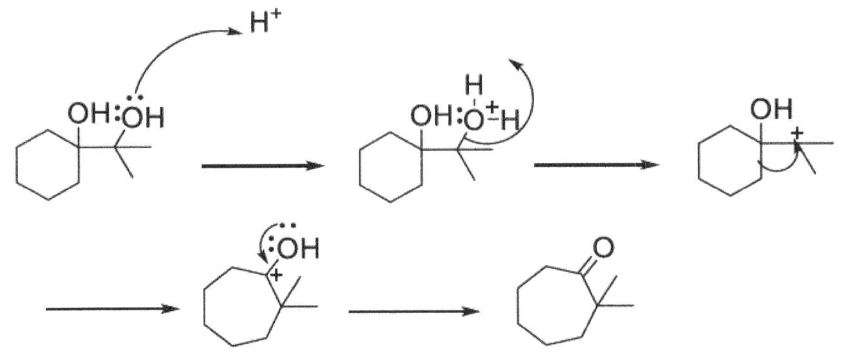
H+
OH
OH
H
OH
O
H
OH
OH
O

472

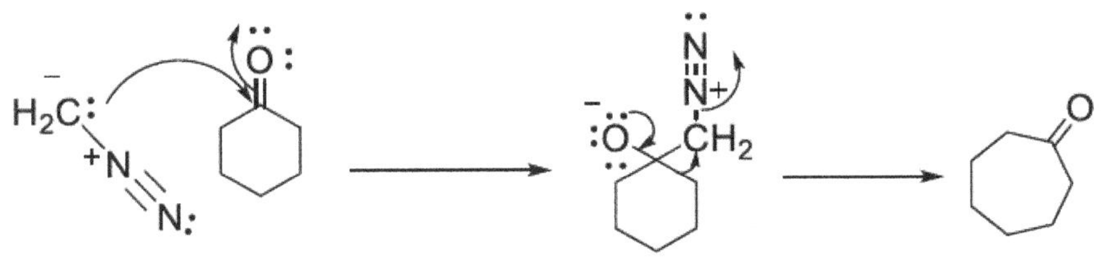
H2C
N
N
O
O
N
N+
CH2
O

473

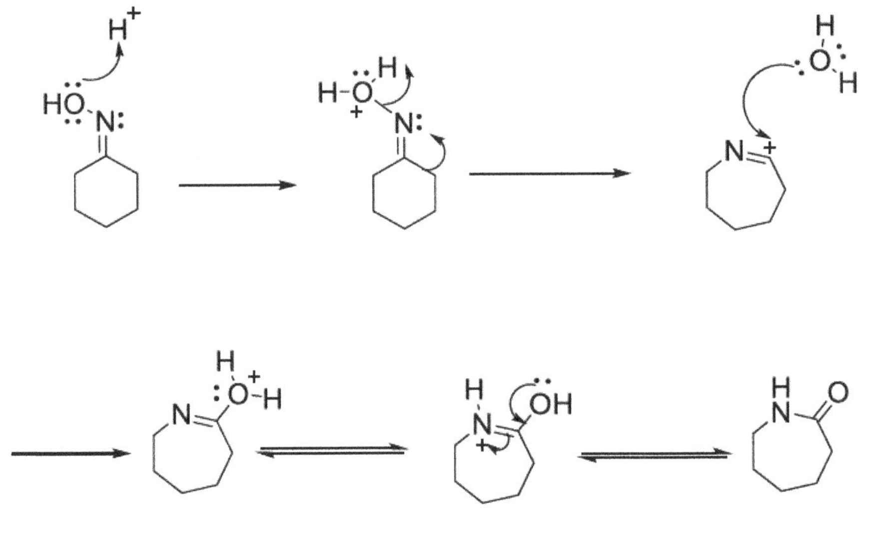
H+
HO
N
H-O
H
N
H
O
H
N
H
O-H
N
H
N
OH
H
N
O

This is an example of the *Beckman rearrangement*

474

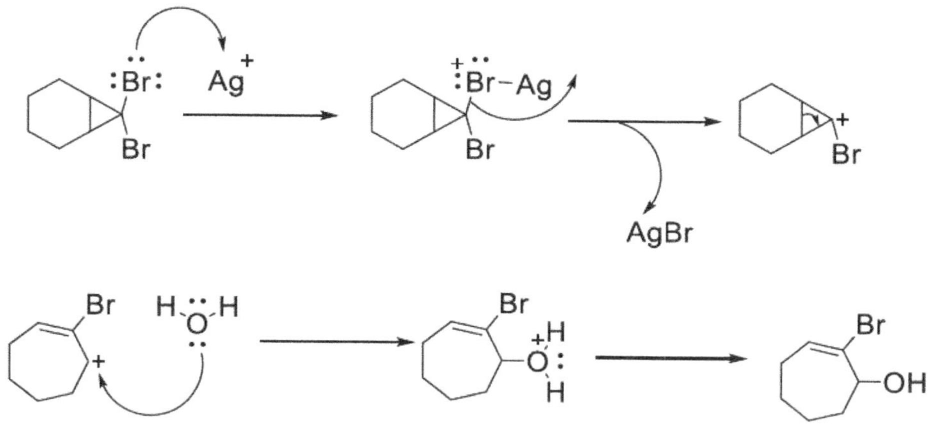

:Br:
Ag+
Br
+:Br–Ag
Br
Br
AgBr
Br
H–O–H
Br
+H
O:
H
Br
OH

475

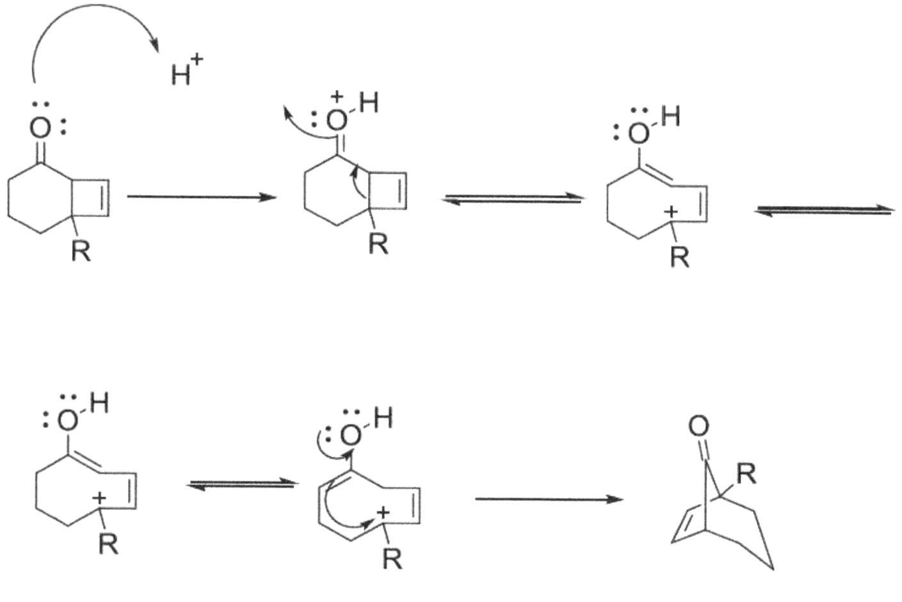

H+
O:
R
:O–H
R
:O–H
R
:O–H
R
:O–H
R
O
R

476

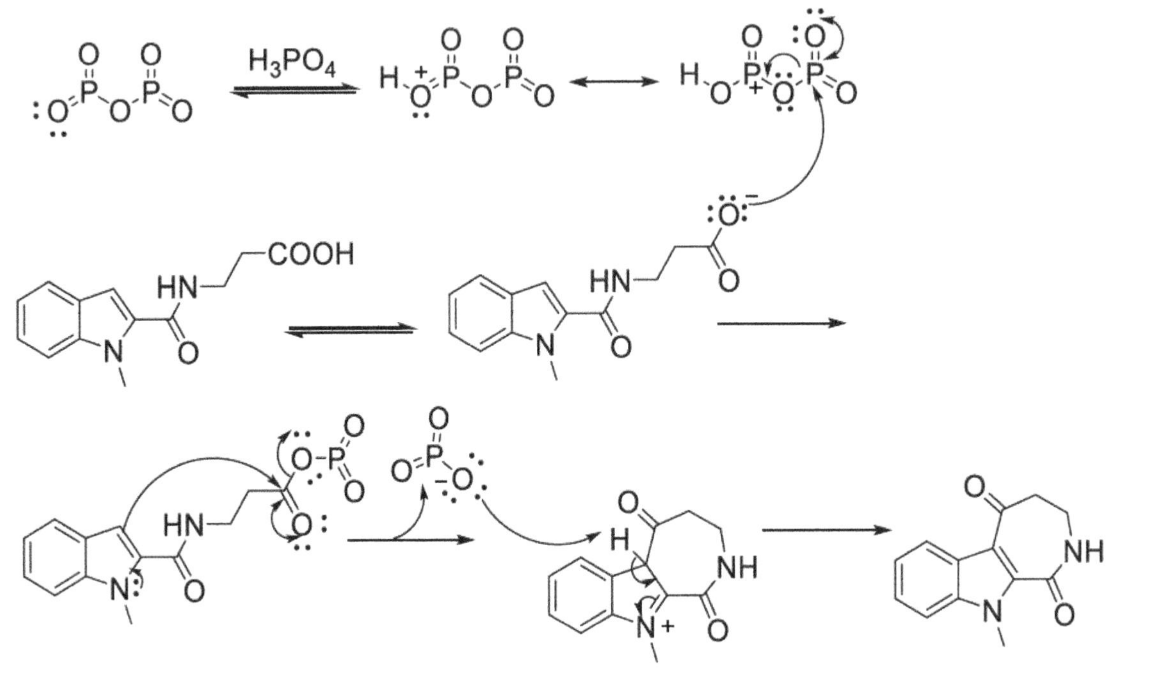
H3PO4
COOH
HN
N
NH
H
R

477

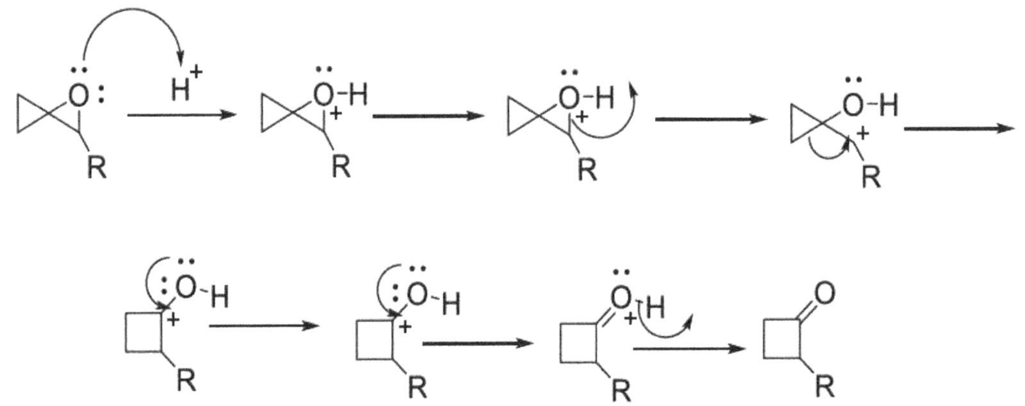
H+
O-H
R

478

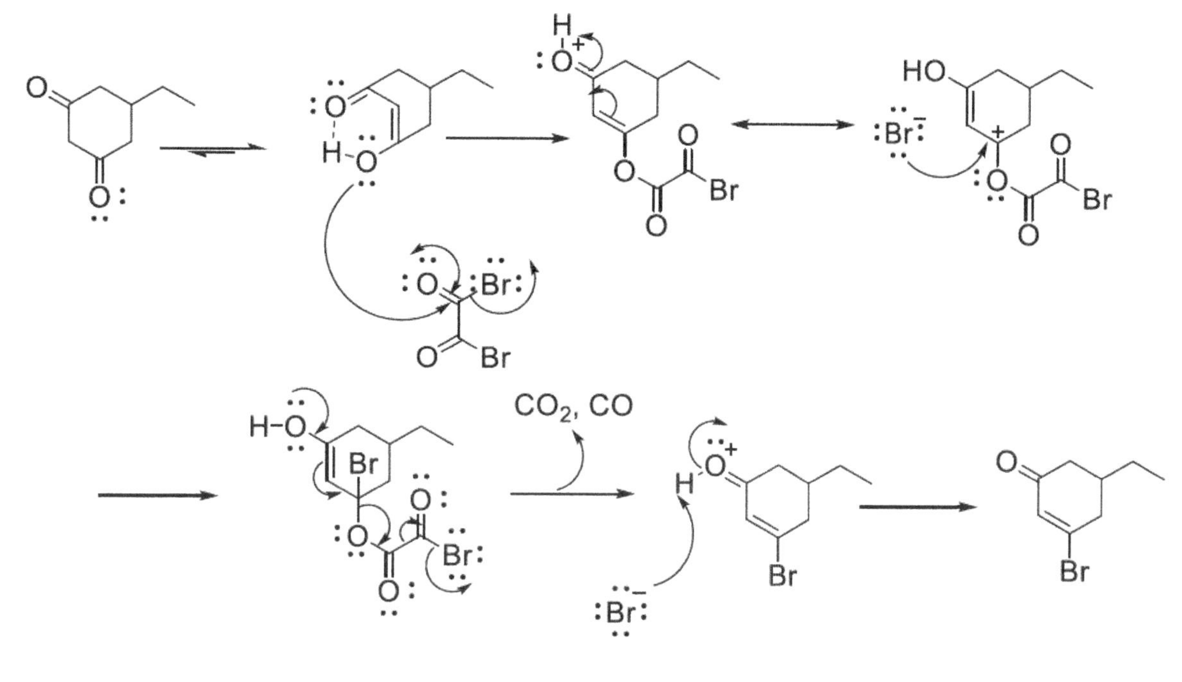

CO2, CO
Br

479

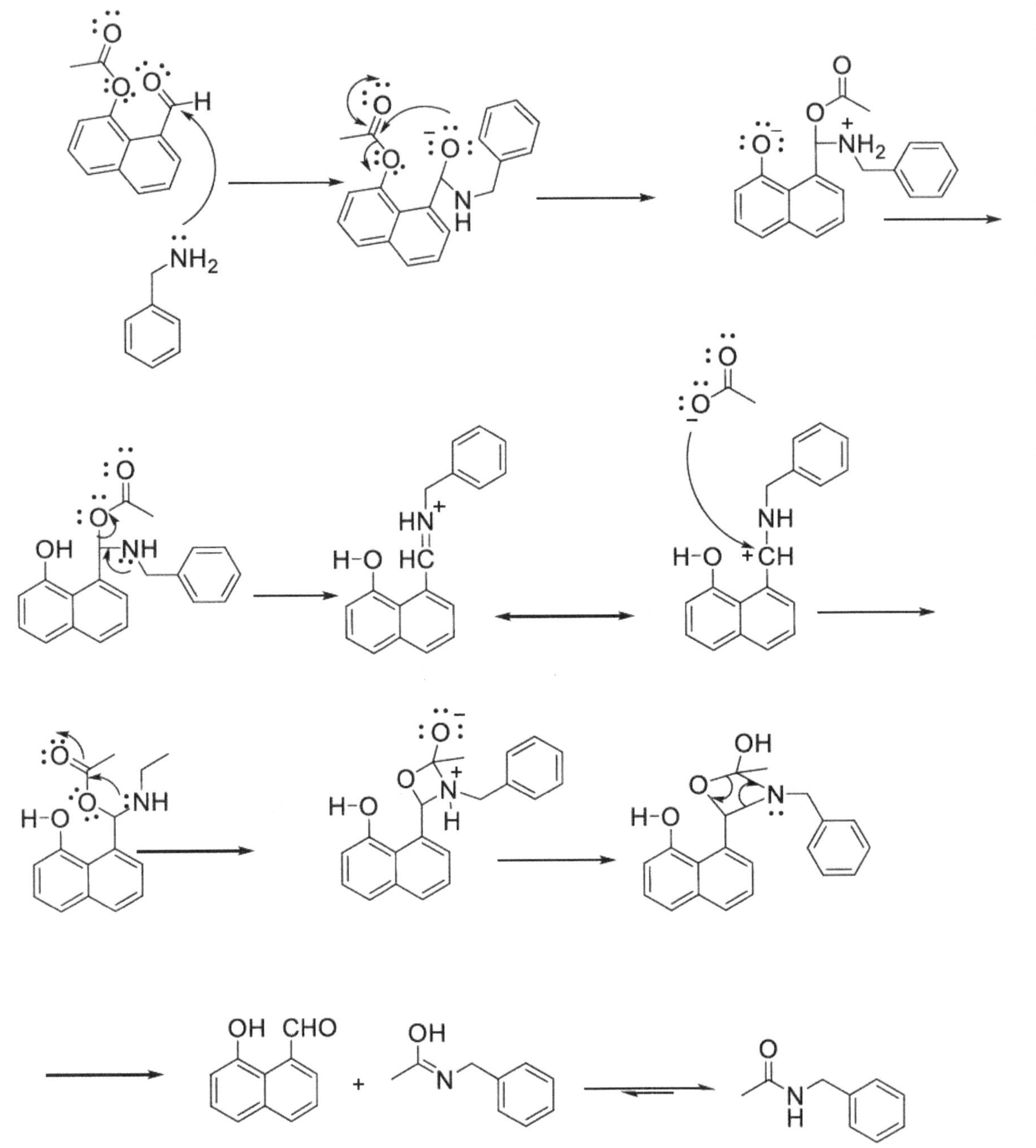

480

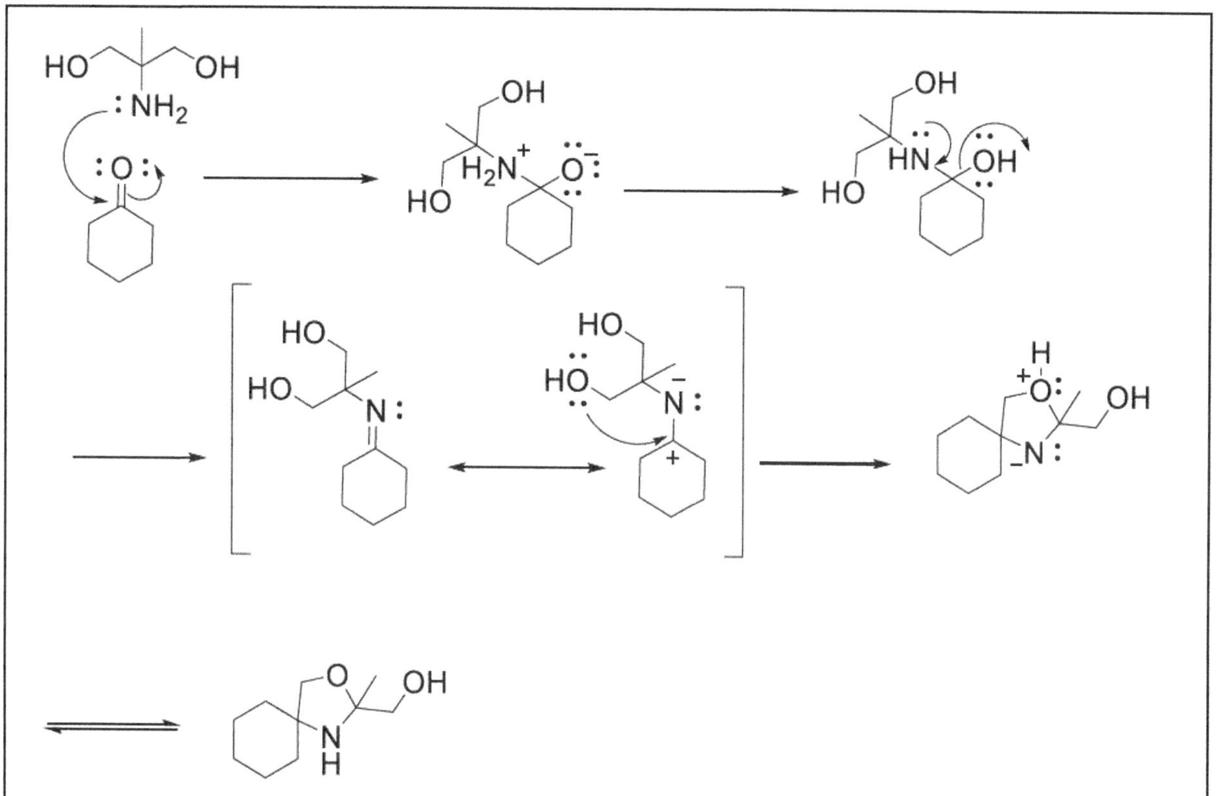
HO
OH
NH2
O
OH
H2N
O
HO
OH
HN
OH
HO
HO
HO
N
HO
HO
N
H
O
OH
N
O
OH
N
H

481

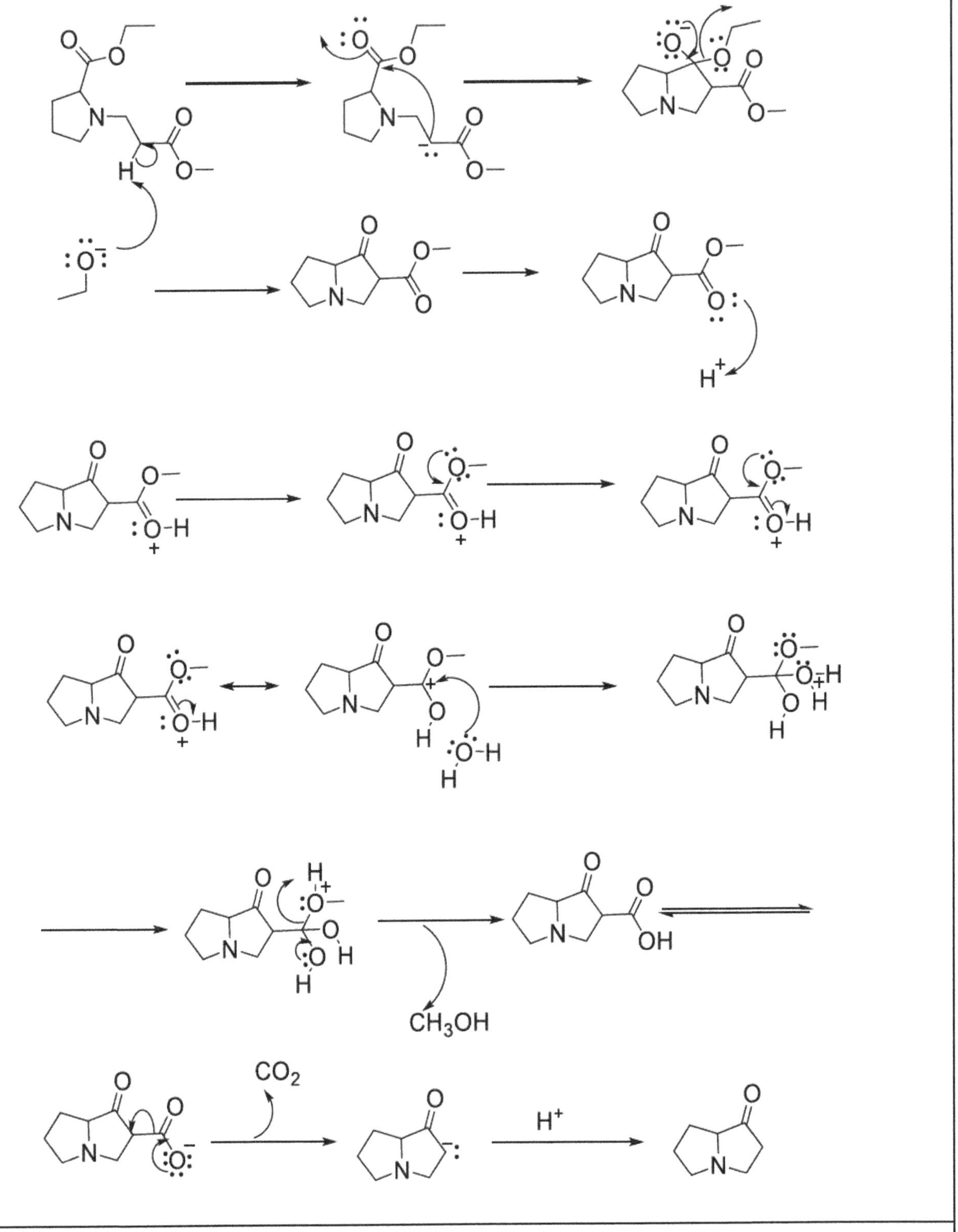
H⁺
CH3OH
CO2
H⁺

482

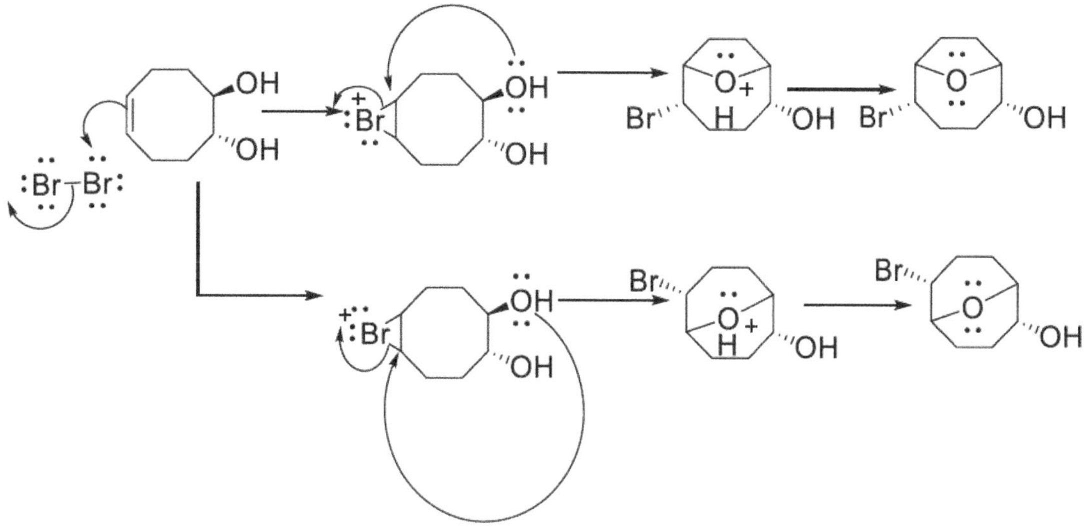

OH
OH
:Br–Br:
Br
OH
OH
O+
Br
H
OH
O
Br
OH
Br
OH
OH
Br
O+
H
OH
Br
O
OH

483

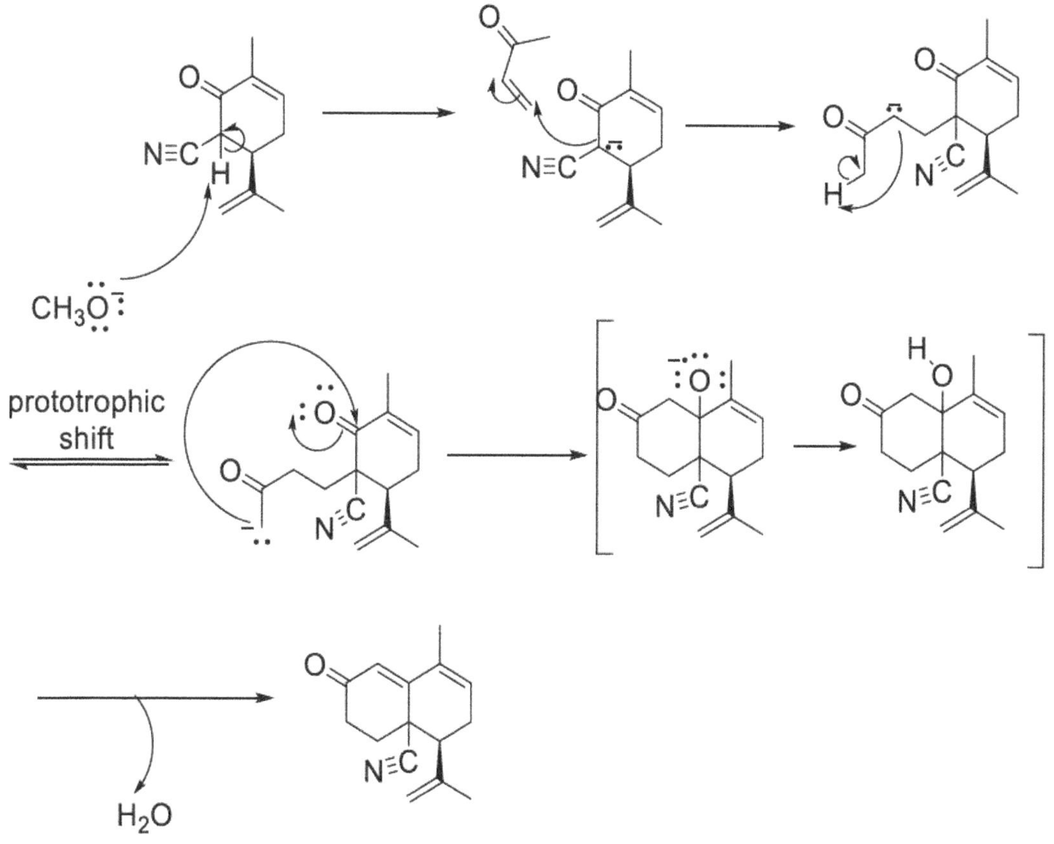

O
N≡C
H
CH_3O^-
O
O
N≡C
O
O
H
N≡C
prototrophic
shift
O
O
N≡C
O
O
N≡C
H
O
O
N≡C
O
N≡C
H_2O

484

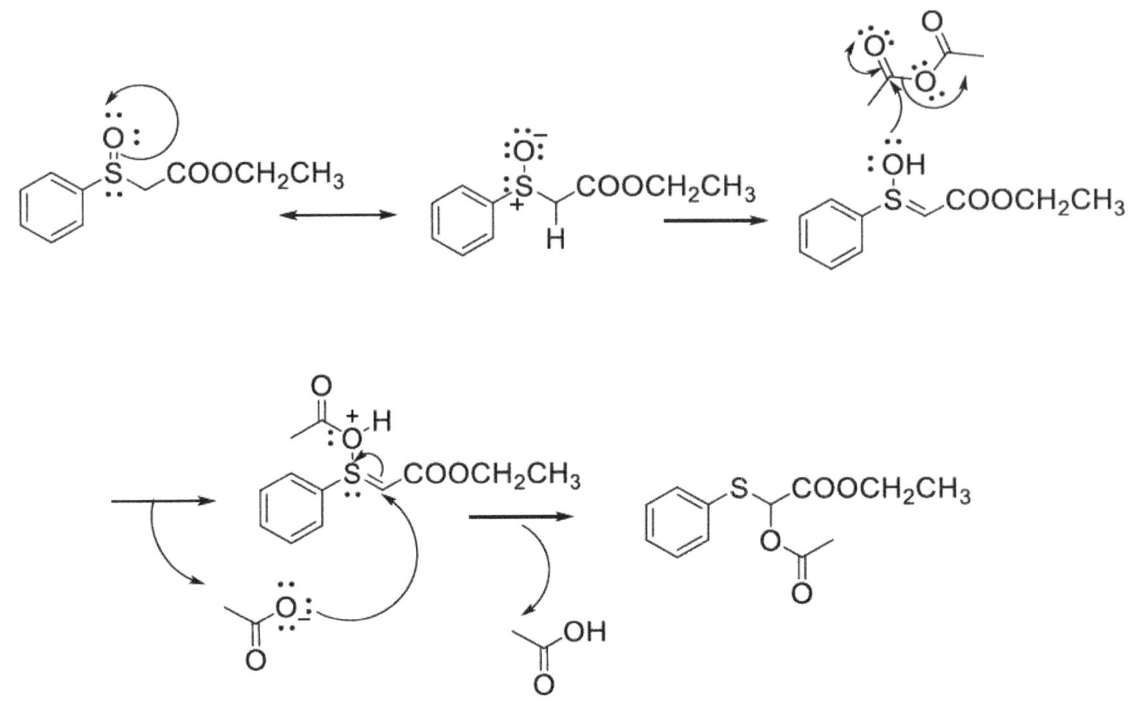
COOCH$_2$CH$_3$
COOCH$_2$CH$_3$
H
COOCH$_2$CH$_3$
OH
COOCH$_2$CH$_3$
H
COOCH$_2$CH$_3$
OH

485

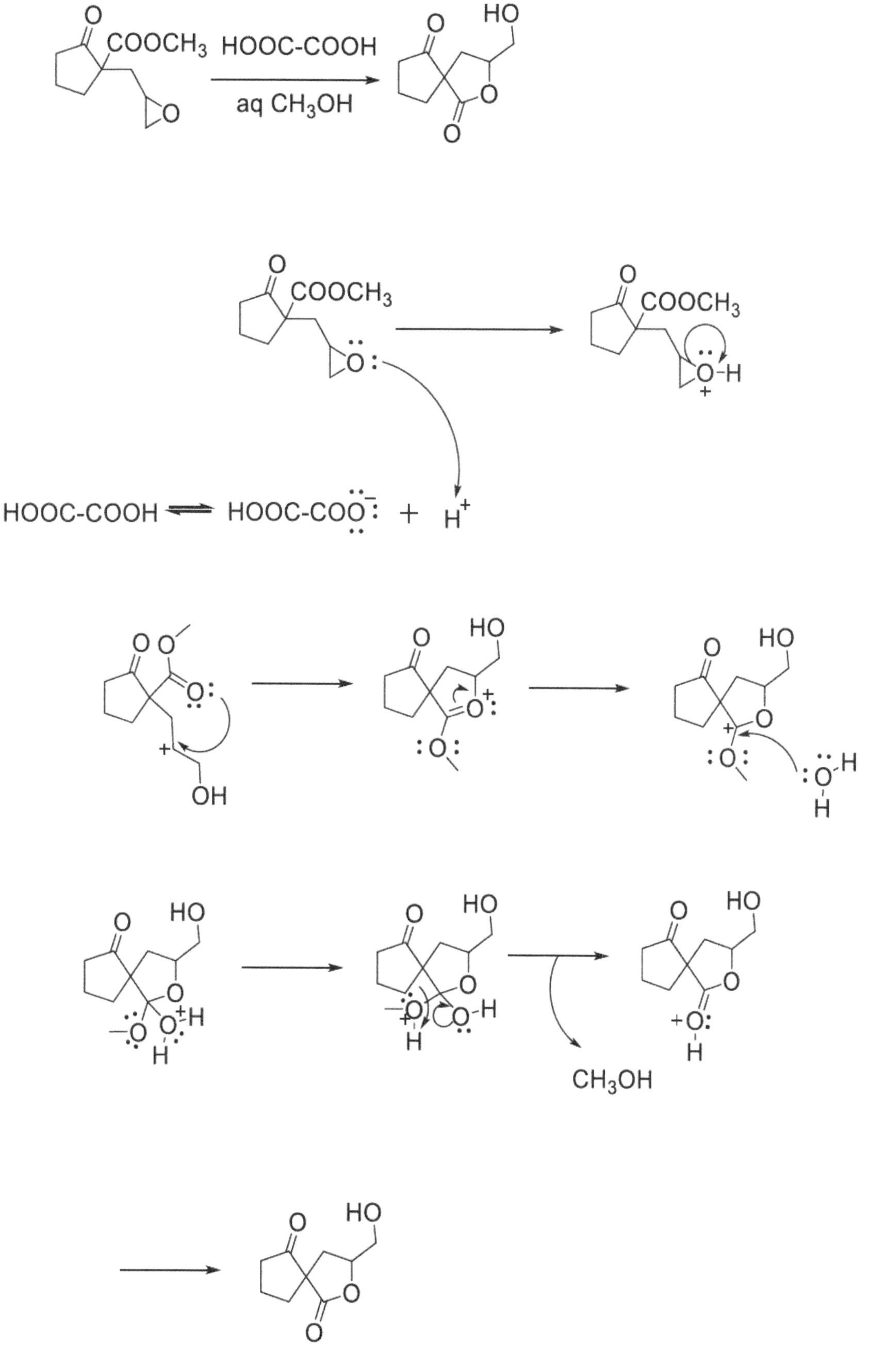
COOCH3
HOOC-COOH
aq CH3OH
HO
HOOC-COOH
HOOC-COO
H+
OH
CH3OH

486

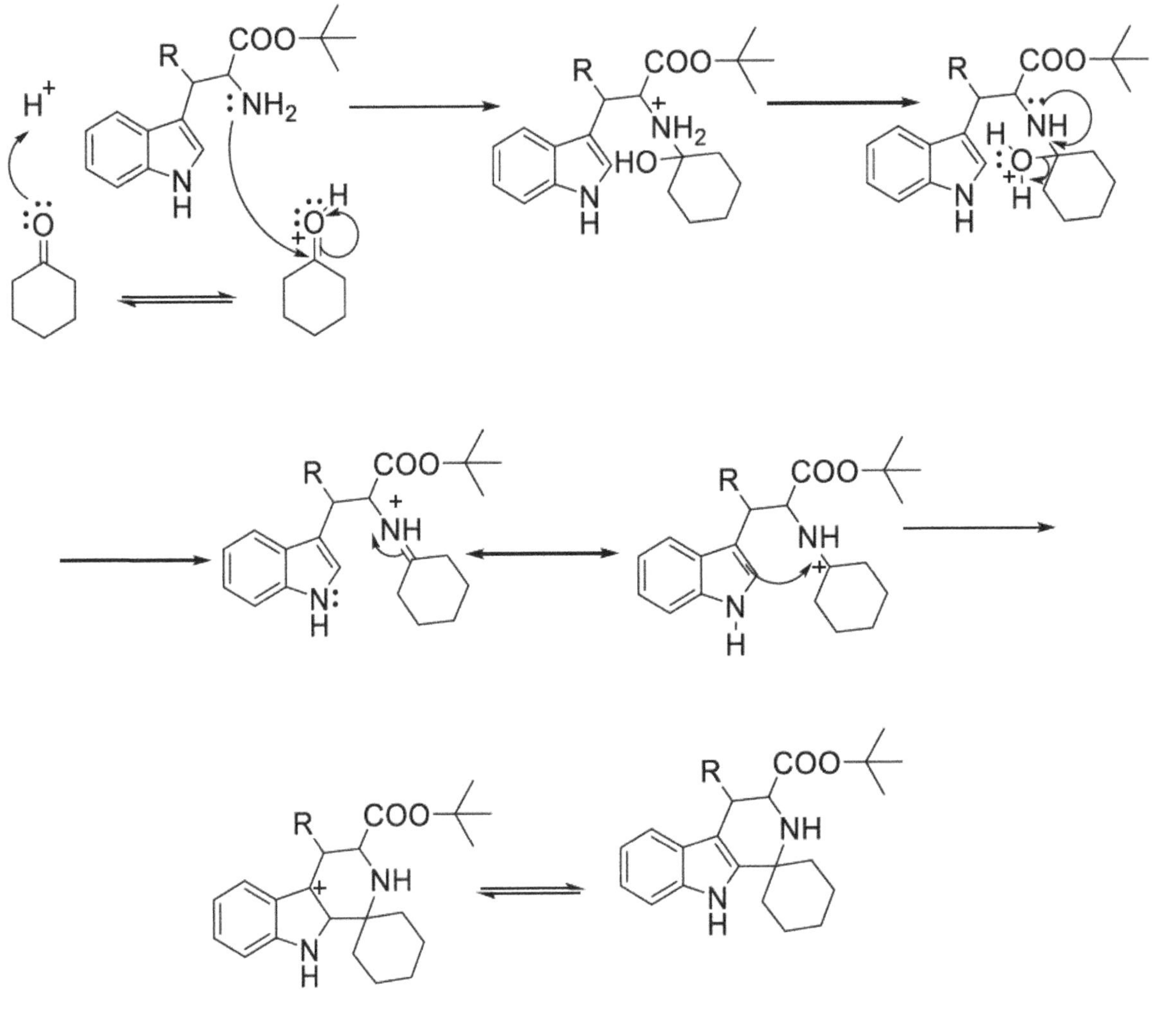

487

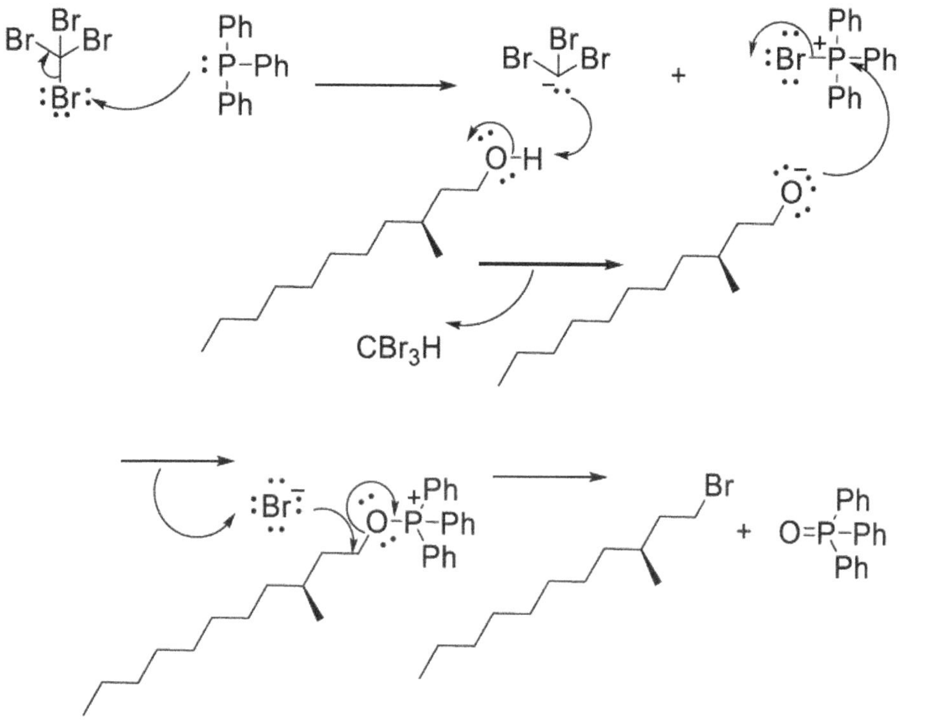
Br
Br
Br
Br
Ph
P
Ph
Ph
Br
Br
Br
+
Br
P
Ph
Ph
Ph
O
H
O
CBr3H
Br
O
P
Ph
Ph
Ph
Br
+
O=P
Ph
Ph
Ph

488

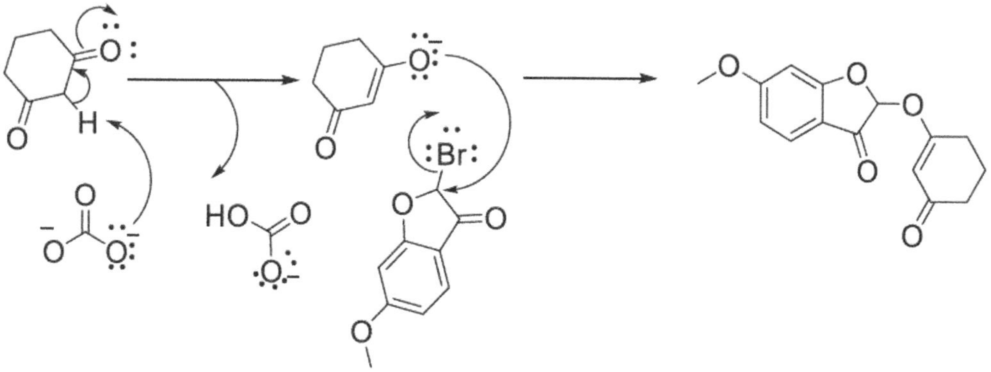
O
O
H
O
O
O
HO
O
O
O
O
Br
O
O
O
O
O
O
O
O
O

489

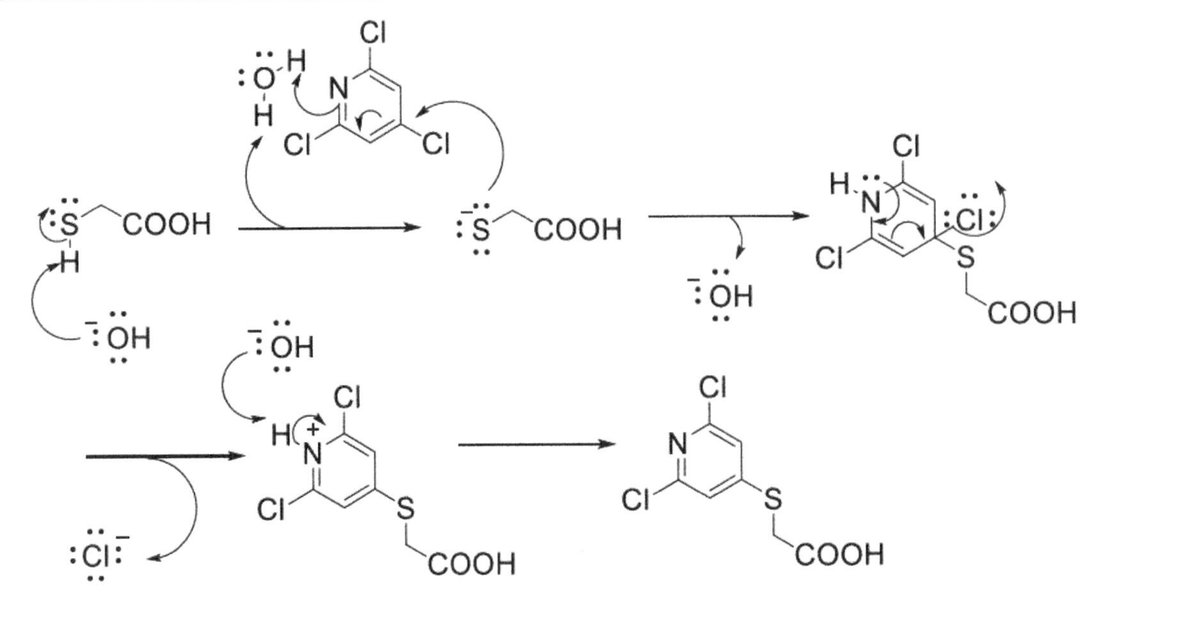

490

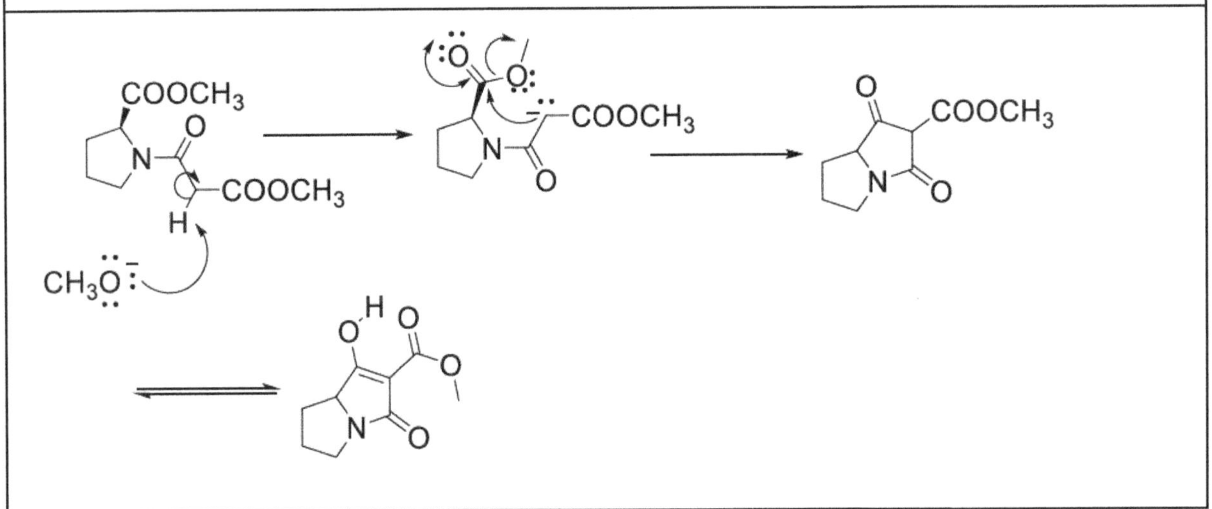

PART C
Questions 491-500

491	When the antibiotic emycin F is treated with acid, it rearranges to emycin E. Write a mechanism for this transformation. 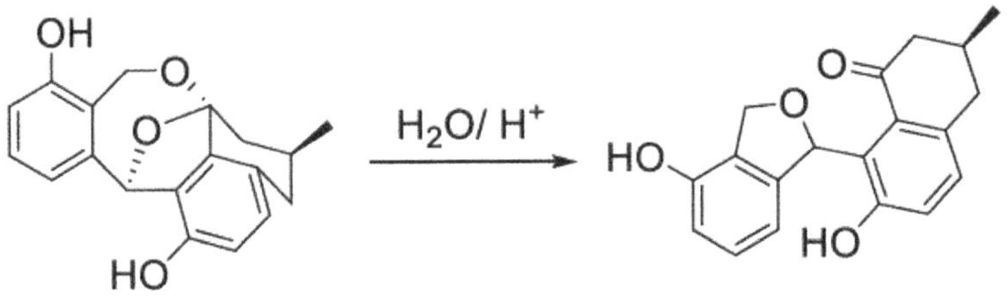 OH O O HO H_2O/ H^+ O HO O HO emycin F emycin E
492	Mechlorethamine hydrochloride (Mustargen) is a drug used in the treatment of Hodgkins disease and lymphosarcoma. Mustargen is administered intravenously, and its action lasts for only a few minutes. Its short duration of action is due to its rapid non-enzymatic hydrolysis. Write a mechanism that accounts for its rapid hydrolysis (recall that ordinary alkyl halides do not react with weak nucleophiles, such as water). 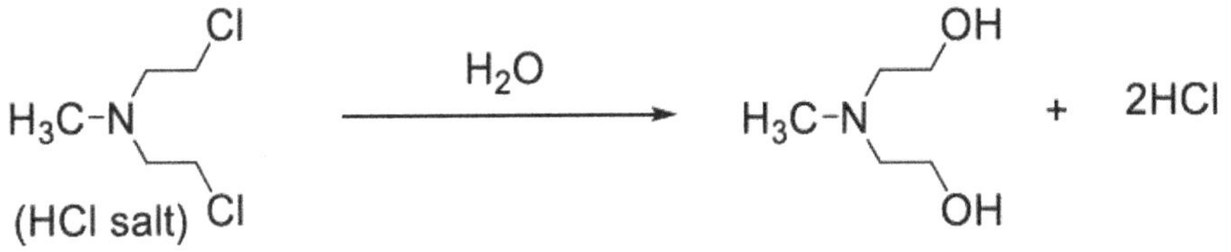 Cl H_3C-N (HCl salt) Cl H_2O OH H_3C-N OH + 2HCl
493	*A prodrug* is a biologically inactive form of a drug that in vivo yields the active form of the drug. The liberation of the drug from the prodrug can be a non-enzymatic or enzyme-catalyzed process. Prodrugs are typically used because of their greater chemical stability, better transport and/or lower toxicity. The antibiotic cycloserine has a tendency to dimerize, forming an inactive dimer (see below). In order to lower the instability of cycloserine the prodrug shown below was synthesized. It was found to be an efficacious prodrug of increased stability. The prodrug releases the active ingredient in phosphate buffer, pH 7.0. Write a mechanism for the hydrolysis of the prodrug to the formation of the cycloserine and a mechanism for the formation of the cycloserine dimer.

	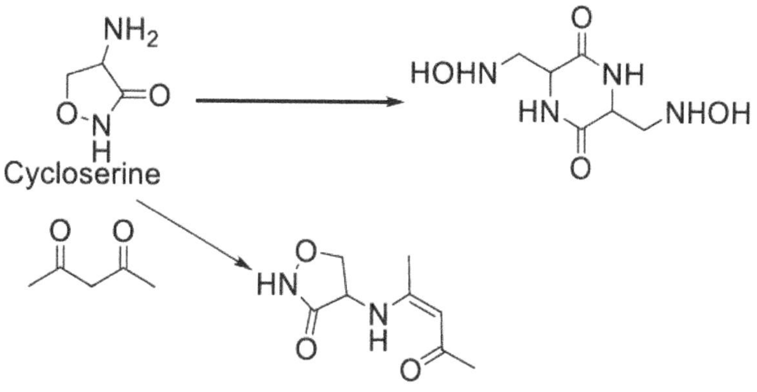
494	Mammalian cells produce a range of bioactive substances, including prostaglandins, prostacyclins, thromboxanes, and leukotrienes. These hormone-like substances act as mediators of inflammation, pain, fever, blood clotting, etc. Prostacyclin is a potent vasodilator and inhibitor of platelet aggregation produced by vascular endothelial cells. Decreased production of prostacyclin is associated with platelet aggregation and the formation of blood clots (thrombosis). The latter is the primary cause of heart attacks and stroke. Prostacyclin is transformed into the compound shown below in vivo. Write a mechanism for this reaction. 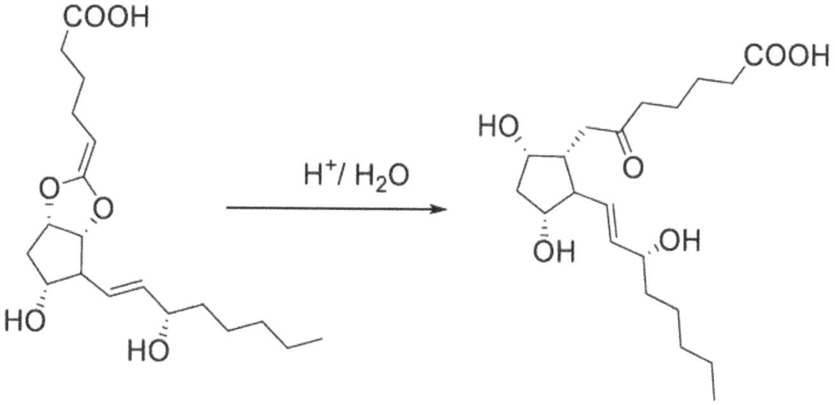 Prostacyclin

495	The antitumor antibiotic leinamycin cleaves DNA, however, it does so only in the presence of added thiols. The precise mechanism of action of leinamycin is not known. It has been suggested that nucleophilic attack of a thiol on the 1,2-dithiolan-3-one 1-oxide heterocycle present in leinamycin triggers DNA cleavage. The thiol-activated DNA-cleavage chemistry of leinamycin was probed using 1,2,-dithiolan-3-one 1-oxide as a model compound and n-propane thiol (see reaction below).Write a plausible mechanism for this reaction. 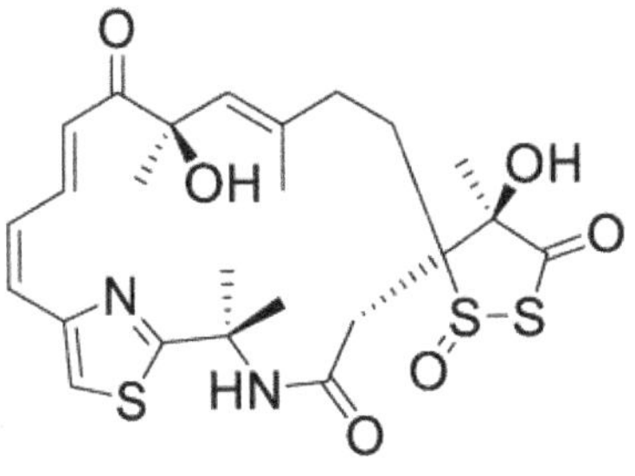 leinamycin 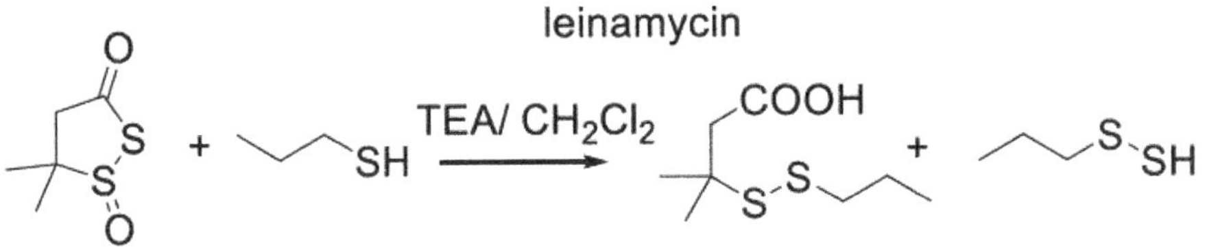 TEA/ CH_2Cl_2 COOH SH +
496	The prodrug shown below yields the active ingredient (chloral, a sedative) in the stomach. Write a mechanism for this transformation. 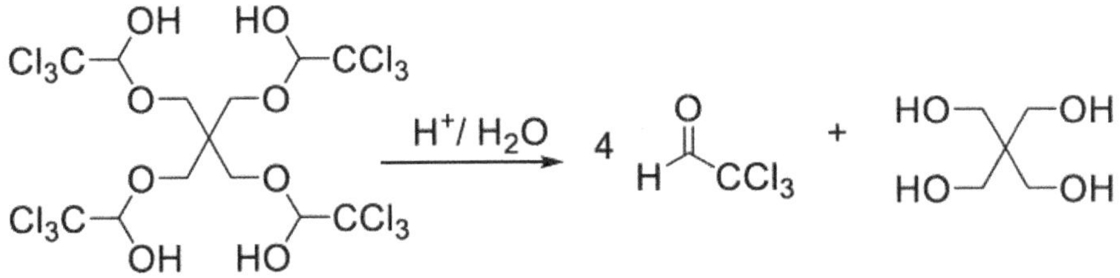 H^+/ H_2O 4 Cl_3C CCl_3 HO OH
497	Solutions of the broad-spectrum antibiotic chlortetracycline hydrochloride lose their therapeutic potency with time. This is due to rapid epimerization at the C-4 position, forming an epimerized product having greatly reduced antibiotic activity. Write a mechanism for the epimerization reaction.

	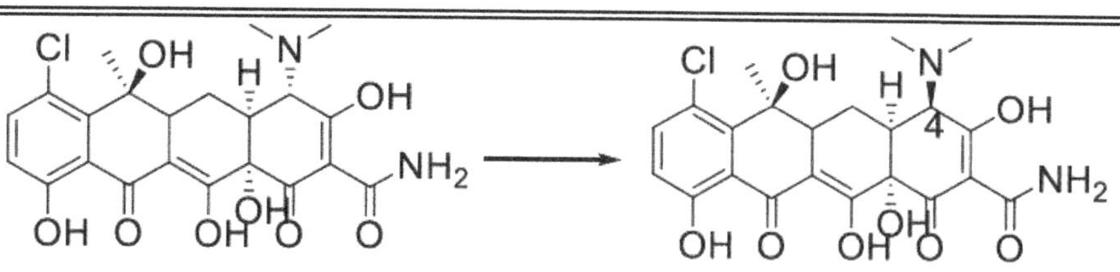 Chlortetracycline hydrochloride (Aureomycin)
498	The intracellular enzyme CMP-KDO synthetase is the key enzyme in the biosynthesis of the lipopolysaccharide (LPS) of gram-negative bacteria. Compounds that inhibit this enzyme are of potential therapeutic value as antibacterial agents. Compound 1 is a potent inhibitor of CMP-KDO synthetase in vitro but is inactive as an antibacterial agent in vivo. In an effort to circumvent this problem a series of simple esters were synthesized and investigated for their antibacterial activity. None of the esters showed any activity. Ultimately, compound 2 was conceived which was found to be a highly effective antibacterial agent. The design of prodrug 2 was based on the following biological and chemical considerations. (a) gram-negative bacteria contain significant quantities of glutathione (gama-glu-cys-gly); (b) in vitro disulfide exchange reactions are facile; (c) the disulfide exchange was expected to be irreversible because of the "peri" effect; (d) when model compound 3 was reacted with n-propanethiol in the presence of triethylamine at room temperature, compound 4 [($C_{11}H_8S$; ^1H NMR: 4.78 (S,2H), 7.20-7.60 (m,4H)] was formed. Write a mechanism for the in vivo conversion of 2 to 1, and comment on the biochemical and chemical rationale underlying the design of compound 2.

	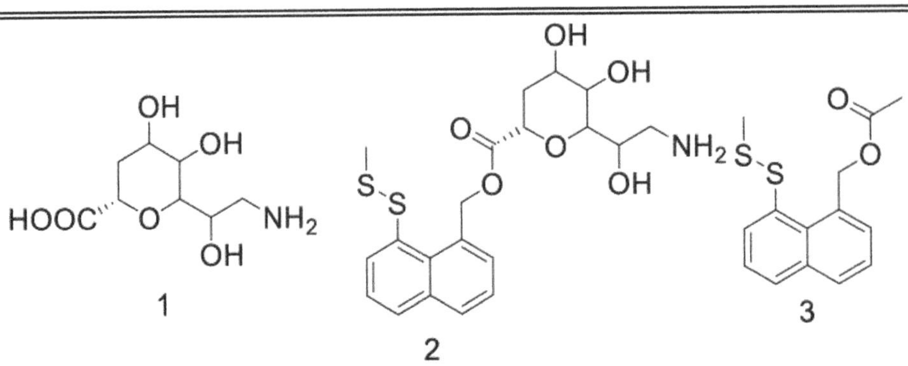 1 2 3
499	Isoilludin M behave as a bifunctional alkylating agent, yielding the aromatic product shown below. Write a plausible mechanism for this reaction. 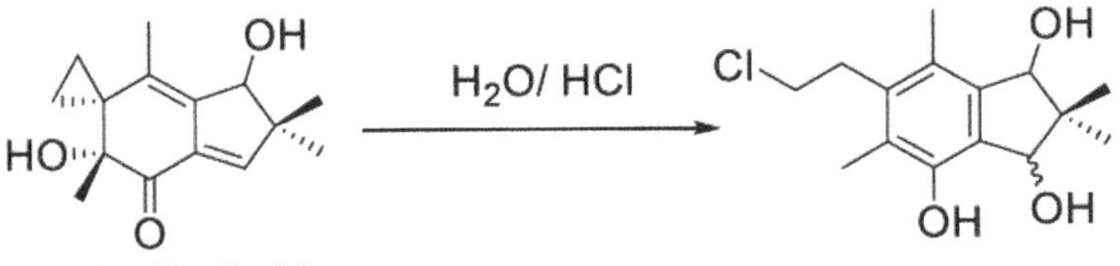 H_2O/ HCl isoilludin M
500	The natural product Bripiodionen is an inhibitor of human cytomegalovirus protease. When dissolved in methanol for a prolonged period of time, bripiodionen undergoes geometric isomerization to compound X. The chemical shift of H-8 in bripiodionen is 7.45 and 7.61 in compound X. What is the structure of X? Write a mechanism for the formation of X. 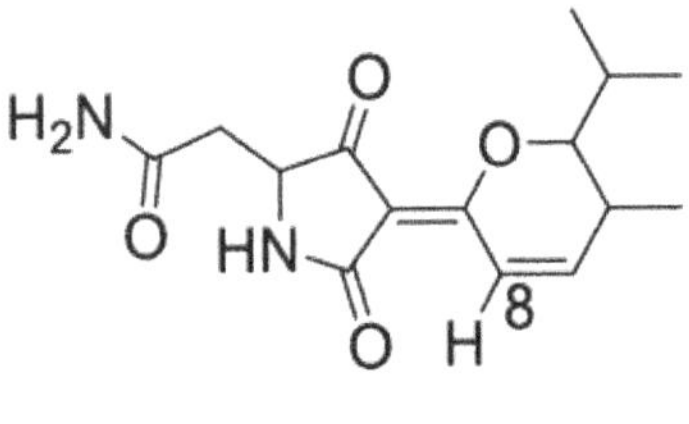 Bripiodionen

Answers to Questions 491-500

491

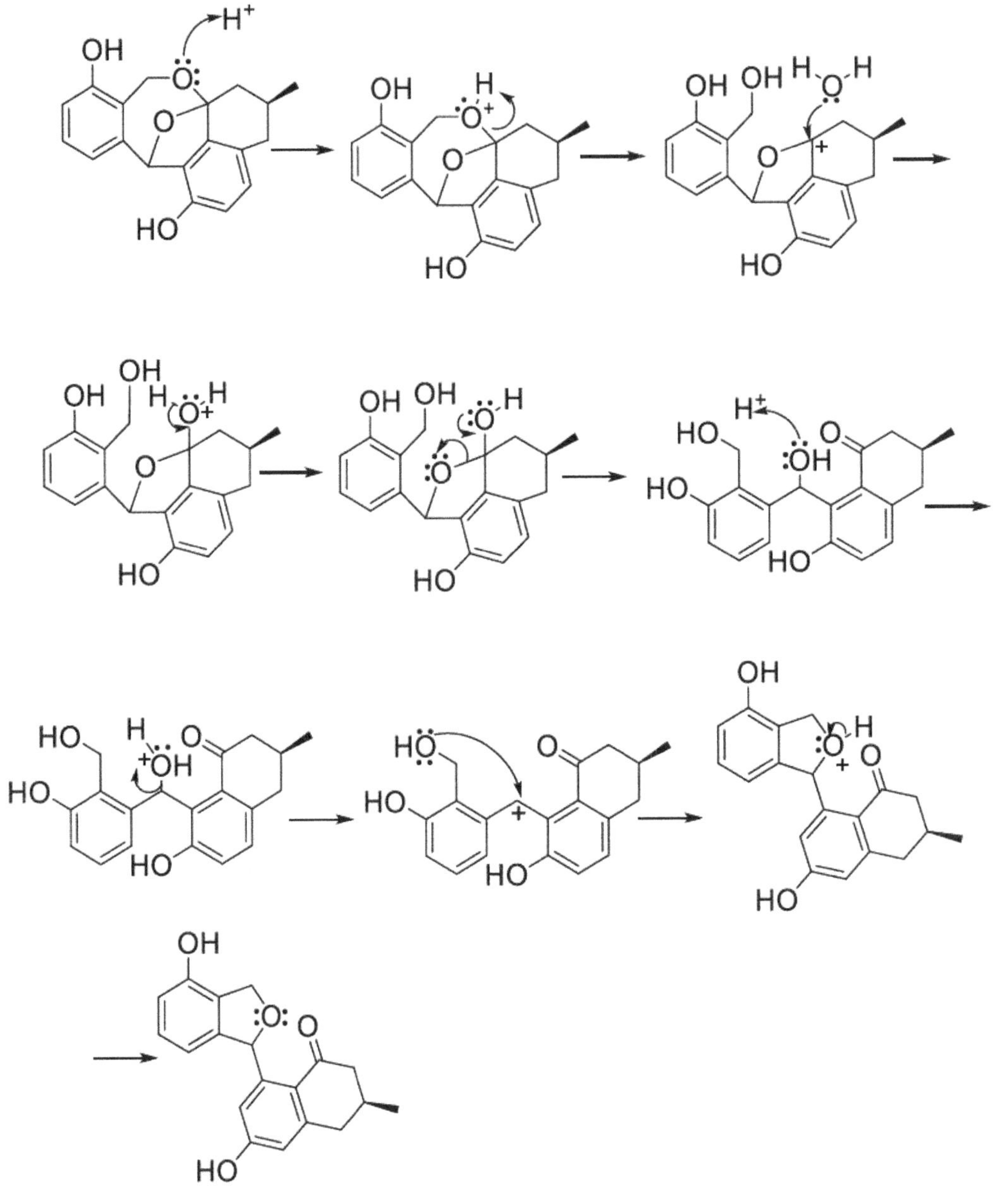

H+
OH
O
HO
H
OH
H
+

492

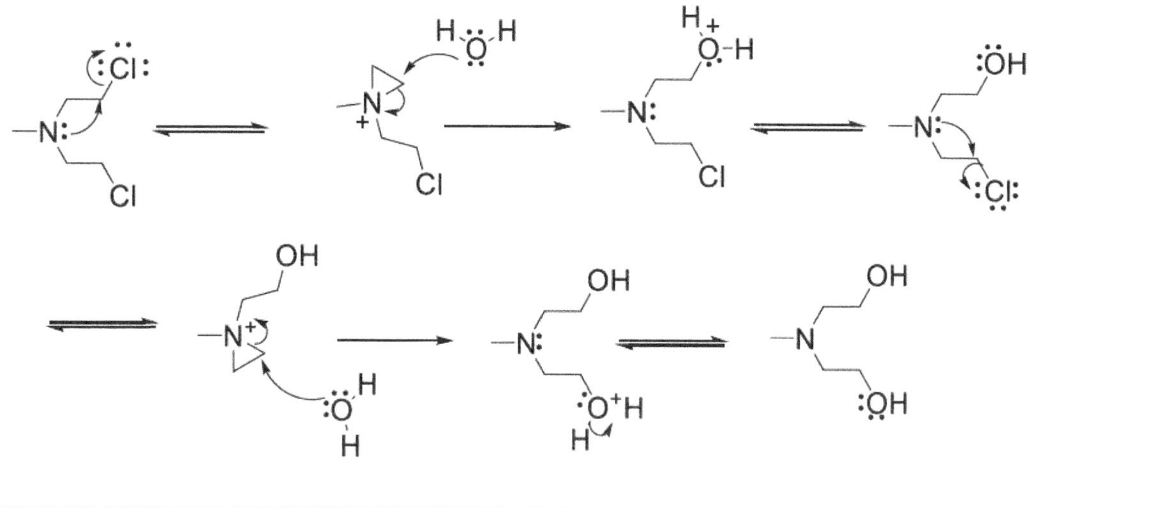

493

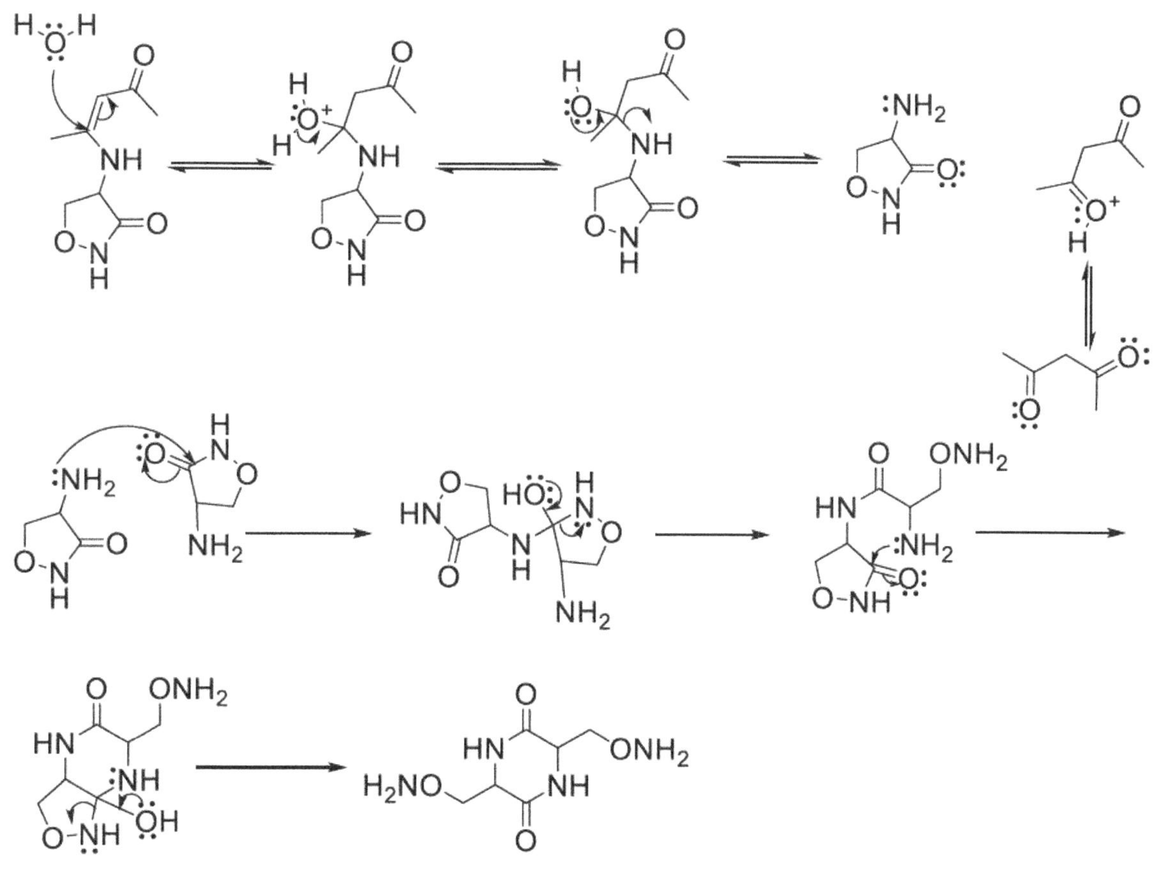

494

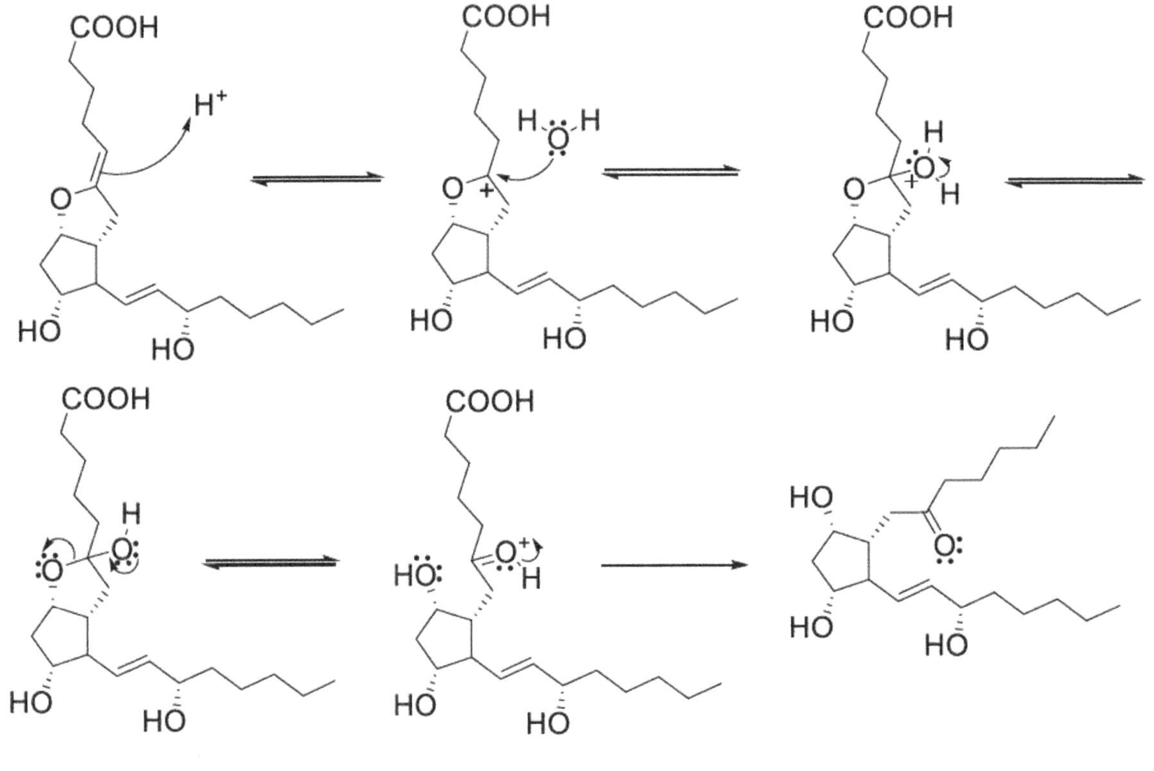

495

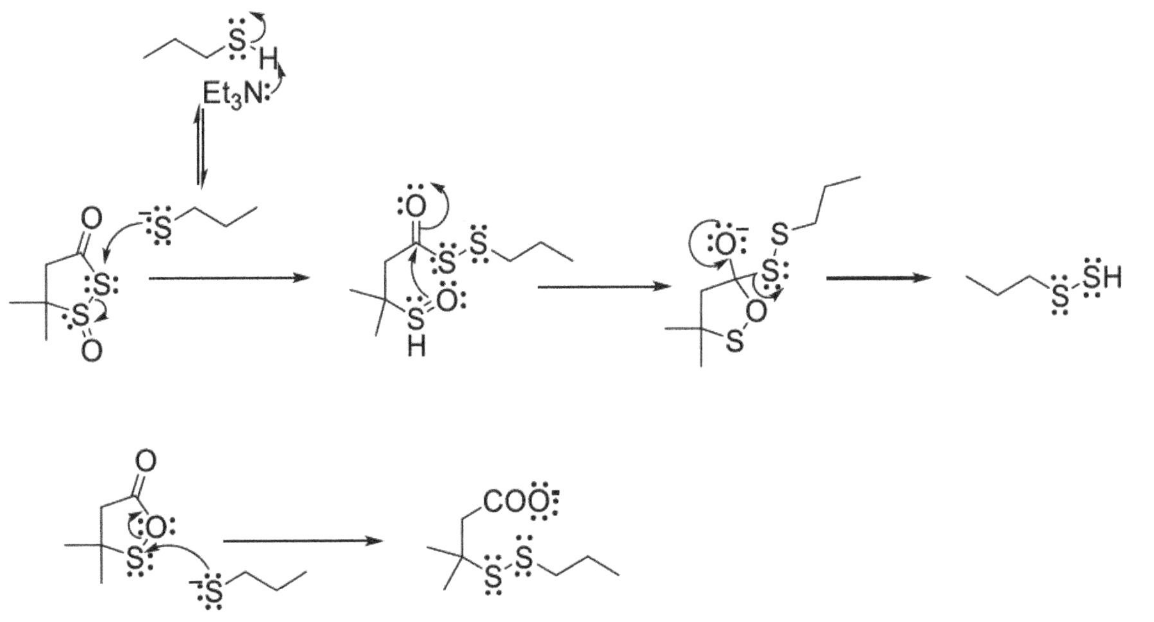

496

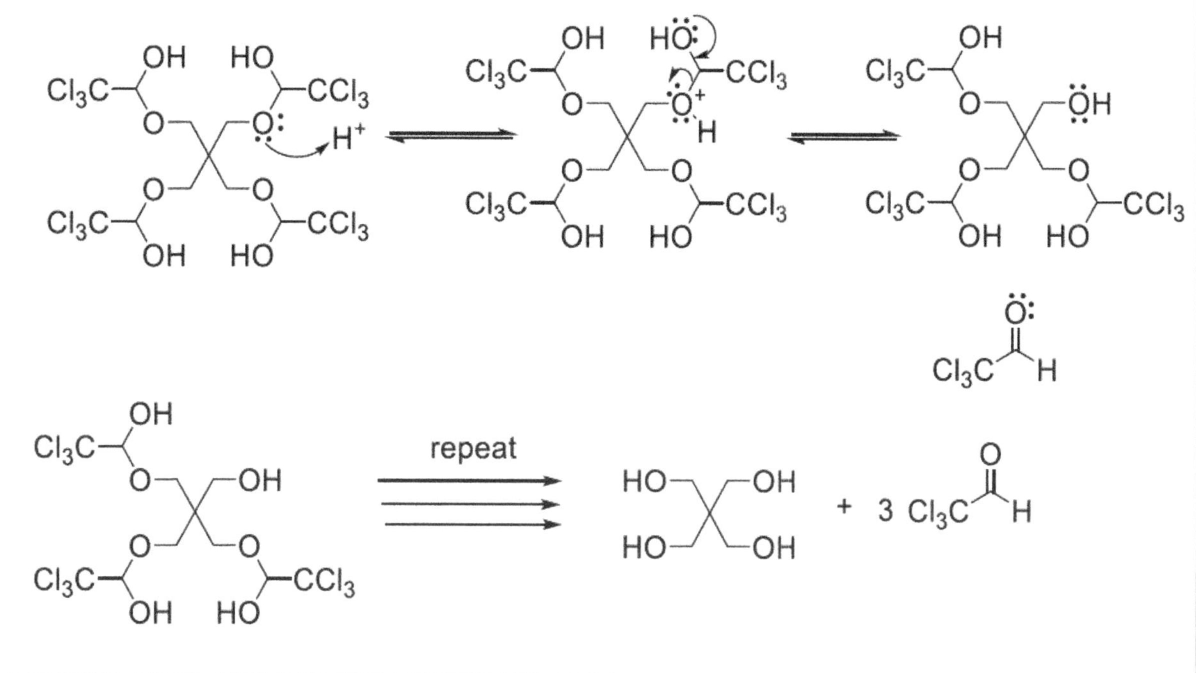

497

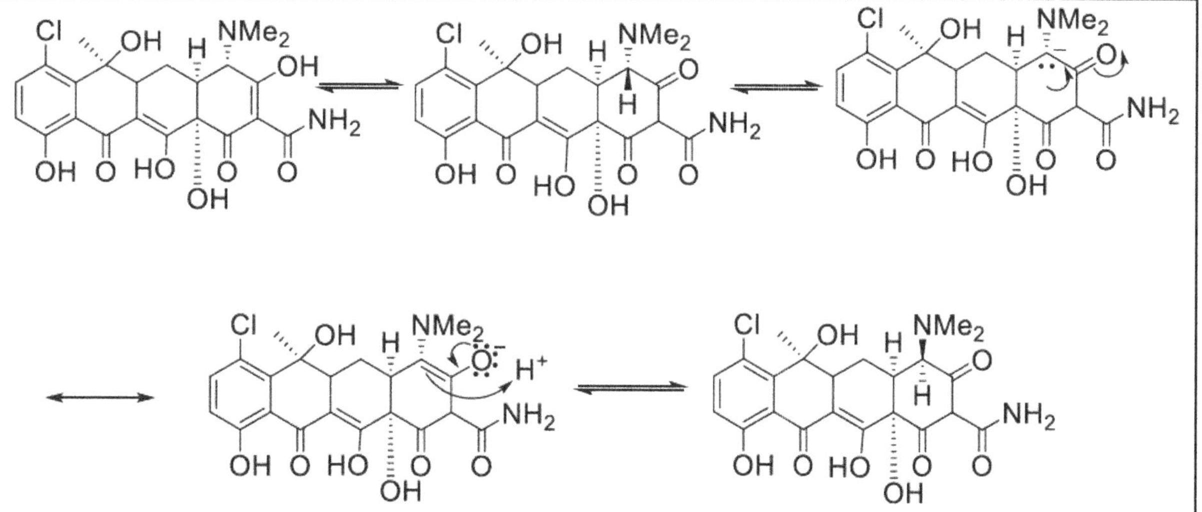

498

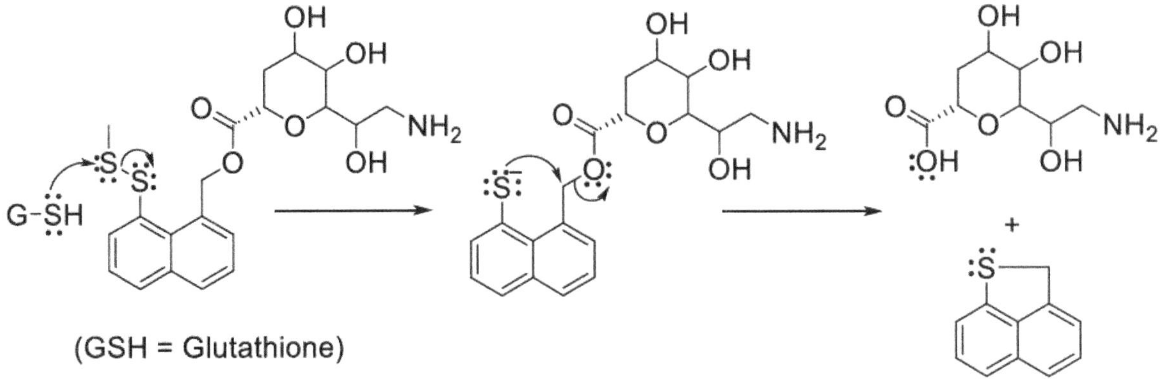

OH
OH
O
O
NH2
OH
S
S
G-SH
(GSH = Glutathione)
+

499

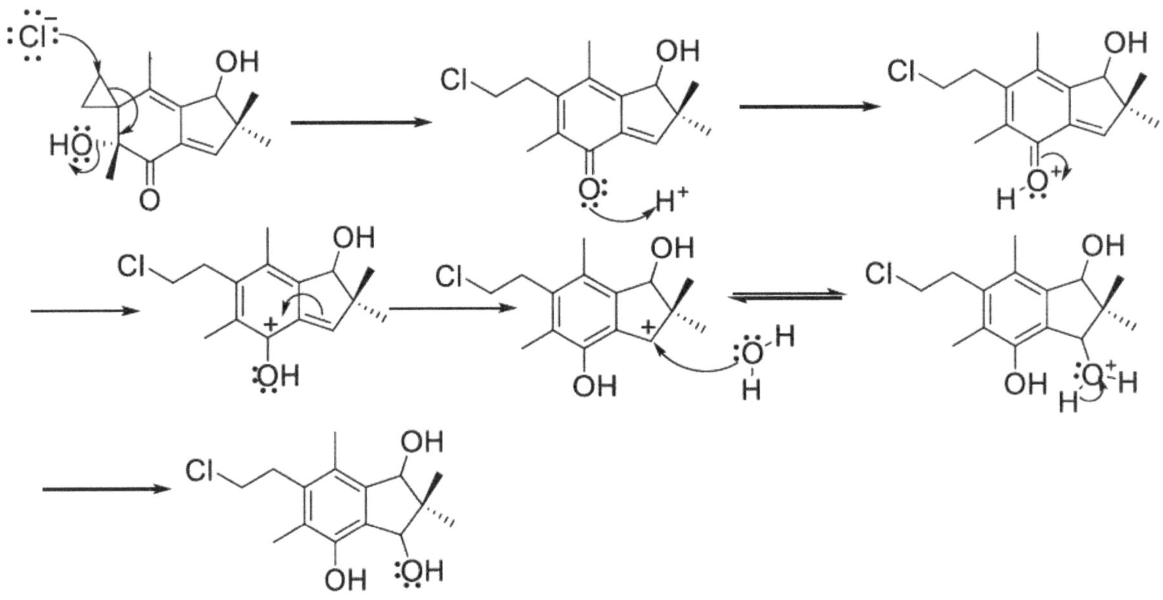

Cl
OH
HO
O
H+

500

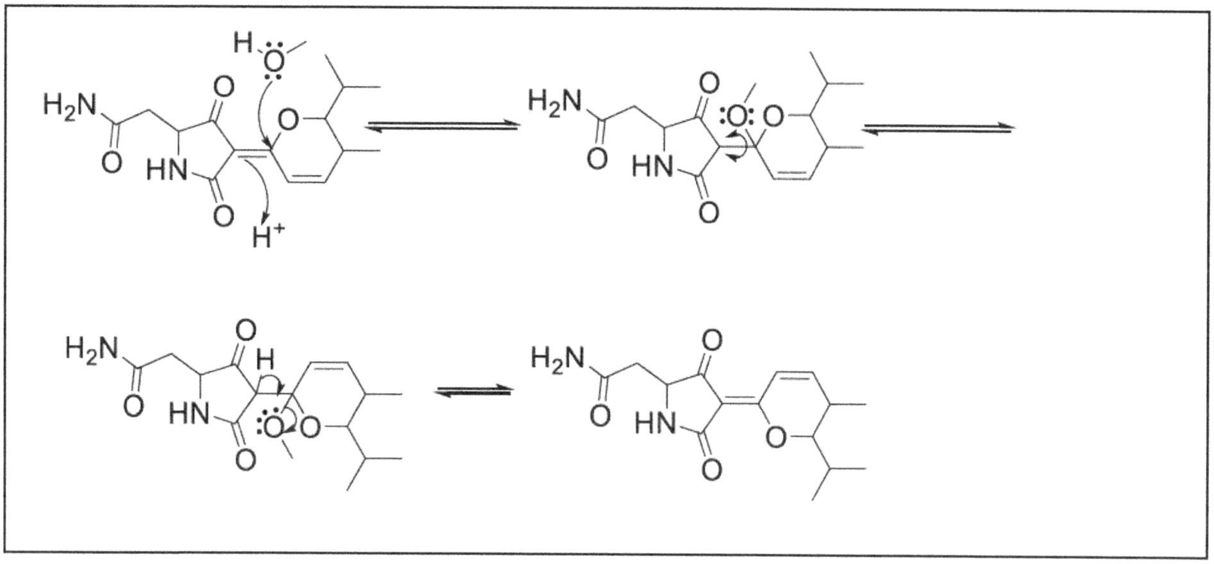

H
H₂N
O
O
HN
O
H⁺
H₂N
O
O
HN
O
H₂N
O
H
O
HN
O
O
H₂N
O
O
HN
O
O

REFERENCES

1. Stoner, E. J.; Cothron, D. A.; Balmer, M. K.; Roden, B. A., Benzylation via tandem Grignard reaction - Iodotrimethylsilane (TMSI) mediated reduction. *Tetrahedron* **1995,** *51* (41), 11043.

2. Azzena, U.; Demartis, S.; Fiori, M. G.; Melloni, G.; Pisano, L., Reductive electrophilic substitution of phthalans and ring expansion to isochroman derivatives. *Tetrahedron letters* **1995,** *36* (44), 8123-8126.

3. KRAATZ, U.; SAMIMI, M. N.; KORTE, F., Direktsynthese von Tetrahydrodibenz [b, d] oxepin-6 (7H)-onen. *Synthesis* **1977,** *1977* (06), 430-431.

4. Taylor, S. K.; May, S. A.; Hopkins, J. A., Cyclizations wherein an epoxide acts as the source of initiation and termination steps. Evidence for an early transition state in biomimetic epoxide cyclizations. *Tetrahedron letters* **1993,** *34* (8), 1283-1286.

5. Mazdiyasni, H.; Konopacki, D. B.; Dickman, D. A.; Zydowsky, T. M., Enzyme-catalyzed synthesis of optically Pure β-sulfonamidopropionic acids. Useful starting materials for P-3 site modified renin inhibitors. *Tetrahedron letters* **1993,** *34* (3), 435-438.

6. Kaminska, J.; Schwegler, M.; Hoefnagel, A.; Van Bekkum, H., The isomerization of α-pinene oxide with Brønsted and Lewis acids. *Recueil des Travaux Chimiques des Pays-Bas* **1992,** *111* (10), 432-437.

7. Almena, J.; Foubelo, F.; Yus, M., Lithium 2-(2-lithiomethylphenyl) ethanolate from isochroman: Easy preparation of substituted benzoxepines and functionalised arenes. *Tetrahedron* **1995,** *51* (11), 3365-3374.

8. Blank, I.; Grosch, W.; Eisenreich, W.; Bacher, A.; Firl, J., Determination of the chemical structure of linden ether. *Helvetica chimica acta* **1990,** *73* (5), 1250-1257.

9. Ferreri, C.; Ambrosone, M.; Chatgilialoglu, C., A Facile Entry to Secondary Cyclopropylcarbinols Further Developments in the Stereospecific Synthesis of (E)-Homoallylic Bromides. *Synthetic communications* **1995,** *25* (21), 3351-3356.

10. Huckin, S. N.; Weiler, L., Aldol Type Condensations of β-Keto Ester Dianions. *Canadian Journal of Chemistry* **1974,** *52* (11), 2157-2164.

11. Chen, J.; Fletcher, M. T.; Kitching, W., (2S, 6S, 8S)-2, 8-Dimethyl-1, 7-dioxaspiro [5.5] undecane: A natural spiroacetal lacking anomeric stabilisation. *Tetrahedron: Asymmetry* **1995,** *6* (4), 967-972.

12. Cohen, T.; Chen, F.; Kulinski, T.; Florio, S.; Capriati, V., Reductive cleavage and ring expansion of thiochromane and benzodihydrothiophene. *Tetrahedron letters* **1995,** *36* (25), 4459-4462.

13. Rossinskii, A. P.; Bunina-Krivorukova, L. I.; Bal'yan, K. V., Condensation of aromatic compounds with allyl type halides. XXI. Alkenylation of p-cresol by 1-chloro-2,3-dimethyl-2-butene and 3-chloro-2,3-dimethyl-1-butene. *Zh. Org. Khim.* **1975,** *11* (11), 2364.

14. Djakovitch, L.; Eames, J.; Jones, R. V.; McIntyre, S.; Warren, S., Rearrangements of phenylthio substituted 1, n-diols with toluene-p-sulfonic acid and with toluene-p-sulfonyl chloride. *Tetrahedron letters* **1995,** *36* (10), 1723-1726.

15. Brandes, B. D.; Jacobsen, E. N., Highly enantioselective, catalytic epoxidation of trisubstituted olefins. *The Journal of Organic Chemistry* **1994,** *59* (16), 4378-4380.

16. Lottaz, P. A.; Edwards, T. R. G.; Mentha, Y. G.; Burger, U., A convenient synthesis of dibenzo[a,c]cyclooctene based on thioenol ether reduction. *Tetrahedron Lett.* **1993,** *34* (4), 639.

17. Hagenbruch, B.; Hünig, S., Ein Beitrag zur Dienon-Phenol-Umlagerung. *Chemische Berichte* **1983,** *116* (12), 3884-3894.

18. Seuring, B.; Seebach, D., Synthese von vier chiralen, elektrophilen C3-und C4-Synthesebausteinen aus Hydroxycarbonsäuren. *Helvetica Chimica Acta* **1977,** *60* (4), 1175-1181.

19. Moffett, R., Cyclopentadiene and 3-Chlorocyclopentene-cyclopentene, 3-Chloro. *Organic Syntheses* **1952,** *32*, 41-44.

20. Masuda, S.; Nakajima, T.; Suga, S., Retentive Friedel-Crafts alkylation of benzene with optically active 2-chloro-1-phenylpropane and 1-chloro-2-phenylpropane. *Bulletin of the Chemical Society of Japan* **1983,** *56* (4), 1089-1094.

21. McCullough, D. W.; Cohen, T., Preparation of 2-vinylcyclobutanones viam-chloroperbenzoic acid oxidation of allylidenecyclopropanes. *Tetrahedron letters* **1988,** *29* (1), 27-30.

22. Guy, A.; Guetté, J.-P.; Lang, G., Utilization of polyphosphoric acid in the presence of a co-solvent. *Synthesis* **1980,** *1980* (03), 222-223.

23. Almena, J.; Foubelo, F.; Yus, M., Lithium 2-(2-lithiomethylphenyl) ethanolate from isochroman: Easy preparation of substituted benzoxepines and functionalised arenes. *Tetrahedron* **1995,** *51* (11), 3365-3374.

24. Prasad, R. S.; Roberts, R. M., Cyclialkylation studies. 3. Acid-catalyzed cyclodehydration of some benzyltetralols, with and without rearrangement, to yield tetracyclic hydrocarbons. *The Journal of Organic Chemistry* **1991,** *56* (9), 2998-3000.

25. Gu, X.-L.; Liu, H.-B.; Jia, Q.-H.; Li, J.-F.; Liu, Y.-L., Design and synthesis of novel miconazole-based ciprofloxacin hybrids as potential antimicrobial agents. *Monatsh. Chem.* **2015,** *146* (4), 713-720.

26. Kagabu, S.; Kojima, Y., A synthesis of indane musk Celestolide. *Journal of Chemical Education* **1992,** *69* (5), 420.

27. Kano, S.; Ebata, T.; Shibuya, S., Formation of 2, 3-dihydro-4 (1 H)-quinolones and related compounds via Fries-type acid-catalysed rearrangement of 1-arylazetidin-2-ones. *Journal of the Chemical Society, Perkin Transactions 1* **1980**, 2105-2111.

28. Morris, J.; Wishka, D. G., Synthesis of novel antagonists of leukotriene B4. *Tetrahedron letters* **1988,** *29* (2), 143-146.

29. Wiberg, K. B.; Matturro, M. G.; Okarma, P. J.; Jason, M. E., Tricyclo[4.2.2.22,5]dodeca-1,5-diene. *J. Am. Chem. Soc.* **1984,** *106* (7), 2194.

30. Fatope, M. O.; Okogun, J. I., A convenient solvent-specific synthesis of 4-[4-acetoxy-3-(3-methylbut-2-enyl) phenyl] butyric acid. *Journal of the Chemical Society, Perkin Transactions 1* **1982**, 1601-1603.

31. Dike, S. Y.; Ner, D. H.; Kumar, A., A new enantioselective chemoenzymatic synthesis of R-(−) thiazesim hydrochloride. *Bioorganic & Medicinal Chemistry Letters* **1991,** *1* (8), 383-386.

32. Constantino, M. G.; Beltrame Jr, M.; de Medeiros, E. F.; da Silva, G.-V. J., Synthesis of an Allylic Spiro-Lactone. *Synthetic communications* **1992,** *22* (19), 2859-2864.

33. Almena, J.; Foubelo, F.; Yus, M., Lithium 2-(2-lithiomethylphenyl) ethanolate from isochroman: Easy preparation of substituted benzoxepines and functionalised arenes. *Tetrahedron* **1995,** *51* (11), 3365-3374.

34. Trost, B. M.; Bogdanowicz, M. J., New synthetic reactions. X. Versatile cyclobutanone (spiroannelation) and. gamma.-butyrolactone (lactone annelation) synthesis. *Journal of the American Chemical Society* **1973,** *95* (16), 5321-5334.

35. Boeckman Jr, R. K.; Ko, S. S., Stereocontrol in the intramolecular Diels-Alder reaction. 1. An application to the total synthesis of (.+-.) marasmic acid. *Journal of the American Chemical Society* **1980,** *102* (23), 7146-7149.

36. Karim, M. R.; Sampson, P., A new and efficient approach to macrocyclic keto lactones. *The Journal of Organic Chemistry* **1990,** *55* (2), 598-605.

37. Koch, K.; Biggers, M. S., General preparation of 7-substituted 4-chromanones: synthesis of a potent aldose reductase inhibitor. *The Journal of Organic Chemistry* **1994,** *59* (5), 1216-1218.

38. Corey, E.; Das, J., A new method for stereospecific cis hydroxylation of olefins. *Tetrahedron Letters* **1982,** *23* (41), 4217-4220.

39. Karikomi, M.; Yamazaki, T.; Toda, T., Synthesis of 2-aminomethyloxiranes. *Chemistry letters* **1993,** *22* (10), 1787-1790.

40. Amlaiky, N.; Leclerc, G.; Carpy, A., Unusual reaction of N-hydroxyphthalimido ethers leading to oxygen-nitrogen heterocycles. *The Journal of Organic Chemistry* **1982,** *47* (3), 517-523.

41. Pinnick, H. W.; Chang, Y.-H., New approaches to the pyrrolizidine ring system: total synthesis of (.+-.)-isoretronecanol and (.+-.)-trachelanthamidine. *The Journal of Organic Chemistry* **1978,** *43* (24), 4662-4663.

42. Lin, Y.; Liang, X.; Chan, Y., Ring-size effect in the reaction of α,ω-dibromoalkanes with ethyl acetoacetate and its application to the synthesis of some disubstituted oxa- and aza-heterocycles. *Chin. J. Chem.* **1990,** (2), 153.

43. Das, S.; Karpha, T. K.; Ghosal, M.; Mukherjee, D., Aryl participated cyclisations involving indane derivatives a total synthesis of (±)-isolongifolene. *Tetrahedron letters* **1992,** *33* (9), 1229-

1232.

44. Benedetti, F.; Berti, F.; Risaliti, A., Cyclization of γ, δ-epoxy-α-cyanosulphones. A simple, diastereoselective route to cyclopropane carboxylic acids. *Tetrahedron letters* **1993,** *34* (40), 6443-6446.

45. Sangwan, N. K.; Rastogi, S. N., Synthesis of novel heterocycles, 2,3,4,11-tetrahydro-3,6-dihydroxy-5H-pyrano[2,3-c][1]benzoxepin-5-one and 4'-(chloromethyl)-6-hydroxyspiro[1-benzoxepin-3(2H),2'-[1,3]dioxolan]-5(4H)-one. *Indian J. Chem., Sect. B* **1984,** *23B* (12), 1284.

46. Greeves, N.; Torode, J. S., Synthesis of Phenylthiomethyl Substitution Furans by Lewis Acid Catalysed Substitution. *Synthesis* **1993,** *1993* (11), 1109-1112.

47. Takeuchi, Y.; Kirihara, K.; Shibata, N.; Kirk, K. L., Synthesis of carbonyl-bridged peptides containing an α-fluoroglycine residue. *Chem. Commun. (Cambridge)* **2000,** (9), 785-786.

48. a) Gokhale, A.; Schiess, P., Regioselectivity of the base-induced ring cleavage of 1-oxygenated derivatives of cyclobutabenzene. *Helvetica chimica acta* **1998,** *81* (2), 251-267. b) Cava, M.; Muth, K., Condensed cyclobutane aromatic compounds. IX. Benzocyclobutenol and benzocyclobutenone. *Journal of the American Chemical Society* **1960,** *82* (3), 652-654.

49. Kim, S.-C.; Kwon, B.-M., A New Convenient Synthesis of Hydantoin Derivatives by the Phase-Transfer Method. *Synthesis* **1982,** *1982* (09), 795-796.

50. Kozikowski, A. P.; Stein, P. D., The INOC route to carbocyclics: a formal total synthesis of (.+-.)-sarkomycin. *Journal of the American Chemical Society* **1982,** *104* (14), 4023-4024.

51. Wong, S. C.; Sasso, S.; Jones, H.; Kaminski, J. J., Stereochemical considerations and the antiinflammatory activity of 6-amino-6, 7, 8, 9-tetrahydro-5H-benzocyclohepten-5-ols and related derivatives. *Journal of medicinal chemistry* **1984,** *27* (1), 20-27.

52. Stevens, R. V.; Lee, A. W., Studies on the stereochemistry of nucleophilic additions to tetrahydropyridinium salts. A stereospecific total synthesis of (±)-monomorine I. *Journal of the Chemical Society, Chemical Communications* **1982,** (2), 102-103.

53. Skinnemoen, K.; Undheim, K.; Ullenius, C.; Kylin, A.; Glaumann, H., Synthesis of 2H-Pyran-3-(6H)-ones. *Acta Chemica Scandinavica, Series B* **1980**, 295-297.

54. Bunce, R. A.; Peeples, C. J.; Jones, P. B., Tandem SN2-Michael reactions for the preparation of simple five-and six-membered-ring nitrogen and sulfur heterocycles. *The Journal of Organic Chemistry* **1992,** *57* (6), 1727-1733.

55. Trigo, G. G.; Avendaño, C.; Santos, E.; Christensen, H. N.; Handlogten, M. E., 3-Aminotropane-3-carboxylic acids. Preparation and properties. *Canadian Journal of Chemistry* **1980,** *58* (21), 2295-2299.

56. Bertram, H. J.; Güntert, M.; Sommer, H.; Thielmann, T.; Werkhoff, P., Synthesis of 1H-Pyrrolo-[2.1-c]-1, 4-thiazine–A Novel Sulfur-Nitrogen Heterocyclic Flavor Constituent. *Journal für Praktische Chemie/Chemiker-Zeitung* **1993,** *335* (1), 101-102.

57. Loevgren, K.; Hedberg, A.; Nilsson, J. L. G., Adrenergic receptor agonists. *Journal of*

Medicinal Chemistry **1980,** *23* (6), 624-627.

58. Monti, S. A.; Yang, Y.-L., Stereoselective total synthesis of racemic stachenone. *J. Org. Chem.* **1979,** *44* (5), 897.

59. Zhang, D.; Closs, G. L.; Chung, D. D.; Norris, J. R., Free energy and entropy changes in vertical and nonvertical triplet energy transfer processes between rigid and nonrigid molecules. A laser photolysis study. *Journal of the American Chemical Society* **1993,** *115* (9), 3670-3673.

60. Gras, J.-L.; Bertrand, M., cis Δ5-octahydro-1-naphtalenones and cis Δ5-decahydro-1-naphthalenes (cis-1-decalones) by stereoselective intramolecular diels-alder reaction. *Tetrahedron Letters* **1979,** *20* (47), 4549-4552.

61. Roxburgh, C. J., Syntheses of medium sized rings by ring expansion reactions. *Tetrahedron* **1993,** *49* (47), 10749-10784.

62. Barco, A.; Benetti, S.; Pollini, G. P.; Baraldi, P. G.; Gandolfi, C., A new, elegant route to a key intermediate for the synthesis of 9 (0)-methanoprostacyclin. *The Journal of Organic Chemistry* **1980,** *45* (23), 4776-4778.

63. Engelhart, J. E.; McDivitt, J. R., Anomalous dimerization of 5, 5-dimethyl-2-cyclohexen-1-one. *The Journal of Organic Chemistry* **1971,** *36* (2), 367-368.

64. Sisido, K.; Kurozumi, S.; Utimoto, K., Synthesis of methyl dl-jasmonate. *The Journal of Organic Chemistry* **1969,** *34* (9), 2661-2664.

65. Gothelf, K.; Thomsen, I.; Torssell, K., A Convenient Synthesis of Flavones. Synthesis of Apigenin. *ChemInform* **1992,** *23* (34), no-no.

66. Loftus, F., The synthesis of some 2-substituted morpholines. *Synthetic Communications* **1980,** *10* (1), 59-73.

67. Curran, D. P., A short synthesis of γ-hydroxycyclopentemones. *Tetrahedron Letters* **1983,** *24* (33), 3443-3446.

68. Cookson, R. C.; Ray, P. S., A new synthesis of macrocyclic keto-lactones via ring expansion of 2-(3-hydroxypropyl)-2-nitrocycloalkanones. *Tetrahedron Letters* **1982,** *23* (34), 3521-3524.

69. Klutchko, S.; Von Strandtmann, M., General rearrangement of 3-substituted chromones. Rearrangement of 1-benzoxepins and 4-chromanones. *Synthesis* **1977,** (1), 61.

70. a) Babler, J. H.; Schlidt, S. A., An expedient route to a versatile intermediate for the stereoselective synthesis of all-trans-retinoic Acid and beta-carotene. *Tetrahedron letters* **1992,** *33* (50), 7697-7700. b) Rosenberger, M.; Jackson, W.; Saucy, G., The Synthesis of β, γ-and α, β-Unsaturated Aldehydes via Polyene Epoxides. *Helvetica Chimica Acta* **1980,** *63* (6), 1665-1674.

71. Zoretic, P.; Barcelos, F.; Jardin, J.; Bhakta, C., Synthetic approaches to 10-azaprostaglandins. *The Journal of Organic Chemistry* **1980,** *45* (5), 810-814.

72. GARDNER, P. D.; HAYNES, G. R.; BRANDON, R. L., Formation of Dieckmann reaction

products under acyloin conditions. Competition of the two reactions. *The Journal of Organic Chemistry* **1957,** *22* (10), 1206-1210.

73. Albright, J.; Moran, D.; Wright Jr, W.; Collins, J.; Beer, B.; Lippa, A.; Greenblatt, E., Synthesis and anxiolytic activity of 6-(substituted-phenyl)-1, 2, 4-triazolo [4, 3-b] pyridazines. *Journal of medicinal chemistry* **1981,** *24* (5), 592-600.

74. Bunce, R. A.; Bennett, M. J., A Tandem Ester Cleavage-Michael Addition Reaction for the Synthesis of Oxygen Heterocycles. *Synthetic communications* **1993,** *23* (7), 1009-1020.

75. Morris, J.; Wishka, D. G., Synthesis of novel antagonists of leukotriene B4. *Tetrahedron letters* **1988,** *29* (2), 143-146.

76. Rousseau, G., Medium ring lactones. *Tetrahedron* **1995,** *51* (10), 2777-2849.

77. Amlaiky, N.; Leclerc, G., Synthesis of the New Heterocyclic Hydroxy Compounds 4-Hydroxyisoxazolidine, 3-Hydroxymethylisoxazolidine, 3-Hydroxymethylhexahydro-1, 2-oxazine, and 4-Hydroxy hexahydro-1, 2-oxazepine. *Synthesis* **1982,** *1982* (05), 426-428.

78. Matsuyama, H.; Miyazawa, Y.; Takei, Y.; Kobayashi, M., Regioselective synthesis of cyclopentenones from 4-thianone. *J. Org. Chem.* **1987,** *52* (9), 1703.

79. Reitz, D. B., Reinvestigation of the reaction of ethyl acetoacetate with styrene oxide. *The Journal of Organic Chemistry* **1979,** *44* (25), 4707-4709.

80. Busqué, F.; Cid, P.; De March, P.; Figueredo, M.; Font, J., Preparation of γ-heterosubstituted α, β-hexenolides and their 1, 3-dipolar cycloaddition to 2, 3, 4, 5-tetrahydropyridine 1-oxide. *Heterocycles* **1995,** *1* (40), 387-399.

81. Xie, Z. F.; Suemune, H.; Sakai, K., A facile ring enlargement. *Synth. Commun.* **1989,** *19* (5-6), 987.

82. Girard, M.; Moir, D. B.; ApSimon, J. W., A simple and efficient synthesis of 5'-(2H3) olivetol. *Canadian journal of chemistry* **1987,** *65* (1), 189-190.

83. Marx, J. N.; Minaskanian, G., Regiospecific synthesis of sarkomycin and some analogs. *The Journal of Organic Chemistry* **1982,** *47* (17), 3306-3310.

84. Sato, K.; Inoue, S.; Ozawa, K.; Kobayashi, T.; Ota, T.; Tazaki, M., Selective ortho-alkylation of phenols with sulphoxides via [2, 3] sigmatropic rearrangement: synthesis of coumarins. *Journal of the Chemical Society, Perkin Transactions 1* **1987**, 1753-1756.

85. Yates, P.; Bichan, D. J., Bridged xanthenes. I. Intermolecular cycloaddition route. *Can. J. Chem.* **1975,** *53* (14), 2045.

86. Urban, F. J., Synthesis of R-and S-3-Oxo-2-oxaspiro [4.4]-nonane-1-carboxylic acid. *Tetrahedron: Asymmetry* **1994,** *5* (2), 211-214.

87. Nicholson, J. M.; Edafiogho, I. O.; Moore, J. A.; Farrar, V. A.; Scott, K., Cyclization reactions leading to β-hydroxyketo esters. *Journal of pharmaceutical sciences* **1994,** *83* (1), 76-78.

88. Seebach, D.; Hoffmann, T.; Kühnle, F. N.; Kinkel, J. N.; Schulte, M., Preparation, structure, and properties of all possible cyclic dimers (diolides) of 3-hydroxybutanoic acid. *Helvetica chimica acta* **1995,** *78* (6), 1525-1540.

89. Kim, D. H., Improved syntheses of 1, 4-benzodiazepine-2, 5-diones. *Journal of Heterocyclic Chemistry* **1975,** *12* (6), 1323-1324.

90. Jernow, J.; Tautz, W.; Rosen, P.; Williams, T. H., Methylenomycin B: revised structure and total syntheses. *The Journal of Organic Chemistry* **1979,** *44* (23), 4212-4213.

91. Hikino, H.; De Mayo, P., Photochemical cycloaddition as a device for general annelation. *Journal Of The American Chemical Society* **1964,** *86* (17), 3582-3583.

92. Barco, A.; Benetti, S.; Pollini, G. P.; Veronesi, B.; Baraldi, P. G.; Guarneri, M.; Vicentini, C. B., Synthesis of 14-hydroxy-8-azaprostanoic acids methyl ester. *Synth. Commun.* **1978,** *8* (4), 219.

93. Fall, Y.; Santana, L.; Uriarte, E., A convenient synthesis of benzofuran-3-acetic acids. *Heterocycles-Sendai Institute of Heterocyclic Chemistry* **1995,** *41* (4), 647-650.

94. McEvoy, F. J.; Albright, J. D., Alkylations and acylations of. alpha.-(aryl)-4-morpholineacetonitriles (masked acyl anion equivalents) and their use in 1, 4-additions. *The Journal of Organic Chemistry* **1979,** *44* (25), 4597-4603.

95. Walker, G. N.; Alkalay, D., New bicyclic enamines and iminium salts. II. Synthesis of 1, 4-dihydro-1, 4-ethanoisoquinolinium salts and 4, 5-dihydro-1H-1, 4-methano-3-benzazepinium salts by reaction of bridged lactams with organometallic reagents. *The Journal of Organic Chemistry* **1971,** *36* (4), 491-500.

96. Suzuki, M.; Orr, G.; Stammer, C., The synthesis of cyclopropyl tyrosine. *Bioorganic Chemistry* **1987,** *15* (1), 43-49.

97. Nicholson, J. M.; Edafiogho, I. O.; Moore, J. A.; Farrar, V. A.; Scott, K., Cyclization reactions leading to β-hydroxyketo esters. *Journal of pharmaceutical sciences* **1994,** *83* (1), 76-78.

98. Baraldi, P. G.; Achille, B.; Simoneta, B.; Piero, P. G.; Vinicio, Z., 2, 3a, 5, 6, 7, 7a-Hexahydro-3h, 4h-benzothiophene-3, 4-dione and cyclopenta [b]-tetrahydrothiophene-3, 4-dione enolate anions as synthetic equivalents to cyclohex-2-enone and cyclopent-2-enone c-2-carbanions. *Tetrahedron letters* **1984,** *25* (38), 4291-4294.

99. Chang, C.-P.; Hsu, L.-F.; Chang, N.-C., Bicyclo [3.2. 1] octenones as Building Blocks in Natural Products Synthesis. 2. Formal Synthesis of (.+-.)-Verbenalol. *The Journal of Organic Chemistry* **1994,** *59* (7), 1898-1899.

100. Mahmoodi, N. O.; Jazayri, M., Direct synthesis of γ-butyrolactones via γ-phenyl substituted butyric acids mediated benzyl radical cyclization. *Synth. Commun.* **2001,** *31* (10), 1467-1475.

101. Nair, V.; Treesa, P., Hetero Diels–Alder trapping of 3-methylene-1, 2, 4-[3H] naphthalenetrione: an efficient protocol for the synthesis of α-and β-lapachone derivatives. *Tetrahedron Letters* **2001,** *42* (27), 4549-4551.

102. Taber, D. F.; Saleh, S. A.; Korsmeyer, R. W., Preparation of cyclohexanones and cyclopentanones of high optical purity. *The Journal of Organic Chemistry* **1980,** *45* (23), 4699-4702.

103. Hutchinson, I. S.; Matlin, S. A.; Mete, A., The synthesis of 3-diazo-2-nitromethylenepiperidine. *Tetrahedron Letters* **2001,** *42* (9), 1773-1776.

104. Kim, S. H.; Kim, K. H.; Kim, H. S.; Kim, J. N., Regioselective synthesis of 1, 2, 4, 5-tetrasubstituted pyridines from Baylis–Hillman adducts via consecutive [3+ 2+ 1] annulation protocol. *Tetrahedron Letters* **2008,** *49* (12), 1948-1951.

105. Sun, Y.; Wang, H.; Prins, R., Synthesis of 4, 6-dimethyl-tetrahydro-and hexahydro-dibenzothiophene. *Tetrahedron Letters* **2008,** *49* (13), 2063-2065.

106. Ramesh, E.; Vidhya, T. K. S.; Raghunathan, R., Indium chloride/silica gel supported synthesis of pyrano/thiopyranoquinolines through intramolecular imino Diels–Alder reaction using microwave irradiation. *Tetrahedron Letters* **2008,** *49* (17), 2810-2814.

107. Piras, L.; Ghiron, C.; Minetto, G.; Taddei, M., Microwave-assisted synthesis of tetrahydroindoles. *Tetrahedron Letters* **2008,** *49* (3), 459-462.

108. Dziedzic, P.; Ibrahem, I.; Cordova, A., Direct catalytic asymmetric three-component Mannich reactions with dihydroxyacetone: enantioselective synthesis of amino sugar derivatives. *Tetrahedron Letters* **2008,** *49* (5), 803-807.

109. Santos, L. S.; Pilli, R. A., Total synthesis of (±)-homopumiliotoxin 223G. *Tetrahedron Letters* **2001,** *42* (40), 6999-7001.

110. Chen, B.-C.; Ngu, K.; Guo, P.; Liu, W.; Sundeen, J. E.; Weinstein, D. S.; Atwal, K. S.; Ahmad, S., A new facile method for the stereoselective synthesis of trans-2-aryl-3, 3-dimethylcyclopropane-1-carboxylic acids. *Tetrahedron Letters* **2001,** *42* (36), 6227-6229.

111. Guz, N. R.; Lorenz, P.; Stermitz, F. R., New coumarins from Harbouria trachypleura: isolation and synthesis. *Tetrahedron Letters* **2001,** *42* (37), 6491-6494.

112. Guz, N. R.; Lorenz, P.; Stermitz, F. R., New coumarins from Harbouria trachypleura: isolation and synthesis. *Tetrahedron Letters* **2001,** *42* (37), 6491-6494.

113. Kiewel, K.; Tallant, M.; Sulikowski, G. A., Asymmetric Heck cyclization route to indolizidine and azaazulene alkaloids: synthesis of (+)-5-epiindolizidine 167B and indolizidine 223AB. *Tetrahedron Letters* **2001,** *42* (38), 6621-6623.

114. Bieniek, A.; Epsztajn, J.; Kowalska, J. A.; Malinowski, Z., Application of organolithium and related reagents in synthesis. Part 25: Novel specific synthesis of the 4-arylisochroman-3-acetic acids via conversion of benzoic acids. *Tetrahedron Letters* **2001,** *42* (52), 9293-9295.

115. Cabell, L. A.; Hedrich, L. W.; McMurray, J. S., Cleavage of the N (1)□ C (4) bond of 4-(4′-hydroxyphenyl)-azetidine-2-ones via quinone methide intermediates. *Tetrahedron Letters* **2001,** *42* (48), 8409-8413.

116. Bernsmann, H.; Gruner, M.; Fröhlich, R.; Metz, P., A shortcut to the smaller fragment of

pamamycin-607. *Tetrahedron Letters* **2001,** *42* (32), 5377-5380.

117. Sato, M.; Sakaki, J.-i.; Sugita, Y.; Yasuda, S.; Sakoda, H.; Kaneko, C., Two lactone formation reactions from 1, 3-dioxin-4-ones having hydroxyalkyl group at the 6-position: Difference in ring opening and closure. *Tetrahedron* **1991,** *47* (30), 5689-5708.

118. Bashford, K. E.; Cooper, A. L.; Kane, P. D.; Moody, C. J., A new protecting-group strategy for indoles. *Tetrahedron letters* **2002,** *43* (1), 135-137.

119. Gallagher, P. O.; McErlean, C. S.; Jacobs, M. F.; Watters, D. J.; Kitching, W., Sub-structure syntheses and relative stereochemistry in the bistramide (bistratene) series of marine metabolites. *Tetrahedron letters* **2002,** *43* (3), 531-535.

120. D'hooghe, M.; Hofkens, A.; De Kimpe, N., A new route towards N-(α-methoxybenzyl) aziridines. *Tetrahedron letters* **2003,** *44* (6), 1137-1139.

121. Vatèle, J.-M., One-pot selective cleavage of prenyl carbamates using iodine in methanol followed by zinc. *Tetrahedron letters* **2003,** *44* (51), 9127-9129.

122. Daia, D. E.; Gabbutt, C. D.; Heron, B. M.; Hepworth, J. D.; Hursthouse, M. B.; Malik, K. A., Conjugate additions of lithium dialkynylcuprates [(RC□ C) 2CuLi] to activated chromones. Unexpected formation of the 6H-bis [1] benzopyrano [2, 3-b: 3′, 4′-e] pyridine system. *Tetrahedron letters* **2003,** *44* (7), 1461-1464.

123. Maier, M. E., Product class 6: lactones. *Sci. Synth.* **2006,** *20b*, 1421-1551.

124. Sikoraiová, J.; Marchalín, Š.; Daïch, A.; Decroix, B., Acid-mediated intramolecular cationic cyclization using an oxygen atom as internal nucleophile: synthesis of substituted oxazolo-, oxazino-and oxazepinoisoindolinones. *Tetrahedron letters* **2002,** *43* (27), 4747-4751.

125. Matesanz, E.; Alcázar, J.; Andrés, J. I.; Bartolomé, J. M.; De Bruyn, M.; Fernández, J.; Van Emelen, K., Synthesis of novel aza analogues of 2-substituted-2, 3-dihydro-1, 4-benzodioxins as potential new scaffolds for drug discovery. *Tetrahedron letters* **2003,** *44* (11), 2275-2277.

126. Matesanz, E.; Alcázar, J.; Andrés, J. I.; Bartolomé, J. M.; De Bruyn, M.; Fernández, J.; Van Emelen, K., Synthesis of novel aza analogues of 2-substituted-2, 3-dihydro-1, 4-benzodioxins as potential new scaffolds for drug discovery. *Tetrahedron letters* **2003,** *44* (11), 2275-2277.

127. Madhusudhan, G.; Reddy, G. O.; Ramanatham, J.; Dubey, P., Stereoselective synthesis of anti-2-oxazolidinones by Ph3P–CCl4–Et3N mediated SN2 cyclization of N-Boc-β-amino alcohols. *Tetrahedron letters* **2003,** *44* (33), 6323-6325.

128. Patel, M. V.; Rohde, J. J.; Gracias, V.; Kolasa, T., An efficient one-pot synthesis of substituted 2-arylbenzo [b] thiophene derivatives. *Tetrahedron letters* **2003,** *44* (35), 6665-6667.

129. Zhao, C.; Lu, J.; Yan, J.; Xi, Z., One-pot four-component synthesis of tetrahydrofuran derivatives involving an alkyne, an ethylene and two aldehydes via CuCl-mediated reactions of oxazirconacyclopentenes with aldehydes. *Tetrahedron letters* **2003,** *44* (36), 6895-6898.

130. Ortiz, A.; Quintero, L.; Hernández, H.; Maldonado, S.; Mendoza, G.; Bernès, S., (S)-4-Isopropyl-5, 5-dimethyl-1, 3-oxazolidinethione as chiral auxiliary for the intramolecular sulfur

transfer in α, β-unsaturated N-acylimides, promoted by NbCl5. *Tetrahedron letters* **2003,** *44* (6), 1129-1132.

131. Liu, H.-G.; Wu, C.-S.; Wang, J.-F.; Yang, D.-Y., Isomerization of enol esters derived from 2-acyl-1, 3-cyclohexanediones: mechanism and driving force. *Tetrahedron letters* **2003,** *44* (15), 3137-3141.

132. Peng, X.; She, X.; Su, Y.; Wu, T.; Pan, X., A novel approach to synthesis of tricyclic diterpenoid. *Tetrahedron letters* **2004,** *45* (16), 3283-3285.

133. Reid, M.; Taylor, R. J., A stannous chloride-induced deacetalisation–cyclisation process to prepare the ABC ring system of'upenamide. *Tetrahedron letters* **2004,** *45* (21), 4181-4183.

134. GowriSankar, S.; Lee, K. Y.; Lee, C. G.; Kim, J. N., Synthesis of methyl 9-phenyl-7H-benzocycloheptene-6-carboxylates from Baylis–Hillman adducts: use of intramolecular Friedel–Crafts alkenylation reaction. *Tetrahedron letters* **2004,** *45* (32), 6141-6146.

135. Vets, N.; Smet, M.; Dehaen, W., Synthesis and thermolysis of a Diels–Alder adduct of pentacene and thiophosgene. *Tetrahedron letters* **2004,** *45* (39), 7287-7289.

136. Ding, K.; Flippen-Anderson, J.; Deschamps, J. R.; Wang, S., An efficient synthesis of optically pure (S)-2-functionalized 1, 2, 3, 4-tetrahydroquinoline. *Tetrahedron letters* **2004,** *45* (5), 1027-1029.

137. Nakamura, M.; Masaki, M.; Maki, S.; Matsui, R.; Hieda, M.; Mamino, M.; Hirano, T.; Ohmiya, Y.; Niwa, H., Synthesis of Latia luciferin benzoate analogues and their bioluminescent activity. *Tetrahedron letters* **2004,** *45* (10), 2203-2205.

138. Le, T. N.; Gang, S. G.; Cho, W.-J., A facile synthesis of benzo [c] phenanthridine alkaloids: oxynitidine and oxysanguinarine using lithiated toluamide–benzonitrile cycloaddition. *Tetrahedron letters* **2004,** *45* (13), 2763-2766.

139. Zenouz, A. M., Synthesis of some substituted 5, 6, 11, 12-tetrahydro-dibenzo [a, e] cyclooctene derivatives through the intermediacy of tricarbonyl (η6-arene) chromium complexes. *Tetrahedron letters* **2004,** *45* (14), 2967-2971.

140. Malachowski, W. P.; Banerji, M., Sequential Birch reduction–allylation and Cope rearrangement of o-anisic acid derivatives. *Tetrahedron letters* **2004,** *45* (44), 8183-8185.

141. Devi, I.; Bhuyan, P. J., Stereoselective intramolecular hetero Diels–Alder reactions of 1-oxa-1, 3-butadienes: a novel approach for the synthesis of complex annulated uracils. *Tetrahedron letters* **2004,** *45* (41), 7727-7728.

142. Lacey, J. R.; Anzalone, P. W.; Duncan, C. M.; Hackert, M. J.; Mohan, R. S., A study of epoxyolefin cyclizations catalyzed by bismuth trifluoromethanesulfonate and other metal triflates. *Tetrahedron letters* **2005,** *46* (49), 8507-8511.

143. Chang, M.-Y.; Pai, C.-L.; Kung, Y.-H., Synthesis of (±)-coerulescine and a formal synthesis of (±)-horsfiline. *Tetrahedron letters* **2005,** *46* (49), 8463-8465.

144. Shawcross, F.; Sard, H., Cyclization of 3-fluorophenol and 3, 3-dimethylacrylic acid: A

structure correction. *Journal of heterocyclic chemistry* **1995,** *32* (4), 1393-1395.

145. Shawcross, F.; Sard, H., Cyclization of 3-fluorophenol and 3, 3-dimethylacrylic acid: A structure correction. *Journal of heterocyclic chemistry* **1995,** *32* (4), 1393-1395.

146. Brown, G. R.; Foubister, A. J.; Wright, B., Hexahydro-1, 4-oxazepines. Part 1. Synthesis by morpholine ring expansion. *Journal of the Chemical Society, Perkin Transactions 1* **1987**, 553-555.

147. Len, C.; Violeau, B., Asymmetric synthesis of thymine nucleoside analogues based on the isochroman core. *Tetrahedron letters* **2005,** *46* (29), 4835-4838.

148. Guy, A.; Guetté, J.-P.; Lang, G., Utilization of polyphosphoric acid in the presence of a co-solvent. *Synthesis* **1980,** *1980* (03), 222-223.

149. Sato, M.; Sakaki, J.-i.; Sugita, Y.; Yasuda, S.; Sakoda, H.; Kaneko, C., Two lactone formation reactions from 1, 3-dioxin-4-ones having hydroxyalkyl group at the 6-position: Difference in ring opening and closure. *Tetrahedron* **1991,** *47* (30), 5689-5708.

150. Gammill, R.; Nash, S.; Mizsak, S., The addition of amines to 3-bromochromone and 6-bromofurochromone. An unexpected ring contraction of the pyrone ring. *Tetrahedron Letters* **1983,** *24* (33), 3435-3438.

151. Curran, D. P., A short synthesis of γ-hydroxycyclopentemones. *Tetrahedron Letters* **1983,** *24* (33), 3443-3446.

152. Gras, J.-L.; Bertrand, M., cis Δ5-octahydro-1-naphtalenones and cis Δ5-decahydro-1-naphthalenes (cis-1-decalones) by stereoselective intramolecular diels-alder reaction. *Tetrahedron Letters* **1979,** *20* (47), 4549-4552.

153. Aida, T.; Legault, R.; Dugat, D.; Durst, T., Cyclization reactions of 4-(3′-butenyl) azetidin-2-one a route to the carbopenam ring system. *Tetrahedron Letters* **1979,** *20* (52), 4993-4994.

154. Taylor, S. K.; May, S. A.; Hopkins, J. A., Cyclizations wherein an epoxide acts as the source of initiation and termination steps. Evidence for an early transition state in biomimetic epoxide cyclizations. *Tetrahedron letters* **1993,** *34* (8), 1283-1286.

155. Azzena, U.; Demartis, S.; Fiori, M. G.; Melloni, G.; Pisano, L., Reductive electrophilic substitution of phthalans and ring expansion to isochroman derivatives. *Tetrahedron letters* **1995,** *36* (44), 8123-8126.

156. Scarborough Jr, R. M.; Toder, B. H.; Smith III, A. B., A stereospecific total synthesis of (.+-.)-methylenomycin A and its epimer,(.+-.)-epimethylenomycin A. *Journal of the American Chemical Society* **1980,** *102* (11), 3904-3913.

157. McCullough, D. W.; Cohen, T., Preparation of 2-vinylcyclobutanones viam-chloroperbenzoic acid oxidation of allylidenecyclopropanes. *Tetrahedron letters* **1988,** *29* (1), 27-30.

158. Stefani, H. A.; Petragnani, N.; Valduga, C. J.; Brandt, C. A., Iodine promoted cyclofunctionalization reaction of 2, 4-dialkenyl-1, 3-dicarbonyl compounds. *Tetrahedron letters*

1997, *38* (28), 4977-4980.

159. Almena, J.; Foubelo, F.; Yus, M., Lithium 2-(2-lithiomethylphenyl) ethanolate from isochroman: Easy preparation of substituted benzoxepines and functionalised arenes. *Tetrahedron* **1995,** *51* (11), 3365-3374.

160. Roxburgh, C. J., Syntheses of medium sized rings by ring expansion reactions. *Tetrahedron* **1993,** *49* (47), 10749-10784.

161. Nesvadba, P., Easy large scale syntheses of 2, 6-di-t-butyl-7-cyano-, 7-carboxy-and 7-methoxycarbonyl quinone methides. *Synthetic Communications* **2000,** *30* (15), 2825-2832.

162. Basavaiah, D.; Hyma, R. S.; Muthukumaran, K.; Kumaragurubaran, N., Stereoselective transformation of Baylis-Hillman adducts into (3E)-3-(alkoxymethyl) alk-3-en-2-ones. *Synthesis* **2000,** *2000* (02), 217-219.

163. Serra, S., Bisabolane sesquiterpenes: Synthesis of (R)-(+)-sydowic acid and (R)-(+)-curcumene ether. *Synlett* **2000,** *2000* (06), 0890-0892.

164. Sgariglia, E. A.; Schopp, R.; Gavardinas, K.; Mohan, R. S., The Discovery-Oriented Approach to Organic Chemistry. 3. Rearrangement of cis-and trans-Stilbene Oxides with Boron Trifluoride Etherate. An Exercise in 1H NMR Spectroscopy for Sophomore Organic Laboratories. *Journal of Chemical Education* **2000,** *77* (1), 79.

165. Lewis, D. B.; Matecka, D.; Zhang, Y.; Hsin, L.-W.; Dersch, C. M.; Stafford, D.; Glowa, J. R.; Rothman, R. B.; Rice, K. C., Oxygenated analogues of 1-[2-(diphenylmethoxy) ethyl]-and 1-[2-[bis (4-fluorophenyl) methoxy] ethyl]-4-(3-phenylpropyl) piperazines (GBR 12935 and GBR 12909) as potential extended-action cocaine-abuse therapeutic agents. *Journal of medicinal chemistry* **1999,** *42* (24), 5029-5042.

166. Maslak, P.; Varadarajan, S.; Burkey, J. D., Synthesis, structure, and nucleophile-induced rearrangements of spiroketones. *The Journal of Organic Chemistry* **1999,** *64* (22), 8201-8209.

167. Adam, W.; Peters, E.-M.; Peters, K.; Schmidt, E.; von Schnering, H. G., Bis(3-phenyl-1,4-benzodioxin-2-yl) Sulfide: Formation and Structure Elucidation. *Chemische Berichte* **1983,** *116* (4), 1686-1689.

168. Miller, S. A.; Gadwood, R. C., Synthesis of Cyclobutanones via 1-Bromo-1-Ethoxycyclopropane:(E)-2-(1-Propenyl) Cyclobutanone: Cyclobutanone, 2-(1-propenyl)-,(E)-. *Organic Syntheses* **2003,** *67*, 210-210.

169. Malpass, J. R.; White, R., 7-Substituted 2-Azabicyclo [2.2.1] heptanes as Key Intermediates for the Synthesis of Novel Epibatidine Analogues; Synthesis of syn-and anti-Isoepiboxidine. *The Journal of Organic Chemistry* **2004,** *69* (16), 5328-5334.

170. Kim, J. H.; Lee, Y. S.; Park, H.; Kim, C. S., Formation of pyrazinoisoquinoline ring system by the tandem amidoalkylation and N-acyliminium ion cyclization: an efficient synthesis of praziquantel. *Tetrahedron* **1998,** *54* (26), 7395-7400.

171. Kamal, A.; Krishnaji, T.; Khanna, G. R., Chemoenzymatic synthesis of enantiomerically enriched kavalactones. *Tetrahedron letters* **2006,** *47* (49), 8657-8660.

172. Moisan, L.; Wagner, M.; Comesse, S.; Doris, E., Ring expansions of a spirocyclohexadienone system. *Tetrahedron letters* **2006,** *47* (51), 9093-9094.

173. Srikrishna, A.; Lakshmi, B. V.; Mathews, M., Construction of spiro [5.5] undecanes containing a quaternary carbon atom adjacent to a spirocentre via an Ireland ester Claisen rearrangement and RCM reaction sequence. Total syntheses of (±)-α-chamigrene,(±)-β-chamigrene and (±)-laurencenone C. *Tetrahedron letters* **2006,** *47* (13), 2103-2106.

174. Manhas, M. S.; Ganguly, S. N.; Mukherjee, S.; Jain, A. K.; Bose, A. K., Microwave initiated reactions: Pechmann coumarin synthesis, Biginelli reaction, and acylation. *Tetrahedron letters* **2006,** *47* (14), 2423-2425.

175. Bhar, S. S.; Ramana, M., Annulation strategy for the biomimetic synthesis of cis-fused diterpenoids. *Tetrahedron letters* **2006,** *47* (44), 7805-7807.

176. Len, C.; Violeau, B., Asymmetric synthesis of thymine nucleoside analogues based on the isochroman core. *Tetrahedron letters* **2005,** *46* (29), 4835-4838.

177. Lin, C.-T.; Shih, J.-H.; Chen, C.-L.; Yang, D.-Y., Synthesis of novel triketone-based acidichromic colorants. *Tetrahedron letters* **2005,** *46* (30), 5033-5037.

178. Reddy, D. S.; Palani, K.; Balasubrahmanyam, D.; Kamath, V. B.; Iqbal, J., The first synthesis of a noreremophilane isolated from the roots of Ligularia przewalskii. *Tetrahedron letters* **2005,** *46* (31), 5211-5213.

179. Drevermann, B.; Lingham, A.; Hügel, H.; Marriott, P., Microwave assisted synthesis of the fragrant compound Calone 1951®. *Tetrahedron letters* **2005,** *46* (1), 39-41.

180. Travert, N.; Martin, M.-T.; Bourguet-Kondracki, M.-L.; Al-Mourabit, A., Regioselective intramolecular N1–C3 cyclizations on pyrrole–proline to ABC tricycles of dibromophakellin and ugibohlin. *Tetrahedron letters* **2005,** *46* (2), 249-252.

181. de Cienfuegos, L. A.; Mota, A. J.; Rodríguez, C.; Robles, R., Highly efficient synthesis of 2′, 3′-didehydro-2′, 3′-dideoxy-β-nucleosides through a sulfur-mediated reductive 2′, 3′-trans-elimination. From iodomethylcyclopropanes to thiirane analogs. *Tetrahedron letters* **2005,** *46* (3), 469-473.

182. Langer, P., Synthesis of functionalized benzopyrans by sequential [3+ 3]-cyclization—Williamson reactions of 1, 3-bis (trimethylsilyloxy)-7-chlorohepta-1, 3-dienes. *Tetrahedron letters* **2005,** *46* (5), 815-817.

183. Banik, B. K.; Chapa, M.; Marquez, J.; Cardona, M., A remarkable iodine-catalyzed protection of carbonyl compounds. *Tetrahedron letters* **2005,** *46* (13), 2341-2343.

184. Kim, S.-G.; Kim, J.; Jung, H., Efficient total synthesis of (+)-curcuphenol via asymmetric organocatalysis. *Tetrahedron letters* **2005,** *46* (14), 2437-2439.

185. Popović-Đorđević, J. B.; Ivanović, M. D.; Kiricojević, V. D., A novel tandem process leading to functionalized glutarimides. *Tetrahedron letters* **2005,** *46* (15), 2611-2614.

186. Klumpp, D. A.; Kindelin, P. J.; Li, A., Superacid-promoted reactions of

pyrazolecarboxaldehydes and the role of dicationic electrophiles. *Tetrahedron letters* **2005,** *46* (16), 2931-2935.

187. Win, T.; Bittner, S., Novel 2-amino-3-(2, 4-dinitrophenylamino) derivatives of 1, 4-naphthoquinone. *Tetrahedron letters* **2005,** *46* (18), 3229-3231.

188. Shao, B., Synthesis of fused bicyclic pyridines with microwave-assisted intramolecular hetero-Diels–Alder cycloaddition of acetylenic pyrimidines. *Tetrahedron letters* **2005,** *46* (19), 3423-3427.

189. Rotzoll, S.; Appel, B.; Langer, P., Synthesis of 2, 3-benzoxepins by sequential cyclopropanation/ring-enlargement reactions of benzopyrylium triflates with diazoesters. *Tetrahedron letters* **2005,** *46* (23), 4057-4059.

190. Rigo, B.; Gautret, P., On the structure of compounds obtained from the reaction of amines with 6, 6-dimethyl-5, 7-dioxaspiro [2.5] octane-4, 8-dione. *Tetrahedron letters* **2006,** *47* (3), 295-298.

191. Torr, J. E.; Large, J. M.; Horton, P. N.; Hursthouse, M. B.; McDonald, E., On the nucleophilic tele-substitution of dichloropyrazines by metallated dithianes. *Tetrahedron letters* **2006,** *47* (1), 31-34.

192. Krülle, T. M.; Barba, O.; Davis, S. H.; Dawson, G.; Procter, M. J.; Staroske, T.; Thomas, G. H., A simple route to 6-and 7-fluoro-substituted naphthalene-1-carboxylic acids. *Tetrahedron letters* **2007,** *48* (9), 1537-1540.

193. Jones, I. L.; Schofield, D. J.; Strevens, R. R.; Horton, P. N.; Hursthouse, M. B.; Tomkinson, N. C., Novel steroid mimics: synthesis of tri-and tetra-substituted oxamides and oxoamides. *Tetrahedron letters* **2007,** *48* (4), 521-525.

194. Goh, W. K.; Black, D. S.; Kumar, N., Synthesis of novel 7-substituted 5, 6-dihydroindol-2-ones via a Suzuki–Miyaura cross-coupling strategy. *Tetrahedron Letters* **2007,** *48* (51), 9008-9011.

195. Leleti, R. R.; Hu, B.; Prashad, M.; Repič, O., Highly selective methanesulfonic acid-catalyzed 1, 3-isomerization of allylic alcohols. *Tetrahedron Letters* **2007,** *48* (48), 8505-8507.

196. Fesenko, A. A.; Shutalev, A. D., Diastereoselective synthesis of 5-benzylthio-and 5-mercaptohexahydropyrimidin-2-ones. *Tetrahedron Letters* **2007,** *48* (48), 8420-8423.

197. Piras, L.; Ghiron, C.; Minetto, G.; Taddei, M., Microwave-assisted synthesis of tetrahydroindoles. *Tetrahedron Letters* **2008,** *49* (3), 459-462.

198. Özcan, S.; Balci, M., The chemistry of homophthalic acid: A new synthetic strategy for construction of substituted isocoumarin and indole skeletons. *Tetrahedron* **2008,** *64* (23), 5531-5540.

199. Mehta, G.; Bera, M. K., A concise approach towards the bicyclo [3.3. 1] nonan-9-one core present in the phloroglucin natural product hyperforin. *Tetrahedron Letters* **2008,** *49* (8), 1417-1420.

200. Schlapbach, A.; Revesz, L.; Koch, G. Heterocyclic compounds useful as MK2 inhibitors and their preparation, pharmaceutical compositions and use in the treatment of diseases. WO2009010488, 2009.

201. Almena, J.; Foubelo, F.; Yus, M., Reductive opening of thiophthalan: a new route to functionalized sulfur-containing compounds. *The Journal of Organic Chemistry* **1996,** *61* (5), 1859-1862.

202. Stefani, H. A.; Petragnani, N.; Valduga, C. J.; Brandt, C. A., Iodine promoted cyclofunctionalization reaction of 2, 4-dialkenyl-1, 3-dicarbonyl compounds. *Tetrahedron letters* **1997,** *38* (28), 4977-4980.

203. Mueller, L. G.; Lawton, R. G., Ring expansion synthesis of fused trans-. alpha. methylene-. gamma.-lactones. *The Journal of Organic Chemistry* **1979,** *44* (25), 4741-4742.

204. Wasacz, J.; Badding, V., A hydration of an alkyne illustrating steam and vacuum distillation. *Journal of Chemical Education* **1982,** *59* (8), 694.

205. Hiyama, T.; Shinoda, M.; Nozaki, H., Regio-and stereoselective cyclopentenone annulation by means of ketone-propargyl alcohol adducts. *Journal of the American Chemical Society* **1979,** *101* (6), 1599-1600.

206. Davis, B. R.; Hinds, M. G., Synthetic, Structural and vibrational spectroscopic studies in bismuth (III) halide/N, N′-aromatic bidentate base systems. IV. Bismuth (III) halide/N, N′-bidentate ligand (1: 1) systems. *Australian Journal of Chemistry* **1997,** *50* (4), 309-320.

207. Cohen, T.; Bhupathy, M.; Matz, J. R., A practical method for using alkoxide-accelerated vinylcyclobutane ring expansions in the synthesis of six-membered rings. Unexpected orbital symmetry allowed and forbidden 1, 3-sigmatropic rearrangements. *Journal of the American Chemical Society* **1983,** *105* (3), 520-525.

208. Apostolopoulos, C. D.; Georgiadis, M. P.; Couladouros, E. A., Ring chain transformation of γ-keto-δ-crotonolactones: A convenient synthesis of 1, 4-piperazinones, 1, 4-thiazinones and 1, 4-diazepinones. *Journal of heterocyclic chemistry* **1996,** *33* (3), 703-708.

209. Abdul-Aziz, M.; Auping, J. V.; Meador, M. A., Synthesis of Substituted 2, 3, 5, 6-Tetraarylbenzo [1, 2-b: 4, 5-b'] difurans. *The Journal of Organic Chemistry* **1995,** *60* (5), 1303-1308.

210. Kantorowski, E. J.; Kurth, M. J., Expansion to seven-membered rings. *Tetrahedron* **2000,** *56* (26), 4317-4353.

211. Williams, E. L., Synthesis and Rearrangement of Dispiro-Epoxy-β-Lactams. *Synthetic communications* **1992,** *22* (7), 1017-1021.

212. Valencia, E.; Freyer, A. J.; Shamma, M.; Fajardo, V., (±)-Nuevamine, an isoindoloisoquinoline alkaloid, and (±)-lennoxamine, an isoindolobenzazepine. *Tetrahedron letters* **1984,** *25* (6), 599-602.

213. Branan, B. M.; Paquette, L. A., Heteroatomic Effects on the Acid-catalyzed Rearrangements of Dispiro [4.0. 4.4] tetradeca-11, 13-dienes. *Journal of the American Chemical*

Society **1994,** *116* (17), 7658-7667.

214. Williams, E. L., Synthesis and Rearrangement of Dispiro-Epoxy-β-Lactams. *Synthetic communications* **1992,** *22* (7), 1017-1021.

215. Stefancich, G.; Artico, M.; Massa, S.; Vomero, S., Research on nitrogen heterocyclic compounds. XII. Synthesis of 5H-pyrrolo [1, 2-b][2] benzazepine derivatives. *Journal of Heterocyclic Chemistry* **1979,** *16* (7), 1443-1447.

216. Nicolaou, K.; Seitz, S.; Sipio, W.; Blount, J., Phenylseleno-and phenylsulfenolactonizations. Two highly efficient and synthetically useful cyclization procedures. *Journal of the American Chemical Society* **1979,** *101* (14), 3884-3893.

217. Taylor, S. K.; May, S. A.; Hopkins, J. A., Cyclizations wherein an epoxide acts as the source of initiation and termination steps. Evidence for an early transition state in biomimetic epoxide cyclizations. *Tetrahedron letters* **1993,** *34* (8), 1283-1286.

218. Dittami, J. P.; Xu, F.; Qi, H.; Martin, M. W.; Bordner, J.; Decosta, D. L.; Kiplinger, J.; Reiche, P.; Ware, R., Preparation of N-alkyl pyrrolidinones via photocyclization of γ-keto-α, β-unsaturated amides. *Tetrahedron letters* **1995,** *36* (24), 4197-4200.

219. Branan, B. M.; Paquette, L. A., Heteroatomic Effects on the Acid-catalyzed Rearrangements of Dispiro [4.0. 4.4] tetradeca-11, 13-dienes. *Journal of the American Chemical Society* **1994,** *116* (17), 7658-7667.

220. Gollnick, K.; Schade, G.; Cameron, A.; Hannaway, C.; Roberts, J.; Robertson, J., The structure of an isocaryophyllene rearrangement product, 1, 5, 9, 9-tetramethyltricyclo [6, 2, 1, 0 4, 11] undec-5-ene: X-ray analysis of the dibromo-derivative. *Journal of the Chemical Society D: Chemical Communications* **1970,** (4), 248-249.

221. Shi, X.; Miller, B., Cyclization and rearrangement processes resulting from bromination of 3-benzylcycloalkenes. *The Journal of Organic Chemistry* **1993,** *58* (10), 2907-2909.

222. Andersen, K. E.; Braestrup, C.; Groenwald, F. C.; Joergensen, A. S.; Nielsen, E. B.; Sonnewald, U.; Soerensen, P. O.; Suzdak, P. D.; Knutsen, L. J., The synthesis of novel GABA uptake inhibitors. 1. Elucidation of the structure-activity studies leading to the choice of (R)-1-[4, 4-bis (3-methyl-2-thienyl)-3-butenyl]-3-piperidinecarboxylic acid (tiagabine) as an anticonvulsant drug candidate. *Journal of medicinal chemistry* **1993,** *36* (12), 1716-1725.

223. Descotes, G.; Missos, D., Synthesis of Δ3-Flavene from Δ2-Chromene. *synthesis* **1971,** *1971* (03), 149-149.

224. Lee, W. Y.; Park, C. H.; Kim, E. H., Orthocyclophanes. 4. Functionalization of [1n] Orthocyclophanes on the Aromatic Rings. *The Journal of Organic Chemistry* **1994,** *59* (16), 4495-4500.

225. Negri, J. T.; Rogers, R. D.; Paquette, L. A., Belted spirocyclic tetrahydrofurans-a new class of preorganized ionophoric polyethers. Molecular structure, conformation, and binding to alkali metal atoms. *Journal of the American Chemical Society* **1991,** *113* (13), 5073-5075.

226. Raner, K. D.; Strauss, C. R.; Vyskoc, F.; Mokbel, L., A comparison of reaction kinetics

observed under microwave irradiation and conventional heating. *The Journal of Organic Chemistry* **1993,** *58* (4), 950-953.

227. Aida, T.; Legault, R.; Dugat, D.; Durst, T., Cyclization reactions of 4-(3′-butenyl) azetidin-2-one a route to the carbopenam ring system. *Tetrahedron Letters* **1979,** *20* (52), 4993-4994.

228. Aleksiuk, O.; Cohen, S.; Biali, S. E., Selective hydroxyl replacement in calixarenes: amino-, azo-, and xanthenocalixarene derivatives. *Journal of the American Chemical Society* **1995,** *117* (38), 9645-9652.

229. Galatsis, P.; Manwell, J. J., Deconjugative aldol-cyclization sequence for the construction of substituted 2-methoxytetrahydrofurans. *Tetrahedron letters* **1995,** *36* (45), 8179-8182.

230. Nath, A.; Mal, J.; Venkateswaran, R. V., Rearrangement of α-hydroxycyclobutanes. Unusual rearrangement of some cyclobutachromanols to a component of the ring system of averufin. *Journal of the Chemical Society, Chemical Communications* **1993,** (18), 1374-1375.

231. Cannone, P.; Bélanger, D.; Lemay, G., A Facile Synthesis of Spirochromanes-3, 4-Dihydrospiro [2H-1-benzopyran-2, 1′-cycloalkanes]. *Synthesis* **1980,** *1980* (04), 301-303.

232. Ren, X.-F.; Turos, E.; Lake, C. H.; Churchill, M. R., Regiochemical and stereochemical studies on halocyclization reactions of unsaturated sulfides. *The Journal of Organic Chemistry* **1995,** *60* (20), 6468-6483.

233. Seebach, D.; Hoffmann, T.; Kühnle, F. N.; Kinkel, J. N.; Schulte, M., Preparation, structure, and properties of all possible cyclic dimers (diolides) of 3-hydroxybutanoic acid. *Helvetica chimica acta* **1995,** *78* (6), 1525-1540.

234. Sido, A. S. S.; Chassaing, S.; Kumarraja, M.; Pale, P.; Sommer, J., Solvent-dependent behavior of arylvinylketones in HUSY-zeolite: a green alternative to liquid superacid medium. *Tetrahedron letters* **2007,** *48* (33), 5911-5914.

235. Allin, S. M.; Hodkinson, C. C.; Taj, N., Neighbouring group assistance in the formation of phthalimidines from o-phthalaldehyde: an intramolecular auxiliary effect. *Synlett* **1996,** (8), 781.

236. Takano, S.; Inomata, K.; Ogasawara, K., A new route to (+)-2, 3-(isopropylidenedioxy)-4-cyclopentenone via the optically active dicyclopentadiene intermediate. *Chemistry Letters* **1989,** *18* (2), 359-362.

237. Martin, P.; Winkler, T., Synthese und Eigenschaften der Furo-und Thieno-Analogen von PQQ-Triester. *Helvetica chimica acta* **1994,** *77* (1), 100-110.

238. Vlattas, I.; Harrison, I. T.; Tokes, L.; Fried, J. H.; Cross, A. D., Synthesis of (+-)-zearalenone. *The Journal of Organic Chemistry* **1968,** *33* (11), 4176-4179.

239. Kocienski, P.; Yeates, C., A new synthesis of 1, 7-dioxaspiro [5.5] undecanes. Application to a rectal gland secretion of the olive fruit fly (dacus oleae). *Tetrahedron Letters* **1983,** *24* (36), 3905-3906.

240. Montagnat, O. D.; Lessene, G.; Hughes, A. B., Synthesis of azide-alkyne fragments for "click" chemical applications. Part 2. Formation of oligomers from orthogonally protected chiral

trialkylsilylhomopropargyl azides and homopropargyl alcohols. *The Journal of Organic Chemistry* **2010,** *75* (2), 390-398.

241. Dong, W.; Zhang, M.; Xiao, F.; Wang, Y.; Liu, W.; Hu, X.; Yuan, Q.; Zhang, S., Gold-Catalyzed Intermolecular Carboalkoxylation of Alkenes. *Synlett* **2012,** *23* (19), 2799-2802.

242. Ong, C. W.; Chen, C. M.; Juang, S. S., Novel Cyclopentenone Synthesis by Base-Catalyzed Cyclization of Dienediones. *The Journal of Organic Chemistry* **1994,** *59* (25), 7915-7916.

243. Padwa, A.; Murphree, S. S.; Yeske, P. E., Use of 2, 3-dibromo-1-(phenylsulfonyl)-1-propene as a reagent for the synthesis of annulated furans. *The Journal of Organic Chemistry* **1990,** *55* (14), 4241-4242.

244. Crimmins, M. T.; Bankaitis, D. M., Addition of 1-methoxy-1-buten-3-yne to lactones: synthesis of substituted spiroketals. *Tetrahedron letters* **1983,** *24* (42), 4551-4554.

245. Carre, M. C.; Gregoire, B.; Caubere, P., Arynic condensation of ketone enolates. 16. Efficient access to a new series of benzocyclobutenols. *The Journal of Organic Chemistry* **1984,** *49* (11), 2050-2052.

246. Mi, X.; Luo, S.; Cheng, J.-P., Ionic liquid-immobilized quinuclidine-catalyzed Morita− Baylis−Hillman reactions. *The Journal of organic chemistry* **2005,** *70* (6), 2338-2341

247. Barco, A.; Benetti, S.; Pollini, G. P.; Baraldi, P. G.; Gandolfi, C., A new, elegant route to a key intermediate for the synthesis of 9 (0)-methanoprostacyclin. *The Journal of Organic Chemistry* **1980,** *45* (23), 4776-4778.

248. Nelson, R. P.; McEuen, J. M.; Lawton, R. G., . alpha.,. alpha.'Annulation of cyclic ketones. Synthesis of bicyclo [3.2. 1] octane derivatives. *The Journal of Organic Chemistry* **1969,** *34* (5), 1225-1229.

249. Baraldi, P. G.; Barco, A.; Benetti, S.; Moroder, F.; Pollini, G. P.; Simoni, D.; Zanirato, V., A novel α-acrylate anion equivalent: a useful synthon for α-substituted acrylic esters. *Journal of the Chemical Society, Chemical Communications* **1982,** (21), 1265-1266.

250. Yoshimura, H.; Nagai, M.; Hibi, S.; Kikuchi, K.; Abe, S.; Hida, T.; Higashi, S.; Hishinuma, I.; Yamanaka, T., A novel type of retinoic acid receptor antagonist: synthesis and structure-activity relationships of heterocyclic ring-containing benzoic acid derivatives. *Journal of medicinal chemistry* **1995,** *38* (16), 3163-3173.

251. Kawano, T.; Ogawa, T.; Islam, S. M.; Ueda, I., A facile synthesis of 5-substituted 2-furylacetates via 6-hydroxy-3-oxo-4-hexenoates. *Tetrahedron letters* **1995,** *36* (42), 7685-7688.

252. South, M. S., An unusual labilization of a 4-(trifluoromethyl) thiazole. *Journal of heterocyclic chemistry* **1991,** *28* (4), 1013-1016.

253. Ley, S. V.; Leslie, R.; Tiffin, P. D.; Woods, M., Dispiroketals in synthesis (part 2): A new group for the selective protection of diequatorial vicinal diols in carbohydrates. *Tetrahedron letters* **1992,** *33* (33), 4767-4770.

254. Bakuzis, P.; Bakuzis, M. L.; Weingartner, T. F., A new β-acyl vinyl anion equivalent. Synthesis of pyrenophorin. *Tetrahedron Letters* **1978,** *19* (27), 2371-2374.

255. Dickens, J. P.; Ellames, G. J.; Hare, N. J.; Lawson, K. R.; McKay, W. R.; Metters, A. P.; Myers, P. L.; Pope, A. M.; Upton, R. M., 2-Chloro-1-(2, 4-dichlorophenyl)-3-(1H-imidazol-1-yl)-2-phenylpropan-1-one hydrochloride, a novel, nonmutagenic antibacterial with specific activity against anaerobic bacteria. *Journal of medicinal chemistry* **1991,** *34* (8), 2356-2360.

256. Koroleva, E. B.; Backvall, J.-E.; Andersson, P. G., Palladium-catalyzed stereocontrolled endo cyclization of 3-hydroxypropyl-1,3-cyclohexadiene leading to versatile fused tetrahydropyrans. *Tetrahedron Lett.* **1995,** *36* (30), 5397.

257. Lawton, R. G.; Dunham, D. J., Synthesis of spiro systems by the. alpha.,. alpha.'-annelation process. *Journal of the American Chemical Society* **1971,** *93* (8), 2074-2075.

258. Richard, J., The synthesis of novel benzomorphan analogues: a new intramolecular acid-catalysed aldol route to benzannelated bicyclo [3.3. 1] nonane derivatives. X-Ray molecular structure of 8-hydroxy-1-methoxytricyclo [7.3. 1.0 2, 7] trideca-2, 4, 6-trien-10-one. *Journal of the Chemical Society, Perkin Transactions 1* **1990,** (7), 2145-2150.

259. Vincek, W. C.; Aldrich, C. S.; Borchardt, R. T.; Grunewald, G. L., Importance of the aromatic ring in adrenergic amines. 5. Nonaromatic analogs of phenylethanolamine as inhibitors of phenylethanolamine N-methyltransferase: role of hydrophobic and steric interactions. *Journal of Medicinal Chemistry* **1981,** *24* (1), 7-12.

260. Bhakuni, D. S.; Kumar, P., Studies on Mannich reaction of 1-benzyltetrahydroisoquinolines. *J. Indian Chem. Soc.* **1988,** *65* (6), 417.

261. Rao, A. R.; Deshmukh, M.; Kamalam, M., A convenient synthesis of 1-triacontanol. *Tetrahedron* **1981,** *37* (1), 227-230.

262. Marfat, A.; Helquist, P., Copper-catalyzed conjugate addition of an acetal-containing grignard reagent. A method for cyclopentene annulation. *Tetrahedron Letters* **1978,** *19* (44), 4217-4220.

263. Schatz, P. F., Synthesis of chrysanthemic acid: A multistep organic synthesis for undergraduate students. *Journal of Chemical Education* **1978,** *55* (7), 468.

264. Lansbury, P. T.; Hangauer, D. G., C-10 epimerization of hydroazulen-7-onesby vinylogus activation. *Tetrahedron Letters* **1979,** *20* (38), 3623-3626.

265. Gammill, R.; Nash, S.; Mizsak, S., The addition of amines to 3-bromochromone and 6-bromofurochromone. An unexpected ring contraction of the pyrone ring. *Tetrahedron Letters* **1983,** *24* (33), 3435-3438.

266. Campaigne, E.; Forsch, R. A., Cyclization of ylidenemalononitriles. 9. Rearrangement of 2-cyano-3-(1-methylcyclopentyl) indenone to 4a-methyl-9-oxo-10-cyano-1, 2, 3, 4, 4a, 9-hexahydrophenanthrene. *The Journal of Organic Chemistry* **1978,** *43* (6), 1044-1050.

267. Rogers, M. E.; Wilkinson, D. S.; Thweatt, J. R.; Halenda, S. P., Some spiro analogs of the potent analgesic ketobemidone. *J. Med. Chem.* **1980,** *23* (6), 688.

268. a) Tramontini, M., Advances in the chemistry of Mannich bases. *Synthesis* **1973,** *1973* (12), 703-775. b) Tramontini, M.; Angiolini, L., Further advances in the chemistry of Mannich bases. *Tetrahedron* **1990,** *46* (6), 1791-1837.

269. Reutrakul, V.; Poochaivatanon, P., Eliminative deoxygenation of substituted-α-halosulfoxides. *Tetrahedron Letters* **1983,** *24* (5), 531-534.

270. Hill, R. K.; Ledford, N. D., Mechanism of the abnormal Michael reaction between ethyl cyanoacetate and 3-methyl-2-cyclohexenone. *J. Am. Chem. Soc.* **1975,** *97* (3), 666.

271. Matsuda, H.; Maruyama, K. Fragrance composition containing 3-(3-hexenyl)-2-cyclopentenone. EP1321508, 2003.

272. Scarborough Jr, R. M.; Toder, B. H.; Smith III, A. B., A stereospecific total synthesis of (.+-.)-methylenomycin A and its epimer,(.+-.)-epimethylenomycin A. *Journal of the American Chemical Society* **1980,** *102* (11), 3904-3913.

273. Löwe, W.; Jeske, P., Umsetzungen von 4-Chromon-3-sulfonsäure-phenylester mit methylenaktiven Verbindungen. *Liebigs Annalen der Chemie* **1987,** *1987* (6), 549-550.

274. Gammill, R.; Nash, S.; Mizsak, S., The addition of amines to 3-bromochromone and 6-bromofurochromone. An unexpected ring contraction of the pyrone ring. *Tetrahedron Letters* **1983,** *24* (33), 3435-3438.

275. Khatana, S. S.; Boschelli, D. H.; Kramer, J. B.; Connor, D. T.; Barth, H.; Stoss, P., Preparation of benzothieno [2, 3-f]-1, 4-oxazepin-and-thiazepin-5 (2 H)-ones and of benzothieno [3, 2-e]-1, 4-diazepin-5-ones. *The Journal of Organic Chemistry* **1996,** *61* (17), 6060-6062.

276. Black, T. H.; Arrivo, S. M.; Schumm, J. S.; Knobeloch, J. M., 4-(Dimethylamino)pyridine as a catalyst for carbon acylation. 2. Control of carbon vs. oxygen acylation in benzofuranones. *J. Org. Chem.* **1987,** *52* (24), 5425.

277. Johnson, W. S.; Daub, G. H., The S tobbe Condensation. *Organic Reactions* **2004,** *6*, 1-73.

278. Queignec, R.; Kirschleger, B.; Lambert, F.; Aboutaj, M., High Yield Synthesis of α Propargylic Acrylic Ester: A General Access to α Substituted Acrylic Esters. *Synthetic Communications* **1988,** *18* (11), 1213-1223.

279. Taber, D. F.; Amedio Jr, J. C.; Jung, K. Y., Phosphorus pentoxide/dimethyl sulfoxide/triethylamine (PDT): a convenient procedure for oxidation of alcohols to ketones and aldehydes. *The Journal of Organic Chemistry* **1987,** *52* (25), 5621-5622.

280. Bagli, J.; Bogri, T.; Voith, K., Troponoids. 6. Troponylpiperazines: a new class of dopamine agonists. *Journal of medicinal chemistry* **1984,** *27* (7), 875-881.

281. Dike, S. Y.; Ner, D. H.; Kumar, A., A new enantioselective chemoenzymatic synthesis of R-(−) thiazesim hydrochloride. *Bioorganic & Medicinal Chemistry Letters* **1991,** *1* (8), 383-386.

282. Arct, J.; Jakubska, E.; Olszewska, G., Conversion of Mannich Phenol Bases; III. Synthesis and Transformations of 3, 4-Dihydro-2H-1, 3-Benzoxazin-2-one Derivatives. *Synthetic*

Communications **1978,** *8* (3), 143-149.

283. Van Hove, F.; Vanwetswinkel, S.; Marchand-Brynaert, J.; Fastrez, J., Synthesis and rearrangment of potential zinc β-lactamase inhibitors. *Tetrahedron letters* **1995,** *36* (51), 9313-9316.

284. Staab, H. A.; Walther, G.; Rohr, W., Synthesis of carboxylic acid anhydrides by the imidazolide method. *Chem. Ber.* **1962,** *95*, 2073.

285. Sato, M.; Sakaki, J.-i.; Sugita, Y.; Yasuda, S.; Sakoda, H.; Kaneko, C., Two lactone formation reactions from 1, 3-dioxin-4-ones having hydroxyalkyl group at the 6-position: Difference in ring opening and closure. *Tetrahedron* **1991,** *47* (30), 5689-5708.

286. Hart, B. P.; Coward, J. K., The synthesis of DL-3,3-difluoroglutamic acid from a 3-oxoprolinol derivative. *Tetrahedron Lett.* **1993,** *34* (31), 4917.

287. Zehnter, R.; Gerlach, H., Synthesis of anacardic acids. *Liebigs Annalen* **1995,** *1995* (12), 2209-2220.

288. Boden, E. P.; Keck, G. E., Proton-transfer steps in Steglich esterification: A very practical new method for macrolactonization. *The Journal of Organic Chemistry* **1985,** *50* (13), 2394-2395.

289. Cooney, J. V., Reissert compounds and their open-chain analogs in organic synthesis. *Journal of Heterocyclic Chemistry* **1983,** *20* (4), 823-837.

290. Bargar, T.; Riley, C., A Rapid and Efficient Method for Dehydration of Primary Amides to Nitriles. Preparation of Acrylonitrile Derivatives. *Synthetic Communications* **1980,** *10* (6), 479-487.

291. Cooney, J. V., Reissert compounds and their open-chain analogs in organic synthesis. *Journal of Heterocyclic Chemistry* **1983,** *20* (4), 823-837.

292. Bayles, R.; Johnson, M.; Maisey, R.; Turner, R., The Smiles rearrangement of 2-aryloxy-2-methylpropanamides. Synthesis of N-aryl-2-hydroxy-2-methylpropanamides. *Synthesis* **1977,** *1977* (01), 31-33.

293. Boeckman Jr, R. K.; Ko, S. S., Stereocontrol in the intramolecular Diels-Alder reaction. 1. An application to the total synthesis of (.+-.) marasmic acid. *Journal of the American Chemical Society* **1980,** *102* (23), 7146-7149.

294. Hauser, F. M.; Rhee, R. P., New synthetic methods for the regioselective annelation of aromatic rings: 1-hydroxy-2, 3-disubstituted naphthalenes and 1, 4-dihydroxy-2, 3-disubstituted naphthalenes. *The Journal of Organic Chemistry* **1978,** *43* (1), 178-180.

295. Huckin, S. N.; Weiler, L., Alkylation of dianions of. beta.-keto esters. *Journal of the American Chemical Society* **1974,** *96* (4), 1082-1087.

296. Paquette, L. A.; Freeman, J. P., Mechanism of thietane formation from the reaction of 1, 3-dioxan-2-ones with thiocyanate ion. Stereochemical investigation. *The Journal of Organic Chemistry* **1970,** *35* (7), 2249-2253.

297. Raina, S.; Singh, V. K., Reaction of epoxides with activated DMSO reagent. General

method for synthesis of α-chlorocarbonyl compounds: Application in asymmetric synthesis of (3S)-2, 3-oxidosqualene. *Tetrahedron* **1995,** *51* (8), 2467-2476.

298. Hughes, D. L., The mitsunobu reaction. *Organic reactions* **1992**.

299. Armstrong, A.; Brackenridge, I.; Jackson, R. F.; Kirk, J. M., A new method for the preparation of tertiary butyl ethers and esters. *Tetrahedron letters* **1988,** *29* (20), 2483-2486.

300. Muxworthy, J. P.; Wilkinson, J. A.; Procter, G., Stereoselective cycloadditions of chiral acyl-nitroso compounds; unexpected formation of an oxazolidinone. *Tetrahedron letters* **1995,** *36* (41), 7539-7540.

301. Statham, F., 44. Anhydro-N-carboxyamino-acids. A Friedel–Crafts type reaction. *Journal of the Chemical Society (Resumed)* **1951**, 213-215.

302. Ley, S. V.; Leslie, R.; Tiffin, P. D.; Woods, M., Dispiroketals in synthesis (part 2): A new group for the selective protection of diequatorial vicinal diols in carbohydrates. *Tetrahedron letters* **1992,** *33* (33), 4767-4770.

303. Muxworthy, J. P.; Wilkinson, J. A.; Procter, G., Stereoselective cycloadditions of chiral acyl-nitroso compounds; unexpected formation of an oxazolidinone. *Tetrahedron letters* **1995,** *36* (41), 7539-7540.

304. Van Hove, F.; Vanwetswinkel, S.; Marchand-Brynaert, J.; Fastrez, J., Synthesis and rearrangment of potential zinc β-lactamase inhibitors. *Tetrahedron letters* **1995,** *36* (51), 9313-9316.

305. Dittami, J. P.; Xu, F.; Qi, H.; Martin, M. W.; Bordner, J.; Decosta, D. L.; Kiplinger, J.; Reiche, P.; Ware, R., Preparation of N-alkyl pyrrolidinones via photocyclization of γ-keto-α, β-unsaturated amides. *Tetrahedron letters* **1995,** *36* (24), 4197-4200.

306. Morris, J.; Wishka, D. G., Synthesis of novel antagonists of leukotriene B4. *Tetrahedron letters* **1988,** *29* (2), 143-146.

307. Hart, D. J.; Kim, A.; Krishnamurthy, R.; Merriman, G. H.; Waltos, A.-M., Synthesis of 6H-dibenzo [b, d] pyran-6-ones via dienone-phenol rearrangements of spiro [2, 5-cyclohexadiene-1, 1′(3′ H)-isobenzofuran]-3′-ones. *Tetrahedron* **1992,** *48* (38), 8179-8188.

308. Baraldi, P. G.; Achille, B.; Simoneta, B.; Piero, P. G.; Vinicio, Z., 2, 3a, 5, 6, 7, 7a-Hexahydro-3h, 4h-benzothiophene-3, 4-dione and cyclopenta [b]-tetrahydrothiophene-3, 4-dione enolate anions as synthetic equivalents to cyclohex-2-enone and cyclopent-2-enone c-2-carbanions. *Tetrahedron letters* **1984,** *25* (38), 4291-4294.

309. Babler, J. H.; Schlidt, S. A., An expedient route to a versatile intermediate for the stereoselective synthesis of all-trans-retinoic Acid and beta-carotene. *Tetrahedron letters* **1992,** *33* (50), 7697-7700.

310. Raina, S.; Singh, V. K., Reaction of epoxides with activated DMSO reagent. General method for synthesis of α-chlorocarbonyl compounds: Application in asymmetric synthesis of (3S)-2, 3-oxidosqualene. *Tetrahedron* **1995,** *51* (8), 2467-2476.

311. Bakuzis, P.; Bakuzis, M. L.; Weingartner, T. F., A new β-acyl vinyl anion equivalent. Synthesis of pyrenophorin. *Tetrahedron Letters* **1978,** *19* (27), 2371-2374.

312. Arrault, A.; Touzeau, F.; Guillaumet, G.; Mérour, J.-Y., A Straightforward Synthesis of 1, 2-Dihydronaphtho [2, 1-b] furans from 2-Naphthols. *Synthesis* **1999,** *1999* (07), 1241-1245.

313. Soós, T.; Timári, G.; Hajós, G., A concise synthesis of furostifoline. *Tetrahedron letters* **1999,** *40* (49), 8607-8610.

314. Valderrama, J. A.; Leiva, H.; Tapia, R., Studies on Quinones. Part. 33.1 Synthetic Approach to Podands Containing Quinone Fragments. *Synthetic Communications* **2000,** *30* (4), 737-749.

315. Potts, D.; Stevenson, P. J.; Thompson, N., Expedient synthesis of (+)-trans-5-allylhexahydroindolizidin-3-one. *Tetrahedron Letters* **2000,** *41* (2), 275-278.

316. Marson, C. M.; Pink, J. H.; Smith, C.; Hursthouse, M. B.; Malik, K. A., A biomimetic synthesis of the pyrrolizidine ring by sequential intramolecular cyclizations. *Tetrahedron Letters* **2000,** *41* (1), 127-129.

317. Ranu, B. C.; Jana, U., A new redundant rearrangement of aromatic ring fused cyclic α-hydroxydithiane derivatives. Synthesis of aromatic ring fused cyclic 1, 2-diketones with one-carbon ring expansion. *The Journal of Organic Chemistry* **1999,** *64* (17), 6380-6386.

318. Black, D. S.; Craig, D. C.; Santoso, M., Mechanism-controlled regioselective synthesis of indolyl benzo [b] carbazoles. *Tetrahedron letters* **1999,** *40* (36), 6653-6656.

319. Trauner, D.; Danishefsky, S. J., Studies towards the total synthesis of halichlorine: asymmetric synthesis of the spiroquinolizidine subunit. *Tetrahedron letters* **1999,** *40* (36), 6513-6516.

320. Statham, F. S., Anhydro-N-carboxyamino acids. A Friedel-Crafts type reaction. *J. Chem. Soc.* **1951**, 213.

321. Hahn, H.-G.; Dal Nam, K.; Mah, H.; Lee, J. J., A New Synthesis of 1, 4-Thiazin-3-ones by a Novel Rearrangement of 1, 4-Oxathiins. *The Journal of organic chemistry* **1996,** *61* (11), 3894-3896.

322. Montes, I. F.; Burger, U., The cyanide catalyzed isomerization of enol esters derived from cyclic 1, 3-diketones. *Tetrahedron letters* **1996,** *37* (7), 1007-1010.

323. Kumar, G. B.; Shah, A. C.; Pilati, T., Formation of a novel ring system: An unexpected intermolecular cyclization. *Tetrahedron letters* **1997,** *38* (18), 3297-3300.

324. Allin, S. M.; Northfield, C. J.; Page, M. I.; Slawin, A. M., A facile and highly stereoselective approach to a polycyclic isoindolinone ring system via an N-acyliminium ion cyclization reaction. *Tetrahedron letters* **1998,** *39* (27), 4905-4908.

325. Nagafuji, P.; Cushman, M., A general synthesis of pyrroles and fused pyrrole systems from ketones and amino acids. *The Journal of Organic Chemistry* **1996,** *61* (15), 4999-5003.

326. Rho, T.; Lankin, C. M.; Lankin, M. E.; Shih, D. H., A Facile Synthesis of an Advanced Glycosylation Endproduct 2-(2′-furoyl)-4 (5)-(2′-furanyl)-1 H-imidazole. *Synthetic communications* **1997,** *27* (24), 4315-4318.

327. Grivas, J. C., Rearrangement of 2-phenacyl-1, 2-benzisothiazolin-3-one to 2-benzoyl-2H-1, 3-benzothiazin-4 (3H)-one. *The Journal of Organic Chemistry* **1976,** *41* (8), 1325-1327.

328. Gerlitz, M.; Udarnoki, G.; Rohr, J., Biosyntheses of Novel Emycins from the Mutant Strain Streptomyces cellulosae ssp. griseoincarnatus 1114-2. *Angewandte Chemie International Edition in English* **1995,** *34* (15), 1617-1621.

329. Rogers, M. E.; Wilkinson, D. S.; Thweatt, J. R.; Halenda, S. P., Some spiro analogs of the potent analgesic ketobemidone. *J. Med. Chem.* **1980,** *23* (6), 688.

330. Rios, R.; Sundén, H.; Ibrahem, I.; Zhao, G.-L.; Eriksson, L.; Cordova, A., Highly enantioselective synthesis of 2H-1-benzothiopyrans by a catalytic domino reaction. *Tetrahedron letters* **2006,** *47* (48), 8547-8551.

331. Kim, I.; Kim, T.-H.; Kang, Y.; Lim, Y.-b., BBr3-promoted cyclization to produce ladder-type conjugated polymer. *Tetrahedron letters* **2006,** *47* (49), 8689-8692.

332. El Azzaoui, B.; Rachid, B.; Doumbia, M. L.; Essassi, E. M.; Gornitzka, H.; Bellan, J., Unexpected opening of benzimidazole derivatives during 1, 3-dipolar cycloaddition. *Tetrahedron letters* **2006,** *47* (50), 8807-8810.

333. Ko, S.; Yao, C.-F., Heterogeneous catalyst: Amberlyst-15 catalyzes the synthesis of 14-substituted-14H-dibenzo [a, j] xanthenes under solvent-free conditions. *Tetrahedron Letters* **2006,** *47* (50), 8827-8829.

334. Prukała, D., New compounds via Mannich reaction of cytosine, paraformaldehyde and cyclic secondary amines. *Tetrahedron letters* **2006,** *47* (51), 9045-9047.

335. Christoforou, A.; Nicolaou, G.; Elemes, Y., N-Phenyltriazolinedione as an efficient, selective, and reusable reagent for the oxidation of thiols to disulfides. *Tetrahedron Letters* **2006,** *47* (52), 9211-9213.

336. Gaddam, V.; Sreenivas, D. K.; Nagarajan, R., Highly diastereoselective synthesis of new chromenylaminoanthraquinones through a one-pot, three-component hetero Diels–Alder reaction. *Tetrahedron letters* **2006,** *47* (52), 9291-9295.

337. Bellur, E.; Langer, P., Synthesis of functionalized pyrroles and 6, 7-dihydro-1H-indol-4 (5H)-ones by reaction of 1, 3-dicarbonyl compounds with 2-azido-1, 1-diethoxyethane. *Tetrahedron letters* **2006,** *47* (13), 2151-2154.

338. Singh, B. K.; Verma, S. S.; Dwivedi, N.; Tripathi, R. P., l-Ascorbic acid in organic synthesis: DBU-catalysed one-pot synthesis of tetramic acid derivatives from 5, 6-O-isopropylidene ascorbic acid. *Tetrahedron letters* **2006,** *47* (13), 2219-2222.

339. Pask, C. M.; Camm, K. D.; Kilner, C. A.; Halcrow, M. A., Synthesis of a new series of ditopic proligands for metal salts: differing regiochemistry of electrophilic attack at 3 {5}-amino-5 {3}-(pyrid-2-yl)-1H-pyrazole. *Tetrahedron letters* **2006,** *47* (15), 2531-2534.

340. Parchinsky, V. Z.; Shuvalova, O.; Ushakova, O.; Kravchenko, D. V.; Krasavin, M., Multi-component reactions between 2-aminopyrimidine, aldehydes and isonitriles: the use of a nonpolar solvent suppresses formation of multiple products. *Tetrahedron letters* **2006,** *47* (6), 947-951.

341. Moulin, A.; Martinez, J.; Fehrentz, J.-A., Convenient two-step preparation of [1, 2, 4] triazolo [4, 3-a] pyridines from 2-hydrazinopyridine and carboxylic acids. *Tetrahedron letters* **2006,** *47* (43), 7591-7594.

342. Kumar, S.; Malik, V.; Kaur, N.; Kaur, K., A simple synthesis of di (uracilyl) aryl methanes and 1, ω-bis [di (uracilyl) methyl] benzenes. *Tetrahedron letters* **2006,** *47* (48), 8483-8487.

343. Nunami, K.; Suzuki, M.; Yoneda, N., Synthesis of heterocyclic compounds using isocyano compounds. 5. One-step synthesis of 1-oxo-1, 2-dihydroisoquinoline-3-carboxylic acid derivatives. *The Journal of Organic Chemistry* **1979,** *44* (11), 1887-1888.

344. Kawasaki, I.; Terano, M.; Yada, E.; Kawai, M.; Yamashita, M.; Ohta, S., Novel cyclo-dimerization of 1-tert-butoxycarbonyl-3-alkenylindole derivatives. *Tetrahedron letters* **2005,** *46* (7), 1199-1203.

345. Yadav, J.; Reddy, B. S.; Vishnumurthy, P., Amberlyst-15® as a novel and recyclable solid acid for the coupling of aromatic aldehydes with homopropargyl alcohol. *Tetrahedron letters* **2005,** *46* (8), 1311-1313.

346. Ceglia, S. S.; Kress, M. H.; Nelson, T. D.; McNamara, J. M., A regioselective synthesis of 2, 4-dialkyl resorcinols. *Tetrahedron letters* **2005,** *46* (10), 1731-1734.

347. Lakner, F. J.; Xia, H.; Pervin, A.; Hammaker, J. R.; Jahangiri, K. G.; Dalton, M. K.; Khvat, A.; Kiselyov, A.; Ivachtchenko, A. V., A novel mesoionic ring system: unusual cyclization of thio-and amino-acid derivatives of 6-azauracil. *Tetrahedron letters* **2005,** *46* (32), 5325-5328.

348. Cook, A. G.; Schering, C. A.; Campbell, P. A.; Hayes, S. S., Pyrolysis of perhydro [1, 2-c][1, 3] oxazines: a green method of synthesizing 2, 3-dehydropiperidine enamines. *Tetrahedron letters* **2005,** *46* (33), 5451-5454.

349. Svetlik, J.; Kettmann, V.; Zaleska, B., A new convenient synthesis of functionalized 2, 3-dihydro-4-pyridones. *Tetrahedron letters* **2005,** *46* (33), 5511-5514.

350. Shi, X.; Miller, B., Cyclization and rearrangement processes resulting from bromination of 3-benzylcycloalkenes. *The Journal of Organic Chemistry* **1993,** *58* (10), 2907-2909.

351. Ech-Chahad, A.; Minassi, A.; Berton, L.; Appendino, G., An expeditious hydroxyamidation of carboxylic acids. *Tetrahedron letters* **2005,** *46* (31), 5113-5115.

352. Li, D.; Shi, F.; Guo, S.; Deng, Y., Highly efficient Beckmann rearrangement and dehydration of oximes. *Tetrahedron Letters* **2005,** *46* (4), 671-674.

353. Banerji, A.; Bandyopadhyay, D.; Prangé, T.; Neuman, A., Unexpected cycloadducts from 1, 3-dipolar cycloaddition of 3, 4-dehydromorpholine N-oxide to N-cinnamoyl piperidines—first report of the novel formation of 2: 1 cycloadducts. *Tetrahedron letters* **2005,** *46* (15), 2619-2622.

354. Kawashima, T.; Kashima, H.; Wakasugi, D.; Satoh, T., A novel synthesis of bicyclo [3.3.

0] oct-1-en-3-ones from cyclobutanones through [chloro (p-tolylsulfinyl) methylidene] cyclobutanes with ring expansion. *Tetrahedron letters* **2005,** *46* (21), 3767-3770.

355. Dhanabal, T.; Sangeetha, R.; Mohan, P., Fischer indole synthesis of the indoloquinoline alkaloid: cryptosanguinolentine. *Tetrahedron letters* **2005,** *46* (26), 4509-4510.

356. Oda, M.; Fukuchi, Y.; Ito, S.; Thanh, N. C.; Kuroda, S., A facile non-oxidative method for synthesizing 1, 3-disubstituted pyrroles from pyrrolidine and aldehydes. *Tetrahedron letters* **2007,** *48* (52), 9159-9162.

357. Nagumo, S.; Mizukami, M.; Wada, K.; Miura, T.; Bando, H.; Kawahara, N.; Hashimoto, Y.; Miyashita, M.; Akita, H., Novel construction of hydro-2-benzazepines based on 7-endo selective Friedel–Crafts-type reaction of vinyloxiranes. *Tetrahedron Letters* **2007,** *48* (48), 8558-8561.

358. Ardes-Guisot, N.; Ouled-Lahoucine, B.; Canet, I.; Sinibaldi, M.-E., A straightforward route to spiroketals. *Tetrahedron Letters* **2007,** *48* (48), 8511-8513.

359. Ying, Y.; Hong, J., Synthesis of brasilibactin A and confirmation of absolute configuration of β-hydroxy acid fragment. *Tetrahedron Letters* **2007,** *48* (46), 8104-8107.

360. Piras, L.; Ghiron, C.; Minetto, G.; Taddei, M., Microwave-assisted synthesis of tetrahydroindoles. *Tetrahedron Letters* **2008,** *49* (3), 459-462.

361. Kowalewska, M.; Kwiecień, H., A new course of the Perkin cyclization of 2-(2-formyl-6-methoxyphenoxy) alkanoic acids. Synthesis of 2-alkyl-7-methoxy-5-nitrobenzo [b] furans. *Tetrahedron* **2008,** *64* (22), 5085-5090.

362. Elgersma, R. C.; Mulder, G. E.; Posthuma, G.; Rijkers, D. T.; Liskamp, R. M., Mirror image supramolecular helical tapes formed by the enantiomeric-depsipeptide derivatives of the amyloidogenic peptide amylin (20–29). *Tetrahedron Letters* **2008,** *49* (6), 987-991.

363. Methot, J. L.; Dunstan, T. A.; Mampreian, D. M.; Adams, B.; Altman, M. D., An unexpected aminocyclopropane reductive rearrangement. *Tetrahedron Letters* **2008,** *49* (7), 1155-1159.

364. Methot, J. L.; Dunstan, T. A.; Mampreian, D. M.; Adams, B.; Altman, M. D., An unexpected aminocyclopropane reductive rearrangement. *Tetrahedron Letters* **2008,** *49* (7), 1155-1159.

365. Yamada, N.; Mizuochi, M.; Takeda, M.; Kawaguchi, H.; Morita, H., A facile and efficient one-pot synthesis of thiirans by the reaction of benzoxazolyl β-ketosulfides with NaBH4/NaOH. *Tetrahedron Letters* **2008,** *49* (7), 1166-1168.

366. Métro, T.-X.; Fayet, C.; Arnaud, F.; Rameix, N.; Fraisse, P.; Janody, S.; Sevrin, M.; George, P.; Vogel, R., Synthesis of 2, 2-disubstituted azaindolines by intramolecular cyclization in acidic media. *Synlett* **2011,** *2011* (05), 684-688.

367. Uchida, H.; Ogawa, S.; Makabe, M.; Maeda, Y. Preparation of heterocyclylidene-N-(heterocyclyl)acetamide derivatives as antagonists of transient receptor potential type I receptor (TRPV1). WO2008091021, 2008.

368. Tarrade-Matha, A.; Pillon, F.; Doris, E., Straightforward Conversion of Alcohols into Nitriles. *Synthetic Communications®* **2010,** *40* (11), 1646-1649.

369. Allin, S. M.; Hodkinson, C. C.; Taj, N., Neighbouring group assistance in the formation of phthalimidines from o-phthalaldehyde: an intramolecular auxiliary effect. *Synlett* **1996,** (8), 781.

370. Kim, H.-O.; Kahn, M., The synthesis of aminoazole analogs of lysine and arginine: the Mitsunobu reaction with lysinol and argininol. *Synlett* **1999,** *1999* (08), 1239-1240.

371. Sepiol, J. J.; Wilamowski, J., New aromatic rearrangement accompanying ring closure of 2-arylpropylidenemalonodinitriles to 1-aminonaphthalene-2-carbonitriles. *Tetrahedron Letters* **2001,** *42* (31), 5287-5289.

372. Paul, S.; Gupta, M.; Gupta, R.; Loupy, A., Microwave assisted solvent-free synthesis of pyrazolo [3, 4-b] quinolines and pyrazolo [3, 4-c] pyrazoles using p-TsOH. *Tetrahedron Letters* **2001,** *42* (23), 3827-3829.

373. Han, G.; Hruby, V. J., A study of conjugate addition to a γ, δ-dioxolanyl-α, β-unsaturated ester. *Tetrahedron Letters* **2001,** *42* (26), 4281-4283.

374. Kajikawa, S.; Nishino, H.; Kurosawa, K., Synthesis of naphthalenes using acid-catalyzed ring-opening and recyclization of 3-acetyl-5, 5-diaryl-2-methyl-4, 5-dihydrofurans. Isolation of intermediates. *Tetrahedron Letters* **2001,** *42* (19), 3351-3354.

375. Risitano, F.; Grassi, G.; Foti, F.; Bilardo, C., A convenient synthesis of furo [3, 2-c] coumarins by a tandem alkylation/intramolecular aldolisation reaction. *Tetrahedron Letters* **2001,** *42* (20), 3503-3505.

376. Epstein, O. L.; Kulinkovich, O. G., Two-step synthesis of (±)-stigmolone, the pheromone of Stigmatella aurantiaca. *Tetrahedron Letters* **2001,** *42* (22), 3757-3758.

377. Hagiwara, H.; Okabe, T.; Hakoda, K.; Hoshi, T.; Ono, H.; Kamat, V. P.; Suzuki, T.; Ando, M., Catalytic enamine reaction: an expedient 1, 4-conjugate addition of naked aldehydes to vinylketones and its application to synthesis of cyclohexenone from Stevia purpurea. *Tetrahedron Letters* **2001,** *42* (14), 2705-2707.

378. Suzuki, H.; Yamazaki, N.; Kibayashi, C., Synthesis of the azatricyclic core of FR901483 by bridgehead vinylation via an anti-Bredt iminium ion. *Tetrahedron Letters* **2001,** *42* (16), 3013-3015.

379. Faure, S.; Piva, O., Application of chiral tethers to intramolecular [2+ 2] photocycloadditions: synthetic approach to (−)-italicene and (+)-isoitalicene. *Tetrahedron Letters* **2001,** *42* (2), 255-259.

380. Lohse, O.; Beutler, U.; Fünfschilling, P.; Furet, P.; France, J.; Kaufmann, D.; Penn, G.; Zaugg, W., New synthesis of oxcarbazepine via remote metalation of protected No-tolyl-anthranilamide derivatives. *Tetrahedron Letters* **2001,** *42* (3), 385-389.

381. Reddy, G. V.; Sreevani, V.; Iyengar, D., A novel nonclassical Wittig reaction of dioxolanones: highly facile and concise enantiospecific synthesis of (3S, 4S)-3-hydroxy-4-phenylbutyrolactone. *Tetrahedron Letters* **2001,** *42* (3), 531-532.

382. Mani, N. S.; Chen, J.-H.; Edwards, J. P.; Wu, M.; Chen, P.; Higuchi, R. I., Efficient synthesis of an androgen receptor modulator. *Tetrahedron Letters* **2008,** *49* (12), 1903-1905.

383. Zhang, D.; Bender, D. M.; Victor, F.; Peterson, J. A.; Boyer, R. D.; Stephenson, G. A.; Azman, A.; McCarthy, J. R., Facile rearrangement of N4-(α-aminoacyl) cytidines to N-(4-cytidinyl) amino acid amides. *Tetrahedron Letters* **2008,** *49* (13), 2052-2055.

384. Faggi, C.; Neo, A. G.; Marcaccini, S.; Menchi, G.; Revuelta, J., Ugi four-component condensation with two cleavable components: the easiest synthesis of 2, N-diarylglycines. *Tetrahedron Letters* **2008,** *49* (13), 2099-2102.

385. Kumar, G. K.; Natarajan, A., Total synthesis of ovalifoliolatin B, acerogenins A and C. *Tetrahedron Letters* **2008,** *49* (13), 2103-2105.

386. Li, J.; Han, Y.; Freedman, T. B.; Zhu, S.; Kerwood, D. J.; Luk, Y.-Y., Utilizing the high dielectric constant of water: efficient synthesis of amino acid-derivatized cyclobutenones. *Tetrahedron Letters* **2008,** *49* (13), 2128-2131.

387. Juma, B.; Adeel, M.; Villinger, A.; Langer, P., Efficient synthesis of 2, 6-dioxo-1, 2, 3, 4, 5, 6-hexahydroindoles based on the synthesis and reactions of (2, 4-dioxocyclohex-1-yl) acetic acid derivatives. *Tetrahedron Letters* **2008,** *49* (14), 2272-2274.

388. Lamberth, C.; Querniard, F., First synthesis and further functionalization of 7-chloro-imidazo [2, 1-b][1, 3] thiazin-5-ones. *Tetrahedron Letters* **2008,** *49* (14), 2286-2288.

389. Yavari, I.; Mirzaei, A.; Moradi, L.; Hosseini, N., Stereoselective synthesis of dialkyl 3-spiroindanedione-1, 2, 3, 3a-tetrahydropyrrolo [1, 2-a] quinoline-1, 2-dicarboxylates. *Tetrahedron Letters* **2008,** *49* (15), 2355-2358.

390. Amere, M.; Blanchet, J.; Lasne, M.-C.; Rouden, J., 4-Toluenesulfonic acid: an environmentally benign catalyst for Nazarov cyclizations. *Tetrahedron Letters* **2008,** *49* (16), 2541-2545.

391. Bikbulatov, R. V.; Stewart, J.; Jin, W.; Yan, F.; Roth, B. L.; Ferreira, D.; Zjawiony, J. K., Short synthesis of a novel class of salvinorin A analogs with hemiacetalic structure. *Tetrahedron letters* **2008,** *49* (6), 937-940.

392. Taylor, S. K.; May, S. A.; Hopkins, J. A., Cyclizations wherein an epoxide acts as the source of initiation and termination steps. Evidence for an early transition state in biomimetic epoxide cyclizations. *Tetrahedron letters* **1993,** *34* (8), 1283-1286.

393. Butin, A. V.; Smirnov, S. K.; Trushkov, I. V., The effect of an N-substituent on the recyclization of (2-aminoaryl) bis (5-tert-butyl-2-furyl) methanes: synthesis of 3-furylindoles and triketoindoles. *Tetrahedron Letters* **2008,** *49* (1), 20-24.

394. Yadav, J.; Reddy, B. S.; Praneeth, K., FeCl3-catalyzed alkylation of indoles with 1, 3-dicarbonyl compounds: an expedient synthesis of 3-substituted indoles. *Tetrahedron Letters* **2008,** *49* (1), 199-202.

395. Raj, T.; Ishar, M.; Gupta, V.; Pannu, A. P. S.; Kanwal, P.; Singh, G., Unusual conversion of substituted-3-formylchromones to 3-(5-phenyl-3H-[1, 2, 4] dithiazol-3-yl) chromen-4-ones: a

facile and efficient route to novel 1, 2, 4-dithiazoles. *Tetrahedron Letters* **2008,** *49* (2), 243-246.

396. Boto, A.; Hernández, D.; Hernández, R., One-pot synthesis of azanucleosides from proline derivatives. *Tetrahedron Letters* **2008,** *49* (3), 455-458.

397. Carreira, E. M.; Fessard, T. C., Four-membered ring-containing spirocycles: synthetic strategies and opportunities. *Chemical Reviews* **2014,** *114* (16), 8257-8322.

398. Bakthadoss, M.; Sivakumar, N.; Sivakumar, G.; Murugan, G., Highly regio-and stereoselective synthesis of tricyclic frameworks using Baylis–Hillman derivatives. *Tetrahedron Letters* **2008,** *49* (5), 820-823.

399. Gil, M.; Román, E.; Serrano, J., Nitrous acid elimination from 4-alkyl-5-formyl-4-nitrocyclohex-1-enes: synthesis of mono and bicyclic benzene and dihydrobenzene derivatives. *Tetrahedron Letters* **2001,** *42* (28), 4625-4628.

400. Omura, Y.; Taruno, Y.; Irisa, Y.; Morimoto, M.; Saimoto, H.; Shigemasa, Y., Regioselective Mannich reaction of phenolic compounds and its application to the synthesis of new chitosan derivatives. *Tetrahedron Letters* **2001,** *42* (41), 7273-7275.

401. Zhou, P.; Li, Y.; Meagher, K. L.; Mewshaw, R. G.; Harrison, B. L., A new efficient synthesis of 3-(4-pyridinyl) methylindoles. *Tetrahedron Letters* **2001,** *42* (42), 7333-7335.

402. Huang, S.-T.; Kuo, H.-S.; Chen, C.-T., Total synthesis of NADH: ubiquinone oxidoreductase (complex I) antagonist pterulone and its analogue. *Tetrahedron Letters* **2001,** *42* (42), 7473-7475.

403. Kise, N.; Ozaki, H.; Terui, H.; Ohya, K.; Ueda, N., A convenient synthesis of N-Boc-protected tert-butyl esters of phenylglycines from benzylamines. *Tetrahedron Letters* **2001,** *42* (43), 7637-7639.

404. Tse, B.; Jones, A. B., A novel approach to the synthesis of 4-aryl-furan-3-ols. *Tetrahedron Letters* **2001,** *42* (37), 6429-6431.

405. Ujjainwalla, F.; Walsh, T. F., Total syntheses of 6-and 7-azaindole derived GnRH antagonists. *Tetrahedron Letters* **2001,** *42* (37), 6441-6445.

406. Martínez-Teipel, B.; Michelotti, E.; Kelly, M. J.; Weaver, D. G.; Acholla, F.; Beshah, K.; Teixidó, J., Solid-phase synthesis of 1-substituted 4, 5-dihydro-1, 2, 4-triazin-6-ones. *Tetrahedron Letters* **2001,** *42* (37), 6455-6457.

407. Guiso, M.; Marra, C.; Cavarischia, C., Isochromans from 2-(3′, 4′-dihydroxy) phenylethanol. *Tetrahedron Letters* **2001,** *42* (37), 6531-6534.

408. Ohmura, H.; Mikami, K., Heterogeneous acid-catalyzed (2, 5) oxonium-ene reaction for eight-membered ring formation. *Tetrahedron Letters* **2001,** *42* (39), 6859-6863.

409. Kai, H.; Nakai, T., A convenient synthesis of 1H-2, 3-benzoxazines by an acid-catalyzed intramolecular Mitsunobu reaction. *Tetrahedron Letters* **2001,** *42* (39), 6895-6897.

410. Snider, B. B.; Shi, B., A novel extension of the Stork–Jung vinylsilane Robinson annelation

procedure for the introduction of the cyclohexene of guanacastepene. *Tetrahedron Letters* **2001,** *42* (52), 9123-9126.

411. Moghaddam, F. M.; Dekamin, M. G.; Ghaffarzadeh, M., FeCl3 as an efficient and new catalyst for the thia-Fries rearrangement of aryl sulfinates. *Tetrahedron Letters* **2001,** *42* (45), 8119-8121.

412. Ciblat, S.; Canet, J.-L.; Troin, Y., A new route to 2-spiropiperidines. *Tetrahedron Letters* **2001,** *42* (29), 4815-4817.

413. Kuduk, S. D.; Ng, C.; Chang, R. K.; Bock, M. G., Synthesis of 2, 3-diaminodihydropyrroles via thioimidate cyclopropane rearrangement. *Tetrahedron letters* **2003,** *44* (7), 1437-1440.

414. Klein, A.; Miesch, M., New cascade reactions starting from acetylenic ω-ketoesters: an easy access to electrophilic allenes and to 1, 3-bridgehead ketones. *Tetrahedron letters* **2003,** *44* (24), 4483-4485.

415. Kandula, S. V.; Puranik, V. G.; Kumar, P., Synthesis of novel chiral spirodione,(6R, 7R)-7-phenyl-1-oxaspiro [5.5] undec-3-ene-2, 5-dione: application to the asymmetric Diels–Alder reaction with high π-facial selectivity. *Tetrahedron letters* **2003,** *44* (27), 5015-5017.

416. Ballini, R.; Bosica, G.; Fiorini, D.; Petrini, M., Unprecedented, selective Nef reaction of secondary nitroalkanes promoted by DBU under basic homogeneous conditions. *Tetrahedron letters* **2002,** *43* (30), 5233-5235.

417. Liao, Y.-X.; Kuo, P.-Y.; Yang, D.-Y., Efficient synthesis of trisubstituted [1] benzopyrano [4, 3-b] pyrrol-4 (1H)-one derivatives from 4-hydroxycoumarin. *Tetrahedron letters* **2003,** *44* (8), 1599-1602.

418. Song, A.; Wang, X.; Lam, K. S., A convenient synthesis of coumarin-3-carboxylic acids via Knoevenagel condensation of Meldrum's acid with ortho-hydroxyaryl aldehydes or ketones. *Tetrahedron Letters* **2003,** *44* (9), 1755-1758.

419. Snow, R. J.; Butz, T.; Hammach, A.; Kapadia, S.; Morwick, T. M.; Prokopowicz III, A. S.; Takahashi, H.; Tan, J. D.; Tschantz, M. A.; Wang, X.-J., Isoquinolinone synthesis by SNAr reaction: a versatile route to imidazo [4, 5-h] isoquinolin-9-ones. *Tetrahedron letters* **2002,** *43* (42), 7553-7556.

420. Shaabani, A.; Teimouri, M. B.; Bijanzadeh, H. R., One-pot three component condensation reaction in water: an efficient and improved procedure for the synthesis of furo [2, 3-d] pyrimidine-2, 4 (1H, 3H)-diones. *Tetrahedron letters* **2002,** *43* (50), 9151-9154.

421. Rogers, J. F.; Green, D. M., Mild conversion of electron deficient aryl fluorides to phenols using 2-(methylsulfonyl) ethanol. *Tetrahedron letters* **2002,** *43* (19), 3585-3587.

422. Basavaiah, D.; Satyanarayana, T., One-pot facile conversion of the acetates of Baylis–Hillman adducts into substituted fused pyrimidones in aqueous media. *Tetrahedron letters* **2002,** *43* (24), 4301-4303.

423. Kele, P.; Orbulescu, J.; Calhoun, T. L.; Gawley, R. E.; Leblanc, R. M., Coumaryl crown ether based chemosensors: selective detection of saxitoxin in the presence of sodium and

potassium ions. *Tetrahedron letters* **2002,** *43* (25), 4413-4416.

424. Ferreira, P. M.; Maia, H. L.; Monteiro, L. s. S., Synthesis of 2, 3, 5-substituted pyrrole derivatives. *Tetrahedron letters* **2002,** *43* (25), 4491-4493.

425. Daia, G. E.; Gabbutt, C. D.; Hepworth, J. D.; Heron, B. M.; Hibbs, D. E.; Hursthouse, M. B., Synthesis and cycloadditions of 9H-furo [3, 4-b][1] benzo (thio) pyran-9-ones: furan ring formation by a novel hydrolytically induced cycloreversion. *Tetrahedron letters* **2002,** *43* (25), 4507-4510.

426. Ram, V. J.; Agarwal, N., Carbanion induced, base-catalyzed, synthesis of highly functionalized 8-aryl-3, 4-dihydro-2 (1H)-naphthalenones from 2H-pyran-2-ones. *Tetrahedron letters* **2002,** *43* (18), 3281-3283.

427. Wasserman, H. H.; Long, Y. O.; Zhang, R.; Carr, A. J.; Parr, J., Vinyl vicinal tricarbonyl esters as trielectrophiles. Reactions with diamines and related trinucleophiles. *Tetrahedron letters* **2002,** *43* (18), 3347-3350.

428. Agami, C.; Beauseigneur, A.; Comesse, S.; Dechoux, L., Synthesis of new enantiomerically pure α, β-unsaturated bicyclic lactams. *Tetrahedron letters* **2003,** *44* (41), 7667-7669.

429. Wong, K.-T.; Hung, Y.-Y., A convenient one-pot synthesis of homoallylic halides and 1, 3-butadienes. *Tetrahedron letters* **2003,** *44* (43), 8033-8036.

430. Blay, G.; Cardona, L.; Collado, A. M.; García, B.; Pedro, J. R., Silicon-guided rearrangement of 10-methyl-4, 5-epoxydecalins. Methyl versus methylene migration. *Tetrahedron letters* **2003,** *44* (44), 8117-8119.

431. e Melo, T. M. P.; Santos, C. I.; Gonsalves, A. M. A. R.; Paixão, J. A.; Beja, A. M.; Silva, M. R., Synthesis of novel tricyclic isoindole derivatives. *Tetrahedron letters* **2003,** *44* (45), 8285-8287.

432. Saikia, P.; Prajapati, D.; Sandhu, J. S., A novel indium-catalysed synthesis of tetra-substituted pyridine derivatives. *Tetrahedron letters* **2003,** *44* (48), 8725-8727.

433. Heinisch, G.; Matuszczak, B.; Mereiter, K., Pyridazines 71. A novel type of 1, 2-diazine→ 1, 2-diazole ring contraction. *Heterocycles* **1994,** *38* (9), 2081-2089.

434. Tasber, E. S.; Garbaccio, R. M., Thermodynamic equilibration of dihydropyridone enolates: application to the total synthesis of (+/−)-epiuleine. *Tetrahedron letters* **2003,** *44* (51), 9185-9188.

435. Portevin, B.; Golsteyn, R. M.; Pierré, A.; De Nanteuil, G., An expeditious multigram preparation of the marine protein kinase inhibitor debromohymenialdisine. *Tetrahedron letters* **2003,** *44* (52), 9263-9265.

436. Kuo, Y.-H.; Chyu, C.-F., Two novel sesquiterpenes from the roots of Taiwania cryptomerioides Hayata. *Tetrahedron letters* **2003,** *44* (38), 7221-7223.

437. Basavaiah, D.; Rao, A. J., 1-Benzopyran-4 (4H)-ones as novel activated alkenes in the

Baylis–Hillman reaction: a simple and facile synthesis of indolizine-fused-chromones. *Tetrahedron letters* **2003,** *44* (23), 4365-4368.

438. Allin, S. M.; Thomas, C. I.; Allard, J. E.; Duncton, M.; Elsegood, M. R.; Edgar, M., Stereoselective synthesis of the indolizinoindole ring system. *Tetrahedron letters* **2003,** *44* (11), 2335-2337.

439. Cuevas-Yañez, E.; García, M. A.; Marco, A.; Muchowski, J. M.; Cruz-Almanza, R., Novel synthesis of α-diazoketones from acyloxyphosphonium salts and diazomethane. *Tetrahedron letters* **2003,** *44* (26), 4815-4817.

440. Carreira, E. M.; Fessard, T. C., Four-membered ring-containing spirocycles: synthetic strategies and opportunities. *Chemical Reviews* **2014,** *114* (16), 8257-8322.

441. Chiang, Y.-M.; Kuo, Y.-H., Two novel α-tocopheroids from the aerial roots of Ficus microcarpa. *Tetrahedron Letters* **2003,** *44* (27), 5125-5128.

442. Mischne, M., 1, 2, 4-Trioxane in organic synthesis. Unusual entry to diverse carbocyclic frameworks derived from β-ionone. *Tetrahedron letters* **2003,** *44* (31), 5823-5826.

443. Díaz-Marrero, A. R.; Brito, I.; Dorta, E.; Cueto, M.; San-Martín, A.; Darias, J., Caminatal, an aldehyde sesterterpene with a novel carbon skeleton from the Antarctic sponge Suberites caminatus. *Tetrahedron letters* **2003,** *44* (31), 5939-5942.

444. Gil, M.; Román, E.; Serrano, J., Nitrous acid elimination from 4-alkyl-5-formyl-4-nitrocyclohex-1-enes: synthesis of mono and bicyclic benzene and dihydrobenzene derivatives. *Tetrahedron Letters* **2001,** *42* (28), 4625-4628.

445. Nagamitsu, T.; Takano, D.; Shiomi, K.; Ui, H.; Yamaguchi, Y.; Masuma, R.; Harigaya, Y.; Kuwajima, I.; Ōmura, S., Total synthesis of nafuredin-γ, a γ-lactone related to nafuredin with selective inhibitory activity against NADH-fumarate reductase. *Tetrahedron letters* **2003,** *44* (34), 6441-6444.

446. Ding, X.; Nguyen, S. T.; Williams, J. D.; Peet, N. P., Diels–Alder reactions of five-membered heterocycles containing one heteroatom. *Tetrahedron letters* **2014**, *55* (51), 7002-7006..

447. Kato, K.; Yamamoto, Y.; Akita, H., Unusual formation of cyclic-orthoesters by Pd (II)-mediated cyclization–carbonylation of propargylic acetates. *Tetrahedron letters* **2002,** *43* (37), 6587-6590.

448. Kim, J. N.; Im, Y. J.; Kim, J. M., Synthesis of ortho-hydroxyacetophenone derivatives from Baylis–Hillman acetates. *Tetrahedron letters* **2002,** *43* (37), 6597-6600.

449. Turet, L.; Markó, I. E.; Tinant, B.; Declercq, J.-P.; Touillaux, R., Novel anionic polycyclisation cascade. Highly stereocontrolled assembly of functionalised tetracycles akin to the middle core of the manzamines. *Tetrahedron letters* **2002,** *43* (37), 6591-6595.

450. Klymchenko, A. S.; Ozturk, T.; Demchenko, A. P., Synthesis of furanochromones: a new step in improvement of fluorescence properties. *Tetrahedron letters* **2002,** *43* (39), 7079-7082.

451. Groutas, W. C.; Venkataraman, R.; Brubaker, M. J.; Tagusagawa, F., Facile redox formation of a 3-substituted maleimide from a 3-substituted N-hydroxysuccinimide. *Tetrahedron Lett.* **1991,** *32* (43), 6093.

452. Zhong, J.; Lai, Z.; Groutas, C. S.; Wong, T.; Gan, X.; Alliston, K. R.; Eichhorn, D.; Hoidal, J. R.; Groutas, W. C., Serendipitous discovery of an unexpected rearrangement leads to two new classes of potential protease inhibitors. *Bioorg. Med. Chem.* **2004,** *12* (23), 6249-6254.

453. Bergmeier, S. C., The synthesis of vicinal amino alcohols. *Tetrahedron* **2000,** *17* (56), 2561-2576.

454. Koltun, E. S.; Kass, S. R., Rearrangement of the trans-Tricyclo [4.2. 0.01, 3] oct-4-enyl Skeleton. *Tetrahedron* **2000,** *56* (17), 2591-2594.

455. Koltun, E. S.; Kass, S. R., Rearrangement of the trans-Tricyclo [4.2. 0.01, 3] oct-4-enyl Skeleton. *Tetrahedron* **2000,** *56* (17), 2591-2594.

456. Rudolph, J., Facile access to N-thiazolyl α-amino acids from α-bromo ketones and α-amino acids. *Tetrahedron* **2000,** *56* (20), 3161-3165.

457. Paleta, O.; Pelter, A.; Kebrle, J.; Duda, Z.; Hajduch, J., Fluorine-containing butanolides and butenolides. Vinylic fluorine displacement in 4, 4-dialkyl-2, 3-difluoro-2-buten-4-olides and a novel rearrangement induced by organolithium addition to a carbonyl group. *Tetrahedron* **2000,** *56* (20), 3197-3207.

458. Deng, B.-L.; Demillequand, M.; Laurent, M.; Touillaux, R.; Belmans, M.; Kemps, L.; Cérésiat, M.; Marchand-Brynaert, J., Preparation of (3S, 4S)-1-Benzhydryl-3-[(5R)-1'-hydroxyethyl]-4-acyl-2-azetidinones from (2R, 3R)-Epoxybutyramide Precursors. *Tetrahedron* **2000,** *56* (20), 3209-3217.

459. Macías, F. A.; Aguilar, J. M. a.; Molinillo, J. M. a. G.; Rodríguez-Luís, F.; Collado, I. G.; Massanet, G. M.; Fronczek, F. R., Studies on the stereostructure of eudesmanolides from umbelliferae: Total synthesis of (+)-decipienin A. *Tetrahedron* **2000,** *56* (21), 3409-3414.

460. Clapham, G.; Shipman, M., Selective Lewis Acid Complexation of 2-Hydroxyethyl Esters using Competitive Diels–Alder Reactions as a Mechanistic Probe. *Tetrahedron* **2000,** *56* (8), 1127-1134.

461. Taylor, S. K., Reactions of epoxides with ester, ketone and amide enolates. *Tetrahedron* **2000,** *56* (9), 1149-1163.

462. Taylor, S. K., Reactions of epoxides with ester, ketone and amide enolates. *Tetrahedron* **2000,** *56* (9), 1149-1163.

463. Mehta, G.; Venkateswaran, R. V., Haller-Bauer reaction revisited: synthetic applications of a versatile C-C bond scission reaction. *Tetrahedron* **2000,** *56* (11), 1399-1422.

464. Mehta, G.; Venkateswaran, R. V., Haller-Bauer reaction revisited: synthetic applications of a versatile C-C bond scission reaction. *Tetrahedron* **2000,** *56* (11), 1399-1422.

465. Mehta, G.; Venkateswaran, R. V., Haller-Bauer reaction revisited: synthetic applications

of a versatile C-C bond scission reaction. *Tetrahedron* **2000,** *56* (11), 1399-1422.

466. Mehta, G.; Venkateswaran, R. V., Haller-Bauer reaction revisited: synthetic applications of a versatile C-C bond scission reaction. *Tetrahedron* **2000,** *56* (11), 1399-1422.

467. Mehta, G.; Venkateswaran, R. V., Haller-Bauer reaction revisited: synthetic applications of a versatile C-C bond scission reaction. *Tetrahedron* **2000,** *56* (11), 1399-1422.

468. Zamri, A.; Abdallah, M. A., An improved stereocontrolled synthesis of pyochelin, siderophore of Pseudomonas aeruginosa and Burkholderia cepacia. *Tetrahedron* **2000,** *56* (2), 249-256.

469. Buon, C.; Chacun-Lefevre, L.; Rabot, R.; Bouyssou, P.; Coudert, G., Synthesis of 3-Substituted and 2, 3-Disubstituted-4H-1, 4-Benzoxazines. *Tetrahedron* **2000,** *56* (4), 605-614.

470. Kantorowski, E. J.; Kurth, M. J., Expansion to seven-membered rings. *Tetrahedron* **2000,** *56* (26), 4317-4353.

471. Kantorowski, E. J.; Kurth, M. J., Expansion to seven-membered rings. *Tetrahedron* **2000,** *56* (26), 4317-4353.

472. Kantorowski, E. J.; Kurth, M. J., Expansion to seven-membered rings. *Tetrahedron* **2000,** *56* (26), 4317-4353.

473. Kantorowski, E. J.; Kurth, M. J., Expansion to seven-membered rings. *Tetrahedron* **2000,** *56* (26), 4317-4353.

474. Kantorowski, E. J.; Kurth, M. J., Expansion to seven-membered rings. *Tetrahedron* **2000,** *56* (26), 4317-4353.

475. Kantorowski, E. J.; Kurth, M. J., Expansion to seven-membered rings. *Tetrahedron* **2000,** *56* (26), 4317-4353.

476. Chacun-Lefèvre, L.; Joseph, B. t.; Mérour, J.-Y., Synthesis and reactivity of azepino [3, 4-b] indol-5-yl trifluoromethanesulfonate. *Tetrahedron* **2000,** *56* (26), 4491-4499.

477. Bernard, A. M.; Floris, C.; Frongia, A.; Piras, P. P., Synthesis of tertiary cyclobutanols through stereoselective ring expansion of oxaspiropentanes induced by Grignard reagents. *Tetrahedron* **2000,** *56* (26), 4555-4563.

478. Krohn, K.; Micheel, J.; Zukowski, M., Total Synthesis of Angucyclines. Part 15: A Short Synthesis of (±)-6-Deoxybrasiliquinone B. *Tetrahedron* **2000,** *56* (27), 4753-4758.

479. Coltart, D. M., Peptide segment coupling by prior ligation and proximity-induced intramolecular acyl transfer. *Tetrahedron* **2000,** *56* (22), 3449-3491.

480. Darabantu, M.; Plé, G.; Maiereanu, C.; Silaghi-Dumitrescu, I.; Ramondenc, Y.; Mager, S., Synthesis and stereochemistry of some 1, 3-oxazolidine systems based on TRIS (α, α, α-trimethylolaminomethane) and related aminopolyols skeleton. Part 2: 1-aza-3, 7-dioxabicyclo [3.3. 0] octanes. *Tetrahedron* **2000,** *56* (23), 3799-3816.

481. Christine, C.; Ikhiri, K.; Ahond, A.; Al Mourabit, A.; Poupat, C.; Potier, P., Synthèse des

1-amidopyrrolizidines naturelles, absouline et laburnamine, de dérivés et d'analogues pyrrolidinoimidazoliques. *Tetrahedron* **2000,** *56* (13), 1837-1850.

482. Takahashi, A.; Aso, M.; Tanaka, M.; Suemune, H., Synthesis of Optically Active 9-Oxabicyclo [3.3. 1] nona-2, 6-diene as a Cycloocta-1, 5-diene Equivalent and the Corresponding Tetrol. *Tetrahedron* **2000,** *56* (14), 1999-2006.

483. Jansen, B. J.; Hendrikx, C. C.; Masalov, N.; Stork, G. A.; Meulemans, T. M.; Macaev, F. Z.; de Groot, A., Enantioselective Synthesis of Functionalised Decalones by Robinson Annulation of Substituted Cyclohexanones, Derived from R-(−)-Carvone. *Tetrahedron* **2000,** *56* (14), 2075-2094.

484. De Lucchi, O.; Miotti, U.; Modena, G., The P ummerer Reaction of Sulfinyl Compounds. *Organic Reactions* **2004,** *40*, 157-405.

485. Santos, M. R. d. L.; Barreiro, E. J.; Braz-Filho, R.; Fraga, C. A. M., Synthesis of functionalized γ-spirolactone and 2-oxabicyclo [3.3. 0] octane derivatives from nucleophilic oxirane ring opening. *Tetrahedron* **2000,** *56* (30), 5289-5295.

486. Nemes, C.; Jeannin, L.; Sapi, J.; Laronze, M.; Seghir, H.; Augé, F.; Laronze, J.-Y., A convenient Synthesis of conformationally Constrained β-substituted Tryptophans. *Tetrahedron* **2000,** *56* (30), 5479-5492.

487. Pempo, D.; Cintrat, J.-C.; Parrain, J.-L.; Santelli, M., Synthesis of [3H2]-(11S, 17R)-11, 17-Dimethylhentriacontane: A useful tool for the study of the Internalisation of communication pheromones of ant Camponotus vagus. *Tetrahedron* **2000,** *56* (30), 5493-5497.

488. Chandrasekhar, B.; Ramadas, S.; Ramana, D., A Convenient and Simple Method for the Synthesis of Condensed γ-Lactams and Substituted Xanthones from Cyclic-1, 3-diketones. *Tetrahedron* **2000,** *56* (32), 5947-5952.

489. D'Andrea, S. V.; Bonner, D.; Bronson, J. J.; Clark, J.; Denbleyker, K.; Fung-Tomc, J.; Hoeft, S. E.; Hudyma, T. W.; Matiskella, J. D.; Miller, R. F., Synthesis and anti-MRSA activity of novel cephalosporin derivatives. *Tetrahedron* **2000,** *56* (31), 5687-5698.

490. Despinoy, X. L. M.; McNab, H., The Synthesis of 1,(7)-Substituted Pyrrolizidin-3-ones. *Tetrahedron* **2000,** *56* (35), 6359-6383.

491. Gerlitz, M.; Udarnoki, G.; Rohr, J., Biosyntheses of Novel Emycins from the Mutant Strain Streptomyces cellulosae ssp. griseoincarnatus 1114-2. *Angewandte Chemie International Edition in English* **1995,** *34* (15), 1617-1621.

492. Golumbic, C.; Fruton, J. S.; Bergmann, M., Chemical reactions of the nitrogen mustard gases. I. The transformation of methylbis(β-chloroethyl)amine in water. *J. Org. Chem.* **1946,** *11*, 518.

493. a) Walker, M.; Pohl, E.; Herbst-Irmer, R.; Gerlitz, M.; Rohr, J.; Sheldrick, G. M., Absolute configurations of Emycin D, E and F; mimicry of centrosymmetric space groups by mixtures of chiral stereoisomers. *Acta Crystallogr., Sect. B: Struct. Sci.* **1999,** *B55* (4), 607-616. b) Jensen, N.; Friedman, J.; Kropp, H.; Kahan, F., Use of acetylacetone to prepare a prodrug of cycloserine. *Journal of medicinal chemistry* **1980,** *23* (1), 6-8.

494. Nicolaou, K. C.; Barnette, W. E.; Magolda, R. L., Synthesis and chemistry of prostacyclin. *J. Chem. Res., Synop.* **1979,** (6), 202.

495. Behroozi, S. J.; Kim, W.; Gates, K. S., Reaction of n-Propanethiol with 3H-1, 2-Benzodithiol-3-one 1-Oxide and 5, 5-Dimethyl-1, 2-dithiolan-3-one 1-Oxide: Studies Related to the Reaction of Antitumor Antibiotic Leinamycin with DNA. *The Journal of Organic Chemistry* **1995,** *60* (13), 3964-3966.

496. Bruce, W. F. Chloral derivatives. US2784237, 1957.

497. Ramachanderan, R.; Schaefer, B., Tetracycline antibiotics. *ChemTexts* **2021,** *7* (3), 1-42.

498. Norbeck, D. W.; Rosenbrook, W.; Kramer, J. B.; Grampovnik, D. J.; Lartey, P. A., A novel prodrug of an impermeant inhibitor of 3-deoxy-D-manno-2-octulosonate cytidylyltransferase has antibacterial activity. *Journal of medicinal chemistry* **1989,** *32* (3), 625-629.

499. Kinder Jr, F. R.; Bair, K. W., Total Synthesis of (.+-.)-Illudin M. *The Journal of Organic Chemistry* **1994,** *59* (23), 6965-6967.

500. Shu, Y.-Z.; Ye, Q.; Kolb, J. M.; Huang, S.; Veitch, J. A.; Lowe, S. E.; Manly, S. P., Bripiodionen, a new inhibitor of human cytomegalovirus protease from Streptomyces sp. WC76599. *Journal of natural products* **1997,** *60* (5), 529-532.